Springer Series in
CHEMICAL PHYSICS 71

Springer
Berlin
Heidelberg
New York
Hong Kong
London
Milan
Paris
Tokyo

Physics and Astronomy ONLINE LIBRARY

http://www.springer.de/phys/

Springer Series in
CHEMICAL PHYSICS

The purpose of this series is to provide comprehensive up-to-date monographs in both well established disciplines and emerging research areas within the broad fields of chemical physics and physical chemistry. The books deal with both fundamental science and applications, and may have either a theoretical or an experimental emphasis. They are aimed primarily at researchers and graduate students in chemical physics and related fields.

63 **Ultrafast Phenomena XI**
Editors: T. Elsaesser, J.G. Fujimoto, D.A. Wiersma, and W. Zinth

64 **Asymptotic Methods in Quantum Mechanics**
Application to Atoms, Molecules and Nuclei
By S.H. Patil and K.T. Tang

65 **Fluorescence Correlation Spectroscopy**
Theory and Applications
Editors: R. Rigler and E.S. Elson

66 **Ultrafast Phenomena XII**
Editors: T. Elsaesser, S. Mukamel, M.M. Murnane, and N.F. Scherer

67 **Single Molecule Spectroscopy**
Nobel Conference Lectures
Editors: R. Rigler, M. Orrit, T. Basché

68 **Nonequilibrium Nondissipative Thermodynamics**
With Application to Low-Pressure Diamond Synthesis
By J.-T. Wang

69 **Selective Spectroscopy of Single Molecules**
By I.S. Osad'ko

70 **Chemistry of Nanomolecular Systems**
Towards the Realization of Molecular Devices
Editors: T. Nakamura, T. Matsumoto, H. Tada, K.-I. Sugiura

71 **Ultrafast PhenomenaXIII**
Editors: D. Miller, M.M. Murnane, N.R. Scherer, and A.M. Weiner

72 **Physical Chemistry of Polymer Rheology**
By J. Furukawa

73 **Organometallic Conjugation**
Structures, Reactions and Functions
of d–d and d–π Conjugated Systems
Editors: A. Nakamura, N. Ueyama, and K. Yamaguchi

Series homepage – http://www.springer.de/phys/books/chemical-physics/

Volumes 1–62 are listed at the end of the book

R.D. Miller, M.M. Murnane
N.F. Scherer, A.M. Weiner (Eds.)

Ultrafast Phenomena XIII

Proceedings
of the 13th International Conference,
Vancouver, BC, Canada,
May 12–17, 2002

With 284 Figures and 69 Tables

Springer

Dr. R. Dwayne Miller
Department of Physics, University of Toronto
60 St. George St., Toronto, ON M5S-1A7, Canada
E-mail: dmiller@lphys.chem.utoronto.ca

Dr. Margaret M. Murnane
JILA, University of Colorado, Boulder
Boulder, CO 80309-0440, USA

Dr. Norbert F. Scherer
Department of Chemistry, University of Chicago
5735 S. Ellis Ave., Chicago, IL 60637, USA
E-mail: nfschere@uchicago.edu

Dr. Andrew M. Weiner
School of Elect. and Comp. Eng., Purdue University
1285 Elec. Eng. Bldg., W.
Lafayette, IN 47907-1285, USA
E-mail: amw@ecn.purdue.edu

Series Editors:

Professor F.P. Schäfer
Max-Planck-Institut für Biophysikalische Chemie
37077 Göttingen-Nikolausberg, Germany

Professor J.P. Toennies
Max-Planck-Institut für Strömungsforschung
Bunsenstrasse 10
37073 Göttingen, Germany

Professor W. Zinth
Universität München,
Institut für Medizinische Optik
Öttingerstr. 67
80538 München, Germany

ISSN 0172-6218

ISBN-13: 978-3-642-63933-3 e-ISBN-13: 978-3-642-59319-2

DOI: 10.1007/978-3-642-59319-2

Library of Congress Cataloging-in-Publication Data applied for

Bibliographic information published by Die Deutsche Bibliothek
Die Deutsche Bibliothek lists this publication in the Deutsche Nationalbibliografie; detailed bibliographic data is available in the Internet at <http://dnb.ddb.de>.

Springer-Verlag Berlin Heidelberg New York
a member of BertelsmannSpringer Science+Business Media GmbH

http://www.springer.de

Softcover reprint of the hardcover 1st edition 2003

Cover concept: eStudio Calamar Steinen
Cover production: *design & production* GmbH, Heidelberg

Printed on acid-free paper SPIN: 10872043 57/3141/XO - 5 4 3 2 1 0

Preface

This volume is a collection of papers presented at the Thirteenth International Conference on Ultrafast Phenomena held at Vancouver, Canada, from May 12–17, 2002. The Ultrafast Phenomena Conferences are held every two years and provide a forum for discussion of the latest results in ultrafast optics and their applications in science and engineering. A total of more than 200 papers were presented, reporting the forefront of research in ultrashort pulse generation and characterization, including new techniques for shortening the duration of laser pulses, for stabilizing their absolute phase, and for improving tunability over broad wavelength ranges, output powers and peak intensities. Ultrafast spectroscopies, particularly time-resolved X-ray diffraction and two-dimensional spectroscopy, continue to give new insights into fundamental processes in physics, chemistry and biology. Control and optimization of the outcome of ultrafast processes represent another important field of research. There are an increasing number of applications of ultrafast methodology in material diagnostics and processing, microscopy and medical imaging. The enthusiasm of the participants, the involvement of many students, the high quality of the papers and the cosmopolitan location made the conference very successful.

Many people and organizations made invaluable contributions. The members of the international program committee reviewed the submissions and organized the program. The staff of the Optical Society of America deserves special thanks for making the meeting arrangements and running the meeting smoothly. We thank the Optical Society of America for sponsorship, and also acknowledge support from the National Research Council of Canada, Air Force Office of Scientific Research, Canadian Light Source, Photonics Research Ontario, and the Canadian Institute for Photonics Innovations. Also, Positive Light, Spectra Physics Lasers, Coherent, Newport Research Corporation and Clark/MXR provided support. We thank Mary Kulberg (Chicago) for the work involved in creating this volume.

Toronto, CA — *R. J. Dwayne Miller*
Boulder, USA — *Margaret Murnane*
Chicago, USA — *Norbert F. Scherer*
West Lafayette, USA — *Andrew M. Weiner*
September 2002

Contents

Part I: Phase Stabilization

Part II: Short Wavelength Science

Part III: Control and Dynamics

Part IV: Ultrafast Sources and Phase Stabilization

Part VI: THz Generation and Applications

Part VII: Dynamics at Interfaces and Solids

Part X: Control and Dynamics of Condensed Phase Processes

Part XIII: Applications of Ultrashort Pulses

Part I

Attosecond and High Field Science

Attosecond pulse generation from high harmonics and polarization gating

Yann Mairesse[1], Milutin Kovacev[1], Enrico Priori[1,3], Ahmed Merdji[1], Pascal Salières[1], Bertrand Carré[1], Pierre Breger[1], Patrick Monchicourt[1], Olivier Tcherbakov[2], Eric Constant[2], Eric Mevel[2], Pierre Agostini[1]

[1] Commissariat à l'Energie Atomique DSM/DRECAM/SPAM
CE Saclay 91191 Gif Sur Yvette France
E-mail: pagostini@cea.fr
[2] CELIA Université de Bordeaux 1, Cours de la Libération, 33405 Talence, France
[3] INFM, Dipartimento di Fisica, Politecnico
Piazza L. da Vinci 32, I-20133 Milano, Italy

Abstract. High harmonic generation by a "long" infrared pulse is shown by a quantum interference method to produce a periodic train of 250 as pulses separated by 1.35 fs. Such a train is a potential source of a single attosecond pulse through polarization gating. Preliminary results of such a gating showing a spectral broadening up to a factor 2 are reported.

1. Introduction

The generation of sub-femtosecond light pulses requires a source with a wide bandwidth and a well-behaved spectral phase. High-harmonic generation obviously satisfies the first requirement and was long predicted to fulfil second one[1]. It is only recently though, that the demonstration of this last property was accomplished. Harmonics 11-19 of a 40 fs, 800nm pulse generated in argon were found to have a linear relationship between their phases. The corresponding time-domain profile is a periodic train of 250 attosecond pulses separated by half the optical cycle 1.35 fs. This is, at the moment the shortest pulse duration ever measured[2].

Although no measurement of the harmonic pulse envelope was carried in this particular experiment, it was known from previous measurements to be about 10 fs. Therefore the train was 7-or 8-pulse long at FWHM. It is not impossible that such a train could be useful for some applications. However, a more ambitious goal is to achieve the generation of a single attosecond pulse. One straightforward, but demanding solution to this challenge is to start from a sub-10fs infrared pulse. Such pulses are indeed produced from Ti:S lasers in a few groups which master the most sophisticated techniques of pulse recompression[3]. The other approach is to retain the "long" 40 fs pulses which are the standard in many laboratories, generate the train, and subsequently select a single attosecond pulse, taking advantage of the high sensitivity of the harmonic generation process to polarization, phase or intensity of the fundamental light. This paper summarizes the measurements of [2] and reports preliminary results in polarization gating..

2. Reconstruction of attosecond beating by interference of two-photon transition (RABITT)

The method is derived from the theory developed some time ago by Véniard et al[4]. It is moreover a logical extension of a series of cross-correlation measurements of ultrashort XUV pulses opened in 1994 with the observation of a Laser-Assisted Auger decay and continued with high harmonics as Laser-Assisted photoeffect or two-photon, two-color ionisation[5]. It utilizes essentially the same setup, and in particular an annular fundamental beam which greatly helps its subsequent separation from the harmonics[6]. The method was named RABITT by Harm Geert Muller in the spirit of FROG and other SPIDER[7]. Based on two-color, two-photon ionization this method was first implemented as shown in Fig.1:

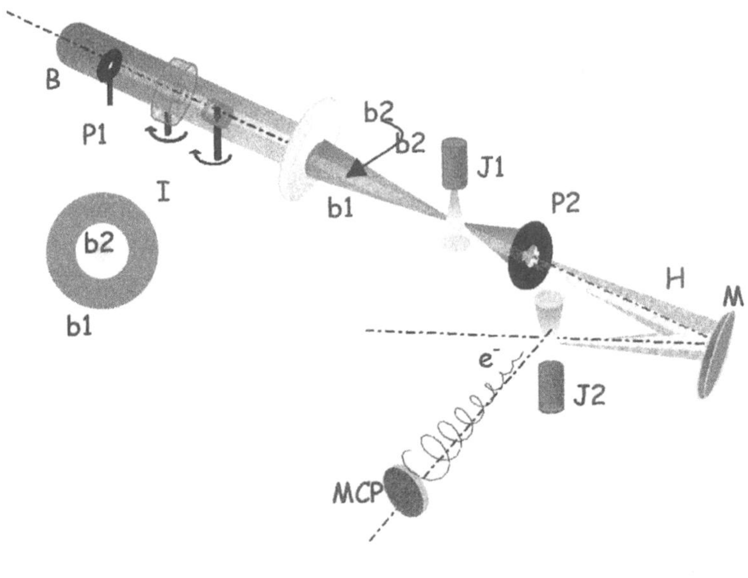

Fig. 1. Experimental implementation of the RABITT method: the Ti:S beam B is split by the pinhole p1 into the collinear annular beam b1 and central beam b2 whose relative phase is controlled by the interferometer I made of two glass plates which can be precisely positioned. b1 generates high harmonics in the atomic jet J1. After elimination of b1 by the pinhole P2, beam b2 and harmonics H are focused by the mirror M into a second atomic jet J2. Photoelectrons issued from two-color, two-photon ionization of atoms in J2 are energy-analyzed in a magnetic bottle, time-of-flight spectrometer and detected on a pair of MCPs. The reflection angle at M is grossly exaggerated on the figure.

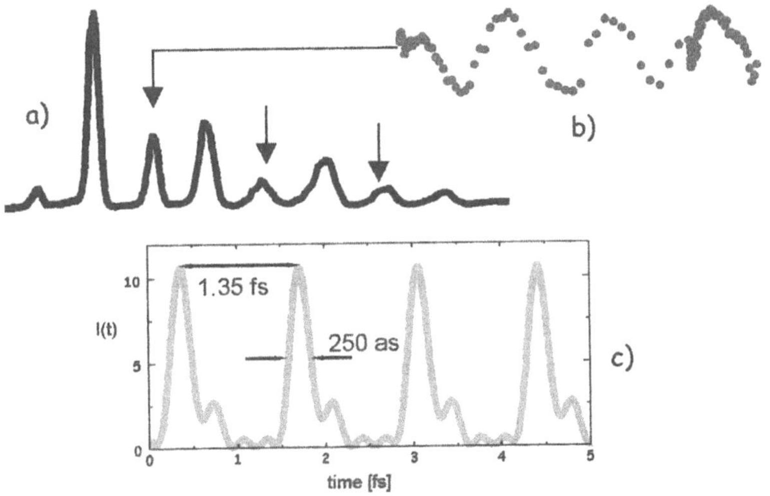

Fig.2 The RABITT I method data (see text).

The photoelectron energy spectrum displays two series of interspersed peaks separated by twice the fundamental photon energy and resulting either from photoionization by one harmonic or from two-photon (harmonic and fundamental) ionization (Fig.2a). Varying the relative phase through rotation of one of the plates of the interferometer while monitoring the amplitude of one of the two-photon peaks yields a sine function with a period equals to half the fundamental optical cycle as shown in (Fig.2b). From the phase of this function determined by fit, the phase difference between the two harmonics on either side of this peak can be retrieved and hence, the time beating of the harmonics reconstructed as the amplitudes of each harmonic peak is known from the photoionization spectrum (Fig.2c). See [2] for more details.
The resulting pulse width depends of course on the number of harmonics locked in phase. In the experiment above this number was limited by the reflectivity of the mirror M Tungsten coating (Fig.1) which dropped beyond harmonic 19.

3. Selection of one attosecond pulse by polarization gating

The idea to take advantage of the rapid dependence of high harmonics generation on the ellipticity of the pump light was proposed by Corkum et al. in 1994[8]. This dependence is intuitively clear since harmonic generation is forbidden in circular polarization by angular momentum conservation. The harmonic yield was measured to depend on the ellipticity ε as the exponential function $\exp(-\beta\varepsilon^2)$ with β ≈41 for harmonic 21 in argon for example[9]. Hence if the pump pulse polarization is changed as a function of time from circular to linear and back to circular the resulting harmonic pulse would last essentially the duration of the linear polarization. The first suggestion[8] was to combine two pulses with slightly different frequencies linearly polarized along perpendicular directions. A variant of this idea was actually implemented by the Lund group[10]. The spectra of the resulting harmonics were expected to show a broadening but, on the contrary, the spectral widths were found to be narrowed. Careful investigations demonstrated that the chirp of the harmonics is in fact responsible for this situation[10]. Both the intensity-dependent dipole phase and the ionization chirp play a role although the first factor was shown to dominate. The finding [2] that harmonics, at least under certain conditions, appear to be phase-locked, as well as recent TDSE calculations [11] and a new experimental approach allowing more control of the various parameters were the motivations for the present attempt.

4. Polarization gating at a single frequency

The scheme described above can be substantially simplified, as proposed by several people[12], by using two properly delayed circularly polarized pulses. An first implementation was attempted at the CELIA laboratory[13] with 35 fs pulses. The changes observed in the harmonic spectra (spectral broadening in the cutoff and spectral squeezing in the plateau) were consistent with a 5 fs temporal confinement of the high harmonic generation. In the present experiment carried out on the LUCA laser at Saclay, the initial linearly polarized 50fs pulse (intensity, FWHM) is fed into a Michelson interferometer. The polarization direction of the pulses in each arm is controlled independently by two quarter-wave plates, equivalent to two half-wave plates due to the double pass. The two pulses leave the interferometer and pass through a quarter-wave plate whose optical axis orientation governs the resulting polarization. If the latter is set at 45° from two mutually orthogonal linearly polarized pulses, a time-dependent ellipticity is determined: the light polarization is first circular, becomes exactly linear during a short time and evolves to circular again ("narrow gate"). The width of the gate depends on the temporal profiles of the pulses and the delay introduced between them by the interferometer. The radiation is then focused inside a 3mm long argon jet where harmonics are generated. Harmonics are spectrally resolved by a grating flat-field spectrometer and detected on a pair of MCPs or a windowless PMT.

The chirp due to SPM during propagation of the pump pulse between the output of the compressor and the harmonic chamber entrance window increased the pulse duration to 80 fs. Experiments were performed either with or without pre-compensation of this chirp by adjusting the compressor length while monitoring the harmonic signal.

Fig. 3 shows preliminary measurements of the spectral width of harmonics in the plateau region without pre-compensation. A clear broadening is seen when the gate is reduced to 8fs FWHM, the shortest duration compatible with the intensity required by harmonic generation (obtained for a delay of 45fs) as compared to the 0 delay case (generation by the 80fs linearly-polarized laser pulse). The maximum broadening recorded varied from about a factor 2 at harmonic 15 to essentially 0 at harmonic 23. It should be noted that the ratio between the two gates is 5-6 (rather than 10 as suggested by the 80fs and 8 fs figures.) if a power law I^3 and β=60 are assumed. Numerical simulations using a simple model for

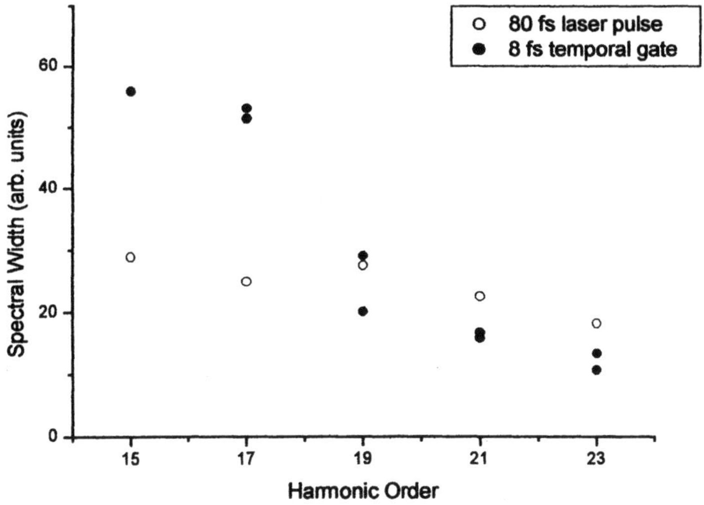

Fig.3 Spectral broadening as a function of the harmonic order.

harmonic generation and the ellipticity dependence discussed above are in progress. However more investigations are needed for a complete understanding of the present results and an evaluation of the role of the harmonic chirp and the ultimate possibilities of the polarization gating method.

References

1 G. Farkas and C. Toth Phys. Lett. A **168**, 447 (1992); S. Harris et al. Opt. Comm. **100**, 487 (1993); Ph. Antoine et al. Phys. Rev. Lett. 77, 1234 (1996).

2 P. M. Paul et al. Science **292**, 1689 (2001).

3 M. Drescher et al. Science **291**, 1923 (2001); M. Hentschel et al. Nature **414**, 509 (2001).

4 V. Véniard, R. Taieb and A. Maquet Phys. Rev. A **54**, 721 (1996).

5 J. M. Schins et al. Phys. Rev. Lett. **73**, 2180 (1994).

6 J. Peatross et al. Opt. Lett. **19**, 942 (1994).

7 H. G. Muller Ultrafast 2001 (Appl. Phys. B to be published)

8 P. B. Corkum et al. Opt. Lett. **19**, 1870 (1994).

9 P. Dietrich et al. Phys. Rev. A **50**, R3585 (1994).

10 C. Altucci et al. Phys. Rev. A **58**, 3934 (1998).

11 M. B. Gaarde and K. J. Schafer Phys. Rev. A **65**, 031406(R) (2002).

12 V.T. Platonenko and V.V. Strelkov Quantum Electronics **28**, 749 (1998); E. Constant PhD thesis (unpublished).

13 E. Constant, E. Mevel and O. Tcherbakov ATTO EU workshop, Milan (2001).

Methods for Complete Characterization of Single Attosecond Pulses

Fabien Quéré[1,2], Jiro Itatani[1], Gennady L. Yudin[1], and Paul B. Corkum[1]

[1] Steacie Institute for Molecular Sciences, National Research Council of Canada, 100 Sussex Drive, Ottawa, ON K1A 0R6, Canada
[2] CEA-CNRS-Ecole Polytechnique, 01128 Palaiseau Cedex, France

Abstract. We propose two methods to completely characterize single attosecond x-ray pulses, using x-ray photoionization of atoms in a strong laser field. (1) Attosecond streak camera. – The laser field is used to spread the energy- or spatial distribution of electron wavepackets. (2) Attosecond SPIDER. – Two identical x-ray pulses produce twin photoelectron wavepackets, one of which is spectrally sheared by the laser field.

1. Introduction

Recent development of attosecond pulse generation opened the route to probing ultrafast electronic dynamics in atoms and molecules [1]. Since the duration of attosecond pulses falls below one period of the laser field, conventional technique (e.g. XUV-laser cross correlation) cannot be applied. We propose two methods based on the modulation of photoelectron wavepackets by the strong laser field.

2. Attosecond Streak Camera [2]

When an atom is simultaneously irradiated by x-ray and laser pulses, the x-ray photoelectron spectrum is modulated by the laser field. If the field is strong and the x-ray photon energy is high, a semiclassical two-step model describes this two- color photoionization process well. We calculated the photoelectron spectra with the semiclassical model or a full quantum-mechanical model with strong field approximation. Both results agree.

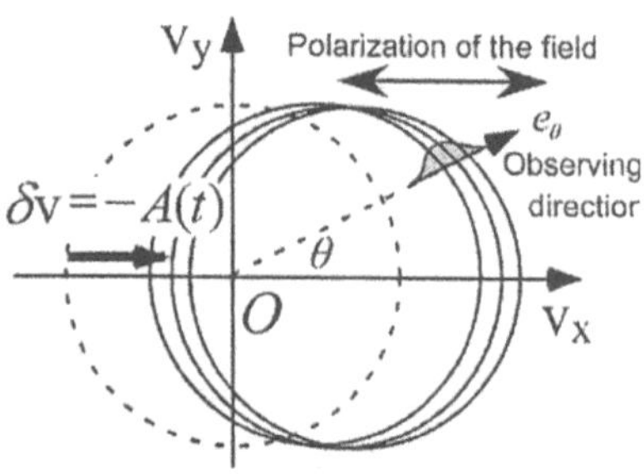

Fig. 1. Effect of a strong laser field on the velocity distribution of X-ray photoelectrons.

At the time of ionization t, the photo-electrons are freed with constant velocity

v_0. The final drift velocity $v_d(t)$ is classically given as $v_d(t) = v_0 - A(t)$ with the vector potential $A(t)$ at the time of ionization. Thus the photoelectrons born at slightly different times will have different final velocities (Fig.1). The laser field $E(t) = E_0(t)\cos\omega_L t$ translates the initial velocity distribution periodically. The final kinetic energy of the electron is then given as:

$$W(t) \approx W_0 + 2U_p \cos 2\theta \sin^2 \omega_L t + \sqrt{8W_0U_p} \cos\theta \sin \omega_L t\,. \qquad (1)$$

Here θ is the observation angle, $W_0 = \Omega(t) - I_p$ is the kinetic energy at t, and $\Omega(t)$ is the instantaneous frequency of the x-ray pulse. By appropriately choosing the phase of the field and the observing angle, we can measure the phase-dependent change of the photoelectron distribution. For transform-limited (TFL) x-ray pulses, this effect results in the spectral broadening (Fig.2a). For chirped x-ray pulses, the photoelectron spectrum is broadened if the signs of streaking $\partial W(t)/\partial t$ and chirp of the x-ray pulse $\partial\Omega(t)/\partial t$ are the same, and it becomes narrower if the signs are opposite (Fig.2b). Thus, we can estimate the chirp as well as the duration of attosecond pulses by measuring the phase-dependent modulation of the photoelectron spectra.

With a circularly polarized field, the angular distribution of photoelectrons at fixed kinetic energy contains equivalent information as angle-resolved photoelectron spectra. In this case, the streaking speed is constant within the laser period. This extends the streaking duration up to one laser period.

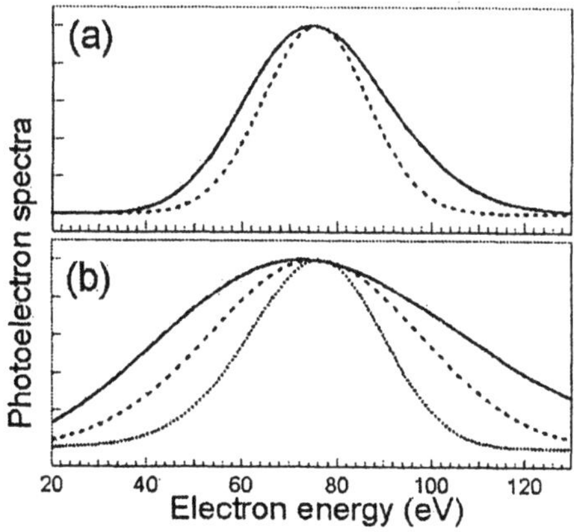

Fig. 2. (a) The photoelectron spectra of He produced by TFL x-ray pulse (τ_p=70 asec, Ω=100 eV). Dashed curve – field-free case, solid curve – with a laser field (I=6x10^{14} W/cm^2, λ=0.8 μm). (b) The photoelectron spectra produced by chirped x-ray pulse (τ_p=70 asec, $\tau_{p,TFL}$=35 asec). Dashed curve – field-free case, solid curve – positive chirp, dotted curve – negative chirp. All the spectra shown are calculated with a SFA model.

3. Attosecond SPIDER [3]

If an atom is ionized by two time-delayed replicas of the attosecond pulse to be characterized, two electron wavepackets are generated in the continuum. They can be shifted in energy with respect to each other by simultaneously applying a strong laser field. The interference of these wave-packets leads to the appearance of fringes in the photoelectron spectrum. The spectrum obtained using the strong field approximation is shown in Fig.3a. As in the spectral phase interferometry for

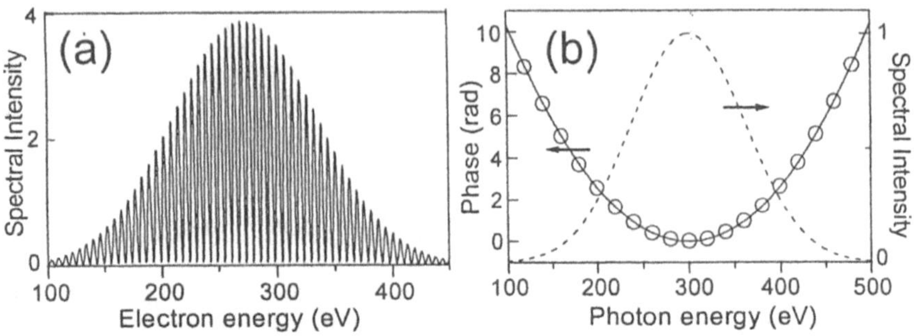

Fig. 3. (a) The simulated photoelectron spectra of hydrogen atoms produced by twin x-ray pulses (τ_p=50 asec, $\tau_{p,TFL}$=12.5 asec, Ω=300 eV) at $\theta = 90°$. The laser field is I=$2\text{x}10^{13}$ W/cm^2, λ=0.8 μm. (b) Comparison between the spectral phases retrieved from the SPIDER spectrum (circles) and the exact phase (solid curve). The dashed curve shows the spectrum of the x-ray pulse.

direct electric field reconstruction (SPIDER) [4], these fringes contain the spectral phase information of the electron wavepacket, and hence the phase of the attosecond x-ray pulse.

Indeed, some phases of the laser field satisfy the condition $\partial W(t)/\partial t = 0$ (i.e., the streaking speed is zero). At these phases, the distortion of photoelectron spectrum is minimized, but its energy is shifted by $\delta W = W(t) - W_0$. Once the photoelectron spectrum is measured, the algorithm of conventional SPIDER can be used to retrieve the spectral phase (Fig.3b). No knowledge of the ionization processes is needed.

5. Discussion and Conclusions

The temporal resolution for the attosecond streak camera is limited by the x-ray bandwidth and the maximum intensity of the laser field. It is estimated to be ~70 asec for 100-eV transform-limited x-ray pulses, and improves for higher photon energy or by iterative deconvolution procedure. As for attosecond SPIDER, it can be applied to arbitrarily short x-ray pulses.

Both methods rely on the phasing of the x-ray pulse to the laser field. In high harmonic generation, the most likely source of attosecond pulses, the x-ray pulses are naturally phase locked to the driving laser pulses. Therefore, the combination of these x-ray and laser pulses is ideal for attosecond pulse measurements.

References

1 M. Drescher *et al.*, Science **291**, 1923 (2001); H. Hentschel *et al.*, Nature **414**, 509 (2001); P.M. Paul *et al.*, Science **292**, 1689 (2001).
2 J. Itatani *et al.*, Phys. Rev. Lett. **88**, 173903 (2002).
3 F. Quéré *et al.*, submitted to Phys. Rev. Lett.
4 C. Iaconis and I.A. Walmsley, Opt. Lett. **23**, 792 (1998).

Strong-field photoionization in the few-optical-cycle regime: the role of the absolute phase

Mauro Nisoli[1], Salvatore Stagira[1], Enrico Priori[1], Giuseppe Sansone[1], Sandro De Silvestri[1], Paolo Villoresi[2], Gerhard G. Paulus[3], Felix Grasbon[3], Herbert Walther[3]

[1] Istituto Nazionale per la Fisica della Materia- I.N.F.M., Politecnico di Milano, Piazza L. da Vinci 32, 20133 Milano, Italy
E-mail: mauro.nisoli@fisi.polimi.it
[2] INFM, Laboratorio di Elettronica Quantistica – D.E.I., Università di Padova, Padova, Italy
[3] Max-Planck-Institut für Quantenoptik, Garching bei München (Germany)

Abstract. Intense-field photoionization is used to investigate the effects of absolute phase of a few-cycle light pulse interacting with a gas medium. Novel features in above-threshold-ionization spectra have been observed.

1. Introduction

Progress in femtosecond laser technology led to the production of laser pulses whose envelope varies on a time scale comparable to that of the electromagnetic field itself [1]. Such pulses consist of less than two optical cycles in FWHM. Under these conditions, the phase of the carrier frequency with respect to the envelope (absolute phase) determines the variation of the laser electric field in time. Since all effects in strong-field laser interaction are driven by the electromagnetic field of the laser, the absolute phase is important for many different topics in laser physics ranging from coherent control of chemical reactions and generation of attosecond pulses to photoionization, production of coherent soft X-rays and particle acceleration. Photoionization using intense laser pulses is characterized by the fact that an atom may absorb more photons than necessary for its ionization. The effect is known as above-threshold ionization (ATI). ATI has found applications for the characterization of few-cycle and attosecond pulses [2-3].

In this paper, we describe a photoionization experiment that provided the first evidence of effects of the absolute phase and we show some relevant features of the photoelectron spectra in the few-optical-cycle regime. We have employed a technique that does not require phase stabilization to demonstrate experimentally the influence of the absolute phase on the photoelectron emission.

2. Experimental Methods

Until now, the only proven way to produce high-peak power light pulses in the 5-fs regime has relied on the hollow-fiber compression technique [1,4]. 25-fs pulses, with an energy of 0.5 mJ, were coupled into an argon-filled (gas pressure 0.5 bar), 60-cm-long hollow fiber, with an inner radius of 250 μm. The pulses are subsequently compressed by specifically designed ultrabroadband chirped mirrors. Typical pulse duration, measured using the spectral phase interferometry for direct electric field reconstruction (SPIDER) technique, ranges from 5 to 7 fs, and the spectral phase is nearly constant over the pulse spectrum. Circularly polarized sub-7-fs light pulses have been generated using an achromatic quarter-wave plate placed between the laser amplifier and the hollow fiber. The compressed pulses have been focused onto a gas jet in a vacuum chamber. Slits with a width of 250 μm allow only photoelectrons originating from a predefinite region in the laser focus to enter the shielded drift tubes of the ATI spectrometer. At the end of the drift tubes, placed in opposite directions with respect to the laser focus (stereo-ATI spectrometer), the electrons are recorded by microchannel plates. The electron time of flight is measured by two computer-hosted multiscalers.

3. Results and Discussion

For linear laser polarization, the appearance of ATI spectra presents an initial steep decrease in electron photon yield for the first few orders followed by a plateau-like structure, i.e. a more or less constant yield over an extended energy range. The plateaus end with a steep roll-off, the so-called ATI-cutoff. These spectra closely resembles those of high-order harmonic generation (HHG)

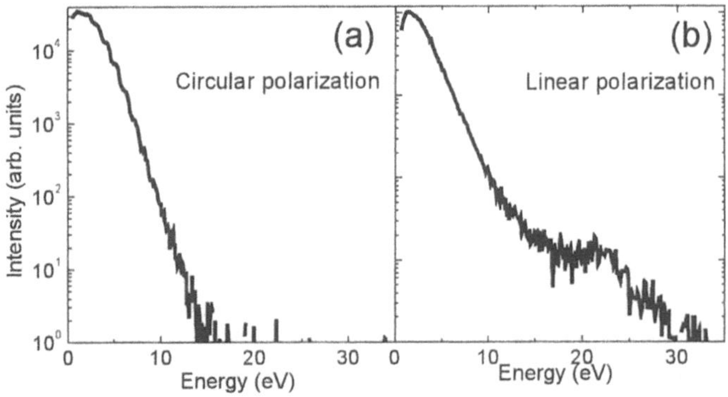

Fig. 1. Typical ATI spectra measured using sub-10-fs excitation pulses with circular (a) and linear (b) polarization.

The major difference is that in ATI the spectra are not affected by propagation, since they originate from single-atom effect. In addition, ATI due to its close relation to HHG can be instrumental also for the investigation of the HHG process. Using few-cycle laser pulses, ATI spectra allow to study processes originating from a single optical cycle. For longer pulses, in contrast, many optical cycles can contribute and strongly alter the appearance of spectra. Here we present for the first time ATI spectra generated with few-cycle laser pulses.

Figure 1 shows typical ATI spectra measured using sub-7-fs pulses with linear and circular polarization. Since the ATI plateau is due to scattering of the photoelectrons, which revisit the ion core, it is absent in the case of circularly polarized excitation pulses. Moreover, no discrete peaks are visible in the measured spectra. The ATI peak structure is due to the periodicity of the ionization process: in every cycle of the laser electric field an electron wavepacket tunnels out of the potential; the wavepackets disperse and interfere each other eventually. In the energy domain this gives rise to the ATI peaks. Using few-optical-cycle pulses the ionization process is not periodically repeated, therefore the corresponding ATI spectrum does not present peaks.

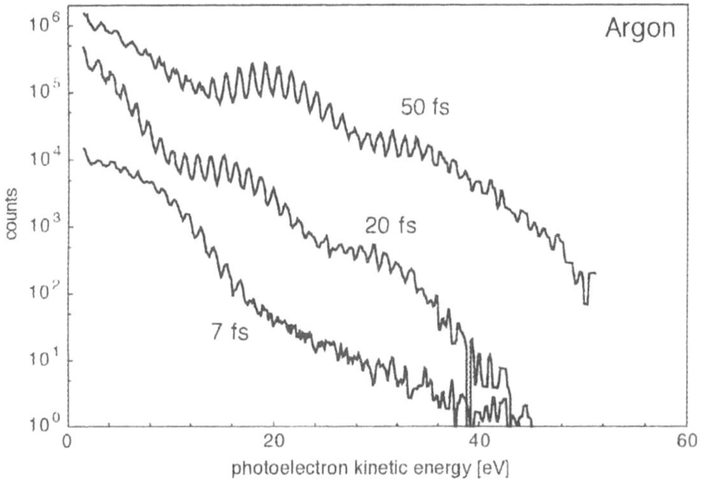

Fig. 2. ATI spectra of argon recorded for different pulse duration. The spectra are separated in the vertical direction for visual convenience.

Figure 2 shows ATI spectra of argon recorded for linearly polarized pulses, with duration ranging from 50 fs to 7 fs, and roughly the same laser intensity. The spectra measured using 50- and 20-fs pulses clearly show a significant enhancement of groups of ATI peaks, within the lower two thirds of the plateau. In the case of 7-fs this effect is completely absent. This can be traced back to the absence of resonance-like effects due to constructive interference of electron wave-packets that return several times to the ion core during the ionization process [5]. In the case of few-cycle pulses there are no electron trajectories that revisit the ion core more than once (and even the first return seems to be attenuated).

We have then used the process of intense-field photoionization to experimentally demonstrate the role of the absolute phase. For a few-cycle laser pulse, depending on the absolute phase, the temporal variation of the electric field becomes asymmetric. Accordingly, the emission of photoelectrons looses inversion symmetry, i.e. a left-right asymmetry appears. Depending on the number of electrons emitted in opposite directions. The strength of the anticorrelation depends on the degree of influence of the absolute phase. The basic idea of our experiment is to place two electron detectors in opposite directions with respect to the laser focus [6]. For each laser pulse, the number of electrons detected with both detectors is recorded. Each laser pulse is characterized by two numbers: namely, the number of electrons detected in the left-hand and that in the right-hand arm of the ATI spectrometer. These pairs of numbers can be interpreted as coordinate in a contingency map. As the signature of an effect from the absolute phase is an anticorrelation, it should appear as a structure inclined at –45° in the contingency map. This approach is expected to be very sensitive to correlations, because, in principle, an arbitrary number of laser shots can be used to make even small (anti)correlations statistically significant. The use of circularly polarized light pulses offers various advantages [7]: (i) since the electric field amplitude varies smoothly with time, the ionization is not modulated by rapidly changing electric field; (ii) it prevents electron rescattering from the parent ions, so that the photoelectron direction is uniquely determined by the absolute phase.

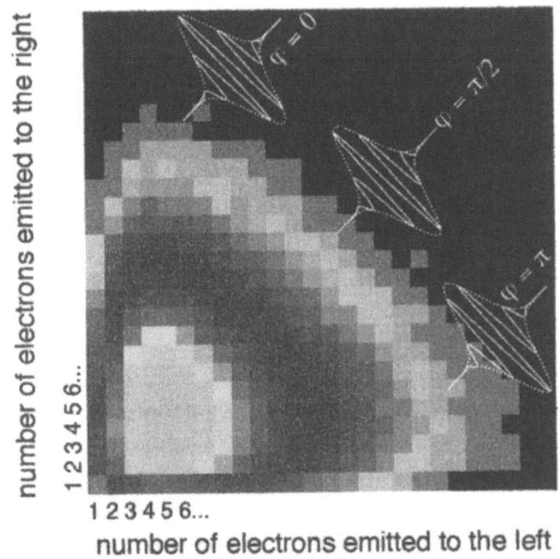

Fig. 3. Contingency map. The color of each pixel of the false color plot represents the number of recorded laser pulses with an absolute phase such that electrons according to the coordinate of the pixel were emitted to the left and the right. The absolute phase of the carrier with respect to the envelope of the pulse is shown for a few specific values.

6-fs, circularly polarized pulses, have been focused at an intensity of 5×10^{13} W/cm^2, onto a Krypton jet. The measured contingency map, shown in Fig. 1, clearly displays a structure inclined by –45°. In order to prove the anticorrelation, and evaluate its strength and significance, we calculated the Kendall τ. A positive value of τ indicates (positive) correlation, while a negative value indicates an anticorrelation. Kendall's τ is the result of a non-parametric algorithm that relies on rank-ordering: essentially, the number of concordant and discordant events is subtracted. It is worth pointing out that non-parametric procedures, in particular value of τ indicates (positive) correlation, while a negative value indicates an anticorrelation. Kendall's τ is the result of a non-parametric algorithm that relies on rank-ordering: essentially, the number of concordant and discordant events is subtracted. It is worth pointing out that non-parametric procedures, in particular Kendall's τ, are conservative in the sense that if they indicate (anti)correlation, then it definitely exists.

Using 6-fs excitation pulses we measured negative values of τ, (typically $\tau \approx -0.05$) with distinct significance (typically more than two standard deviations of the null hypothesis of no correlation). This was compared with the result of the same procedure for longer pulses (>7 fs). There, Kendall's τ had a positive sign with a significance of typically six standard deviations or more. This is expected because of the inevitable laser pulse fluctuation. For this reason, the observed anticorrelation is a rather strong effect.

The correlation technique presented here is important for its potential applications. These include: detailed study of the influence of the absolute phase on strong-field effects, without phase stabilization; investigation of phase stability of a pulse train; measurement of the absolute phase of a few-cycle light pulse.

4. Conclusions

In this work we have reported the first experimental demonstration of the influence of the absolute phase of a few-cycle light pulse on intense-field photoionization. We have then evidenced novel features in above-threshold-ionization spectra generated by few-cycle laser pulses.

References

1 M. Nisoli, S. De Silvestri, O. Svelto, R. Szipöcs, K. Ferencz, Ch. Spielmann, S. Sartania, and F. Krausz, Opt. Lett. **22**, 522, 1997.

2 P.M. Paul, E.S. Toma, P. Breger, G. Mullot, F. Augé, Ph. Balcou, H.G. Muller, and P. Agostini, Science **292**, 1689, 2001.

3 M. Hentschel, R. Kienberger, Ch. Spielmann, G.A. Reider, N. Milosevic, T. Brabec, P. Corkum, U. Heinzmann, M. Drescher, and F. Krausz, Nature **414**, 509, 2001.

4 M. Nisoli, S. De Silvestri, and O. Svelto, Appl. Phys. Lett. **68**, 2793, 1996.

5 G.G. Paulus, F. Grasbon, H. Walther, R. Kopold, and W. Becker, Phys. Rev. A **64**, 021401(R), 2001.

6 G.G. Paulus, F. Grasbon, H. Walther, P. Villoresi, M. Nisoli, S. Stagira, E. Priori, and S. De Silvestri, Nature **414**, 182, 2001.

7 P. Dietrich, F. Krausz, and P.B. Corkum, Opt. Lett. **25**, 16 (2000).

Ultrafast EUV Lithography

Stefan Düsterer, Heinrich Schwoerer, Wolfgang Ziegler, Roland Sauerbrey

Institut für Optik und Quantenelektronik, Friedrich-Schiller-Universität,
Max-Wien-Platz 1, D-07743 Jena, Germany
Tel. +49 (0) 3641 / 947218, Fax: +49 (0) 3641 / 947202
email: duesterer@ioq.uni-jena.de

Abstract: Using two 2 ps laser pulses, separated hundreds of ps, rather than longer pulses we could improve the conversion of laser radiation into EUV radiation from water droplet targets by a factor of four.

1. Introduction

The industrial application of EUV lithography demands hundreds of watts of EUV average power in an ~ 2.5% bandwidth region around a center wavelength of 13 nm. The current approach to get the desired power is to use large scale long pulse (hundreds of ps to several nanoseconds) lasers with high average power heating a target which subsequently emits EUV radiation. But as we could demonstrate, the conversion efficiency (CE) from laser radiation into EUV radiation is much higher for two short pulses of about 2 ps duration than for longer (or shorter) ones. This discovery suggests that it is more efficient to use smaller short pulse lasers for EUV production which can produce the same EUV average power than long pulse lasers due to the higher CE.

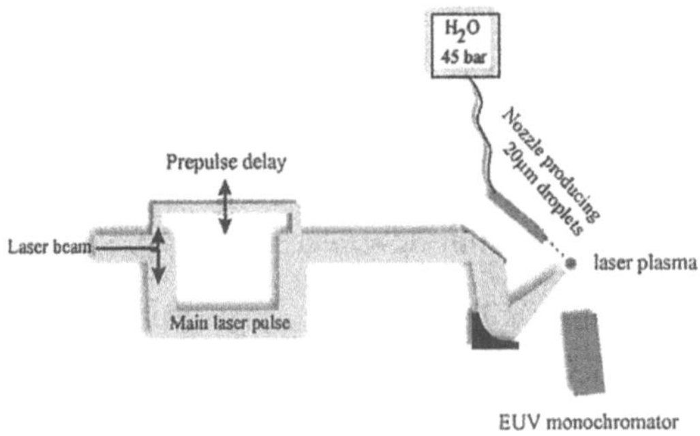

Fig 1: Scheme of the experimental setup. The laser pulse duration ranges from 200 fs up to 120 ps with a laser pulse energy of 20 mJ to 100 mJ. The whole setup was placed in vacuum better 10^{-3} mbar.

In our experiments water droplets with 20 μm diameter were irradiated with laser pulses, focussed to a 20 μm diameter focal spot, from a Ti:sapphire laser system (λ= 800nm) [1]. The measured EUV radiation at 13 nm, produced by the laser plasma interaction, resulted from fivefold ionized oxygen ions undergoing the transition 4d $\rightarrow$ 2p. The duration of the laser pulses varied from 200 fs to 120 ps for laser pulse energies of 20mJ to 100 mJ. In addition we introduced a 2 mJ prepulse, with the same pulse duration as the main pulse, reaching the target up to 700 ps before the main pulse. The EUV conversion efficiency was measured for each pulse duration, energy and prepulse delay. The result of the measurement for 20 mJ laser pulses is shown in a compact form in fig. 2. Obviously the conversion efficiency was enhanced with increasing prepulse delay, within the measured region, for each pulse duration. Independent of the laser pulse length, the introduction of a prepulse increases the CE, whereas the effect of the prepulse is much more dominant for short pulses.

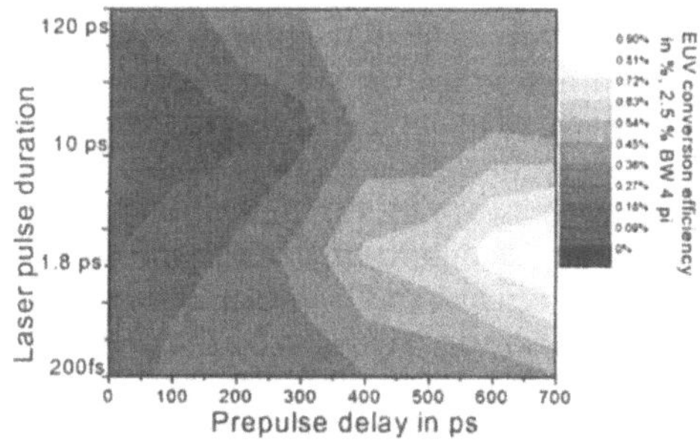

Fig 2: *Measured* conversion efficiency plotted as function of the laser pulse duration and the prepulse delay. The shown data was measured for 20 mJ main pulse energy and 2 mJ prepulse energy.

While the optimum CE without prepulse was indeed found for long pulses with a pulse duration of 120 ps at about 0.2 % in 2.5 % bandwidth and 4 π (fig. 2 and more detailed in [1]), the best CE was measured for 2 ps short pulses preceded by a weak prepulse. A CE of 0.9 % in 2.5 % bandwidth and 4 π and was obtained for 2 ps laser pulses with a 10 % prepulse hitting the droplet 700 ps before the main pulse. Fig. 2 clearly shows that 2 ps pulses with prepulses yield the highest CE for the experimental conditions and the CE of either longer or shorter pulses is less.
To understand the underlying physics of this promising result simulations were performed. The dependence of the EUV radiation on the plasma density and the plasma temperature was calculated by an atomic physics code. In addition plasma dynamic simulations of the laser-plasma interaction and the subsequent plasma evolution were done with the simulation code MEDUSA. The result of the simulations, using the parameters of the experiment shown in fig. 2, can be seen in fig. 3. A good agreement between the simulation and the experiment was achieved.

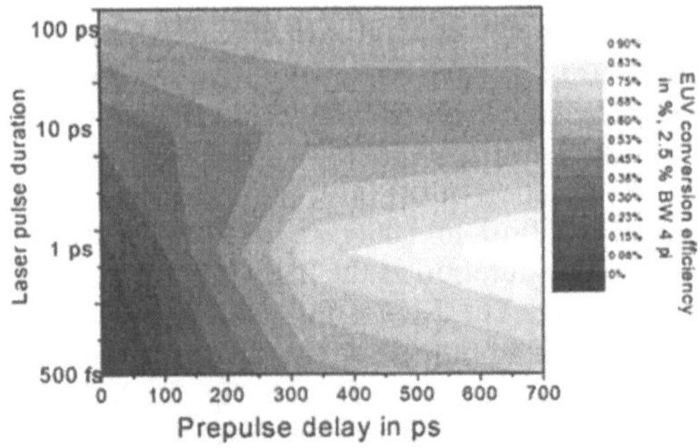

Fig 3: *Simulated* conversion efficiency plotted as function of the laser pulse duration and the prepulse delay. The simulation was performed for the same laser parameters as were used for the experiment shown in Fig. 2 .

The EUV emissivity is strongly dependent on the plasma temperature and electron density and there is only a narrow range for these parameters in which EUV is produced efficiently. To achieve the optimum plasma conditions in the laser produced plasma, collisional as well as collective absorption mechanisms play an important role. The best balance between the different absorption processes in the preplasma was found at a laser pulse duration of 2 ps. For either longer or shorter pulses only one of the absorption mechanisms is dominant and the CE decreases. The absorption mechanisms are also very sensitive on the preplasma. To produce optimum preplasma conditions, one single prepulse has only two degrees of freedom (energy and delay) and as seen in the experiment enhances the EUV efficiency already by a factor of four. Novel pulse shaping techniques, which are currently installed in our terawatt laser system, enable much more degrees of freedom. Complicated pulse forms, including multiple prepulses, can be tailored to optimize the absorption in a way that optimum plasma conditions for EUV production can be achieved and the CE increases even more. Since the mechanisms do not strongly depend on the atomic structure, the results are valuable for other target materials.

In conclusion, the introduction of shaped pulses is a novel method to enhance the EUV yield substantially, especially for ultrashort laser pulses. EUV lithography steppers would require an ultrashort laser source with an average power of ~10 kW which constitutes a mayor challenge for future ultrashort pulse laser technology.

References

[1] S. Düsterer, H. Schwoerer et al.: „ Optimization of EUV radiation yield from laser-produced plasma", Appl. Phys. B 73, 693 (2001)

Distortion-free guiding of 0.2×10^{18} W/cm^2 femtosecond pulses through a fully-ionized 1.5 cm He plasma channel

E. W. Gaul, F. Grigsby, N. H. Matlis, B. Shim, R. Zgadzaj, and M. C. Downer

Department of Physics, University of Texas at Austin, Austin, Texas 78712-1081

Abstract. We couple *80fs* laser pulses into preformed Helium plasma channels with ~70% efficiency, and guide them at near-relativistic intensity over 60 Rayleigh ranges with no detectable distortion of their spectrum or 16μm-diameter TEM_{00} mode structure.

Plasma waveguides capable of guiding relativistically intense fs laser pulses *without optical distortion* are essential to developing GeV-scale Laser Wakefield (LWF) Accelerators [1] and coherent short-wavelength sources. We report, to our knowledge, the highest intensity-length product (0.2×10^{18} *W/cm*^{2} *x1.5 cm*) to date for distortion-free guiding. We generated our channels in a backfill of He gas (*300-700 Torr*) using a modification of a method developed by Milchberg *et al.* [2], who line-focused *~100 ps, ~0.3 J Nd:YAG* laser pulses to intensities of ~10^{13} *W/cm*^{2} into a gas using an axicon (see Fig. 1a). This intensity readily field-ionizes high-Z gases (e.g. Ar, N_2), thus initiating the channel-forming process, but forms an *incompletely-ionized* channel, in which relativistically intense pulses suffer ionization-induced phase distortion during guiding and ionization-induced defocusing when entering the channel. Indeed we observe distorted exit mode and spectrum (see Fig. 1c) and only *5%* throughput when we attempt to guide a pulse at $\geq 10^{17}$ *W/cm*^{2} through a nitrogen channel.

To overcome these problems, we formed channels in a He backfill. A backfill provides more *uniform* gas density than a jet [2], essential for obtaining uniform channel diameter. Use of He gas minimizes ionization-induced distortions *inside* the channel, where (like H_2) it can be fully–ionized, *and* in the entrance/exit regions, where (unlike H_2) its high field-ionization threshold (~10^{15} W/cm^2) suppresses ionization until the focusing *fs* pulse is nearly inside the channel. Unfortunately, this high threshold also prevents the channel-forming YAG pulse from field ionizing neutral He. However, we succeeded in forming excellent He channels with 3 modifications of the procedure of Ref. [1]: (i) pre-ionizing He with a discharge to provide seed electrons; (ii) lengthening the channel-forming pulse to 400 ps to increase Inverse Bremsstrahlung heating; (iii) amplifying the channel-forming pulse to *1.1 J* to maintain line-focused intensity ~10^{13} *W/cm*^{2} [3]. Transverse interferometry (Fig. 1b) confirms that the resulting channels are fully ionized. Most importantly, we now report that *800nm, 80 fs, 50 mJ* pulses focused to beam diameter *16 μm* at the channel entrance by an *f/8* off-axis parabolic mirror, exit the channel with ~*50%* of their incident energy, undistorted transverse mode and undistorted spectrum (Fig. 1c). Best results are obtained by coupling ~*3*

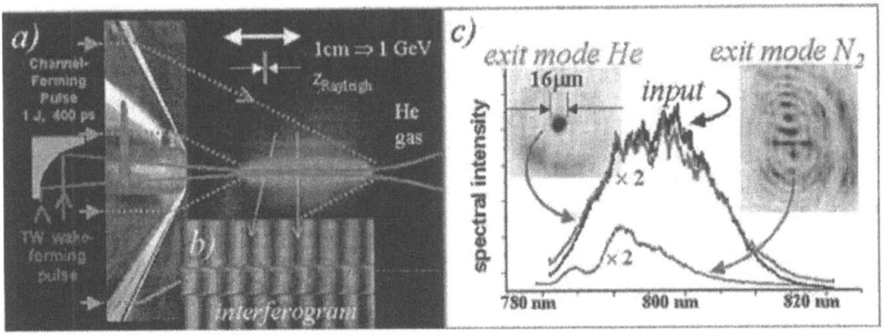

Fig. 1. Demonstration of distortion-free guiding of 0.2×10^{18} W/cm^2 through a fully-ionized He plasma channel. **(a)** photo of axicon (left) and fluorescencing filament (right) from plasma heated by line-focused *1 J, 400 ps* Nd:YAG pulse. The Ti:Sapphire pulse (red lines) is coupled into the channel through the hole in the center of the axicon. **(b)** transverse interferogram of plasma channel *3 ns* after line focus. Abel inversion of this interferogram reveals a density depression on axis which guides the Ti:sapphire pulse over *1.5 cm* . **(c)** mode and spectra of intense Ti:S pulse exiting channel. Mode exiting He channel (left) is TEM_{00}, spectrally indistinguishable (red line) from input; mode exiting N_2 channel (right) is spatially distorted, spectrally blue-shifted (solid green line) and severely attenuated (~0.05) by poor coupling.

ns after the Nd:YAG line-focus, to enable formation of a channel of axial density $n_e(r=0) = 4 \times 10^{18}\ cm^{-3}$, radius $40\ \mu m$.

Since, at $I_{guided} > 10^{17}\ W/cm^2$, He ionization fronts occur well in the leading edge of the guided pulse, they might blue-shift only a few % of its energy even if present. We therefore also performed pump-probe experiments, with $\lambda_{probe} = \lambda_{pump} = 800\ nm$ to avoid group-velocity walk-off and $E_{probe} \perp E_{pump}$ to discriminate transmitted probe from pump spectra. The probe thus "rode" the pump ionization front through the length L of the channel, and characterized it more sensitively through its cumulative blue-shift $\Delta\omega \sim \int^L (dn_e/dt)dz$. Fig. 2a shows differential spectra (i.e. an undistorted reference spectrum was subtracted) of probe pulses that co-propagated with a $0.1 \times 10^{18}\ W/cm^2$ pump pulse through channels formed in *700Torr* He at several representative time delays in the interval $-1000\ fs < \Delta t < +333\ fs$. Frequency-domain interference (FDI) fringes ~ $E_{probe}E_{scatt.pump}\cos\omega\Delta t$ evident in the data are caused by the probe and depolarized scattered pump light of comparable intensity co-propagating through the channel and interfering at the CCD detector of the spectrometer. The variable fringe period $(\Delta t)^{-1}$ calibrated Δt, while the fringe phase recorded probe-reference phase shifts. Small blue-shifts were indeed observed for $\Delta t < 0$ (2nd–4th spectra) that were not detectable in the transmitted spectrum of the main pulse. The shift did not vary with channel length, proving that it originated from ionization in the entrance/exit regions, and became undetectable for channels generated in $p < 500\ Torr$ He gas. On the other hand, strong probe blue-shifts were observed at $\Delta t \sim -300\ fs$ in channels formed in N_2 gas of only $p \sim 100\ Torr$ (see Fig. 2b).

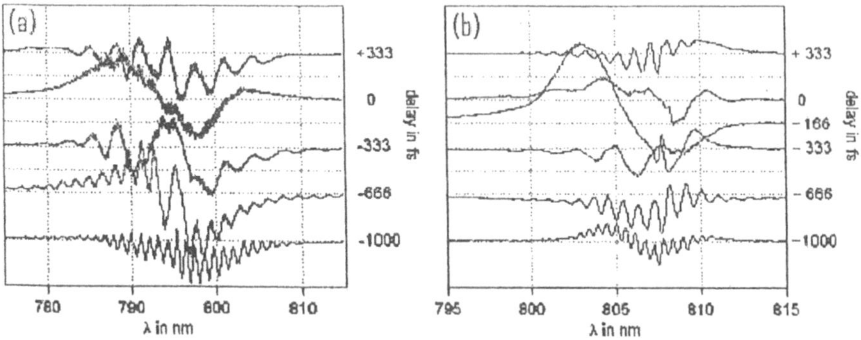

Fig. 2. Spectra of probe pulses after propagation through 1.5 cm plasma channels formed in (a) *700 Torr* He or (b) *100 Torr* N_2, at time delays $-1000 < \Delta t < +333$ fs from peak of intense, orthogonally-polarized guided pulse, normalized to undistorted pulse spectrum. Oscillations with period $\omega\Delta t$ are produced by frequency domain interference between probe pulse and depolarized scattered pump light.

We are currently upgrading our *1 TW* Ti:S system to *~4 TW* as one step toward guiding pulses at *~3 x* higher intensity ($\sim 0.6 \times 10^{18}$ W/cm^2) that can generate high-amplitude plasma wakefields ($\Delta n_e/n_e \sim 0.5$, $E_z \sim 1$ GV/cm) capable of accelerating electrons to *1 GeV* in *1 cm* [1]. However, such pulses will ionize the back-fill gas before reaching the channel, even in Helium. We have therefore also built and characterized a differentially-pumped cell, which still enables the channel to be generated in a region of uniform density, but leaves an exponentially decaying neutral gas density profile $n(z) = n(0)exp(-\beta z)$ at the entrance/exit regions, for which we have measured scale length $\beta^{-1} \approx 0.5$ *mm*. In parallel with these measurements, we have simulated the propagation of intense focusing pulses through this inhomogeneous ionizing gas profile, using coupled ADK ionization, relativistic cold fluid and Maxwell equations [4]. These simulations predict that focused intensity 10^{18} W/cm^2 can be coupled into H_2 or He channels and guided with minimal distortion if straightforward constraints on the focusing geometry are satisfied [4].

References

1 T. Tajima and J. M. Dawson, Phys. Rev. Lett. 43, 267-270 (1979) .

2 C. G. Durfee, III and H. M. Milchberg, Phys. Rev. Lett. 71, 2409 (1993); J. Fan, T. R. Clark, and H.M. Milchberg, Appl. Phys. Lett. 73, 3064 (1998).

3 E. W. Gaul, S.P. LeBlanc, A.R. Rundquist, R. Zgazaj, H. Langhoff, and M.C. Downer, Appl. Phys. Lett. 77, 4112 (2000).

4 N. Andreev, M.V. Chegotov, M.C. Downer, E.W. Gaul, N.H.Matlis, A.A. Pogosova, and A.R. Rundquist, IEEE Trans. Plasma Science 28, 1090 (2000).

Observation of Townes-soliton formation

K.D. Moll and Alexander L. Gaeta

School of Applied and Engineering Physics, 159 Clark Hall, Cornell University, Ithaca, NY 14853, USA

Abstract. We investigate the spatial dynamics of an ultrashort pulse undergoing self-focusing collapse. For both elliptically- and irregularly-shaped input beams, the spatial profile of the pulse is observed to evolve to a circularly-symmetric profile known as a Townes soliton.

1. Introduction

Nonlinear wave collapse is an inherent phenomenon to several areas of physics including hydrodynamics, plasma physics, and nonlinear optics. In the latter case, the process responsible for collapse is the self-focusing of light in a medium with an intensity-dependent refractive index. The nature of the collapse has been theoretically analyzed using a variety of techniques including the variational principle [1], modulation theory [2], and numerical simulations [2,3]. Although the technique based on the variational principle is simple to apply, the other techniques have demonstrated that the collapse dynamics are too complicated to be fully captured by the ansatz that the beam maintains a Gaussian profile. In particular, modulation theory and numerical simulations predict that regardless of the shape of the input beam profile a collapsing beam transforms its radial profile into a focused *Townes soliton*. In this report, we present experimental evidence demonstrating that the spatial profile of a collapsing femtosecond laser pulse evolves into the cylindrically symmetric profile described by the Townes soliton for the case in which the input beam is either elliptically or irregularly shaped. These results could lend further understanding to recent experiments [4] that have observed extended light filaments in air with femtosecond laser pulses.

2. Numerical Simulations

An example of the evolution of an input beam into the Townes profile is the case in which the input profile is not smooth, but is noisy; this case might correspond to the profile that emerges from an amplified femtosecond laser system. As seen in Figs. 1(a) and 1(b), simulations show that the pulse evolves into a smooth, circularly symmetric profile. Another important case is that of an elliptically-shaped profile. Figure 1(c) and 1(d) shows results from our numerical simulations in which an elliptical beam with a power above the threshold for collapse transforms into a circularly symmetric beam as it approaches the collapse point. Figure 1(d) shows lineouts along both directions near the collapse point for the

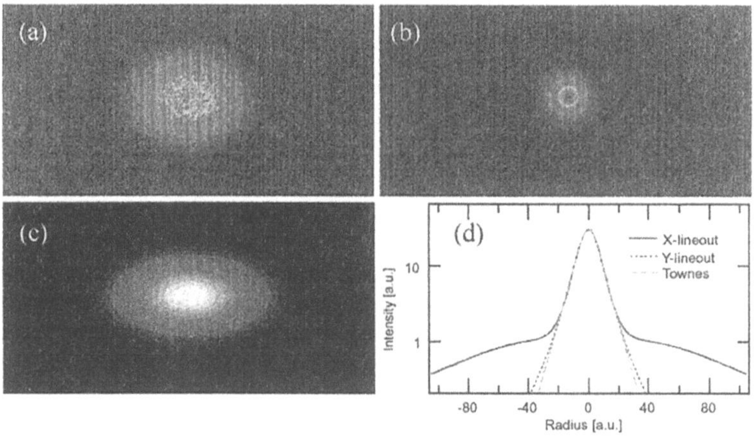

Fig. 1. Simulations for (a) a randomly distorted input beam and (c) an elliptically shaped input beam above the critical power for self-focusing. Near the points of collapse, the radial profiles (b) and (d) evolved to circularly-symmetric profiles. Lineouts of the collapsing elliptical beam (d) show that the on-axis portion of the beam transforms into the theoretically predicted Townes profile.

initial profile shown in Fig. 1(c). As is clearly seen, the on-axis portion overlaps exactly with the theoretically calculated Townes profile. Part (d) also shows the partial-beam self focusing inherent to this process. Since the profile near the location of collapse is insensitive to the initial profile, we expect that after higher-order nonlinear processes have halted the collapse, the profile will also be circularly symmetric and insensitive to the input profile.

3. Experimental Results

We verified these predictions by using a beam from an amplified femtosecond Ti:sapphire laser system and passing it through a 30-cm BK7 glass sample at powers close to the threshold power for collapse P_{cr}. The output face was blocked by a piece of frosted glass such that a digital camera imaged the beam or by a thin slit such that a linear photodiode array could record lineouts of the beam. As can be seen from the photograph in Fig. 2(a), the beam profile from our system was noisy. However, after passing through the glass at powers near P_{cr}, the output became smooth and circularly symmetric. A separate beam was imaged with a linear photodiode array at both low and high powers [See Fig. 2(c)]. The on-axis component is clearly seen to evolve into the theoretically prediced Townes profile, while the low-intensity wings are seen to undergo nearly linear propagation.

In a second experiment, an elliptically-shaped input beam was created by passing the input beam through a series of cylindrical lenses. At low powers, an elliptically shaped input beam was observed at the output face of the sample, but as the power was increased, circularly-symmetric supercontinuum generation (SCG) was observed and imaged. The observed circularly symmetric SCG

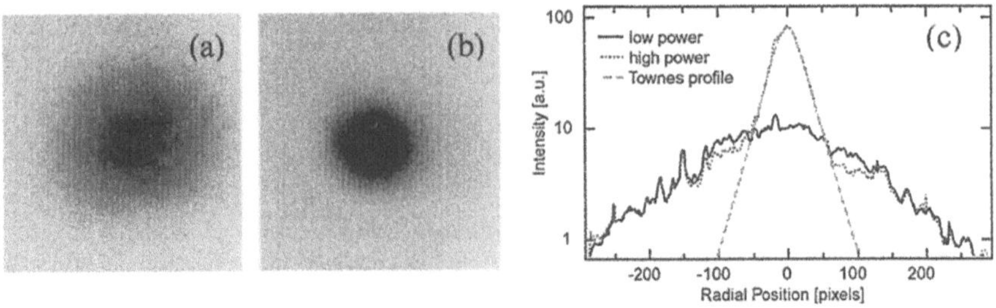

Fig. 2. A noisy input beam (a) with low power is passed through a BK7 glass sample. As the power is increased, the output beam becomes smooth and circularly symmetric. In a separate experiment, (c) a linear photodiode array captures lineouts of the low and high intensity beams. The lineouts show that the on-axis component precisely matches the Townes profile as predicted by theory.

provides further evidence that verifies the theoretical predictions for the generation of the Townes profile when using highly irregular input beams.

In addition to these single-beam experiments, we investigated the interactions between two pulses separated transversely in space, and we observed that when the pulses are allowed to overlap in space and time within the medium, mutually-induced collapse of the pulses can occur at peak powers such that collapse of the pulses would not occur in the non-interacting case.

4. Conclusions

In conclusion, we have demonstrated that a laser beam with either an elliptically- or irregular-shaped spatial profile will collapse to a circularly symmetric profile, as predicted by theoretical simulations. We have also made a direct measurement of the radial profile and have observed remarkable agreement between the experimentally measured beam and the theoretically predicted Townes profile. We have also demonstrated that it is possible to induce a weak pulse to catastrophically collapse by interaction with a second pulse.

References

1. M. Desaix, D. Anderson, and M. Lisak, J. Opt. Soc. Am. B **8**, 2082-2086 (1991).
2. G. Fibich and B Ilan, J. Opt. Soc. Am. B **17**, 1749-1758 (2000).
3. B. Gross and J. T. Manassah, Phys. Lett. A **169**, 371-378 (1992).
4. A. Braun *et al.*, Opt. Lett. **20**, 73-75 (1995).

Probing wave packet motion of D_2^+ with sub femtosecond resolution

H. Niikura[1], F. Légaré[2], R. Hasbani[1], M. Yu. Ivanov[1], D. M. Villeneuve[1] and P. B. Corkum[1]

[1] National Research Council of Canada, 100 Sussex Dr., Ottawa, Ontario, Canada K1A0R6
E-mail: Paul.Corkum@nrc.ca

[2] Universite de Sherbrooke Sherbrooke PQ. Canada

Abstract. In order to achieve better time resolution in pump-probe experiments, it is generally assumed that one must reduce the time duration of two laser pulses. This assumption is not correct. We show that sub-femtosecond time resolution can be achieved even using 50 fsec laser pulses. The key idea is the correlated formation of electron and vibrational wave packet by strong laser field. Here the electron probes the motion of vibrational wave packet. We demonstrate this technique by measuring the vibrational wave packet motion of D_2^+, on the time scale of 1.5 – 4.5 fs.

1. Introduction

Upon strong field ionization, the electron and vibrational wavepacket are produced simultaneously. The vibrational wave packet is produced mostly the D_2^+ X $^2\Sigma_g$ surface and propagate on it. This is the pump stage in a sense of preparing and staring the wave packet.

The ionized electron wave packet is going away from the core in laser field, but when the electric field changes its sign, it returns to the core resulting in collision with it. This process is referred to re-collision, and we use it as a probe source.

The kinetic energy of the returning electron depends on the strength of laser field and its wavelength. At 800 nm, it is around 30eV, and extending to a few hundred electron volts if using the infrared light. The current density of electron seen from the ion core and its time structure has been characterized elsewhere [1].

The vibrational wave packet moves on the D_2 ions surface until the electron comes back. When the electron re-collides with the ion core, the D_2^+ ion is promoted onto the excited states or other higher states, leading to dissociation.

The kinetic energy of fragment D^+ depends on the internuclear separation of the parent molecule at the time of re-collision. Measuring the kinetic energy measures the wave packet position.

The time between ionization and when the electron returns to the ion core depends on the laser wavelength. The highest probability of recollision occurs at ~ 2/3 of laser period. For 800 nm it is ~1.7fs after its birth. Increasing laser wavelength delays the re-collision time. That is, changing the laser wavelength allows us to change the time of electron's birth and its re-collision. This is like changing the pump-probe delay time. Therefore, by measuring the kinetic energy of D^+ as a function of laser wavelength, we can follow the motion of the vibrational wave packet on X $^2\Sigma_g$ surface of D_2^+.

2. Experimental

The kinetic energy spectrum of D^+ is measured with the time-of-flight (TOF) mass spectrometry. We use 50 fs duration laser pulses. We shift the 800 nm output from a Ti: sapphire regenerative amplifier to longer by optical parametric amplifier. The polarization of the laser is perpendicular to the TOF axis. The small hole in the electrode selects only the fragment dissociating from the parent ion aligning parallel to the TOF axis, in other words, perpendicular to laser polarization. That configuration ensures that the vibrational wave packet moves on the field free potential energy surface of D_2^+, without any laser induced coupling between other excited states.

To confirm that the fragments are due to electron re-collision, we use linear and elliptically polarized light. Elliptically polarized light pushes the electron away from the core, resulting in no-re-collision. The difference of kinetic energy spectra at between linear and elliptical polarization is due to re-collision.

3. Results and discussion

We measure the kinetic energy spectra of D^+ as a function of laser wavelength, in other words, re-collision time. We use linear and elliptically polarized light, and take the difference between them in order to obtain the kinetic spectra only associated with re-collision.

As increasing laser wavelength we found that the peaks of the spectra shift towards lower kinetic energy. This agrees with the general trend that the kinetic energy of fragment should decrease as the vibrational wave packet moves on X $^2\Sigma_g$ surface, because all excited states of D_2^+ are repulsive.

Among all excited states, we select the excitation to $A^2\Sigma_u$ as a reference state. The reason is that it has the largest contribution on the kinetic energy spectra of D^+ because of its largest inelastic cross section from the ground state of D_2^+. This allows us to clear identification of the observed peaks in kinetic energy spectra.

We reflect the kinetic energy spectra at each wavelength onto the Σ_u surface, and back to the Σ_g surface. We take the peak of the reflected wave packet on Σ_g as a internuclear separation at a time of the electron re-collision. Fig. 1 plots the data points of internuclear separation as a function of the re-collision time (squares). We represent the re-collision time as the time when the recollision current density of electron is at its maximum [1]. As a reference we have also calculated the vibrational wave packet motion by solving time dependent Shroendinger equation. The straight line in Fig. 1 is the calculated one. Only the data point at 4.4 fs is not so good agreement. The raw data at this wavelength is noisy. It is clear from Fig. 1 that our time resolution is sub-femtosecond.

4. Conclusion

Recollision electrons are energetic, sub-Angstrom wavelength probes, slaved to the laser field. They can be used almost as easily as photons. We illustrate their potential for measuring molecular structure and dynamics by resolving D_2^+ vibrational wave packet motion with sub-femtosecond and sub-Angstrom precision.

References

1 H. Niikura, F. Légaré, R. Hasbani, A. D. Bandrauk, M. Yu. Ivanov, D. M. Villeneuve and P. B. Corkum, *Nature*, to be published.

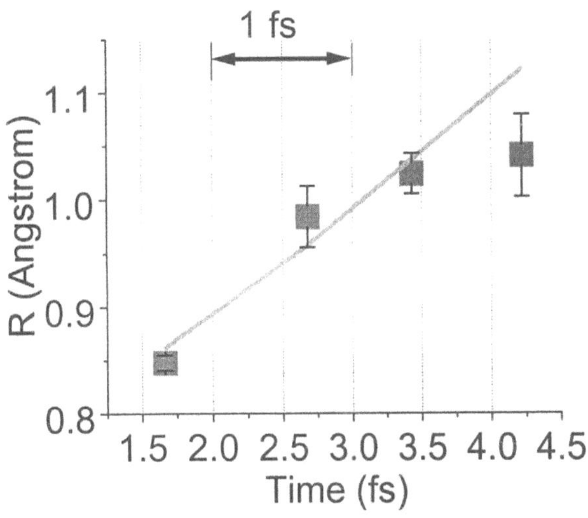

Fig. 1. The vibrational wave packet motion of D_2^+. Squares are obtained from experimental results. We calculate the vibrational wave packet motion quantum mechanically (straight line).

Evolution of pulse shapes during compressor scans in a CPA system and control of electron acceleration in plasmas

Csaba Tóth, Joeri de Groot, Jeroen van Tilborg, Cameron G. R. Geddes, Jerome Faure, Palma Catravas, Carl Schroeder, B. A. Shadwick, Eric Esarey, and Wim Leemans

L'OASIS Group, Accelerator and Fusion Research Division Lawrence Berkeley National Laboratory, MS 71-259, 1 Cyclotron Rd., Berkeley, CA 94720, USA
E-mail: ctoth@lbl.gov

Abstract. The skewness of the envelope function of 50-100 femtosecond Ti:sapphire laser pulses has been controlled by appropriate choice of the higher order spectral phase coefficients, and used for optimization of a plasma wake-field electron accelerator.

1. Introduction

The delicate details of the envelope function of ultrashort light pulses ('pulse shape') play an important role in most nonlinear processes. Two pulses with the same full width half maximum (FWHM), but having different rising edge and tail characteristics or amplitude modulation, could lead to strongly different results, especially in cases where certain cumulative processes are involved in the first part of the interactions. For example, in laser wake-field acceleration [1], a driving pulse with a sharp leading edge will result in faster growth of the plasma wake-field, and consequently, higher energy electrons after acceleration.

There are several means of active control of the shape of ultrashort pulses, such as frequency-domain filtering and envelope transfer to the time-domain by using liquid crystal phase modulators [2] or acousto-optic phase and amplitude control [3]. In this report we investigate the intrinsic pulse shaping behavior of the most widely used optical pulse compressor, the grating pair – with special emphasis on the pulse skewness and its effect on a strongly nonlinear process, the laser wake-field generation.

2. Envelopes, chirp and skew

The properties of an ultrashort pulse depend on the higher order phase characteristics. In addition to the simplest case of linear 'chirp' (when the phase of an optical pulse changes quadratically as a function of time or frequency), the presence of higher order components (such as 'cubic' third order, or 'quartic' fourth order) will modify the shape of the pulse, resulting in modulation, and even pre- or post pulses [4,5]. In every chirped pulse amplification (CPA) system, the final pulse-forming device is the grating pulse compressor. In contrast to the most simplistic view and practice, by scanning the grating separation the experimenter changes not only the linear chirp, but also modifies all the higher order terms [6]. This effect becomes negligible far from the shortest compression settings, i.e., where the pulse duration is much longer than the shortest possible. However, if the pulse duration is in the range of a few times the shortest one, or the compressor angle is set slightly differently from the optimum, or the pulse contains uncompensated higher order terms, then a typical scan around the shortest pulse will produce significant changes to the pulse shape. For example, when compressing stretched pulses with a positive bias third order phase component, the pulse shape in the course of a compressor separation scan initially will be skewed toward the head of the pulse (s<1), then flips to become a skewed pulse at the tail (s>1), and again flips back to the original asymmetric one (s<1). The skewness parameter (s) is conveniently defined here as the ratio of the 'head-width-half-max' (HWHM) and the 'tail-width-half-max' (TWHM), respectively. We note here, that these changes are fundamentally different from the well known and widely used mapping of the spectral amplitude to the time domain at strongly chirped pulses, where only the dominant, second order phase coefficient plays any role.

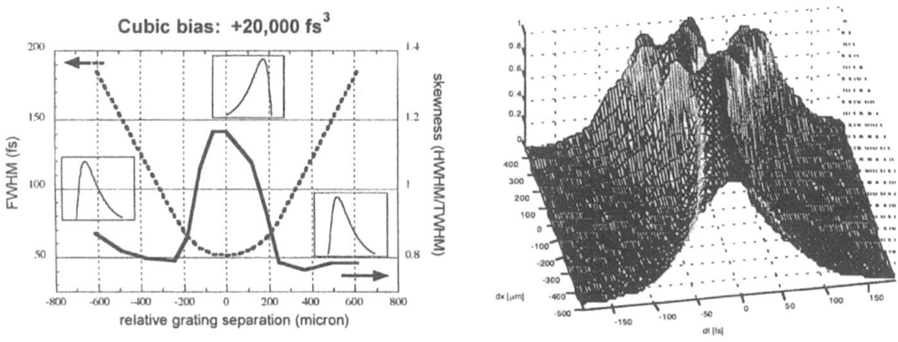

(a) (b)

Fig. 1. The variation of skewness and pulse width (a), and global shape of the envelope (b), in the function of grating separation.

The described qualitative behavior can be clearly seen in Figure 1(a), which shows the pulse shape dependence of a ~19 nm bandwidth pulse during an experimental compressor scan around the shortest pulse. On the small inserts time flows from left to right. The grating parameters are: groove density: 1480 1/mm; angle of incidence: 50.0 deg; grating separation at the shortest pulse: 323 mm; l=795 nm; third order bias: +20,000 fs^3. In addition to the conventional autocorrelation measurement – which usually washes out the details of the pulse shape – we measured the real pulse envelope by using a PG-FROG device [8], showing the same flip in characteristics of pulse skewness around the shortest pulse. A more detailed simulation of the pulse shape evolution using the experimentally retrieved parameters of the pulse can be seen on Figure 1(b). On the time-axis time flows from the right to the left, and the front region of the plot represents the positive chirp cases, while the back corresponds to negatively chirped pulses.

3. Option for control of wakefield accelerators

In laser plasma wake-field electron acceleration experiments in the L'OASIS Lab of LBNL we observed strong asymmetry in the yield of neutrons produced by accelerated electrons of multiple-tens of MeV as a function of compressor scans [8,9]. There is a strong indication that the interplay of pulse shapes and sign of chirp leads to enhanced growth of the plasma wake in the standard wake-field accelerator regime at early parts of the interaction. This process eventually results in optimal acceleration conditions different from the shortest, highest intensity pulse, favoring pulse shapes with sharp leading edges. The detailed knowledge of the evolution of the pulse shapes provides us an additional path to control the parameters of plasma wake-field accelerators even in a conventional grating compressor scan by intentionally biasing the higher order phase components. The presented method of higher order phase control is a simple alternative to the recently developed acousto-optic phase modulation devices such as the DAZZLER [3], and can conveniently be used at arbitrarily high power and energy levels limited only by the damage threshold of the compressor gratings.

Acknowledgements. This work has been supported by the DOE contract DE-AC-03-76SF0098. C. Geddes is also supported by the Hertz Foundation.

References

1 W. P. Leemans, C. W. Siders, E. Esarey, N. E. Andreev, G. Shvets, and W. B. Mori, IEEE Transactions on Plasma Science, **24**, 331-342, 1996.

2 A. M. Weiner, D. E. Leaird, J. S. Patel, and J. R. Wullert, IEEE Journal of Quantum Electronics, **28**, 908-920, 1992.

3 F. Verluise, V. Laude, Z. Cheng, C. Spielmann, and P. Tournois, Optics Letters, **25**, 575-577, 2000.

4 D. N. Fittinghoff, B. C. Walker, J. A. Squier, Cs. Tóth, C. Rose-Petruck, and C. P. J. Barty, IEEE J. Sel. Top. Quant. Elec. **JSTQE-4**, 430-440, 1998.

5 S. Backus, C. G. Durfee, III, M. M. Murnane, and H. C. Kapteyn, Rev. Sci. Inst. **69**, 1207-1223, 1998.

6 Cs. Tóth, D. N. Fittinghoff, B. C. Walker, J. A. Squier, C. P. J. Barty, in *Ultrafast Phenomena XI*, Edited by T. Elsaesser, J. G. Fujimoto, D. A. Wiersma, W. Zinth (Springer, Berlin, 1998) pp. 109-111.

7 D. J. Kane and R. Trebino, Optics Letters, **18**, 823-825, 1993.

8 W. P. Leemans, D. Rodgers, P. E. Catravas, C. G. R. Geddes, G. Fubiani, E. Esarey, B. A. Shadwick, R. Donahue, and A. Smith, Physics of Plasmas, **8**, 2510-2516, 2001.

9 W. P. Leemans, P. Catravas, E. Esarey, C.G.R. Geddes, C. Tóth, R. Trines, C. B. Schroeder, B. A. Shadwick, J. van Tilborg, and J. Faure, submitted to Phys. Rev. Lett., 2002.

Part II

Short Wavelength Science

Femtosecond X-ray Diffraction Measurement of a Solid-Solid Phase Transition in VO_2

A.Cavalleri[1], Cs. Tóth[1], C.W. Siders[2], J.A. Squier[3], F. Ráksi[4], P. Forget[5], J.C. Kieffer[5]

[1]Lawrence Berkeley National Laboratory MS 2-300, 1 Cyclotron Rd. Berkeley, CA 94720 Tel: 510 495 2536; Fax: 510 486 5530; email:acavalleri@lbl.gov

[2]School of Optics, CREOL. University of Central Florida

[3]University of California San Diego, La Jolla, California 92093-0339, U.S.A.

[4]Opt-X, Inc. Lake Forest, CA 92630

[5]Université du Québec, INRS énergie et matériauxt, Varennes, Québec

Abstract: Femtosecond x-rays were for the first time used to probe a photo-induced solid-solid phase transition in VO_2. The fast timescale observed suggests that, in this regime, the structural distortion may not be thermally initiated.

Vanadium Dioxide (VO_2) exhibits both an insulator-to-metal and a solid-solid phase transition upon heating (T_c=340 K)[1]. The intriguing role of correlations between structural and electronic degrees of freedom[2] in transition metal oxides[3] makes this problem amenable to combined ultrafast optical and x-ray studies[4].
Firstly, optical measurements were performed using time-resolved microscopy.

The reflectivity of the pumped-probed surface was recorded for different time delays with spatial resolution and thus with sensitivity to the local pump fluence (figure

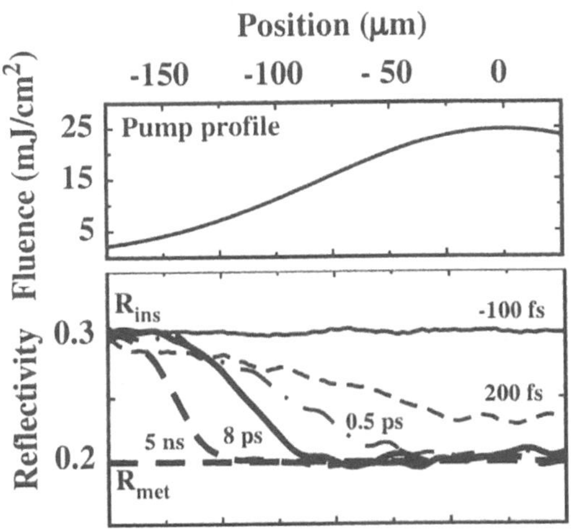

Figure 1. Space and time dependent optical reflectivity, measured using time resolved microscopy. Reflectivity evolutions sampled at three different positions of the photo-pumped spot are shown in the respective insets.

1). For excitation at 15-25 mJ/cm^2, the reflectivity was observed to reach the equilibrium value of the metallic phase within hundreds of femtoseconds, indicative of a sub-picosecond insulator-to-metal transition. The edges of the photo-excited area (7-15 mJ/cm^2), were on the contrary observed to follow a slower pathway toward the same reflectivity of the high temperature, metallic phase.

Secondly, in a parallel optical pump, x-ray probe experiment, ultrafast structural probing was achieved by using bursts of 8-keV, Cu-K_a radiation from a femtosecond plasma source[5]. The measured diffraction profiles were, at negative time delays, identical to those measured from the unperturbed sample, evidencing no significant cumulative or pre-pulse effects.At positive time delays, a shoulder originating from the new crystallographic phase appeared at about 0.1° degrees from the center of the unperturbed curve (figure 2a). Because the high T phase was initially formed over a small fraction of the depth probed by the x-rays (≈3 mm), the peak of the shoulder was only a few percents of the main lines. Figure 2b displays normalized diffraction curves at early times, obtained by dividing the measured time resolved signals by those from the unperturbed sample.

Thus, optical and x-ray data point toward approximately simultaneous electronic and structural transitions, occurring within 500 fs or less. This timescale is comparable to that for energy transfer from the photo-excited electrons to the

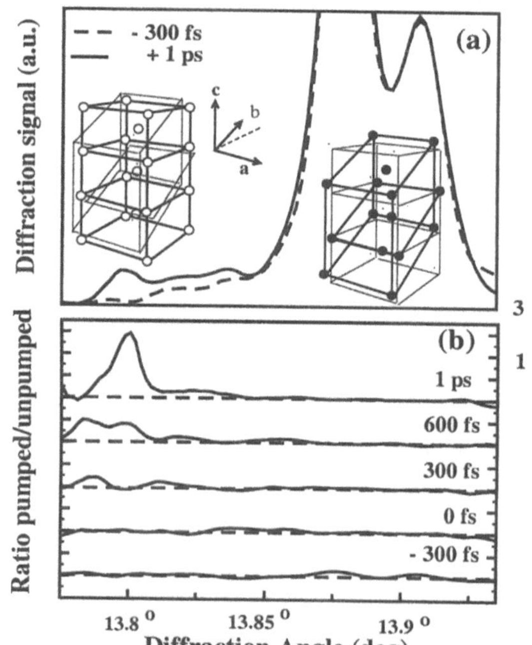

Figure 2: Angle and time dependent diffraction of a bulk VO_2 sample, measured in an optical pump, x-ray probe configuration, with excitation limited to where no cumulative damage was observed (15 mJ/cm^2).

lattice and it is shorter than the typical internal thermalization time of a non-equilibrium phonon distribution. Therefore, the transition may occur along a non-equilibrium pathway, although it is still not clear what microscopic mechanism originates it (e.g. Raman excitation along the ground state potential energy surface or re-shaping of the potential energy surface in the excited states).

In summary, we have reported on the first conjunct optical and x-ray measurements of an optically driven solid-solid and insulator-to-metal phase transition in the correlated oxide VO_2. We find direct evidence for a sub-picosecond structural transition from the low-T monoclinic phase to the high-T rutile phase. Our results challenge the thermal model for a photo-induced solid-solid transition in this excitation regime. As opposed to previous studies of non-thermal order-disorder phase transitions[6], this is the first experiment where the product state of a transformation is structurally identified. Based on our experimental results, however, we also cannot conclusively establish whether the metallic phase appears primarily due to the structural distortion or to destruction of correlations giving rise to a Mott-Hubbard bandgap.

---

[1] F. Morin *Phys. Rev. Lett.* **3**, 34 (1959).

[2] A. Zylberstein, N. Mott *Phys. Rev. B* **11**, 4383 (1975); R.M. Wentzcovic, et al. *Phys. Rev. Lett.* **72**, 3389 (1994).

[3] M. Imada, A. Fujimori, Y. Tokura *Rev. Mod. Phys.* **70**, 1034 (1998).

[4] A. Cavalleri et al. *Phys. Rev. Lett.* **87**, 237401 (2001).

[5] C. Rischel, *et al. Nature* **390**, 490 (1997); C. Rose Petruck et al. *et al. Nature* **398**, 310 (1999).

[6] C.W. Siders *et al. Science* **286**, 1340 (1999); A. Chin et al. *Phys. Rev. Lett.* **83**, 336 (1999); K. Sokolowski-Tinten et al. *Phys. Rev. Lett.* **87** 225701 (2001); A. Rousse et al. *Nature* **410**, 65 (2001).

Time-resolved X-ray diffraction study of ultrafast structural dynamics in laser-excited solids

K. Sokolowski-Tinten[1], C. Blome[1], C. Dietrich[1], A. Tarasevitch[1], M. Horn-von-Hoegen[1], D. von der Linde[1], A. Cavalleri[2], J.A. Squier[3], and M. Kammler[4]

[1] Institute for Laser and Plasma Physics, University of Essen, 45117 Essen, Germany, E-mail: kst@ilp.physik.uni-essen.de
[2] Materials Science Division, Lawrence Berkeley Nat.l Lab., Berkeley, CA 94720, USA
[3] Dep. of Chem. and Biochem., Univ. of California at San Diego, La Jolla, CA 92093, USA
[4] Insitute for Solid State Physics, University of Hannover, 30167 Hannover, Germany

Abstract. Time-resolved X-ray diffraction using sub-300 fs, multi-keV x-ray pulses from a laser-produced plasma has been used to study ultrafast melting of Germanium.

1. Introduction and Experimental Methods

Ultrashort x-ray pulses offer a unique combination of *atomic-scale* spatial and temporal resolution, which permits direct measurements of structural transients on a femtosecond time scale. Using femtosecond, multi-keV x-ray pulses from a laser-produced plasma we have performed time-resolved diffraction experiments during ultrafast phase transitions in optically excited germanium.

Short bursts of Ti-K_α-radiation at 4.51 keV were produced by focusing laser pulses of about 100mJ in energy and a pulse duration of 120 fs onto the surface of a moving titanium wire. Part of the emitted radiation was collected by a toroidally bent Si (311)-crystal to form a monochromatic, 1:1 image of the plasma that was generated on the surface of the Ti-wire. Single-crystalline thin films of germanium, grown by surfactant-mediated heteroepitaxy on crystalline silicon substrates, were placed in the image plane under the appropriate Bragg-angle, and the transient changes of the x-ray diffraction after optical excitation were measured.

2. Non-Thermal Melting of Germanium

In covalent semiconductors there is extensive evidence from time-resolved optical spectroscopy [1-3] that a solid-to-liquid phase transition occurs *non-thermally* in a few hundred femtoseconds, as a result of strong optical excitation. This process represents an ideal case to be studied by time-resolved x-ray diffraction [4-6]. The data displayed in the left viewgraph of Fig. 1, obtained on a 170 nm thick, (111)-oriented Ge-film, show the angular integrated diffraction efficiency as a function of pump-probe time delay for two different excitation fluences.

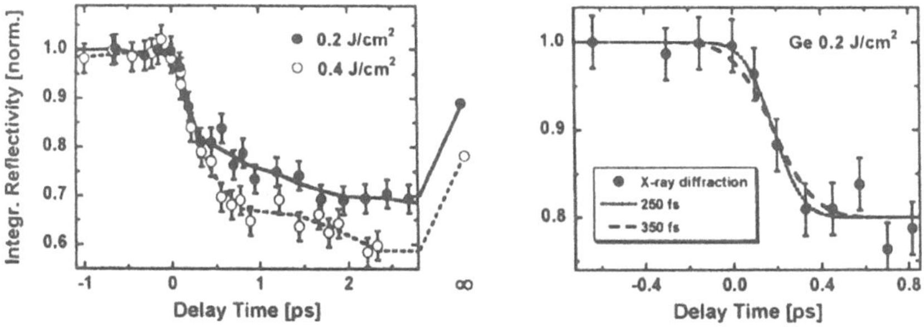

Fig. 1. Left: angular integrated X-ray reflectivity of a 170 nm Ge-film (111-reflection) as a function of time delay for two different pump fluences. Right: fits of the diffraction data (0.2 J/cm^2) assuming a step-like material response and Gaussian-shaped X-ray pulses.

For both fluences the diffraction efficiency exhibits an fast drop within a few hundred femtoseconds, followed by a more gradual decrease on a picosecond time-scale. Analysis of the data shows that a loss of order over a depth of 40 nm has occurred in less than 300 fs, a clear indication of *non-thermal* melting. Explanations of the observed decrease of diffraction as being due to plasma formation and/or ablation can be ruled out, mainly because of the nearly complete recovery of the diffraction efficiency after long times (data points "∞" in Fig. 1).

These results permit also a reliable estimate of the duration of the Ti-K_α-emission. As is demonstrated by the right part of Fig. 1, deconvolution of the data assuming an *instantaneous,* step-like material-response (which *overestimates* the X-ray pulse duration) leads to an x-ray pulse width of about (300 ± 50) fs.

3. Transient Thermo-Acoustic Effects

Additional insight into the physical processes following ultrafast melting can be obtained from an analysis of the angular dependence of diffraction, the so-called *rocking curves*. Examples for different delay times measured at a pump fluence of 0.2 J/cm^2 are displayed in Figure 2. The solid and dashed curves represent the transient rocking curves and the rocking curves of the unexcited material, respectively.

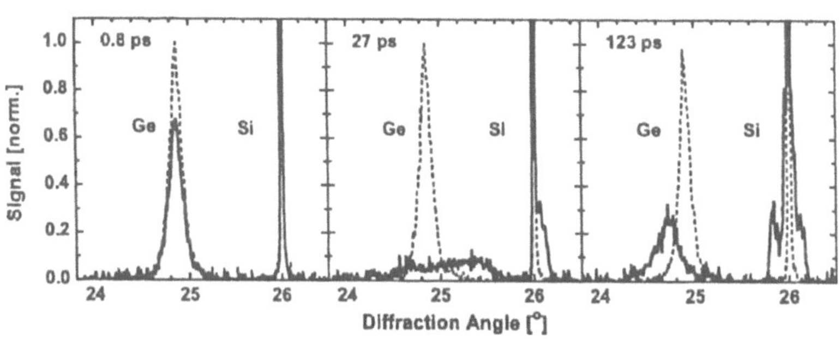

Fig. 2: Rocking curves of the laser-excited Ge/Si-herterostructure for different delay times. Pump fluence: 0.2 J/cm2, solid: transient, dashed: reference of the unpumped sample.

The x-rays, incident from the crystal mirror onto the sample, cover an angular range of about 2.5 degrees. Therefore, we are able to simultaneously measure diffraction from the Ge overlayer as well as of the Si substrate. At early time delays (0.8 ps) only a drop of the diffraction efficiency of the Ge line without any change in shape is observed, because this time-scale is too short for any volume changes to occur.

At later time delays the non-thermally, impulsively generated hot and pressurized liquid layer expands, leading to a compression of the non-molten material (Ge and Si) underneath. Such compressive strain is indicated by the appearance of diffraction contributions at larger angles with respect to the original Bragg-angles (Ge: 24.88^0; Si: 26^0). Measured maximum compressive strains of about 1.5-2 % in Ge and 0.5 % in Si correspond to transient pressures of about 2-3 GPa in Ge and 1 GPa in Si, respectively. Simple acoustic considerations allow then to roughly estimate the maximum pressure in the liquid layer at the surface to about 5-10 GPa.

At even later times (123 ps) the acoustic perturbations of the Ge layer have damped out, leaving behind the relaxed, thermally expanded material, as indicated by a shift of the Ge line to smaller angles. The measured expansive strain of 0.8-1% corresponds to a temperature of approximately 1.200 K, close to the melting temperature of Ge, suggesting that liquid and solid have come to an equilibrium near the melting point.

References

1 C.V. Shank, R. Yen, and C. Hirlimann, Phys. Rev. Lett. **50**, 454 (1983); Phys. Rev. Lett. **51**, 900 (1983)

2 P. Saeta, J.K. Wang, Y.N. Siegal, N. Bloembergen, and E. Mazur, Phys. Rev. Lett. **67**, 1023 (1991).

3 K. Sokolowski-Tinten, J. Bialkowski, and D. von der Linde, Phys. Rev. B **51**, 14186 (1995).

4 L. Huang, J. P. Callan, E. N. Glezer, and E. Mazur, Phys. Rev. Lett. **80**, 185 (1998).

5 C.W. Siders, A. Cavalleri, K. Sokolowski-Tinten, Cs. Tóth, T. Guo, M. Kammler, M. Horn-von-Hoegen, K.R. Wilson, D. von der Linde, and C.P.J. Barty, Science **286**, 1340 (1999).

6 A. Rousse, C. Rischel, S. Fourmaux, I. Uschmann, S. Sebban,.G. Grillon, Ph. Balcou, E. Förster, J.P. Geindre, P. Audebert, J.C. Gauthier, and D. Hulin, Nature **410**, 65 (2001).

7 K. Sokolowski-Tinten, C. Blome, C. Dietrich, A. Tarasevitch, M. Horn-von-Hoegen, D. von der Linde, A. Cavalleri, and M. Kammler, Phys. Rev. Lett. **87**, 225701 (2001).

Properties of liquid silicon and carbon studied by ultrafast time-resolved x-ray absorption spectroscopy

S.L. Johnson,[1] P.A. Heimann,[2] A.M. Lindenberg,[1] A.G. MacPhee,[1] R.W. Falcone,[1] H.O. Jeschke,[3] M. Garcia,[3] J.J. Rehr,[4] R.W. Lee[5] and Z. Chang[6]

[1]Department of Physics,University of California at Berkeley, Berkeley, CA 94720, USA
[2]Advanced Light Source, Lawrence Berkeley National Laboratory, 1 Cyclotron Rd., Berkeley, CA 94720, USA
PAHeimann@lbl.gov
[3]Institute for Theoretical Physics, Free Universitat Berlin, Arnimallee 14, 14195 Berlin, Germany
[4]Department of Physics, University of Washington, Seattle, WA 98195, USA
[5]Lawrence Livermore National Laboratory, Livermore, CA 94550, USA
[6]Department of Physics, Kansas State University, Manhattan, KS 66506, USA

Abstract. Ultrafast time-resolved x-ray absorption spectroscopy has been developed based on synchrotron radiation and a dispersive spectrometer. Liquid silicon and carbon have been studied.

Time-resolved x-ray absorption spectroscopy provides information about the structure and electronic structure of a transient state induced by a laser pulse. The x-ray absorption near edge structure (XANES) is especially sensitive to the electronic structure of the material. The extended x-ray absorption fine structure (EXAFS) is an accurate probe of bond distances and numbers of nearest neighbors. Picosecond time resolution allows the spectroscopic measurement of a material at high temperature and near solid density before it undergoes significant expansion.

Bend magnet beamline 5.3.1 at the Advanced Light Source (ALS) provides a broad spectrum of x-rays from 100 eV to 10 keV. The x-rays are produced at high repetition rate with about 70 ps pulse duration. Excited by a femtosecond laser pulse, a sample foil is probed in transmission by the focused broad bandwidth x-rays. Between laser pulses the sample is translated to a new position. A dispersive spectrometer measures an entire x-ray absorption spectrum simultaneously. The grating spectrograph, built by Hettrick Scientific, Inc., employs a varied line spacing grating and two focusing mirrors to provide an erect image of the x-ray spectrum. High photon energies are suppressed with a pair of plane mirrors having an appropriate coating. Two detectors are used: a streak camera and microchannel plates (MCP). The time resolution using the two

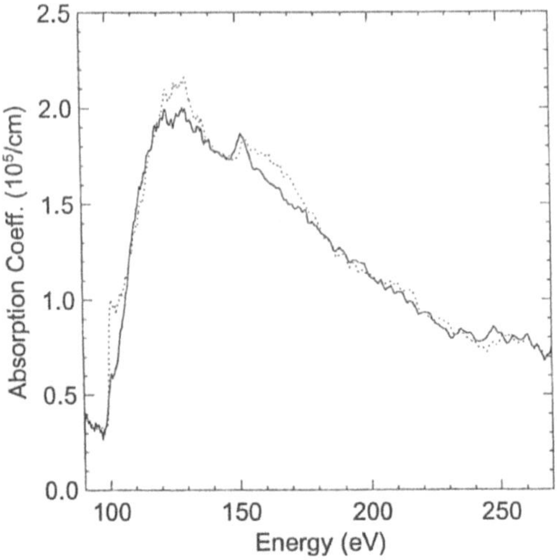

Fig. 1. Measured L-edge absorption spectrum of silicon. Dotted line: unheated spectrum. Solid line: heated spectrum 100 ps after laser pulse.

detectors is ~ 2 ps with the streak camera [1] and the x-ray pulse duration,~ 70 ps, with the MCP detector.

Solid covalently bonded semiconductors are known from molecular dynamics calculations to form liquids, which have both metallic and covalent bonding [3]. However, there has been little experimental confirmation of this description of the structure and bonding of melted semiconductors. With a femtosecond laser pulse fluence of 0.7 J/cm^2, silicon foils, 50 to 100 nm thick, are brought well into its liquid phase. In figure 1, the dashed curve shows the absorption spectrum before excitation, and the solid line shows the spectrum 100 ps after laser excitation. The $L_{II,III}$ edge is located at 100 eV while at 150 eV the smaller L_I edge appears. The melting induced by the laser causes a dramatic reduction at the $L_{II,III}$ edge and an energy shift in p-d absorption maximum. In addition, the L_I edge is shifted to lower energies by 1.6 $\pm$ 0.2 eV and the amplitude of the EXAFS oscillations is greatly reduced. With the streak camera detector, the changes in the $L_{II,III}$ edge are observed to be instantaneous within the temporal resolution of 5 ps. Thereafter the absorption spectrum is unchanged for about 10 ns.

Molecular dynamics simulations have been used to determine atomic positions in liquid silicon. The simulated atoms were impulsively heated beyond the melting temperature and permitted to equilibrate for several picoseconds. The L-edge x-ray absorption was calculated with the ab-initio x-ray scattering code FEFF 8.1 [2]. These calculations show good agreement with the experimentally measured x-ray absorption spectrum of liquid silicon. From the model calculations, it can be understood that the lower absorption at the $L_{II,III}$ edge reflects a lower unoccupied s density of

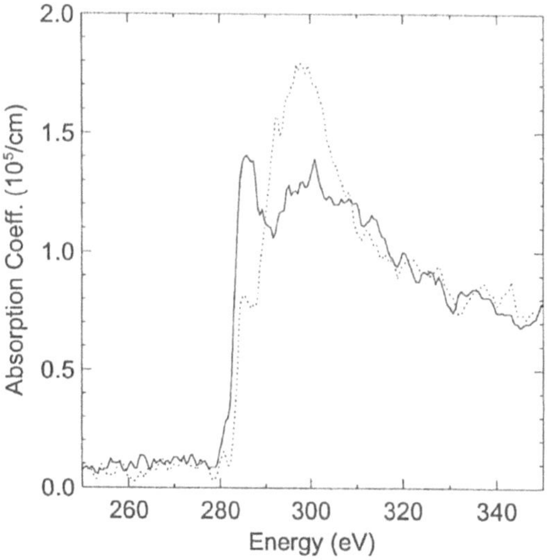

Fig. 2. K-edge absorption spectrum of carbon. Dotted line: unheated spectrum. Solid line: heated spectrum 100 ps after laser pulse.

states resulting from the collapse of the semiconductor band gap. The shift in the p-d absorption maximum results from an increase in the nearest neighbor distance of 0.14 ± 0.08 A. Finally, the energy shift of the L_I edge is a consequence of both the metallic nature and density of liquid silicon.

The properties of liquid carbon are interesting for astrophysical reasons because carbon is an important constituent of some planets. Also, liquid carbon is difficult to study expect by time-resolved methods because of its high melting temperature, ~ 3800 K. Measurements on foils of amorphous carbon, summarized in figure 2, show large differences between the unheated and heated spectra at the carbon K-edge. After 100 ps, the laser-induced melting causes an increase in the size of the π^* resonance at 285 eV accompanied by a decrease and asymmetric broadening in the σ^* resonance at 296 eV. These changes indicate a transition in bonding from the predominant sp^2 bonding of soft amorphous carbon to the sp bonded chains recently predicted by molecular dynamics simulations for the low-density liquid carbon [4].

References

1 J. Larsson, Z. Chang, E. Judd, R.W. Falcone, P.A. Heimann, H.A. Padmore, P.H. Bucksbaum, M. Murnane, H. Kapteyn, R.W. Lee and J.S. Wark, Optics Letters **22**, 1012 (1997).
2 A.L. Ankudinov, B. Ravel, J.J. Rehr and S.D. Conradson, Phys.Rev. B **58**, 7565 (1998).
3 I. Stich, R. Car and M. Parrinello, Phys. Rev. Lett. **63**, 2240 (1989).
4 J.N. Glosgli and F.H. Lee, Phys. Rev. Lett. **82**, 4659 (1999).

Kinetics of Cluster Formation During Femtosecond Laser Ablation

T.E. Glover, G.D. Ackermann, A. Belkacem, P.A. Heimann, Z. Hussain , H.A. Padmore, C. Ray, R.W. Schoenlein, W.F. Steele

E. O. Lawrence Berkeley National Laboratory, Berkeley, California 94720, USA
Email : teglover@lbl.gov

Abstract. We report, to our knowledge, the first use of pump-probe photoemission spectroscopy to study core-level dynamics during a laser-driven phase transition. Synchrotron x-rays probe the kinetics of cluster formation during the initial moments of femtosecond laser ablation.

In laser ablation a material is rapidly heated to a point where the system's thermal energy exceeds its cohesive energy. Material ejection occurs leading to rapid cooling and associated synthesis of small scale-length particles. While laser ablation has been applied to a wide range of materials and is of considerable importance in the synthesis of organic and inorganic micro-particles and films, the dynamical processes by which micro-particles form within an ablation plume are complex and poorly understood. While recent optical experiments have shed some light on these processes [1], ambiguities exist regarding even basic qualitative features. Here we report the first application of xray photoemission spectroscopy for probing rapid atomic-scale dynamics during the initial moments of femtosecond laser ablation. Core-level photoemission spectroscopy (CPS) is extended to the picosecond time domain and enables us to follow rapid metal-insulator phase transitions as a highly pressurized, supercritical fluid expands into vacuum.

Laser pump and x-ray probe photoemission experiments are performed at the Advanced Light Source (ALS) synchrotron. Femtosecond laser pulses (200 fs, 800 nm, 1 KHz) are focused to high fluence (~10 J/cm^2) and strike a silicon wafer. Photoemission by time-delayed (400 eV, 80 ps) x-ray pulses allows one to follow a fixed set of near-surface atoms; initially compressed at solid density and later dispersed upon expansion into vacuum. X-ray photoemission spectra are collected using a hemispherical analyzer and the spectral window of interest is that corresponding to Si 2p core-level (~100 eV binding energy). The synchrotron is operated in 'cam-shaft' mode whereby a single electron bucket is isolated (> 20 ns) from the remaining bunches in the storage ring; electronic gating is then used to collect photoelectrons produced by synchrotron pulses temporally overlapped (or nearly overlapped) with laser pulses.

A sequence of photoemission spectra covering a range of pump-probe time delays is shown in Fig. 1. The spectra illustrate a transient shift of the Si 2p photoelectron peak to lower binding energy. The dynamics of the shift are emphasized in the inset to Figure 2. Here electrons counts (from 97-99 eV) are plotted as a function of pump-probe time delay. The curve (80 ps FWHM) is a convolution between the synchrotron pulse duration (80 ps) and the duration of the transient photoemission peak shift. The curve indicates that the duration of the photoemission peak-shift is shorter than the synchrotron pulse duration and we put an upper limit of 50 ps on the peak-shift duration.

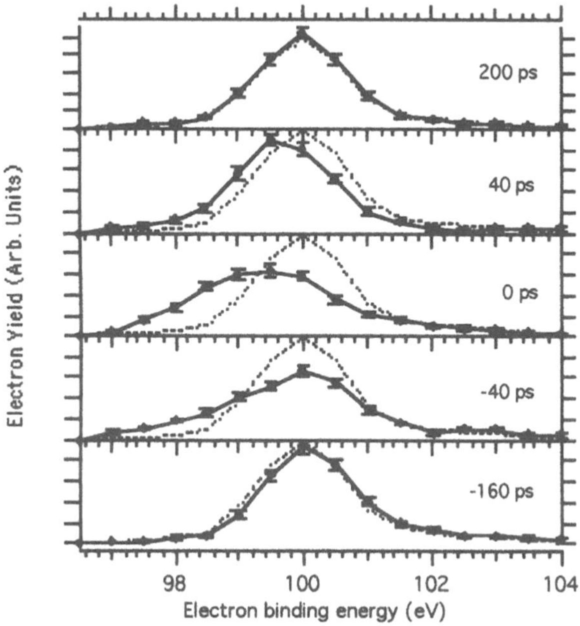

Figure 1. Silicon 2p photoelectron spectra obtained at various time delay between laser pump pulse and x-ray probe pulse. The relevant time delay is shown with negative time corresponding to laser exciting before the x-ray probe.

The maximum spectral shift is emphasized in Figure 2. Spectra corresponding to x-ray probe pulses arriving either well before or well after laser excitation are similar while the spectrum corresponding to simultaneous arrival of xray and laser pulses is lowered, broadened and shifted. The distorted spectrum has been deconvolved into two gaussians : one at the original peak location and a second whose position is adjustable. A good fit to the data is obtained for a photoemission peak shift of 1.3 eV.

The observed shift is consistent with the core-level shift expected upon a solid-to-fluid phase transition in Si. [2]. The current laser fluence is well above the melting threshold in Si. Upon melting silicon undergoes an insulator (solid) to metal (liquid) phase transition. The mobile charge associated with the metallic phase efficiently screens an x-ray core-hole thus producing a photoelectron peak shift. The recovery of the photoemission peak is consistent with rapid (< 50 ps) solidification of the ejected material. Core-level photoemission spectroscopy is sensitive to the local valence charge density and we find that for times > ~100 ps after laser excitation the photoelectron spectrum in the ejected material is similar to that of the initial, unperturbed material. We take this as evidence that solid-phase clusters with local diamond structure are obtained within 100 ps after ablation. Importantly we note that the residual bulk surface remains liquid for a time >> 100 ps so that we are not observing the residual bulk surface. Further, the transient spectra show no evidence of isolated silicon atoms since they would be shifted outside of our spectral window; instead we observe constant photoemission yield (to within 10%). This indicates that isolated atoms comprise < 10% of the ejected material. Solid phase clusters therefore dominate the ejected material soon after laser excitation.

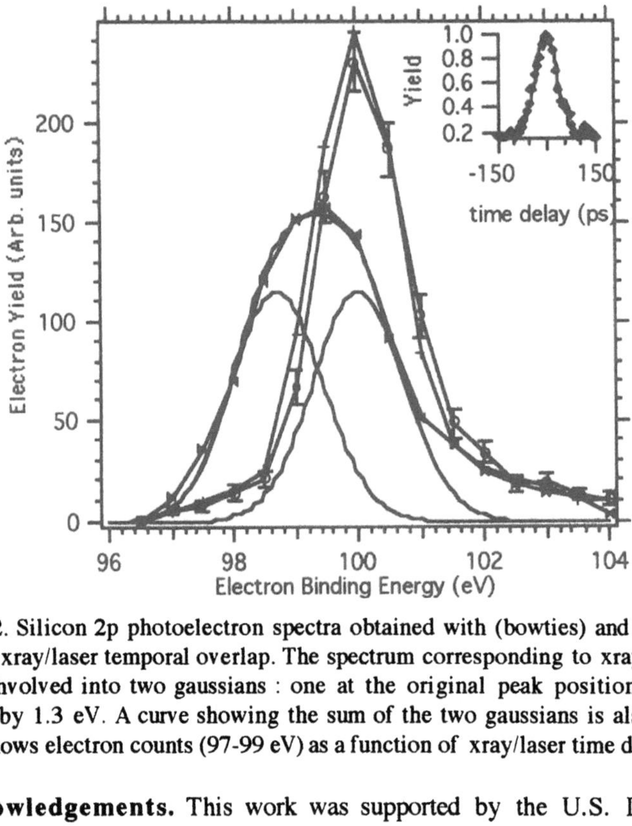

Figure 2. Silicon 2p photoelectron spectra obtained with (bowties) and without (stars, circles) xray/laser temporal overlap. The spectrum corresponding to xray/laser overlap is deconvolved into two gaussians : one at the original peak position and a second shifted by 1.3 eV. A curve showing the sum of the two gaussians is also shown. The inset shows electron counts (97-99 eV) as a function of xray/laser time delay.

Acknowledgements. This work was supported by the U.S. Department of Energy under Contract No. AC0376SF00098.

References

1. Sokolowski-Tinten, K. *et al.* Transient States of Matter during Short Pulse Laser Ablation. Phys. Rev. Lett. **81**, 224-227 (1998).

2. H.C. Gerritsen *et al.*, "A time-resolved x-ray absorption study of amorphous Si during pulsed laser irradiation," J. Appl. Phys. **60**, 1774-1781 (1986).

Femtosecond Si-Kα pulses from laser produced plasmas

Andreas Morak, Ingo Uschmann, Thomas Feurer, Eckhart Förster, and Roland Sauerbrey

Friedrich-Schiller Universität Jena, Max-Wien-Platz 1, 07743 Jena, Germany

Abstract. We have measured the strain induced by a fs laser in an InSb crystal temporally resolved using well characterized Si-Kα x-ray pulses. The generation of electron-hole-pairs is identified as the dominant strain inducing process.

In several experiments [1–4], Kα radiation was used to monitor ultrafast structural changes in matter. An open question in all of these pump-probe experiments was, however, the exact temporal duration of the probing x-rays. This ultimately determines the temporal resolution of the experiment. The determination of the temporal duration of sub-picosecond x-ray pulses is very challenging. Previously we reported on the measurement of the temporal profile of the Si-Kα line emission from a femtosecond laser produced plasma [5]. The measurement was performed by cross-correlating the Si-Kα pulse with the ultrafast response of a laser pumped MBE-grown thin CdTe crystal layer. The pump laser induces an ultrafast change of the CdTe crystal structure and causes a rapid decrease of the x-ray reflectivity. In contrast to InSb bulk measurements [3], the x-ray penetration depth in this measurement is definitely limited by the thickness of the crystalline layer. The Si-Kα pulse duration is (450 ± 150) fs.

The experimental setup used in this measurement is described in reference [6].

Using these ultrashort Kα pulses we have studied the propagation of shock waves with unprecedented temporal resolution in particular for early times ($\leq$ 5 ps), by measuring the elastic deformation of an InSb-crystal lattice by time-resolved x-ray diffraction. The impulsive excitation following laser irradiation launches a shock wave which propagates through the crystal volume. In front of the shock wave the crystal is compressed (larger Bragg-angle) and after the shock wave the crystal experiences an expansion (smaller Bragg-angle). Signatures of both regions were detected in time-resolved experiments (see figure 1). Due to the high temporal resolution of the x-ray pulse, this is the first measurement were the compression of the crystal lattice is directly observed. To get a sufficiently high signal-to-noise ratio, photon accumulation over 55 laser shots was necessary. The simulation reproduces the experiments with high accurancy.

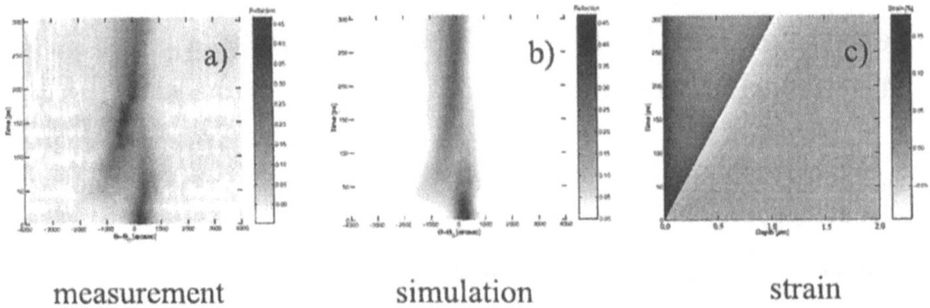

Fig. 1. Transient Bragg reflection of the laser excited InSb crystal lattice. The elastic relaxation of the lattice starts with a compression followed by an expansion. The experiment is compared with thermoelastic simulations combined with x-ray diffraction theory. As a result the time dependent strain is presented as a function of crystal depth.

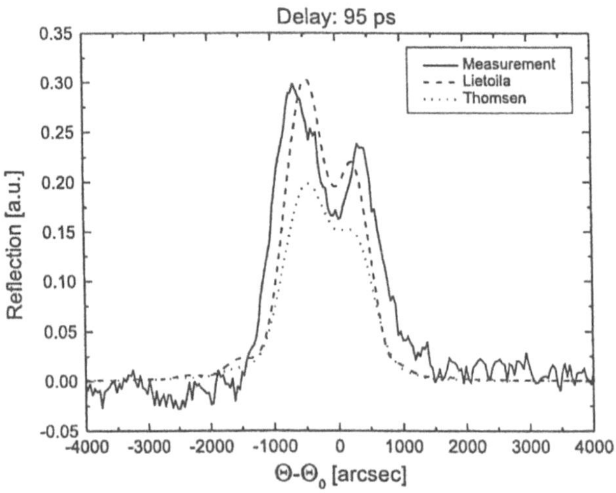

Fig. 2. Comparison of the different models and the measured transient Bragg reflection of the laser excited InSb crystal lattice.

The results are compared with two different models. The model of Thomsen et al. [7] calculates the strain in the thick InSb crystal for an instantaneous and exponentially decaying stress due to short-pulse laser absorption. This model reproduces many of the features of the data, but there is a clear discrepancy in the temporal window from 0 to 100 ps. The model by Lietoila et al. [8] considers the build up of strain due to electron-hole pair production and electron-lattice dynamics. The absorbed laser pulse energy creates a dense electron-hole plasma with large excess energy, which quickly thermalizes via carrier-carrier-scattering. The Auger recombination is the dominant carrier recombination mechanism. The carrier energy is transferred to the lattice through the emission of longitudinal-optical phonons. In comparison with the experiment it turns out, that the electron-hole contribution to the stress is the significant part, while the thermal phonons do not play an important role. In figure 2 the measured transient Bragg reflection at a pump-probe delay of 95 ps is compared with the two different models. The Lietoila model [8] reproduces the shape of the measured rocking curve significantly better than the Thomsen model.

In conclusion the femtosecond x-ray temporal resolution allows the unique identification of an electron-hole plasma as the primary source of strain in a femtosecond laser irradiated semiconductor.

References

1. C. Rose-Petruck et al.; Nature **398**, 310 (1999)
2. C.W. Siders et al.; Science **286**, 1340 (1999)
3. A. Rousse et al.; Nature **410**, 65 (2001)
4. D.A. Reis et al.; PRL **86**, 3072 (2001)
5. T. Feurer et al.; Phys. Rev. E **65**, 16412 (2001)
6. T. Feurer et al.; Appl. Phys. B **72**, 15 (2000)
7. C. Thomsen et al.; Phys. Rev. B **34**, 4129 (1986)
8. A. Lietoila et al.; Appl. Phys. Lett. **40**, 624 (1982)

Ultrafast x-ray radiation from intense femtosecond laser systems

K. Ta Phuoc[1], L. Notebaert[1], M. Pittman[1], J.P. Rousseau[1], V. Malka[1], S. Sebban[1], R. Marquès[2], P. Balcou[1], S. Fritzler[1], P.G. David[2], D. Hulin[1], and A. Rousse[1]

[1] Laboratoire d'Optique Appliquée, ENSTA, CNRS UMR7639, Ecole Polytechnique, Chemin de la Hunière,91761 Palaiseau, France.
E-mail: rousse@enstay.ensta.fr

[2] Laboratoire pour l'Utilisation des Lasers Intenses, UMR7605, CNRS, Ecole Polytechnique, CEA, Universitée Paris VI, Ecole Polytechnique 91128 Palaiseau, France.

Abstract. Polychromatic X-ray radiation have been produced during the relativistic interaction of a 50 TW femtosecond laser with an Helium gas jet. All the measurements indicate that the most plausible process at the origin of the observed X-ray emission is the Larmor radiation. Such femtosecond laser-based Larmor radiation should mark the emergence of a new generation of ultrafast x-ray bursts produced from the relativistic motion of particles, with the challenge to extend x-ray absorption probing techniques (EXAFS) on femtosecond time-scales.

1. Introduction

Electrons in the electromagnetic field experience strong accelerations and produce elementary Larmor-type radiation [1]. The emission generated by this interaction between electrons and photons is also called non linear Thomson scattering and was observed few years ago by the detection of the second and third harmonic light of the laser [2]. The x-ray spectral range was anticipated at relativistic quiver energies provided by the powerful laser fields. This electron-photon interaction can either be described by an undulator model. In synchrotrons, highly energetic particles (few GeV) from the storage rings are launched inside periodic magnetic structures - the undulators - to produce the x-ray radiation. The laser can be seen as an electromagnetic structure with a period much shorter than that of the conventional of undulators. The use of intense femtosecond lasers offers two main advantages. First, the spectral range can be extended to the x-ray wavelengths despite the significantly lower electron energies achieved in the laser field [3]. Second, the x-ray pulse duration is expected to remain in the order of magnitude of the duration of the laser pulse (30 fs).

2. Experimental Methods

The experiment was conducted at the Laboratoire d'Optique Appliquée (LOA) where we used the 30 fs and 1 J (33 TW) Ti:Sa laser system. The schematic of the setup is displayed in Figure 1. The laser beam, 55 mm in diameter, was focused

with an f/5.45 off-axis parabolic mirror onto the front edge of a supersonic gas jet of helium or argon. The diameter of the focal spot was 6 μm at full width half maximum (FWHM) in intensity and contained 60 % of the laser energy. At such laser intensities, the helium gas was fully ionized early in the pedestal of the laser pulse. The x-ray radiation was collected with grazing incidence metallic mirrors (nickel or gold) over a solid angle of 1.5 10^{-2} sr, and was focussed onto a back illuminated CCD camera. Filters were positioned between the plasma and the detector to block the laser infrared light.
We selected spectral bands of the incoming polychromatic x-rays by combining the spectral absorption of the filters and the spectral reflectivity of the mirrors. Our x-ray spectrometer was sensitive to x-rays in a bandwidth from 20 eV to 2 keV. The angular dependence of the x-ray emission was obtained by rotating the whole spectrometer around the laser focal spot.

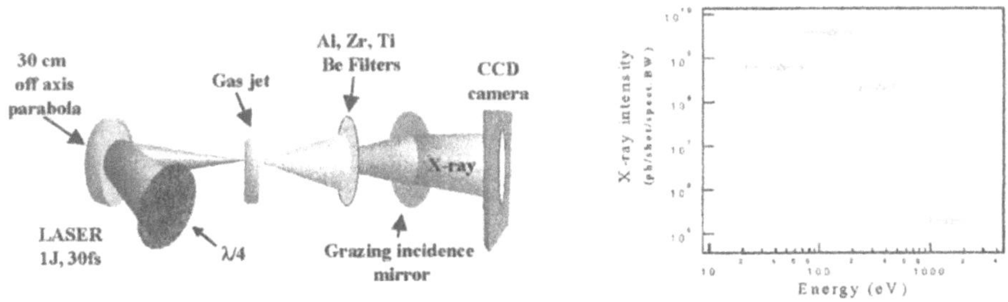

Fig. 1. Experimental setup and experimental spectrum of the observed X-ray radiation

3. Results and Discussion

The spectral distribution obtained on-axis (the direction of the laser propagation) is presented in Figure 1. The spectral bandwidths were selected with aluminum, zirconium, titanium and beryllium filters which respectively provide spectral ranges centered on 45 eV, 150 eV, 350 eV and 1500 eV. We can see that the output spectrum is peaked on the bandwidth centered at 0.15 keV with a significant tail up to 2 keV. This spectral distribution was recorded where the produced x-ray intensity was the highest.

Many radiative processes are expected during the interaction between the intense ultrafast laser and the matter. The generation of high order harmonics of the fundamental laser frequency from atomic ensembles can occur in the wings of the focal spot where the laser intensity is sufficiently weak. In case of low atomic weight gases, a fully ionized plasma is generated at the focal spot of the tightly focussed beam and Bremsstrahlung radiation can be produced following the inelastic collision of the electrons with the nucleus. Electrons oscillating inside the

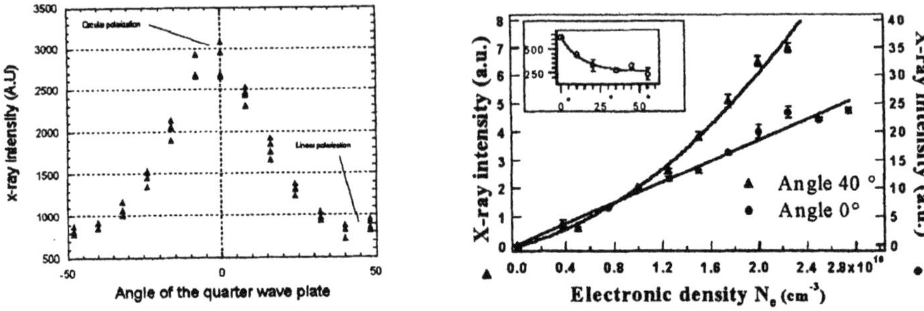

Fig. 2. Behavior of the observed x-ray signal as a function of the polarization of the laser, the electronic density of the Helium plasma and its spatial distribution compared to the axis of laser propagation (angle 0°)

laser field, trapped in the plasma wave by self modulated laser wake field acceleration (SMLWF) or accelerated directly by the ponderomotive force, and the bath of thermal electrons of the hot plasma releasing their energy a long time after the laser, could participate in the overall x-ray emission through this collisional process. We have found that the experimental features of the x-ray emission do not match the characteristics of all these radiative processes except the case of the Larmor radiation: peaked spectrum (Figure 1), collimated angular distribution (Figure 2), higher efficiency for circularly polarized laser light (Figure 2) and linear increase with the plasma electronic density (Figure 2).

4. Conclusions

We have demonstrated that the x-ray signal follows the expected features of the Larmor radiation: variation with the polarization of the laser, spectral distribution, anisotropic angular distribution, scaling with the electronic density. The Larmor radiation is the most plausible process at the origin of the observed X-ray signal.

Acknowledgements. This work was supported by the European Community under contracts HPRI-CT-1999-00086 and HPRI-CT-2000-40016.

References

1 J. D. Jackson, in *Classical Electrodynamics (3rd ed)*, edited by John Wiley and Sons, New York, 1998
2 S. Y. Chen, M. A. and D. Umstadter, Nature **396**, 653, 1998.
3 E. Esarey, S. K. Ride and P. Sprangle, Phys. Rev. E **48**, 3003, 1993.

EUV Photonics: Quasi-phasematching at short wavelengths

Randy A. Bartels[1], Ariel Paul[1], Sterling Backus[1], Raanan Tobey[1], Henry Kapteyn[1], Margaret Murnane[1]

[1] JILA, University of Colorado, 440 UCB, Boulder CO 80309-0440, USA
E-mail: bartels@colorado.edu

Abstract. We experimentally demonstrate quasi phase matched generation of EUV high harmonic radiation using an engineered hollow waveguide. A periodic structure allows for phase-matching of higher-order harmonics compared with an unstructured waveguide.

1. Introduction

High harmonic generation (HHG) is an excellent source of extreme ultraviolet (EUV) radiation that upconverts an intense laser pulse to much shorter wavelengths. Over the past decade, many experiments have studied this process, providing detailed insights into the physics of HHG. Phase matching the HHG process, either using a hollow waveguide,(*1*) or using a near-plane-wave geometry, (*2, 3*) increases the brightness of the source by orders of magnitude and generates and EUV beam with full spatial coherence.(*4*) In this work, we experimentally demonstrate that phase-matching of HHG radiation can be extended to shorter wavelengths and higher pulse energies using a structured hollow waveguide. Moreover, this photonic structure results in the generation of a broad continuum that may support an isolated 250 attosecond EUV pulse, as has been previously predicted theoretically.(*5*) This work is a compelling demonstration of the applicability of photonics technology concepts to a new region of the spectrum.

High harmonic emission results when atoms subjected to an intense short pulse of light emit EUV radiation as a result of their electric field-induced ionization. This HHG process can be understood in a "recollision" model—after ionization, the newly-freed electron begins to oscillate in the oscillating electric field of the driving laser.(*6*) If this electron re-encounters its parent ion, it can recombine and release its impact kinetic energy as a photon in the EUV region of the spectrum. In this picture, repeated ionization and recollision events result in emission as a series of attosecond-duration "bursts," separated in time by 1/2 optical cycle, or ~1.2 femtoseconds. This periodic emission results in a spectrum with a series of peaks separated in energy by twice the fundamental photon energy.

The conversion efficiency of the laser light into EUV can be increased by matching the phase velocities of the fundamental and harmonic radiation. The phase velocity of EUV radiation in a gas is higher than for the infrared fundamental, but the dispersion of a hollow-core fiber increases the phase velocity of the fundamental pulse. The pressure of the gas filling the hollow fiber can be adjusted to balance the phase velocity of the fundamental and EUV fields, dramatically increasing the conversion efficiency.(*1*) Increasing the driving pulse energy increases the level of ionization, and thus the electron density in the fiber. This also increases the phase velocity of the fundamental pulse.(*1, 2*) When a critical fraction of gas in the fiber is ionized, the phase velocity of the fundamental laser pulse is too high for phase matching to be possible - which limits the conversion efficiency. Since the level of ionization is increasing along with the laser intensity on the leading edge of the pulse, this ionization-induced phase mismatch limits the highest energy photon that can be generated in a phase-matched configuration.

2. Quasi-phasematching of HHG radiation

In this work we show that this phase mismatch at higher photon energies can be overcome by engineering the hollow fiber. By modulating the inner radius of the fiber, the on-axis intensity of the driving laser pulse is also modulated periodically. Since HHG is highly nonlinear, the effect is to strongly modulate the generation of the highest harmonic orders. In this way, HHG can be restricted to regions of the fiber in which the HHG radiation adds in phase, thereby quasi phase matching the generation process.(*5*) Such a modulated fiber can be fabricated using glassblowing techniques; a photo of the QPM fiber we used for HHG is shown in Fig. 1.

Fig. 1. Modulated fiber for quasi phase matching of high harmonic generation. This fiber had an initial inner diameter of 150 μm, and was periodically modulated with a period of 1mm. These experiments used a fiber with 10 modulation periods.

The HHG spectra obtained both from the modulated fiber, and from a conventional hollow fiber with the same inner radius, are shown in Fig. 2 for the case of argon gas. At low pressure, (Fig 2(a)) for the modulated fiber we see an increase in the brightness of 6 higher harmonic orders (shorter wavelengths) compared to the spectrum obtained from a conventional fiber. At high pressures (Fig. 2(b)), the shift is even more dramatic. Here, the QPM creates a single broad-bandwidth emission peak with minor spectral modulations. The overall bandwidth of this peak, ~3 eV, corresponds to a time-bandwidth limit of ~250 attoseconds.

Previous theoretical calculations for QPM harmonic generation have indicated the possibility that such a geometry will allow for the generation of single, isolated, attosecond pulses, even using ~20 fs driving pulses (as were used in this experiment). Therefore, there is excellent agreement between theory and our observations.

Future work will extend these advances to higher photon energies by using shorter laser pulses, higher laser intensities, better fidelity and tapered fibers, and higher-ionization potential gases such as helium or even ions.

We acknowledge support for this work from the National Science Foundation.

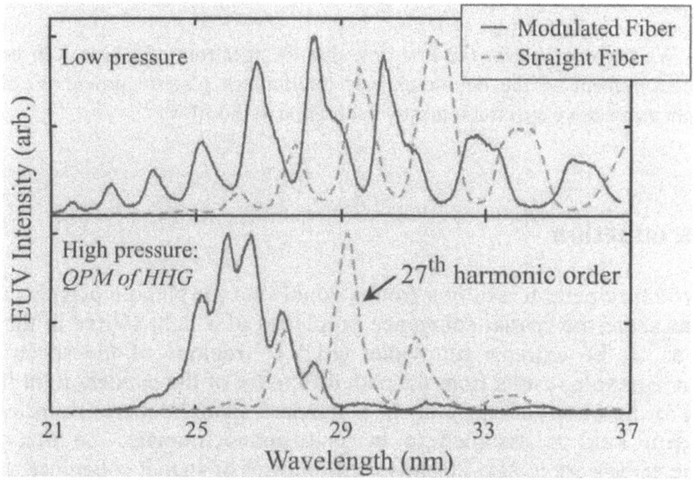

Fig. 2. HHG spectra measured from a hollow-core fiber with constant radius (dashed) and a modulated fiber (solid). At low pressure (3 Torr), the fiber structure slightly reduces the phase mis-match for cutoff harmonics, resulting in an increased intensity at shorter wavelengths. At higher pressures, (35 Torr) the cutoff harmonics are quasi phase matched, and substantially enhanced relative to lower harmonic orders. Furthermore, the QPM spectrum supports a single, isolated, 250 attosecond light pulse.

References

1 A. Rundquist *et al.*, *Science* **280**, 1412 (1998).
2 A. Rundquist, Ph.D., Washington State University (1998).
3 Y. Tamaki *et al.*, *Phys. Rev. A* **6206**, 063802 (2000).
4 R. A. Bartels *et al.*, *Science*, in press (2002).
5 I. P. Christov *et al.*, *Optics Express* **7**, 362-367 (2000).
6 M. Lewenstein *et al.*, *Phys. Rev. A* **49**, 2117 (1994).

Determination of HHG spectrum by measuring the mutual coherence

Randy A. Bartels[1], Sterling Backus[1], Ariel Paul[1], Henry Kapteyn[1], Margaret Murnane[1], Yanwei Liu[2], David Attwood[2]

[1] JILA, University of Colorado, 440 UCB, Boulder CO 80309-0440, USA
E-mail: bartels@colorado.edu
[2] Center for X-ray Optics, LBNL,EECS, University of California, Berkeley CA, USA

Abstract. We demonstrate for the first time that the spectrum of a beam can be obtained from a measurement of the double-pinhole interference pattern, providing an absolute wavelength and relative spectral intensity calibration in the EUV.

1. Introduction

The interference pattern resulting from a double slit (or pinhole pair) is commonly used to measure the spatial coherence properties of a light source in the visible[1], as well as in the extreme ultraviolet (EUV)[2,3] regions of the spectrum. This interference pattern results from the path difference of the incident light field from each slit to the observation point. In a standard spatial coherence measurement, the incident field is assumed to be quasi-monochromatic, so that temporal coherence effects are isolated from a measurement of spatial coherence. However, under conditions where a broad bandwidth source illuminates a pinhole pair, the interference pattern will contain both temporal and spatial coherence information of the source, as well as the power spectrum. Here, we show that the spectrum of a beam can be extracted from a measurement of a double-pinhole interference pattern and the geometry of the set-up[4]. Such a spectral measurement is unique in that it determines the absolute wavelength and relative spectral intensity of the source, which is a great advantage in the EUV region where calibration is very difficult.

2. Young's double slit as a single-shot field autocorrelation

The exact relationship between the interference pattern and power spectrum can be derived as follows. The intensity distribution in the observation plane after illuminating two pinholes equally can be written as

$$I(x) = 2I^{(0)}(x)\left[1+\gamma_{12}(x)\cos\left(2\pi\frac{d}{\lambda_0 z}x\right)\right] \tag{1}$$

where $I^{(0)}(x)$ is the Airy distribution due to diffraction through a pinhole of width δ, d is the pinhole separation, z is the distance from the pinhole pair to the observation plane, λ_0 is the central wavelength of the light field, and γ_{12} is the complex degree of mutual coherence.

Equation (1) is best analyzed in the spatial frequency domain. A Fourier transform of Eq. (1) gives

$$2T(\nu)\otimes\left\{\delta(\nu)+\frac{1}{2}\hat{S}(\nu)\mu_{12}(\nu)\otimes\left[\delta\left(\nu-\frac{zc}{d}f_0\right)+\delta\left(\nu+\frac{zc}{d}f_0\right)\right]\right\} \tag{2}$$

where $\otimes$ is the convolution operator, $T(f_x)=\Im\{I^{(0)}(x)\}$ is a "dc" spike, $\delta(f_x)$ is the Dirac delta function, $f_0 = d/(z\lambda_0)$ is the carrier spatial frequency due to the pinhole-pair interference pattern, and $\hat{S}(f_x)\mu_{12}(f_x)=\Im\{\gamma_{12}(x)\}$. Thus, a Fourier transform of the interferogram produced from a Young's pinhole pair measurement should yield three terms – a dc term corresponding to a "spike" at zero (or dc) frequency and two terms containing information on the power spectrum convolved with the dc spike and weighted by the spatial coherence function at that frequency. As a result, the resolution of a spectral measurement is limited by the width of the dc spike.

Figure 1 shows the interferogram (a) and its Fourier transform (b) for a double pinhole of diameter 20 μm and separation 575μm, while Fig. 2 shows a comparison between the deconvolved spectrum and spectrum measured by a grating spectrometer. The broadening in the pinhole measurement is due to the geometry of the pinhole pair and is evident from the comparison of the spectrum measured by the x-ray spectrometer (dashed line of Fig. 2). However, the width of the spectrum obtained from the x-ray spectrometer after being convolved with the dc spike is identical to that obtained from the pinhole pair, indicating that we are limited only by geometry. Further improvement in resolution is possible by either shrinking the pinhole diameter or increasing the pinhole separation. The difference in the intensity of the two spectra is due to the varying efficiency of spectrometer response, assuming constant spatial coherence across the harmonic spectrum as is the case for this experiment.

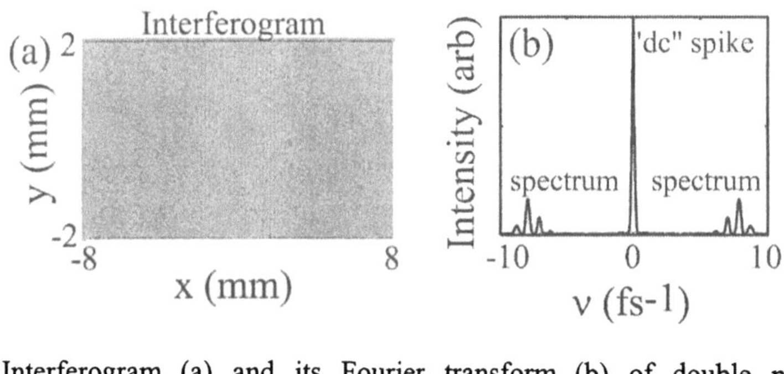

Fig. 1. Interferogram (a) and its Fourier transform (b) of double pinhole experiment for 20 μm x 575 μm pinhole.

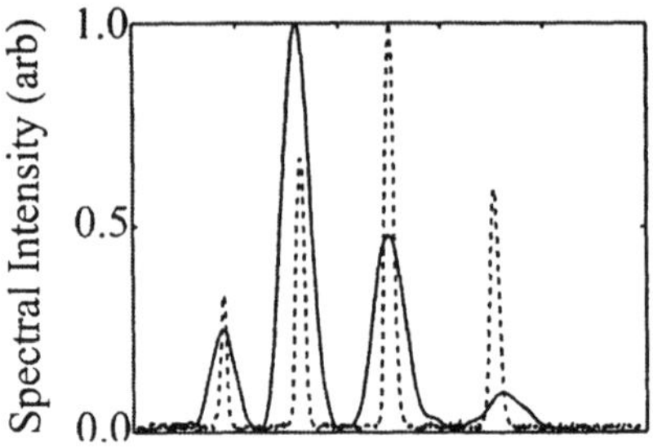

Fig. 2. Comparison of spectra obtained from the interferogram (solid line) and a grating spectrometer (dashed line).

3. Conclusions

In summary, we have experimentally demonstrated a robust and accurate technique that allows the absolute wavelength and spectrum of a light field to be determined from the far-field interferogram produced by a pinhole pair. Furthermore, this technique provides a convenient method of absolute calibration in the EUV region of the spectrum where few calibrated tunable sources currently exist. The spectrum obtained is verified with the spectrum measured by a conventional grating spectrometer. The resolution of pinhole pair spectrum can be improved by a simple change in geometry. The quality of the data is extremely high due to the implicit averaging of the 2-D Fourier transform. Finally, this experiment provides verification of an important link between the spatial coherence properties of an optical field and its power spectrum, and can be extended to other wavelengths.

References

1 B Thompson and E. Wolf, J. Opt. Soc. Am. **46**, 895 (1957).
2 C. Chang *et al.*, Opt. Comm. **182**, 25 (2000).
3 Y. Lui *et al.*, Phys. Rev. A **63**, 033802 (2001).
4 R. Bartels *et al.*, *Optics Letters* **27**, 707 (2002).

Adaptive Optimization of High Order Harmonic Generation in a Free Propagation Geometry

David Reitze[1], Frederick Weihe[2], Sophie Kazamias[2], Geneviève Mullot[2], Denis Douillet[2], Frederika Augé[2], Jean Paul Chambaret[2], Danièle Hulin[2], and Philippe Balcou[2]

1 Physics Department, University of Florida, Gainesville, FL 32611 USA
E-mail: reitze@phys.ufl.edu
2 Laboratoire d'Optique Appliquée, École Nationale Supérieure des Techniques Avancées– École, Polytechnique, Palaiseau, France
E-mail: balcou@enstay.ensta.fr

Abstract. An adaptive learning loop is used to optimize the wavelength and amplitude of soft x-rays via high order harmonic generation in a free propagation geometry. We demonstrate that it is possible to tune and enhance the amplitude of the harmonics in the 25 – 40 nm range at intensities exceeding the saturation intensity. Unlike simple pulse chirping, we achieve a broader tuning range. The complex nature of the optimizing spectral phase indicates that ionization dynamics and propagation play an important role.

1. Introduction

Adaptive control has become a powerful method for optimizing the outcome of ultrashort pulse laser-matter interactions. A variety of experiments demonstrating 'closed loop' have been reported for controlling laser pulse shape [1,2] as well as physical and chemical processes. Recently, it has been shown that in a waveguide geometry, specific harmonic orders near the cutoff can be significantly enhanced by optimizing the spectral phase of the driving pulse [3]. It is interesting to consider if adaptive learning can be effective for high order harmonic generation in a phase-matched geometry in the absence of a waveguide and at high intensities.

2. Experimental Methods

In these experiments, our goal was to optimize the wavelength and amplitude of harmonics in the 25-40 nm range. Our experiments were conducted using a Ti:sapphire chirped pulse amplifier producing 6.5 mJ, 35 fs pulses at a 1 kHz repetition rate. Prior to each experiment, the pulse spectral phase was measured using SPIDER. Pulses were apertured and focused using a 1 m focal length lens into a gas cell of varying cell length containing up to 15 torr of Argon. Harmonics were analyzed using a spectrometer consisting of a transmission grating, a gold parabolic mirror, and a CDD camera. A genetic algorithm (GA) was used to optimize the spectral phase of the driving pulse. The 'actuator' was an acousto-optic programmable dispersive filter [4], capable of modifying the spectral phase of the pulse prior to amplification. The spectral phase was parameterized by 10 arbitrary values across the spectrum summed with $\Phi^{(2)}$ and $\Phi^{(3)}$, the second and third order Taylor series spectral phase coefficients. The cost function was determined by first fitting the experimentally measured spectral peak to a Gaussian and comparing to a target peak with respect to peak location, peak height, and peak width using appropriate weights and then adding a term which maximized the total integrated harmonic output energy.

3. Results and Discussion

Figure 1 displays an optimized spectrum (solid line) to position the wavelength of the 25th harmonic to 33.6 nm. For comparison, the spectrum for a transform limited pulse is shown (dashed line) prior to optimization. In this case, the algorithm was able to converge exactly to the desired wavelength. Figure 2 shows the tuning of the 25th harmonic in the range 24hν - 26hν as indicated by the arrows. With the exception of a forbidden band extending 0.6 nm below 32 nm, essentially complete coverage is obtained. Similar behavior was found for other harmonic orders.

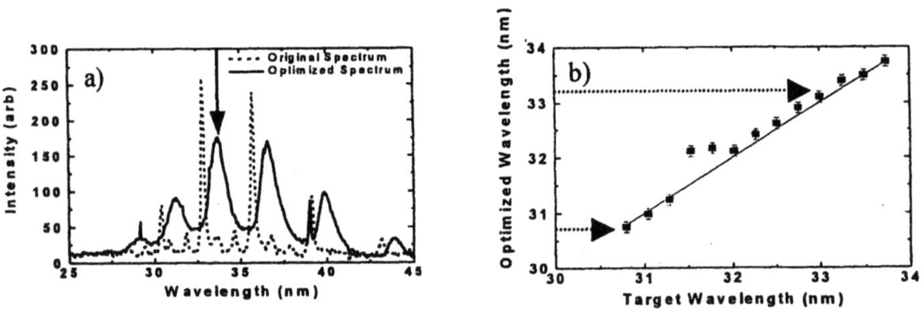

Fig. 1. a) Spectrum before (dashed line) and after optimization (solid line). The arrow indicates the target wavelength. The attenuation in peak amplitudes at longer wavelengths arises from Argon absorption. b) Optimized wavelength vs. target wavelength of the 25th harmonic (32 nm). With the exception of a 0.6 nm 'energy gap' below 32 nm, the harmonic peak can be tuned in the range 25hν +/- hν (indicated by arrows), allowing for essentially complete tuning. The solid line represents perfect tuning.

For comparative purposes, harmonic spectra were taken by changing the chirp parameter, $\Phi^{(2)}$. It has been shown previously that pulse chirping can tune harmonic peaks [5,6,7]. However, in the 25-40 nm band, less tuning bandwidth is obtained using this method (Figure 2a). Moreover, substantially greater peak amplitudes were obtained via adaptive optimization. A clue to this behavior can be found in Fig. 2b, which displays the optimizing spectral phase for each of the tuned peaks in 1b. In most cases, the spectral phase displays a complicated structure, suggesting that the interaction of the laser field with the atom is more precisely controlled by taking advantage of the interplay between the dynamic laser phase and intensity and the atomic phase. It is also important to consider both the ionization process and the propagation of the pulse during high order harmonic generation [8]. The peak intensity reached in our experiments is 4 x 10^{14} W/cm^2, above the saturation threshold for Argon. The formation of a plasma and its subsequent blueshifting of the fundamental at the leading edge of the pulse leads to a shift of the harmonic spectrum as the pulse propagates through the gas. The combined action of the laser spectral shift combined with the propagation potentially plays a significant role in these experiments.

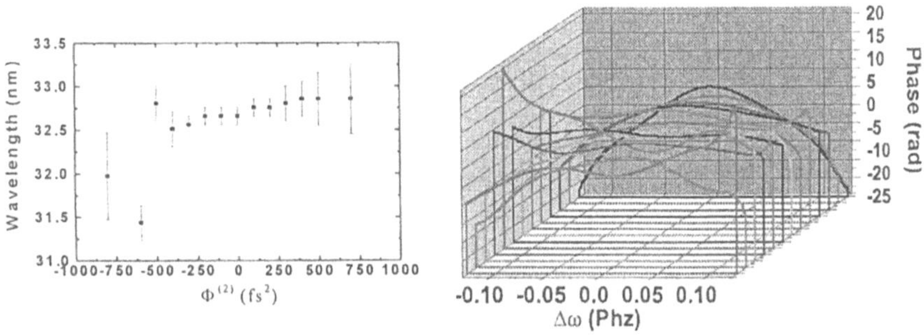

Fig. 2. a) Peak position of the 25th harmonic vs. chirp parameter $\Phi^{(2)}$. Compared with adaptive optimization, a much more limited tuning range is found. The error bars reflect uncertainty due to small amplitudes b) The spectral phase of the optimizing pulse for each of the harmonic peaks shown in Figure 1b. In most cases, the spectral phase displays a complicated structure as opposed to a simple quadratic form.

4. Conclusions

We have shown that adaptive optimization can be used to tune and enhance the amplitude of specific harmonic peaks at intensities well above the saturation intensity in argon. The complex nature of the pulse spectral phase indicates that ionization dynamics combined with propagation effects play an important role.

Acknowledgements. David Reitze gratefully acknowledges the financial support of École Polytechnique.

References

1 D. Yelin, et al., Opt. Lett. **22,** 1793, 1997.
2 A. Efimov, et al., Opt. Lett. **23**, 1915, 1998.
3 R. Bartels, et al., Nature **406,** 164, 2000.
4 F. Verluise, et al., Opt. Lett. **25**, 575, 2000.
5 P. Salieres, et al., Phys. Rev. Lett. **81,** 5544, 1998.
6 Z..Chang, et al., Phys. Rev. A **58,** R30-33, 1998.
7 D. Lee, et al., Phys. Rev. Lett. 87,243902, 2001.
8 M. Nisoli, et al., Appl. Phys..B **70**,S215, 2000.

Few-optical-cycle regime for high-order harmonic generation: enhanced conversion and brightness

P. Villoresi[1], M. Pascolini[1], L. Poletto[1], G. Tondello[1], M. Nisoli[2], E. Priori[2], G. Sansone[2], S. Stagira[2], S. De Silvestri[2]

[1] INFM; Laboratorio di Elettronica Quantistica - D.E.I., Università di Padova, Via Gradenigo 6/a, 35131 Padova, Italy
E-mail: paolo.villoresi@unipd.it
[2] INFM, Centro di Elettronica Quantistica e Strumentazione Elettronica – C.N.R. Dipartimento di Fisica, Politecnico, P.za L. da Vinci 32, 20133 Milano, Italy

Abstract. We show how a particular scheme of laser-gas interaction may lead to enhanced brightness and sustained conversion-efficiency from infrared to extreme-ultraviolet and soft-X-ray photons. Moreover, smooth spatial-profiles and few-fs-duration are appealing features for ready utilization of the source.

The interaction of a few-optical-cycle pulse with matter have shown novel phenomena [1]. Moreover, the high intensity which can be realized for a few semi-cycles of the laser optical-period may induce the tunnel ionization or even ionization due to barrier suppression. The resultant photoelectrons are driven in a coherent motion which influences both the above-threshold-ionization electron spectrum as well as the photon spectrum emitted due to recombination.

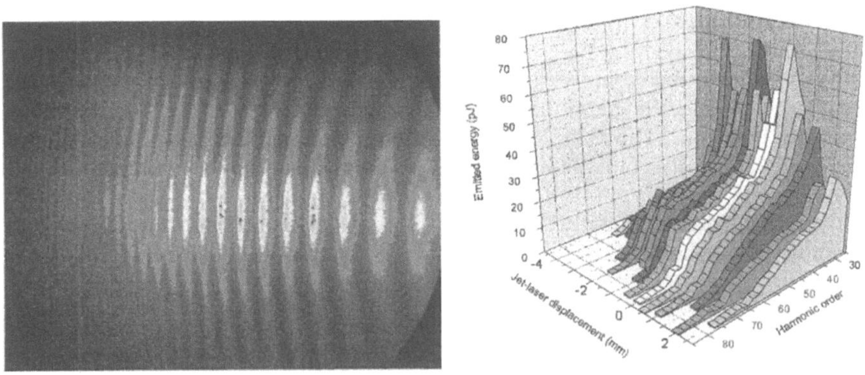

Fig. 1. Left panel: image of the spectrometer focalplane showing the spectral and spatial profile of the harmonics generated using Ne as target and a pulse of 7 fs and 0.14 mJ; horizontal direction: spectral dispersion direction, the 23th harmonic is the first on the right; vertical direction: projection of the spatial profile. Right panel: emitted energy per harmonic order per laser pulse in different interaction geometry. The jet displacement is positive if it is located downstream the laser focus.

In particular, in this work we analyze the condition of enhanced generation efficiency and brightness of extreme ultraviolet (XUV) radiation in the case of a 7 fs pump pulse on a jet of different gases.

The optimization of the harmonic-generation conversion efficiency and of the spatial properties of the harmonic radiation are very important issues in the various applications of the coherent harmonic radiation. The phase matching mechanism that governs HHG has been studied by several researchers. When a Gaussian-profile beam is used as pump, the efficiency at the beam focus is depressed by the mismatch between the intensity-dependent dipole phase and the Gouy phase [2]. Good phase matching can be achieved before the focus, but with annular emission and so degraded brightness, or after the focus with a low divergence but with a minor yield. In both cases the highest intensity region at focus cannot be exploited for efficient harmonic generation. In this work, we report on a novel scheme for the generation of high brightness harmonic radiation using different noble gases (He or Ne) by employing sub-10 fs laser pulses with truncated Bessel radial amplitude profile, generated by the hollow fiber compression technique [3]. Left panel of Fig. 1 shows a typical harmonic spectrum obtained in Neon, for a laser peak intensity of ~9×10^{14} W/cm^2 at the laser focus. The harmonic emission was measured by a absolute-calibrated x-ray spectrometer, based on a chirped grating and a high-resolution CCD detector [4].

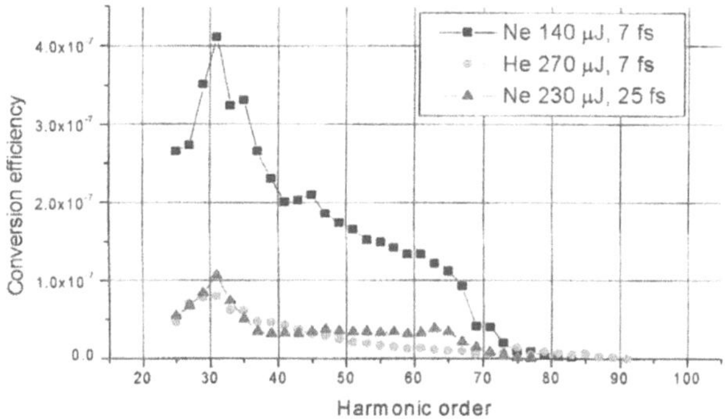

Fig. 2. Comparison of the conversion efficiency for best cases relative to different gases and duration. Same efficiency was obtained in Ne also for double input laser energy.

By using this setup we have measured the total energy irradiated as well as the full width at half maximum of the spatial profile, and an example is reported in the right panel of Fig. 1. From this latter quantity, we have estimated the divergence angle of each harmonic order, for the different gases, laser parameters and interaction geometry. The conversion efficiency and the source brightness can then be derived from the measurements. The reported measurements clearly demonstrate novel spatial properties of the harmonic emission: regular harmonic spectral profiles (see left panel of Fig. 1), with small divergence, can be generated in a broad range of positions around the laser focus. This result is particularly significant for applications where high XUV brightness is required [5].

The conversion efficiency of infrared in XUV and soft X-ray radiation results almost flat in the final plateau of the harmonics spectra, with values in the range 10^{-8} to a few 10^{-7}. In the case of He, similar results were observed by using 25 fs and 7 fs pulses. By using 7 fs pulses with Ne, the conversion efficiency is 3×10^{-7} at 23 nm and of 1×10^{-7} at 13 nm. This finding indicates a relevant enhancement of the efficiency with respect to the case of He as well as to the case of longer pulses with the same gas, as reported in Fig 2. Moreover, we have observed no saturation of this efficiency even by doubling the input energy. The harmonic beams are very regular in the profile, with FWHM angle of about 3 mrad. The emission occurs during the three more intense semi-cycle of the laser pulse, and so the duration of the XUV envelope may be estimated as 3 fs. The corresponding brightness is about 5×10^{13} W/srad cm^2. The remarkably different spatial behavior of the harmonic emission observed using a fundamental beam with Bessel profile instead of one with Gaussian profile, can be understood in terms of the peculiar properties of a truncated-Bessel beam. A significant difference between the two, both focused by a spherical mirror, is that the former shows an axial intensity which, in the focal region, oscillates, with respect to the propagation coordinate, around an almost constant value. The extension of such oscillatory behavior is roughly equal to half of the confocal parameter of the best Gaussian fitting of the Bessel profile. Due to the peculiar properties of a truncated Bessel beam, the polarization phase on the propagation axis presents an oscillatory behavior around an almost constant value, in a ~2-mm-wide region across the laser focus. We propose that such phase oscillations, which lead to a series of stationary phase points, provide a novel phase-matching mechanism, which favors the on-axis harmonic emission when the gas jet is located near the focus. The two different results were also predicted by a numerical simulation [6] of the generation process. By using the actual beam parameters, the different far field patterns are in agreement with our observations as well with those of the gaussian beam.

In conclusion, we have shown that a truncated-Bessel beam generates high-order harmonics with enhanced characteristics with respect to a Gaussian beam. The underlying mechanism was understood as a different phase matching in the laser focal region. By using neon gas, a relevant efficiency is observed in the soft X-ray part of the spectrum. The resultant high brightness is very interesting looking in the perspective of the harmonics utilization.

1 T. Brabec and F. Krausz, "Intense Few-Cycle Laser Fields: Frontiers of Nonlinear Optics," Rev. Mod. Phys **72**, 545 (2000).

2 P. Salieres, A. L'Huillier, M. Lewenstein, "Coherent Control of High-Order Harmonics," Phys. Rev. Lett. **74**, 3776 (1995); Ph. Balcou et al., "Generalized phase-matching conditions for high harmonics: The role of field-gradient forces," Phys. Rev. A **55**, 3204 (1997).

3 M. Nisoli, S. De Silvestri and O. Svelto, "Generation of high energy 10 fs pulses by a new pulse compression technique," Appl. Phys. Lett. **68**, 2793 (1996).

4 L. Poletto, G. Tondello, P. Villoresi, "High-Order Laser Harmonics Detection in the EUV and Soft X-Ray Spectral Regions," Rev. Sci. Instrum. **72**, 2868 (2001).

5 M. Nisoli et al., "High-Brightness High-Order Harmonic Generation by Truncated Bessel Beams in the Sub-10-fs Regime," Phys. Rev. Lett. **88**, 033902 (2002).

6 P. Villoresi et al., 'Spectral Features and Modeling of High-Order Harmonics Generated by Sub-10-fs Pulses," Phys. Rev. Lett. **85**, 2494 (2000).

Coherent Femtosecond X-ray Raman Spectroscopy of Molecules

Satoshi Tanaka[1,2] and Shaul Mukamel[2]

[1] College of Integrated Arts and Sciences, Osaka Prefecture University, Sakai 599-8531, Japan
E-mail: stanaka@ms.cias.osakafu-u.ac.jp
[2] Departments of Chemistry and of Physics and Astronomy, University of Rochester, Rochester, New York, 14627, USA
E-mail: mukamel@chem.rochester.edu

Abstract. By formulating the nonlinear response in terms of correlation functions of charge and current densities we show that coherent x-ray Raman Spectroscopy could provide an excellent local probe for valence excitation dynamics in molecules.

1. Introduction

Recent progress in ultrafast x-ray pulse generation makes attosecond spectroscopy possible [1]; Electron dynamics in atoms, molecules, and solid state materials may then be directly investigated by x-ray nonlinear spectroscopies. Since x-ray $1 \sim 100\,\text{Å}$ wavelength is comparable to a molecular size, the dipole approximation which imposes strict selection rules in optical spectroscopy does not apply, allowing the direct probing of the entire manifold of electronic transitions and inducing a strong dependence of the signal on wavevectors. Nonlinear x-ray spectroscopies thus provide detailed information on spatial coherence in excited states. By tuning the x-ray frequencies across the core excitations of various atoms it may become possible to investigate in detail the nonlocal nature of valence electronic excitations.

We have simulated time-resolved coherent x-ray Raman Scattering (TCXRS) for a model molecular system, and demonstrated its utility as a novel probe for electronic motions. TCXRS is a four wave mixing process involving three incoming x-ray beams $\mathbf{k}_1$, $\mathbf{k}_2$, and $\mathbf{k}_3$, with two coincident pulses $\mathbf{k}_1$ and $\mathbf{k}_2$, and the x-ray signal at $\mathbf{k}_s=\mathbf{k}_1-\mathbf{k}_2+\mathbf{k}_3$ is detected as a function of the variable delay τ of $\mathbf{k}_3$ with respect to $\mathbf{k}_1$ and $\mathbf{k}_2$.

2. Model

We consider a one-dimensional 8 atom molecular wire, each having a core, an occupied valence, and an unoccupied conduction-orbital. The Hamiltonian includes the transfer of the conduction electron (t_c) and valence hole (t_v), but neglects core hole transfer. It further contains the on-site attractive Coulomb interactions between a core hole and a conduction electron (core-exciton effect) (U_{ac}), between the valence hole and a conduction electron (valence-exciton effect) (U_{vc}), and the on-site repulsive Coulomb interaction between conduction electrons (U_{cc}) and between core holes (U_{aa}).

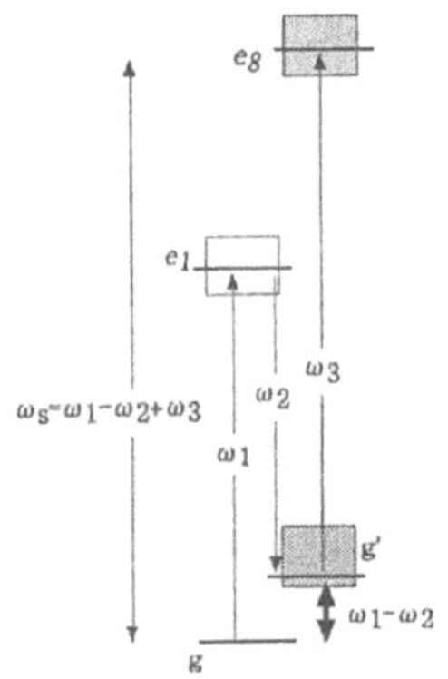

We have calculated a three beam TCXRS where the central frequencies of the three incoming beams ω_1, ω_2, and ω_3, and the signal $\omega_s = \omega_1 - \omega_2 + \omega_3$ are resonant with the core excited states and $\omega_1 - \omega_2$ is tuned in the optical valence excitation. The correlation function approach for the nonlinear x-ray response [2,3] was used in these calculations. In Fig.1 we depict the TCXRS process, where e_1 and e_8 are the core excited states at site 1 and 8, respectively, and g' denotes the valence excited states.

3. Results

In Fig.2 we display the TCXRS signal calculated using rectangular 5fs pulses. ω_1 is tuned at the core exciton at site 1, while $\omega_s = \omega_1 - \omega_2 + \omega_3$ is held fixed at the core exciton at site 1 (a, b, c) or site 8 (d, e, f). $\omega_1 - \omega_2$ is fixed at the resonance with the strongly bound valence exciton states (pair creation of the valence hole and the conduction electron): $\omega_1 - \omega_2 = 6.7\text{eV}$ (panels a, d), $\omega_1 - \omega_2 = 7.4\text{eV}$ (panels b, e), and $\omega_1 - \omega_2 = 8.9\text{eV}$ (panels c, f). Shown is the Fourier transform of the signal with respect to the time delay τ. The time resolved TCXRS signal is displayed in the inset.

Comparison with the calculated optical absorption spectrum (not shown) reveals that the peak positions in the frequency domain coincide with the optical absorption peaks, *i.e.* valence exciton energies. The intensity distribution, however, is completely different; the intensities of the Raman-active valence exciton modes are strong, but they are very weak in the optical absorption. This is because they are induced by the two x-ray photons ($\mathbf{k}_1$ and $\mathbf{k}_2$). The TCXRS selection rules thus complement the optical absorption. The interaction with the $\mathbf{k}_1$ and $\mathbf{k}_2$ beams induces a polarization associated with the valence exciton at site 1, and this polarization is probed by the $\mathbf{k}_3$ beam at site 1(left panel) or site 8 (right panel). Therefore these intensities depend strongly on $\omega_1 - \omega_2$, reflecting the wavepacket motion for each valence exciton. We also see a clear quantum beat in the time resolved TCXRS resulting from interference among the strong valence exciton modes (See Fig.2 (b), (e), and (f)).

The right panels in Fig.2 show that the temporal profile of the TCXRS is greatly changed when we tune the x-ray energies of $\mathbf{k}_3$ and $\mathbf{k}_s$ to the core resonance of site 1, even for the same detuning of $\omega_1 - \omega_2$. In this case we detect the valence exciton motion at site 1. Our calculations illustrate that TCXRS is an atomic scale selective local probe most suitable for investigating the valence exciton wavepacket motion.

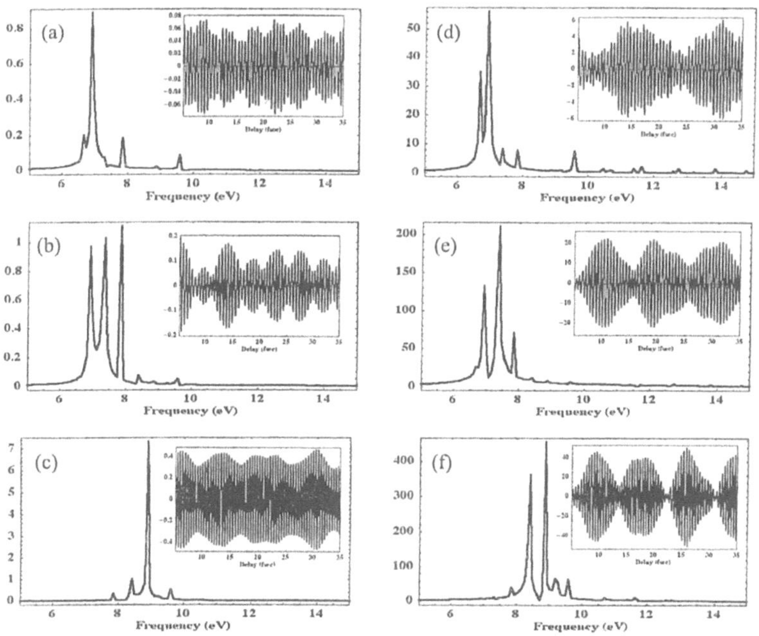

Fig. 2. The calculated TCXRS. ω_1 is fixed at the core exciton at site 1, while ω_s is fixed at the core exciton at site 1 (left panels) or at site 8 (right panels). $\omega_1 - \omega_2$=6.7eV (a,d), 7.4eV (b,d), and 8.9eV (c,f). The results of the Fourier analysis of the temporal behavior of TCXRS with the time delay (inset) are shown.

When the pulses are so short that their bandwidths span the strongly bound exciton states, but are still long enough to resolve the core excited states between the sites 1 and 8, the TCXRS does not depend on $\omega_1 - \omega_2$. Tuning the x-ray energies of $\mathbf{k}_3$ and $\mathbf{k}_s$ gives rise to very different TCXRS profiles.

In addition to the temporal profile shown here, the TCXRS has a strong wavevector dependence, which provides detailed information on the *spatial coherence* of the valence excitations. TCXRS should thus be an ideal technique for investigating the real-time dynamics of valence excitons.

Acknowledgements

The support of the National Science Foundation Grant no. CHE 0132571 and the National Institutes of Health Grant no. 1 RO1 GM59230-01A2 is gratefully acknowledged.

References

1. M. Hentschel *et al.*, Nature Vol. 414, 519, 2001; M. Drescher *et al.*, Science Vol. 291, 1923, 2001; R. W. Schoenlein *et al.*, Science Vol. 287, 2237, 2000.
2. S. Tanaka, V. Chernyak, and S. Mukamel, Phys. Rev. A, Vol. 63, 63405, 2001.
3. S. Tanaka and S. Mukamel, J. Chem. Phys. Vol. 116, 1877, 2002.
4. "Coherent X-Ray Raman Spectroscopy", S. Tanaka and S. Mukamel, Phys. Rev. Lett. (In Press) (2002).

Fully spatially coherent EUV source

Randy A. Bartels[1], Sterling Backus[1], Ariel Paul[1], Henry Kapteyn[1], Margaret Murnane[1], Yanwei Liu[2], David Attwood[2], and Chris Jacobsen[3]

[1] JILA, University of Colorado, 440 UCB, Boulder CO 80309-0440, USA
E-mail: bartels@colorado.edu
[2] Center for X-ray Optics, LBNL,EECS, University of California, Berkeley CA, USA
[3] Department of Physics, SUNY at Stony Brook, Stony Brook NY, USA

Abstract. We demonstrate full spatial coherence of an EUV source, which is used to perform coherent EUV imaging.

1. Introduction

We present measurements of the coherence of EUV light generated using the process of high-harmonic upconversion (HHG). We show that, by using a phase-matched hollow-fiber geometry,[1] the EUV light exhibits the highest inherent spatial coherence of any source in this region of the spectrum to date. This source generates approximately 10^{12} photons/sec in a fraction of an eV FWHM bandwidth around 31eV. Since this source exhibits full spatial coherence at very short wavelength, this light source represents the smallest inherent effective source-size of any light source yet created. Although EUV radiation can also be produced from undulators, extensive spatial filtering is required to produce a spatially coherent EUV beam. Such filtering is lossy and inefficient.

HHG is a useful way of generating coherent light throughout the UV and EUV. In HHG, extremely short duration pulses can be produced by focusing a high-intensity femtosecond laser pulse into a gas. Odd harmonics of the exciting laser are produced in a directed, narrow-divergence beam, with photon energies that can extend up to > 500eV. This process thus coherently up-shifts a femtosecond pulse from the visible into the EUV region of the spectrum. However, experiments to date have shown that the HHG radiation generated in gas jets is only partially spatially coherent, and does not retain the full coherence of the fundamental driving beam, due to multiple harmonic trajectories and plasma dynamics. In contrast, HHG from hollow fibers[1] selects a single electron trajectory and limits ionization, dramatically improving the spatial coherence.[2]

2. Spatial coherence of HHG radiation

We measure the spatial coherence of the EUV beam by placing a pinhole pair in the beam and recording the diffraction pattern. The intensity distribution of the interference from two equally illuminated pinholes can be written as -

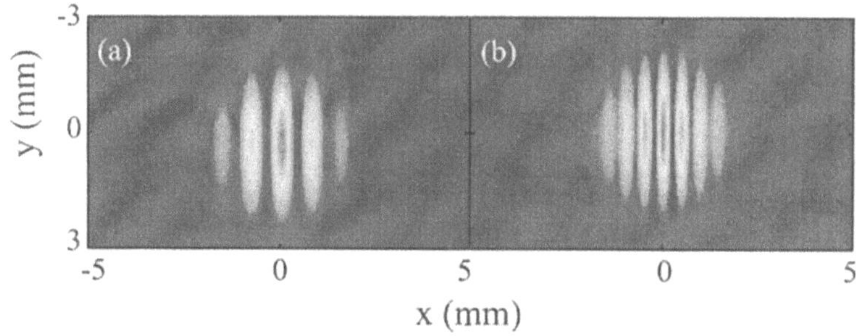

Fig. 1. Interferogram of a (a) 50 x 150 micron and (b) 50 x 250 micron pinhole pair.

$$I(x) = 2I^{(0)}(x)\left[1 + \gamma_{12}(x)\cos\left(2\pi \frac{d}{\lambda_0 z} x\right)\right] \quad (1)$$

where $I^{(0)}(x)$ is the Airy distribution due to diffraction through a pinhole of width δ, d is the pinhole separation, z is the distance from the pinhole pair to the observation plane, λ_0 is the central wavelength, and $\gamma_{12}(x)$ is the degree of mutual coherence. When the incident field spectrum is broad, $\gamma_{12}(x)$ represents the *field autocorrelation* of the incident field, modified by the spatial coherence of each spectral component. The field autocorrelation uniquely determines the power spectrum, thus the incident field spectrum can be determined [3].

The Fourier transform of Eqn. (1) is written as

$$2T(\nu) \otimes \left\{\delta(\nu) + \frac{1}{2}\hat{S}(\nu)\mu_{12}(\nu) \otimes \left[\delta\left(\nu - \frac{zc}{d} f_0\right) + \delta\left(\nu + \frac{zc}{d} f_0\right)\right]\right\} \quad (2)$$

where $\otimes$ is the convolution operator, $T(\nu)$ is a "dc" spike, $\delta(\nu)$ is the Dirac delta function, f_0 is the carrier spatial frequency, $S(\nu)$, and $\mu_{12}(\nu)$ is the coherence of each wavelength in the bandwidth.

The spectrally averaged fringe visibility of a broadband optical field is obtained by the sum of the integral of the sidebands and normalized by the "dc" term in

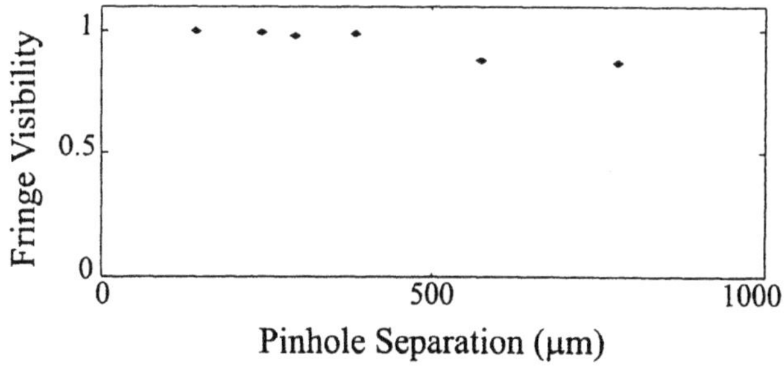

Fig. 2. Fringe visibility across a 1-mm diameter EUV from a phase-matched hollow-core fiber source demonstrating full spatial coherence [2]

Eqn. (2) [2]. Figure 2 shows the fringe visibility for a 1mm diameter EUV HHG beam phase-matched in a hollow-core fiber, and demonstrates full spatial coherence over the majority of the EUV beam.

3. Gabor holography with HHG EUV radiation

We demonstrate the utility of this spatially coherent table-top source of EUV radiation by performing Gabor holography. Figure 3(a) shows a hologram recorded on an EUV CCD camera (Andor Technology, Inc.) of a ~ 10 micron diameter water jet. The reconstruction of the hologram is shown in Figure 3(b). Note that the resolution is determined by the geometry since there is a geometric magnification of in the recording process [2].

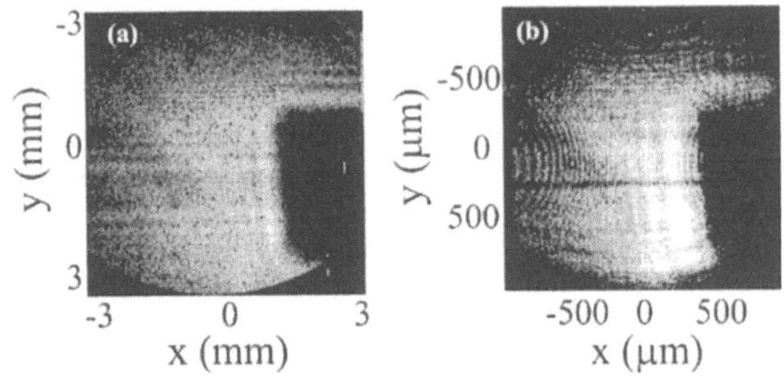

Fig. 3. Gabor hologram (a) and reconstruction (b) of a ~10 micron diameter water jet taken using a < 5 fs HHG EUV pulse. [2]

4. Conclusions

Increasing the magnification of the recording process will increase the resolution to <40 nm, to enable ultrafast imaging in the EUV with high spatial resolution. We acknowledge support for this work from the Department of Energy and the National Science Foundation.

References

1 A. Rundquist *et al.*, *Science* **280**, 1412 (1998).
2 R. Bartels *et al.*, *Science* **In Press** (2002).
3 R. Bartels *et al.*, *Optics Letters* **27**, 707 (2002).

Part III

Control and Dynamics

Time-Resolved Photoelectron Spectroscopy: Charge & Energy Flow in Molecules

James P. Shaffer[1], Thomas Schultz[1], Jonathan G. Underwood[1], Carl C. Hayden[2] and Albert Stolow[1]

[1] Steacie Institute for Molecular Sciences, National Research Council, 100 Sussex Dr. Ottawa ON Canada K1A 0R6
E-mail: albert.stolow@nrc.ca
[2] Combustion Dynamics Facility, Sandia National Laboratories, Livermore CA U.S.A

Abstract: The coupled flow of charge and energy in molecules lies at the heart of photochemistry, photobiology and active molecular electronics. Time resolved photoelectron spectroscopy characterizates these dynamics by disentangling the electronic and vibrational aspects. Femtosecond Coincidence-Imaging, with angle and energy resolved detection of photoions and photoelectrons in coincidence, provides even greater levels of detail.

The excited state dynamics of polyatomic molecules are dominated by the non-adiabatic coupling of vibrational and electronic degrees of freedom, inducing both charge and energy redistribution in molecules. These dynamics are the primary step in the photochemistry of many polyatomic molecules, in photobiological process such as vision and underlie many concepts in molecular electronics. The initially prepared electronically excited zeroth order state mixes with a manifold of vibronic levels of a lower electronic state, leading to the radiationless 'decay' of the initial state. This 'decay' represents a conversion of electronic to vibrational energy and is thus often the first step in the ensuing photochemical dynamics. Non-adiabatic coupling often leads to complex, broadened absorption spectra due to the high density of nuclear states and strong variations of transition dipole with nuclear coordinate. The investigation of these problems is very difficult when the state density becomes high and multi-mode vibronic couplings are involved. The case of greatest importance to photochemistry, when the zeroth order excited states are directly or indirectly coupled to a true continuum, leading to non-adiabatic photodissociation dynamics, is especially challenging. In this paper, we discuss the utility of femtosecond time-resolved photoelectron spectroscopy (TRPES)[1-4] and time-resolved photoelectron-photoion coincidence imaging spectroscopy[5] in providing new views of excited state electronic relaxation and dissociation dynamics in polyatomic molecules.

Photoelectron spectroscopy is particularly well suited to non-adiabatic coupling problems because it is a universal energy and angle resolved detection scheme that is sensitive to both electronic configurations and vibrational dynamics. In the description of photoionization, an elementary and useful picture is that emission of an independent outer valence electron occurs without simultaneous electronic reorganization of the ion core (the Koopmans' picture). This leads to simple correlation rules which indicate the cation state expected to be formed upon single photon, single electron ionization of a given molecular orbital. Partial ionization probabilities into specific ion electronic states can differ drastically with respect to the molecular orbital nature of the neutral electronic state. In Figure 1, we show a cartoon of excited state polyatomic wavepacket dynamics probed via TRPES. A zeroth order bright state α is coherently prepared with a femtosecond pump pulse. According to the molecular orbital picture, it ionizes into the α^+ continuum, the electronic state of the ion obtained upon removal of the outermost valence electron.

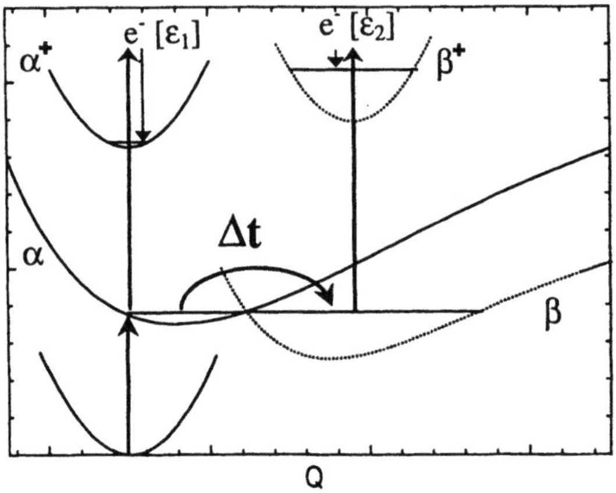

Fig. 1. Time-resolved photoelectron spectroscopy. The non-adiabatically coupled electronic states α and β correlate to different ionization continua. The photoelectron bands $[\varepsilon_1]$ and $[\varepsilon_2]$ reveal the vibrational dynamics in the α and β states.

This process produces a photoelectron band $[\varepsilon_1]$. Any vibrational dynamics in the α state should be reflected in the form of the $[\varepsilon_1]$ band. We now consider a non-adiabatic coupling process which transforms the zeroth order bright state α into a lower lying zeroth order dark state β. The β state should ionize into the β^+ ionization continuum (here an electronically excited state of the ion), producing a photoelectron band $[\varepsilon_2]$. Therefore, for a sufficiently energetic probe photon, we expect a *switching* of the electronic photoionization channel from $[\varepsilon_1]$ to $[\varepsilon_2]$ due to the non-adiabatic process. Any vibrational dynamics in the β state is reflected in the $[\varepsilon_2]$ band. This simple picture illustrates how we can directly monitor the evolving excited state electronic configurations during non-adiabatic processes while simultaneously following the coupled nuclear dynamics via the vibrational structure within each photoelectron band. This disentangling of electronic from vibrational dynamics via TRPES was first demonstrated in a study of internal conversion dynamics in a linear polyene[6]. The TRPES method is quite general and can be applied to a range of excited state problems provided that the sample can be volatilized either by heating or laser ablation/liftoff. For example, we have applied TRPES to the study of excited state intramolecular proton transfer dynamics, revealing time scales for the proton transfer step, and the subsequent excited state dynamics after transfer[7].

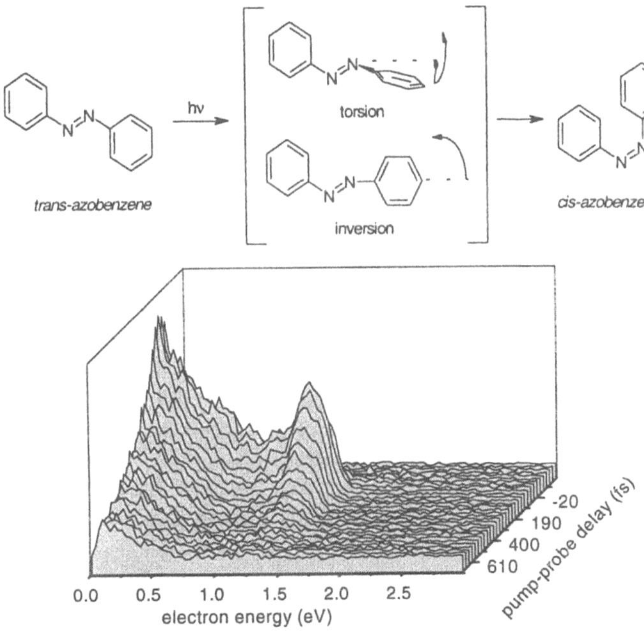

Fig. 2. Time-resolved photoelectron spectroscopy of isomerization dynamics in the model molecular switch azobenzene. Our results lead us to a new, more complete model of azobenzene isomerization involving two overlapping ππ* excited states which exhibit differing decay dynamics.

In Fig.2 (upper) we show a picture of photoisomerization in the model molecular switch azobenzene. Azobenzene is one of the most important chromophores in molecular electronics and has been employed in diverse applications such as switches, triggers for protein folding and data storage, to name but a few. Two isomerization pathways, torsion and inversion, have been discussed at length in the literature. A puzzling aspect of the azobenzene photophysics is that the isomerization quantum-yield decreases with increasing photon energy, contrary to expectations. The origin of this phenomenon has been a matter of debate for over 20 years.

Using the TRPES method, we studied the excited state dynamics of azobenzene in the 285-350 nm range, probing by photoionization with 200 or 207 nm photons. Typical pump-probe cross-correlation FWHM were 140 fs, but FWHM as short as 90 fs were achieved with slightly longer probe wavelengths. In Fig. 2 (lower) we show a typical photoelectron spectrum of count rate *vs.* electron kinetic energy and pump-probe time delay. Two energy-resolved photoelectron bands were easily distinguished by their characteristic decay dynamics along the pump-probe delay axis. It should be noted that the identification of these two bands would be nearly impossible in an integrated measurement (*i.e.* mass resolved ion detection which corresponds to the integral over all photoelectron energies). The detection of the dispersed signal with reasonable resolution of the photoelectron energies was essential for the separation of these two bands. We were able to assign the bands to two different ππ* excited states by investigating (1) Koopmans' ionization correlations, (2) the decay dynamics in each band and (3) the wavelength dependence of the spectra. The lower energy state essentially corresponds to an excitation of the N=N double bond, while the higher energy state is $\pi\pi^*$ excitation largely localized on the phenyl rings. The electronic absorption spectra of these two $\pi\pi^*$ states are completely overlapping. It is surprising that this higher energy $\pi\pi^*$ state has been essentially ignored in the published literature to date. In combination with high level *ab initio* calculations, our results allowed us to develop a new model for the azobenzene photoisomerization mechanism. In our model, the lower $\pi\pi^*$ state undergoes free rotation around the N=N bond (torsional coordinate) and can relax directly to the initial structure. The higher $\pi\pi^*$ state internally converts to the lowest excited state without large amplitude motion and subsequently undergoes photoisomerization along the inversion coordinate. Our two $\pi\pi^*$ state model offers a simpler and more intuitive explanation for the wavelength dependence of the isomerization quantum yield. Furthermore, this model resolves conflicting experimental reports about the existence of a torsional coordinate: due to selection rules in absorption and fluorescence spectroscopy, only part of the excited state dynamics was observed in previous experiments, leading to apparently incompatible descriptions of the isomerization pathway.

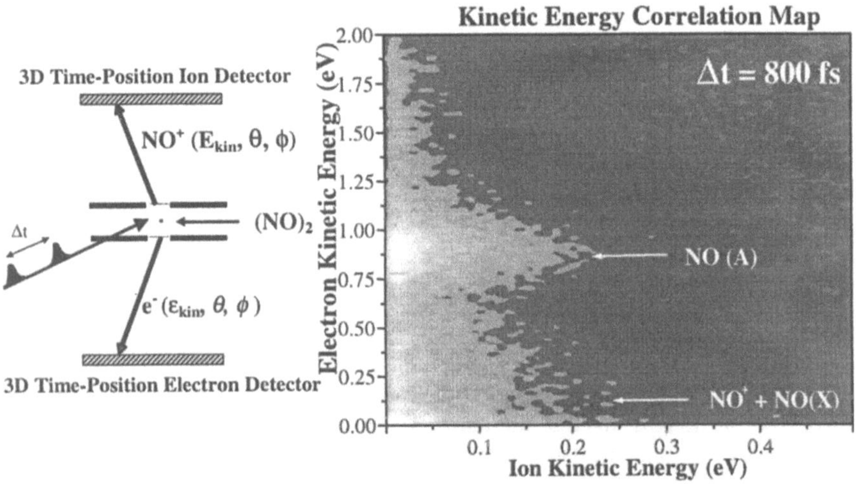

Fig. 3. Time-resolved photoelectron-photoion Coincidence-Imaging spectroscopy of non-adiabatic photodissociation dynamics in the nitric oxide dimer. This method allows for the coincident determination of fs time-resolved full 3D recoil momentum vector correlations.

Although we believe TRPES to be a powerful tool for the study of excited state non-adiabatic dynamics, as the dynamics becomes more complex and dissociative processes occur, it becomes a serious challenge to "follow" the dynamics from the initially prepared state all the way through to products. The only remaining possibility is to make multiply differential measurements of the product state attributes (*e.g.* energy & angular distributions, vector correlations) as a function of time and try to "work backwards" from the products towards the excited state dynamics. In Fig.3 (left) we depict the coincidence imaging method[5], here applied to the photodissociation dynamics of the nitric oxide dimer. In these experiments the full 3D momentum recoil vectors of the ion and electron are recorded *in coincidence* using a pair of *time and position* sensitive MCP detectors and coincidence electronics. The NO dimer molecules are excited with a femtosecond pulse at 210 nm, and the $(NO)_2$ undergoes a non-adiabatic dissociation into NO(A) + NO(X) with a decay time of 0.7ps [3]. Either the excited dimer molecule or the NO(A) photofragment may be single photon ionized by the probe laser at time Δt. A small amount of ionization to the dimer cation is observed. This signal decays with a 0.7ps time constant. A signal due to the formation of NO(A) product grows in. A kinetic energy correlation map at 800fs time delay shown in Fig. 3. The photoelectron kinetic energy is plotted vs. the kinetic energy of the *coincident* fragment ion. The projection onto the vertical axis yields the energy-resolved photoelectron spectrum, while the projection onto the horizontal axis yields the ion kinetic energy release spectrum. Due to the full 3D detection scheme, the ion and electron angular distributions were *simultaneously* obtained but are not shown here. At short delays, the NO(A) channel is absent and the recoil energy of the ion identifies it as coming from dissociative ionization of the dimer into NO^+ + NO(X), the relatively low electron kinetic energy shows that the ionization occurs to an excited state of the dimer cation which undergoes direct dissociation. Thus the excited state of the neutral dimer correlates upon ionization with a dissociative electronically excited state of the cation. Ionization of the NO(A) state product emerges at longer delay times (as the dissociation proceeds) and is clearly identified by the feature at ~0.9 eV photoelectron energy. It can be seen that although the two ion fragment kinetic energy distributions are *completely* overlapped, the correlation map allows the desired NO(A) product state kinetic energy distribution to be cleanly extracted. These observations confirm that the dissociation is not direct but instead occurs from a intermediate state which rapidly evolves into a configuration that precludes ionization to the bound cation state[3]. These time-resolved correlation maps provide a novel way to follow chemical reaction dynamics from the initially excited state through to the final products and allow for unique time-resolved measurements of product state distributions and their various scalar and vector correlations.

References

1 I.Fischer, D.M.Villeneuve, M.J.J.Vrakking & A.Stolow, J.Chem.Phys.**102**, 5566 (1995).
2 D.R.Cyr & C.C.Hayden J. Chem. Phys., **104**, 771 (1996).
3 V. Blanchet and A. Stolow, J. Chem. Phys. **108**, 4361 (1998).
4 C.C.Hayden & A.Stolow, in *Adv.Series Phys. Chem. Vol.10: Photoionization and Photodetachment*, C.Y.Ng, Ed. (World Scientific, Singapore, 2000).
5 J. A. Davies, J. E. LeClaire, R.E.Continetti & C. C. Hayden, J.Chem.Phys.**111**, 1 (1999).
6 V.Blanchet, M.Z.Zgierski, T.Seideman & A.Stolow, Nature **401**, 52 (1999).
7 S.Lochbrunner, T.Schultz, M.Schmitt, J.P.Shaffer, M.Z.Zgierski & A.Stolow, J.Chem.Phys. **114**, 2519 (2001).

Molecular quantum computing: Implementation of global quantum gates applying optimal control theory

Regina de Vivie-Riedle[1] and Carmen M. Tesch[2]

Max-Planck-Institut für Quantenoptik, Hans-Kopfermann-Strasse 1, D-85748 Garching, Germany
[1]E-mail: rdv@mpq.mpg.de, [2]E-mail: crt@mpq.mpg.de

Abstract. Vibrational modes of a polyatomic molecule are selected as qubit basis for a new concept of a molecular quantum computer. Optimal control theory is applied to find the optimized laser fields to realize the quantum gates between these qubits.

1. Introduction

Moving beyond observation to the control of molecular motion with shaped laser pulses is subject of intense current theoretical [1,2] and experimental efforts [3]. In our theoretical work we extended the optimization algorithms (OCT) to handle complex systems and to obtain experimentally feasible laser pulses [4,5] which can directly serve as input pulse. The mask function of the calculated pulses provides the interface to the experimental closed loop setup. These tools at hand we are able to design realistic and robust laser pulses for any given control objective. The possibility to control vibrational population transfer in the electronic groundstate up to 100% is the basis for our newly proposed concept for molecular quantum computing.

2. Results and Discussion

Molecular vibrations are defined as basis states of a qubit system. Our molecular model system is acetylene, but we want to point out that in principle any molecular system and any optically addressable vibrational mode can be used. For acetylene the two IR active modes are selected as basis for the qubit system. Each IR-active mode and its excitation define two separate Hilbert spaces, with a ground state $|0\rangle$ defined by n quanta and an excited state $|1\rangle$ defined by $n' \geq n+1$ quanta of the same vibrational mode. The complete, in this case, two-qubit system is described as the direct product of the individual Hilbert spaces. Two possible examples for basis states within the complete set of the five normal modes $(n_1 n_2 n_3 n_4 n_5)$, are given below:

$$\begin{array}{ll} |00\rangle = |00000\rangle & |00\rangle = |00101\rangle \\ |10\rangle = |00100\rangle & |10\rangle = |00201\rangle \\ |01\rangle = |00001\rangle & |01\rangle = |00102\rangle \\ |11\rangle = |00101\rangle & |11\rangle = |00202\rangle \end{array} \quad (1)$$

The first entry refers to the asymmetric CH-stretch n_3 and the second entry to the cis-bending mode n_5. The qubit basis which corresponds to the second row is shown as possible input and output states in figure 1.

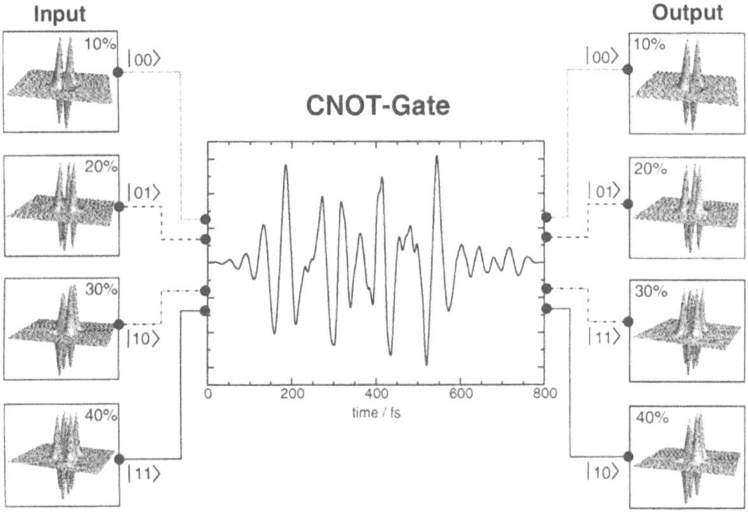

Figure 1. CNOT gate operating on the second qubit, i.e. the cis bending mode, only when the control quibt, the asymmetric stretch mode, is in |1>

For state preparation and implementation of quantum logic operations according to Deutsch's requirements OCT is applied, i.e. the quantum gates are realized by modulated femtosecond laser pulses. In our previous study [6], we could successfully demonstrate the implementation of 1-qubit quantum gates within each qubit space separately. However, it became obvious that in a molecular multi-qubit system it is important to differentiate between active and passive qubits. If the quantum gate operates on the asymmetric CH-stretch, the cis-bending mode, i.e. the second entry, is the passive qubit and vice versa. The influence of the passive qubit plays a relevant role for the construction of global gates. The pulse representing a global quantum gate must transform the qubit input states into the corresponding output states regardless in which quantum state the passive qubits are. This requires flexible laser fields which can only be found with the help of OCT.

Optimizing k transitions simultaneously using the following formulation of the generalized multi target OCT functional:

$$L\ [\Psi_i(t).\Psi_f(t).\varepsilon(t)] =$$

$$\sum_k \left(\left| \langle \Psi_i^k(T) | \Phi_f^k \rangle \right|^2 - 2\,\mathrm{Re} \left\{ \langle \Psi_i^k(T) | \Phi_f^k \rangle \cdot \int_0^T \langle \Psi_f^k(t) | \frac{i}{\hbar} [H^k - \mu^k \varepsilon(t)] + \frac{\partial}{\partial t} | \Psi_i^k(t) \rangle dt \right\} \right) \tag{2}$$

$$- \alpha_0 \int_0^T \frac{|\varepsilon(t)|^2}{s(t)} \cdot dt \quad .$$

enables us to create global one- and two-qubit logic gates [7].

In case of construction of CNOT gates the passive qubit adopts the function of a control qubit. The time evolution of a possible field for a CNOT gate operating on the second qubit is shown in Figure. 1. Here, the first qubit is selected as control qubit and the CNOT gate switches the second qubit from $|0\rangle$ to $|1\rangle$ or vice versa only if the control qubit is in $|1\rangle$. Starting from an incoherent distribution in our qubit system we demonstrate the effect of the CNOT gate. The first two input states do not fulfill the specified condition and are left unchanged by the applied laser field while the third and forth input state meet the condition and their second qubits are switched, respectively. The amount of the two active qubit states are interchanged, but in principle the relative initial distribution of the input qubits is mapped to the output states without any error propagation.

3. Conclusions

The realization of the CNOT gate demonstrates the ability to implement global 2 qubit quantum gates in a molecular system using OCT algorithms and signifies a step ahead on the way to realize a molecular quantum computer

References

1 W. Zhu, J. Botina, and H. Rabitz, J. Chem. Phys., 108 (1998) 1953.

2 H. Rabitz, R. de Vivie-Riedle, M. Motzkus, K. Kompa, Science, 288 (2000) 824-828.

3 A. Assion, T. Baumert, T. Brixner, B. Kiefer, V. Seyfried, M. Strehle, and G. Gerber, Science, 282 (1998) 919.

4 T. Hornung, M. Motzkus, and R. de Vivie-Riedle, J. Chem. Phys, 115 (2001) 3105-3110.

5 T. Hornung, M. Motzkus, and R. de Vivie-Riedle, Phys. Rev. A 65, 0114XX (R) (2002.

6 C.M. Tesch, L. Kurtz, and R. de Vivie-Riedle, Chem. Phys. Lett., 343 (2001) 633-641.

7 C.M. Tesch and R. de Vivie-Riedle, submitted.

Control of Photoinduced Processes by Optimally Shaped Laser Pulses in $MnCp(CO)_3$: Recovering the Information Content Coded in the Optimal Pulse Form

Štefan Vajda, Cosmin Lupulescu, Albrecht Lindinger, Andrea Merli, and Ludger Wöste

Institut für Experimentalphysik, Freie Universität Berlin, Arnimalle 14 , D-14195 Berlin, Germany
E-Mail: vajda@physik.fu-berlin.de

Abstract. Optimal laser control of ultrafast photofragmentation vs. ionization processes leading to a maximal yield of the desired ionic product has underlined the robustness of the closed-loop self-learning experiments. The structure of optimal pulse forms which emerge from these experiments are generally very complex and are expected to reflect the character of the involved molecular states and processes in a coded way. The richness of information contained in pulse forms is demonstrated on the example of interpretation of the pulse shape producing the highest ionic yield of $MnCp(CO)_3^+$.

1. Introduction

The use of closed loop experiments based on the combination of a pulse shaper and a self-learning algorithm has opened new perspectives for controlling ultrafast molecular processes in real time. Such an experiment allows a generation of system specific tailored pulses, which can drive photoinduced processes at a maximum ion yield for a desired product along the most efficient path(s) on the involved potential energy surface(s).

The first control experiment in this regard was performed by Gerber et al. [1]. In our earlier optimization experiments on the highly photofragmenting $MnCp(CO)_3$ molecule, the ion yield of the parent molecule and its fragment ion $MnCp(CO)^+$ was successfully manipulated [2]. The pulse shape retrieved from the SHG FROG trace of the laser pulse producing the highest yield of $MnCp(CO)_3^+$ was a triple pulse sequence consisting of two intense pulses followed by a pulse of very low intensity. A very important issue remained the information content coded in the profile of the acquired pulse. In the case of the described triple pulse, a detailed interpretation based on first pump&probe experimental and available theoretical data was not possible at that time. As observed in power dependent measurements, the relative peak intensities of the first and second pulse of the pulse train (2:3) reflects intuitively the number of photons involved in the pump and probe step, namely the two photonic pump and three photonic probe step, respectively. In order to get deeper insight into the underlying molecular dynamics and to find a direct correlation of the pulse form to these processes, a new extended series of time-resolved experiments with improved time-resolution monitoring the ultrafast fragmentation and ionization process taking place in this organometallic complex was performed.

2. Experimental Methods

The laser system consists of a Kapteyn-Murnane Titanium:Sapphire oscillator, which is pumped by a Nd:YVO_4 laser (Spectra Physics Millennia V). The laser is tuned to a wavelength of 805 nm, where it produces pulses of about 40 fs duration for seeding the multipass amplifier (Quantronix Odin). The amplifier produces 40 fs pulses of up to 1.3 mJ energy at a repetition rate of 1 kHz. In the pump&probe experiment the light pulses were split up and the pump pulse was frequency doubled. The two pulses were recombined in a Mach-Zender-type arrangement, allowing to generate pump&probe sequences of a variable delay. The molecular beam is produced in an adiabatic beam expansion apparatus consisting of a heated oven, from which the evaporated $CpMn(CO)_3$ substance is coexpanded with argon carrier gas across a nozzle of 70-80 μm diameter into the vacuum. The molecular beam passes between the electrical lenses of a quadrupole mass spectrometer, where the molecules are exposed to laser light. The emerging ionic photoproducts are monitored on the quadrupole mass spectrometer.

3. Results and Discussion

In Figure 1, the transients of the parent ion signal $MnCp(CO)_3^+$ and the fragment ions $MnCp(CO)_2^+$ and $MnCp(CO)^+$ are shown. The progressive shift of the signal peaks of the fragmentation intermediates to larger delays reflects the consecutive population of precursors leading to the formation of the ionic products. A modulation superimposed on the fast decays is present in the parent ion signal and the fingerprints of these oscillations are resolved in the time resolved spectra of the

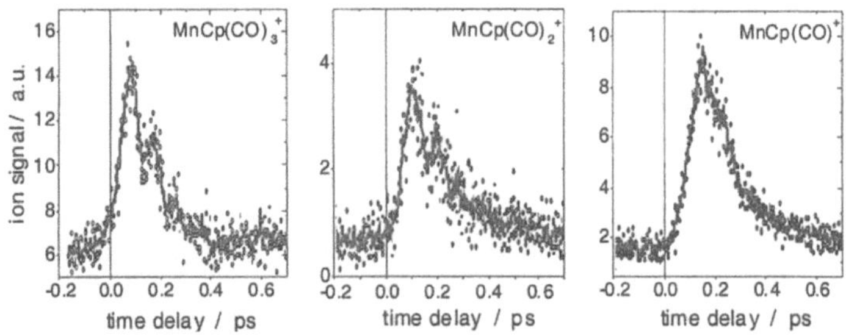

Fig. 1. Transient ion signals of $MnCp(CO)_3^+$, $MnCp(CO)_2^+$ and $MnCp(CO)^+$ obtained by using 402.5 nm pump and 805 nm probe wavelength (circles). In order to visualize the revealed oscillations, smoothed curves are inserted into the individual plots (lines).

emerging fragments as well. The oscillation period is about 80 fs - in a good agreement with the vibration period of the excited parent molecule obtained from preliminary theoretical calculations [3].

The optimized pulse producing the highest yield of $MnCp(CO)_3^+$ is shown in Figure 2, together with the $MnCp(CO)_3^+$ transient. It is clearly visible, that the

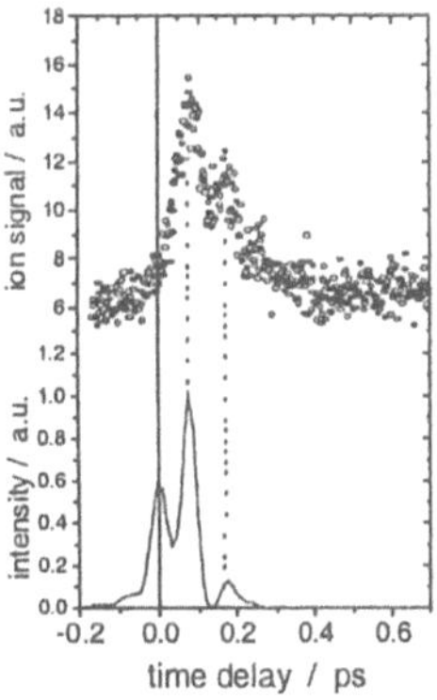

Fig. 2. Optimal pulse producing the maximum $MnCp(CO)_3^+$ yield after phase modulation of an initially unchirped approximately 100 fs laser pulse with a central laser wavelength of 800 nm (bottom) compared with the transient spectrum of $MnCp(CO)_3^+$ (top).

timing of the pulses corresponds exactly to the vibration period of the excited parent molecule. Additionally, the intensity of the second pulse is about 60% higher than the intensity of the first one, reflecting the two-photon excitation and three-photon ionization step involved in the production of $MnCp(CO)_3^+$.

4. Conclusions

This paper was devoted to the interpretation of pulse forms acquired from the control experiments leading to the production of the highest yield of $MnCp(CO)_3^+$. Our experimental results based on pump-probe data with high resolution gave a clear evidence that the obtained pulse shapes are very rich in system-specific information reflecting the intrinsic dynamics and energetics of the involved molecular states and processes.

Acknowledgements. C. L. acknowledges the financial support of his PhD studies at the Freie Universität Berlin by the Gottlieb Daimler und Karl Benz Stiftung. Support by the Deutsche Forschungsgemeinschaft in the frame of the Sonderforschungbereich 450 is gratefully acknowledged.

References

1 A. Assion, T. Baumert, M. Bergt, T. Brixner, B. Kiefer, V. Seyfried, M. Strehle, and G. Gerber, Science 282, 919, 1998

2 C. Daniel, Jürgen Full, L. González, C. Kaposta, M. Krenz, C. Lupulescu, J. Manz, Sh. Minemoto, M. Oppel, P. Rosendo-Francisco, Š. Vajda and L. Wöste, Chem. Phys. 267, 265, 2001.

3 L.Gonzales et al., to be published

Controlling the Dephasing of Vibrational Wavepackets in Potassium Dimers

Ian Walmsley[1], Pablo Londero[1], Michael Fitch[2], Sascha Wallentowitz[3], Leon Waxer[4] and Czeslaw Radzewicz[5]

[1]Clarendon Laboratory, University of Oxford, Parks Rd., Oxford OX1 3PU, UK
[2]Johns Hopkins University Applied Physics Laboratory, 11100 Johns Hopkins Rd., Laurel MD 20723, USA
[3]Fachbereich Physik, Universität Rostock, Universitätsplatz 3, D-18051 Rostock, Germany
[4]Laboratory for Laser Energetics, University of Rochester, Rochester, NY 14627, USA
[5]Institute for Experimental Physics, University of Warsaw, ul. Hoza, Warsaw, Poland

Abstract. We have manipulated vibrational wave packet dephasing in potassium dimers using adaptively-shaped optical pulses controlled by a learning algorithm. We identify several wave packet shapes that are resistant to decoherence due to coupling to the rotational degree of freedom of the molecule. A system-reservoir model of the rotational-vibrational dynamics is presented, and used to interpret the experimental results.

1. Introduction

The objectives of quantum control are predicated on the ability to manipulate the system coherently over a period of time. Any coupling of the system of interest to an external system apart from the control field necessarily induced decoherence. Whether this takes the form of dephasing or dissipation, it is a significant barrier to non-classical control, and is the limiting effect in applying quantum mechanics to technologies such as information processing.

A number of recent experiments have demonstrated that shaped ultrafast laser pulses allow one to excite a broad range of states in atoms and molecules while controlling their relative amplitudes and phases. Some of these small systems are ideal models for studying the loss of coherence when the controlled system is coupled to an ancillary degree of freedom.

Decoherence, the destruction of quantum interference, arises when distinguishing information resides in an unmeasured part of the entity under study. The standard model of decoherence is formulated in terms of a system, defined by the observables that can be measured, and a reservoir, defined by those degrees of freedom that are not accessed by observation. The vibrational and rotational degrees of freedom of diatomic molecules form such a system. One may design measurements that are sensitive only to the vibrational the rotational degree of freedom acts as a dephasing reservoir.[1] The evolution of a vibrational wavepacket in the the A-K_2 system is given by its Wigner distribution $W_s(r,\theta,t)$

$$W_s(r,\theta,t) = \int_0^{2\pi} P(\theta - t,\varphi) W_s(r,\theta,0) d\varphi \,, \qquad (2)$$

where the kernel describes the action of the thermal rotational reservoir, and can be derived analytically and in terms of a single parameter, the dephasing rate •.[1]

We have measured directly the dynamics of the phase-space quasi-probability density by means of Molecular Emission Tomography. [2,3] This enables us to reconstruct the quantum state of the vibrational mode from time-resolved fluorescence spectra [4]. From the reconstructions we were able to confirm that the dephasing dynamics matched our model, and that decoherence half-life was 6.7 vibrational periods at this temperature.

The Hamiltonian for this system admits of no decoherence-free subspaces (DFS). Therefore there are no vibrational wave packets that are free of dephasing. [5] This is, in fact, likely to be the case for the majority of physically realizable Hamiltonians. Moreover, if the Hamiltonian is not known exactly, one cannot determine analytically whether it possesses DFS – this is an experimental question. The concept of a DFS can be generalized, to consider subspaces which are more or less resistant to dephasing- decoherence-resistant subspaces (DRS), of which DFS are the most extreme example. Determining the properties of DRS is clearly an experimental question, since the very definition is most usefully formulated in terms of particular experimental operations.

2. Experimental results

To begin to address this issue for our toy system, we set up a learning control loop in which the fluorescence quantum beats from the outer turning point of the wave packet's trajectory were used as a measure of the dephasing. The shaper was controlled by a learning algorithm, which searched for the longest-lived wavepacket, as measured by the depth of modulation of the quantum beats at a prescribed time after excitation. Examples of the fluorescence beats, exhibiting the characteristic 500 fs period, is shown in Fig. 1.

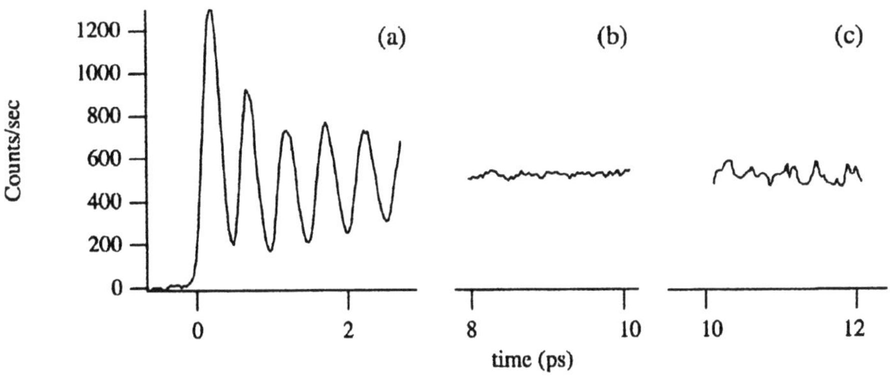

Figure 1. Flourescence quantum beats from the A-X K_2 transition. (a), (b) The impulsively excited wavepacket generates beats that decay in about 7 periods. (c) Learning control feedback results in a longer-lived wavepacket.

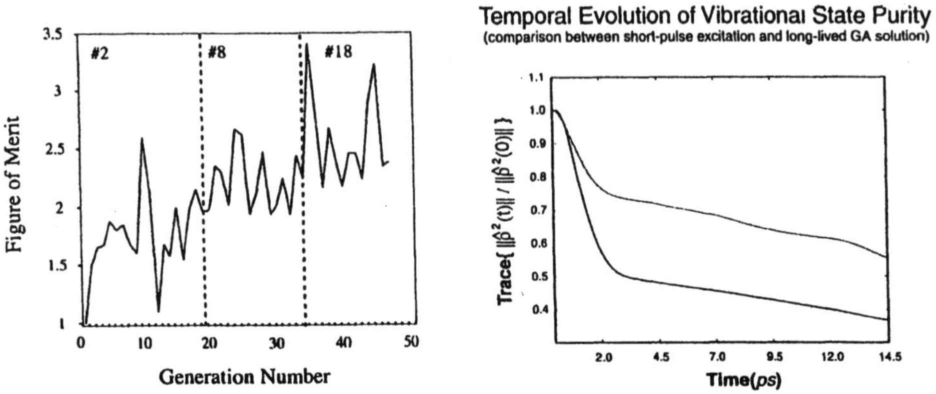

Figure 2. (Left) The generation chart for the genetic algorithm showing improvements in the quantum beat modulation at different vibrational periods. (Right) The trace of the square norm of the vibrational density matrix as a function of time for the impulse-excited (upper) and the optimized (lower) vibrational wavepacket.

The figure of merit controlling the GA is the quantum beat modulation visibility relative to that excited by the initial pulse. There is an improvement of nearly about 80% in the modulation near 12 ps after excitation, as shown in Fig. 2. Our experiments show that a number of interesting decoherence-resistant excitations exist. The nature of the dephasing was examined theoretically. Using measured optical pulse shapes,[6] we simulated numerically the molecular wavepacket in the limit of perturbation theory. The model calculates the complete ro-vibrational density matrix for a thermal initial state, taking into account the rotational-vibrational coupling, and tracing over the rotational degree of freedom. We use the trace of the square of the vibrational density matrix as a measure of the purity of the vibrational wavepacket.

Note that the state generated by the learning control loop exhibits a much less rapid decoherence than the state generated by a Fourier-transform limited pulse with the same bandwidth. Furthermore, this behavior can be controlled with a closed-loop approach in a fashion that can be measured and modeled. We conclude that wavepackets resistant to dephasing do exist, and that learning algorithms provide an efficient means for finding them.

References

1 S. Wallentowitz, I. A. Walmsley, L. J. Waxer and Th. Richter, *J. Phys. B: At. Mol. Opt. Phys.*, 35, 1967, 2002.
2 T. Dunn, I. A. Walmsley and S. Mukamel, *Phys. Rev. Lett.*, 74, 884, 1995.
3 I. A. Walmsley and L. Waxer, *J. Phys. B: At., Mol. Opt. Phys.*, 31, 1825, 1998.
4 T. Dunn, J. Sweetser, I.A. Walmsley and C. Radzewicz, *Phys. Rev. Lett.*, 70, 3388, 1993.
5 C. Brif, H. Rabitz, S. Wallentowitz and I. A. Walmsley, *Phys. Rev. A*, 63, 063404, 2001.
6 C. Iaconis and I. A. Walmsley, *Opt. Lett.*, 23, 792, 1998; L. Lepetit, G. Cheriaux and M. Joffre, *J. Opt. Soc. Am. B*, 12, 2467, 1995.

Ultrafast Two-Photon Free Induction Decay

Dorine Keusters, Peifang Tian, Warren S. Warren

Dept. Chemistry and Electrical Engineering, Princeton University, Princeton, NJ, 08544
E-mail: wwarren@princeton.edu

Abstract. Phase sensitive measurement of two-photon induced coherence is demonstrated, using phase coherent pulse pairs. By using an acousto-optic pulse shaper the different coherence transfer pathways in the two-photon absorption process can be separated.

1. Introduction

Direct optical analogs of (2D) NMR experiments require detection of the phase of a coherence induced by a (series of) laser pulses, which is not trivial. [1]. One method to obtain this phase information is by using a phase coherent pulse pair created by an acousto-optic pulse shaper (AOPS). [2] This technique was demonstrated on the one-photon transitions in Rubidium [2]. The same technique can be used to detect a coherence that is induced by two photons simultaneously [3], as we demonstrate experimentally by detecting the two-photon free induction decay (TPFID) of rubidium and coumarin. Since the two-photon transition in Rb is resonance enhanced, several types of coherences are induced in the two-photon process. The use of an AOPS allows a separation of these types of coherences.

2. Theory

The technique employs two phase coherent pulses with a delay between them. The first pulse introduces a coherence. If the phase relation between the second pulse and the coherence is correct, the detection pulse will transfer the coherence into a population in the excited state. But with the opposite phase relation, the coherence will be transferred to the ground state. Thus, the population in the excited state directly reflects the phase of the coherence at the arrival time of the second pulse. Population in the excited state can easily be detected, e.g. by the fluorescence.

The pulse sequence is generated by an AOPS, using two RF sine waves with different frequencies. This creates two phase coherent laser pulses. [2]

$$E(t) = E_1 + E_2 = A_1(t)e^{-i\omega t} + A_2(t+\tau)e^{-i\omega t}e^{i\phi} + c.c. \tag{1}$$

where A_1 and A_2 are the envelopes of the pulses, ω is the center frequency, τ is the delay, which is proportional to beat frequency $\Delta\omega_{RF}$ of the RF waves. ϕ is the phase difference of the two pulses, and is related to the phase difference of the RF waves. The laser creates a pulse every 12.5 ns. If the first pulse hits the AOPS when the two RF waves are in phase, the resulting laser pulses are in phase too. But when the next pulse arrives after 12.5 ns, the RF waves have propagated through the crystal, and now the phase-difference is $\Delta\omega_{RF}$*12.5 ns, which creates laser pulses with $\phi=\Delta\omega_{RF}$*12.5 ns. The next pulse sees again a different phase-relation. Thus, ϕ is automatically modulated with the beat frequency of the RF waves: $\phi(t_{RF})= \Delta\omega_{RF}*t_{RF}$. The population created by this detection pulse will also be modulated at this beat frequency, which allows lock-in detection. The amplitude and phase of the oscillating population are directly related to the amplitude and phase of the induced coherence. To demonstrate that this technique can also be used to detect the phase of a coherence that is induced by two photons simultaneously, the TPFID of rubidium and coumarin were measured.

Fig. 1a shows the level diagram of Rb. The two-photon transition from 5S to 5D is resonantly enhanced by the 5P states. The 5D population is detected by the 420 nm fluorescence emitted from the 6P levels. This population is proportional to $I^2=|E|^4$, and it requires four interactions with the electric field to reach it. There are several pathways to create a population in the 5D states:

1. Two interactions with the first pulse create a coherence between 5S and 5D. Then two interactions with the second field create a population in the 5D states. The population in the 5D state is proportional to:

$$P_1 \propto A_1^2 A_2^{*2} \cos(2\phi) \propto \cos(2\Delta\omega_{RF} t_{RF}) \tag{2}$$

2. One interaction with pulse 1 creates a 5S-5P coherence. Three interactions with pulse 2 create the 5D population:

$$P_2 \propto A_1 A_2 A_2^{*2} \cos(\phi) \propto \cos(\Delta\omega_{RF} t_{RF}) \tag{3}$$

3. Three interactions with pulse 1 create a 5P-5D coherence. One interaction with pulse 2 creates the 5D population:

$$P_3 \propto A_1^2 A_1^* A_2 \cos(\phi) \propto \cos(\Delta\omega_{RF} t_{RF}) \tag{4}$$

For pathway 1, the signal is modulated at $2\Delta\omega_{RF}$ (2f), and reflects evolution of a two-photon coherence; for pathways 2 and 3, it is modulated at $1\Delta\omega_{RF}$ (1f), and reflects evolution of a one-photon coherence. Thus, the phase modulation introduced by the AOPS allows separation of the different types of coherences.

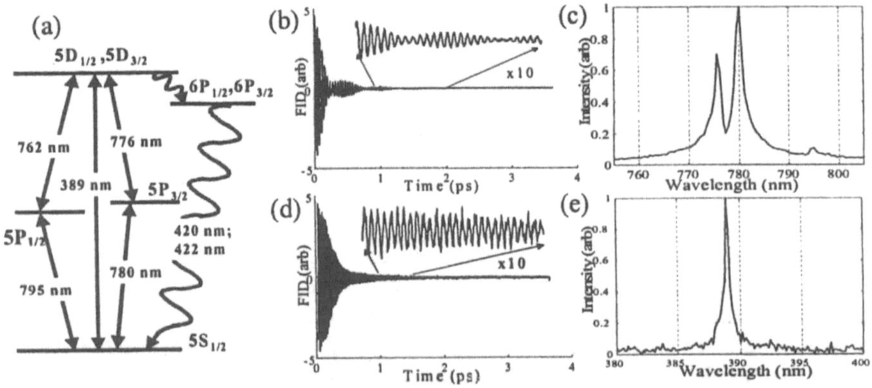

Fig. 1. a) Energy level diagram Rb b) TPFID detected at 1f. Inset: TPFID between 1 and 2 ps. c) Fourier transform of b. d) TPFID detected at 2f. Inset: TPFID between 1 and 1.5 ps e) Fourier transform of d. Note fourier transforms are taken from 1 to 3.6 ps.

3. Results

Pulses from a Ti:Sapph laser are shaped and passed through a sample (a Rb cell at 200°C or a 0.5 mM solution of coumarin in ethanol). The fluorescence is detected at right angle by a PMT, and the 1f and 2f components are selected by a lock-in amplifier. The experiment is repeated with a series of RF beat frequencies ranging from 25 kHz to 16 MHz, which gives the TPFID from 5 fs to 3.6 ps.

Figs. 1b and d show the real part of the TPFID. Both 1f and 2f components decay much faster than expected for the coherence lifetime of Rb. Two factors contribute to this decay. (1) At early times, the two pulses overlap, resulting in an autocorrelation-type signal. (2) The fluorescence lifetimes of the 5D and 6P states are hundreds of nanoseconds. Thus, when the next pulse sequence arrives after 12.5 ns, fluorescence from the previous sequence is still detected. At high RF frequencies (= long delays), the modulation is washed away, and the signal is smaller than expected. For a fluorescence lifetime of 600 ns the amplitude of the signal is expected to drop to 1% of its initial value within 1 ps.

Fig. 1c shows the fourier transform of the 1f TPFID signal. The two main peaks correspond to the coherences between 5S-5$P_{3/2}$ at 780 nm (Eq. 3) and 5$P_{3/2}$-5D at 776 nm (Eq. 4). A small peak is also seen at 795 nm (5S-5$P_{1/2}$). However, since there is no 762 component in the spectrum of the shaped pulse, this peak is not expected. Further investigations in the origin of this peak are underway. Fig. 1e shows the fourier transform of the 2f-TPFID. The single peak at 389 nm corresponds to the 5S-5D-coherence (Eq. 2).

Fig. 2 shows the TPFID of coumarin. Since there are no intermediate levels in coumarin, only pathway 1 is possible, and no 1f component is expected. The two-photon absorption spectrum is much broader than the pulse bandwidth, so the coherence lifetime is expected to be shorter than the pulse. As a result both the 1f and 2f TPFID-signal are dominated by the autocorrelation signal from the two overlapping pulses. Information in this system would be extracted by deconvolution and comparison of the two components.

References

1 S.M.Gallagher, A.W.Albrecht, T.D.Hybl, B.L.Landin, B.Rajaram, D.M.Jonas, "Heterodyne detection of the complete electric field of femtosecond four-wave mixing signals" J. Opt. Soc. Am. B **15** 2338-2345 (1998)

2 D.Keusters, P.Tian, W.S.Warren, "Femtosecond polarization detection using high-speed pulse shaping" in Ultrafast phenomena XII R. Elsaesser, S. Mukamel, M.M. Murnane, N.F. Scherer, eds. (Springer, Berlin, 2001) 165-167

3 H.Kawashima, K.A.Nelson, "Real-time probing of two-photon absorption with phase related pulses" J. Chem. Phys. 100, 6160-6165, (1994).

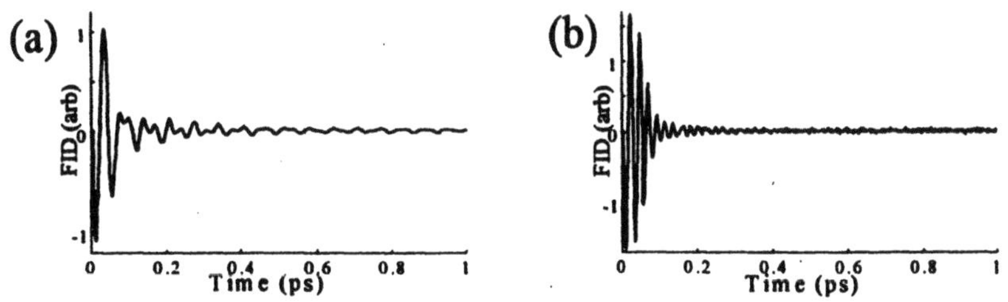

Time-Resolved Coincidence Imaging of the Dissociative Ionization in CF_3I

Anouk M. Rijs[1], Carl C. Hayden[2], and Maurice H.M. Janssen[1]

[1] Laser Center and Chemistry Department, Vrije Universiteit, de Boelelaan 1083, 1081 HV Amsterdam, The Netherlands
E-mail: rijs@chem.vu.nl
[2] Combustion Research Facility, Sandia National Laboratories, P.O. Box 969, Livermore, California 94550-0696, USA

Abstract. The real-time dynamics of the photodissociative ionization of CF_3I has been studied by femtosecond photoelectron-photoion coincidence imaging. With femtosecond coincidence imaging time-resolved photoelectron angular distributions in the molecular frame of the dissociating CF_3I molecule can be obtained. Various stages of the dissociative ionization process are probed at different wavelengths and polarizations.

1. Introduction

Photodissociation dynamics provide a wealth of information on molecular interactions. With the introduction of femtosecond lasers to chemical dynamics it is nowadays possible to study these photon-induced dissociation dynamics in real time [1]. Most of the femtosecond gas phase experiments use sensitive electron [2-4] or ion detection [1,5]. However, it is advantageous to detect the photoelectron and photoion in coincidence, since both particles contain important information about the photodissociation process.

Photoelectron angular distributions (PADs) contain detailed information on the symmetry, electronic configuration and orientation of the molecule as the electron is removed. However, much of this information is lost when the PAD is measured in the lab-frame, with randomly oriented molecules. However, in the molecular frame, where the velocity vector of the electron is correlated with a dissociation axis of the parent molecule, the PAD may be a sensitive probe of photodissociation and dissociative photoionization dynamics [6]. This contribution presents results from a new technique [7,8] that combines energy- and angular resolved photoion-photoelectron coincidence technique with femtosecond pump-probe spectroscopy. With this new femtosecond coicidence imaging we can obtain time-resolved PADs in the frame of the dissociating molecule. The present study reports the multi-photon induced dynamics of CF_3I in real time.

2. Experimental Set-Up

The femtosecond energy and angle resolved photoelectron-photoion coincidence imaging experiments were performed at Sandia National Labs, Livermore [7-9].

Linearly polarized laser pulses from a 1 kHz regeneratively amplified Titanium sapphire laser are frequency doubled to 396 nm and tripled to 264.5 nm. The 264.5 nm pump (~8 μJ) and the 396 nm probe (~25 μJ) beams are combined and mildly focused (focal length 1000 mm) into an ultra high vacuum chamber, where they intersect the pulsed molecular beam. For all the data presented here, the polarization of both the pump and probe pulse was along the direction of the molecular beam velocity and in plane of the two detectors (horizontal polarization). In these experiments the pump beam two-photon excites the neutral CF_3I molecule and the probe dissociatively ionizes the excited molecule to produce CF_3^+ and I^+ fragments.

The electrons and ions are accelerated and extracted in opposite directions towards time- and position-sensitive detectors. The acceptance solid angle of the detectors is 4π, such that full three-dimensional angular distributions are measured. To suppress false coincidences the laser pulse energies were adjusted such that typically the total coincidence rate was less than ~0.01 events per laser shot. Coincidence data was taken at various pump-probe delay times.

3. Time-resolved Molecular Frame PAD

In addition to lab frame angular distributions, with the femtosecond coincidence imaging technique we can also obtain time-resolved angular distributions in the frame of the dissociating molecule. This means that we can correlate the angular direction of the electron with the recoil direction of the ionic fragment. Figure 1 shows the 3D photoelectron angular distribution coincident with I^+ obtained at 132 fs pump-probe delay for electrons with a kinetic energy of 0.3-1.0 eV. The distance of the surface from the center of the plot shows the electron intensity as a function of the polar angle away from the axis. Comparing Fig 1a and 1b, we can conclude that the photoelectron angular distribution corresponding to I^+ is essentially uncorrelated with the ion recoil direction.

Figure 2a and 2b shows the similar electron angular distributions coincident with CF_3^+ for the high kinetic energy channel (0.5-1.3 eV). In this case the molecular frame PAD has strong forward-backward asymmetry with the electron

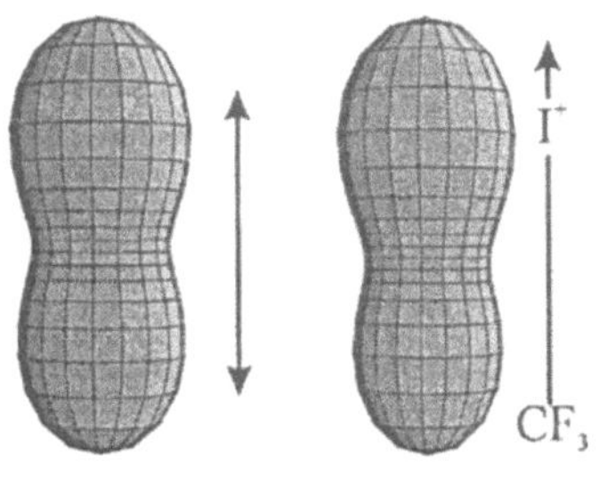

Figure 1: a) Lab frame electron angular distribution referenced to the laser polarization (arrow) for electrons coincident with I^+ at 132 fs pump-probe delay, b) Molecular frame electron angular distribution referenced to the direction of recoiling I^+ for ions recoiling within 30 degrees of the laser polarization.

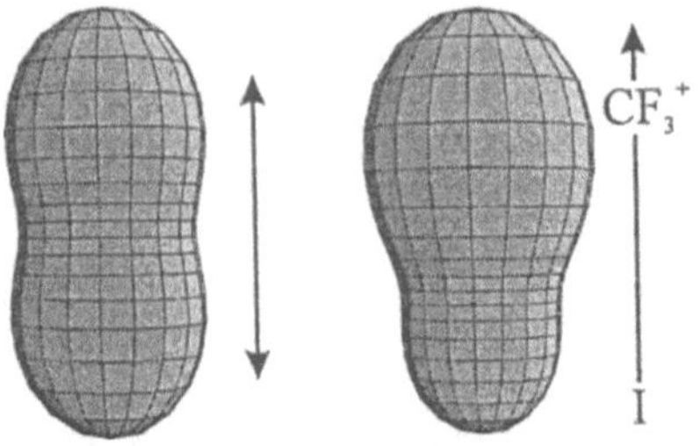

Figure 2: a) Lab frame electron angular distribution referenced to the laser polarization (arrow) for electrons coincident with CF_3^+ at 132 fs pump-probe delay, b) Molecular frame electron angular distribution referenced to the direction of recoiling CF_3^+ for ions recoiling within 30 of the laser polarization.

preferentially recoiling in the direction of the departing ion. These results suggest that CF_3^+ is produced by simultaneous dissociation and ionization, which is energetically accessible with one probe photon. In contrast I^+ can not be produced by absorption of one probe photon so the ionization and dissociation may occur in separate steps, consistent with the lack of correlation between the electron and I^+ recoil velocity vectors.

Acknowledgements. The authors thank Mike Gutzler for his excellent technical support, and Engelene Chrysostom and Khai Luong for help in the experiments. AMR and MHMJ acknowledge the council for Chemistry Sciences of the Netherlands Organization for Scientific Research (CW-NWO) for travel support. AMR gratefully acknowledge the Holland Research School of Molecular Chemistry for a PhD fellowship. The research of CCH is supported by Division of Chemical Sciences, Geosciences and Biosciences, the Office of Basic Energy Sciences, the U.S. Department of Energy.

References

1 A.H. Zewail, *J. Phys. Chem. A.* **104** (2000), 5660.
2 V. Blanchet, M.Z. Zgierski, T. Scheideman, and A. Stolow, *Nature* **401** (1999), 52.
3 L. Wang, H. Kohguchi, and T. Suzuki, *Faraday Discussions* **113** (1999), 37.
4 A. Assion, M. Geisler, J. Helbing, V. Seyfried, and T. Baumert, *Phys. Rev. A* **54** (1996), R4605.
5 W.G. Roeterdink and M.H.M. Janssen, *Phys. Chem. Chem. Phys.* 4 (2002), 601..
6 Y, Arasaki, K. Takatsuka, K. Wang and V. McKoy, *J. Chem. Phys.* **112** (2000), 8871.
7 J.A. Davies, R.E. Continetti, D.W. Chandler, and C.C. Hayden, *Phys. Rev. Lett.* **84** (2000), 5983.
8 J.A. Davies, J.E. Leclaire, R.E. Continetti, and C.C. Hayden, *J. Chem. Phys.* **111** (1999), 1.
9 A.M. Rijs, C.C. Hayden, and M.H.M. Janssen, in *Femtochemistry and Femtobiology: Ultrafast Dynamics in Molecular Science*, A. Douhal and J. Santamaria, World Scientific Publishing Co.Pte.Ltd. (Singapore) (2002), 91.

Making and Measuring Vibrational Wave Packets in Small Molecules through non-Resonant Impulsive Stimulated Raman Scattering

D.A. Samuels[1], T.C. Weinacht[1], O.L.A. Monti[1], E. Gershgoren[1], S.R. Leone[1], H.C. Kapteyn[1], M.M. Murnane[1], R.A. Bartels[1]

[1] JILA, University of Colorado, 440 UCB, Boulder CO 80309-0440, USA
E-mail: david.samuels@colorado.edu

Abstract. We demonstrate the creation and measurement of vibrational wavepackets in small polyatomic molecules using nonresonant impulsive excitation. Modeling reveals features due to vibrational dephasing and rephrasing, rotational dephasing, and dissociation.

1. Introduction

Many chemical reactions take place in the ground electronic state. We present results of experiments featuring the creation, control, and measurement of vibrational wave packets in the ground electronic states of small polyatomic molecules and their photodissociation products. The wave packets are non-resonantly excited using broad bandwidth laser pulses, allowing for study of a large class of molecules. A simple theoretical model of the measurement allows for a comprehensive understanding of the interaction between the laser pulse and the molecular sample. We obtain information on vibrational dephasing and rephasing, intermode coherences, rotational dephasing and dissociation.

2. Experiments and Results

The experiments consist of a series of pump-probe measurements. A shaped pump pulse propagates through a gas sample, creating a vibrational coherence. The spectrum of a time-delayed probe pulse is measured after propagation through the vibrationally excited gas as a function of pump-probe delay and pump pulse shape, yielding detailed information about the vibrational wave packet created by the pump pulse. Several modes can be monitored simultaneously by the probe pulse.

The pump pulse is 20 fs in duration, centered at 800 nm, with an energy of approximately 400 μJ. The probe pulse is approximately 500 fs in duration, centered around 400 nm, and with an energy of approximately 1 μJ. For positive pump-probe delays, the probe pulse develops frequency sidebands whose offsets from the fundamental frequency label which modes are excited, and whose intensities are proportional to the strength of the excitation. Figure 1 shows the probe pulse spectrum as a function of pump-probe delay in CCl_4, along with lineouts at two different frequencies: 460 cm^{-1} (Fig. 1b) and 556 cm^{-1} (Fig. 1c). Note that there are excitations at frequencies not attributable to CCl_4. We attribute these peaks to the decomposition of CCl_4 into Cl_2, as described below.

In order to understand the beats in the CCl_4 wave packet, in addition to the new sidebands, we performed a Filter Diagonalization decomposition of the signals (Fig. 2, dotted lines). Filter Diagonalization is a powerful method for discerning small signals in spectra [4]. The pair of peaks in both data sets (8.2 cm^{-1} and 3.3 cm^{-1} for CCl_4 and 7.8 cm^{-1} and 5.5 cm^{-1} for Cl_2) can be explained in terms of an isotopic mixture of Cl in the CCl_4. Beating between vibrational levels with an anharmonic energy spacing. 556 cm^{-1} is consistent with the vibrational frequency of $^{35}Cl_2$. The two peaks in Fig. 2b agree exactly with the measured anharmonicity of Cl_2 and the vibrational energy differences between molecules composed of different isotopes.

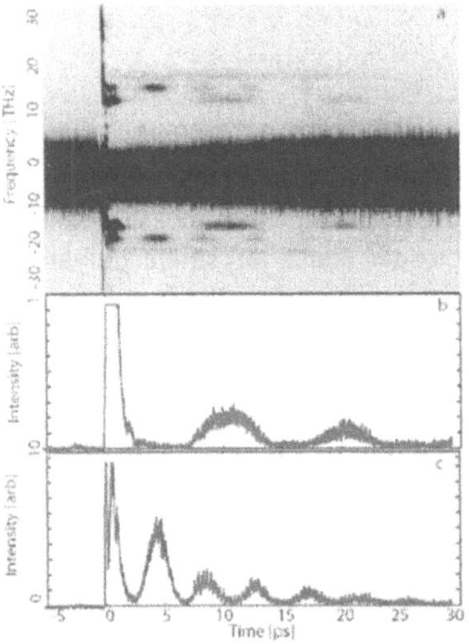

Figure 1 Probe spectrum and sideband intensities as a function of pump-probe delay in CCl_4. Shading is proportional to probe intensity.

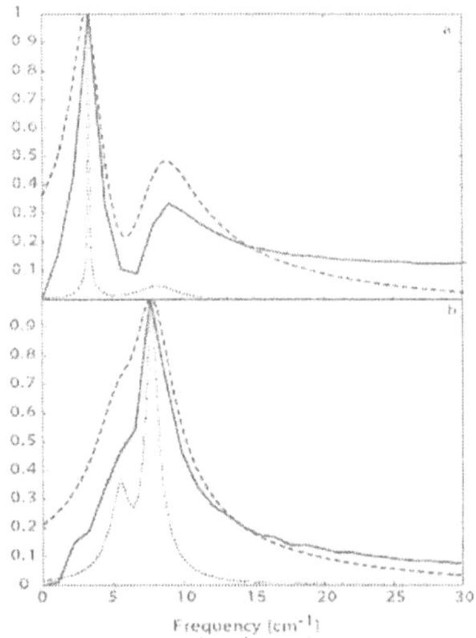

Figure 2 Spectra of sideband intensities vs. pump-probe delay shown in figures 1b and 1c.

One final piece of evidence that points strongly towards Cl_2 formation is the fact that the new modes only appear after several pump laser pulses have passed through the sample. Figure 3 shows the buildup of the new Raman peak at 560 cm^{-1} and the decay of the signal at 460 cm^{-1}, as measured by the intensities of the sidebands at those frequencies. Further studies are under way to improve our understanding of the dissociation process, and also to determine whether the Cl_2 formation is a concerted process or instead occurs through formation of isolated Cl atoms and later recombination into Cl_2.

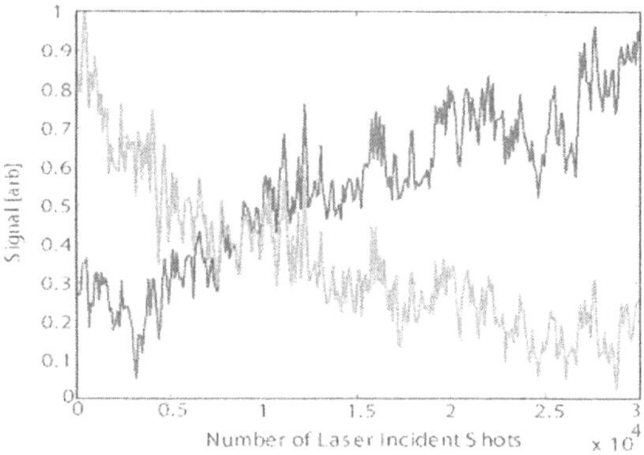

Figure 3 Probe sideband intensities as a function of the number of incident pump pulses in CCl_4

4. Conclusions

In conclusion, we have shown how Impulsive Stimulated Raman Scattering can be used to excite multimode wave packets in the electronic ground state of small molecules. By making use of non-resonant excitation, the technique is general and can be extended to a large class of molecules. We can monitor the development of the wave packet using a time delayed probe pulse.

References

1 T.C. Weinacht *et. al,.Chem. Phys. Lett.* **344**, 333 (2001)
2 R.A. Bartels, *et. al., Phys. Rev. Lett.*, **88**, 033001 (2002)
3 T.C. Weinacht, *et. al., Phys. Rev. A*. Accepted for Publication.
4 G. Hochenbleicher, H.W. Schrotter, *Applied Spectroscopy* **25**, 360 (1971)

Breaking bonds with mid IR pulses: Femtosecond photolysis of gas-phase metal carbonyls

L. Windhorn, T. Witte, J.S. Yeston, D. Proch, M. Motzkus, K.L. Kompa, and W. Fuß

Max-Planck-Institut für Quantenoptik, D-85748 Garching, Germany
E-mail: Lars.Windhorn@mpq.mpg.de

Abstract. Resonant infrared multiphoton dissociation of $Cr(CO)_6$, $Mo(CO)_6$, $W(CO)_6$, $Fe(CO)_5$, and $Ni(CO)_4$ was effected by mid IR femtosecond laser pulses. This requires an excitation to at least v = 7 or 8 of the CO stretch vibration.

1. Introduction

Efficient and selective preparation of highly excited vibrational states in molecules is a challenging objective in physical chemistry. One means of achieving high ground-state vibrational excitation, possibly even leading to bond scission, was infrared multiphoton excitation (IRMPE) by ns lasers [1]. However, in this approach the role of intramolecular vibrational relaxation (IVR) as a process competing with or assisting excitation and possibly being mandatory for populating high vibrational levels was unclear [2,3]. Using femtosecond lasers resolves this uncertainty by restricting excitation to timescales faster than IVR. We have therefore applied fs mid IR laser technology toward the dissociation of transition metal carbonyls (Fig. 1), which had previously been dissociated by nanosecond IR pulses [4]. The fs pulse duration allows detailed modeling of the excitation mechanism. Specifically, analysis of the experimental data affords the number of vibrational modes involved in the excitation.

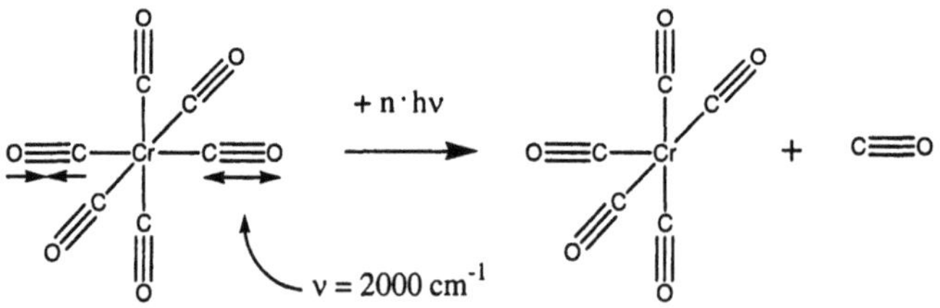

Fig. 1. Resonant excitation of the T_{1u} CO-stretching manifold of $Cr(CO)_6$ with femtosecond pulses at $\lambda = 5\ \mu m$ leads to elimination of a CO.

2. Experiment and Results

Mid IR pulses around 5 μm with energies up to 15 μJ (bandwidth ~140 cm^{-1}, pulse duration ~120 fs, rep rate 1 kHz) were focused with CaF_2 and Ge lenses of f = 100 and 50 mm to a radiation energy density $\Phi \leq 300$ mJ cm^{-2}. $Cr(CO)_6$, $Mo(CO)_6$, $W(CO)_6$, $Fe(CO)_5$, and $Ni(CO)_4$ were filled into a small glass cell with pressures ranging from 3 to 30 Pa. The concentration of metal carbonyls in the cell was determined by their IR absorption spectrum. It was monitored by a FTIR spectrometer on a timescale of minutes. We thus accumulated microscopic dissociation in the focal volume over a few minutes of irradiation to detect a macroscopic depletion of concentration in the cell.

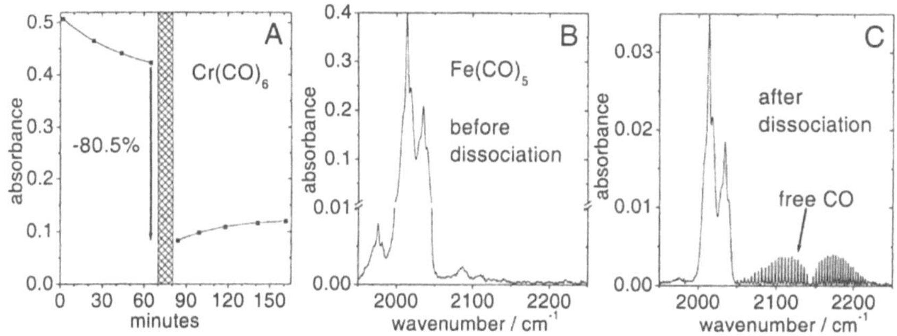

Fig. 2. Part A shows the change in absorbance of $Cr(CO)_6$ after 10 min irradiation (shaded area) with 11μJ at $\lambda = 5$ μm. Part B and C show two absorption spectra of $Fe(CO)_5$, before and after 30 min irradiation with 11μJ at $\lambda = 4950$ nm. Note the different scales in B and C.

We have found that all investigated metal carbonyls can easily be dissociated by resonant multiphoton excitation of their CO stretch band with femtosecond pulses at $\lambda = 5$ μm. This requires an excitation to at least $v = 7$ or 8 as inferred from bond energies for cleaving a single M-CO bond [5].

Macroscopic quantities of metal carbonyls could be dissociated within minutes: The 80% decrease of $Cr(CO)_6$ concentration in Fig. 2A corresponds to 3×10^{10} dissociations per shot. After the dissociation rovibrational lines of free CO appear (Fig. 2C). Dissociation is only observed as long as the excitation wavelength is in resonance with the fundamental CO stretch vibration. The dissociation probability is independent of the laser repetition rate. Hence, dissociation of an individual molecule is a single shot process and not due to consecutive absorption of energy.

The dissociation yield was measured to be a steep function of the incident energy density Φ. To investigate how many degrees of freedom are involved in the excitation process, we modeled the dissociation probability by assuming a stepwise absorption process and solving the resulting rate equations analytically [6,7]. The model provides two parameters, n, which in this case equals the number of degrees of freedom involved in the excitation, and Φ_0^*, a molecule-specific parameter corresponding to a threshold energy density for dissociation. We consider the total conversion U, which is the number of dissociated molecules per pulse divided by the gas density. Fig. 3 shows the theoretical curves for $U(\Phi_{max})$ for different parameters n. Φ_{max} is the maximum energy density in the center of the focal volume. In the plot, both axes have been normalized: the conversion U by the focal volume V_{focal} and Φ_{max} by Φ_0^*. Also shown in the plot are the measured

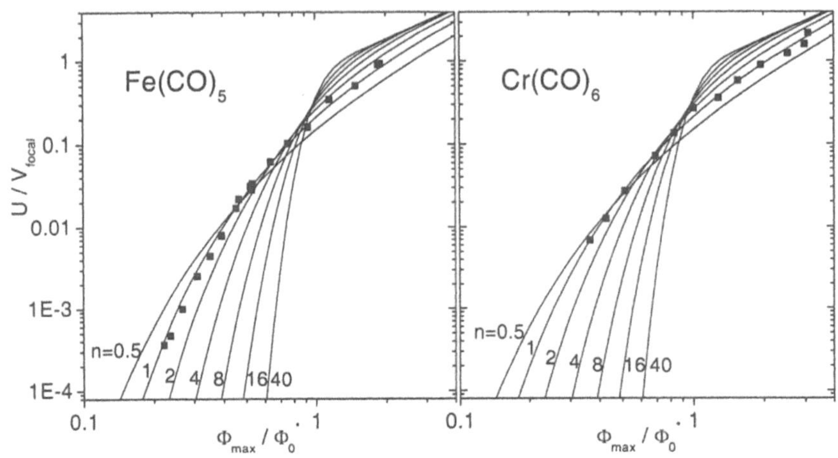

Fig. 3. Theoretical curves of the conversion *U(Φ)* calculated for different parameters *n*. The experimental data shown are for the dissociation of $Fe(CO)_5$ and $Cr(CO)_6$.

data points for the dissociation of $Fe(CO)_5$ and $Cr(CO)_6$. From comparison with the analytical curves, we conclude that only 1 to 2 degrees of freedom participate in the excitation process.

3. Conclusion

Resonant infrared multiphoton dissociation of molecules with femtosecond laser pulses has been demonstrated for the first time. MIR pulses of moderate energies E ~ 15 μJ focused to a Φ up to ~300 mJ cm^{-2} were sufficient to dissociate macroscopic quantities of $Cr(CO)_6$, $Mo(CO)_6$, $W(CO)_6$, $Fe(CO)_5$, and $Ni(CO)_4$. This requires an excitation to at least v = 7 or 8 of the CO stretching manifold. Comparison with an analytical model indicates that only 1 to 2 degrees of freedom are involved in the excitation process.

References

1 R. V. Ambartzumian and V. S. Letokhov, in *Chemical and Biochemical Applications of Lasers*, Edited by C. B. Moore, Vol. 3, 167-316, 1977.
2 W. Fuss and K. L. Kompa, *Progress in Quantum Electr.* **7**, 117, 1981.
3 W. Fuss, *Chem. Phys.* **76**, 343, 1983.
4 M. K. Au, P. A. Hackett, M. Humphries and P. John, *Appl. Phys. B* **33**, 43, 1984.
5 K. E. Lewis, D. M. Golden and G. P. Smith, *J. Am. Chem. Soc.* **106**, 3905, 1984.
6 C. D'Ambrosio, W. Fuss, K. L. Kompa and W. E. Schmid *Opt. Soc. Am. B* **5**, 1540, 1988.
7 W. Fuss, *Chem. Phys.*, **36**, 135, 1979.

Femtosecond four-wave mixing for molecule based computation

Vadim V. Lozovoy, Igor Pastirk, Matthew Comstock, Katherine A. Walowicz, and Marcos Dantus.

Department of Chemistry and Department of Physics and Astronomy, Michigan State University, East Lansing, MI 48824;
E-mail: lozovoy@msu.edu, dantus@msu.edu

Abstract. Analog and digital information can be encoded as spectral amplitude and phase of coherent laser pulses. Nonlinear interactions between light and molecules can be used for performing basic linear algebraic transformations between information in the shaped laser pulses and in the quantum eigenstates. The manipulation of a molecular rovibronic manifold using nonlinear optical processes such as stimulated photon echo is discussed from an information technology point of view. Preliminary experimental results on molecular iodine are presented.

1. Introduction

Massive parallelism is the key to the computer of future. Here we propose to use three-pulse four-wave mixing (FWM), in particular the stimulated photon echo (SPE) and virtual echo (VE) pulse sequences, to perform computational tasks in parallel through the preparation and control coherent ensembles of vibrational states [1]. The transformations are achieved with femtosecond amplitude and phase shaped laser pulses. Results from our group and others indicate that FWM signals from gas phase molecules have a great potential for carrying out molecule-based computation [2-6].

The main idea is to write information on the system with a shaped pulse (as quantum amplitudes on selected eigenstates) and to manipulate the information with subsequent shaped coherent light pulses. The shaped pulses contain specific spectral components that address individual v'-v'' transitions. Each specific vibrational eigenstate is a component of a vector in Hilbert space and the manipulation with complex amplitudes of these vector components allows us to perform a computation with a rate of 10^{13} bits per second.

2. Experimental results

Here we present data obtained from molecular I_2 at ~ 90 °C using the VE arrangement with a 595 nm pump, shaped 800 nm Stokes pulse and 800 nm probe pulse and homodyne detection (additional details can be found in [7]). The FWM signal was dispersed to achieve spectral resolution. The experimental results show

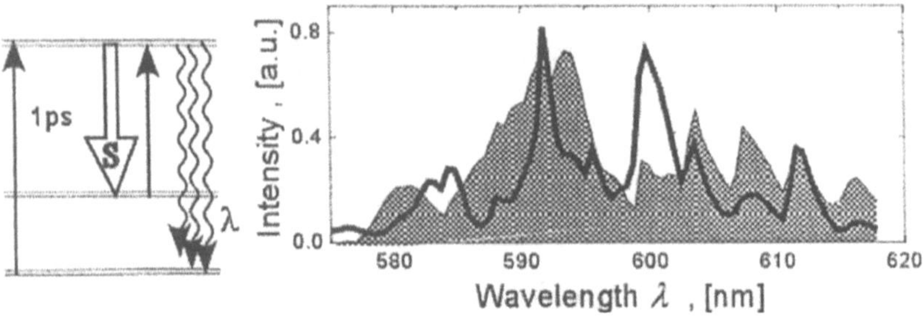

Fig. 1. Schematic of the pulse sequence and FWM spectra for unshaped (thick line) and shaped (shaded) Stokes pulses.

the spectral output obtained when the Stokes pulse was (left) near transform-limited and (right) when it was shaped (amplitude of bluer wavelengths blocked). The differences between the two spectra demonstrated the dependence of the output FWM spectrum from the input laser spectra.

3. Conclusion

From a mathematical point of view, the proposed VE and SPE processes can perform the following functions: write, store, read, sum, scalar product, direct products, matrix product between matrices and vectors, and can be used to construct circuits and networks. Technologically the proposed setup has a number of advantages. The operation time is $\sim 10^{-12}$ s, the coherence dephasing time is approx. $10^{-6} - 10^{-9}$ s, the quanta are 10^2 greater than kT, the information is stored in $\sim 10^{10}$ replicas, and the output is a coherent beam that could be used to build a more complex circuit.

Acknowledgements

This research was funded by Chemical Sciences, Geosciences and Biosciences Division, Office of Basic Energy Sciences, Office of Science, U.S. Department of Energy. MD is a Camille Dreyfus Teacher-Scholar.

References

1 V.V. Lozovoy, B.I. Grimberg, I. Pastirk, M. Dantus, Chem. Phys. Vol. 267, 99, 2001.
2 V.V. Lozovoy, I. Pastirk, et al. Int. Rev. Phys. Chem. Vol. 19, 531, 2000.
3 I. Pastirk, V.V. Lozovoy, M. Dantus, Chem. Phys. Lett. Vol. 333, 76, 2001.
4 T. Chen, V. Engel, M. Heid, et al, J. Mol. Struct. Vol. 481, 33, 1999.
5 G. Knopp, I. Pinkas, Y. Prior, J. Raman Spectr. Vol. 31, 51, 2000.
6 R. Zadoyan, D. Kohen, D. A. Lidar, V. A. Apkarian, Chem. Phys. Vol. 266, 323, 2001.
7 V. V. Lozovoy, Dantus, Chem. Phys. Lett, Vol, 351, 213, 2002.

Preparation of oriented molecules for ultrafast spectroscopic studies

Hirofumi Sakai, Shinichirou Minemoto, Hiroshi Nanjo, Haruka Tanji, and Takayuki Suzuki

Department of Physics, Graduate School of Science, The University of Tokyo, 7-3-1, Hongo, Bunkyo-ku, Tokyo, 113-0033, Japan
E-mail: hsakai@phys.s.u-tokyo.ac.jp

Abstract. We demonstrate that common molecules such as OCS can be oriented with combined electrostatic and pulsed, nonresonant laser fields. The degree of orientation can be increased by increasing the magnitude of electrostatic field and the peak intensity of the laser field or by decreasing the rotational temperature of the molecules.

1. Introduction

The manipulation of atoms and molecules is one of the most important and intriguing subjects in modern physics, chemistry, and biology. The control of the spatial direction of molecules is especially important for studies of orientational effect in chemical reaction dynamics [1]. The alignment of neutral molecules by a strong linearly-polarized laser field has already been demonstrated [2,3]. At present, strong electrostatic field orientation and electric hexapole focusing are used to control the orientation of polar molecules. However, strong electrostatic field orientation is effective only for certain molecules possessing large permanent dipole moments and electric hexapole focusing works only for symmetric top molecules. The experimental realization of molecular orientation to arrange *any* polar molecules in a "head-versus-tail" order is a next challenging subject and should greatly expand the range of applications in stereodynamical studies of chemical reaction dynamics.

Recently, Friedrich and Herschbach suggested exploiting combined electrostatic and nonresonant induced dipole forces to enhance the orientation of polar molecules [4,5]. Here we report the experimental evidence of molecular orientation of *common* molecules such as OCS based on their scheme. Since our experiment is performed in an adiabatic regime where the orientation proceeds slowly compared to the rotational period of the molecule, we can keep a sample of oriented molecules for a few nanoseconds if we employ a laser pulse with the duration of ~10 ns. Femtosecond or picosecond time-resolved studies including femtosecond timed Coulomb explosion studies [6,7] can be performed on such a sample of oriented molecules. Furthermore, oriented molecules can be used to enhance or suppress some photochemical events whose efficiencies depend on the angle between a specific molecular axis and the polarization direction of the laser field.

2. Experimental

We performed a Coulomb explosion experiment using intense femtosecond Ti:sapphire laser pulses (wavelength: ~ 800 nm, peak intensity: 3 x 10^{14} W/cm^2) and a time-of-flight (TOF) mass spectrometer. To achieve molecular orientation, we apply fundamentals (wavelength: 1064 nm, pulse width: ~ 12 ns, peak intensity: 2.6 x 10^{12} W/cm2) from an injection-seeded Nd:YAG laser and utilize an extraction field (380 V/cm) also as an electrostatic field of the above scheme. Therefore, the polarization direction of the YAG laser is also parallel to the TOF axis. We irradiate the Ti:sapphire laser pulses at the peak of the YAG laser pulses to see the degree of orientation. Here we use OCS molecules as a sample.

3. Results and Discussions

When the polarization direction of the intense femtosecond laser pulses is parallel to the TOF axis, we usually observe a pair of peaks consisting of so-called "forward" and "backward" fragments whose initial velocities are directed towards and away from the detector, respectively. This is the result of enhanced ionization [8-10] by which molecules initially aligned along the polarization direction are much more likely to undergo multiple ionization. Figure 1 shows the typical TOF spectra obtained (a) with and (b) without Nd:YAG laser pulses. Here we focus on the S^{3+} signals. When the YAG laser is not applied (b), the "forward" and "backward" signals look almost symmetric, indicating that the molecules are randomly oriented. However, when we apply the YAG laser (a), they look asymmetric, which we interpret as the result that more than half the OCS molecules are oriented with their S atoms directed toward the detector. This is consistent with our expectation based on a simple consideration about electronegativity. (The enhancement of "backward" signal is the result of combined effect of molecular alignment and enhanced ionization, i.e., some molecules are not oriented but just aligned.) We can confirm our interpretation by observing the change of the signal magnitude of counterpart fragment, e.g., CO^+. As we expected, the "backward" signal was larger than the "forward" signal.

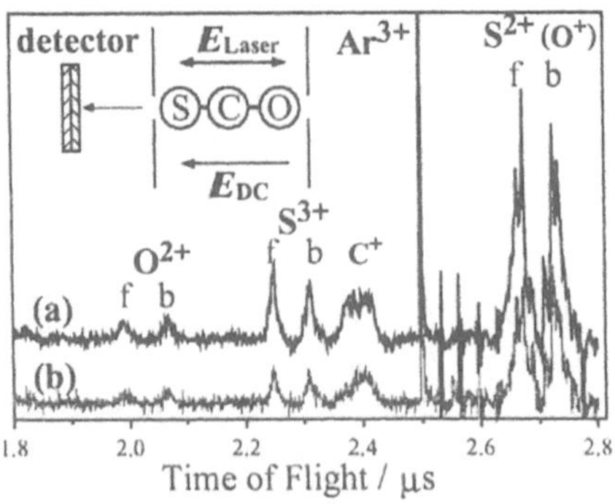

FIG. 1. Typical TOF spectra of OCS molecules (a) with and (b) without Nd:YAG laser pulses. The labels **f** and **b** denote fragment ions coming from the "forward" and "backward" initial emission directions, respectively.

We use the quantity of $F = I_f/(I_f+I_b)$ as an order parameter of orientation, where I_f and I_b are the integrated signals of "forward" and "backward" fragments, respectively. We measured the dependence of the F parameter on the peak intensity of the YAG pulse, the magnitude of electrostatic field, and the backing pressure in the pulsed gas valve (i.e., the rotational temperature of the molecules), which are the three key parameters that determine the degree of orientation. As expected theoretically, the F parameter of S^{3+} (CO^{+}) became larger (smaller) for the higher intensity of the YAG pulse, the larger magnitude of electrostatic field, and the higher backing pressure in the pulsed gas valve (i.e., the lower rotational temperature), supporting our interpretation that OCS molecules are oriented. Since the intensity of the YAG pulse and the magnitude of electrostatic field used in our experiments are still modest, we can easily increase the degree of orientation.

4. Conclusion

We have demonstrated clear evidence of the orientation of OCS molecules. All the results obtained are consistent with one another and compatible with our interpretation that OCS molecules are oriented. Our results show the efficacy of the present method [4,5] and the usefulness of our approach to confirm the molecular orientation.

Acknowledgements. The authors are grateful to Bretislav Friedrich for his helpful discussions. They also appreciate financial support by Japan Society of the Promotion of Science.

References

1 J. Phys. Chem. A **101**, 41, 1997, special issue on Stereodynamics of Chemical Reactions.
2 H. Sakai, C. P. Safvan, J. J. Larsen, K. M. Hilligsφe, K. Hald, and H. Stapelfeldt, J. Chem. Phys. **110**, 10235, 1999.
3 J. J. Larsen, H. Sakai, C. P. Safvan, I. Wendt-Larsen, and H. Stapelfeldt, J. Chem. Phys. **111**, 7774, 1999.
4 B. Friedrich and D. Herschbach, J. Chem. Phys. **111**, 6157, 1999.
5 B. Friedrich and D. Herschbach, J. Phys. Chem. A **103**, 10280, 1999.
6 Ch. Ellert, H. Stapelfeldt, E. Constant, H. Sakai, J. Wright, D. M. Rayner, and P. B. Corkum, Phil. Trans. R. Soc. Lond, A, **356**, 329, 1998.
7. H. Stapelfeldt, E. Constant, H. Sakai, and P. B. Corkum, Phys. Rev. A **58**, 426, 1998.
8. T. Seideman, M. Yu. Ivanov, and P. B. Corkum, Phys. Rev. Lett. **75**, 2819, 1995.
9. T. Zuo and A. D. Bandrauk, Phys. Rev. A **52**, R2511, 1995.
10. E. Constant, H. Stapelfeldt, and P. B. Corkum, Phys. Rev. Lett. **76**, 4140, 1996.

3/2 Harmonic Generation by Femtosecond Laser Pulses on Solid Targets

A. Tarasevitch, C. Dietrich, C. Blome, K. Sokolowski-Tinten, and D. von der Linde

Institute for Laser and Plasma Physics, University of Essen, 45117 Essen, Germany

Abstract. 3/2 harmonic radiation is generated by femtosecond pulses in a very steep-gradient plasma on solid surfaces. A well-collimated harmonic emission with high conversion efficiency (5×10^{-4}) and an unusually broad spectrum is observed.

1. Introduction

The development of high-power femtosecond lasers has enabled experiments on laser-plasma interactions in the relativistic regime. Experiments at this intensity level are of great interest because of a number of applications such as fast ignition [1], particle acceleration [2-4], hard x-ray [2-4] and high order harmonic generation [5], etc. Plasma instabilities can play an important role in these interactions. Moreover, completely new regimes of electronic instabilities not evident at lower intensities can be expected.

The two-plasmon decay (TPD) instability has the lowest threshold with respect to plasma inhomogeneity. This type of instability manifests itself in the emission at 3/2 times the fundamental frequency ω. The TPD and 3/2 ω generation in plasmas have been studied both experimentally and theoretically for more than three decades (see [6,7] and references therein). These studies have dealt with relatively long spatial and temporal scales and with moderate intensities, typical of nano- and subnanosecond laser-plasma interaction.

We report here on the observation of the TPD and 3/2 ω generation in a weakly relativistic regime in plasmas produced by the interaction of high-contrast-ratio femtosecond laser pulses with solid targets. The plasma scale length was changed in a controlled way in order to explicitly investigate its influence on the instability threshold. Strong 3/2 ω emission with an unusually broad spectrum and high directionality was already observed at $L \sim \lambda$, where λ is the laser wavelength.

2. Experimental Results

In the experiments we used a titanium sapphire laser producing 200 mJ pulses of 120 fs duration at a wavelength of 800 nm. The ratio of the pulse peak intensity to the intensity at ~ 1 ps (intensity contrast) was determined to be about 10^5. The ASE intensity contrast was approximately 10^9. The incident laser beam was focused onto a glass target with the help of an off-axis parabolic mirror. The beam was p-polarized and the angle of incidence was 38^0. The peak intensity on the target surface was 7×10^{17} W/cm^2 corresponding to a normalized laser wave amplitude $a_0 = eA/mc^2 \sim 0.6$. The beam providing the controlled "pre-pulse" was split from the "main" beam with the help of small mirrors. The intensity of the "pre-pulse" on the target surface was 10^{15} W/cm^2. The plasma scale length L as a function of delay time, was estimated using the "Medusa" hydro-code.

We observed the angular distribution of the reflected light on a screen with the help of a CCD-camera. The reflected radiation at the fundamental frequency was attenuated with suitable optical filters. At short delays only the collimated beam of the specularly reflected fundamental and second harmonic was present. However, starting from a certain delay green 3/2 ω emission can be seen (Fig. 1).

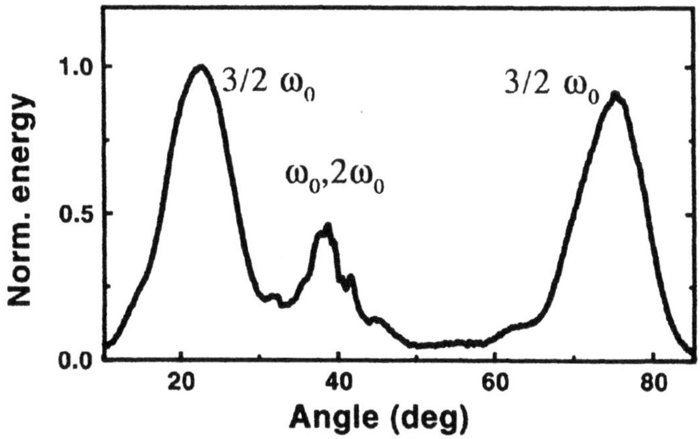

Fig. 1: Angular distribution of the radiation emitted from the target surface for a delay time $\tau = 25$ ps, $L \sim 1.1\lambda$.

It is interesting that in a particular range of delay times (see below) the 3/2 ω emission is well collimated and directed at $\sim 25^0$ and $\sim 75^0$ to the target normal. At all delay times, no harmonic radiation in the backward direction was observed.

In Fig. 2 the 3/2 ω energy is plotted as a function of delay time (lower scale) and scale length (upper scale). The energy was measured with two calibrated photodiodes. The dashed vertical lines mark the range where the 3/2 ω was emitted as two collimated beams. It can be seen that the TPD instability already develops at a very steep plasma-vacuum interface. The energies of the 3/2 ω at the angles of $\sim 25^0$ and $\sim 75^0$ were approximately equal. The total energy of the 3/2 ω emission was about 10 μJ at a delay time of 30 ps ($L \sim 1.2\lambda$), resulting in a conversion efficiency of 5×10^{-4}.

The measured harmonic spectra were much broader than classical dispersion relations would allow. This discrepancy may be connected with high intensity/relativistic effects such as nonlinear saturation of the instability, merging of the TPD and Raman instabilities, etc. [8,9].

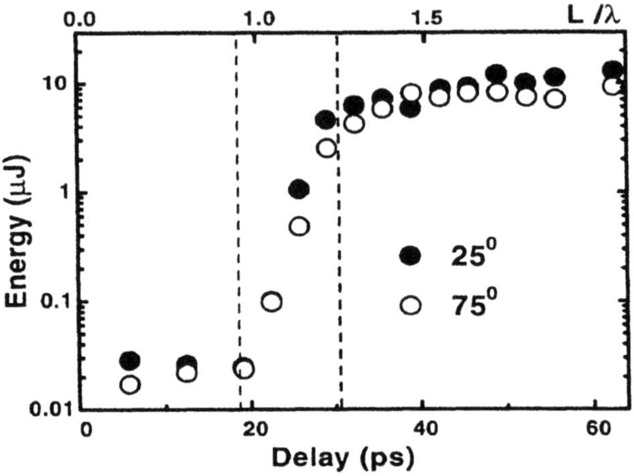

Fig. 2: Energy of the 3/2 ω for the angles of 25^0 and 75^0 as a function of delay time and L.

3. Conclusions

In conclusion, we have observed the onset of the TPD plasma instability and 3/2 ω emission with femtosecond pulses in the weakly relativistic intensity regime (7×10^{17} W/cm^2, $a_0 \sim 0.6$). Using plasmas with controlled scale lengths L we found the threshold for TPD at $a_0 \sim 0.6$ to be $L \sim \lambda$. At this scale length the 3/2 ω was emitted in the form of two well-collimated beams. A very high conversion efficiency of 5×10^{-4} to the 3/2 ω was reached. A broad spectrum of harmonic emission indicated an unusually broad spectrum of TPD.

References

1 M. Tabak et al., Phys. Plasmas **1**, 1626 (1994).
2 D. Umstadter, Phys. Plasmas **8**, 1774-1785 (2001).
3 K. Kruschelnick et al., Phys. Plasmas 7, 2055 (2000).
4 S. P. Hatchett et al, Phys. Plasmas 7, 2076 (2000).
5 A. Tarasevitch et al., Phys. Rev. A **62**, 023816 (2000).
6 W. L. Kruer, *The Physics of Laser Plasma Interactions* (Addison-Wessly, Redwood City 1988).
7 N. G. Basov et al., *Heating and compression of thermonuclear targets by laser beam* (Cambridge University Press, 1986).
8 B. Quesnel et al., Phys. Plasmas, **4**, 3358 (1999).
9 H. C. Barr et al. Phys. Plasmas 7, 2604-2615 (2000).

Fast electron and ion emission from metal targets in intense femtosecond laser fields

Yasuaki Okano[1], Yoichiro Hironaka[1], Yuji Oishi[2], Takuya Nayuki[2], Koshichi Nemoto[2], Kazutaka G. Nakamura[1], and Ken-ichi Kondo[1]

[1] Materials and Structures Laboratory, Tokyo Institute of Technology, 4259 Nagatsuta, Midori, Yokohama 226-8503, Japan
E-mail: okano@kkhp8.rlem.titech.ac.jp
[2] Central Research Institute of Electric Power Industry, 2-11-1 Iwado kita, Komae, Tokyo 201-8511, Japan

Abstract. Electron and ion emission has been investigated by infrared 50-fs laser irradiation on copper targets at power density up to 10^{18} W/cm^2. Energies of electrons and ions are measured directly using an electron magnetic spectrometer and a charge collector, respectively. Fast electron temperature agrees with the ponderomotive scaling in laser power density of 10^{18} W/cm^2, and copper ion energies are comparable to ponderomotive potential in the power density between 10^{16} to 10^{17} W/cm^2.

1. Introduction

Interactions between intense femtosecond laser pulses and metal targets have been widely studied in connection with generation of pulsed x-rays and energetic charged particles in recent years. Among them, the electron generation is the most basic phenomenon because x-rays and other particles, such as ions and electrons, are generated by the interaction of electrons with matter. In this work, we have studied electron and ion emission from copper targets induced by femtosecond laser irradiation where hard x-ray emission takes place.

2. Experimental Methods

The used laser system (B.M.industries, α10us) consists of a Ti:sapphire oscillator and amplifiers based on a chirped pulse amplification technique [1]. The amplified laser beam has a pulse duration of 50 fs in full-width at half-maximum, a maximum energy of 200 mJ/pulse at a fundamental wavelength of 790 nm, and a repetition rate of 10 Hz. The contrast ratio of the main pulse to the prepulse that precedes it by 8 ns is greater than 10^6:1. In a vacuum chamber ($\approx 10^{-5}$ Torr), the femtosecond pulse is focused on a metal target by an off-axis parabola in the power density range between 10^{16} and 10^{18} W/cm^2 to generate energetic electrons and ions. The used samples are copper targets in the shape of a disk (70 mm-$\phi \times$ 5 mm-t) or a tape-like foil (30 μm-t). Because of damages on the target after laser irradiation, the target is rotated or translated to expose a fresh surface for each laser shot.

The energy spectrum of electrons is measured with a magnetic spectrometer with an imaging plate (Fuji photo film, FDL-UR-V), which is used to monitor the electron flux [2]. The spatial resolution of the imaging plate is 25 μm and the continuous spectrum is measured with an intensity resolution of 18 bits. Charged particles from laser-produced plasma are directly detected using a charge collector consisting of an electrostatic probe in the shape of a Faraday cup. This method is one of the simplest diagnostic tools for ions of laser plasma. The kinetic energy of positive ions is obtained from a time-of-flight spectrum of emitted ions. The ion signal is amplified up to 30 dB with a high-bandwidth (350 MHz) DC voltage amplifier (Analog Modules, 353A), and is recorded by a digital-storage-oscilloscope (Tektronix, TDS684B).

3. Results and Discussion

Figure 1(a) shows typical energy spectrum of electrons emitted perpendicularly from the backside of a copper film target. The spectrum is obtained with 20 laser shots. The laser pulse is focused at an incident angle of 45° with respect to the surface normal in p-polarization and the power density is approximately 2.8×10^{18} W/cm^2. The electron temperature is estimated from fitting to the Boltzmann distribution in the range between 500 keV and 1 MeV. The estimated temperature is 300 keV. The laser intensity is on the order of the relativistic region where the ponderomotive force is dominant. The effective electron temperature is close to the temperature of 260 keV estimated from the simulated ponderomotive scaling of $k_B T_h = m_0 c^2\ [(1 + I\lambda^2 / 1.37 \times 10^{18}\ \mathrm{W\ \mu m^2/cm^2})^{1/2} - 1]$ [3,4] in the power density region between 10^{18} and 10^{19} W/cm^2.

On the other hand, figure 1(b) shows typical kinetic-energy spectra of copper ions emitted from a front side of a copper disk target at an angle of 10° to the surface normal. Incident laser pulses are focused at 60° to the normal in p-polarization in the power density up to 10^{17} W/cm^2. Each plot is an average for

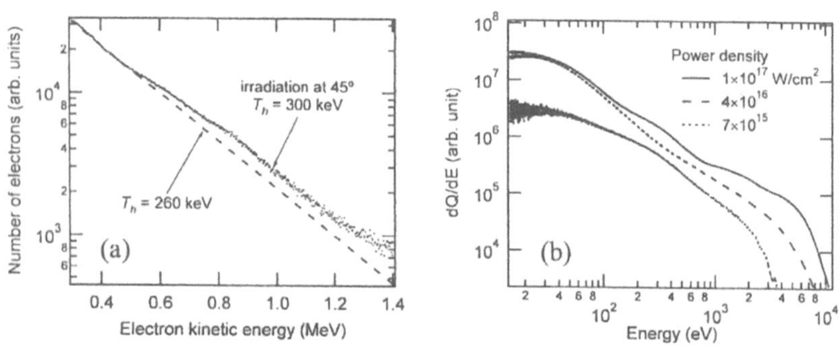

Fig. 1. (a) Energy spectrum of electrons penetrate the 30 μm thick copper target by femtosecond laser irradiation at power densities approximately 2.8×10^{18} W/cm^2. The dashed line represents ponderomotive scaling. (b) Energy spectra of ions emitted from a front side of a copper disk target by femtosecond laser irradiation at power densities between 10^{15} and 10^{17} W/cm^2.

500 laser shots. The energy spectrum of ions shows two features, which are an energy cutoff and a low-energy peak. The cutoff energy is strongly dependent on the incident laser energy and they increase in the energy range between 2 and 6 keV of copper ions, which are close to the ponderomotive potential in the laser field described by $U_p(\mathrm{eV}) = 9.33 \times 10^{-14}\, I\lambda^2\ \mathrm{W\,\mu m^2/cm^2}$ [4,5]. For example, the cutoff energy of copper ions at approximately 10^{17} W/cm^2 is estimated to be 5 keV, and ponderomotive potential is about 6 keV. Ion acceleration is commonly considered to be due to energy transfer from the electron to ion subsystem, which occurs due to the locally induced space-charge separation fields and electron-ion collisions [6]. Therefore, this kind of ion generation is considered to be dominated by interactions with the electrons, which are accelerated by ponderomotive potential at the present power density. On the other hand, the ion charge yield of the low-energy peak around 20 eV is rapidly increase above laser power density of 4×10^{16} W/cm^2. The low-energy ions are considered to be emitted by thermal ablation [7]. The deep holes in the depth of several μm are also observed on the laser-irradiated surface above laser power density of 4×10^{16} W/cm^2.

3. Summary

In summary, we have directly measured electron and ion emission from copper targets by infrared 50-fs laser irradiation. Fast electron temperature is obtained to be 300 keV at the laser power density of 2.8×10^{18} W/cm^2, and cutoff energies of copper ions are estimated to be several keV at the power density between 10^{16} to 10^{17} W/cm^2.

Acknowledgements. We would like to thank M. Hasegawa for constructing the experimental setup.

References

1 M. Yoshida, Y. Fujimoto, Y. Hironaka, K. G. Nakamura, K. Kondo, M. Ohtani, and H. Tsunemi, Appl. Phys. Lett. **73**, 2393, 1998.
2 Y. Oishi, T. Nayuki, K. Nemoto, Y. Okano, Y. Hironaka, K. G. Nakamura, and K. Kondo, Appl. Phys. Lett. **79**, 1234, 2001.
3 S. C. Wilks, W. L. Kruer, M. Tabak, and A. B. Langdon, Phys. Rev. Lett. **69**, 1383, 1992.
4 T. E. Cowan *et al.*, Phys. Rev. Lett. **84**, 903, 2000.
5 P. H. Bucksbaum, R. R. Freeman, M. Bashkansky, and T. J. McIlrath, J. Opt. Soc. Am. B **4**, 760, 1987.
6 C. Momma, B. N. Chichkov, S. Nolte, F. Alvensleben, A. Tünnermann, H. Welling, and B. Wellegehausen, Opt. Commun. **129**, 134, 1996.
7 S. Nolte, C. Momma, H. Jacobs, A. Tünnermann, B. N. Chichkov, B. Wellegehausen, and H. Welling, J. Opt. Soc. Am. B **14**, 2716, 1997.

Reaction dynamics of hydrogen bonded aniline-ammonia clusters in intense laser fields by tandem mass spectroscopy

Ryuji Itakura, Jun Watanabe, Taiki Asano, and Kaoru Yamanouchi

Department of Chemistry, School of Science, The University of Tokyo, 7-3-1 Hongo, Bunkyo-ku, Tokyo 113-0033, Japan
E-mail: ita@chem.s.u-tokyo.ac.jp

Abstract. Mass-selected aniline cations and [aniline-$(NH_3)_n$]$^+$ (n = 1 and 2) cluster cations are exposed to the femtosecond laser fields (λ ~ 395 nm, I ~ 4×10^{15} W/cm^2) using a tandem-type time-of-flight spectrometer. It is found that a bare aniline cation decomposes into $C_5H_6^+$, while the decomposition is suppressed when one or two ammonia molecules are attached to the aniline cation. This suppression was interpreted in terms of an intermolecular energy flow through the hydrogen bonding.

1. Introduction

In the last decade, characteristic molecular responses to intense laser fields such as alignment along the laser polarization direction, and multiple ionization leading to the Coulomb explosion, have stimulated researchers to investigate dynamical behavior of molecules in intense laser fields. Along with the expansion of this research field, it has been revealed that the ionization and fragmentation mechanisms of organic compounds in intense laser fields are influenced significantly by the resonant coupling between electronic states of a parent molecule with a specific charge number [1].

In the present study, we focus our attention on an effect of hydrogen bonding on the response of clusters containing an aromatic molecule to intense laser fields, and chose aniline-$(NH_3)_n$ cations as target clusters.

2. Experiment

Aniline cations (AN^+) and $[AN(NH_3)_n]^+$ produced by irradiating with a fourth harmonics of Nd:YAG laser output are mass-selected in the first stage of a tandem-type time-of-flight (TOF) mass spectrometer. The mass-selected AN^+ and $[AN(NH_3)_n]^+$ (n = 1 - 2) are exposed to the intense laser fields generated either by the fs laser pulse (Δt ~ 50fs, λ ~ 395 nm, I ~ 10^{15}W/cm^2) or by the ns laser pulse (Δt ~ 6 ns, λ = 532 nm, I ~ 10^{10}W/cm^2). The resultant fragment cations are mass-separated in the second stage of the tandem-type TOF mass spectrometer, and are detected by an MCP detector.

3. Results and Discussion

When aniline cations are exposed to the fs intense laser fields ($\lambda \sim 395$ nm), the fragmentation into $C_5H_6^+$ and HNC proceeds dominantly without being ionized into the multiply charged cations. This fragmentation is considered to occur through the resonant coupling between the electronically excited $^2B_1$ and ground $^2B_1$ states. However, in the case of the hydrogen bonded $[AN(NH_3)_n]^+$ (n = 1, 2) clusters, the decomposition of the aniline cation moiety is suppressed significantly, and instead the evaporation of NH_3 proceeds. As shown in figure 1, the yield ratios of $C_5H_6^+$ from the $[AN(NH_3)_n]^+$ (n = 1, 2) cluster with respect to the corresponding parent ions are only $5.6(1)\times 10^{-4}$ and $2.2(2)\times 10^{-3}$, respectively, which are 11 and 44 % of the corresponding ratio, $5.2(3)\times 10^{-3}$, for the bare AN^+.

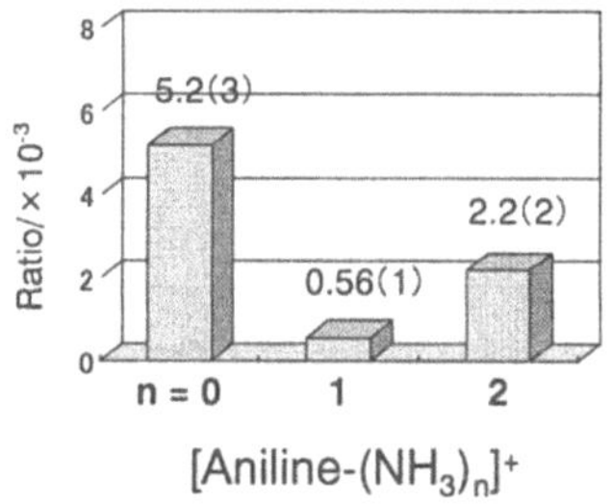

Fig. 1. The yield ratios of $C_5H_6^+$ from the bare aniline cation and the $[AN(NH_3)_n]^+$ (n = 1 and 2) cluster cations with respect to the corresponding parent ions.

The formation of $C_5H_6^+$ from the aniline cation is expected to proceed through the transition state in which the geometrical structure around the amino group is substantially deformed, since one of the two H atoms in the amino group needs to migrate into the ortho-carbon site by simultaneously carrying a positive charge. This suppression of the decomposition of an aniline cation moiety in the cluster indicates that the NH_3 attachment to the amino group hinders largely the migration of H atom in the amino group.

In view of the intra-cluster energy flow, it is possible that a part of the energy gained from the photon fields is distributed to the NH_3 moiety through the intermolecular hydrogen bonding, and eventually, the small portion is consumed for the breaking of one or two intermolecular bonds. Such an energy flow may lower the energy kept in the AN^+ moiety below the threshold energy for its decomposition.

Reference

1 R. Itakura, J. Watanabe, A. Hishikawa, and K. Yamanouchi, J. Chem. Phys. **114**, 5598, 2001.

Indole$(NH_3)_n$ Clusters: Hydrogen Atom Transfer Initiated by Femtosecond Laser Pulses

H. Lippert, V.Stert, L. Hesse, C. P. Schulz, I. V. Hertel, and W. Radloff

Max–Born–Institut für Nichtlineare Optik und Kurzzeitspektroskopie
Max–Born–Str. 2A, D–12489 Berlin, Germany
E-mail: radloff@mbi-berlin.de

Abstract. The H atom transfer reaction in excited indole–ammonia clusters is characterized by successive processes on the femtosecond and picosecond time scale, which are analyzed by time–resolved photoelectron spectroscopy.

1 Introduction

The femtochemistry of the indole molecule which represents the chromophore of the amino acid tryptophan is of particular interest because of its biological relevance. Recent studies of indole in aqueous solution have shown, that the photoexcitation leads to a fast charge separation process characterized by the formation of a solvated electron [1]. The microscopic mechanism of the solvation process can be elucidated by gas phase studies of indole-solvent clusters with polar molecules such as water or ammonia. Recent ab initio calculations of indole–water clusters have stressed the crucial role of the non-adiabatic coupling between the optically excited $\pi\pi^*$and a low-lying dark $\pi\sigma^*$ state for the electron transfer process [2]. In first pump-probe experiments on indole$(NH_3)_n$ clusters with ns laser pulses a H atom transfer reaction has been stated [3].

We have studied the H–transfer reaction in indole$(NH_3)_n$ clusters in pump–probe experiments with fs laser pulses by detection of the photoions and the coincident photoelectrons. The clusters formed by adiabatic expansion through a pulsed nozzle were excited by a pump photon at 263 nm (4.71 eV) well above the $S_1(\pi\pi^*)$ state. A probe photon at 395 nm (3.14 eV) was applied to ionize the parent clusters and the reaction products. The width of the laser pulses was about 140 fs. By irradiating the molecular beam pulse in its front region where the cluster formation is just starting the width of the indole$(NH_3)_n$ cluster distribution interacting with the laser field has been varied, e.g. from a broad ($n \leq 6$) to a narrow ($n \leq 3$) distribution.

2 Results and Discussion

In Fig. 1 the time–dependent ion signals are shown for indole$(NH_3)_1^+$ obtained with a narrow distribution and for indole$(NH_3)_4^+$ with a broad one. The

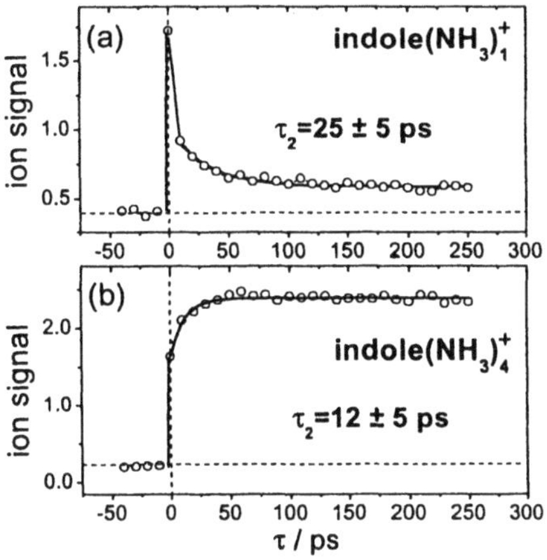

Fig. 1. Ion signal vs. delay time τ

ion signals of small clusters ($n = 1\ldots3$) were fit by a single–exponential decay (see Fig. 1a) with a time constant τ_2 growing with the clusters size ($25\ldots90$ ps). These signals are preceded by fast transients on the fs time scale (e.g. 300 fs for $n = 1$) [4]. For the larger clusters ($n \geq 4$) we observed a relatively fast rise of the ion signals on the ps time scale followed by a constant signal at long delay times (see e.g. Fig. 1b for $n = 4$).

For the broad cluster distribution ($n \leq 6$) we observed the product ions $(NH_3)_{n-1}NH_4^+$ only for such radicals with an ionization potential below the probe photon energy. This is a clear indication of a dissociative H atom transfer reaction in the electronically excited indole-ammonia clusters. The rise times of the product signals were determined to be about 140 ps for $n \leq 4$ and about 80 ps for $n \geq 5$ [5]. We found no product ions $(NH_3)_{n-1}NH_4^+$ for $n > 6$, because for larger clusters the H–transfer reaction proceeds in the electronic ground state [3].

To elucidate further the processes of the H–transfer reaction proceeding on the different time scales we have analyzed the photoelectron spectra measured in coincidence to the photoion signals. In Fig. 2 the photoelectron spectra are represented obtained for the two typical cluster sizes given in Fig. 1. The electron spectra of indole$(NH_3)_1^+$ show a strong peak at low energies ($E_{el} \approx 0.1$ eV) for $\tau = 0$ ps, which rapidly decreases for $\tau = 0.25$ ps and vanishes for $\tau = 2$ ps (cf. Fig. 2a). Tentatively, this ultrafast signal decay with a time constant of 300 fs is explained by the internal conversion from the initially excited $\pi\pi^*$ to the $\pi\sigma^*$ state (probably followed by the ultrafast H–transfer). In the electron spectra of the larger cluster indole$(NH_3)_4^+$ this fast contribution is not observable (see Fig. 2b) due to Franck–Condon reasons. However, at longer delay times a strong electron signal grows up in the low energy region $E_{el} \leq 0.6$ eV (see Fig. 2b) and stays nearly constant above $\tau = 60$ ps, in agreement with the time–dependent ion signal (Fig. 1b). As the growing electron spectra reflect a more efficient Franck–Condon overlap for

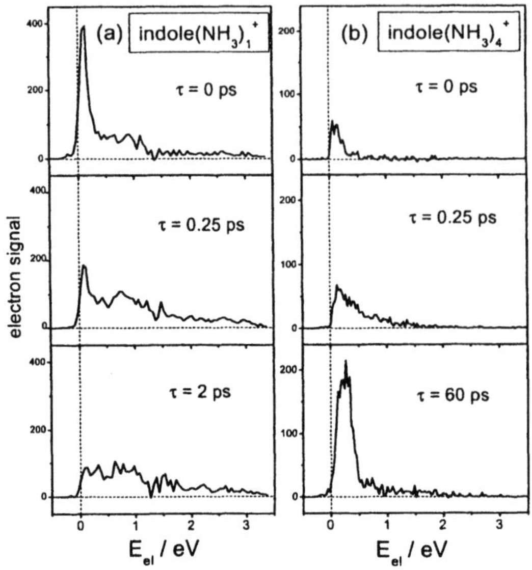

Fig. 2. Electron spectra at different delay times τ

the transition from the excited electronic to the ion state we conclude that the increasing transition probability reflects the structural rearrangement of the clusters on the ps time scale after the H–transfer.

In conclusion, the time–resolved photoelectron spectra confirms our preliminary reaction scheme [4] describing the H–transfer reaction in indole–ammonia clusters to proceed in the following steps: 1) internal conversion from the initially excited $\pi\pi^*$ to the $\pi\sigma^*$ state followed by the H–transfer on the fs time scale (a few 100 fs), 2) structural rearrangement of the clusters after the H–transfer on the ps time scale (a few 10 ps) and 3) vibrational relaxation leading to a partial dissociation of the clusters after the H-transfer in significantly longer times (80 ... 140 ps).

Acknowledgement. This project is financially supported by the DFG through Sonderforschungsbereich 450.

References

1. see e.g. J. Peon, G.C. Hess, J.-M. L. Pecourt, T. Yuzawa and B. Kohler, J. Phys. Chem. A **103**, 2460, 1999.
2. A.L. Sobolewski and W. Domcke, Chem. Phys. Lett. **329**, 130, 2000.
3. C. Dedonder-Lardeux, D. Grosswasser, C. Jouvet and S. Matrenchard, Phys. Chem. Chem. Phys. **4**, 1, 2001.
4. V. Stert, L. Hesse, H. Lippert, C.P. Schulz and W. Radloff, J. Phys. Chem. A, in press.
5. H. Lippert, V. Stert, L. Hesse, C.P. Schulz, W. Radloff and I.V. Hertel, Eur. Phys. J. D, submitted.

Theoretical models predicting the outcome of Raman type control experiments

Thomas Hornung, Marcus Motzkus, and Regina de Vivie-Riedle

Max-Planck-Institut für Quantenoptik,
Hans-Kopfermann-Strasse 1, D-85748 Garching / Germany

1 Introduction and FWM control experiments

In this contribution the control experiments on the four-wave mixing (FWM) response of Ref. [1] are compared and analyzed with three different theoretical models. Our reference system was the potassium dimer. The experimental observation is that it is possible to control the contributions in the FWM signal originating from the vibrational motion in the electronic ground X and first excited state A potential by suitably modulating the interacting pulses. Feedback-loop optimization of the A state peak using, e.g. chirped excitation represented by a polynomial phase modulation, leads to the pulse shown in Fig. 1. The central question arising in all FWM based control experiments

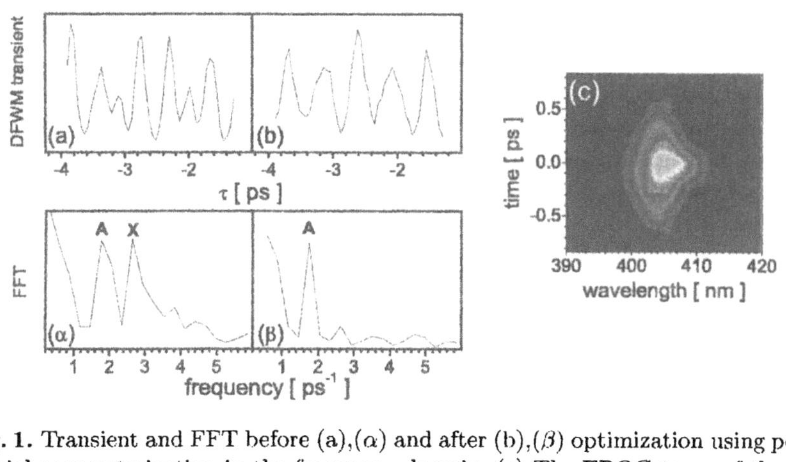

Fig. 1. Transient and FFT before (a),(α) and after (b),(β) optimization using polynomial parameterization in the frequency domain. (c) The FROG trace of the optimal pulse. The A optimizing pulse has negative linear and quadratic chirp and a pulse duration of about 560 fs.

is whether the observed control is due to a controlled population transfer or whether it is due to an interference of the different Liouville space pathways, that constitute the DFWM signal.

2 Theoretical models clarifying FWM experiments

First an analytical model was derived, which leads to very good agreement with experimental results [2]. However it is restricted to predict the FWM signal from modulated pulses consisting of subpulses with constant phase. Therefore in this contribution two more sophisticated models, that allow a comparison with experimental results using chirped pulses as in Fig. 1 are presented.

A perturbational approach. This theory is based on a perturbational calculation of the FWM polarization [3] and is able to predict the FWM signal for arbitrarily shaped pulses. Within this theory the FWM polarization generated by the pulse sequence SS-U (the two equally shaped pulses SS interacting before the unshaped pulse U) is given by three double-sided Feynman diagrams [3]. In order to account for a thermal ensemble of molecules the incoherent sum of all polarization contributions of different vibrational states within the Boltzmann distribution must be evaluated. The total DFWM signal $I(\tau)$ is then given by integrating the sum of all polarization contributions from the thermal ensemble over two times the *fwhm* (full width at half maximum) duration of the last interacting pulse (U).

A model system is chosen based on the characteristic properties of the potassium dimer with some modifications to reduce the necessary computing time. The model system consists of an harmonic potential as ground state X, the real anharmonic excited state potential $A^1\Sigma_u$ and the K-K separation dependent dipole moment between X and A potentials. It is observed, that the A state peak increases with increasing negative linear chirp (see Fig. 2).

Optimal control theory (OCT). In order to find the shaped pulse that would maximize A state contribution in the FWM signal due to a complete population transfer the usual OCT functional [4] is extended to N initial and final states in the system and is then

$$\bar{J} = J - 2\mathrm{Re}\sum_{k=1}^{N}\int_0^T \mathrm{d}t\left\langle\lambda^{(k)}\middle|\partial_t + \mathrm{i}H\middle|\psi^{(k)}\right\rangle \quad \text{with}$$

$$J = \sum_{k=1}^{N}\left|\left\langle\psi^{(k)}(T)\middle|O\middle|\phi^{(k)}(T)\right\rangle\right|^2 - \alpha\int_0^T \mathrm{d}t\frac{\epsilon(t)^2}{s(t)} \tag{1}$$

This functional is used to find a single optimal field connecting several initial states to their respective desired target states or maximizing an operator O starting from an incoherent ensemble of initial states. In our case we consider a Boltzmann distribution of five vibrational states. OCT then obtains the pulse of Fig. 3, that displays a non linear positive chirp. This result could not be reproduced yet in an experimental optimization limited to polynomial phases.

In summary, all theoretical models suggest that the observed control is not due to an optimal population transfer, but due to a controlled interplay between population dynamics and interference of Liouville space pathways.

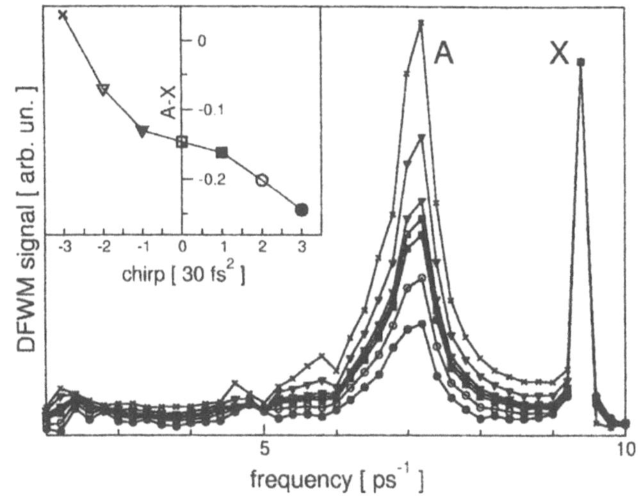

Fig. 2. The Fourier transform of the DFWM signal obtained through excitation with a SS-U sequence, where the two time coincident pulses are shaped with a linear chirp. Each dotted line corresponds to a different linear chirp. Clearly visible the X and A peak corresponding to the vibrational recurrence time in the respective potential. The data is normalized to have the same X state peak magnitude. The peak height difference A-X is plotted as a function of chirp in the inset and summarizes the information of the main graph.

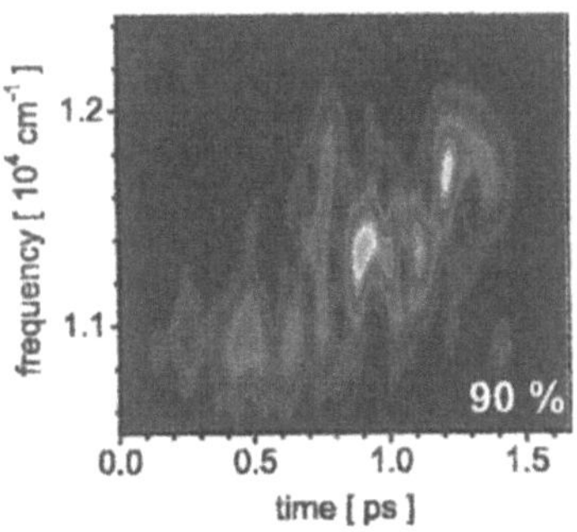

Fig. 3. Molecular π-pulse transferring an initial Boltzmann distribution of five vibrational ground states $v'' = 0 \ldots 5$ into A $^1\Sigma_u$.

References

1. T. Hornung, R. Meier, M. Motzkus: Chem. Phys. Lett. **326**, 445 (2000)
2. T. Hornung, R. Meier, R. de Vivie-Riedle, M. Motzkus: Chem. Phys. **267**, 261 (2000)
3. S. Meyer, M. Schmitt, A. Materny, W. Kiefer, V. Engel: Chem. Phys. Lett. **281**, 332 (1997) and Chem. Phys. Lett. **287**, 753 (1998)
4. W. Zhu, J. Botina, H. Rabitz: J. Chem. Phys. **108**, 1953 (1998)

Part IV

Ultrafast Sources and Phase Stabilization

Generation and characterization of few-cycle mid-infrared pulses from 2800 to 4600 nm

Julie A. Gruetzmacher and Norbert F. Scherer

Department of Chemistry, James Franck Institute, and Institute for Biophysical Dynamics, University of Chicago, Chicago, IL 60637, USA
E-mail: nfschere@uchicago.edu

Abstract. Sub-five-cycle pulses in the spectral region from 2800 to 4600 nm are generated using a potassium niobate optical parametric amplifier employing a germanium pulse compression scheme. The sub-65 fs mid-infrared pulses are fully characterized by XFROG.

Generation of ultrashort pulses in the mid-infrared (mid-IR) spectral region is a goal that has been pursued intensely, with the major interest being the extension of traditional nonlinear optical measurements to the study of molecular vibrations. While much activity has focused on the region of carbonyl stretching vibrations near 5000 nm, the 3000 to 4000 nm wavelength range is of great importance due to the hydrogen-bond-sensitive O-H and N-H stretching modes in this region. While several groups have studied the dynamics of the O-H stretching vibration in isoptopically-diluted water, the pulses used in these experiments have been >100 fs in duration; such experiments would benefit from improved time-resolution as mid-IR echoes on water have suggested a pure dephasing time $\leq$ 90 fs [1]. Furthermore, KTP-based OPA schemes commonly used for mid-IR pulse generation typically have been limited to the region near 3000 nm, requiring the addition of new parametric stages to study the deuterated analogs of these vibrations [2]. The availability of a sub-100-fs source of pulses spanning the full 3000 to 4000 nm region therefore would be a great advance toward the study of hydride stretching dynamics in a wealth of samples, particularly aqueous systems.

Potassium niobate ($KNbO_3$) has shown promise as a nonlinear crystal for ultrashort mid-IR pulse generation due to its large second-order nonlinearity, transparency beyond 4000 nm, and broad phase matching bandwidths for pump wavelengths near the Ti:sapphire fundamental. However, previous mid-IR generation from this crystal has shown that while the pulses generated have broad spectral bandwidths, they typically are not transform limited. Combining a short-pulse-seeded $KNbO_3$ OPA design with a mid-IR pulse compression scheme should allow the generation of sub-100 fs mid-IR pulses from 3000 nm to beyond 4000 nm. This paper presents a $KNbO_3$ optical parametric amplifier (OPA) that achieves this goal. The OPA generates mid-IR pulses containing five or fewer optical cycles with center wavelengths from 2800 nm to 4600 nm in a single crystal device. These pulses are suitable for probing the ultrafast vibrational responses of X-H and X-D modes (X = C, N, O). The pulses are fully characterized in intensity and phase using cross-correlation frequency-resolved optical gating (XFROG) [3].

The OPA was pumped with 800 nm light from a home-built Ti:sapphire regenerative amplifier producing pulses ~ 42 fs in duration. Near-IR signal pulses

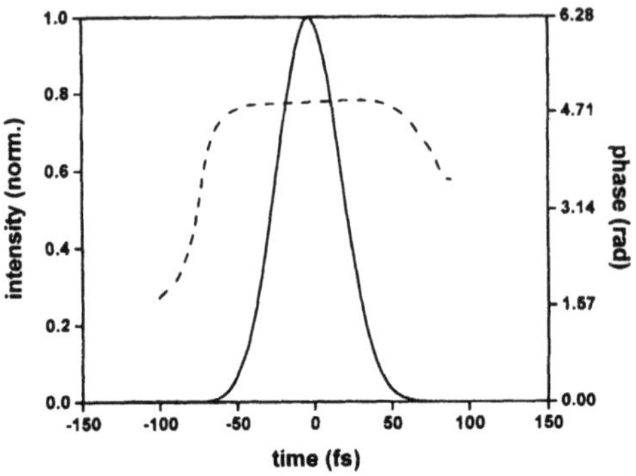

Fig. 1. Intensity (solid) and phase (dashed) of a four-cycle pulse at 3600 nm (48 fs).

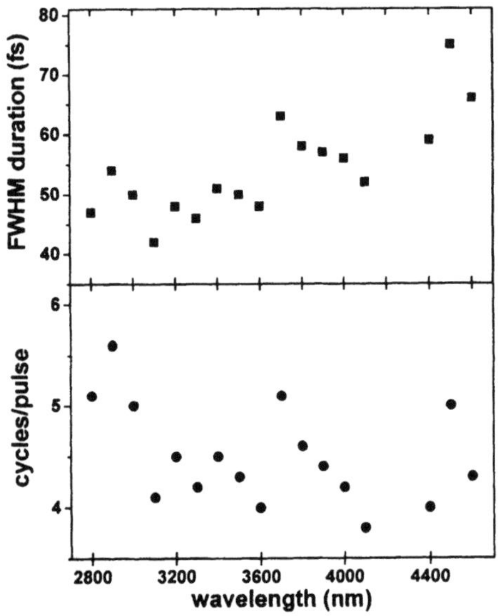

Fig. 2. FWHM duration and cycles per pulse for compressed pulses.

were obtained from a white light continuum generated in 2 mm of [0001]-cut sapphire. The pump and near-IR signal beams were combined with a dichroic mirror for collinear type-I parametric amplification in the xz-plane of a 1 mm $KNbO_3$ crystal (VLOC; θ = 40.5°off b toward c, ϕ = 0°). The mid-IR output of the OPA was tuned from 2800 to 4600 nm by adjusting the temporal overlap of the pump and signal pulses; minor tuning of the θ–angle was also required. The design is similar to that of Kafka and Watts [4] and Cussat-Blanc, et al. [5], but uses shorter pump pulses and a thinner OPA crystal.

Mid-IR pulses were characterized by sum-frequency generation (SFG) XFROG. The SFG cross-correlation was generated in a thin (100 or 300 μm) $KNbO_3$ crystal cut at θ = 37.6°. The signal was imaged onto the entrance slit of a 0.32 m monochromator, and XFROG spectrograms were obtained using a CCD array to acquire spectra at each delay. Compensation of the negative group velocity dispersion (GVD) of the $KNbO_3$ crystal and CaF_2 optics was accomplished using germanium disks of varying thicknesses, which have positive GVD in this spectral region [6,7]. The gate pulse was characterized via second-harmonic generation autocorrelation and FROG. The intensity and phase of the mid-IR pulses were extracted from spectrograms using FROG 3.06 software (Femtosoft Technologies). All traces were corrected for the spectral response of the detection system prior to analysis. The intensity and phase of a mid-IR pulse at 3600 nm is shown in Figure 1. The FWHM duration of this pulse is 48 fs, corresponding to four optical cycles. The performance of the system over the full tuning range explored is given in Figure 2; pulses are shorter than 65 fs in all but one case, equivalent to five or fewer optical cycles per pulse. To our knowledge, these are the shortest pulses reported to date from a femtosecond mid-IR OPA throughout this entire spectral range; the duration of these pulses is sufficient to time-resolve the vibrational dephasing of hydrogen-bonded modes in condensed phase systems.

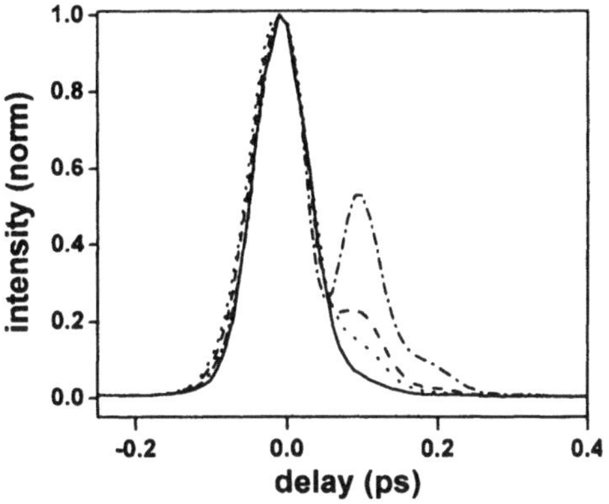

Fig. 3. FID-modified pulse profiles

We have used this system and the XFROG technique to examine coherent pulse propagation in optically dense samples of HDO in liquid D_2O [7]. Figure 3 shows the wavelength-integrated profiles of pulses resonant with the OH-stretching vibration of HDO (~ 2950 nm) after traversing samples with optical densities of 0.5 (dotted), 0.9 (dashed), and 1.7 (dot-dashed) at the absorption maximum (solid = instrument response). Severe distortions are observed due to interference between the finite-duration initial pulse and the free-induction decay field that is 180° out of phase with the driving field. These features are chirp-dependent, and the associated phase shifts between them have been determined using XFROG. Such distortions should be considered when interpreting linear and nonlinear optical temporal measurements. Simulations of these effects using a semiclassical implementation of the finite-difference time-domain technique are in progress.

References

1 J. Stenger, D. Madsen, P. Hamm, E. T. J. Nibbering and T. Elsaesser, Phys. Rev. Lett. **87**, 027401, 2001.

2 M. F. Kropman, H.-K. Nienhuys, S. Woutersen and H. J. Bakker, J. Phys. Chem. A **105**, 4622, 2001.

3 S. Linden, H. Giessen and J. Kuhl, Phys. Stat. Sol. B **206**, 119, 1998.

4 J. D. Kafka and M. L. Watts, in *Ultrafast Phenomena X*, edited by P. F. Barbara, J. G. Fujimoto, W. H. Knox and W. Zinth, Springer-Verlag, 38, 1996.

5 S. Cussat-Blanc, A. Ivanov, D. Lupinski and E. Freysz, Appl. Phys. B **70**, S247, 2000.

6 N. Demirdöven, M. Khalil, O. Golonzka, and A. Tokmakoff, Opt. Lett. **27**, 433, 2002.

7 J. A. Gruetzmacher and N. F. Scherer, Rev. Sci. Instrum. **73**, 2227 (2002).

1-GHz repetition-rate femtosecond optical parametric oscillator

X. P. Zhang [1], J. Kuhl [2], J. Hebling [3], A. Bartels [4,5], D. Nau [4], W. W. Rühle [1], and H. Giessen [4]

[1] Department of Physics, Philipps University, D-35032 Marburg, Germany
[2] Max-Planck-Institut für Festkörperforschung, D-70569 Stuttgart, Germany
[3] Department of Experimental Physics, University of Pecs, H-7624 Pecs, Hungary
[4] Institute of Applied Physics, University of Bonn, D-53115 Bonn, Germany
[5] GigaOptics GmbH, Louis-Raverat Str. 23, 63165 Mühlheim am Main, Germany

Abstract. We achieved 1-GHz-repetition-rate and low-pump-threshold operation of a femtosecond optical parametric oscillator (OPO) based on periodically poled $LiNbO_3$ (PPLN) using higher-order synchronous pumping with a 1GHz Ti:sapphire laser.

Introduction. High-repetition-rate femtosecond OPOs operating at about 1.3 μm and 1.55 μm are potentially of great importance for optical communication systems to achieve high data rates and high capacity. High-repetition-rate operation of femtosecond laser pulses also means a high signal to noise ratio, which is very useful for time-resolved spectroscopy and pump-probe measurements. However, femtosecond OPOs reported so far operated below 400 MHz. [1]

In our work, we pumped a PPLN OPO with a 1 GHz Kerr-lens mode-locked Ti:sapphire laser, so that we achieved 1 GHz operation of the PPLN OPO that was tunable from 1160 nm to 1320 nm. The signal pulses of the OPO were as short as 65 fs. This is to our knowledge the highest repetition rate of a femtosecond OPO and the first OPO directly pumped with a GHz oscillator.

Experimental Methods. The ring cavity of the PPLN OPO is illustrated in Fig. 1. Synchronous pumping at high repetition rate means a very short OPO cavity, e.g., the cavity length should be as short as 30 cm for 1 GHz operation. According to our theoretical analysis, the signal beam waist (D_0) in the nonlinear crystal increases quickly with shortening the OPO cavity length (L_{OPO}) at a given curvature radius (r) of the curved mirrors. For perfect mode matching between the signal and the pump beams, the focus of the pump beam in the crystal should be as large as the signal beam waist, reducing the intensities of both the signal and the pump waves in the interaction region. Consequently, the efficiency of the OPO is reduced and the pump threshold is increased. Although a smaller D_0 can be obtained by reducing r, sizes of the mirror holders, astigmatisms resulting from the curved mirrors and stable operation conditions of the OPO limit the reduction of r. Furthermore, too small r results in spatial inhomogeneity of the reflectivity specifications of the curved mirrors.

We can solve above problems using higher order synchronous pumping, namely choosing a longer cavity length for the OPO if compared with the pump laser with $L_{OPO}=N \cdot L_P$, where N=2, 3, 4,... and L_P is the cavity length of the pump laser. Fig. 2 demonstrates the 4th order synchronous pumping at 1 GHz. The pump pulses excite four 250 MHz signal pulse trains (S1-S4), each delayed sequentially by one repetition period (1 ns). At the output, a 1 GHz signal pulse train (S) will be created. For Nth order synchronous pumping, the OPO repetition rate is always as high as that of the pump. Different from lower-order synchronous pumping method, where $L_{OPO}=L_P/N$ with N=2, 3, 4,..., [1, 2] higher-order synchronous pumping configuration generates an equal-intensity signal pulse train.

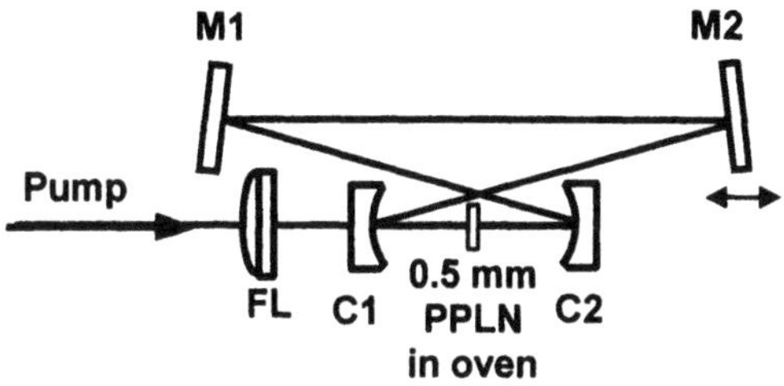

Fig. 1. Ring cavity of the PPLN OPO. M1, 2% output coupler; M2, high reflector; C1, C2, curved mirrors (r=100 mm); FL, focusing lens (f=60 mm). PPLN grating period is around 21 μm.

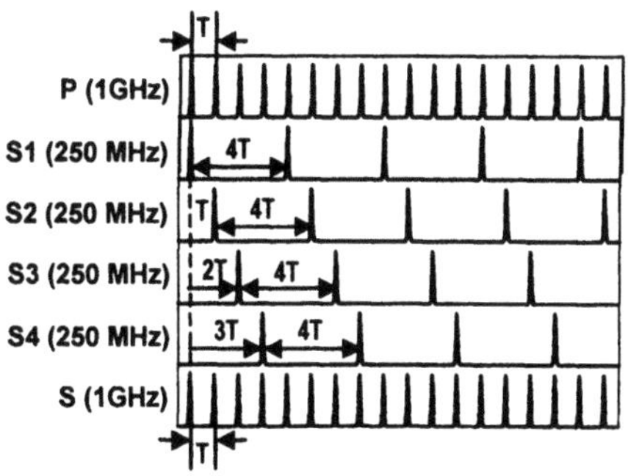

Fig. 2. 4th order synchronous pumping. P, the pump; S1-S4, the 4 generated signal pulses trains; S, the output signal pulse train; T, cavity round-trip time of the pump laser.

Results and Discussion. In our experiment, we pumped the PPLN OPO with 900 mW pump power at 1 GHz and used 4th order synchronous pumping (L_{OPO}=120.8 cm, L_P=30.2 cm). Fig. 3 (a) presents the output characteristics of the OPO when tuning the cavity length. The highest signal output power was approximately 18 mW, and the pump threshold of the OPO was measured to be lower than 790 mW. The signal wavelength was tunable from 1160 nm to 1235 nm. Nearly transform-limited 65 fs signal pulses were measured at 1220 nm with 34 nm bandwidth (see Fig.3(b)), creating a time-bandwidth product of 0.445.

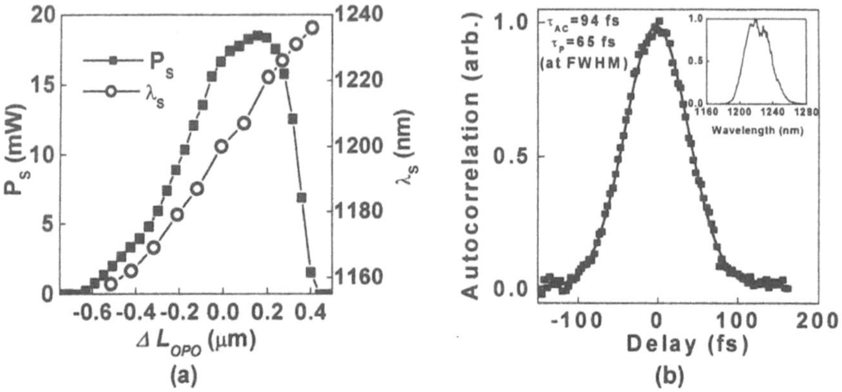

Fig. 3. (a) Signal output power (P_S) and wavelength (λ_S) versus change of the cavity length (ΔL_{OPO}) of the PPLN OPO. (b) Intensity autocorrelation measurement (filled squares) of the signal pulses, the pulse duration and the autocorrelation width are denoted by τ_P and τ_{AC}. The solid line is a Gaussian fit to the data. Inset: corresponding spectrum.

Replacing *FL* (see Fig. 1) by a f = 50 mm achromat and choosing 6th order synchronous pumping with $L_{OPO} = 6L_P \approx 181.2$ cm, we were able to greatly increase the OPO efficiency. More than 43 mW signal power was measured for 950 mW pump power and the pump threshold was lowered to approximately 580 mW at 1 GHz. Raising the crystal temperature from 100°C to 125°C increased the efficiency by 20 %. Replacing M2 by a chirped mirror (GDD≈-100 fs^2, R>99.8% from 1100 nm to 1300 nm) did not shorten the signal pulses but reduced the spectral bandwidth, resulting in a time-bandwidth product of 0.36. The tuning range of the signal wavelength was shifted to 1230 nm – 1320 nm.

Conclusions. In conclusion, we have demonstrated 1 GHz operation of a femtosecond PPLN OPO. Using higher-order synchronous pumping, we optimized the mode matching between the pump and signal beams and achieved low-threshold operation of the PPLN OPO. Using different cavity mirror sets, generation of 1.55 μm signal pulses at 1 GHz should also be possible.

References

1 P. J. Phillips, S. Das, and M. Ebrahimzadeh, Appl. Phys. Lett. **77**, 469 (2000).
2 B. Ruffing, A. Nebel, R. Wallenstein, Appl. Phys. B **67**, 537 (1998).

Hybrid Chirped Pulse Amplification

Igor Jovanovic and C. P. J. Barty

Lawrence Livermore National Laboratory, Mail Code L-490, 7000 East Avenue, Livermore, CA 94550
E-mail: jovanovic1@llnl.gov, barty1@llnl.gov

Abstract. We present a novel chirped pulse amplification method which combines optical parametric amplification and laser amplification. We have demonstrated this hybrid CPA concept with a combination of beta-barium borate and Ti:sapphire. High-efficiency, multi-terawatt compatible amplification is achieved without gain narrowing and without electro-optic modulators using a simple commercial pump laser.

1. Introduction

Optical parametric chirped pulse amplification (OPCPA) [1, 2] is an attractive short pulse amplification technology because of its simplicity, broad bandwidth and large gain. Furthermore, OPCPA does not intrinsically produce pre-pulses and does not require electro-optic switching to either select a single pulse from an oscillator pulse train prior to multipass amplification or to inject/eject a pulse from a regenerative amplifier. Finally, the absence of quantum defect in OPCPA allows scaling to high average power. The major drawback of OPCPA, however, is pump conversion efficiency. The spatio-temporal evolution and relatively long pulse width of commercial Q-switched lasers limits conversion. The highest OPCPA extraction efficiency reported to date from a system pumped by a commercial Q-switched Nd:YAG laser is 6%. [3] It is expensive and complicated to produce an ideal pump pulse with a top hat spatial and temporal profile. However, a unique feature of optical parametric amplification that distinguishes it from amplification in laser gain media is the fact that no pump absorption and energy storage occurs in the optical parametric amplifier (OPA). If a small temporal slice of the pump pulse is converted in OPCPA, essentially all of the remaining pump energy is available after OPCPA to pump another laser amplifier. Since a laser amplifier acts as an integrator of the residual pump energy, it is essentially insensitive to the temporal modulation of the pump pulse. In this way, hybrid chirped pulse amplification (HCPA) combines a high-gain amplifier stage based on OPCPA with a high-efficiency amplifier stage based on laser amplification. Pump-to-signal conversion efficiencies of >50% or 90% of the maximum theoretical quantum efficiency have been demonstrated from chirped-pulse, Ti:sapphire-based laser amplifiers pumped by commercial Nd:YAG lasers [4].

The HCPA scheme can be applied to a number of nonlinear materials/laser gain media combinations (see Table 1). The requirements that dictate the selection of materials in HCPA are the existence of pump bands at pump laser wavelength, existence of laser transition at the desired operating wavelength, and the broad bandwidth of selected OPA nonlinear crystal at the operating wavelength.

Table 1. Some potential combinations of OPA nonlinear crystals and laser amplifiers for use in HCPA

Center wavelength	Pump laser	OPA nonlinear crystal	Laser amplifier
700-900 nm	527 or 532 nm	BBO noncollinear/ LBO noncollinear/ KTP noncollinear/ $LiNbO_3$ noncollinear/ PPLN	Ti:sapphire
720-840 nm	527 or 532 nm	BBO noncollinear/PPLN	Cr:LiCAF
780-920 nm	527 or 532 nm	BBO noncollinear/LBO noncollinear/PPLN	Cr:LiSAF
350-380 nm	266 nm	BBO noncollinear/ PPLN	Ce:LiSAF
700-818 nm	527 or 532 nm	BBO noncollinear/PPLN	Alexandrite
1540 nm	946 nm	KTP/ $LiIO_3$/ KTA/ $LiNbO_3$/ PPLN	Er:glass
1053 nm	527 or 532 nm	BBO/LBO/KDP/KTP/$LiNbO_3$/PPLN	Nd:YLF
1054 nm	527 or 532 nm	BBO/LBO/KDP/KTP/$LiNbO_3$/PPLN	Nd:glass
550 nm	351-355 nm	KDP noncollinear/PPLN	Dye

2. Experimental demonstration of HCPA at 800 nm

We have experimentally demonstrated HCPA at 800 nm. Our system is presented in Figure 1. A mode-locked Ti:sapphire oscillator produces 26.5-nm FWHM pulses centered near 820 nm, which are stretched to a duration of 600 ps. The resulting 0.5-nJ pulses are used to seed the OPCPA preamplifier, consisting of a 20-mm and a 15-mm BBO crystal. Noncollinear OPCPA is pumped by a single 225-mJ pump pulse from a commercial frequency-doubled, Q-switched Nd:YAG pump laser operating with single longitudinal mode. We obtained up to 3-mJ amplified pulses from the OPCPA preamplifier, with a gain of $6x10^6$.

After pump conversion to signal and idler in the OPCPA, and after some losses on unoptimized coatings of dichroic beamplitters, the remaining pump energy available is 195 mJ. This energy is relay-imaged onto a Ti:sapphire crystal bow-tie amplifier, which had a single pass absorption of 90%. The output of this amplifier was 65 mJ, or 37% conversion efficiency of the absorbed pump energy. The measured beam quality of thee amplified pulses was M^2(horizontal)=1.4, and M^2(vertical)=1.7. The measured shot- to-shot energy stability of the four-pass

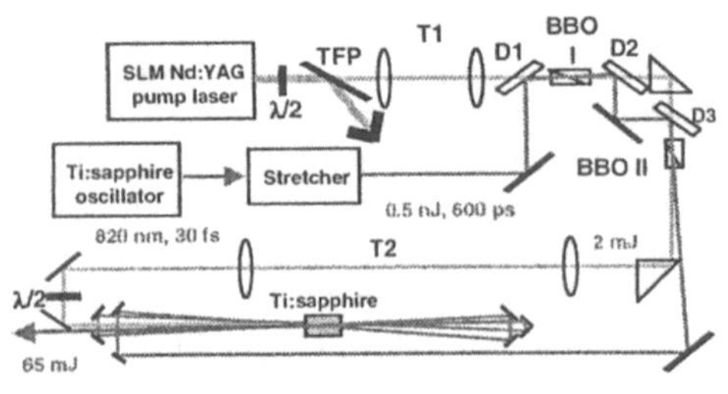

Fig. 1. Experimental setup. TFP-thin film polarizer, T-telescope, D-dichroic.

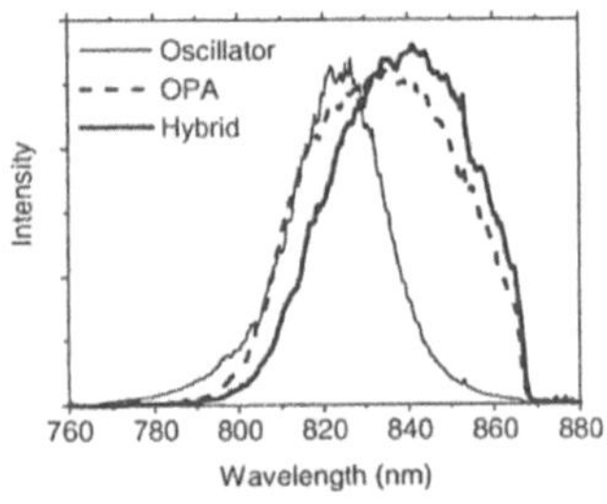

Fig. 2. Seed and amplified signal spectra

amplifier was 3% (1 std. deviation in 100 shots). The measured prepulse contrast level of our system is set by the total gain and was better than 10^8.

In Figure 2 we show the measured spectra of the stretched oscillator pulse, OPCPA, and four-pass amplifier. Oscillator pulse has a FWHM bandwidth of 26.5 nm, while the spectrum of the pulses amplified in OPCPA is broadened to 49 nm FWHM as a result of nonuniform temporal saturation. After saturated amplification in Ti:sapphire, the center wavelength of the amplified pulse shifts to 840 nm. The FWHM bandwidth of the 65-mJ output was 44.5 nm. Transform-limited pulse width related to this spectrum is 30 fs.

3. Conclusions

In summary, we have demonstrated a novel hybrid architecture combining OPCPA and laser amplification, which eliminates the need for electro-optic modulators, produces gain "broadened" output capable of supporting multiterawatt pulses, and intrinsically provides high prepulse contrast. This scheme exhibits superior energetic performance when compared to a system based on OPAs alone and superior bandwidth and simplicity when compared to a system based on laser amplification alone. The demonstrated 37% absorbed energy-to-signal conversion efficiency in BBO-Ti:sapphire HCPA is highest to date in an OPA-based CPA system pumped by a tabletop commercial pump laser. Scaling of this scheme can produce >10 TW pulses using a single commercial pump laser.

Acknowledgements. We would like to thank Chris Ebbers, Rick Cross, John Crane, David Gibson, Al Erlandson, Steve Payne, Bill Krupke, and Mark Hermann for their help and suggestions.
This work was performed under the auspices of the U. S. Department of Energy by Lawrence Livermore National Laboratory under Contract No. W-7405-Eng-48.

References

1 A. Dubietis, G. Jonusauskas, and A. Piskarskas, Opt. Commun. **88,** 437, 1992.
2 I. N. Ross, P. Matousek, M. Towrie, A. J. Langley, and J. L. Collier, Opt. Commun. **144**, 125, 1997.
3 I. Jovanovic, B. J. Comaskey, C. A. Ebbers, R. A. Bonner, D. M. Pennington, and E. C. Morse, Appl. Opt. **41**, 2923, 2002.
4 B. Walker, C. Toth, D. Fittinghoff, T. Guo, D.-E. Kim, C. Rose-Petruck, J. A. Squier, K. Yamakawa, K. R. Wilson, and C. P. J. Barty, Opt. Exp. **5**, 196, 1999.

High average system power, 10kHz, ultrafast laser

Sterling Backus, Randy A. Bartels' Ariel Paul, Henry Kapteyn, Margaret Murnane

JILA, University of Colorado, 440 UCB, Boulder CO 80309-0440, USA
E-mail: sbackus@colorado.edu

Abstract. We demonstrate a simple and practical, single-stage, ultrafast laser amplifier system that operates at a repetition frequency from 1 to 10kHz, with millijoule pulse energy, and up to 15 W average power.

1. Introduction

The development of compact, high-intensity, ultrafast lasers has enabled new experiments in high field science. Many applications of these sources, including laser-synchrotron experiments, coherent x-ray generation, time-resolved holography, ultrafast surface science, and metrology for EUV optical systems, are limited by the average flux available for experiments, rather than the available peak power or pulse energy. These and other considerations motivate the development of simple, high repetition-rate (>>1kHz) laser systems capable of generating ≈ 1 millijoule pulse energies, with good ($M^2 < 1.5$) beam quality. The ability to continuously vary the repetition rate of the laser, so that initial alignment of an experiment can be performed at modest average powers, is also very desirable.

Past work has demonstrated ultrafast laser systems that operate at TW peak powers for 1 kHz repetition-rate systems, and at average powers of up to 25W for 5-10kHz repetition rate systems[1,2]. However, previous high average-power systems that have been demonstrated to date have been large (requiring two or more optical tables), complex, and expensive, typically requiring several stages of amplification and >100W of 532nm pump light from multiple pump lasers. In this talk, we will describe a compact, single-stage, diode-pumped ultrafast laser system delivering millijoule pulse energies, with 24fs duration pulses, at a repetition-rate that can be varied from 1 to 10kHz[3].

2. Thermal management

This laser system employs well-established Ti:sapphire oscillator designs that routinely generate 11fs pulses, as well as chirped-pulse amplification technologies, and a liquid-nitrogen-cooled 12-pass single-stage amplifier. By cooling the crystal to 77K, thermal lensing effects are essentially eliminated, even when pumped continuously at a repetition-rate of 10kHz and 75W of average power. Elimination of the thermal lens has two benefits - it allows us to use a simple 12-pass ring amplifier design, and also to obtain an extremely high conversion efficiency (26%) from pump light to amplified output pulses. Previous high average-power systems have relied on compensating for the thermal lens with a static optical element, one example of which is the "thermal eigenmode" amplifier[4]. This approach can partially eliminate the effect of the lens, but does not allow for moderate changes in the average power of the pump laser that occur as the repetition rate is changed, and does not compensate for higher-order aberrations in beam focusing resulting from the thermal lens. Thus, in the case of the thermal eigenmode design, the system must be designed for a specific repetition rate. Furthermore, for the very high (>100W) pump powers needed for multi-kHz amplification, the induced thermal lens is simply too strong to be compensated.

In contrast, by removing the thermal lens by cooling the amplifier crystal, our laser system is capable of running at widely-varying repetition rates and average powers. At temperatures of 300K for example, thermal lensing can be extremely severe, varying from 11cm at 10W, to 1cm at 100W. However, at temperatures of 77 K, the power of the lens varies from 5000cm at 10W, to 500cm at 100W. Therefore, there is a 500-fold decrease in the strength of the thermal lens by cooling the crystal from 300K to 77K. This decrease is the result both of the large increase in thermal conductivity, and the large decrease in dn/dT of sapphire. At 300K, the thermal conductivity is 0.26 W/cm*K, while at 77K, it increases to 9.8 W/cm*K. The corresponding values of dn/dT are 1.3×10^{-5}/K at 300K, and 1.8×10^{-6}/K at 77K respectively.

3. Performance

In our amplifier system, the 11fs pulses from the oscillator are stretched to 206ps by the pulse stretcher. Pulses of the desired repetition rate are then injected into the liquid-nitrogen-cooled, 12-pass, single-stage amplifier. The B-integral calculated for this system is 0.33 radians, which takes into account the dynamically changing pulse width in the amplifier. The amplifier is pumped by a diode-pumped frequency-doubled Nd:YAG laser (Coherent Inc. Corona). The performance of the amplifier is shown in Fig. 1. The lower curve (stars) shows that for ~8mJ of pump energy at repetition rates varying from 1 to 5kHz, we obtain 1.4mJ of compressed output. At 6 and 7kHz, the output drops to ~1.3mJ, and for 8, 9, and 10kHz the output drops to 1.0 mJ, 700μJ, and 600μJ respectively. Much of this energy drop is due to decreasing pump pulse energy (from 8.6mJ to 6.34mJ), due to both expansion of the pump beam, and roll-off in pump energy at higher repetition rates. The maximum average power from the amplifier system is 15 W at 5 kHz - data at higher repetition rates between 10 and 20kHz is in progress.

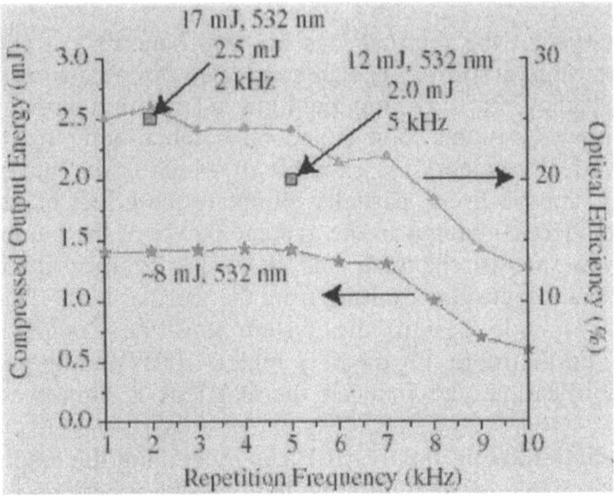

Fig. 1. Performance graph. Left axis is compressed energy (stars), and right axis is amplifier optical to optical efficiency (triangles). Squares are performance points taken under higher energy pumping conditions.

4. Conclusion

In conclusion, we have demonstrated a simple and practical, single-stage, high average power, all-solid-state, ultrafast laser system, that generates the highest average power from a single-stage, amplifier system to date. We acknowledge support for this work from the Department of Energy and the National Science Foundation.

References

1 Brent Stuart (personal communication) 2000.
2 Y. Nabekawa *et al.*, *Appl. Phys. B*, **70,** 171 (2000).
3 S. Backus *et al.*, *Optics Letters* **26**, 465 (2001).
4 F. Salin *et al.*, *Optics Letters* **23**, 718 (1998).

Continuum generation in a prism-less Ti:sapphire laser

T. R. Schibli[1], L. M. Matos[2], F. J. Grawert[1], J. G. Fujimoto[1], and F. X. Kärtner[1]

[1]Department of Electrical Engineering and Computer Science,
[2]Department of Physics and Research Laboratory of Electronics
Massachusetts Institute of Technology, Cambridge, MA 02139, USA
phone: (617) 452 5093, fax: (617) 253 9611, email: schibli@mit.edu

Abstract. Continuum generation in a prism-less Ti:sapphire laser is demonstrated. The spectral output extends from 700 to 950 nm at –3dB and from 600 – 1240 nm at –60dB. Eliminating the prisms generally used to control the optical dispersion in a laser cavity leads to an extremely stable and reproducible generation of ultra short Kerr lens modelocked pulses.

Kerr lens mode-locked Ti:sapphire lasers are capable of ultra short pulse and continuum generation, and, therefore, have enabled many novel experiments in nonlinear optics, optical coherence tomography and frequency metrology [1,2,3] over the last years. The key issue in sub-10 fs laser systems is the precise compensation of the optical dispersion caused by the laser crystal and the air residing within the laser cavity. Therefore, dispersion compensating laser optics, i.e. various forms of chirped mirrors and multi-GTI-mirrors have been developed [4,5,6]. To date, extremely stable, all mirror-dispersion controlled Ti:sapphire lasers emitting 8 fs pulses with spectra extending over 105 nm full width at half-maximum (FWHM) have been demonstrated [7]. Additional use of prism pairs for dispersion compensation results in an increased flexibility in the mirror design and therefore, in pulses as short as 5 fs with octave spanning spectra directly from the laser [8,9]. Recently, it has been found, that the fluctuations in the intracavity beam-pointing angle translates into undesired dispersion fluctuations in lasers with prisms, which is detrimental in experiments concerned with carrier-envelope phase stabilization. In this paper, we present an all mirror-dispersion controlled Ti:sapphire laser, which generates sub-8fs pulses and most notably, emits over a spectral width as large as 250 nm (FWHM) and even generates significant spectrum over 400 nm on a logarithmic scale suitable for carrier-envelope phase stabilization based on interference of second and third harmonic light [3]. Due to the reduced intracavity-losses, an optical to optical efficiency of more than 10% in the mode-locked state is achieved allowing for pump-powers below 3 W with a total output power of typically 300 mW.

Fig. 1. shows a schematic diagram of the prism-less Ti:sapphire laser. All cavity mirrors except the output coupler are double-chirped mirrors (DCMs) [5]. The DCM-design is such that there is high transmission of the pump light at 532 nm

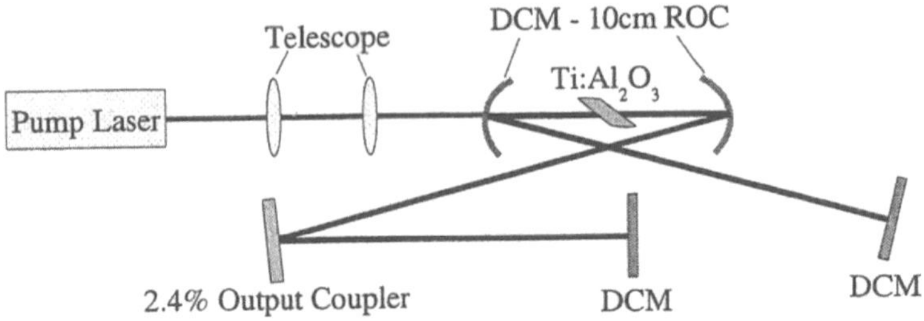

Fig. 1. Schematic diagram of the astigmatically compensated, prism-less Ti:sapphire laser. All mirrors except the output coupler mirror are double-chirped mirrors (DCMs), which compensate the second and third order dispersion of the laser crystal and the air, which resides inside the laser cavity.

and high reflectivity greater than 99.9% from 700-1000 nm. On average, the DCMs generate within six bounces per roundtrip the necessary dispersion to compensate the positive second and third order dispersion of the laser crystal and the 4.5 meters of air which are present in this 75 MHz laser cavity, see Fig. 2. The laser crystal has a path length of 2.3 mm and absorbs approximately 70% of the pump light, which is emitted from a frequency doubled, diode pumped Nd:YVO_4 laser.

Fig. 3. shows a typical spectrum and the corresponding interferometric autocorrelation of the pulses emitted by the laser shown in Fig. 1. A fit to the autocorrelation is obtained by using the phase retrieval algorithm published in [10], resulting in about 8 fs pulses. (The Fourier transform limit of the spectrum shown in Fig. 3 assuming constant spectral phase is 6.9 fs). The interferometric autocorrelator used to record the trace is based upon the dispersion balanced setup published in [11]. The optical nonlinearity used in the autocorrelator is based on two-photon absorption in an edge cleaved multiple quantum well laser diode from Sanyo (DL3038-011, 635 nm emission wavelength). To avoid the material-dispersion of the laser diode the light is not coupled into the diode's waveguide but the quantum wells are rather illuminated under grazing incidence. The dispersion of the output coupler mirror substrate (3.2 mm fused silica) and the air between the laser and the autocorrelator is compensated by three additional bounces on DCMs as the ones used in the laser cavity. No further fine-tuning of the extra-cavity dispersion has been performed, which explains the residual chirp of the pulses.

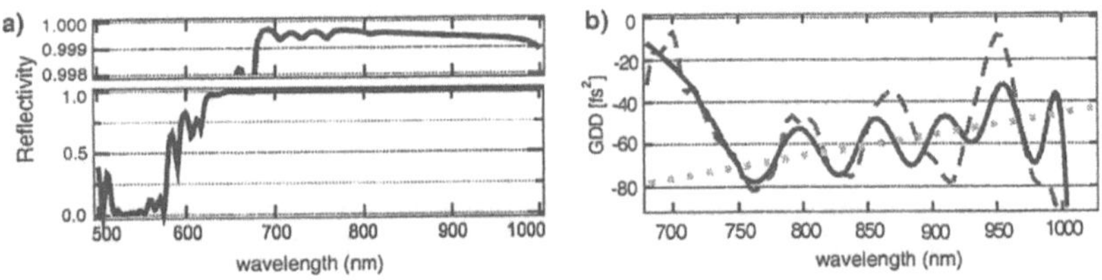

Fig. 2: a) Reflectivity of the DCMs. b) Desired (dotted line), designed (solid line) and measured (dashed line) group delay dispersion of the DCMs.

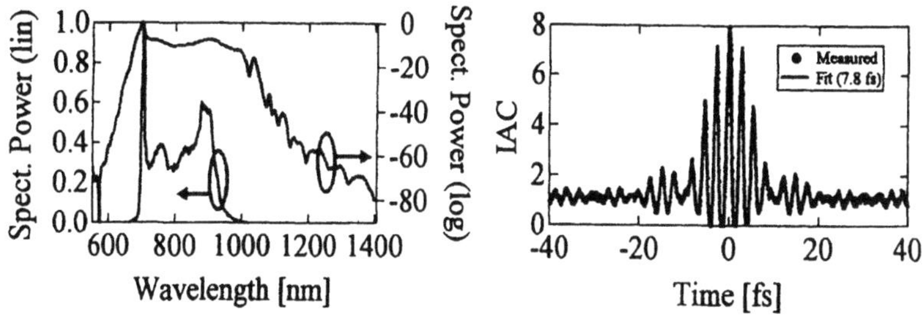

Fig. 3. Typical spectrum and interferometric autocorrelation generated by the laser shown in Fig.1.

In conclusion, we have shown that all DCM-compensated Ti:sapphire lasers have the ability to generate ultra-broad spectra with high conversion efficiencies. Since the resonator dispersion is independent of the alignment of the cavity, stable mode-locking is observed over many hours, and since no fine-tuning of the dispersion is needed, the operation of such a laser is significantly easier compared to a laser with additional prism pairs for-dispersion compensation.

References

1 W. Drexler, U. Morgner, F. X. Kärtner, C. Pitris, Y. Chen, S. A. Boppart, X. D. Li, E. P. Ippen, J. G. Fujimoto, Opt. Lett., **24**, 1224, (1999)

2 S. A. Diddams, D. J. Jones, J. Ye, S. T. Cundiff, J. L. Hall, J. K. Ranka, R. S. Windeler, R. Holzwarth, T. Udem, T. W. Hänsch, Phys. Rev. Lett., **84**, 5102, (2000)

3 U. Morgner, R. Ell, G. Metzler, T. R. Schibli, F. X. Kärtner, J. G. Fujimoto, H. A. Haus, E. P. Ippen, Phys. Rev. Lett., **86**, 5462, (2001)

4 R. Szipöcs, K. Ferencz, C. Spielmann, F. Krausz, Opt. Lett., **19**, 201, (1994)

5 F. X. Kärtner, N.Matuschek, T. Schibli, U. Keller, H. A. Haus, C. Heine, R. Morf, V. Scheuer, M. Tilsch, T. Tschudi, Opt. Lett., **22**, 831, (1997)

6 B. Golubovic, R. R. Austin, M. K. Steiner-Shepard, M. K. Reed, S. A. Diddams, D. J. Jones, A. G. V. Engen, Opt. Lett., **25**, 275, (2002)

7 A. Stingl, M. Lenzner, Ch. Spielmann, F. Krausz, R. Szipöcs, Opt. Lett., **20**, 602, (1995)

8 D.H. Sutter, G. Steinmeyer, L.Gallmann, N.Matuschek, F. Morier-Genoud, U. Keller, V. Scheuer, G. Angelow, T. Tschudi, Opt. Lett **24,** 631, (1999)

9 R. Ell, U. Morgner, F. X. Kärtner, J. G. Fujimoto, E. P. Ippen, V. Scheuer, G. Angelow, T. Tschudi, Opt. Lett., **26**, 373, (2001)

10 A. Baltuska, A. Pugzlys, M. Pshenichnickov, D. Wiersma, B. Hoenders, H. Ferwerda, in Proceedings of Ultrafast Optics 1999, paper Th10, Ascona, Switzerland, (1999)

11 C. Spielmann, P. Curley, T. Brabec, F. Krausz, Appl. Opt. **36**, 2523, (1997)

Octave spanning femtosecond laser at GHz repetition rates

A. Bartels and H. Kurz

Institut für Halbleitertechnik, RWTH Aachen, Sommerfeldstr. 24, 52074 Aachen, Germany

Abstract. A 5-element Ti:sapphire femtosecond ring oscillator with an optical spectrum ranging from 570 nm to 1140 nm at –47 dB below its maximum, thus spaning one octave, is demonstrated at a repetition rate of 1 GHz. The key element is a slightly convex cavity mirror that serves to increase the self amplitude modulation of a pulse inside the resonator.

1. Introduction

Utilizing their frequency comb properties modelocked femtosecond lasers with repetition rates in the GHz range [1] have served as optical clockworks to phase-coherently link optical and microwave frequencies in frequency metrology applications and new atomic clocks based on transitions in the PHz range [2-4]. Aside from the easily measurable repetition rate f_r the crucial task has been to access the comb's frequency offset from 0 Hz, f_{CEO}. To accomplish this, in current schemes an octave spanning spectrum has been necessary. Microstructure fibers were used for extracavity spectral broadening of the laser output to meet this requirement [5]. However, these fibers have difficulties with regard to operation on a timescale of several hours. Specifically, it is difficult to maintain efficient coupling of the broadband light into the sub 2-μm fiber core and ensure long-term stability of the output spectrum which is extremely important for the stability of the f_{CEO} signal. Favorably, a directly octave spanning femtosecond laser has been realized by Ell et al. [6]. Although there, f_{CEO} could be measured directly, with only 120 mW output power and 65 MHz repetition rate the complicated device does not yet appear optimal for frequency comb applications. We report a new approach to achieve octave spanning spectra directly from femtosecond oscillators in a compact and powerful 5-element resonator operating at 1 GHz repetition rate.

2. New approach to increased bandwidth

The standard ring resonator to achieve GHz repetition rate femtosecond operation is shown in Figure 1.. It consists of two concave mirrors M1 and M2 with focal length of 15 mm, flat M3 and and output coupler OC. A Ti:sapphire crystal is located at Brewster's angle between M1 and M2. The factor $\delta = w_0^{-1}\,\partial w_0/\partial P$ describes the change of beam waist radius inside the laser crystal w_0 with respect to instantaneous power P due to the Kerr lens effect and is a measure for the self gain modulation produced by a short pulse in combination with the finite gain volume [7]. In the limit of pure solitonlike pulse propagation a negative δ is only necessary to stabilize modelocking in the cavity, the pulse duration is determined by group velocity dispersion (GDD) and self phase modulation. Equivalently, the self gain modulation can be considered as an effective fast saturable absorber. In a laser operated in the limit of pure fast saturable absorber modelocking the pulse duration is inversely proportional to the modulation depth. Replacing the flat mirror M3 by a convex mirror with 1 m focal length, aside from that given equal conditions, theoretically results in an increase of the achievable negative amount of δ by a factor of approximately five, i.e. an increase of modulation depth of the effective saturable absorber. Although the calculations are only valid for peak powers below the critical power for self-focusing (960 kW in sapphire) and our lasers are usually well above this threshold, we pursued the route of using a convex mirror M3 in order to enhance the contribution of the effective saturable absorber to the pulse shaping in the resonator in an effort to achieve shorter pulses and broader spectra.

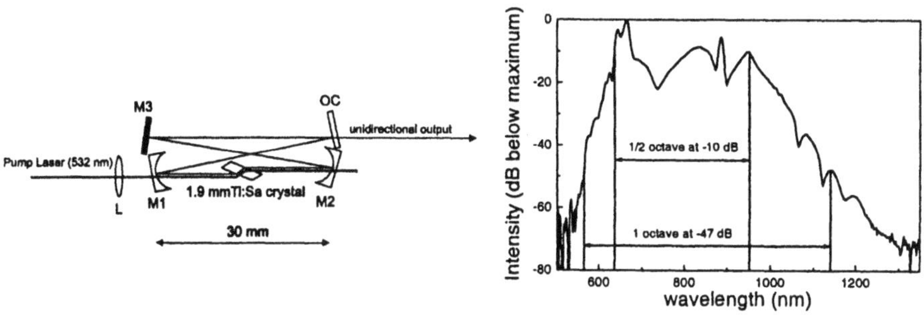

Fig. 1. Left panel: The cavity consists of the two concave chirped mirrors M1 and M2 (15 mm focal length), mirror M3 (see text for details) and the output coupler OC. The 1.9 mm thick Ti:sapphire crystal is located at Brewster's angle in between mirror M1 and M2 and is pumped through lens L (30 mm focal length). Right panel: Laser output spectrum on logarithmic scale. The spectrum spans one octave from 570 nm to 1140 nm at –47 dB and half an octave from 636 nm to 954 nm at –10 dB.

3. Octave spanning spectrum from a 5-element ring oscillator

The experimental resonator, shown in Fig. 1 contains a 1.9 mm thick Ti:sapphire crystal that is pumped through lens L (focal length 30 mm) by 10.5 W from a Coherent Verdi. M3 is a convex mirror with 1 m focal length. The output coupler has a transmission of 2% between 700 nm and 850 nm with losses greater than 20% out of this band. Mirrors M1-M3 have a chirped coating with a GDD of approximately –45 fs^2 from 750 nm to 850 nm. Outside this band they have positive GDD with third order dispersion of several 1000 fs^3. To initiate unidirectional modelocking M2 is translated towards M1 to achieve the highest negative value of δ. The output power in pulsed operation is 1.2 W whereas it drops by more than 90% when pulsed operation is interrupted by shortly intersecting the pump beam. This is a strong indication for the presence of an effective saturable absorber. The output spectrum is shown in Fig. 2. It ranges from 570 nm to 1140 nm at –47 dB below its maximum. The pulse duration measured after compression with chirped mirrors with the same coating as the cavity mirrors is 12 fs. We believe that the intracavity pulse duration is shorter but phase distortions due to the output coupler can only be compensated for partly by our external chirped mirrors. Whereas the full octave is spanned at a relatively low power level, half an octave is spanned from 636 nm to 954 nm at –10 dB. In a scheme where a low frequency part of the spectrum is frequency tripled and beat against the second harmonic of a high frequency part, this has been shown to be sufficient to access f_{CEO} [8]. In fact an optical clock using the presented laser has been demonstrated according to this [9]. It is important to note that the resonator with a flat M3 behaves equivalent to the laser decribed in reference [1]. This suggests that indeed the enhanced effective fast saturable absorber is responsible for the broadband emission and considering the mirror properties that the requirements for accurate control of the net cavity GDD for the achievement of an ultrabroadband spectrum are greatly relaxed compared to earlier approaches [6].

4. Conclusion

In conclusion we demonstrated a powerful and compact Ti:sapphire femtosecond laser spanning one octave at –50 dB and half an octave at –10 dB below its maximum. We expect that the device will make future optical clockworks for frequency metrology applications and new optical atomic clocks simpler and more reliable.

References

1 A. Bartels, T. Dekorsy, and H. Kurz, *Optics* Letters, Vol. 24, 996, 1999.
2 Th. Udem, J. Reichert, R. Holzwarth, and T.W. Hänsch, in *Physical Review Letters*, Vol. 82, 3568, 1999.
3 R. Holzwarth, Th. Udem, T.W. Hänsch, J.C. Knight, W.J. Wadsworth, and P. St. J. Russell, in *Physical Review Letters*, Vol 85, 2264, 2000.
4 S.A. Diddams, Th. Udem, J.C. Bergquist, E.A. Curtis, R.E. Drullinger, L. Hollberg, W.M. Itano, W.D. Lee, C.W. Oates, K.R. Vogel, and D.J. Wineland, in *Science*, Vol. 293, 825, 2001..
5 J.K. Ranka, R.S. Windeler, and A.J. Stentz, *Optics* Letters, Vol. 25, 25, 2000.
6 R. Ell, U. Morgner, F.X. Kärtner, J.G. Fujimoto, E.P. Ippen, V. Scheuer, G. Angelow, T. Tschudi, M.J. Lederer, A. Boiko, and B. Luther-Davies, in *Optics Letters*, Vol 26, 373, 2001.
7 we calculate δ according to V. Magni, G. Cerullo, and S. de Silvestri, *Optics* Communications, Vol. 96, 348, 1993.
8 U. Morgner, R. Ell, G. Metzler, T.R. Schibli, F.X. Kärtner, J.G. Fujimoto, H.A. Haus, and E.P. Ippen, *Physical Review* Letters, Vol. 86, 5462, 2001.
9 T.M. Ramond, S.A. Diddams, L. Hollberg, and A. Bartels, submitted to Optics Letters, 2002.

Compact ultrashort hard-x-ray source driven by high-repetition rate femtosecond laser pulses

A. Egbert, B. Mader, B. Tkachenko, C. Fallnich, and B. N. Chichkov

Laser Zentrum Hannover e.V., Hollerithallee 8, D-30419 Hannover, Germany
Phone:+49-511-2788-218, fax: +49-511-2788-100
E-mail: ae@lzh.de

Abstract. A compact, high-repetition rate, ultrashort-pulse laser-driven hard-x-ray source based on the combination of a femtosecond laser system with an x-ray diode is demonstrated. A comparison with available laser-plasma sources is presented. A numerical modeling is performed which proves that picosecond and sub-picosecond hard-x-ray pulses can be produced with this source.

There is a strong demand for the development of ultrashort hard-x-ray sources for applications in time-resolved diffraction studies, control of chemical reactions, medical imaging, etc. In this paper, we address a possibility to develop a compact, high-repetition rate hard-x-ray source. The investigated concept is based on the combination of low-intensity femtosecond laser pulses with an x-ray diode. The femtosecond laser pulses induce photoemission of electrons from a metal cathode which then are accelerated towards a high-Z target material by applying an external high-voltage electric field. Hard-x-rays are produced by Bremsstrahlung of accelerated electrons and characteristic line emission in the target. This concept has been "floating around" for a decade [1,2] but has not been experimentally realized so far. For the realization of this kind of x-ray source, the experimental requirements are significantly reduced as compared to the currently demonstrated ultrashort hard-x-ray sources.

In our work commercial Ti:sapphire laser systems delivering 120 fs and 160 fs laser pulses at 780 nm with pulse repetition rates of 1 kHz and 250 kHz, respectively, are used to produce photoelectrons from metal cathodes. The laser pulses are focussed onto the surface of a metal cathode to a spot size of $A \sim 5 \cdot 10^{-4}$ cm^2. The cathode consists of a quartz substrate coated on the backside with a 100-nm thick metal film. The copper anode (target) has positive potential which can be varied continuously up to 100 kV, whereas the cathode is set to ground potential. The distance between the metal cathode and the anode is 1 cm. The generated x-rays leave the diode through a 1 mm thick beryllium window with a diameter of 40 mm. The average diode current is measured with an ampere meter providing accuracy of 0.1 nA (Keithley, mod. 195 A digital multimeter). The hard-x-ray spectra are recorded using a Si-photodiode-based energy dispersive x-ray detector (AMPTEK, XR-100 CR) operating in a single-photon counting regime. This allows to measure the absolute number of hard-x-ray photons.

In Fig. 1, a hard-x-ray spectrum obtained with a copper target at an accelerating voltage of U = 50 kV is shown. The x-ray detector was equipped with a 2 mm

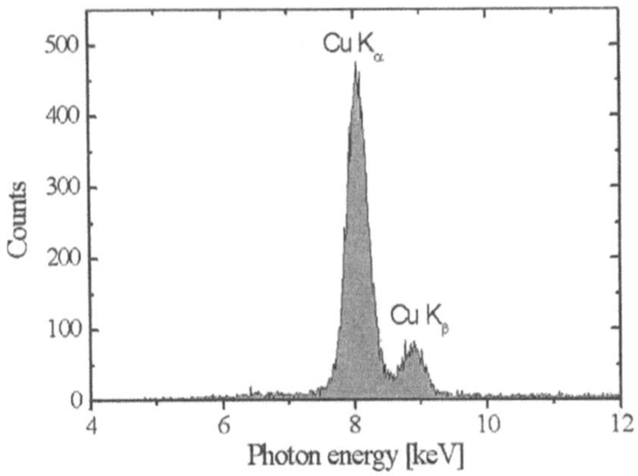

Fig. 1. Hard-x-ray spectrum obtained with an copper target at an accelerating voltage of 50 kV.

lead diaphragm positioned 46 cm away from the x-ray source to avoid pile-up effects in the detector.

A comparison of the x-ray flux corresponding to the copper K-lines generated with our laser-driven x-ray diode and that obtained with laser-plasma x-ray sources is illustrated in Fig. 2 as a function of laser peak power. Our results are marked by stars. As can be seen, at a repetition rate of 250 kHz more than 10^{10} hard-x-ray photons/s (emitted in 4π) are generated with a laser pulse energy as small as 3 μJ. These results are comparable with that obtained with much more complex high-power laser systems.

The evolution of the electron pulse during the propagation from the cathode to the target has been modeled using a space-charge tracking algorithm (ASTRA),

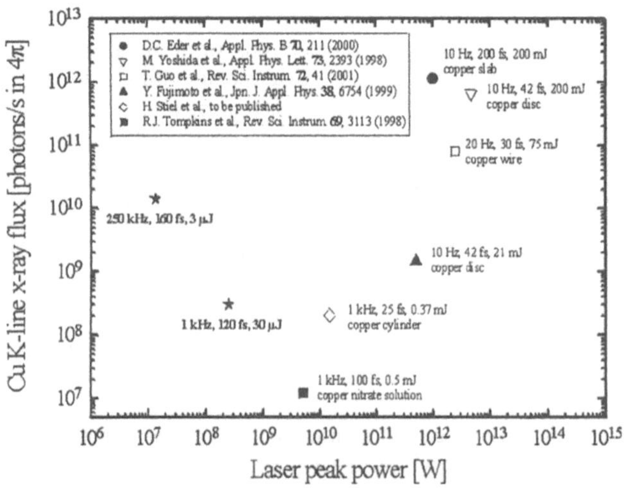

Fig. 2. Comparison between the x-ray fluxes corresponding to copper K-lines generated in this work (shown by stars) and that obtained with available laser-plasma x-ray sources.

which has been developed at DESY (German electron synchrotron facility, Hamburg) by K. Flöttmann to simulate the propagation of charged particles through external fields. The code takes into account pulse broadening due to the initial energy distribution and space-charge effects. In Fig. 3, the calculated temporal duration of the electron pulse, after propagation of 1 cm distance, is presented as a function of the total pulse charge. The calculations are performed for 100 fs laser pulses, a focal spot size of $A = 5 \cdot 10^{-4}$ cm^2 at the cathode surface, and an accelerating voltage of $U = 50$ kV. For the total electron pulse charge of $Q \sim 0.1 - 1$ pC, which corresponds to our experimental conditions, the generated electron and hard-x-ray pulses should have a temporal duration below 5 ps. Measurements of the x-ray pulse duration using an ultrafast x-ray streak camera in combination with a laser-triggered accumulation system are currently under preparation.

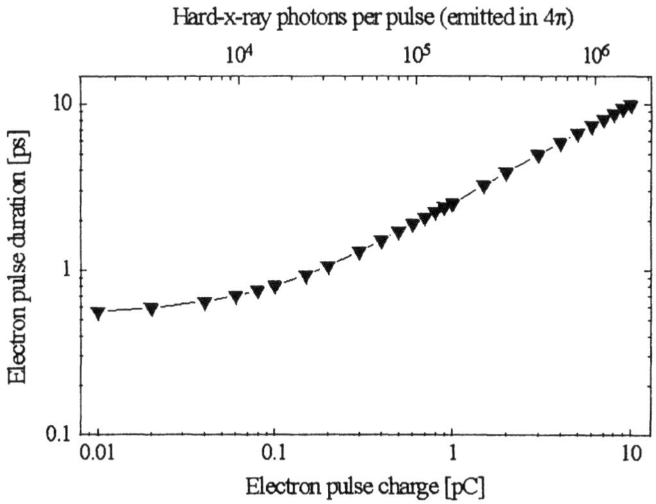

Fig. 3. Calculated electron pulse duration versus the total electron pulse charge and the number of hard-x-ray photons per pulse.

In conclusion, a novel femtosecond laser-driven hard-x-ray source has been demonstrated. This source is based on the combination of a high-repetition rate femtosecond laser system with an x-ray diode. Electron pulses are produced due to femtosecond laser-induced photoemission and accelerated by a high-voltage electric field. Electron current, x-ray spectrum, and absolute hard-x-ray yield have been measured. A comparison with available laser-plasma hard-x-ray sources has been performed. It has been demonstrated via numerical modeling that with our source picosecond and sub-picosecond hard-x-ray pulses can be generated.

References

1 T. Anderson, I.V. Tomov, and P.M. Rentzepis, J. Appl. Phys. **71**, 5161, 1992.

2 J.-P. Girardeau-Montaut, B. Kiraly, C. Girardeau-Montaut, and H. Leboutet, Nucl. Instrum. Meth. Phys. Res. A **452**, 361, 2000.

Synchronously Pumped Optical Parametric Oscillators with Anomalous Wavelength Tuning Behaviour

P. Loza-Alvarez[1], M. B. Flynn[1], W. Sibbett[1], P. Kinsler[2] and G. H. C. New[2]

[1] School of Physics and Astronomy, University of St Andrews, North Haugh, St. Andrews, Fife KY16 9SS, UK
E-mail: pl6@st-andrews.ac.uk
[2] Blackett Laboratory, Imperial College, London SW& 2BW, UK
E-mail: g.new@ic.ac.uk

Abstract. We present unusual cavity length tuning behaviour of PPLN-based femtosecond OPOs. Computer simulations relating to the experimental observations and their interpretation are also described.

1. Introduction

High repetition rate, synchronously pumped optical parametric oscillators (OPOs) have been widely used to extend the tuning wavelength range that can be accessed by conventional picosecond and femtosecond lasers [1]. The oscillation frequency of an OPO can be controlled in several different ways. One technique involves tuning the pump wavelength, while others are based on changing the crystal parameters such as the temperature, the angle (birefringent phase-matching), or poling period (quasi-phase matching). In a third method (perhaps the simplest of all), tuning is affected by adjusting the OPO cavity length. In general, this is possible through the conjunction of synchronous pumping and the inherent normal dispersion of the nonlinear crystal. The result is that shorter (longer) signal wavelengths oscillate for shorter (longer) OPO cavities. In this paper, we report the characterisation of OPOs that exhibit a distinctive (and anomalous) tuning behaviour when the cavity length is varied.

2. Experimental Methods and Results

The OPO was synchronously pumped by a Kerr-lens mode-locked Ti:sapphire laser delivering 1.7 W average power at a repetition rate of 84 MHz. The pump pulses were centred at 820 nm, and had a duration of $\Delta\tau = 110$ fs and a time-bandwidth product $\Delta\nu\Delta\tau = 0.41$. They were focused with a 50 mm focal length lens into a periodically-poled lithium niobate (PPLN) crystal with a period of 21.5 μm. The crystal temperature was maintained at 150°C to avoid photorefractive damage. The OPO cavity consisted of two high-reflectivity mirrors, and an output coupler with 20% transmission in the resonant signal wavelength range.

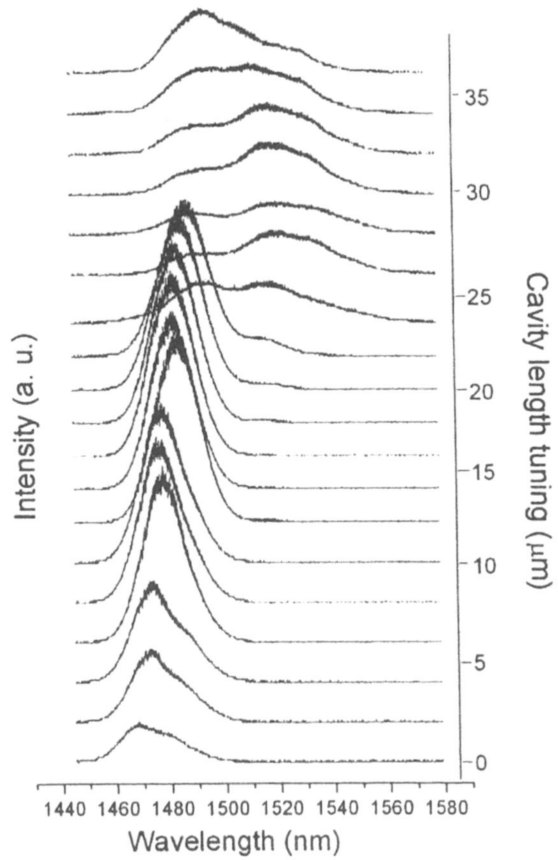

Fig. 1. Cavity length tuning behaviour of the signal pulses of the OPO

The properties of the OPO were studied in two different wavelength bands around 1300 nm and 1500 nm respectively. In the latter case, the OPO cavity included two SF10 prisms with an apex separation of 30 cm. Under these conditions, we recorded the signal pulse spectra shown in figure 1 as the cavity length was varied at regular intervals of 2 µm. It is clear from the figure that the central wavelength of the signal pulse does not follow the cavity length displacement in the way one would expect. Starting from the short cavity end, the wavelength shift is initially slow, and the pulse spectral bandwidth remains essentially constant at $\Delta\lambda \cong 20$nm. After reaching a critical point (at 24 μm displacement), the pulse spectrum experiences an abrupt change of position and the spectral bandwidth suddenly increases to $\Delta\lambda \cong 57$ nm. Thereafter, the cavity length tuning behaviour is different and, as the change in length is increased beyond 30 μm, the sign is reversed with the signal pulse spectrum moving progressively to shorter wavelengths. Broadly similar results were obtained around 1300 nm, although the anomalies were somewhat less pronounced.

In parallel with the experimental work, we have developed a sophisticated computer code to model the OPO. The code is presently based on a plane wave approximation, but it includes all other known features of the device, including dispersion (to all orders), dispersion compensation (if applicable), pulse chirping, and stochastic background noise. The results show good agreement with the experimental tuning characteristics around 1300 nm, as well as many of the temporal and spectral characteristics of the output pulses. The theoretical tuning range is narrower than that of the laboratory device, and we believe that this is likely to be due to the finite angular spread of the focused beam in the PPLN crystal.

We believe that a complete understanding of such a simple cavity setup will lead to interesting insights in the generation of ultrashort pulses for synchronously pumped femtosecond OPOs.

References

1 See, for example, the Special Issue, *Optical parametric processes and their applications,* (M. Ebrahimzadeh, R. Eckardt, M. H. Dunn, Eds.), J. Opt. Soc. Am. B, **16**, 1475 (1999).

Optical Modulator for Passive Synchronization and Active Modelocking of Lasers

Wolfgang Seitz[1], Richard Ell[1], Thomas R. Schibli[2], Uwe Morgner[1], Franz X. Kärtner[1,2], Christoph H. Lange[3], Wolfgang Richter[3], Bernd Braun[4]

[1] High-Frequency and Quantum Electronics Laboratory, University of Karlsruhe Engesserstraße 5, D-76128 Karlsruhe, Germany
E-mail: W.Seitz@ihq.uni-karlsruhe.de
[2] Research Laboratory of Electronics, Massachusetts Institute of Technology, Cambridge, MA 02139, USA, E-mail: kaertner@mit.edu
[3] Institute of Solid State Physics, Friedrich-Schiller-University, D-07743 Jena, Germany
[4] JENOPTIK Laser, Optik, Systeme GmbH, D-07739 Jena, Germany

Abstract. We present an all optical intracavity modulator based on a nonlinear semiconductor Fabry-Perot mirror. It is used for passive synchronization of two independent laser oscillators and for active mode-locking.

Modulation of intracavity losses is an important tool for controlling laser dynamics [1], for active mode-locking [2], and for synchronization of different lasers (see also [3]). In this paper, we present an optical device that allows for fast modulation of the intracavity losses by external illumination and its application as an intracavity device for active mode-locking and laser synchronization [4].

The loss modulation device is a nonlinear semiconductor Fabry-Perot mirror [5, 6] that consists of a 5-μm-thick InGaAs layer on top of a GaAs/AlAs Bragg reflector centered at 1064 nm. The In content is chosen such that the bandgap energy is 1.28 eV (λ_{gap} = 970 nm). Therefore, due to the Bragg mirror the device is highly reflecting for wavelengths longer than λ_{gap}. The Fresnel reflection at the air-InGaAs interface together with the Bragg-mirror constitutes a nonlinear Fabry-Perot resonator. Illumination of the InGaAs layer with light at shorter wavelengths than λ_{gap} leads to an index change due to free carrier generation. The index change is converted into a change in reflectivity by shifting the resonance of the Fabry-Perot structure to shorter wavelengths (see Fig. 1(a), (b)). By varying the intensity of the pump light, the reflectivity at the operation wavelength is externally controlled.

To characterize the properties of the optically controlled Fabry-Perot-Modulator (FPM), we performed a spectrally resolved two color pump-probe measurement using the light of an octave spanning Ti:sapphire laser [7] as the pump at 800 nm and continuum probe beyond 1000 nm. Fig.1(c) shows the spectrally resolved differential reflectivity of the structure near a resonance. The resonances are about 30 nm apart. It shows a change in reflectivity up to 0.2 %, when pumped with only 450 W/cm^2 of light from the modelocked Ti:sapphire laser from 700-900 nm, which is already sufficient for intracavity control of the laser dynamics and other applications.

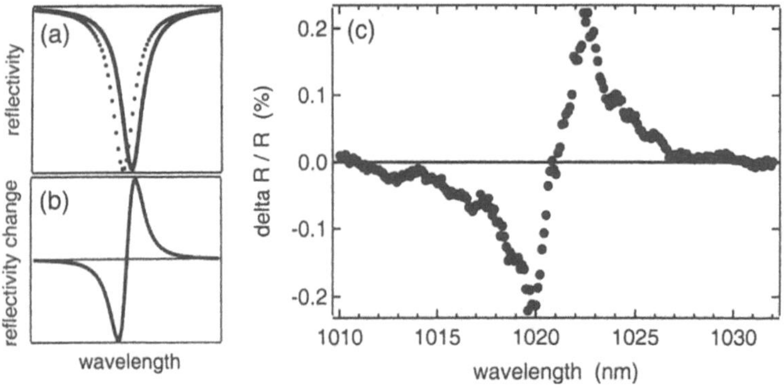

Fig. 1. (a) Schematic drawing of the reflectivity of the FPM near a resonance before carrier generation (solid curve) and after carrier generation (dotted curve). (b) Reflectivity change as a result of the resonance shift in Fig. (a). (c) Measured reflectivity change of the FPM used in the experiments.

One possible application of the FPM is the passive synchronization of two independently mode-locked laser oscillators in a master-slave configuration [4]. As an example, the scheme is used to synchronize a picosecond mode-locked $Nd:YVO_4$ laser (at 1064 nm) as the slave to a femtosecond mode-locked Ti:sapphire laser (around 850 nm) as the master oscillator [4]. For synchronization, one highly reflecting mirror of the $Nd:YVO_4$ laser cavity (dashed box) is replaced by the FPM (see Fig. 2). For external optical modulation, a fraction of the Ti:sapphire laser output power (approximately 100 mW) is focused on the FPM in a spot overlapping with the $Nd:YVO_4$ laser spot. By controlling the temperature of the FPM, one of the resonances is tuned just below 1064 nm. With incident Ti:sapphire radiation, the resonance frequency shifts to shorter wavelengths, resulting in a loss reduction in the $Nd:YVO_4$ cavity modulated at the repetition rate of the Ti:sapphire master oscillator. To achieve synchronization of the two pulse trains, the round-trip time of the driven oscillator has to be adjusted close to the modulating pulse period (about 82.6 MHz). To verify the synchronism of the two pulse trains a cross correlation measurement between the two laser pulse trains was taken (Fig. 3(a)). By analyzing the intensity fluctuations in the falling/rising edge of the cross correlation the timing jitter can be estimated conservatively well below 1 ps, which is much smaller than the duration of the $Nd:YVO_4$ pulses.

The optical modulation of the intracavity losses of the $Nd:YVO_4$ laser is deep enough and fast enough to actively modelock the $Nd:YVO_4$ oscillator in the picosecond range. To verify this, we replaced the semiconductor saturable absorber mirror (SESAM) in the cavity by a highly reflecting mirror and measured again the cross correlation trace between the two synchronized pulse trains (s. Fig. 3(b)). Due to the fact that the modulation depth of the FPM is comparable to the modulation depth of the SESAM the actively mode-locked laser pulses (τ_{FWHM} = 13 ps) are only slightly longer than the passively generated

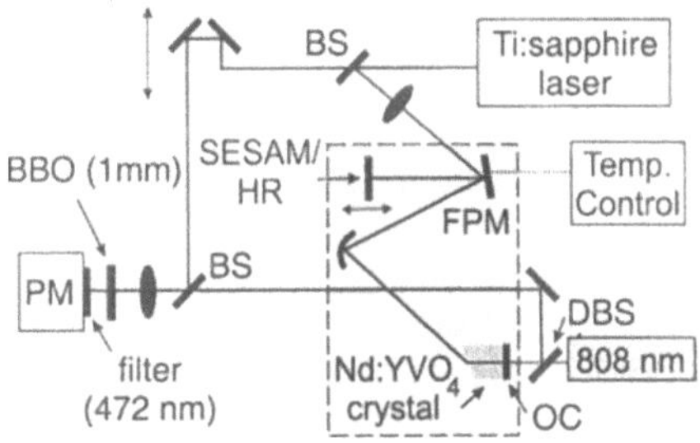

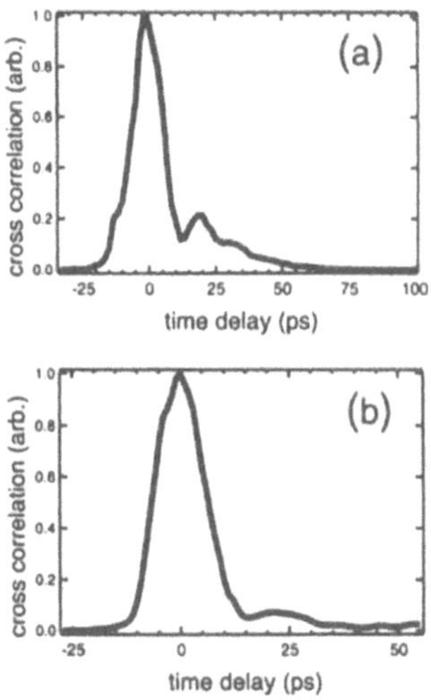

Fig. 2: Experimental setup: PM, Photomultiplier; SESAM, Semiconductor Saturable Absorber Mirror; HR, High Reflector; FPM, Fabry-Perot-Modulator; OC, Output Coupler; DBS, Dichroic Beam Splitter; BS, Beam Splitter

Fig. 3: (a) Measurement of the cross correlation between the synchronized pulse trains with an additional SESAM in the $Nd:YVO_4$ cavity. (b) Measured cross correlation without the SESAM in the $Nd:YVO_4$-cavity

pulses (τ_{FWHM} = 10 ps). The difference in pulse width is due to the additional losses of the FPM. The pulse duration of the actively mode-locked laser can be further reduced by increasing the modulation depth of the FPM and/or by reducing the carrier lifetime of the InGaAs layer.

In conclusion, we have presented a rather general method for intracavity control of the laser dynamics by optically modulating the reflectivity of a semiconductor Fabry-Perot mirror (Fabry-Perot-Modulator, FPM). The capabilities of the FPM were demonstrated by active mode-locking and by synchronizing two independent laser oscillators.

References

1 N. H. Bonadeo, W. H. Knox, J. M. Roth, and K. Bergman, Opt. Lett. **25**, 1421 (2000).

2 L. R. Brovelli, M. Lanker, U. Keller, K. W. Goossen, J. A. Walker, and J. E. Cunningham, Electron. Lett. **31**, 381-2 (1995).

3 R. K. Shelton, L.-S. Ma, H. C. Kapteyn, M. M. Murnane, J. L. Hall, and J. Ye, Science **293**, 1286 (2001).

4 W. Seitz, T. R. Schibli, U. Morgner, F. X. Kärtner, C. H. Lange, W. Richter, and B. Braun, Opt. Lett. **27**, 454 (2002).

5 J. Paye and D. Hulin, Appl. Phys. Lett. **62**, 1326 (1992).

6 R. Quintero-Torres and M. Thakur, Appl. Phys. Lett. **66**, 1310 (1995).

7 R. Ell, U. Morgner, F. X. Kärtner, J. G. Fujimoto, E. P. Ippen, V. Scheuer, G. Angelow, T. Tschudi, M. J. Lederer, A. Boiko, and B. Luther-Davies, Opt. Lett. 26, 373 (2001).

Generation of sub-10-fs pulses from a mode-locked Cr^{3+}:LiCAF laser

Philipp Wagenblast[1], Uwe Morgner[1], Felix Grawert[2], Volker Scheuer[3], Gregor Angelow[3], Max J. Lederer[4], and Franz X. Kärtner[1,2]

1 High-Frequency and Quantum Electronics Laboratory, University of Karlsruhe, Engesserstr. 5, D-76128 Karlsruhe, Germany
E-mail: wagenblast@ihq.uni-karlsruhe.de
2 Dept. of Electrical Engineering and Computer Science, Research Laboratory of Electronics, Massachusetts Institute of Technology, Cambridge 02139, USA
3 Nanolayers Optical Coatings, Rheinbreitbach, Germany
4 Laser Physics Centre, Australian National University, Canberra

Abstract. We demonstrate sub-10-fs-pulses with 220 mW average output power from a mode-locked Cr^{3+}:LiCAF laser employing double-chirped mirrors for second and third order dispersion compensation and Ti:sapphire pumping at 693 nm. .

1. Introduction

Cr^{3+}:LiCAF is a very promising material for compact femtosecond laser sources because of its low quantum defect, a broadband emission range around 800 nm, and an absorption band in a spectral range where high-brightness laser diodes are available. For applications such as seeding, ultra high resolution optical coherence tomography, or metrology, a compact replacement for the Ti:sapphire laser would be highly desirable. Although the gain bandwidth of Cr^{3+}:LiCAF supports pulses substantially shorter than 10 fs, the shortest pulse width so far demonstrated from a mode-locked Cr^{3+}:LiCAF-Laser is 19 fs [1]. It has been suggested that the strong Raman-effect occurring in Colquiriite crystals [1, 2], leading to a red shift of the spectrum, is limiting short pulse generation. We demonstrate that broadband dispersion compensation is a key to sub-10 fs pulse generation in low-gain, diode pumped Colquiriite lasers. As a first step in this direction, we demonstrate broadband laser emission over more than 200 nm and pulses as short as 8.9 fs from a Ti:sapphire pumped Cr^{3+}:LiCAF-oscillator with specially designed laser mirrors.

2. Experimental Set-up

The mirrors in our set-up are dispersion-compensating double-chirped mirrors (DCM [3]), and low-dispersive pump mirrors with a pump hole around 680 nm. To demonstrate the potential of these mirrors in a Cr^{3+}:LiCAF-laser for short pulse generation without being limited by imperfections of the pump source, we use a properly tuned cw-Ti:sapphire laser for pumping.

The set-up is shown in Fig. 1a. A 100 MHz standard z-folded cavity with 75-mm-ROC folding mirrors is used. The cw-Ti:sapphire pump laser is tuned to 693 nm and emits up to 2.1 W into a TEM_{00}-beam, which is focussed down to a diameter of 22×31 μm in the laser crystal, resulting in a maximum of 1.3 W of absorbed pump power in the 5 mm long, 5% Cr-doped crystal.

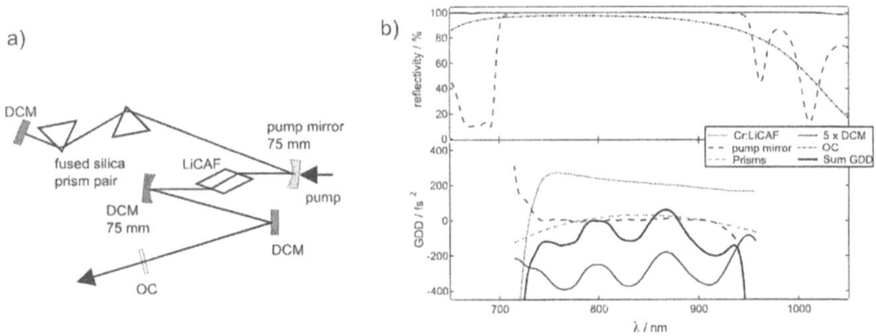

Fig. 1. a) Laser set-up. b) Measured reflectivity and GDD of the resonator elements. Bandwidth and GDD compensation range are limited by the pump mirror.

Fig.1b shows the group delay dispersion (GDD) and reflectivities of the resonator elements. The DCMs are designed to compensate for the second and third order dispersion of Cr:LiCAF, whereas the prism sequence is used for dispersion fine adjustment, which allows us to reach the regime of dispersion managed mode-locking [4]. Below 730 nm, the intracavity dispersion is strongly distorted by higher order, anomalous dispersion contributions resulting from the absorption band of Cr^{3+}:LiCAF. The pump mirror has a high-reflectivity range extending from 700 nm to 950 nm. The contribution of the 2.4% broadband output coupler to the net intracavity dispersion is negligible between 700 and 1000 nm.

3. Results

With soft aperture Kerr-lens mode-locking, we obtain pulse trains with an average output power of 220 mW. The resulting spectrum is shown in Fig. 2a. It extends from 700 to almost 1000 nm, filling the bandwidth of the limiting pump mirror. The spectrum shows a modulation, which originates from GDD oscillations of the DCMs caused by spurious reflections at the mirror-to-air matching structure, leading to echo pulses. The simultaneously measured interferometric autocorrelation trace (IAC) is shown in Fig. 2b. Here, the echo pulses are clearly shown at a delay of 25 fs. To calculate the pulse duration, we reconstruct the IAC using a phase-retrieval algorithm [5]. This is done by an optimization of the spectral phase, minimizing the deviation of the measured IAC and the IAC calculated from the assumed phase and the measured spectrum. Assuming a Fourier-limited pulse, the calculated pulse width would be 7.9 fs, and the pulse width obtained with the phase-retrieval algorithm is 8.9 fs. The measured and reconstructed IAC are in good agreement.

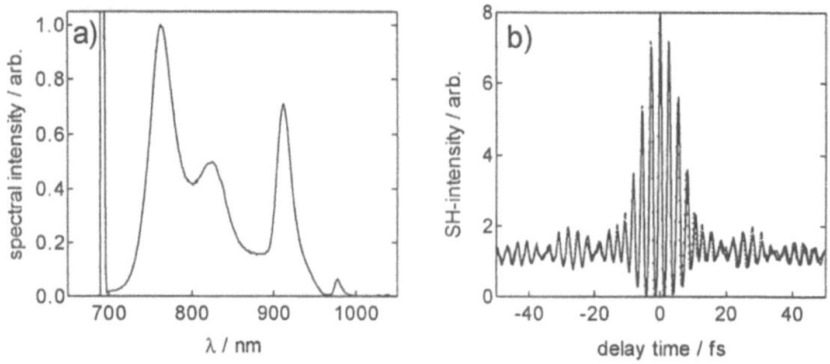

Fig. 2. a) Mode-locked spectrum. The peak at 693 nm is from the pump laser.
b) Interferometric autocorrelation. Line: measured, dots: reconstructed. The reconstruction results in a pulse duration of 8.9 fs.

4. Conclusions

In conclusion, we demonstrate 9-fs pulses from a Cr^{3+}:LiCAF laser. These are to our knowledge the shortest pulses ever generated from a Colquiriite laser. This result is achieved by the use of broad-band second and third order dispersion compensating optics. A further reduction in pulse duration seems possible with optics showing smoother dispersion characterisitics, and an extended high reflectivity range into the infrared region.

Acknowledgements. This work has been supported by the Deutsche Forschungsgemeinschaft within the Center for Functional Nanostructures (CFN) at the University of Karlsruhe (TH) and the National Science Foundation by contract ECS-0119452.

References

1 K. M. Gäbel, P. Rußbüldt, R. Lebert, and A. Valster, Opt. Commun. **157**, pp. 327-334, (1998).
2 V. L. Kalashnikov, E. Sorokin, I. Sorokina, J. Opt. Soc. Am. B, **18**, pp. 1732-1741 (2001).
3 F. X. Kärtner, N. Matuschek, T. Schibli, U. Keller, H. A. Haus, C. Heine, R. Morf, V. Scheuer, M. Tilsch and T. Tschudi, Opt. Lett. **22**, pp. 831-833 (1997).
4 Y. Chen, F. X. Kärtner, U. Morgner, S. H. Cho, H. A. Haus, E. P. Ippen, and J. G. Fujimoto, J. Opt. Soc. Am. B **16**, pp. 1999-2004 (1999).
5 R. Ell, U. Morgner, F. X. Kärtner, J. G. Fujimoto, E. P. Ippen, V. Scheuer, G. Angelow, T. Tschudi, M. J. Lederer, A. Boiko, and B. Luther-Davies, Opt. Lett. **26**, pp. 373-375 (2001).

One Step Mid-infrared Generation from a Dual-wavelength Ti:sapphire System and its Multi-pass Amplifier

Jie Song, Jiangfan Xia, Donna T. Strickland

Department of Physics, Guelph-Waterloo Physics Institute, University of Waterloo, 200 University Ave. W., Waterloo, Ontario N2L 3G1, Canada

Abstract. Intense mid-infrared pulses were generated by direct difference frequency mixing two pulses from a dual-wavelength Ti:sapphire system. Pulse energy of 1.6 μJ and 7.4 μJ is achieved respectively with and without its multi-pass amplifier at 9-11 μm.

1. Introduction

An area of research receiving considerable attention at present is coherent control of molecules. One of the pursuits in this field is excitation of molecules into high lying vibration states by using chirped excitation. To date, the only experimental work on chirped mid-infrared excitation has been carried out with a free electron laser, which could generate pulse energies of 2 μJ at a wavelength of 5 μm with 350 fs duration [1]. In this paper, we describe a high intensity, dual-wavelength laser system that can be applied to carry out these types of experiments.

2. Dual-wavelength Laser System and Gain Control

Our dual-wavelength oscillator is identical to that described by Dykaar [2] except that we use different mirror coatings in order to have the two wavelengths widely separated [3]. After being stretched by a dual-wavelength pulse stretcher, the two beams are then combined and injected into a regenerative amplifier, which is pumped by 32 mJ of 532 nm radiation from a 10 Hz, Q-switched Nd:YAG laser that can generate ~ 300 mJ pulses. The remaining ~ 260 mJ energy is used to pump a multi-pass amplifier. The regenerative amplifier outputs a total energy of 1.5 mJ, evenly split between the two wavelengths of 800 and 870 nm. After propagating through a compressor, the beams have a total energy of 0.8 mJ.

To further increase the pulse energy, a multi-pass amplifier can be added between the regenerative amplifier and the compressor. The pump beam for the multi-pass is down collimated to ~ 6mm diameter. The total output energy of the multi-pass amplifier is 30 mJ when being evenly split between the two wavelengths of 800 and 870 nm. After the compressor, the total energy remaining is 15 mJ. Auto-correlation measurement shows the pulse duration is ~ 180 fs.

There were two methods available to vary the energy ratio of the two outputs. First, because the Ti:sapphire crystals are cut at Brewster's angle, the two colors experience different angular dispersion, causing the beams coming from the regenerative amplifier to be not perfectly collinear. The energy ratio can then be varied, by changing the relative overlap of each beam with the single green pump beam. However, this method of spectral gain control does affect the beam quality. Secondly, we have control over the gain competition in the regenerative amplifier to control the input energy ratio to the multi-pass. This method can maintain the good beam quality. We measured the output pulse energies for different input ratios as shown in Table 1. Since we are not operating in the saturated regime, the multi-pass amplifier is sensitive to the seeded energy. To obtain an equal energy split at the output of the multi-pass, the regenerative amplifier was operated with 4 times more energy in the long wavelength for a wavelength of 870 nm and about 6 times more energy when the long wavelength was 885 nm.

Table 1. Output energies for different input ratios in two wavelengths of 870 and 800 nm

Input (mJ:mJ)	1.2:0.1	1.2:0.2	1.2:0.3	0.9:0.35	0.8:0.4
Output (mJ:mJ)	9.1:0.6	12.2:10.1	14.9:13.8	13.5:16.5	6.1:17.9

3. Mid-infrared Generation

In order to avoid two-photon absorption (TPA), mid-infrared radiation is typically generated through an intermediate step of optical parametric amplification (OPA) [4]. However, mid-infrared short pulses have been generated by direct mixing two wavelengths, each shorter than 1 µm [5,6]. The conversion efficiency was low because both or one of the pump beams had low intensity. In our experiment, we increased the conversion efficiency with a type II $AgGaS_2$ crystal, by two orders of magnitude, by having both pump beams at the maximum intensity imposed by TPA. We generated 1.5 µJ pulses at 9-11 µm, with total pump energy of 0.84 mJ evenly split between the two color pulses. Although limited by TPA, the overall efficiency of ~ 0.2 % from pump energy to mid-infrared energy is comparable to the best results using the OPA step [7]. Recently, Fraser difference frequency mixed the output of a two-color regenerative amplifier in a GaSe to achieve 600 nJ pulses at wavelengths longer than 15 µm [8]. GaSe has a higher nonlinear coefficient than $AgGaS_2$ but higher TPA at wavelengths below 1 µm.

By introducing the multi-pass amplifier, the mid-infrared energy can be increased, but not the conversion efficiency. The beam diameter incident on the crystal is increased from 3 mm to 6-7 mm, corresponding to a total pump energy limitation of ~ 5 mJ. We achieved a mid-infrared output of 7.4 µJ at wavelength of ~ 10 µm with the two pump pulses centered at 800 nm and 870 nm and the total pump energy of 4.7 mJ, evenly split between the two pulses, This corresponds to a total photon conversion efficiency of 1.8 %. By increasing the beam area by a factor of three and using a 2-cm-diameter crystal, 23 µJ of mid-infrared could be generated with the pump energy available. Radiation at a different wavelength of 8.3 µm was also achieved with the two pump pulses centered at 800 nm and 885 nm. The achieved mid-infrared output is 6.6 µJ under the total pump energy of 3.0 mJ, corresponding to a total photon conversion efficiency of 2.1 %. Figure 1 shows the spectra of the generated mid-infrared pulses using the two different

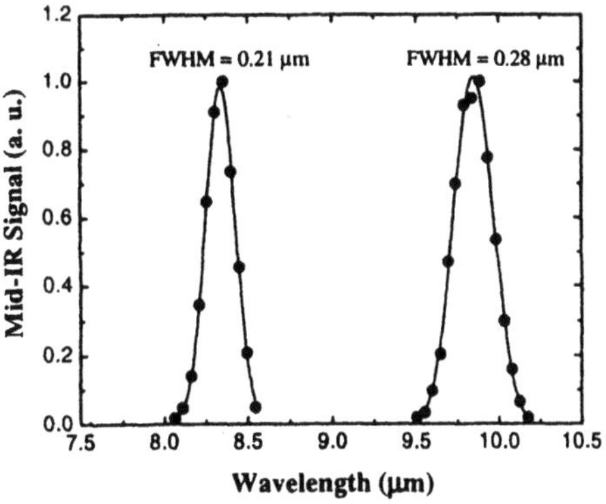

Fig. 1. Mid-infrared spectra generated by the difference frequency mixing of 800 nm short wavelength radiation with 870 or 885 nm long wavelength radiation

long pump wavelengths. We used a single shot cross-correlator to measure the temporal profile of the mid-infrared pulses. A cross-correlation width of 650 fs was measured for the 9.85 μm pulses. By de-convoluting the input pulse width along with the group velocity mismatch to the single-shot cross correlation width, a best fit pulse duration of about 500 fs for the mid-infrared pulse is derived.

4. Conclusions

We have demonstrated the amplification of two tunable wavelength pulses from a dual-wavelength, Ti:sapphire laser system. The mid-infrared from 8-11 μm was generated by direct different frequency mixing the output from the two wavelength tunable pulses. Mid-infrared pulse energy as high as 1.6 μJ and 7.4 μJ was achieved respectively with and without the additional multi-pass amplifier.

Acknowledgements. The authors gratefully acknowledge funding support from Photonics Research Ontario, the National Sciences and Engineering Research Council (Canada), the Canada Foundation for Innovation, the Ontario Innovation Trust, and the Alfred P. Sloan Foundation. Donna Strickland is a Cottrell scholar of the Research Corporation.

References

1 D.J. Maas, D.I. Duncan, R.B.Vrijen, W.J. van der Zande and L.D.Noordam, in *Chem. Phys. Lett.*, Vol. 290, 75, 1998.
2 D.R. Dykaar, and S.B. Darack, in *Opt. Lett.*, Vol. 18, 634, 1993.
3 Z. Zhang, A. M. Deslauriers and D. Strickland, in *Opt. Lett.*, Vol. 25, 581, 2000.
4 F. Seifert, V. Petrov, and M. Woerner, in *Opt. Lett.*, Vol. 19, 2009, 1994.
5 M.R.X. de Barros, R.S. Miranda, T.M. Jedju, and P.C. Becker, in *Opt. Lett.*, Vol. 20, 480, 1995.
6 P. Hamm, C. Lauterwasser, and W. Zinth, in *Opt. Lett.*, Vol. 18, 1943, 1993.
7 F. Rotermund, V. Petrov, and F. Noack, in *Opt. Commun.*, Vol. 185, 177, 2000.
8 J.M.Fraser, I.Cheung, F.Legare, D.M.Villeneuve, J.-P. Likforman, M.Joffre, P.B.Corkum, in Proceedings of Ultrafast Optics 2001, 208, Canada, 2001.

Convenient tunability of sub-10 fs-pulses in the visible range

Erik Baigar[1], Markus Braun[1], Andre Peine[1], Vladimir Konjaev[1], and Wolfgang Zinth[1]

[1] Lehrstuhl für BioMolekulare Optik, Sektion Physik, Ludwig-Maximilians-Universität München, Oettingenstr. 67, 80538 München, Germany
E-mail: markus.braun@physik.uni-muenchen.de

Abstract. We present a system for tunable laser pulses with a duration of about 7 fs. The pulses are generated by compression of the output of a tunable non-collinear optical parametric amplifier (NOPA) via a fiber compressor.

1. Introduction

The first ultra-short pulses below 10 fs were generated by compressing pulses of a CPM laser in a fiber compressor [1]. Since then the progress in solid-state laser techniques has led to the more reliable Ti:Sapphire-based systems as sub-ps laser source. In combination with nonlinear optical mixing processes, these lasers provide tunable pulses in the 20-30 fs range [2, 3]. As a standard method the angular dispersion of prisms and gratings is used to compensate the phase dispersion of ultra-short pulses. By using a combination of prisms and gratings it is also possible to compensate or apply even higher orders of the chirp. Another method to manipulate short pulses is pulse shaping in the Fourier space. Here the use of masks, LCD displays or deformable mirrors in the Fourier plane was tested and is promising for the use as a standard tool [4-6]. Particularly LCD displays are used in coherent control experiments to generate pulses of nearly every possible shape, but only few groups use them for optimal pulse compression [7].
In this work we describe a hybrid compressor system consisting of prism compressor, single-mode fiber and grating-prism compressor. We combine the handy spectral tunability of a NOPA set-up with the generation of well defined, broad spectra in a single-mode fiber to obtain tunable sub-10 fs pulses in the visible range. The stability of the set-up leads to reliable operation with non-varying pulse characteristics and allows spectroscopic applications.

2. Experimental Methods

The tunability of the center wavelength in the visible range is achieved by a NOPA set-up [2, 3]. The output of the Ti:Sapph oscillator/regenerative amplifier system (80 fs, 1 kHz, 500 μJ) is used to pump two NOPAs (Figure 1).

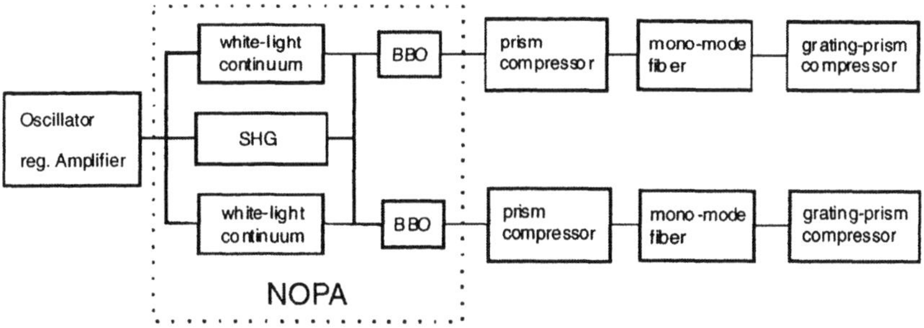

Fig. 1. Schematic sketch of the laser system.

For this process a small fraction of the 800 nm fundamental pulse is focussed on a 2 mm thick sapphire plate to generate a weak white-light seed continuum, which is slightly chirped. This seed pulse is overlapped with the frequency doubled main part of the fundamental wave in a BBO crystal. Phase matching is achieved by focussing pump and seed pulse under a non-collinear geometry in the BBO crystal. The center wavelength of the optical pulse is tuned by choosing the BBO angle and the delay between the white-light seed and the 400 nm pump pulse.

The visible pulses delivered by the NOPA are compressed to about 30 fs in a standard quartz prism compressor. This compressor can be optimized to produce a broad or most of all unstructured spectrum of the pulses in the mono-mode fiber by self-phase modulation. Depending on the application, spectrally broad (for shortest pulses) or unstructured pulses can be generated. The resulting output of the fiber is then compressed by a hybrid grating-prism compressor and yields pulses of about 7 fs. An autocorrelation trace is shown in Figure 2.

In order to adjust the grating-prism compressor a cross-correlation FROG trace of the initial pulse after the mono-mode fiber with spectral narrow pulses from a second NOPA is recorded in a 30 μm thick KDP crystal. From that trace the

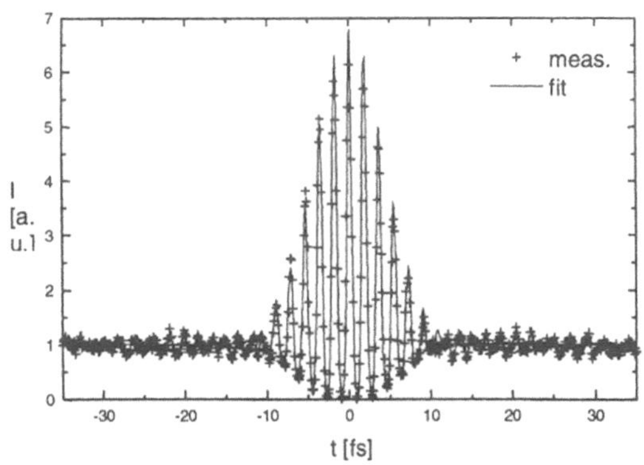

Fig. 2. Interferometric autocorrelation trace of a 7 fs pulse (center wavelength 640 nm) in a BBO crystal (25μm).

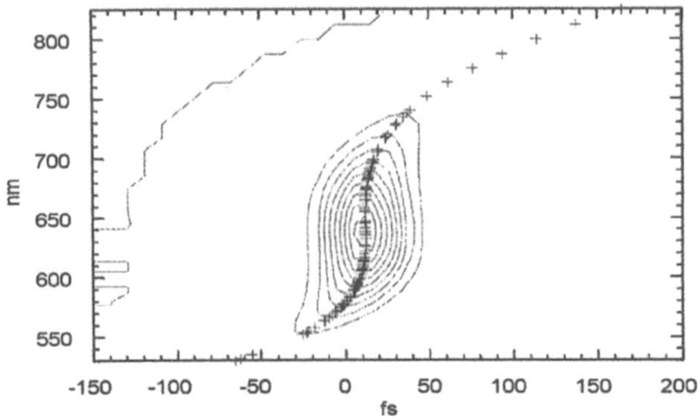

Fig. 3. Cross-correlation FROG of a 7 fs pulse taken at the sample position giving information on the phase of the pulse. A slit in the grating-prism compressor can be used to suppress the spectral wings of the pulse.

optimal compressor parameters are calculated and the set-up is tuned accordingly (mainly by adjusting the grating distance and the prism position of the grating-prism compressor). Then a FROG trace (Figure 3) after passing the compressor system, is taken and compared to the theoretical prediction. If necessary, an additional optimization loop for the parameters can be applied by evaluating the FROG trace.

For fast and reproducible operation of the system the compressor set-up was mechanically optimized to get direct access to those compressor parameters, which are also used in calculations for the optimal parameter set. The completely parallel arrangement of the pump and the probe path, guarantee minimum sensibility on thermal fluctuations and a good long term stability.

The combination of a NOPA and fiber-grating-prism compressor system allows to generate pulses in the 7 fs range where a fast change of the center wavelength and optimal compression of the pulses is obtained within about 1 hour.

References

1 R. L. Fork, C. H. Brito Cruz, P. C. Becker, and C. V. Shank, Opt. Lett. **12**, 483, 1987.
2 T. Wilhelm, J. Piel, and E. Riedle, Opt. Lett. **22**, 1494, 1997.
3 E. Riedle, M. Beutter, S. Lochbrunner, J. Piel, S. Schenkl, S. Spörlein, W. Zinth, Appl. Phys. B **71**, 457, 2000.
4 J. P. Heritage, A. M. Weiner, and R. N. Thurston, Opt. Lett. **10**, 609, 1985.
5 A. M. Weiner, D. E. Leaird, J.S. Patel, and J. R. Wullert, IEEE J. Quantum Elect. **28**, 908, 1992.
6 E. Zeek, K. Maginnis, S. Backus, U. Russek, M. Murnane, G. Mourou, and H. Kapteyn, Opt. Lett. **24**, 493, 1999.
7 T. Brixner, N.H. Damrauer, P. Niklaus, and G. Gerber, Nature **414**, 57 (2001).

High-energy Femtosecond Stretched-pulse Fiber Laser With A Nonlinear Optical Loop Mirror

F. Ömer Ilday, and Frank W. Wise

Department of Applied Physics, Cornell University, Ithaca NY 14583, USA
E-mail: foi1@cornell.edu

Abstract. A stretched-pulse fiber laser with a nonlinear optical loop mirror (NOLM) that produces 100-fs pulses with 1 nJ energy is demonstrated. These results constitute a 30-fold increase in pulse energy over previously reported femtosecond fiber lasers with a NOLM. Compared to previous stretched-pulse lasers, this laser offers a cleaner spectrum and improved stability, with comparable pulse duration and energy. Implications for the construction of truly environmentally-stable lasers are discussed.

Fiber lasers offer an attractive alternative to solid state lasers for generation of femtosecond pulses at near-infrared wavelengths with their compact size and simplicity of operation. The primary drawbacks of fiber lasers are insufficient environmental-stability for use outside of basic research laboratories, limited by uncontrolled birefringence and lack of high pulse energies directly from an oscillator, limited by excessive nonlinear phase shifts. The best performance in fiber lasers is currently provided by stretched-pulse lasers [1,2] employing nonlinear polarization evolution for amplitude modulation (NPE-SPL): 100-fs long, 1-nJ pulses are routinely achievable.

For maximum environmental stability, fiber lasers can be constructed of polarization-maintaining (PM) fiber. However, this approach precludes the use of NPE. NOLM can be implemented with PM fiber and it is capable of fast and deep modulation in a simple scheme. In addition, a NOLM operates with nonlinear

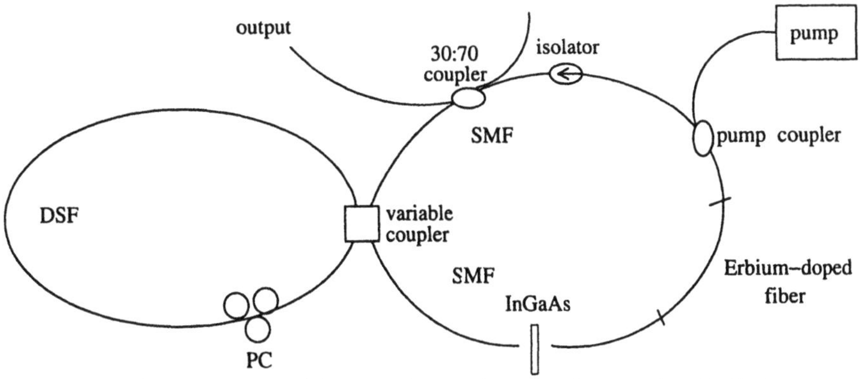

Fig. 1. Experimental setup of the stretched-pulse fiber laser with a NOLM.

phase shift accumulated only in the nonlinear loop, whereas the entire laser cavity contributes in the case of NPE. Therefore, overdriving NOLM for high-energy operation can be prevented simply through the use of a shorter nonlinear loop. In this paper, we consider the use of a NOLM to stabilize a stretched-pulse laser. The stretched-pulse technique is implemented to suppress spectral sidebands and for increased pulse energy.

The experimental setup is depicted in Figure 1. The erbium-doped fiber, which has normal GVD of 75 ps^2/km, is 93 cm long. It is pumped by a 980-nm diode with total power of approximately 250 mW. The rest of the linear loop is composed of SMF28 and a small amount of Corning Flexcor 1060. The total length of SMF is 2.6 m (β_2 = - 23 ps^2/km) and the Flexcor is 0.5 m (β_2 = - 7 ps^2/km). The nonlinear loop is made of 3 m of DSF (β_2 ~ - 3.5 ps^2/km). Total dispersion of the cavity is calculated to be ~ - 0.002 ps^2. The repetition rate is 23 MHz. A polarization-insensitive isolator provides unidirectional operation in the linear loop. Pulses are extracted from the cavity with a 30% output coupler. A variable-ratio coupler is used in the NOLM, which permits fine adjustment of the peak switching intensity. A polarization controller (PC) in the NOLM sets the linear bias. A slow (~ 100 ps response time) InGaAs SA is placed between the NOLM and the gain fiber to assist initial pulse formation from noise fluctuations. Without the use of the InGaAs SA, similar pulsed operation is achieved, but an external perturbation is necessary to initiate mode-locking.

Mode-locked operation is obtained by adjusting the variable coupler for given intra-cavity energy and setting the linear phase bias with the PC. Long-term stability of the laser over several days of continuous operation is excellent, similar to that of the NPE-SPL. However, the laser is significantly more stable against environmental effects (mechanical perturbations and temperature variations) than the NPE-SPL. The increased stability is presumably due to the fact that NOLM operation does not depend on polarization directly. Positively-chirped pulses are extracted and are de-chirped in 0.8 m of the fiber coupler.

The benefits of operating at small and anomalous dispersion can be seen by comparison to the NPE-SPL. Autocorrelations and the spectra of the pulses from the two lasers are compared in Figure 2. The NOLM-SPL produces pulses with a clean, smooth spectrum much narrower than that of the NPE-SPL. Despite the significantly smaller bandwidth, pulses from the NOLM-SPL are only slightly longer (125-fs, assuming a *sech*(t) pulse shape) than those of the NPE-SPL (100-fs, assuming a Gaussian pulse shape). The undesirable secondary structure in the autocorrelation trace is noticeably smaller for the NOLM-SPL. The maximum pulse energies are comparable, 1.0 nJ for the NOLM-SPL, and 1.3 nJ for the NPE-SPL.

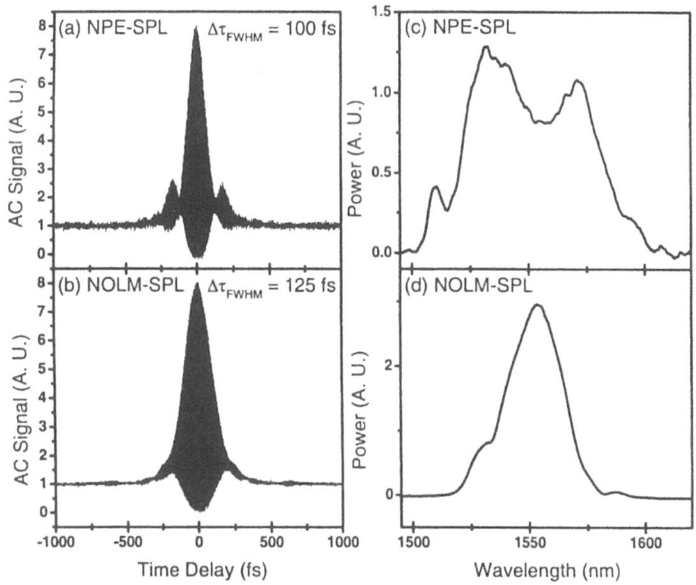

Fig. 2. Comparison of the two lasers: autocorrelation trace of (a) the NPE-SPL and (b) the NOLM-SPL, spectrum of (c) the NPE-SPL and (d) NOLM-SPL.

In conclusion, we have demonstrated a stretched-pulse laser stabilized by a NOLM that produces 1-nJ and 100-fs pulses. The pulse energy is 30 times higher than previously-reported soliton fiber lasers with a NALM [3,4]. By arranging the total cavity dispersion to be small and anomalous while maintaining the same dispersion contrast as in a NPE-SPL, a much cleaner spectrum is obtained at the expense of a small decrease in pulse energy. We identify a route to an environmentally stable, all-PM fiber laser producing high-energy, femtosecond pulses in a simple scheme.

References

1 M. Hofer, M. H. Ober, and M. E. Fermann, in *IEEE J. Quan. Electron.* Vol.28, 720, 1992.

2 K. Tamura, E. P. Ippen, H. A. Haus, and L. E. Nelson, in *Opt. Lett.* Vol.18, 1080, 1993.

3 D. J. Richardson, R. I. Laming, D. N. Payne, M. W. Phillips, V. J. Matsas, in *Elec. Lett.* Vol.27, 730, 1991.

4 M. Nakazawa, E. Yoshida, Y. Kimura, in *Appl. Phys. Lett.* Vol.59, 2073, 1991.

Passively mode-locked, self-frequency doubled, diode-pumped $Yb:YAl_3(BO_3)_4$ laser

M. J. Lederer[1], M. Hildebrandt[1], V. Z. Kolev[1], B. Luther-Davies[1], B. Taylor[2], J. Dawes[2], P. Dekker[2], J. Piper[2], H. H. Tan[3] and C. Jagadish[3].

[1] Laser Physics Centre, Research School of Physical Sciences and Engineering, The Australian National University, ACT 0200, Australia
[2] Centre for Lasers and Applications, Macquarie University, New South Wales 2109, Australia
[3] Electronic Materials Engineering, Research School of Physical Sciences and Engineering, The Australian National University, ACT 0200, Australia

Abstract: We report a self-frequency doubled $Yb:YAl_3(BO_3)_4$ laser mode-locked by an ion-implanted SESAM. Smooth tuning from the sub-picosecond to picosecond pulsewidth regime was achieved by varying the crystal angle with a maximum average green power of 270 mW.

1. Introduction. The broad fluorescence of Ytterbium doped lasers allows tunability and/or sub-picosecond pulse generation at wavelengths around 1040 nm. As a resul there is a growing range of mode-locked Yb^{3+}-doped lasers, including the 40 mW, 90 fs pulses from an $Yb^{3+}:Ca_4GdO(BO_3)_3$ laser[1]. In particular, $Yb:YAl_3(BO_3)_4$ (Yb:YAB) has a large second-order nonlinearity (d_{eff} >1.4 pm/V) making it well suited to visible subpicosecond pulse generation. CW and Q -switched Yb:YAB lasers[2,3] have previously been demonstrated with 1.1 W of green output. We report, to our knowledge, the first semiconductor saturable absorber mirror (SESAM) mode-locked Yb:YAB laser.

2. Experimental Set Up. A 2.5 mm long, 5.6% doped, anti-reflection coated Yb:YAB crystal was placed at the focus between the curved mirrors M1 and M2 (see Fig. 1)

We calculated $1/e^2$ mode-radii of $\omega_x \approx 50$ µm and $\omega_y \approx 40$ µm inside the crystal, while the crystal-mount temperature was 20°C. With an upper state lifetime of 0.68 ms

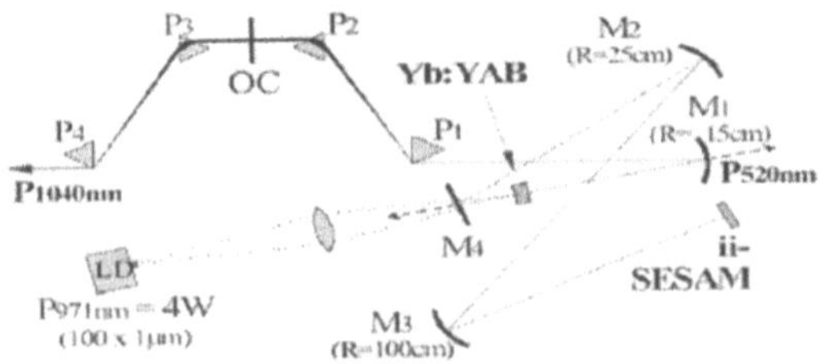

Fig 1. Cavity layout of SESAM mode-locked Yb:YAB laser. P1-4 (SF14 prisms), M1 -4 (mirrors), OC (output coupler), ii (ion-implanted)[4].

and emission cross-section of 0.8×10^{-20} cm^2, Yb:YAB displays a tendency to Q-switch or Q-switch mode-lock in the presence of intra-cavity saturable absorption. Thus a picosecond SESAM recovery time is critical for stable cw mode-locking of Yb:YAB. The SESAM contained three 9.5nm In0.26Ga0.74As quantum-wells in an antiresonant structure that was implanted with 40keV oxygen ions to a dose of 2x1014cm-2 and was annealed at 600°C for 20min under arsine ambien [5]. This resulted in a saturation fluence between 20-30μJcm-2, picosecond recovery time and modulation depth of up to 1.5%. To accommodate the quasi three -level energy scheme, high brightness pumping was achieved by a 4W, 100 μm×1 μm aperture InGaAs laser diode with λ_{pump}=971 nm and $\Delta\lambda$=3 nm. The crystal absorbed a max pump power of 2.7 W and the pump spot size in the crystal was 110 μm × 80 μm. Finally, dispersion compensation was realized through a pair of minimum deviation SF14 prisms. Since the crystal was cut for Type-I phase matching it was initially tilted in the angle-tuning plane by 20° away from phase matching to avoid efficient second harmonic generation (SHG) and strong phase modulation due to $\chi^{(2)}$ -cascading. The crystal was then gradually rotated towards the phase matching angle.

3. **Results.** Far off phase matching, we obtained transform-limited nfrared pulses with pulsewidths of 198 fs with average power of 440 mW. Figs. 2(a) and (b) depic the intensity autocorrelation and visible spectrum respectively of the shortest pulses from the laser.

Assuming $sech^2$-pulses we deduce a pulsewidth of 198 fs and $\Delta\nu\cdot\tau$FWHM = 0.315 from $\Delta\lambda_{ir}$=5.75nm. With an average infrared output power of 440 mW in TEM00 we calculate an energy fluence on the SESAM of <600 $\mu J\cdot cm^{-2}$, a peak output power of 28 kW and an optical efficiency of 16%. The average intracavity green power in this case was 270 mW. As the crystal was rotated towards phase matching the laser produced more 520 nm radiation. The pulse length increased as the green average power increased, however, the green average power increased at a faster rate than the pulsewidth. As a result there is an area around 1-2 degrees off phase matching where the laser produced up to 60 mW (intracavity) of stable green 200 fs pulses.

When the crystal was aligned close to phase matching a total of 270 mW of average green power resulted from stable strongly chirped pulses. Figs. 3(a) and (b) depict the intensity autocorrelation and visible spectrum respectively of the highest

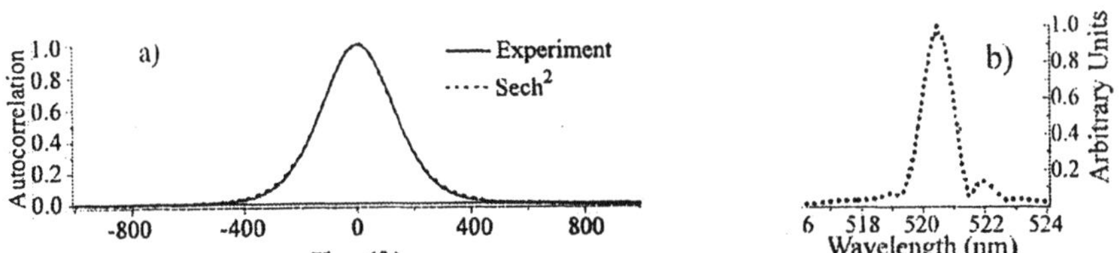

Fig 2. (a) Autocorrelation of the infrared shortest pulses (far off phase matching) together with that of an ideal $sech^2$–pulse (dotted line). (b) Corresponding visible spectrum

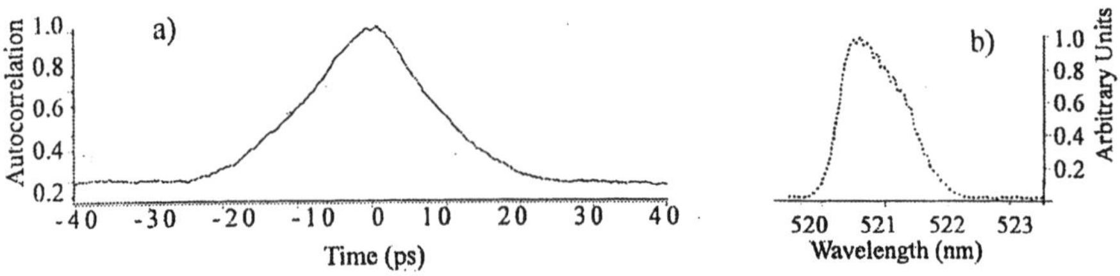

Fig. 3. (a) Autocorrelation of the longest infrared pulse (close to phase matching). (b) Corresponding visible spectru

power green pulses from the laser. Mode-locking was maintained smoothly throughout these adjustments except for perfect phase matching where the laser Q-switched.

4. Summary. We built a self-frequency doubled, passively mode -locked, diode pumped Yb:YAB laser. We were able to tune the pulse width and visible power by changing the angular distance of the laser crystal from phase matching. The highest intracavity green power obtained was 270 mW with 20 ps pulse width at 0.1 ° from phase matching. The shortest pulse was 198 fs with 20 mW of green power at 12 ° from phase matching. Changing the crystal angle in the phase matching plane resulted in smooth tuning of the pulse width and visible power between these two points. Pulse width was found to change at a different rate to green power, this allows for an optima angle from phase matching to be determined. This optimal angle was 1 -2° from phase matching where green power of ~60 mW was obtained with pulse widths as short as 200 fs.

References

1. F. Druon, F. Balembois, P. Georges, A. Brun, A. Courjaud, C. Hönninger, F. Salin, A. Aron, F. Mougel, G. Aka, and D. Vivien, "Generation of 90 -fs pulses from a mode-locked diode-pumped $Yb^{3+}:Ca_4GdO(BO_3)_3$ laser," *Opt. Lett.* 25, 423-425 (2000)
2. P. Wang, P. Dekker, J. Dawes, J. Piper, Y. Liu, J. Wang, "Efficient continuous-wave self-frequency-doubling green diode -pumped $Yb:YAl_3(BO_3)_4$ lasers" *Opt. Lett.*, 25, 731-733, (2000)
3. P. Dekker, J. M. Dawes, J. A. Piper, Y. Liu, J. Wang, "1.1 W CW self-frequency-doubled diode-pumped Yb:YA13(BO3)4 laser." *Opt. Commun.* 195, 431-436 (2001)
4. M. J. Lederer, V. Z. Kolev, B. Luther-Davies, B. Taylor, J. Dawes, P. Dekker, J. Piper, H. H. Tan, and C. Jagadish, "Passive mode-locking of a self-frequency-doubling Yb:YA13(BO3)4 laser", *Opt. Lett.*, 27, 436-438, (2002
5. M. J. Lederer, V. Z. Kolev, B. Luther-Davies, H. H. Tan, and C. Jagadish, "Ion-implanted InGaAs Single Quantum Well Semiconductor Saturable Absorber Mirrors for Passive Mode-locking,", *J. Phys. D: Appl. Phys.* 34, 2455-2464 (2001)

High-order self-dispersion-compensation in a Ti:sapphire laser oscillator

H. Crespo[1], M. A. Cataluna[2], A. Guerreiro[1], and J. T. Mendonça[2]

1 Departamento de Física, Faculdade de Ciências, Universidade do Porto, R. do Campo Alegre, 687, 4169-007 Porto, Portugal
E-mail: hcrespo@fc.up.pt
2 GoLP/Centro de Física de Plasmas, Instituto Superior Técnico, Av. Rovisco Pais, 1049-001 Lisboa, Portugal

Abstract. We describe a new regime of operation where sub-10-fs ultrashort laser pulses with symmetric spectra centered at 800 nm can be directly generated from a prism-dispersion-controlled modelocked Ti:sapphire oscillator that only uses commercially available standard optics.

1. Introduction

Ever since the discovery of self- or Kerr-lens modelocking in Ti:sapphire over a decade ago [1,2], much work has been done in order to stretch the boundaries of ultrashort pulse generation in solid-state lasers. In early systems employing low-dispersion fused-silica prisms as a primary source of negative group-delay dispersion (GDD), it was found that residual third-order dispersion (TOD) was the limiting factor that prevented further reduction in pulse length [3]. Additional compensation of TOD [4,5] proved somewhat insufficient for the generation of pulses shorter than 10-fs, since higher-order dispersion would now come into play, with the resulting spectra becoming asymmetric, as well as red-shifted with respect to the maximum of the gain-profile of Ti:sapphire [5]. The invention of chirped mirrors [6] provided engineered linear dispersion for the first time, which combined with thin (2-mm), highly-doped crystals, has resulted in the direct generation of sub-10-fs pulses centered at 800-nm [7]. Hybrid lasers that use both prisms and ultra-broadband double-chirped mirrors (DCM's) can produce pulses down to the two-cycle regime [8,9], which combined with low-dispersion CaF_2 prisms has resulted in the generation of the shortest laser pulses to date, setting the present world-record of 5-fs sub-two-cycle pulses [10, 11].

Due to the linear dispersive characteristics of both standard optical materials and commercial ultrafast laser optics, it is assumed [12,13] that sub-10-fs pulses can only be obtained from standard prism-dispersion-controlled oscillators at the expense of spectral asymmetry (which may compromise pulse quality) and spectral detuning (from 800 to around 850-nm), which makes the pulses less suitable for applications such as subsequent amplification in Ti:sapphire. We show here that sub-10-fs pulses having broadband (>140-nm) symmetric spectra centered at 800-nm can be obtained from such a system, by using a tight-focusing geometry to promote strong self-phase and amplitude modulation in the crystal.

2. Experimental Methods

Our home-built Ti:sapphire oscillator comprises a relatively thick (4.5-mm) crystal in a standard z-fold sub-cavity (composed of two 10-cm radius mirrors), fused-silica prisms, and commercially available standard ultrafast laser mirrors (Spectra-Physics and Melles Griot). Output power is 150 mW for a pump power of 4 W (all-lines Ar ion). Modelocking is started by means of a mobile cavity mirror and the pulse repetition rate is 85 MHz [14]. One of the major advantages of this type of starting mechanism is that, as the intracavity parameters are changed, the laser will self-modelock whenever possible, which is particularly useful for probing unusual regimes, in particular those occurring for large negative/positive linear cavity GDD and/or extreme geometric stability conditions. In order to maximize self-phase modulation and spectral broadening required for ultrashort pulse generation, we placed the crystal near the geometrical center of the sub-cavity, well in the instability region [see, e.g., ref. 15]. The resulting modelocked regime is nevertheless very stable and reproducible, and can last for several hours.

3. Results and Discussion

For tight-focusing geometries and overall negative linear GDD, the laser can be made to operate in a stable regime where the emitted spectrum extends from 710 to 920 nm (Fig. 1c), as opposed to the 60 nm bandwidth which is otherwise typical of this system (Fig. 1a). The near-rectangular shape of the spectrum evidences strong self-phase modulation occurring in the laser crystal, and the spectral width is mostly limited by the finite reflectivity of the dielectric mirrors; hard-aperturing within the cavity allows for further spectral shaping (Fig. 1c). The smooth spectral distribution centered at 800-nm evidences the balance achieved in the total dispersion. The interferometric autocorrelation of the generated sub-10-fs laser pulses (corresponding to the spectrum of Fig. 1c) is shown in Fig. 1d, where the pulses were first sent through an extra-cavity prism line prior to measurement, in order to compensate for GDD in the laser output coupler and autocorrelator beamsplitter. The pulses are not transform-limited, mostly due to residual TOD in the prism compressor.

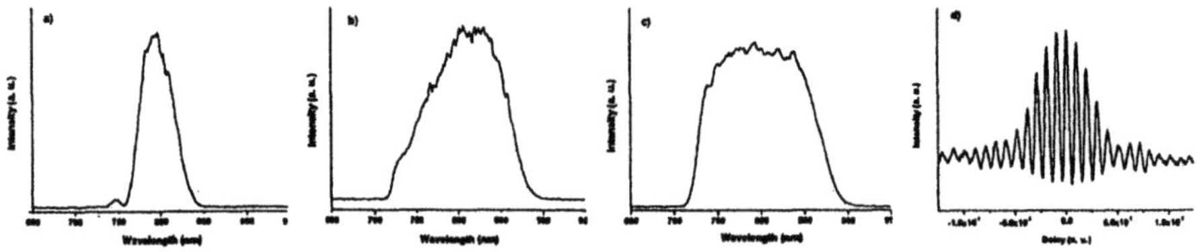

Fig . 1. a), b), c) Evolution of the pulse spectrum for increasing values of GDD; d) Interferometric autocorrelation trace of the shortest generated pulses, obtained using two-photon absorption in a GaAsP photodiode.

4. Conclusions

We describe for what is believed to be the first time, a new regime of operation where sub-10-fs ultrashort laser pulses can be produced at 800-nm using a standard prism-dispersion-controlled modelocked Ti:sapphire oscillator. The combination of the active high-order dispersion due to self-phase modulation and the linear dispersion of all the optical elements used in the cavity results in the broadest symmetric spectrum centered at 800 nm ever to be directly generated from a laser that only uses commercially available standard optics. Further study of this regime, combined with the use of broadband optics, may lead to more simple, reliable and efficient methods for ultrashort pulse generation in self-modelocked laser oscillators.

Acknowledgements. We thank Armindo dos Santos for very helpful suggestions.

References

1. D. E. Spence, P. N. Kean, and W. Sibbett, Opt. Lett. 16, 32 (1991).
2. U. Keller et al., Opt. Lett. 16, 1022 (1991).
3. M. T. Asaki, C. Huang, D. Garvey, J. Zhou, H. C. Kapteyn, M. Murnane, Opt. Lett. 18, 977 (1993).
4. B. E. Lemoff and C. P. J. Barty, Opt. Lett. 18, 57 (1993).
5. J. Zhou, G. Taft, C. Huang, M. Murnane, H. C. Kapteyn, I. P. Christov, Opt. Lett. 19, 1149 (1994).
6. R. Szipöcs, K. Ferencz, Ch. Spielmann, and F. Krausz, Opt. Lett. 19, 201 (1994).
7. A. Stingl, M. Lenzner, Ch. Spielmann, F. Krauzs, and R. Szipöcs, Opt. Lett. 20, 602 (1995).
8. I. D. Jung, F. X. Kärtner, N. Matuschek, D. H. Sutter, F. Morier-Genoud, G. Zhang, U. Keller, V. Scheuer, M. Tilsch, and T. Tschudi, Opt. Lett. 22, 1009 (1997).
9. D. H. Sutter, G. Steinmeyer, L. Gallmann, N. Matuschek, F. Morier-Genoud, U. Keller, V. Scheuer, G. Angelow, and T. Tschudi, Opt. Lett. 24, 631 (1999).
10. U. Morgner, F. X. Kärtner, S. H. Cho, Y. Chen, H. A. Haus, J. G. Fujimoto, E. P. Ippen, V. Scheuer, G. Angelow, and T. Tschudi, Opt. Lett. 24, 411 (1999).
11. R. Ell, U. Morgner, F. X. Kärtner, J. G. Fujimoto, E. P. Ippen, V. Scheuer, G. Angelow, T. Tschudi, M. J. Lederer, A. Boiko, and B. Luther-Davies, Opt. Lett. 26, 373 (2001).
12. M. Lenzner, Ch. Spielmann, E. Wintner, F. Krausz, and A. J. Schmidt, Opt. Lett. 20, 1397 (1995).
13. Ch. Spielmann, M. Lenzner, F. Krausz, R. Szipöcs, and K. Ferencz, "Chirped dielectric mirrors improve Ti:sapphire lasers", Laser Focus World, Vol. 31, No. 12, p. 55 (December 1995).
14. M. A. Cataluna, H. Crespo, A. Guerreiro, and J. T. Mendonça, Physica Scripta T98, 134-136 (2002).
15. T. Beddard, W. Sibbett, D. T. Reid, J. Garduno-Mejia, N. Jamasbi, M. Mohebi, Opt. Lett. 24, 163 (1999).

IR Linac Driven by a High-Intensity TM_{01} Diffraction-Limited Laser Beam

Charles Varin, and Michel Piché

Centre d'optique photonique et laser, Département de physique, Pavillon Vachon, Université Laval, Sainte-Foy, Québec, G1K 7P4, Canada
E-mail: cvarin@phy.ulaval.ca

Abstract. It is widely believed that the future of particle accelerator technology relies upon the development of accelerator schemes based on the use of powerful laser pulses. We describe here a linear accelerator driven by a TM_{01} diffraction-limited laser beam propagating in free space. Under extreme conditions the energy gain experienced by the accelerated electrons can be well in the multi-GeV range.

1. Introduction

We consider IR linear accelerators (IR linacs) driven by laser beams whose wavelength is in the 1-μm range. To achieve energy gains much larger than those obtained with RF linacs, beam intensities must exceed the damage threshold of most optical materials, thus inhibiting the use of wave-slowing structures as done with RF linacs. Laser beams propagating in free space are characterized by a phase velocity exceeding the speed of light; hence most laser beams cannot efficiently transfer their energy to charged particles. In this paper we show how this fundamental limitation can be overcome with TM_{01} laser beams.

2. IR Linac driven by a TM_{01} diffraction-limited laser beam

The longitudinal electric field associated to the TM_{01} (transverse magnetic) Laguerre-Gauss laser beam in free space has a Gaussian distribution. A complete description of this beam can be found in [1-2]. We recall that its intensity profile exhibits a donut-like pattern (see Fig. 1); at the center, the longitudinal electric field is maximum and the transverse fields are zero.

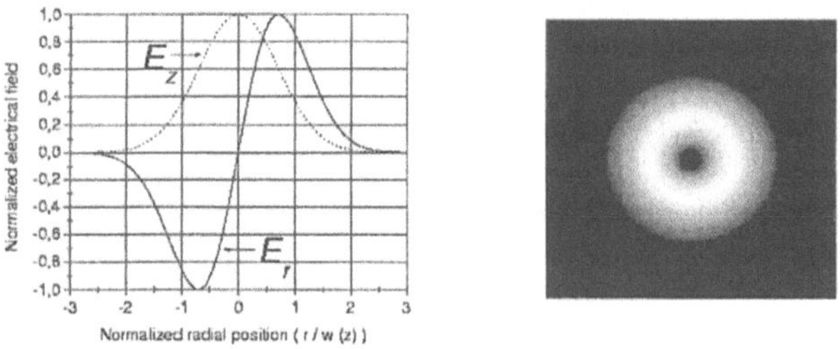

Fig. 1. Electric fields and intensity profile of a TM_{01} Laguerre-Gauss beam

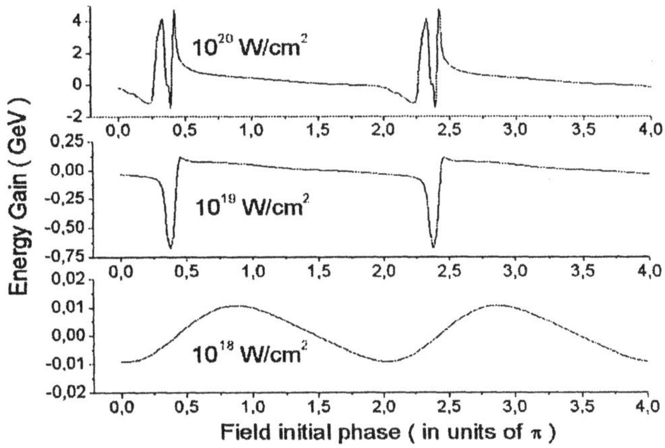

Fig. 2. Phase dependence of the energy gain for a 1.5-GeV electron after a 20-meter on-axis interaction. The beam wavelength is 0.8 μm, the focus is at z = 10 m (the interaction begins at z = 0), and the beam waist at focus is 10 μm. The beam intensity is given on each curve.

An ultra-relativistic electron moving on axis will only experience the effects of the longitudinal electric field. Numerical solutions of the equations of motion (given in [2]) for high electron energies (>1 GeV) and high laser intensities ($>10^{18}$ W/cm^2) give striking evidence of substantial energy gains.

The top curve of Fig. 2, showing energy gain as a function of the phase of the laser beam, clearly exhibits two peaks with energy gains in the multi-GeV range. These peaks correspond to positions in a field cycle for which there is an effective phase matching between the longitudinal field of the laser beam and the electrons. Such phase matching is generally believed to be impossible, since laser beams propagating in free space have phase velocities higher than c, the speed of light in vacuo. We interpret the origin of acceleration as a consequence of the Gouy phase shift experienced by the TM_{01} laser beam about its focus. In fact, the longitudinal electric field of the TM_{01} beam has a phase velocity equal to c, plus a weak correction due to the Gouy phase shift. This correction produces a phase advance of π from $-\infty$ to $+\infty$. As a result, the phase slippage experienced by an ultra-relativistic electron can be such that its trajectory is contained within a half-cycle of the field, for the last part of its transit after beam focus. This behavior is clearly seen in Fig. 3; electrons are decelerated just before focus, and then they experience a violent acceleration beyond focus. The acceleration benefits also from the fact that the longitudinal field falls off as $1/(z - z_f)$ away from focus; this behavior produces a sustained acceleration throughout the entire electron trajectory.

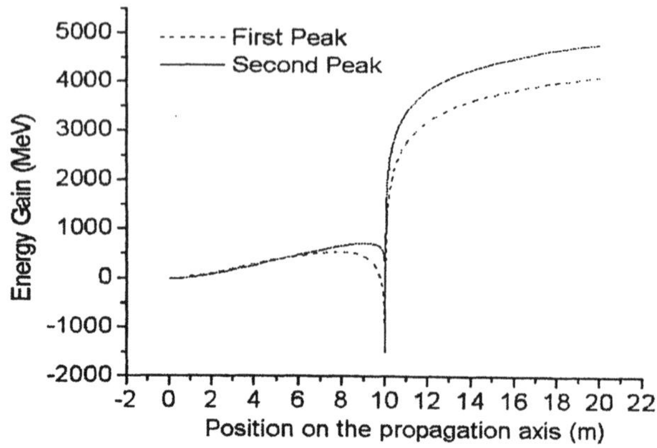

Fig. 3. Energy gain along the z axis for a 1.5-GeV electron. The beam wavelength is 0.8 μm, the focus is at z = 10 m (the interaction begins at z = 0), and the beam waist at focus is 10 μm. The beam intensity is 10^{20} W/cm^2.

3. Conclusions

The dynamics of charged particles accelerated by TM electromagnetic fields is well known from the RF linac technology [3]. The results presented here are consistent with it, although the IR laser beams we have examined are propagating in free space. The novel feature revealed by our results is that the acceleration proceeds in two steps: the electrons are first slowed by the field before focus, in such a way as to move to the next half-cycle of the field where they are accelerated to GeV energies. We are now investigating how this process could be cascaded to obtain even larger energy gains. We are also investigating the effect of transverse fields on electrons that are traveling at off-center positions.

Acknowledgements. We would like to thank La Fondation J.-Arthur Vincent, the Natural Sciences and Engineering Research Council (Canada), Les fonds pour la formation de chercheurs et l'aide à la recherche (Québec), and the Canadian Institute for Photonic Innovations.

References

1 A.E. Siegman, *Lasers* (University Science Books, 1986)
2 C. Varin & M. Piché, *Acceleration of ultra-relativistic electrons using high intensity TM_{01} laser beams*, to be published in Appl. Phys. B : Lasers and Optics, Ultra-Fast Optics 2001.
3 T. Wangler, *RF Linear Accelerators*, (John Wiley & Sons, 1998)

Efficient Frequency Conversion From a Novel QPM Semiconductor Waveguide Crystal

E. U. Rafailov, P. Loza-Alvarez, D. Artigas, W. Sibbett

School of Physics and Astronomy, University of St Andrews, North Haugh, St. Andrews, Fife KY16 9SS UK
E-mail: er8@st-andrews.ac.uk

Abstract. We demonstrate SHG at 980 nm from a novel first-order QPM semiconductor GaAs/AlGaAs waveguide crystal. Our calculations imply that the SHG conversion efficiency from this crystal exceeds significantly that from PPLN for wavelengths exceeding 4 μm.

1. Introduction

Recent progress in the development of electric-field poling techniques for the patterning of the domain structure of ferroelectric materials has enabled the implementation of distinctive quasi-phase-matched (QPM) structures for efficient nonlinear optical interactions [1]. Such QPM devices have been fabricated in $LiNbO_3$, $LiTaO_3$ and in the KTP-family of crystals. Semiconductor materials which have larger nonlinear coefficients (e.g. $\chi^{(2)}$) are also promising candidates for efficient frequency up/down conversion. The technology employed at present to fabricate semiconductor QPM devices suffers from limitations such as difficulty in producing and stacking thin plates for first-order QPM. We report here a first-order QPM semiconductor device based on a novel methodology that involves a "periodically-switched-nonlinearity (PSN)" [2]. In these PSN-structures, the $\chi^{(2)}$-coefficient in the material is changed periodically using a re-growth process. To achieve this change in $\chi^{(2)}$ we have engineered a structure of $GaAs/Al_xGa_{1-x}As$, where $\chi^{(2)}_{GaAs} >> \chi^{(2)}_{AlGaAs}$ [3]. The PSN-design is illustrated in Fig.1.

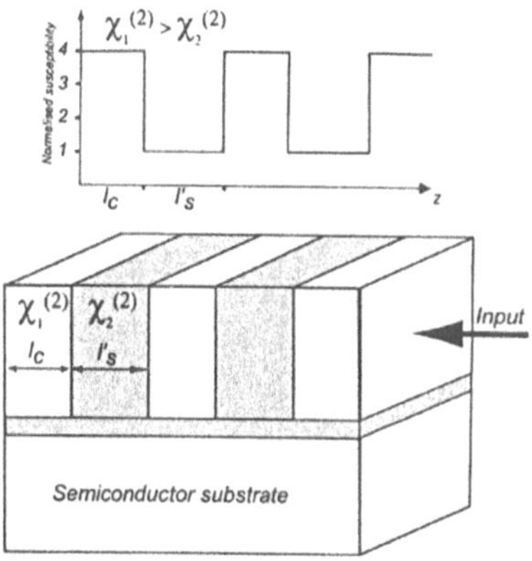

Fig. 1. A simplified schematic of periodically-switched nonlinearity in a PSN crystal.

2. Experimental Results and Discussion

The coherence lengths in the original and re-grown material (l_c and l'_s, respectively) are different due to the unequal refractive indices. The period Λ of the PSN gratings is given by $\Lambda = l_c + l'_s$.

The semiconductor structure was fabricated by growing three different layers on a GaAs <100> substrate: 2 μm of $Al_{0.6}Ga_{0.4}As$, ~0.2 μm of $Al_{0.4}Ga_{0.6}As$, and 1.15 μm of undoped GaAs. The wafer was then processed by etching a 1.1 μm deep grating with a period of $\Lambda = 2.97 + 4.73$ μm into the GaAs layer. This was followed by re-growing two new layers over the grating: 2 μm of $Al_{0.4}Ga_{0.6}As$ and 0.2 μm of $Al_{0.6}Ga_{0.4}As$, terminating with a 10 nm GaAs protective cap. Finally, waveguide ridges 15 μm widths were etched perpendicular to the grating, and the sample was cleaved into ~1.1 mm long devices.

The SHG intensity as a function of the fundamental wavelength power, from a PPLN-based femtosecond (<200 fs) OPO launched into the 15 μm wide waveguide, is plotted in **Fig. 2 (a)** (dots).

These data are in good agreement with a linear fit having a slope of 2, as expected for a QPM quadratic process. We also observed that the SHG signal disappeared when the polarisation at the fundamental wavelength was rotated by 90 degrees, which is consistent with the behaviour expected. The spectra of the fundamental and SHG wavelengths measured following the sample are reproduced in Fig.2 b and c.

To further investigate this technique, a detailed numerical analysis was undertaken. This model was based on the slowly varying envelope approximation that describes pulse evolution in a collinear SHG configuration [4]. Our numerical results showed that efficient SHG conversion in PSN device is limited mainly by the large group velocity mismatches at wavelengths shorter than 4.0 μm (femtosecond regime) and interface reflection losses (in both femtosecond and CW regimes). Fig.3 shows the change of SHG output power for optimal crystal length as a function of fundamental wavelength for (a) the femtosecond and (b) CW regimes and its comparison with a conventional PPLN crystal. For the femtosecond regime, our calculations showed that in the wavelength range longer

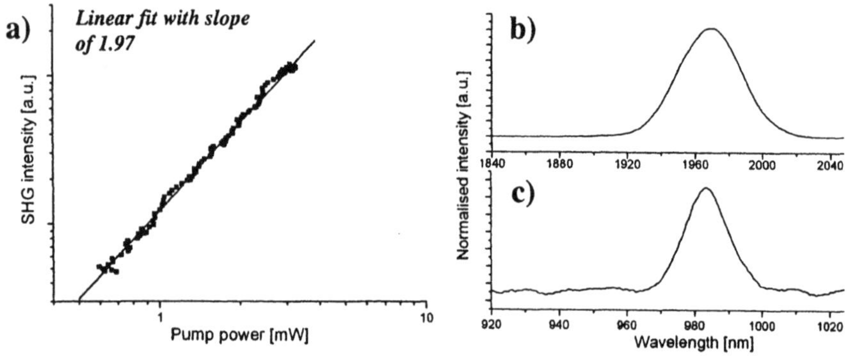

Fig. 2. a) The SHG intensity as a function of fundamental power launched into the waveguide (dots) and spectra of the fundamental (b) and SHG (c) pulses.

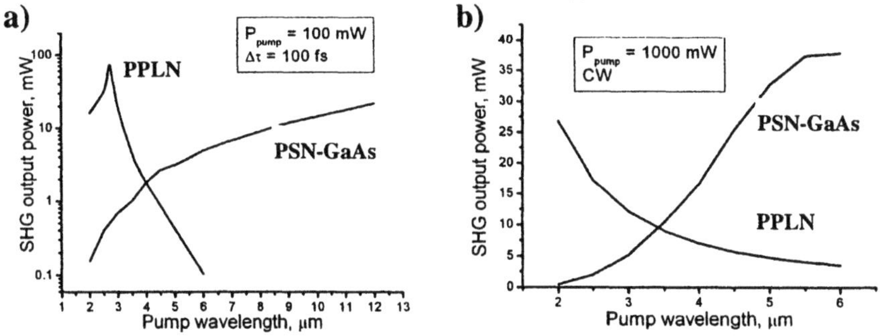

Fig. 3. Calculated output SHG power versus pump wavelength for optimal crystal length with (a) a femtosecond pump beam and (b) a CW pump beam.

than 4.2 μm, the conversion efficiency becomes better than that of a PPLN crystal. For the CW case, shown in Fig.3(b), the conversion efficiency from the PSN device increased significantly in the wavelength range beyond 3.4 μm. Fig.3 also shows that the semiconductor PSN device may cover a wavelength range from 6 to 12 μm that cannot be reached with a conventional periodically-poled nonlinear crystal such as PPLN.

Further experimental investigations will include the use of sources having a narrower spectral bandwidths and the implementation of this technique in OPO-related applications where it has particular potential for efficient light generation in the mid-infrared (>3.4 μm) region.

4. Conclusions

We have reported the first successful demonstration of frequency conversion in first-order QPM devices that utilise a periodically-switched-nonlinearity in a novel design of GaAs/AlGaAs structure. Our calculations show that SHG conversion efficiencies from such devices at wavelengths beyond ~4 μm are higher than those in the 1-3 μm range. For this PSN device the conversion efficiency beyond ~4 μm exceeds significantly that offered by PPLN crystals.

Acknowledgements. The authors wish to thank Prof. R. De La Rue, Dr. D. Yanson (Glasgow University, EE Department) for sample preparation. This work was funded by EPSRC.

References

1 M. M. Fejer, G. A. Magel, D. H. Jundt, R. L. Byer, IEEE J. Quantum Electron., 28, 2631 (1992).

2 E.U.Rafailov, P.Loza-Alvarez, C.T.A. Brown, W.Sibbett, R.M.DeLaRue, P.Millar, D.A.Yanson, J.S.Roberts, P.A.Houston, Opt. Lett., 26, 1984 (2001).

3 M.Ohashi, T.Kondo, R.Ito, S.Fukatsu, Y.Shiraki, K.Kumata, S.S.Kano, J. of Appl. Phys. 74, 596 (1993).

4 P. Loza-Alvarez, M. Ebrahimzadeh, W Sibbett, D. Reid, D. Artigas, M. Missey, J.Opt.Soc.Am., 18, 1212 (2001).

Atomic clocks of the future: using the ultrafast and ultrastable*

Leo Hollberg, Scott Diddams, Chris Oates, Anne Curtis, Sébastien Bize, and Jim Bergquist

National Institute of Standards and Technology, 325 Broadway, Boulder, CO 80305
E-mail: hollberg@boulder.nist.gov

Abstract. The application of ultrafast mode-locked lasers and nonlinear optics to optical frequency metrology is revolutionizing the field ofatomic clocks. The basic concepts and applications are reviewed using our recent results with two optical atomic frequency standards based on laser cooled and trapped atoms.

1. Introduction

In contrast to today's atomic clocks that are based on electronic oscillators locked to microwave transitions in atoms, the next generation of atomic clocks will most likely employ lasers and optical transitions in laser-cooled atoms and ions. Optical atomic frequency standards use frequency-stabilized CW lasers with good short-term stability that are stabilized to narrow atomic resonances using feedback control systems. At NIST we are developing two optical frequency standards, one based on laser cooled and trapped calcium atoms and the other on a single laser cooled trapped Hg+ ion. Similar research is being done at labs around the world.

Moving from microwave frequencies, where one cycle corresponds to 100 ps, to optical frequencies where one cycle is a femtosecond long, allows us to divide time into smaller intervals and hence gain frequency stability and timing precision. Steady progress over the past 30 years has improved the performance of optical frequency standards to the point that they are now competitive with, and even moving beyond, their microwave counterparts.[1-3] A major impediment to the development of optical clocks was that it was extremely difficult to count optical frequencies and to make the coherent connection between laser oscillators and countable microwave sources. To reconfigure an optical frequency standard into an atomic clock we need to provide a mechanism to count and accumulate the optical cycles of the atomically stabilized laser.

2. Femtosecond laser-based optical frequency combs

A breakthrough occurred when Udem et al. proved in 1999 that femtosecond mode-locked lasers could be used to count large optical frequencies intervals (tens of THz) with high accuracy.[2] The method is straightforward and was first clearly described by Hänsch in the '70s, but it was not rigorously tested until recently. Those ideas, along with new techniques from the Hall and Hänsch groups, now provide femtosecond laser-based systems that can phase-coherently connect stable lasers in the visible with frequencies that are accessible to electronics.

Optical frequency combs spanning the entire visible are now readily generated using Ti:Sapphire lasers and nonlinear microstructure fibers. For example, a 1 GHz repetition rate Ti:sapphire laser can be spectrally broadened in a microstructure fiber to produce discrete modes separated by 1 GHz throughout the broadened spectrum from ≈ 500 to 1100 nm. These modes can connect to, and measure, essentially any optical frequency. The frequency of any mode of the comb can be written as $f_N = Nf_{rep}+f_0$, where Nf_{rep} is a large integer N times the pulse repetition rate f_{rep}, and f_0 is the offset of the entire comb extrapolated back to zero frequency. The origin of the offset frequency has a clear physical interpretation in terms of the difference between the group and phase velocities for the pulse in the laser cavity. The "self-referencing" method [1-3] is used to measure the offset frequency by frequency doubling the IR end of the comb and subtracting the existing blue end of the comb from the second harmonic signal, obtaining $2(Nf_{rep}+f_0) - (2Nf_{rep}+f_0) = f_0$. The basic clockwork is shown in Fig. 1.

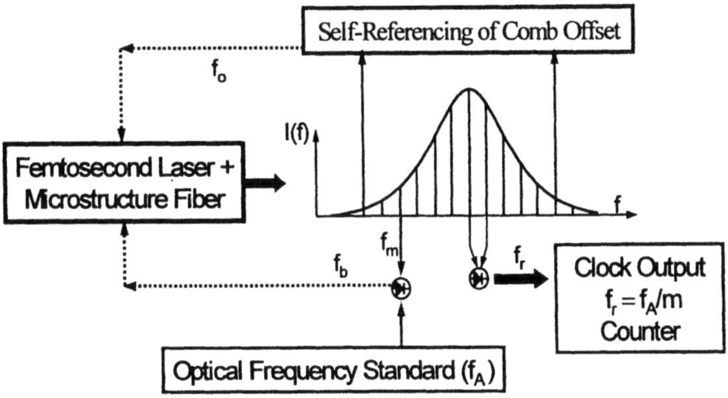

Fig. 1. A simplified diagram of the optical comb shows the counting mechanism used for optical clocks. The beatnote f_b between the optical frequency standard and one tooth of the comb is used to phase-lock that comb tooth to the atomic frequency f_A. This can be done using a PZT on the cavity of the mode-locked laser. The "self-referencing" method generates the offset frequency f_0, which can be controlled by changing the pump power to the femtosecond laser. When the two servos are locked the frequency of all the modes of comb are stabilized relative to the narrow optical atomic reference. The optical clock output then appears at the repetition rate of the modelocked laser. For an appropriate choice of locking parameters the output frequency at the repetition rate is an integer sub harmonic of the optical atomic frequency f_A[3]

The development of convenient optical frequency measuring systems and clockworks based on femtosecond lasers has sparked renewed interest in optical frequency standards and optical atomic clocks. While the comb technology is a powerful and enabling simplification, the optical clocks also rely heavily on advances in laser cooling and trapping of atoms, and the development of highly stabilized cw lasers. Optical frequency combs have been used to measure the frequencies of our two optical standards [f(Ca) = 455 986 240 494 158 +/-26 Hz, and f(Hg+) =1 064 721 609 899 143 +/-10 Hz] relative to the Cs atomic fountain NIST-F1 that realizes the definition of the second.[3]

The performance of optical standards raises obvious questions about the precision, stability, reproducibility, and accuracy of optical frequency metrology based on femtosecond lasers and microstructure fibers. To test for possible errors or limitations of the combs we have been making intercomparisons between two femtosecond combs combined with the two optical standards and a microwave reference derived from a high-stability hydrogen maser. The results show that the combs perform exceptionally well in terms of short-term stability between optical frequencies (< 6.3×10^{-16} at one second), and in frequency reproducibility across the entire visible spectrum (< 4×10^{-17}), see Fig. 2.[4] Though more than adequate for now, the combs will require improvement to reach the accuracy and stability anticipated for the next generation of optical frequency standards. They are presently at ≈ 7×10^{-15} at 1 second, but the projected instabilities and inaccuracies should approach ≈ 1×10^{-18}.[3]

Some technical challenges that we face with femotsecond laser-based frequency combs are instabilities in coupling into the small core of the microstructure fibers, excess broadband noise in the output of the fiber, and extraction of the microwave signals at the repetition rate with high fidelity. Transferring the exceptional short-term stability of the optical frequency standards to the microwave signal from the pulses has proven particularly challenging. In converting the stable optical comb

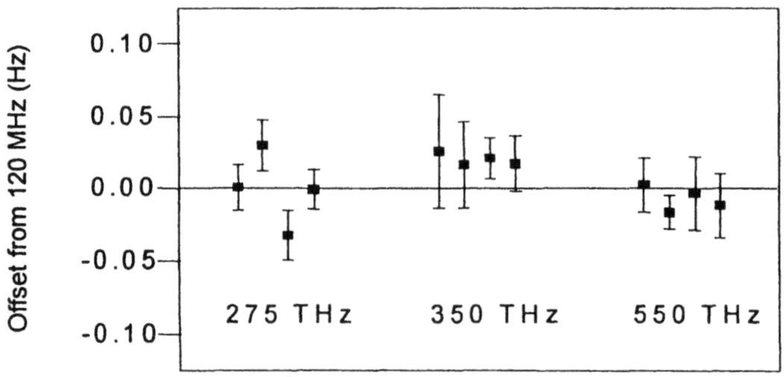

Fig. 2. In this comparison of two octavespanning combs, two independent femtosecond laser systems were operated as optical clockworks referenced to the same 657nm frequency standard. The repetition rates were set equal but theyhad offset frequencies that differed by 120 MHz. Using different bandpass filters and photodetectors we examined the integrity of the comb in three different wavelength regions as indicated. Within the measurement uncertainty we found no systematic differences between the two combs [5].

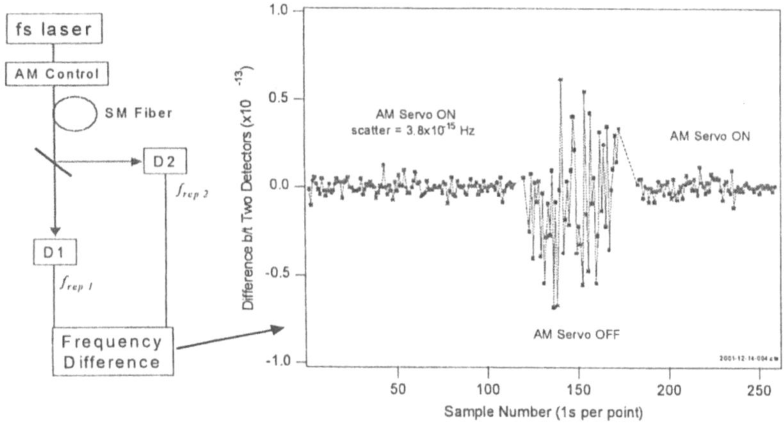

Fig. 3. Two photodetectors measure a stable 1 GHz train of pulses from a stabilized mode-locked laser. The fluctuations on the frequency difference between the two detectors results from phase-noise in the detection process due to time varying differential time delays in the detectors and amplifiers. Removing the fluctuations in pointing, power, and polarization of the input beam using a single mode fiber and a power control servoreduces this noise. With the power control activated the apparent fractional fluctuations in the repetition rate are $\approx 4\times10^{-15}$ in one second, whereas the apparent fluctuations can be $>3\times10^{-14}$ without the AM control, as seen in the central region of the trace.

to electrical pulses, and hence a microwave comb, we find that significant phase-noise is added by the photodetectors and electronics. These effects are illustrated in Fig. 3, which shows the apparent difference in frequency measured by two photodiodes detecting the same femtosecond pulse train at 1GHz.

3. Applications and Discussion

As optical frequency standards and systems develop, we are beginning to access regimes of measurement that were not previously possible. In addition to better clocks, we anticipate improved tests of the symmetries of space-time as understood by relativity, new searches for time variation of fundamental "constants", and even applications to advanced communication and navigation systems. One such application is to use the superior stability of the optical atomic clocks to realize a secure communication link. The idea is simply that one can encode information on the laser frequency (say by a small frequency modulation, or by the absolute optical frequency itself) that would be impossible to detect without a frequency reference with comparable stability and accuracy. For example, adding square-wave FM at 1 Hz with a frequency excursion of 25 Hz on the stable optical clock laser at 10^{15} Hz could not be detected using the world's best quartz oscillators or even a hydrogen maser. This capability is demonstrated in Fig 4.

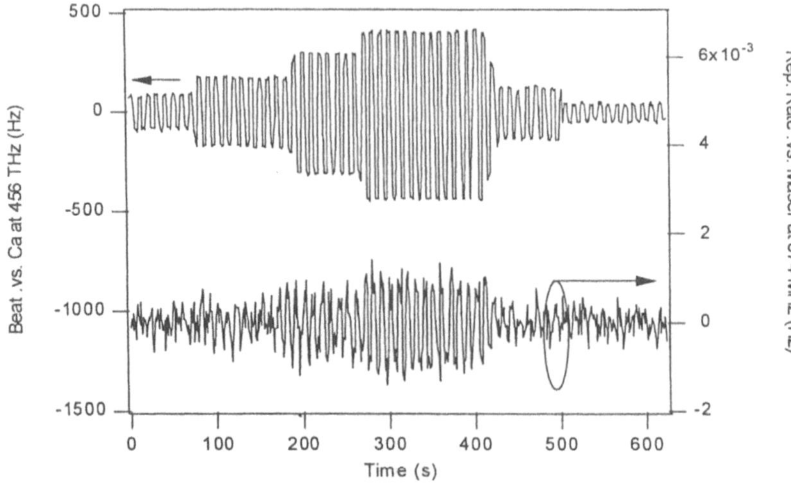

Fig. 4. A small square-wave FM signal with a period of 10 s and varying amplitude was applied to the Hg+ stabilized laser (563 nm, 532 THz) and was easily detected against the Ca stabilized laser (657 nm, 456 THz) once the gap between them is spanned by a self-referenced optical frequency comb locked to the Hg + optical standards. The amplitude of FM detected at 456 THz ranged from +/- 420 to +/- 42 Hz. The lower trace is data taken simultaneously against a hydrogen maser referenced synthesizer at 874 MHz. By locking the fs-comb to the modulated 532 THz laser the repetition rate has the same fractional modulation. Thus, when the modulation is $\lesssim$ 200 Hz at 456 THz the H-maser instability dominates and the FM signal (0.4 mHz at 874 MHz) is not visible against the maser.

Acknowledgements. We gratefully acknowledge important contributions to this work by Th. Udem, E. Ivanov, D. Jones, S. Cundiff, A. Bartels, J.Ye, J. Hall, W. Itano, and D. Wineland.

References

1 J.L. Hall, J.Ye, S.A. Diddams, L.-S. Ma, S.T. Cundiff and D.J. Jones, IEEE J. Quantum Electronics, 37, 1482-1492 (2001).
2 R. Holzwarth, M. Zimmermann, T. Udem and T.W. Hänsch, IEEE J. Quantum Electronics, 37, 1493-1501 (2001).
3 L. Hollberg, C.W. Oates, E. A. Curtis, E. N. Ivanov, S.A. Diddams, T. Udem, H.G. Robinson, J.C. Bergquist, R.J. Rafac, W.M. Itano, R.E. Drullinger, and D.J. Wineland, IEEE J. Quantum Electronics, 37, 1502-1513 (2001) and references therein.
4 Scott A. Diddams, L. Hollberg, Long-Sheng Ma and Lennart Robertsson, Optics Letts., 17, 58-60 (2002).

Carrier-envelope offset dynamics and stabilization of femtosecond lasers

F. W. Helbing[1], G. Steinmeyer[1], and U. Keller[1]

Swiss Federal Institute of Technology (ETH) Physics Department / Institute of Quantum Electronics, ETH Hönggerberg HPT, CH-8093 Zürich, Switzerland

Abstract. We analyze physical effects governing the carrier-envelope offset (CEO) dynamics of femtosecond mode-locked lasers. The insight on the underlying mechanisms enabled us to phaselock the carrier-envelope offset (CEO) frequency of our laser to an RF reference oscillator with an unprecedented 20 mrad rms phase jitter (0.01 Hz - 10 kHz bandwidth)

With the recent progress of femtosecond pulse generation yielding few-cycle pulses, the relative phase between carrier and envelope of an optical pulse becomes an interesting parameter. Control of the carrier-envelope offset (CEO) phase enables optical waveform synthesis [1] and is of considerable importance in optical precision frequency metrology [2]. Meanwhile, measurement of the CEO phase based on the scheme originally proposed in [3] has been reported by many groups. Many experiments, however, require control of the CEO phase to a residual jitter much smaller than π. This turns out to be a demanding problem, requiring a deeper understanding of the physical mechanisms giving rise to CEO phase noise. In this paper we identify several CEO noise contributions. These noise mechanisms fall into two categories: Environmental contributions, e.g., caused by air pressure changes or crystal temperature changes can cause a large CEO phase drift and dominate the low-frequency part of the CEO phase noise. Nonlinear optical contributions act via amplitude-to-phase coupling and dominate the high-frequency part of the noise spectrum. The latter is a much more severe problem for stabilization of the CEO phase. Spectral shifting and dispersion, nonlinear refraction, and beam pointing variations in combination with geometrical dispersion have been considered as a case of a fast coupling between intracavity intensity fluctuations and CEO phase [4–6].

Here we concentrate on dynamic contributions to the CEO phase stemming directly from nonlinear refraction. This mechanisms is most suitably explained considering the difference between phase velocity $v_\phi = c_0/n$ and group velocity $v_g = c_0/n_g = c_0/\frac{\mathrm{d}n}{\mathrm{d}\omega}$ (group-phase offset). A pulse inside the cavity accumulates a CEO phase shift of

$$\Delta\phi_{\mathrm{CEO}} = \left(\omega^2/c \int_0^L \frac{\partial n(z)}{\partial \omega} \mathrm{d}z \right) \bmod 2\pi$$

per round-trip time T, corresponding to $\omega_{\mathrm{CEO}} = \Delta\phi_{\mathrm{CEO}}/T$. It is important to note that therefore only a change of the first-order dispersion term $\frac{\mathrm{d}n}{\mathrm{d}\omega}$ can induce CEO phase noise. Formally, a dynamic contribution to the CEO phase is therefore to be expected from the first-order dispersion of nonlinear refraction $\frac{\mathrm{d}}{\mathrm{d}\omega}n_2(\omega)I$.

We use the model for the dispersion of nonlinear refraction proposed in [7] for estimating the magnitude of this effect. The computation yields a coupling coefficient of 0.4×10^{-8} Hz CEO frequency change per intracavity intensity change of 1 W/m^2 inside the Ti:sapphire crystal. Experimentally, this particular coupling mechanism can be easily explored by modulating the intracavity power and by simultaneously measuring the CEO frequency. For this measurement, we used a prismless laser set-up to exclude additional effects from nonlinear beam steering. Using a wide range of modulation frequencies we mapped out the transfer function of intensity fluctuations into CEO phase noise. The lowest observed value of 0.9×10^{-8} Hz/m^2 at about 20 kHz agrees well with the theoretical estimate given above. We attribute the increased coupling coefficient at lower frequencies to additional thermal mechanisms inside the laser crystal.

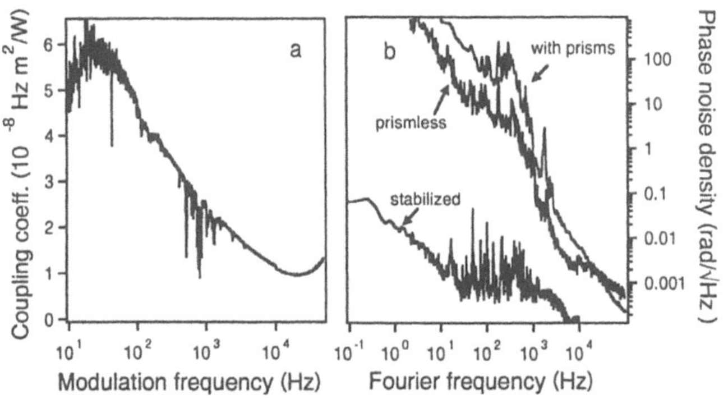

Fig. 1. a: Coupling between CEO frequency change (in hertz) and intracavity intensity modulation (W/cm^2) for a range of modulation frequencies in a prismless laser. **b**: Phase noise spectra of the measured CEO signal for an unstabilized laser with intracavity prisms and a prismless laser, with and without active phase stabilization.

More insight into this coupling mechanism is provided by measurements of the spectral CEO phase noise density. In Figure 1b, the measured phase noise density is depicted for a prismless femtosecond Ti:sapphire laser and a similar laser with intracavity prisms for dispersion compensation. Clearly, the prismless laser exhibits about 10 times reduced noise over most of the spectral range shown. This makes the prismless laser the most suitable set-up for an active stabilization of the CEO-frequency. Such a servo loop uses the very mechanism that couples power noise into phase noise to control the CEO phase of the laser. Our CEO stablilization servo loop modulates the pump power in such a way that a phase lock to an external reference oscillator is enforced. The lower trace in Fig. 1b clearly confirms the drastic reduction of phase noise compared to the unstabilized cases. Our argument that most of the phase noise at kHz frequencies is caused by amplitude-to-phase coupling is confirmed by a reduction of laser-power noise above 2 kHz on establishing the lock. This was also observed in [8]

Integration of the phase noise density from 0.01 Hz to 10 kHz yields an rms value of 20 mrad residual phase jitter. This is about two orders of magnitude better than the best measurements reported so far [1,8]. This low value corresponds to a relative timing jitter between laser carrier and envelope of 10 attoseconds (rms) during the full experimental integration time of about one minute. Integration times of at least a few minutes are an important prerequisite for a direct experimental observation of a cycle-sensitive nonlinear optical effect and the breakdown of the slowly-varying envelope approximation.

Stabilization of the carrier envelope offset phase of a femtosecond laser is a very demanding task. Our analysis disseminates CEO noise mechanisms into beam pointing, environmental and nonlinear optical contributions. Some of these contributions can be reduced or avoided by proper design of the oscillator. It has to be emphasized that the reported values cannot be reached without prior understanding of the mechanisms ruling the excessive phase fluctuations in unstabilized lasers. The unprecedented control of the CEO phase with residual timing jitters in the few-attosecond range enables many new experiments in extreme nonlinear optics, attosecond pulse generation, and frequency metrology.

References

1. R. K. Shelton, L. S. Ma, H. C. Kapteyn, M. M. Murnane, J. L. Hall, J. Ye: Science **293**, 1286 (2001)
2. A. Apolonski, A. Poppe, G. Tempea, C. Spielmann, T. Udem, R. Holzwarth, T. W. Hänsch: Phys. Rev. Lett. **85**, 740 (2000)
3. H. R. Telle, G. Steinmeyer, A. E. Dunlop, J. Stenger, D. H. Sutter, U. Keller: Appl. Phys. B **69**, 327 (1999)
4. L. Xu, C. Spielmann, A. Poppe, T. Brabec, F. Krausz, T. W. Hänsch: Optics Letters **21**, 2008 (1996)
5. H. A. Haus, E. P. Ippen: Optics Letters **26**, 1654 (2001)
6. F. W. Helbing, G. Steinmeyer, U. Keller, R. S. Windeler, J. Stenger, H. R. Telle: Optics Letters **27**, 194 (2002)
7. M. Sheik-Bahae, D. C. Hutchings, D. J. Hagan, E. W. Van Stryland: IEEE J. Quant. Electron. **27**, 1296 (1991)
8. A. Poppe, R. Holzwarth, A. Apolonski, G. Tempea, C. Spielmann, T. W. Hänsch, and F. Krausz: Appl. Phys. B **72**, 373 (2000)

Long term (>300 seconds) carrier-envelope phase coherence of a train of ultrashort pulses

Tara M. Fortier[1], David J. Jones[1], Jun Ye[1], Steven T. Cundiff[1], and Robert S. Windeler[2]

[1] JILA, University of Colorado and National Institute of Standards and Technology, Boulder CO 80309-0440 Tel:(303) 492-4841, fax: (303) 492-5235 fortier@jilau1.colorado.edu
[2] OFS Fitel Laboratories, Murray Hill, New Jersey 07974

Abstract. The carrier-envelope phase of the pulse train emitted by a 10-fs mode-locked laser has been stabilized such that phase coherence is maintained for at least 300 seconds as measured both in-loop and out-of-loop.

Following the recent demonstrations [1-4] of stabilizing the relative pulse-to-pulse phase between the carrier wave and the pulse envelope (carrier-envelope phase) generated by mode-locked femtosecond (fs) lasers, effort has focused on measuring and increasing the coherence time of this phase stability. Carrier-envelope phase stabilization is an important advance for fs technology, as the advent of few-cycle pulses makes it possible to study processes that are directly sensitive to the electric field of each pulse, rather than just the intensity envelope [5,6]. In order for most of these physical processes to be experimentally investigated it is of interest to extend the time period (i.e., the coherence time) over which phase coherence of the carrier-envelope phase is maintained. In this paper we present measurement of the lifetime of the carrier-envelope phase > 300s on the output of a phase stabilized 100 MHz, 10-fs mode-locked laser.

The frequency spectrum emitted by a mode-locked laser is a series of lines with optical frequencies $\nu_n = nf_{rep} + \delta$, where n is a large integer, f_{rep} is the laser repetition rate, and δ is the comb-offset frequency. The offset frequency is the result of the difference between group and phase velocities in the cavity and is thus related to the carrier-envelope phase by $\Delta\phi_{CE} = 2\pi\delta/f_{rep}$ modulo 2π. Stabilization of the ratio δ/f_{rep} allows for coherence of the carrier-envelope phase between adjacent pulses to be established.

The comb-offset frequency is measured using an f-to-$2f$ interferometer as discussed in prior publications (e.g. [1]). The input to the interferometer is a continuum spectrum, spanning an optical octave, generated via external broadening of Ti:sapphire pulses in microstructure fiber [7]. The output of the interferometer is the interference between comb lines on the high frequency end of the continuum with frequency doubled light on the low frequency end of the spectrum. Given that spectral overlap exists, this yields a heterodyne signal which is proportional to $\cos(2\pi\delta t + \phi_{CE}(t))$, where $\phi_{CE}(t)$ represents jitter in the carrier-envelope phase. Once determined the offset frequency is stabilized to an rf

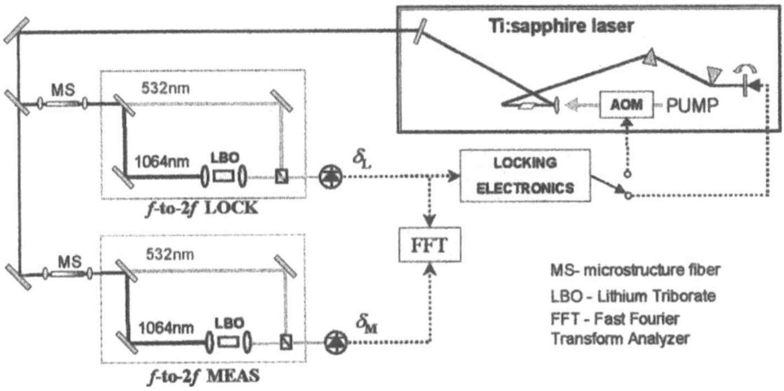

Fig. 1 Schematic of the two *f*-to-2*f* interferometers used for phase stabilization and measurement of the comb-offset frequencies. δ_L and δ_M are the in-loop and out-of-loop comb-offset frequencies, respectively, whose phase power spectral densities and linewidths are measured on a fast Fourier transform spectrum analyzer (FFT).

signal derived from the laser repetition rate, f_{rep}, as measured on a fast Si photodetector. Stabilization of δ is then obtained via negative feedback to the laser by tilting the laser end mirror using a fast (~30 kHz bandwidth) piezo-electric actuator [refs. within 1]. Feedback in this work can also be realized using an acoustic-optic modulator (AOM) to modulate the pump laser [3]. Both schemes yield similar performance.

Once stabilized the coherence time of the carrier-envelope phase is determined from the comb-offset frequency's linewidth. This relationship exists because jitter in the carrier-envelope phase results in phase noise sidebands on δ. Integration of these phase noise sidebands, otherwise known as the phase power spectral density (PSD) of the noise, $S_\phi(\nu)$ in rad^2/Hz, reveals the RMS phase jitter in the carrier-envelope phase:

$$\Delta\phi_{RMS}\,|_{\tau_{obs}} = \sqrt{2\int_{-\infty}^{1/\tau_{obs}} S_\phi(\nu)\,d\nu} \qquad (1)$$

In the Eqn. 1 τ_{obs} is the observation time and ν is an offset frequency relative to the carrier. In the above expression only the lower sideband is integrated. The observation time over which $\Delta\phi_{RMS}$ accumulates ~1 radian is generally taken to define the coherence time, $\tau = 1/(2\pi\tau_{obs})$. $\Delta\phi_{RMS}$ is obtained via measurement of the phase PSD both within the stabilization loop (in-loop), and out-of-loop using a second *f*-to-2*f* interferometer as shown in Fig.1. The in-loop measurement determines the effectiveness of the stabilization circuitry while the out-of-loop measurement determines the effect of the extra-cavity phase noise on the useful output of the laser. This latter step is crucial as it gives an indication of the coherence of the pulses emitted directly by the laser (rather than the pulses exiting the fiber used for the *f*-to-2*f* lock).

a) b)

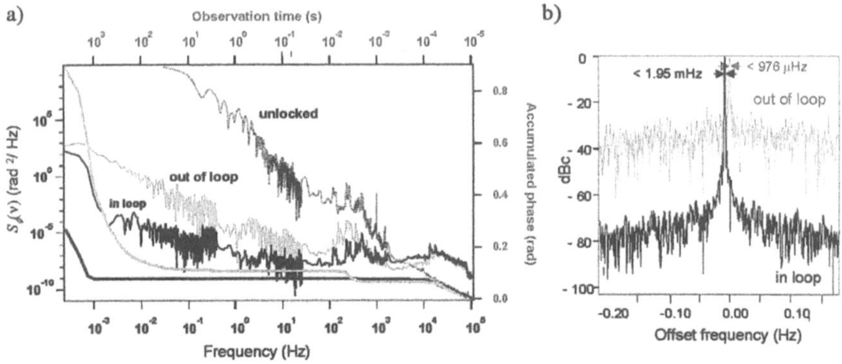

Fig. 2 a) Phase PSD, $S_\phi(\nu)$, of the carrier-envelope phase (left axis) vs. frequency and the accumulated phase jitter, $\Delta\phi_{RMS}$ (right axis), as the upper integration limit $1/\tau_{obs}$ approaches zero. An unlocked spectrum of δ is included to indicate the effectiveness of the stabilization loop. b) Comb-offset frequency linewidths (offset in frequency for clarity).

Using the above procedure we measure an accumulated phase jitter of 0.12 rad (0.80 rad) present in-loop (out-of-loop) via integration of the phase PSD displayed in Figure 2 a). This phase was accumulated over the frequency range 102kHz to 488 μHz, indicating a minimum coherence time of >300s (measurement limited), both in and out-of-loop. For the out-of-loop case, we suspect the sidebands in frequency range 100 Hz - 1 kHz which contribute the most toward $\Delta\phi_{RMS}$, are mechanical (fiber coupling) resonances common to both interferometers. We next examine the linewidth of δ which presents a more intuitive representation of the noise on the carrier. Shown in Figure 2 b), the linewidth measurements indicate the comb-offset linewidths are unresolved below 976 μHz (1.95 mHz) out-of-loop (in-loop). Prior measurements using a prismless Ti:saphire laser only achieved an in-loop resolution of 10 mHz [8] while the out-of-loop coherence time has hitherto not been measured.

Acknowledgements: This project was funded by the National Science Foundation and the National Institute for Science and Technology.

References

1 D. J. Jones *et.al.*, in *Science*, Vol. 288, 635-639, 2000.
2 A. Apolonski *et.al.*, in *Physical Review Letter*, Vol. 85, 740-743, 2000.
3 A. Poppe *et.al.*, in *Applied Physics B*, Vol. 73, 373-376, 2001.
4 U. Morgner *et.al.*, in *Physical Review Letters*, Vol. 86, 5462-5465, 2001.
5 P. Dietrich, F. Krausz, and P. B. Corkum, in *Optics Letters*, Vol. 25, 16-18, 2000.
6 T. Brabec and F. Krausz, in *Reviews of Modern Physics*, Vol. **72**, 545-591, 2001.
7 J. K. Ranka *et. al.*, in *Optics Letters*, Vol. **25**, 25, 2000.
8 F. W. Helbing *et. al.*, in *Optics Letters*, Vol. **27**, 194, 2002.

All-Optical Self-Stabilization of Carrier-Envelope Phase Offset in Few-Cycle Pulses by Optical Parametric Amplifiers

Andrius Baltuška[1,2], Nobuhisa Ishii[1], and Takayoshi Kobayashi[1]

[1] Department of Physics, Faculty of Science, University of Tokyo, 7-3-1 Hongo, Bunkyo, Tokyo 113-0033, Japan
[2] Technical University of Vienna, Institute of Photonics, Gusshasstrasse 27/378 A-1040, Vienna, Austria
E-mail: baltuska@tuwien.ac.at

Abstract. We present versatile schemes of white-light-seeded optical parametric amplifiers (OPA), which permit generation of extremely short visible and infrared pulses with precisely controlled oscillations of the electric field. Passive locking of carrier-envelope phase is demonstrated without generation of octave-spanning bandwidths and active phase tracking. Next, we show that a modified OPA can also be employed for a high-precision measurement of the pulse-to-pulse drift of the carrier-envelope phase.

Optical processes based on extreme nonlinearities are typically sensitive to the electric field of a laser pulse rather than to its envelope. The outcome of such high-order interactions [1] depends on the arrangement of the optical fringes in a few-cycle laser pulse, which is described by the so-called carrier-envelope phase (CEP). In general, the value of CEP in a femtosecond laser drifts from pulse to pulse [2]. Therefore, development of high-energy CEP-stable laser systems is a prerequisite for applications of field-sensitive nonlinear optics, such as high-order harmonic generation and production of coherent soft X-rays with attosecond durations [3]. To date, active CEP stabilization of Ti:sapphire oscillators operating at MHz repetition rates was achieved by employing feedback servo loops [4, 5]. However, this method is impractical for laser amplifiers due to their low, typically 1-10 kHz, repetition rate. We present an alternative approach based on *all-optical* phase stabilization that can be achieved by employing parametric frequency conversion and amplification.

Non-collinear OPAs driven by the second harmonic (SH) of the Ti:sapphire laser provide an extremely broad amplification bandwidth and allow generation of two-cycle light fields [6]. Remarkably, the phase properties of these devices can be applied for passive CEP stabilization of intense laser pulses. We demonstrate a number of ways in which OPAs turn out ultrashort pulses with an automatically determined shot-to-shot deviation of CEP or, alternatively, even maintain the value of the latter constant.

The phase relationship among the pump, signal, and idler waves in a white-light seeded optical parametric amplifier can be used to render an idler pulse with a CEP that is independent of the CEP fluctuation of the input pulse. The link between the phases of all three pulses taking part in a parametric interaction can be described by a simplified phase equation $\psi_I = -\pi/2 + \psi_P - \psi_S$, where ψ stands for a pulse-to-pulse fluctuation of CEP, and I, S, and P denote idler, signal, and pump, respectively. Therefore, self-stabilization of the CEP is achieved in an OPA seeded with a white light that is derived from the pump pulse. Alternatively, if a frequency-broadened version of the fundamental pulse is employed as a seed in combination with a SH pump such an OPA transfers the CEP fluctuation of the fundamental pulse to both the idler and the signal waves (phase-repeating OPA).

The experimental proof of passive phase stabilization is presented in Fig.1. In a noncollinear OPA (Fig.1a) we took advantage of the fact that the SH generation of the radially dispersed idler is also phase-matched in the same parametric crystal (1 mm Type 1 BBO). Here we employed spectral interferometry to study shot-to-shot changes of the interference between i) the idler and the fundamental, and ii) the idler and the idler-SH. In the first case corresponds to a nearly random shot-to-shot CEP fluctuation of the fundamental pulse, as is clearly seen from the rapidly changing interferograms in Fig.1d and irregular relative phase in Fig.1f. In a stark contrast, the pulse-to-pulse phase offset in the second case, in which the CEP of both the idler and idler- SH does not depend on the input pulse, falls within $\pm\pi/10$ limits (Fig.1g, dots). The remaining CEP fluctuation is attributed to the laser intensity noise enhanced through multiple 2nd- and 3rd-order nonlinear processes.

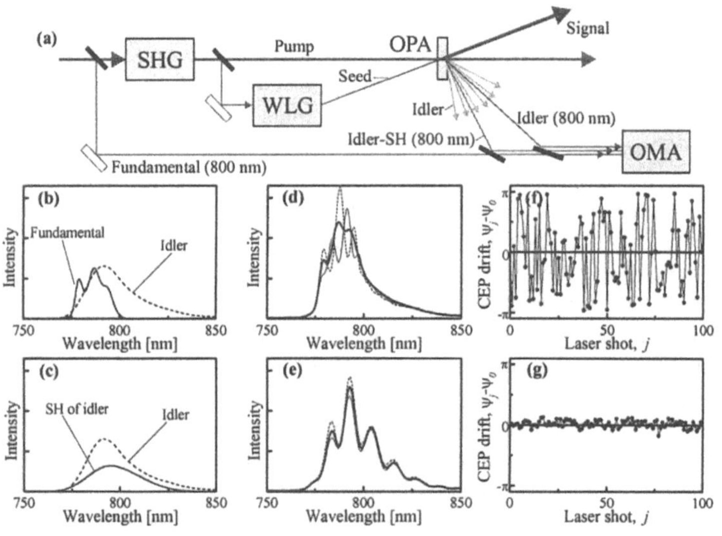

Fig. 1. Experimental results of CEP self-stabilization measurement. (a) OPA and three-beam interferometer. WLG, white-light generation; SHG, second harmonic generation; OMA, spectrometer with an Optical Multichannel Analyzer. (b,c) Overlapped spectral fragments of residual fundamental, idler, and its SH beams. (d,e) Solid curves show interference pattern averaged over 1000 shots, whereas dotted curves represent single-shot interferorgams. (f,g) Dots show relative CEP jumps wrapped on a $\pm\pi$ interval. Note that a stable phase pattern obtained from the idler–idler-SH interference is a direct proof of CEP self-stabilization.

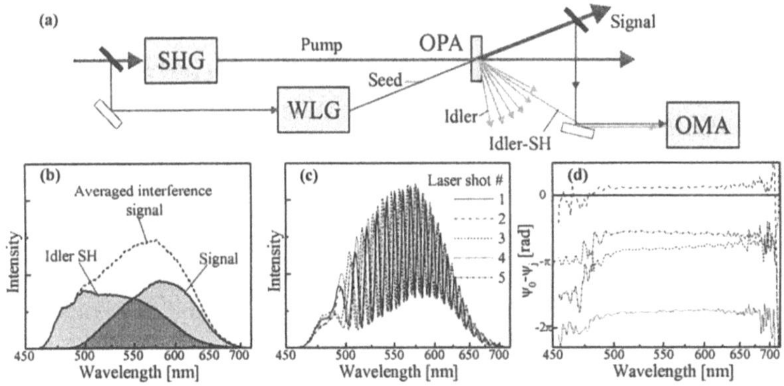

Fig. 2. Experimental results of CEP drift measurement. (a) Apparatus. (b) Spectral fragments of signal and idler-SH waves, and mean intensity of the interference averaged over 300 laser shots. (c) Single-shot interferograms for 5 successive laser pulses. (d) Phase drift relative to the first pulse. Note that nearly constant with frequency phase jumps indicate high precision of the CEP drift measurement.

Whereas a *CEP-stabilizing OPA* can be used as a final or an intermediate amplification stage in a chain of laser amplifiers, a *phase-repeating OPA* provides a convenient tool to track shot-to-shot CEP slip in an experiment based on a phase-unstable laser source. An example of this diagnostic scheme is presented in Fig.2.

Unlike other CEP drift measurement schemes that rely on the generation of octave-spanning bandwidths and harmonic synthesis from *peripheral* frequencies of a white-light spectrum [7], the OPA method utilizes *central* spectral components of the parametric waves. This dramatically reduces the sensitivity of the described technique to the input intensity noise and increases its trustworthiness.

In conclusion, we have demonstrated previously unexplored capabilities of white-light-seeded parametric amplifiers, such as passive CEP stabilization that enables subsequent high-power low-repetition-rate pulse amplification, and an attractive scheme for the measurement of CEP drift in random-phase amplified sources. Both these techniques pave the way to sophisticated experiments in field-sensitive nolinear optics in which control over CEP is critical.

References

1 G.G. Paulus, F. Grabson, H.Walther, *et al.*, *Nature*, **414**, 182, 2001.
2 L. Xu, C. Spielmann, A. Poppe, *et al.*, *Opt. Lett.*, **21**, 2008, 1996.
3 T. Brabec and F. Krausz, *Rev. Mod. Phys.*, **72**, 545, 2000.
4 A. Apolonski, and F. Krausz, *Phys. Rev. Lett.*, **85**, 740, 2000.
5 D.J. Jones, and S.A. Cundiff, *Sience*, **288**, 635, 2000.
6 A. Baltuška, T. Fuji, and T. Kobayashi, *Opt. Lett.*, **27**, 306, 2002.
7 M. Kakehata, H. Takada, Y. Kobayashi, *et al.*, *Opt. Lett.*, **26**, 1436, 2001.

Coherent anti-Stokes Raman scattering microscopy with near-infrared ps pulses synchronized within 20 fs

David J. Jones[1], Eric Potma[2], Ji-Xin Cheng[2], Jun Ye[1], and X. Sunney Xie[2]

[1] JILA, University of Colorado and National Institute of Standards and Technology, Boulder, CO 80309-0440 USA
[2] Department of Chemistry and Chemical Biology, Harvard University, 12 Oxford Street, Cambridge, Massachusetts 02138 USA

Abstract. By employing a new method of stabilizing the relative timing jitter between two, tunable ps lasers to <50 fs we significantly improve the vibrational imaging capability of coherent anti-Stokes Raman scattering microscopy.

Coherent anti-Stokes Raman Scattering (CARS) microscopy [1] has enjoyed a resurgence following a series of recent improvements such as co-linear excitation geometry [2], and epi-direction detection [3]. An additional refinement has been the use of synchronized ps Ti:Sapphire lasers [4]. Due to their wide tunability, ability of access the fingerprint region (1000 to 2000 cm^{-1}) and near-infrared center wavelengths, these lasers are almost the ideal light source for CARS imaging. However, the relative timing jitter between the two lasers has remained a fundamental limitation. As CARS is inherently a multi-photon (four-wave mixing) microscopy technique, the presence of timing jitter is directly mapped to fluctuations in the CARS signal itself. In this work, we utilize a new locking method to reduce the timing jitter to below 50 fs. At this level of jitter, stability of the CARS signal is nearly shot-noise limited.

The experimental setup is shown in Fig. 1. Two picosecond lasers tunable from 700-1000 nm (Coherent Miras individually pumped by two 5 W Coherent Verdis) are employed to generate the pump and Stokes signals. To tightly synchronize the two pulse trains we slave the pump laser to the Stokes laser using a recently reported high harmonic locking technique [5]. Briefly, the fundamental repetition rate (80 MHz) and the 175th harmonic (14 GHz) are simultaneously detected from each laser. The pump laser is first synchronized to the Stokes laser using only the error signal generated from the 80-MHz loop. Next, after temporally overlapping the two pulses trains by adjusting the 80-MHz phase shifter, the feedback loop is gradually switched over to the 14-GHz loop (via an analog switch). By switching to the 175th harmonic, the sensitivity of the feedback loop to any relative timing jitter is greatly enhanced, thereby enabling an extremely tight lock between the two pulse trains. Using this stabilization scheme, the amount of timing jitter

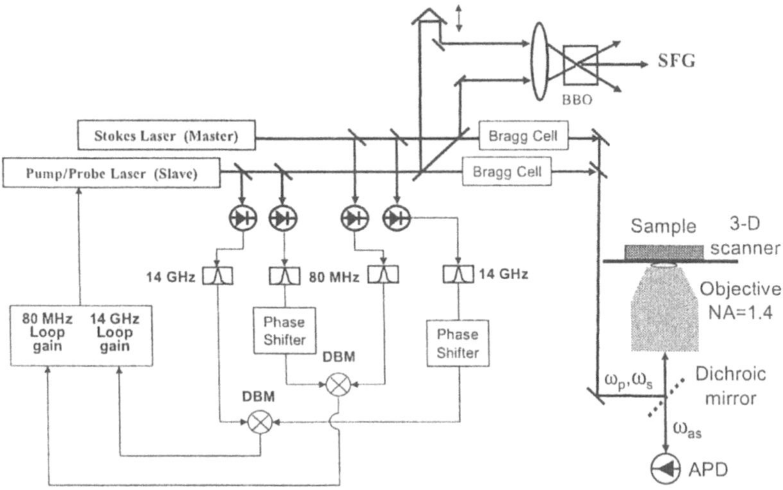

Fig. 1. Schematic of CARS experimental setup. Two ps Ti:sapphire lasers are synchronized together via a dual phase-locked-loop scheme, one operating at the fundamental repetition frequency (80 MHz) and one at the 175th harmonic (14 GHz). In this case the pump laser is slaved to the Stokes laser. Also shown is a fringe-unresolved cross-correlator to characterize the relative pulse timing jitter. For CARS imaging, the repetition rates of the two pulse trains are reduced to 250 kHz with two respective Bragg cells and then sent to the CARS microscope. BBO: β-barium borate; DBM: Double balanced mixer; SFG: sum frequency generation.

can be accurately controlled by changing the amount of 100 MHz and 14 GHz loop gains. A portion of each laser is tapped off to monitor the jitter via sum frequency generation (SFG). To measure the jitter, the pulses are temporally offset in the cross-correlator by one-half of a pulsewidth. Then, amplitude fluctuations recorded in the SFG signal can be converted to timing fluctuations via the width and shape of the cross-correlation pulse. With the pump (Stokes) laser tuned to 753 (857) nm, an autocorrelation measurement yielded 3.4 (4.1) ps, while a cross-correlation width of 3.8 ps was measured.

To evaluate the effect of jitter on the CARS signal and imaging capability, the 1600 cm^{-1} transition in polystyrene beads was investigated. Figure 2(a) displays both the SFG signal (when the pulses are offset as described above) and the CARS signal at two different jitter levels. From the SFG fluctuations (the lower two traces), we measure the RMS jitter in a 160 Hz bandwidth to be 32 fs and 770 fs over a 10 second record length. The upper two traces show the CARS signal, recorded with a 1 ms bin-width, at these two jitter levels, clearly showing the enormous improvement with reduced jitter. In fact, at the lower jitter level, fluctuations of the CARS signal become shot-noise limited as shown in Fig. 2(b) where a linear relationship exists between the

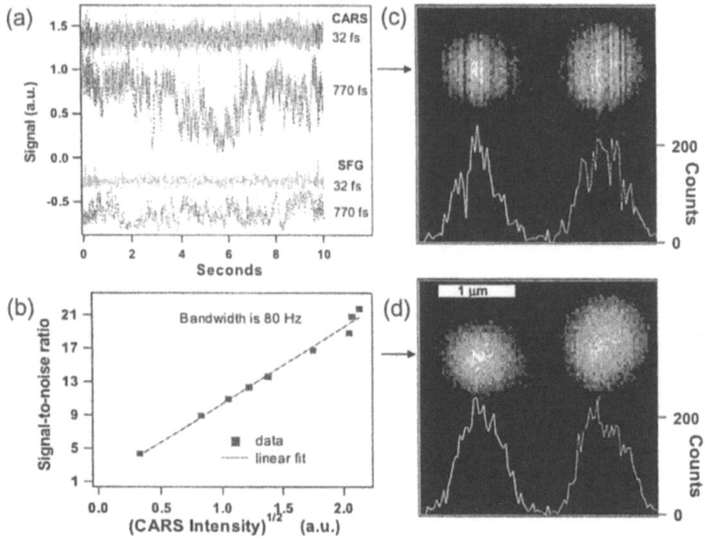

Fig. 2. (a), CARS and SFG signals recorded at two different jitter levels. (b), Plot showing nearly shot-noise limited stability (signal-to-noise ratio $\propto \sqrt{I}$) of the CARS signal. (c) and (d), Images of two polystyrene beads taken with RMS jitter levels of 770 and 32 fs. A horizontal slide taken at the arrow for each image is also shown

signal-to-noise ratio of the CARS signal and the square-root of the CARS intensity.

At low jitter levels, the enhanced stability of CARS signal significantly improves the imaging capability of CARS. Figure 2(c) and (d) shows images of polystyrene beads at jitter levels of 770 and 32 fs, respectively. Also shown is a horizontal slice of the image, taken at the coordinate indicated by the arrow. Clearly, there is a significant improvement in the image quality when the jitter level is reduced below 50 fs, which is sufficiently stable to reveal fundamental shot noise.

References

1. M.D. Duncan, J. Reintjes, and T.J. Manuccia, Opt. Lett. **7**, 350 (1982).
2. A. Zumbusch, G.R. Holtom, and X. S. Xie, Phys. Rev. Lett. **82**, 4142 (1999).
3. A. Volkmer, J.X. Cheng, and X. S. Xie, Phys. Rev. Lett. **87**, 023901 (2001).
4. J.X. Cheng, A. Volkmer, L.D. Book, and X. S. Xie, J. Phys. Chem B **105**, 1277 (2001).
5. R.K. Shelton, L.S. Ma, H.C. Kapteyn, M.M. Murnane, J.L. Hall, and J. Ye, Science **293**, 1286 (2001).

Detection of carrier-envelope phase shift in picosecond lasers

Richard Ell[1], Wolfgang Seitz[1], Thomas R. Schibli[2], Uwe Morgner[1], Franz X. Kärtner[1,2]

[1] High-Frequency and Quantum Electronics Laboratory, University of Karlsruhe Engesserstraße 5, D-76128 Karlsruhe, Germany, E-mail: r.ell@ihq.uni-karlsruhe.de
[2]Research Laboratory of Electronics, Massachusetts Institute of Technology, Cambridge, MA 02139, USA, E-mail: kaertner@mit.edu

Abstract. The carrier-envelope phase shift in a picosecond $Nd:YVO_4$ laser passively synchronized to an octave spanning Ti:sapphire laser is measured. This opens up a route to control the carrier-envelope phase in picosecond lasers.

Recently, coherent superposition of two independently mode-locked, but spectrally overlapping Ti:sapphire laser oscillators has been demonstrated by means of external active control [1]. This concept opens up new perspectives in metrology with femtosecond mode combs (see e.g. [2, 3]), and might lead to phase coherent emission over much broader bandwidth as is possible from a single mode-locked laser (e.g. [4]). The corresponding pulses may reach the single-cycle regime.

In this paper, we demonstrate a possibility to measure the field repetition rate of a picosecond laser with only 0.1 nm bandwidth by passive synchronization of the picosecond laser to an octave spanning Ti:sapphire oscillator. Phase controlled picosecond lasers might become important for particle accelerators and in metrology.

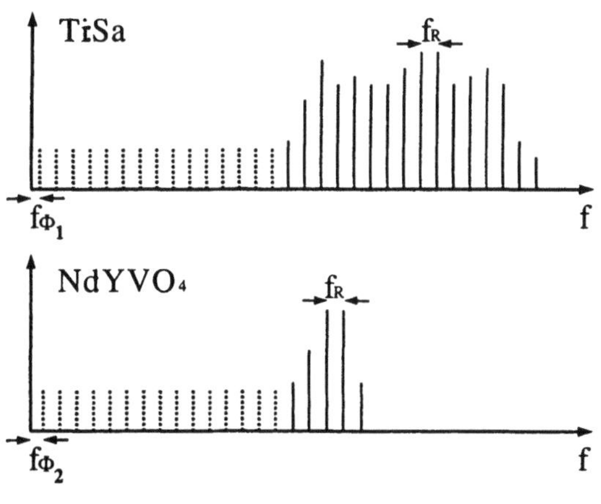

Fig.1 Mode combs of two lasers with equal pulse repetition rate f_R and independent field repetition rates $f_{\phi 1}$ and $f_{\phi 2}$.

The optical spectrum of a mode-locked laser is characterized by two frequencies, the pulse repetition rate f_R, which is the repetition rate of the pulse envelope and the field repetition rate f_ϕ that describes the pulse-to-pulse carrier-envelope phase shift (see Fig. 1). Recently, the measurement of f_ϕ in pulse trains with octave spanning spectra has been demonstrated [5-7].

Here, we demonstrate a possibility to measure the field repetition rate of a picosecond laser with only 0.1 nm bandwidth by synchronizing the picosecond laser to a broadband Ti:sapphire oscillator.

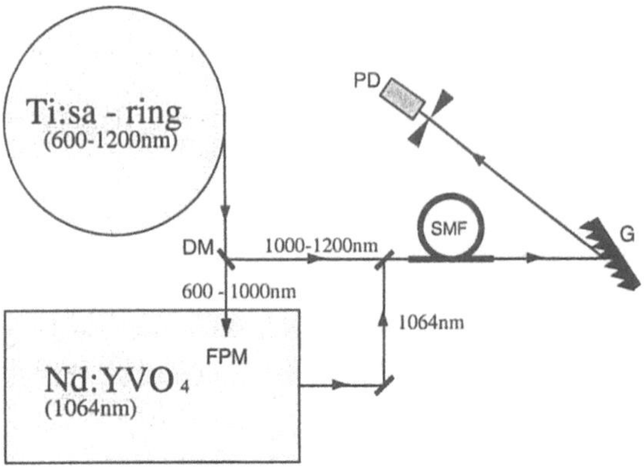

Fig.2 Experimental set-up for the measurement of the difference field repetition rate $f_{\phi 1}$ - $f_{\phi 2}$. DM: dichroic mirror, FPM: Fabry-Perot-Modulator, SMF: single mode fiber, G: grating, PD: photodiode

We use an all-optical approach to synchronize the two lasers [8], where the pulse train of the master laser – a few-cycle-Ti:sapphire ring laser emitting around 800 nm – modulates the losses in a Nd:YVO_4 oscillator with 10-ps-pulses at 1064 nm via an intracavity nonlinear Fabry-Perot semi-conductor mirror.

The set-up is shown in Fig.2: A dichroic mirror splits up the output light of the master laser, a Ti:sapphire ring laser emitting octave spanning spectra from 600 to 1200 nm. The short wavelength part below 1 μm is used to drive the intracavity Fabry-Perot-Modulator (FPM) in the Nd:YVO_4-slave-oscillator to provide synchronism. Within a few micron of cavity length detuning the lasers have the same pulse repetition rate f_R An upper boundary for the timing jitter between the two pulse trains has been estimated by cross-correlation measurements to be well below 1 ps. The long wavelength part of the Ti:sapphire spectrum is spatially superimposed with the output of the Nd:YVO_4-laser in a single-mode fiber and temporally by means of an optical delay line. A PIN photodiode detects the beat signal behind a narrow band spectral filter. Fig.3 shows the measured RF-spectrum which indicates the same pulse repetition rate of both lasers f_R and the position of the difference field repetition rate $f_{\phi 1}$ - $f_{\phi 2}$ between the two lasers. This peak can be shifted by changing the intracavity dispersion in one of the two lasers. The measurement has been done with a resolution bandwidth of 100 kHz which approximately corresponds to the width of the beat signal. The finite width of this line indicates residual phase fluctuations between the two lasers. Nevertheless, the observation of the beat signal proofs in fact that the phase jitter of both lasers is much smaller than one optical cycle of about 3 fs during the observation time of 10 μs. This is longer than the time constant for timing restoration between the two lasers due to the passive synchronization on one hand, on the other hand it is shorter than the bandwidth of typical fluctuations in the carrier-envelope phase of both lasers [7].

To ultimately determine the carrier-envelope phase shift frequency $f_{\phi 2}$ of the Nd:YVO_4 laser we used an experimental setup analogous to that in [7]. We generated the second harmonic of the red part of the Ti:sapphire spectrum around 710 nm and let it interfere with the third harmonic of the Nd:YVO_4 laser. This interference around 355 nm leads to a beat signal similar to those shown in Fig. 3, where the signal frequency is in this case $3f_{\phi 1} - 2f_{\phi 2}$. Together with the fundamental interference shown in Fig.3 , we are able to completely characterize and control the mode combs of the two independent lasers.

In conclusion, a new method for the control of the carrier-envelope phase in picosecond lasers by using passive synchronization of the picosecond laser to an octave spanning Ti:sapphire laser is demonstrated. Such schemes open up possibilities to control the mode combs of independent lasers with different pulse durations and different center wavelengths which is important for the further development of metrology using pico- and femtosecond laser sources.

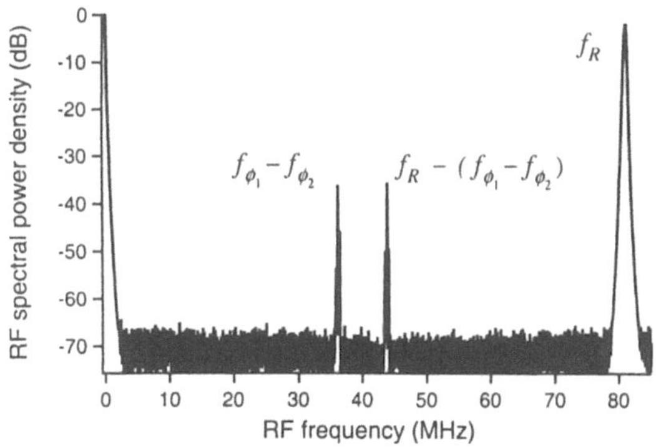

Fig.3 RF spectrum of the beat signal between the two different lasers indicating the difference field repetition rate $f_{\phi 1} - f_{\phi 2}$. The resolution bandwidth is 100 kHz.

References

[1] R. K. Shelton, L.-S. Ma, H. C. Kapteyn, M. M. Murnane, J. L. Hall, and J. Ye, *Science* **293**, 1286 (2001).

[2] J. Reichert, M. Niering, R. Holzwarth, M. Weitz, T. Udem, and T. W. Hänsch, " *Phys. Rev. Lett.* **84**, 3232-5 (2000).

[3] J. Stenger, C. Tamm, N. Haverkamp, S. Weyers, and H. R. Telle, *Opt. Lett.* **26**, 1589-91 (2001).

[4] R. Ell, U. Morgner, F. X. Kärtner, J. G. Fujimoto, E. P. Ippen, V. Scheuer, G. Angelow, T. Tschudi, M. J. Lederer, A. Boiko, and B. Luther-Davies, *Opt. Lett.*, **26**, 373 (2001).

[5] D. J. Jones, S. A. Diddams, J. K. Ranka, A. Stentz, R. S. Windeler, J. L. Hall, and S. T. Cundiff, *Science* **288**, 635-9 (2000).

[6] A. Apolonski, A. Poppe, G. Tempea, C. Spielmann, T. Udem, R. Holzwarth, T. W. Hänsch, and F. Krausz, *Phys. Rev. Lett* **85**, 740-43 (2000).

[7] U. Morgner, R. Ell, G. Metzler, T. R. Schibli, F. X. Kärtner, J. G. Fujimoto, H. A. Haus, and E. P. Ippen, *Phys. Rev. Lett.* **86**, 5462-5 (2001).

[8] W. Seitz, T. R. Schibli, U. Morgner, F. X. Kärtner, C. H. Lange, W. Richter, and B. Braun, *Opt. Lett* **27**, 454 (2002).

Part V

Compression, Shaping and Characterization

Cross-correlation frequency-resolved optical-gating measurements of ultrabroadband continuum from microstructure optical fiber

Xun Gu[1], Mark W. Kimmel[1], Erik Zeek[1], Patrick O'Shea[1], Rick Trebino[1], and Robert S. Windeler[2]

[1] School of Physics, Georgia Institute of Technology, Atlanta, GA 30332-0430, USA
E-mail: xg7@prism.gatech.edu
[2] OFS Fitel Laboratories, 700 Mountain Ave., Murray Hill, NJ 07974, USA

Abstract. Cross-correlation frequency-resolved optical gating (XFROG) using an angle-dithered nonlinear-optical crystal allows the measurement of the intensity and phase of ultrabroadband continuum generated using recently developed microstructure optical fiber. Our measurements reveal fine-scale structure in the continuum spectrum, which is confirmed by separate, single-shot measurements and simulations.

Supercontinuum generation in the microstructure fiber has attracted tremendous interest in the community since the first report in 2000 [1]. All experiments done so far have shown a smooth spectrum covering the entire visible spectrum and the near-infrared region.

In order to fully characterize the intensity and phase of the continuum pulses, we performed a cross-correlation frequency-resolved optical gating (XFROG) measurement. To accommodate the enormous bandwidth the continuum pulse possesses, we used sum-frequency generation in an angle-dithered crystal [3]. When a 1-mm-thick BBO crystal was rapidly dithered by 20°, all wavelengths in the continuum could be effectively phase-matched, and the thicker-than-normal crystal length also gave greater signal strength. Results are shown in Figure 1.

Although the retrieved XFROG trace excellently matches the measured trace in the gross parabolic shape, there is very fine structure in the retrieved trace and spectrum, while the measured trace and the independently measured spectrum are quite smooth.

A possible explanation to the discrepancy is that such fine spectral structure does exist in the continuum pulses. It is not shown in the measured XFROG trace and the independently measured spectrum because the structure is unstable and varies from shot to shot. Because these measurements involve integrating as many as 10^5 shots, the structure will wash out and be unobservable. However, the fine structure in spectrum also will be in slow-varying forms in the time domain, and the time domain information is detectable by XFROG. This redundant information is sufficient to lead FROG toward a qualitatively correct solution, which is a highly-structured spectrum and a cubic spectral phase.

We did numerical simulations explicitly imitating this process by creating artificial structure in a theoretical supergaussian spectrum and its XFROG trace. Averaging 100 similar but different highly-structured theoretical traces washed out the structure, but retrieval brought back similar pattern we saw in the retrieved trace of the continuum measurement.

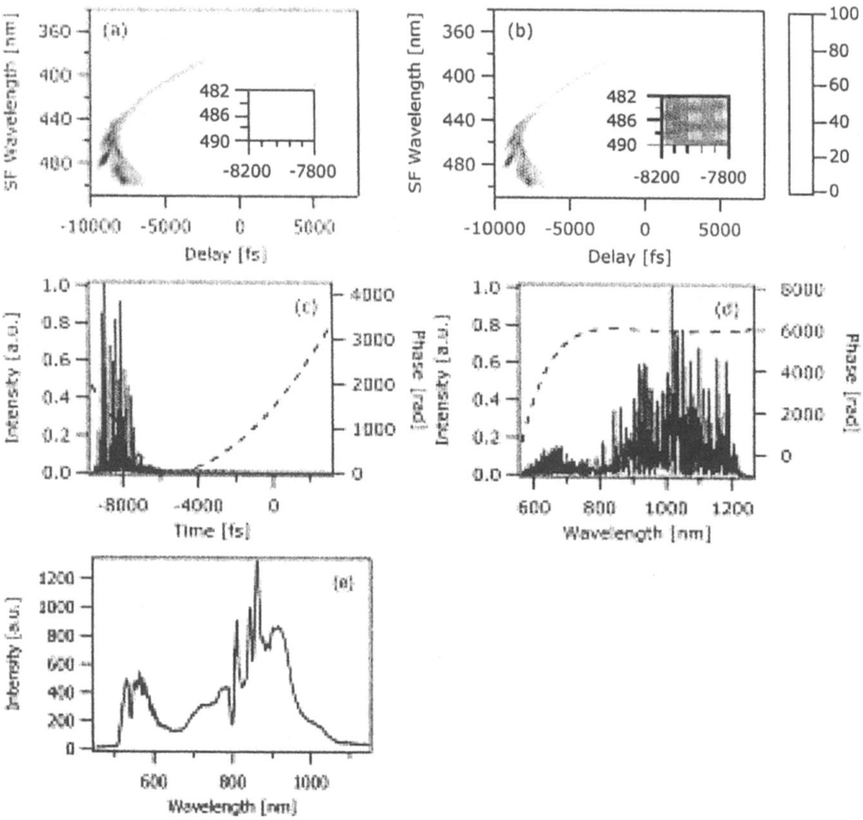

Fig. 1. XFROG measurement results of the microstructure-fiber continuum. (a) Meaured trace, (b) retrieved trace, (c) retrieved temporal intensity and phase, (d) retrieved spectral intensity and phase, and (e) independently measured spectrum.

The most direct confirmation of the fine structure in the continuum spectrum came from a single-shot measurement. We used an adjustable-repetition-rate regenerative amplifier, and attenuated its output to pump the microstructure-fiber continuum. Using different combinations of amplifier repetition rate and camera exposure time, one can capture any number of shots in a single exposure of the camera. We measured a 120-nm section of the continuum spectrum, and the results are in Figure 2.

As fewer shots are integrated in the measurement, fine structure begins to show in the continuum spectrum, and finally, when only one single shot is measured, significantly structure is evident. We took four consecutive single-shot measurements, and the four spectra were all different (see Figure 2e), and, when they were averaged, the amplitude of oscillations became less pronounced. This measurement strongly proves that the fine structure does exist in the continuum spectrum, and it is unstable, and varies from shot to shot.

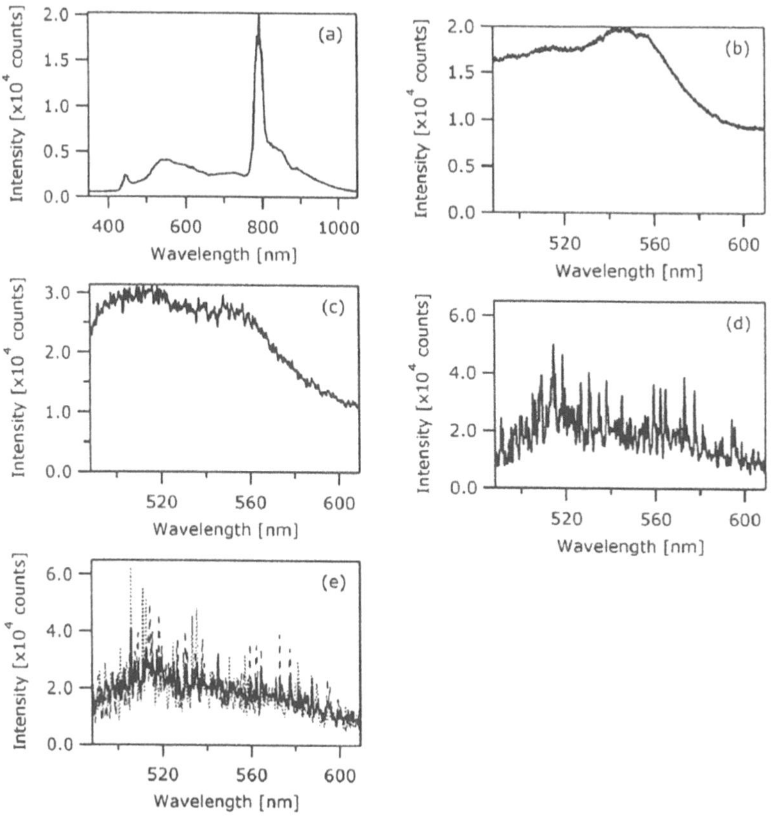

Fig. 2. (a) Entire continuum spectrum, averaged over 10,000 shots, (b) a spectral section exposed for 10,000 shots, (c) exposed for 100 shots, (d) exposed for one single shot, (e) average of consecutive single-shot measurements, where the solid line is the average of four single-shot spectra, of which two are shown as dash and dot lines in the plot. Spectra (a)-(d) were attenuated; (e) was not.

Both figures were excerpted from Xun Gu, Lin Xu, Mark Kimmel, Erik Zeek, Patrick O'Shea, Aparna P. Shreenath, Rick Trebino, and Robert S. Windeler, "Frequency-resolved-optical-gating and single-shot spectral measurements reveal fine structure in microstructure–fiber continuum," *Optics Letters*, in press.

References

1 J. K. Ranka, R. S. Windeler, and A. J. Stentz, *Opt. Lett.* **25**, 25-27, 2000.
2 S. Linden, H. Giessen, and J. Kuhl, *Phys. Status Solidi B* **206**, 119-124, 1998.
3 P. O'Shea, M. Kimmel, X. Gu, and R. Trebino, *Opt. Express* **7**, 342-349, 2000.

Complete Spatio-Temporal Characterization of Ultrashort Optical Pulses using Two-Dimensional Shearing Interferometry

C. Dorrer, [1] E. Kosik[2] and I. A. Walmsley[2]

The Institute of Optics, University of Rochester, Rochester, NY, 14627, USA
Present addresses:
[2]Bell Laboratories, Lucent Technologies, Holmdel, NJ 07733, USA
[3]Clarendon Laboratory, University of Oxford, Parks Rd., Oxford OX1 3PU, UK
E-Mail: walmsley@physics.ox.ac.uk

Abstract. We present a novel method for the complete spatio-temporal characterization of the electric field of an ultrashort light pulse. The method uses shearing interferometry in the spectral and spatial directions, and allows rapid analysis of space-time coupling in ultrashort pulse optical systems.

1. Introduction

It is now straightforward to characterize both the wavefront of a quasi-monochromatic optical beam and the time-dependent electric field of an ultrashort light pulse. However, measurements of the spatio-temporal dependence (i.e. $E(\omega,x)=\sqrt{I(\omega,x)}\exp(i\varphi(\omega,x))$) of an extended pulse have proven difficult They are however, crucial, because space-time coupling is common in ultrafast optics. It arises both from linear propagation (for example, focusing of a pulse by a lens) or by nonlinear processes (such as self-focusing). We have recently developed a method for the complete characterization of a pulsed field by extending shearing interferometry into the space-frequency domain.[1]

Interferometric pulse characterization techniques such as Spectral Phase Interferometry for Direct Electric-field Reconstruction (SPIDER) [2], allow characterization of a one-dimensional field from a one-dimensional experimental data set.

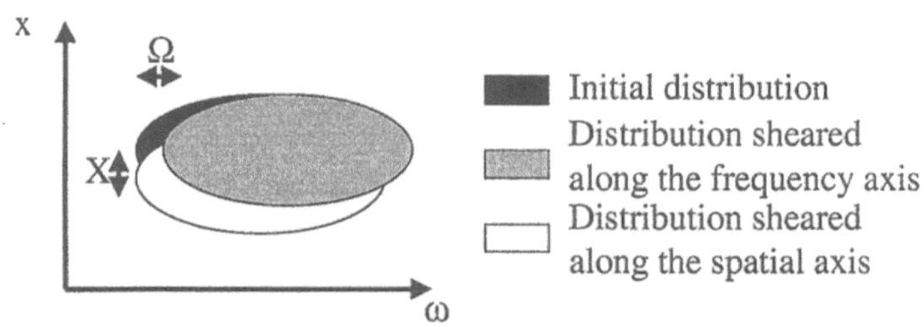

Fig. 1: implementation of shearing interferometry in the (ω,x) domain.

Figure 1 shows the configuration of spatio-spectral fields needed to measure the phase gradients $\partial\varphi/\partial\omega\,(\omega,x)$ and $\partial\varphi/\partial x\,(\omega,x)$, from which the spatio-spectral electric field can be reconstructed. Spectrographic or tomographic techniques for the characterization of a pulse at a single spatial point x_0 are based on the acquisition of a two-dimensional experimental trace, from which a one-dimensional field is reconstructed. SPIDER can be adapted to make use of the extra degree of freedom.[3]

2. Experimental implementation of spatio-spectral shearing interferometry

In SPIDER, a spectral shear between two replicas of the pulse is generated by non-linear interaction of the replicas with a chirped pulse. The difference in upconversion frequencies for the two replicas results in a spectrally-sheared interferogram. The phase gradient $\partial\varphi/\partial\omega$ can be extracted from the one-dimensional spectral interferogram by one of several robust inversion algorithms, and the phase function itself reconstructed to within a constant phase and an arbitrary delay. Together with a measurement of the pulse spectrum (which may be made simultaneously with the interferogram), this provides a complete characterization of the spectral or temporal field of the pulse.

If the nonlinear interaction is performed correctly, then the spatial amplitude and phase distribution of the field is preserved. Then resolving the interferogram both spectrally and spatially using a two-dimensional imaging spectrometer contains a specification of the spatio-spectral phase gradient $\partial\varphi/\partial\omega\,(\omega,x)$. In our experiment, a measurement of $\partial\varphi/\partial x\,(\omega,x)$ is performed using a white-light compensated Michelson interferometer with independent adjustment of the shear, tilt and delay between the two beams. Spatially and spectrally resolving the output of the interferometer using the two-dimensional imaging spectrometer yields a two-dimensional interferogram. Fourier transformation of this interferogram in each of the two orthogonal directions allows the measurement of the two orthogonal phase gradients, $\partial\varphi/\partial x\,(\omega,x)$ and $\partial\varphi/\partial\omega\,(\omega,x)$. These two perpendicular gradients therefore allow the reconstruction of the phase of an arbitrary distribution of the electric field. Together with the spatio-spectral distribution of intensity that is also available from the measured interferogram, the complete spatio-temporal field along one section of the beam can be reconstructed. This contains a specification of all forms of space-time coupling that are present in the beam.

3. Experimental results

This implementation of shearing interferometry has been tested in several important and commonly encountered experimental situations. The electric field of a femtosecond-duration optical pulse has been characterized in the (ω,x) domain after focusing or defocusing by a telescope (as shown, for example, in Fig. 2). The dispersive properties of several prisms have also been studied. In a similar manner, we have measured the distortion of the spatio-temporal phase and amplitude of a pulse propagating in a nonlinear medium, and used this to

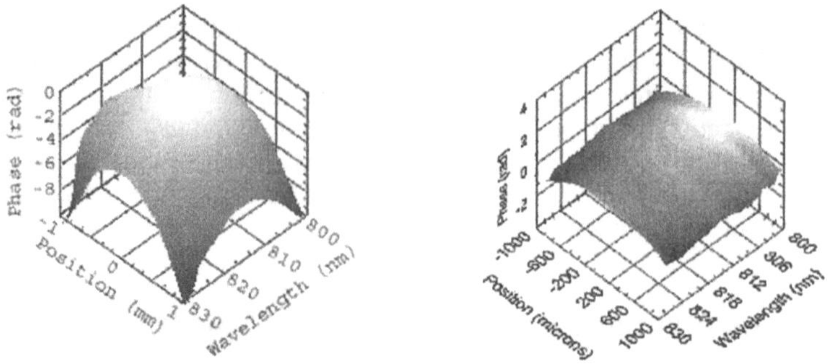

Fig.2: spatio-spectral phase of the electric-field for a slightly defocused and collimated 50 fs pulse (from left to right).

measure the nonlinear refractive index of the material. We believe that application to the non-linear propagation of intense ultrashort optical pulses will be one of the strengths of this technique.

4. Conclusions

By extending shearing interferometry to the (ω, x) domain, we have demonstrated the first self-referencing method for the complete characterization of the electric field of ultrashort optical pulses in space and time. Its simplicity of operation, design and implementation, together with a rapid and robust algorithm for data inversion opens the way to many new experiments. For example, characterization of the output of Chirped Pulse Amplifier systems in this way will lead to a better control (and therefore understanding) of laser-matter interaction, because the distribution of the intensity and phase in time and space on a target is now accessible. Also, real-time optimization of the output of laser systems (for instance, alignment of stretchers and compressors and the choice of focusing elements) will enable easier interpretation of experimental results. One can imagine also that ultrafast technology will benefit from a tool that allows one to precisely optimize the field. Such a tool will also be useful in pulse-shaping applications involving space-time coherent control, as well as in the study of non-linear interactions.

5. References

1 C. Dorrer E. Kosik and I. A. Walmsley, Opt. Lett., **27**, 548 (2002)
2 C. Iaconis and I.A. Walmsley, Opt. Lett., **23**, 792-794 (1998) ; L. Gallmann *et al*, Opt. Lett., **24**, 1314 (1999) ; C. Dorrer *et al*, Opt. Lett., **24**, 1644 (1999).
3 D. Malacara, *Optical shop testing* (John Wiley and Sons, 1978), Chap. 4.

Self-Compression of Ultrafast Optical Pulses using Molecular Phase Modulation

Nick Wagner[1], Tom Weinacht[1], Mark Baertschy[1], Chris Greene[1], Henry Kapteyn[1], Margaret Murnane[1], Randy A. Bartels[1]

[1] JILA, University of Colorado, 440 UCB, Boulder CO 80309-0440, USA
E-mail: bartels@colorado.edu

Abstract. We demonstrate that the phase modulation induced by a molecular rotational wave packet can spectrally broaden and induce a negative chirp on an ultrafast pulse, allowing it to be compressed >9x by propagation through a window.

1. Introduction

Ultrafast optical pulses are useful for a range of scientific applications from spectroscopy to surface science. Since ultrafast lasers are limited in pulse duration to 5-7fs [1], and high-energy laser amplifiers are limited to ~15fs [2], shorter pulses must be created through external pulse compression techniques. The most common configuration is to use self-phase modulation (SPM) to broaden the pulse spectrum, followed by a negative dispersion element to compress the pulse [3,4]. However, these techniques are limited in obtainable pulse duration (particularly at short-wavelengths) by high-order dispersion aberrations. In recent years, molecular modulation of light has been considered as an alternative source of phase modulation [5,6]. In this work, we demonstrate a new technique that uses the time-dependent index modulation of quantum revivals of molecular rotational wave packets in a molecular gas (CO_2). This technique has an inherent advantage over SPM techniques in that the spectral broadening can be induced with a *negative* chirp, allowing for simple pulse compression by propagating through a dispersive medium. This technique differs from Raman techniques previously demonstrated in that a *wavepacket* consisting of many *rotational* levels is used to obtain a *single, self-compressed,* pulse.

When a linear molecule interacts with a linearly-polarized electric field, the molecule experiences a torque as it tries to align itself to the direction of polarization. An ultrashort pulse induces a quantum mechanical wave packet composed of a large number of coherent rotational levels, which evolve in time and periodically dephase and rephase. These quantum revivals lead to a large time-dependent variation in the index of refraction, that can broaden the spectrum of a probe pulse passing through it. Model calculations of this time-dependent index are shown in Fig. 1.

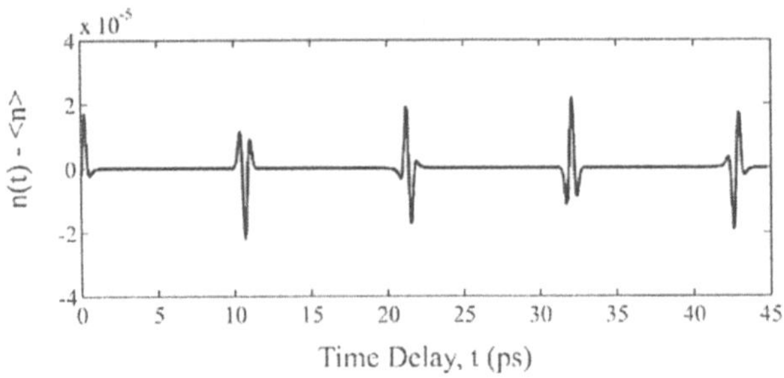

Fig. 1. Calculated index of refraction, n(t), as a function of time for rotating CO2 molecules at room temperature. The plots show deviations from the average value, <n>, where <n>-1 = 1.7x10-4.

For our experiments, we used 20fs laser pulses generated by an amplified Ti:sapphire laser system, at a 1kHz repetition rate, a wavelength of 800nm, and an energy of ~1mJ [7]. A portion of the pump pulse (~ 30%) is split off for use as a probe pulse. This probe pulse is frequency doubled in a nonlinear crystal whose length we vary depending on the desired probe pulse duration. The pump and probe pulses are combined with an adjustable delay and focused into a 60cm long, 450μm diameter, hollow-core fiber filled with CO_2 gas at pressures of 100-500 torr.

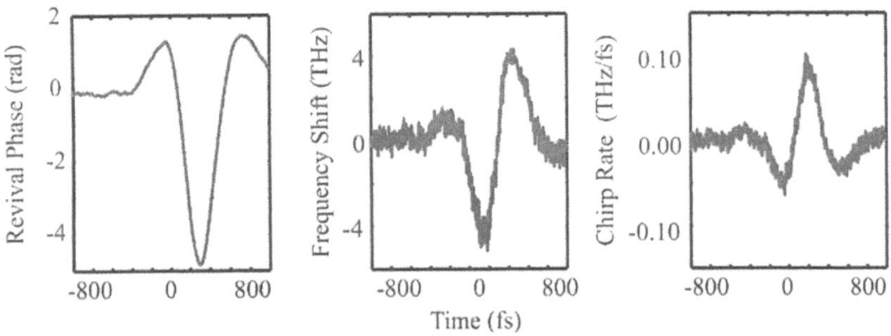

Fig. 2. Measured phase (a), frequency shift (b), and chirp rate (c) for the first observable partial rotational revival.

For our first experiment, the short probe pulse sampled the instantaneous index variation, and we measured the frequency shift (proportional to *dn/dt*) of the temporal phase modulation due to the 1st observed partial rotational revival, shown in Fig. 2(b). Integrating this frequency shift gives the time-dependent phase modulation, while the derivative gives the chirp rate, which is proportional to the spectral broadening that can be experienced by a probe pulse.

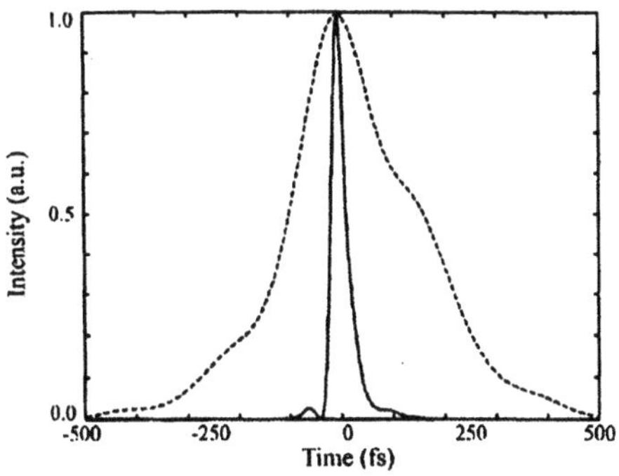

Fig. 3. FROG Temporal profiles of the injected 270 fs pulse (dotted) and the self-compressed 30 fs pulse (solid).

Note that in Fig. 1, the sign of curvature of the time-dependent index modulation varies with delay between the pump and probe pulses. By adjusting the probe pulse to the appropriate delay time, we can broaden the spectrum of the probe while imposing a *negative chirp on the pulse,* which can then be easily compressed in material. Fig. 3 shows the results of such an experiment [8]. A 270fs input pulse with a central wavelength of 400nm is phase modulated in rotationally excited CO_2 gas, resulting in an order of magnitude increase in the bandwidth of the probe. The pulse is then compressed by propagating through a 1" fused silica window. Another advantage of this scheme is that a single, shortened, output pulse is generated.

3. Conclusions

In summary, we demonstrate self-compression of ultrafast optical pulses. New bandwidth is generated with an intrinsic negative chirp that can be compressed by simple propagation though material. This technique is applicable over the entire transparency range of the gas. In particular, modeling predicts that realistic parameters can be found that would lead to compression of a 100fs pulse at 248nm to < 2fs duration, and without loss of pulse energy during the compression process. We acknowledge support for this work from the Department of Energy.

References

1 R. Ell *et al.*, *Opt. Lett.* **26**, 373 (2001).
2 E. Zeek *et al.*, *Opt. Lett.* **25**, 587 (2000).
3 D. Grischkowsky *et al.*, *Appl. Phys. Lett.* **25**, 566 (1974).
4 M. Nisoli *et al.*, *Appl. Phys. Lett.* **68**, 2793 (1996).
5 S. E. Harris and A. V. Sokolov., *Phys. Rev. Lett.* **81**, 2894 (1998).
6 M. Wittmann *et al.*, *Phys. Rev. Lett.* **84**, 5508 (2000).
7 S. Backus *et al.*, *Opt. Lett.* **22**, 1256 (1997).
8 R. A. Bartels *et al.*, *Phys. Rev. Lett.* **88**, 030972 (2002).

Infrared Pulse Shaping by Parametric Amplification

Howe-Siang Tan, Elmar Schreiber, and Warren S. Warren

Department of Chemistry and Center for Ultrafast Laser Applications
Princeton University, Princeton NJ 08544, USA
E-mail: wwarren@princeton.edu

Abstract. We create arbitrarily shaped ultrashort pulses in the mid infrared (~3.0μm). The amplitude and phase profile of a shaped broadband near infrared signal pulse is transferred to the mid infrared idler pulse in a $KNbO_3$ optical parametric amplification system pumped with a 805nm narrow bandwidth pump pulse.

1. Introduction

High resolution pulse shaping capabilities exist for wavelengths ranging from the visible to the near-infrared (NIR) [1]. Extensions of high-power pulse shaping to the mid-infrared (MIR) and ultraviolet (UV) regions would have immediate applications in coherent control, spectroscopy and imaging. However, direct phase and amplitude pulse shaping in these regimes is difficult. Liquid crystal modulators absorb in the UV and MIR; acousto-optic materials such as quartz (for UV) and TeO_2 and GaP (for mid-IR), though transparent, have very low diffraction efficiency.

One solution to the problem of limited modulator transparency is to shape pulses in the visible or NIR wavelengths, and then to transfer the shape to a pulse of another wavelength by a nonlinear optical process. There have been some recent successes in this approach [2]. In all these cases, the shaped pulses are the result of nonlinear optical process between two pulses with similar bandwidth. However, the resolution of the final pulse shape may be restricted as it is a non-trivial convolution between the two input pulses. We have recently shown that the

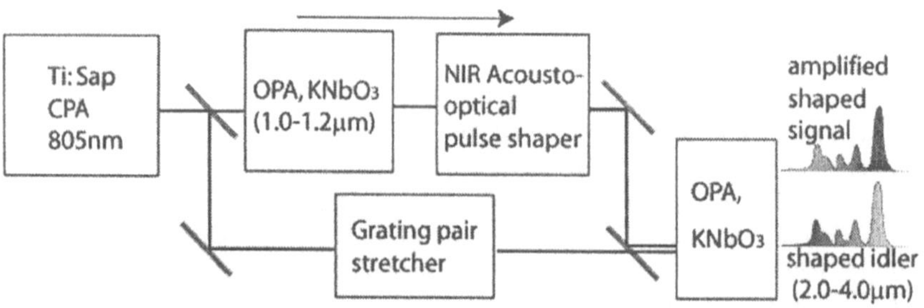

Fig. 1. Experimental setup

optical parametric amplification of a broadband visible shaped signal pulse, pumped by a narrow spectral bandwidth pulse will parametrically force the NIR idler pulse to acquire the shape of the shaped signal with high fidelity [3]. In this paper, we report that complex (amplitude and frequency modulated) MIR pulses can be manufactured from the amplification of NIR shaped pulses using a $KNbO_3$ based OPA process.

2. Experimental Setup

The setup of pulse shaping system is depicted in Fig.1. A first stage $KNbO_3$ OPA system seeded by a white light continuum and pumped by a 805nm regenerative amplifier [4] provides NIR pulses (tunable from 1.0 to 1.2μm) with a spectral bandwidth corresponding to a transform limited pulse of ~50fs. A acousto-optic pulse shaper (TeO_2) is then used to shape the NIR pulse. A second $KNbO_3$ OPA, pumped by a 805nm pulse (50μJ, stretched from ~170nm to ~1.2ps) amplifies the NIR shaped pulse from <0.1μJ to ~1.5μJ. The simultaneously produced shaped idler pulse (~0.5μJ, 2.4-4.0μm) is the time domain replica and spectral mirror image of the signal pulse shape. We characterize both the shaped idler and signal pulses by the spectrally and temporally resolved upconversion technique [5]. Sum frequency generation spectra (0.3mm BBO) between a reference pulse centered at 805nm (6nm spectral bandwidth) and the shaped pulses are obtained as a function of delay between them. From the resultant time-wavelength correlation trace, the group delay versus wavelength of the shaped pulses can be recovered.

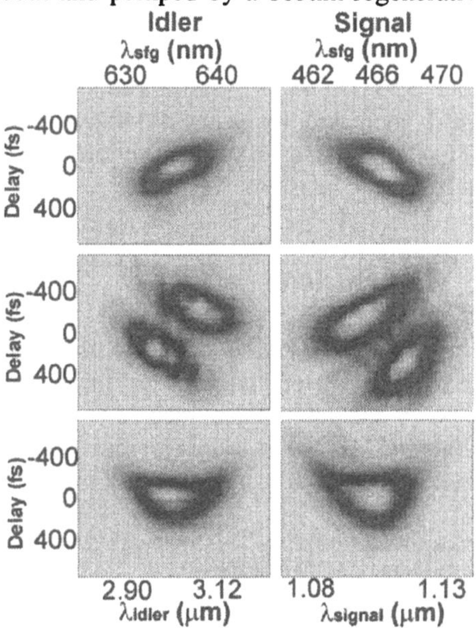

Fig. 2. Time-wavelength correlation trace (Sum frequency spectrum vs. delay between shaped pulse and reference pulse centered at 805nm) of shaped idler pulses (left column) and the corresponding shaped signal pulses (right column). First row, positive linear frequency chirped idler pulse. Second row, idler two pulse train with negative linear frequency chirp. Third row, shaped pulse with quadratic frequency chirp.

3. Results

We amplify arbitrarily shaped pulses centered near 1.1μm to create shaped MIR pulses centered near 3μm using our setup. Fig.2 illustrates a series of NIR shaped signal pulses (right) of different phase modulations and the corresponding MIR shaped idler pulses (left) with the expected spectral mirror image. Fig.3 shows the spectra of an amplified NIR amplitude modulated shaped pulse and the corresponding MIR idler shaped pulse. The amplification gain bandwidth of the OPA (1mm $KNbO_3$) system at 1.1μm is ~10THz while the spectral bandwidth of pump pulse is about 2.5THz. The 4:1 ratio can be further increased by using a thinner nonlinear crystal to effect higher fidelity shape transfer.

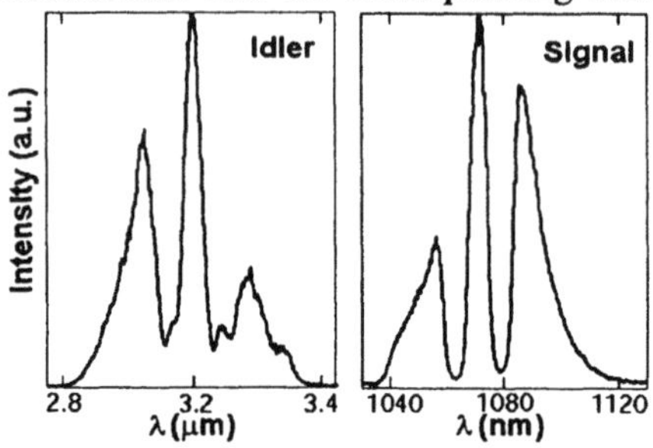

Fig. 3 Spectra of an amplitude modulated signal pulse (right) and the corresponding idler shaped pulse (left).

4. Conclusions

We have demonstrated a system that creates arbitrarily shaped MIR (~3μm) pulses. This system is based on the parametrically transfer of the signal pulse shape to the idler pulse shape by a optical parametric process. This indirect method of MIR pulse shaping avoids the problem of the lack of direct modulation materials in the MIR.

Acknowledgements. This research was supported by the New Jersey Commission on Science and Technology.

References

1 A. M. Weiner, Rev. Sci. Instr. **71**, 1929, 2000; J. X. Tull, M. A. Dugan, and W. S. Warren, Adv. Magn. Opt. Res. **20**, 1, 1997.

2 F. Eickemeyer, R. A. Kaindl, M. Woerner, T. Elsaesser, and A. M. Weiner, Opt. Lett. **25**, 1472, 2000; N. Belabas, J. -P. Likforman, L. Canioni, B. Bousquet, and M. Joffre, Opt. Lett. **26**, 743, 2001; T. Witte, D. Zeidler, D. Proch, K. L. Kompa, M. Motzkus, Opt. Lett. **27**, 131, 2002.

3 H.-S. Tan, E. Schreiber, W. S. Warren, Opt. Lett. **27**, 439, 2002.

4 S. Cussat-Blanc, A. Ivanov, D. Lupinski., and E. Freysz, Appl. Phys. B **70**, S247, 2000.

5 J. –K. Rhee, T. S. Sosnowski, T. B. Norris, J. A. Arns, and W. S. Colburn, Opt. Lett. **19**, 1550 1994.

Mm-wave arbitrary waveform generation using a direct space-to-time optical pulse shaper

J. D. McKinney[1], D. E. Leaird, and A. M. Weiner[2]

School of Electrical and Computer Engineering, Purdue University, 1285 Electrical Engineering Bldg., West Lafayette, IN 47907-1285, USA
[1] E-mail: mckinnjd@ecn.purdue.edu
[2] E-mail: amw@ecn.purdue.edu

Abstract. We have achieved the first mm-wave arbitrary waveform generation at center frequencies approaching 50 GHz by using tailored pulse sequences from a novel, 1.5μm direct space-to-time pulse shaper to drive a high-speed photodetector.

1. Introduction

Fourier transform optical pulse shaping [1] is now a well established technology for generating nearly arbitrarily shaped optical pulses, with applications ranging from coherent control to high-speed communications. On the other hand, arbitrary waveform generation (AWG) capabilities for mm-wave (mm-wave) electromagnetic signals are quite limited. Mm-wave arbitrary waveform generators could be important for a number of applications, including RF test and measurement instrumentation, secure and multiple-access wireless communications, and radar. Commercial AWG instrumentation is limited to the range below 2 GHz. The use of shaped optical pulses to drive a high-speed optical-to-electronic (O/E) converter offers the possibility to dramatically improve the state-of-the-art in mm-wave AWG. In previous work in our group, we used pulse sequences from a Fourier transform (FT) pulse shaper to drive a THz photoconductive antenna, resulting in generation of freely propagating phase- and amplitude-modulated THz radiation signals [2]. Another group recently used a wavelength-division multiplexing technique together with a series of optical fiber delay lines to produce a series of pulses suitable for generating a phase-modulated waveform in the low (~6) GHz range [3]. Here we demonstrate for the first time to our knowledge mm-wave arbitrary waveform generation at center frequencies up to ~50 GHz, thereby filling in the void between the previous results. Our work takes advantage of a novel direct space-to-time (DST) pulse shaper [4], which provides pulse shaping over a 100 psec temporal window, as required for waveform generation at tens of GHz.

2. Experiment and Results

The block diagram for our experiments is shown in Fig. 1. The output from a passively modelocked fsec fiber laser is spectrally filtered to yield ~300 fs pulses, passed through a fiber amplifier, and then directed to a DST pulse shaper, which generates tailored pulse sequences. These pulse sequences are then reamplified and directed onto a 60 GHz photodiode, resulting in shaped mm-wave signals, which are measured on a 50 GHz bandwidth oscilloscope.

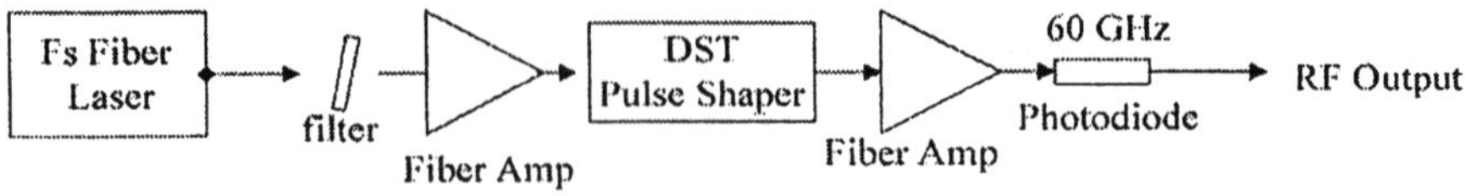

Fig. 1. Schematic diagram of the experimental setup.

In contrast to the well-known FT pulse shaper, the temporal output of a DST shaper is a directly scaled version of the spatial profile of a spatially patterned fsec input beam. We previously demonstrated operation of a DST shaper using femtosecond pulses at 850 nm from a Ti:S laser [4]. Here we use a DST pulse shaper constructed for the first time to operate in the 1.55 μm lightwave communications band. In order to realize the required spatial patterning at the input, we introduce the use of a diffractive optical element (DOE), which splits a single input beam into multiple, nearly equal intensity, spatial spots (8-21 with our DOEs), essentially without loss. Additional novel features include operation with fiber input and output, and the use of a low polarization-dependent-loss grating, which makes the DST pulse shaper insensitive to input polarization (important, since polarization state is usually not preserved in fibers).

We now present several compelling examples of mm-wave arbitrary waveform generation. Figure 2 (a) shows a 48 GHz mm-wave burst and the corresponding optical cross-correlation. The shape of the mm-wave signal is a good sinusoid, unlike the optical drive signal which consists of isolated pulses. This is explained by the limited bandwidth of the O/E converter and electrical measurement, which suppresses harmonics in the mm-wave signal.

Figure 2 (b) shows examples of phase modulation at 48 GHz. The top trace is the same 48 GHz sinusoidal burst as in Fig. 2 (a). However, in the lower traces, the mm-wave waveforms contain a π phase shift, which occurs at different places within the waveform in the different traces. The phase shifts are clearly seen by comparing the positions of peaks and nulls between the top and lower waveforms. In these examples, the mm-wave phase shift is induced by controlling the timing of the optical pulses. For example, in the lowest trace, the optical pulses occur at approximately -60, -40, -20, 0, 30, and 50 ps. Then ~10 ps extra delay between the fourth and fifth yields the π phase shift.

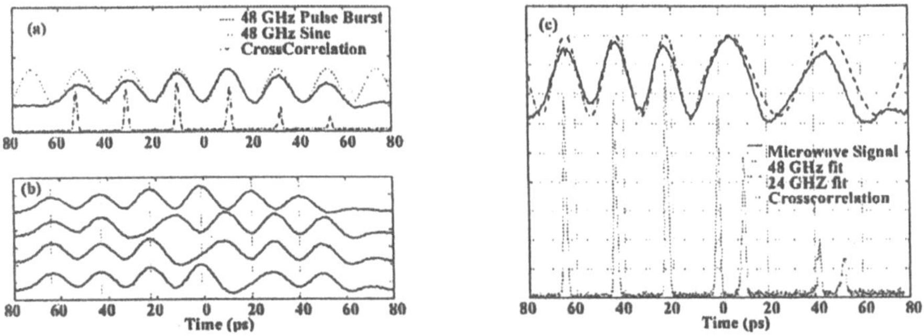

Fig 2. (a) 48 GHz Mm-wave burst. The mm-wave signal is found to be a good sinusoid, in contrast to the optical pulse sequence driving the photodiode (bottom trace). (b) Mm-wave phase modulation. The phase of the mm-wave waveform can be controlled by adjusting the relative *position* of pulses in the driving optical pulse sequence. (c) Mm-wave frequency modulation. Frequency modulation of the mm-waveform is achieved by controlling the relative spacing of pulses in the driving optical pulse sequence.

Figure 2 (c) shows an example of frequency modulation. The first cycles of the waveform occur at 48 GHz, which then changes abruptly to 24 GHz for the last two cycles. Here the frequency modulation is controlled by varying the repetition rate of the driving optical pulse sequence in concert with the electrical bandwidth limitations. In particular, the trick to obtain the lower frequency is to use pulse pairs spaced too closely (~10 ps) to be resolved by the electrical system. The responses from each of the pulses in the pair smear together, yielding a single mm-wave cycle twice as long as for single pulse excitation. This example illustrates how shaped optical pulse sequences enable mm-wave waveform synthesis one cycle at a time.

3. Summary

In summary, we have exploited optical pulse shaping technology to demonstrate mm-wave arbitrary waveform generation up to ~50 GHz. This material is based upon work supported by, or in part by the U.S. Army Research Office under contract DAAD19-00-0497, and by the National Science Foundation, and Tektronix.

References

1 A. M. Weiner, "Femtosecond pulse shaping using spatial light modulators," Rev. Sci. Instr. **71**, 1929 (2000).

2 Y. Liu, S.-G. Park ad A. M. Weiner, "Terahertz waveform synthesis via optical pulse shaping," IEEE J. Sel. Topics Quant. Electron. **2**, 709 (1996).

3 B. Jalali, P. Kelkar, and V. Saxena, "Photonic arbitrary waveform generator," in Proceedings of the 14th Annual Meeting of the IEEE. **1**, IEEE Lasers and Electro-Optics Society, (2001).

4 D.E. Leaird and A. M. Weiner, "Femtosecond direct space-to-time pulse shaping," IEEE J. Quant. Electron. **37**, 494 (2001).

Time-to-space conversion of ultrafast optical pulses via spectral nonlinear optics in the communications band

J.-H. Chung[1], and A. M. Weiner[2]

School of Electrical and Computer Engineering, 1285 EE Building, Purdue University, West Lafayette, IN 47907-1285, USA
Center for Education and Research in Information Assurance and Security, Purdue University, West Lafayette, IN 47907-1315, USA
[1] E-mail: jchung@ecn.purdue.edu
[2] E-mail: amw@ecn.purdue.edu

Abstract. We present the first spectral nonlinear optics time-to-space converter operating in the optical communications band. Results on multi-bit conversion and efficiencies large enough for real-time measurements of a single laser pulse without averaging are reported.

1. Introduction

In ultrahigh-speed optical communication networks, time-to-space conversion, which maps the serial time-domain ultrafast optical signal into its spatial replica, is important both for demultiplexing and to enable spatial or parallel processing for various higher-layer functionalities, such as header recognition and data encryption [1]. Time-to-space conversion is also interesting as a new approach for ultrafast optical waveform measurement. Several groups have reported recently on time-to-space converters [1-4]. From the viewpoint of applications for real-world optical communication networks, however, several requirements will have to be fulfilled simultaneously. For example, the operating wavelength must fall within the 1.55 μm lightwave communications band, the response time of the nonlinear medium must be fast, the efficiency must be high, and the temporal processing window must be large enough to accommodate parallel processing of many optical bits. In this paper we report the first spectral nonlinear optics time-to-space converter operating in the lightwave band. The use of second harmonic generation (SHG) as the spectral nonlinear optical process [1,3,5] provides ultrafast response. Furthermore, drawing on our analysis of the conversion efficiency of a SHG time-to-space converter [5], we have optimized the sensitivity of our converter to realize the first direct, single-shot measurements of a single laser pulse with no averaging.

2. Experiment and Results

Our time-to-space converter, sketched in Fig. 1(a), employs the common 4-f pulse shaper arrangement [1], together with two noncollinear input beams at 1.56 μm

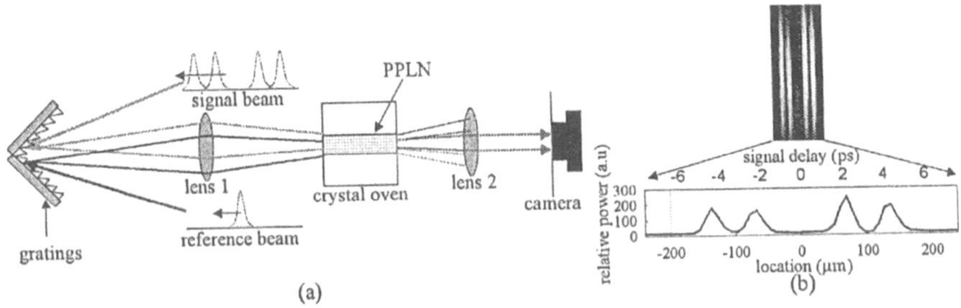

Fig. 1. (a) The schematic of the time-to-space converter with an input data sequence of 11011 indicated. (b) The resultant output image and corresponding transverse power profile.

wavelength and 80 MHz repetition rate from a femtosecond optical parameter oscillator. One input is an ~200-fs, 1-nJ unshaped reference pulse with a broad and smooth spectrum; the second constitutes the signal pulse which may be shaped in time and frequency. Both beams are spectrally dispersed and interact in a 2-cm long Periodically Poled Lithium Niobate (PPLN) crystal to produce a quasi-monochromatic second harmonic (SH) output with a spatial profile determined by the complex spectrum of the signal pulse. Propagation through a subsequent Fourier transform lens yields a spatial output pattern equal to a scaled version of the signal pulse input temporal profile [1,3,5].

A further demonstration of the time-to-space mapping characteristic can be obtained by measuring the transverse power distribution of the red SH output spot as the delay of a single unshaped signal pulse is varied with respect to the reference. The resultant data in Fig. 2(a) show that the position of the output spot varies linearly with delay, with a time-to-space conversion of 31.3μm/ps, while the output intensity rolls off for large delays, with a FWHM temporal window of 24.9ps. Both values are in good agreement with theoretical values [1,5] estimated based on the parameters of our experiment (32.1μm/ps and 29.1ps, respectively).

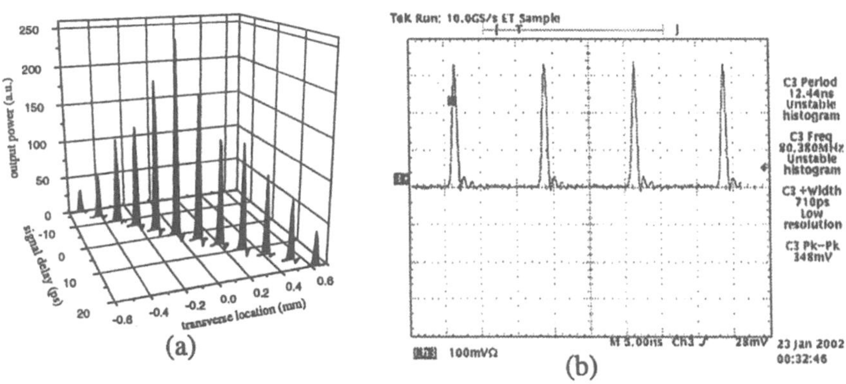

Fig. 2. (a) The relative power of the output red spot as a function of the transverse location and the delay of the signal pulse with respect to the reference pulse. (b) Oscilloscope trace of an output spot as measured on an APD.

Finally, we measured a time-to-space SHG conversion efficiency of 0.8%, close to that predicted based on the formation [5] and our experimental parameters. This efficiency is actually quite good, given the low pulse energy (1 nJ) and the spectrally dispersed geometry of the time-to-space converter, which clearly reduces the peak intensity at the nonlinear crystal compared to conventional SHG experiments. Importantly, the experimental efficiency is already sufficient for observation in real time on a single shot basis without signal averaging. This is demonstrated by Fig. 2(b), which shows the time-to-space output signal as directly measured on an avalanche photodiode (APD) and digital oscilloscope. The 80 MHz repetition rate of the source laser is clearly visible. These results demonstrate that the nonlinear conversion efficiency of the time-to-space converter is coming close to the regime needed for real-time information processing in real-world optical communications applications. Furthermore, this goes well beyond what is required in typical pulse measurement applications (e.g., FROG), where averaging over millions of pulses is permitted.

3. Conclusion

In summary, we have implemented what we believe is the first spectral nonlinear optics time-to-space converter operating in the 1.55 μm band. Further details of our experiments will be covered in the talk.

Acknowledgements. This material was supported in part by the U.S. Army Research Office, by Intel Corporation, by National Science Foundation and by the sponsors of the Center for Education and Research in Information Assurance and Security.

References

1 A. M. Weiner and A. M. Kan'an, "Femtosecond pulse shaping for synthesis, processing, and time-to-space conversion of ultrafast optical waveforms," IEEE J. Sel. Top. Quantum Electron. **4**, 317-331 (1998).

2 M. C. Nuss, M. Li, T. H. Chiu, A. M. Weiner, and A. Partovi, "Time-to-space mapping of femtosecond pulses," Opt. Lett. **19**, 664-666 (1994).

3 P. C. Sun, Y. T. Mazurenko, and Y. Fainman, "Femtosecond pulse imaging: Ultrafast optical oscilloscope," J. Opt. Soc. Amer. A **14**, 1159-1170, (1997).

4 J. Ishi, H. Kunugita, K. Ema, T. Ban, and T. Kondo, "Time-to-space conversion of Tbits/s optical pulses using a self-organized quantum-well material," Appl. Phys. Lett. **77**, 3487-3489, (2000).

5 A. M. Kan'an and A. M. Weiner, "Efficient time-to-space conversion of femtosecond optical pulses," J. Opt. Soc. Amer. B **5**, 1242-1245 (1998).

6 C. W. Siders, J. L. W. Siders, A. J. Taylor, S.-G. Park and A. M. Weiner, "Efficient high energy pulse train generation using a 2^n-pulse Michelson interferometer," Appl. Opt. **37**, 5302-5305 (1998).

A simple theory of the acousto-optic programmable dispersive filter used for femtosecond laser pulse shaping

Daniel Kaplan and Pierre Tournois[1]
[1] Fastlite, XTEC Bâtiment 404, Ecole Polytechnique, 91128 Palaiseau, France
E-mail: tournois@fastlite.com

Abstract. A first order theory allows to address the acousto-optic filter spectral resolution, angular aperture, number of programming points as a function of angular crystal cut and diffraction efficiency as a function of acoustic power.

1. Introduction

We discuss the operation of a newly introduced system for phase and amplitude programmation of ultrafast laser pulses operating from a few femtoseconds to a few picoseconds.The heart of the system is an acousto-optic birefringent device , the Acousto-Optic Programmable Dispersive Filter (AOPDF) which performs for low values of acoustic power density a convolution between the amplitude of an input optical signal and a signal $S(t/\alpha)$ where $S(t)$ is the acoustic signal proportional to the electrical signal applied to the piezoelectric transducer of the device and α a scaling factor related to the ratio between the speed of sound and the speed of light in the material. The small value of α of the order of 10^{-7} enables the control of optical signals in the hundreds of Terahertz range with electrical signal within the tens of Megahertz range [1,2].

2. AOPDF interaction geometry in Paratellurite

We consider an acousto-optic interaction in Paratellurite crystal Te02. The propagation directions of the optical and acoustical waves are in the P-plane which contains the [110] and [001] axis of the crystal. The acoustic wave vector K makes an angle θ_a with the [110] axis. The polarization of the acoustic wave is transverse, perpendicular to the P-plane, along the $[\bar{1}10]$ axis. Because of the strong elastic anisotropy of the crystal, the K vector direction and the direction of the Poynting vector are not collinear. The acoustic Poynting vector makes an angle β_a with the [110] axis. When an incident ordinary optical wave polarized along the $[\bar{1}10]$ direction with a wave vector k_o which makes an angle θ_o with the [110] axis, interacts with the acoustic wave, an extraordinary optical wave polarized in the P-plane with a wave vector k_d is diffracted with an angle θ_d relative to the [110] axis. To maximize interaction length for a given crystal length, in order to decrease the acoustic power, we choose to align the incident ordinary beam with the Poynting vector of the acoustic beam, .i.e. $\beta_a = \theta_o$. Figure 1 shows the k-vector

geometry related to the acoustical and optical slowness curves, V_{110} and V_{001} are the phase velocities of the acoustic shear waves along the [110] axis and along the [001] axis respectively, n_o and n_e are the ordinary and extraordinary indices on the [110] axis and n_d is the extraordinary index associated with the angle θ_d.
The optical anisotropy $\Delta n = (n_e - n_o)$ being generally small as compared to n_o, the following relations can be obtained to first order in $\Delta n / n_o$

$$\theta_d - \theta_o = -\frac{\Delta n}{n_o}\tan(\theta_o - \theta_a)\cos^2\theta_o \text{ and } K = k_o\frac{\Delta n}{n_o}\frac{\cos^2\theta_o}{\cos(\theta_o - \theta_a)} \Rightarrow \alpha = \frac{f}{\nu} = \Delta n\frac{V(\theta_a)}{c}\cdot\frac{\cos^2\theta_o}{\cos(\theta_o - \theta_a)} \quad (1)$$

where α is the ratio between f the acoustic frequency and ν the optical frequency, c being the speed of light.

3. AOPDF interaction efficiency

The single frequency solution of the coupled mode theory allows to relate the diffracted light intensity to the incident light intensity and to the acoustic power density P(f) present in the interaction area by the formula:

$$I_d(\nu) = I_o(\nu)\frac{P(f)}{P_o}\frac{\pi^2}{4}\operatorname{sinc}^2\left[\frac{\pi}{2}\sqrt{\left(\frac{P(f)}{P_o} + \delta\phi^2\right)}\right] \text{ with } P_o = \frac{1}{2M_2}\left[\frac{\lambda}{L\cos(\theta_o - \theta_a)}\right]^2 \quad (2)$$

$\delta\phi$ is an asynchronous factor proportional to the product of the departure δk from the phase matching condition (1) and of the interaction length along the acoustic wave vector K:

$$\pi\delta\phi = \delta k.L\cos(\theta_o - \theta_a) = \frac{\Delta n}{n_o}k_o L\cos\theta_o\left[-\frac{\delta\nu}{\nu} + \delta\theta_o[2\tan\theta_o - \tan(\theta_o - \theta_a)]\right] \quad (3)$$

L being the interaction length along the optical wave vector k_o, λ the wavelength of the light in a vacuum, ρ the density of the Te02 crystal, p an elasto-optic coefficient, and M_2 the merit factor given by:

$$M_2 = \frac{n_o^3.[n_d(\theta_o)]^3.[p(\theta_o,\theta_a)]^2}{\rho.[V(\theta_a)]^3} \text{ with } p = -0{,}17\sin\theta_a\cos\theta_o + 0{,}09\sin\theta_o\cos\theta_a \quad (4)$$

4. AOPDF spectral resolution, angular aperture and number of programming points

From (2), the ratio between the diffracted and incident light equals 0,5 for $\delta\phi = \pm 0{,}8$ when $P(f) = P_o$. So the spectral resolution and angular aperture become:

$$\left(\frac{\delta\lambda}{\lambda}\right)_{1/2} = \left(\frac{\delta\nu}{\nu}\right)_{1/2} = \frac{0{,}8}{\Delta n\cos^2\theta_o}\cdot\frac{\lambda}{L} \text{ and } (\delta\theta_o)_{1/2} = \left(\frac{\delta\lambda}{\lambda}\right)_{1/2}\cdot\frac{1}{[2\tan\theta_o - \tan(\theta_o - \theta_a)]} \quad (5)$$

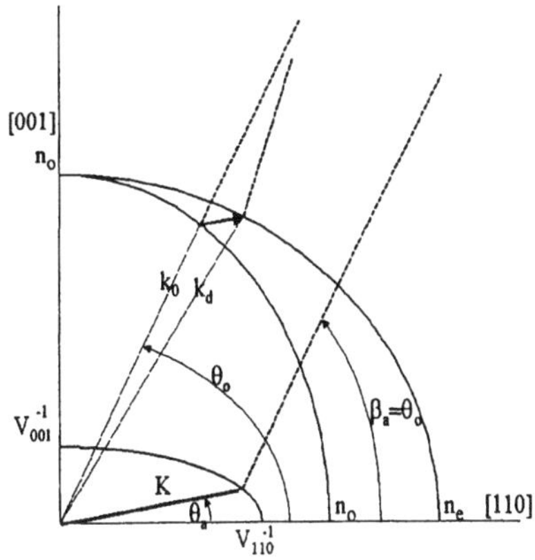

Fig.1. Acoustic and optic slowness curves and k-vector diagram

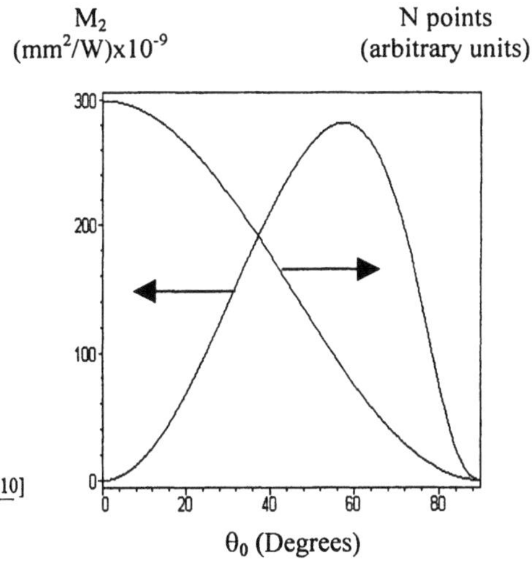

Fig.2. Merit factor and number of points vs. θ_o

If $\Delta\lambda$ is the incident optical bandwidth, the number of programming points N and the estimation P_N of the acoustic power density to maximally diffract the whole bandwidth will be:

$$N = \frac{\Delta\lambda}{(\delta\lambda)_{1/2}} = \frac{\Delta n}{0{,}8} . L\cos^2\theta_0 . \frac{\Delta\lambda}{\lambda^2} \text{ and } P_N = NP_o = \frac{1{,}25.\Delta n\cos^2\theta_o}{2M_2\cos^2(\theta_o - \theta_a)} . \frac{\Delta\lambda}{L} \quad (6)$$

Fig 2 shows the variation of M_2 and N with θ_o when $\beta_a = \theta_o$ for the Te02 around $\lambda = 0{,}8\mu m$ for which n_o=2,226 and n_e=2,374 and $\rho = 5{,}99 g/cm^3$

5. Conclusion: Wide Band and High Resolution AOPDFs

In the design of AOPDFs two different cut optimizations can be made. One for large bandwidth applications optimizes the M_2 factor hence the diffraction efficiency ($\theta_o = 58°, M_2 = 280.10^{-9} mm^2/W, \delta\lambda_{1/2} = 0{,}6nm$ for L=25mm). The other for smaller bandwidth doubles the spectral resolution for the same crystal length ($\theta_o = 38°, M_2 = 200.10^{-9} mm^2/W, \delta\lambda_{1/2} = 0{,}3nm$ for L=25mm).

6. References

1 P.Tournois, Opt.Commun. **140**, 245, 1997.

2 F.Verluise et al., Opt. Lett. **25**, 575-577, 2000.

Automated Two-Dimensional Femtosecond Pulse Shaping and Phased-Array THz Generation

Joshua C. Vaughan, T. Feurer, Keith A. Nelson

Department of Chemistry, Massachusetts Institute of Technology, Cambridge, MA 02139, USA

Abstract. We report the automated generation of high-fidelity spatiotemporally shaped femtosecond pulses using an optically addressed, two-dimensional (2D) liquid crystal spatial light modulator (SLM). The application of 2D pulse shaping to a phased-array source of terahertz (THz) radiation is demonstrated.

Our interest in 2D pulse shaping stems from efforts to control propagating excitations in crystalline solids [1]. The difficulty in using conventional temporal-only (one-dimensional) pulse shaping for controlling propagating responses is that the material response quickly leaves the excitation region [2]. A 2D pulse shaper, which can produce independent temporally-shaped pulses along a transverse spatial coordinate, is able to irradiate a sample with different pulse shapes at different locations, thereby enabling the manipulation of propagating excitations.

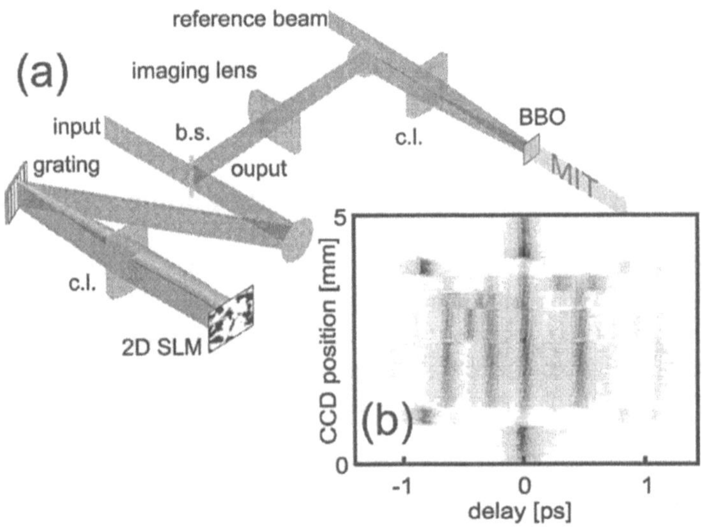

Figure 1. **(a)** Setup for real-space shaping. The shaped pulse was cross-correlated with a reference pulse and the signal was recorded by a CCD camera. **(b)** Cross-correlation of a complex user-defined waveform.

The phase-only pulse shaping setup is similar to conventional 4f spectral filtering setups. However, the incoming beam was expanded in one dimension and a 2D SLM in a reflection geometry was employed [see Fig. 1 (a)]. The optically-addressed SLM used in these experiments, Hamamatsu model SLMM X7550-800, was carefully characterized with regard to: pixelation effects; phase modulation as a function of illumination intensity; and nonlinear dispersion of the spectrum [3].

Real-space shaping was accomplished by imaging (1:1) the SLM onto a beta barium borate (BBO) crystal with a horizontally-aligned 30 cm focal length cylindrical lens [see Fig. 1 (a)]. The shaped pulse was measured by second harmonic generation cross-correlation with a vertically expanded unshaped reference pulse on a motorized delay stage. The second harmonic radiation created by each of the incident beams was blocked, and the intensity of the cross-correlation signal was recorded by imaging (1:1) the BBO crystal onto a CCD camera. Note that this arrangement allowed for a simultaneous cross-correlation measurement of all shaped regions in the vertical direction for a given delay of the reference pulse.

A demonstration of real-space shaping is shown in Fig. 1 (b). The phase pattern necessary to generate the "MIT" waveform was determined by a fast iterative Fourier transform algorithm developed by Hacker et. al [4]. The new Fourier transform algorithm is especially important for 2D pulse shaping because the standard simulated annealing and genetic algorithms are prohibitively time consuming. Sophisticated user-defined waveforms can be generated within a maximum temporal window of approximately ±4.5 ps. Wavevector shaping, in which the spatiotemporal properties of the shaped waveform are controlled at the Fourier plane of a lens, may also be performed with the setup [3].

An immediate application of 2D pulse shaping is the phased-array generation of THz frequency phonon-polaritons by spatiotemporal shaping of an ultrafast excitation waveform. The principle of operation behind the phased-array THz source is similar to those used in radar and ultrasound technologies. Phonon-polaritons are mixed polar lattice-vibrational and electromagnetic modes which are generated with ultrafast laser pulses via impulsive stimulated Raman scattering [1]. They travel at light-like speeds over macroscopic distances, and can be coupled out of their host crystals in the form of freely-propagating electromagnetic radiation. A laser pulse arriving at a single spot on a crystal will generate a phonon-polariton ripple that propagates outwards in all directions. By using the 2D pulse shaper to control the relative timing of many spots along a line, the phonon-polariton response can be steered, focused, or amplified. The setup for phased-array generation of THz radiation is nearly identical to the setup shown in Fig. 1 (a), except that a $LiTaO3$ crystal is used instead of a BBO crystal, and that a large probe beam is used to record snapshots of the propagating phonon-polaritons onto a CCD camera [5]. Phonon-polariton responses for 7 different 2D excitation waveforms are illustrated in Fig. 2.

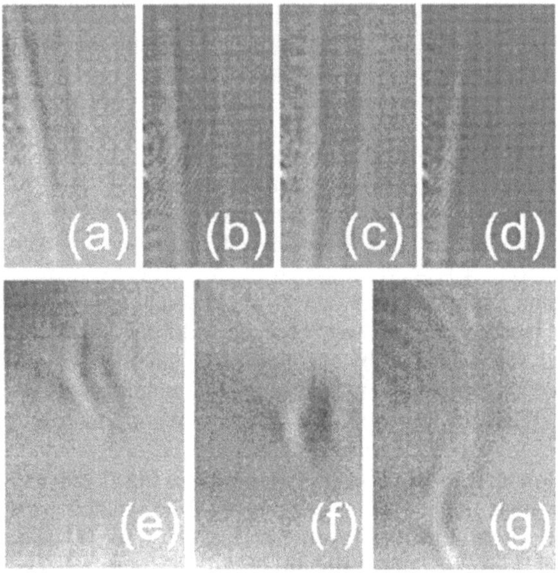

Figure 2. Phonon-polariton responses to several different two-dimensionally shaped excitation waveforms. The excitation waveforms, consisting of approximately 20 independent regions, were focused along a line to the left of the images shown. **(a)-(d)** Tilted phonon-polariton wavefronts resulting from pulses arriving at the spots at progressively later times, with a linear relation between temporal delay and spatial location of the spot. The wavefront and propagation direction depend on the slope of temporal delay vs. position. **(e)-(g)** Steered and focused phonon-polariton responses specified through a combination of linear and parabolic delays.

In conclusion, reconfigurable 2D pulse shaping has been employed to produce high-fidelity spatiotemporally shaped femtosecond laser pulses. The two-dimensionally shaped waveforms have been applied to a phased-array source of THz frequency phonon-polaritons.

Acknowledgments. This work was supported in part by Army Research Office Grants No. DAAD10-01-1-0674. T. Feurer acknowledges financial support by the Max Kade Foundation.

References

1 G.P. Wiederrecht et al., Phys. Rev. B **51**, 916-931 (1995).
2 R.M. Koehl, K.A. Nelson, J. Chem. Phys. B **114**, 1443-1446 (2001).
3 J.C. Vaughan, T. Feurer, K.A. Nelson, JOSA B, in press (2002).
4 M. Hacker, G. Stobrawa, T. Feurer, Opt. Express **9**, 191-199 (2001).
5 R.M. Koehl, S. Adachi, K.A. Nelson, Phys. Chem. A **103**, 10260-10267 (1999).

Jitter of 3-cycle Laser Pulses from a gas-filled capillary – Single Shot SPIDER Pulse Characterization at 1 kHz

J. Biegert[1], W. Kornelis[1], G. Steinmeyer[1], J. W. Tisch[1], U. Keller[1], M. Nisoli[2], G. Sansone[2], S. Stagira[2], C. Vozzi[2], and S. De Silvestri[2]

[1] Swiss Federal Institute of Technology (ETH Zürich), Physics Department, Institute of Quantum Electronics, ETH Hönggerberg-HPT, CH-8093 Zürich, Switzerland,
E-mail: biegert@phys.ethz.ch
[2] Instituto di Fotonica e Nanotecnologie (CNR), Istituto Nazionale per la Fisica della Materia, Dipartimento di Fisica, Politecnico, Milano, Italy

Abstract. Sub-10-fs millijoule-level pulses are usually generated in gas-filled capillaries with subsequent compression. Is the phase-structure preserved in this process? We present results from a preliminary investigation, and observe remarkably low mean shot-to-shot pulse duration variations (21 as).

1. Introduction

Up to now, the only technique capable of generating millijoule-level sub-10-fs pulses is based on spectral broadening in a gas-filled capillary with subsequent compression. Even though this technique is routinely being used, the exact nature of the processes involved in white-light generation processes is not fully understood. We present a measurement approach that opens up a new window to understanding femtosecond continuum generation processes. Aside from yielding a better understanding of the underlying physical principles, this has direct implications in how the pulses are subsequently compressed. More importantly, it might provide the answer to the question of whether a phase-locked pulse keeps its lock during this process.

2. Experiment

We have therefore developed a complete pulse characterization system, based on the SPIDER (spectral interferometry through direct electric field reconstruction) technique, which is able to operate single-shot at a rate of 1 kHz. The SPIDER setup is based on the one outlined in Ref. [1]. The system is capable of measuring the pulse fundamental spectrum and the spectral interferogram in absolute synchronicity at 1 kHz, thus enabling for the very first time a viable approach to understanding pulse-to-pulse fluctuations in ultrashort pulse laser systems, and pave the way to pinpointing and ultimately controlling fluctuations in CEO-phase [2].

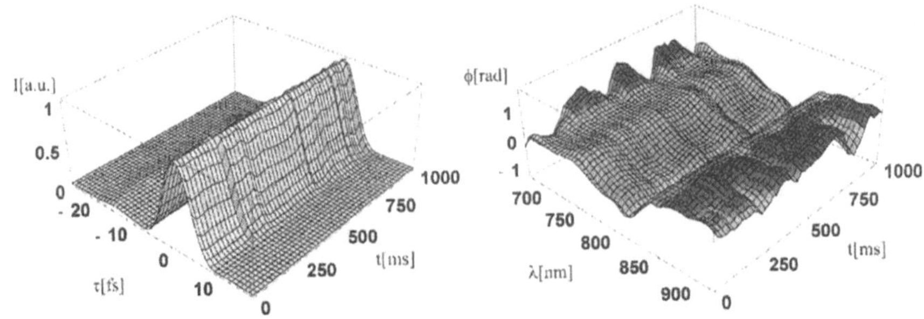

Fig. 1. Temporal pulse shape (to the left) and spectral phase (to the right) as a function of time for a period of 1 s

The laser system delivered amplified, ca. 21-fs (FWHM) long compressed pulses which are sent into a gas-filled hollow fiber, spectrally broadened and compressed through chirped mirrors, yielding pulse durations of about 7-fs FWHM [3,4]. Pulse measurements are then carried out in single-shot mode over a period of 10 s hence 10000 successive pulses are recorded. The measurement is then repeated with the hollow fiber evacuated. Without spectral broadening in the hollow fiber this allows for characterization of the CPA front-end only.

Average pulse durations are measured to be (20.63 ± 0.43) fs for the CPA front-end and (7.18 ± 0.06) fs from the hollow fiber compressor. Mean shot to shot pulse duration variations of 340 as (no gas) and 21 as (with gas) are measured. Figure 1 shows the full data set over a period of 1 s with the pulse intensity profile shown to the left and the corresponding spectral phase to the right. The grid size has been chosen in intervals of 20 ms to highlight the "long-term" drift components. Apart from the generation of sub-10-fs mJ-pulses, continuum generation processes are frequently used for measurement of the carrier-envelope phase in frequency metrology. Hollow-fiber continua have been shown to exhibit remarkable spectral coherence properties [2]. Our measurements allow for an alternative approach to quantifiying spectral coherence properties. This may also prove new insight in understanding the nature of spectral broadening mechanisms in different parts of the spectrum. Random noise - imparted on the phase structure - would destroy or weaken the correlation between neighboring "phase-points" (meaning the value of the spectral phase at one frequency vs. another frequency). Therefore, from a measurement of 10000 successive pulses, we investigated the correlation between each of the individual "phase-points". Such a correlation analysis removes the temporal dependence of the measurements - compared to a time series analysis -, and allows for an investigation to find structural dependencies. This is analogous to finding the principle modes of a system.

Figure 2 shows the results of this analysis: (a) shows the possible spectral phases for the case of a gas-filled capillary. (b) shows a simulated result assuming that the spectral phase is completely random. Comparing the two plots, it becomes clear that the processes in the capillary preserve the correlation between the "phase points".

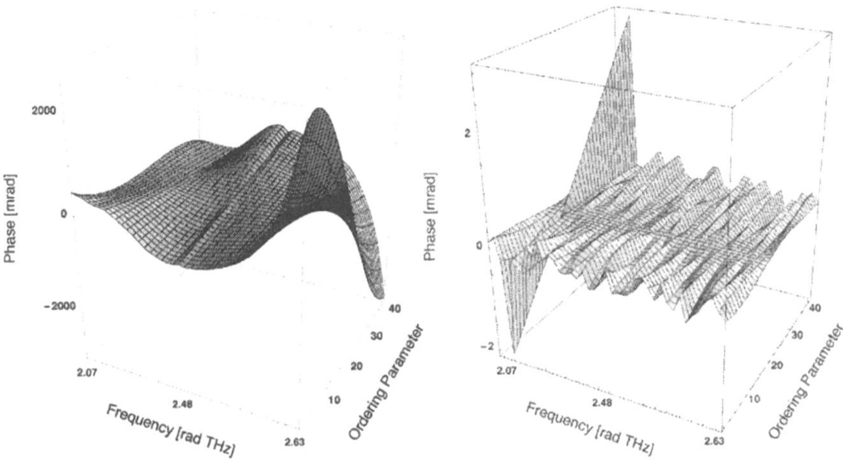

Fig. 1. Plots obtained from a correlation analysis of the individual "phase-points" of the spectral phase. The analysis filters out a coordinated movement of the spectral phase and rejects stochastic noise components. (a) shows these "principle" spectral phases as obtained from a measured data-set of 10000 successive laser pulses after a gas-filled capillary. For comparison, (b) shows the result of the same correlation analysis applied to random spectral phase noise. Comparing (a) and (b) clearly reveals a coordinated movement of the spectral phase indicative of a wide-ranging spectral coherence.

3. Results and Discussion

To summarize, we have for the very first time demonstrated true single-shot total pulse characterization at a 1-kHz rate with 3-cycle laser pulses. The system was used to launch a preliminary investigation into pulse-to-pulse fluctuation and phase preservation inside a gas-filled capillary.

Remarkably small (21 as) pulse-to-pulse fluctuations were measured. Furthermore, an analysis showed that the phase structure of an ingoing pulse is not randomized and that a correlation between the individual phase-points persists.

References

1 L. Gallmann, D. H. Sutter, N. Matuschek, G. Steinmeyer, U. Keller, C. Iaconis, and I. A. Walmsley, Opt. Lett. 24, 1314, 1999.

2 M. Bellini and T. W. Hansch, Opt. Lett. 25, 1049, 2000.

3 M. Nisoli, S. De. Silvestri, and O. Svelto, Appl Phys. Lett. 68, 2793, 1996.

4 M. Nisoli, S. De. Silvestri, O. Svelto, R. Szipocs, K. Ferencz, C. Spielmann, S. Sartania, and F. Krausz, Opt. Lett. 22, 522, 1997.

Time-domain interferometry for direct electric-field reconstruction by use of an acousto-optic programmable dispersive filter

T. Oksenhendler[1], A. Monmayrant[2], R. Herzog[1], D.Kaplan[1], P.Tournois[1] and M. Joffre[2]
[1] Fastlite, XTEC Bâtiment 404, Ecole Polytechnique, 91128 Palaiseau, France
[2] Laboratoire d'Optique et Biosciences, CNRS UMR7645-INSERM U451-Ecole Polytechnique - ENSTA, 91128 Palaiseau France
E-mail: manuel.joffre@polytechnique.fr

Abstract. We demonstrate a new version of SPIDER based on time-domain interferometry. The experiment is implemented using a remarkably simple setup consisting of an acousto-optic programmable dispersive filter (AOPDF) and a two-photon detector.

We demonstrate a new version of spectral phase interferometry for direct electric-field reconstruction (SPIDER) [1]. This new phase-measurement method, hereafter referred to as time-domain SPIDER, is in fact a variant of HOT SPIDER [2] where frequency-domain interferometry is replaced here with time-domain interferometry.

As shown in Fig. 1, time-domain SPIDER consists of the detection using a two-photon detector of the superposition $E_0(t-\tau)+E(t)\pm E_{QM}(t)$ between

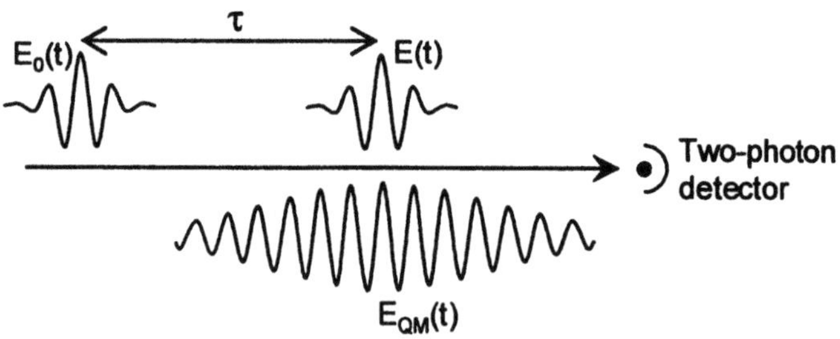

Fig. 1 : Principle of time-domain SPIDER

a reference pulse, the unknown pulse and a quasi-monochromatic pulse $\pm E_{QM}(t)$. The $\pm$ sign means that the phase of this pulse can be shifted by π. The reference pulse, whose exact phase is unimportant, is usually a replica of the unknown pulse itself. In time-domain SPIDER, the time delay τ between the reference pulse and the superposition of the other two pulses must be scanned in order to record the interferometric second-order correlation function :

$$S_{\pm}(\tau) = \int_{-\infty}^{+\infty} \left(E_0(t-\tau) + E(t) \pm E_{QM}(t)\right)^4 dt \tag{1}$$

The Fourier transform of this quantity exhibits peaks at frequencies 0, $\pm\omega_0$ and $\pm 2\omega_0$, where ω_0 is the pulse center frequency. By selecting in $S_{\pm}(\tau)$ the term $S_{2\omega,\pm}(\tau)$ corresponding to the peak at $2\omega_0$, we extract the quantity

$$g(\tau) = S_{+}(\tau) - S_{-}(\tau) \propto \int_{-\infty}^{+\infty} \mathcal{E}_0^{*2}(t-\tau)\left(\mathcal{E}(t)\mathcal{E}_{QM}(t)\right)dt \tag{2}$$

which is the correlation function between the frequency-doubled reference pulse and the sum-frequency mixing between the quasi-monochromatic field and the unknown field. In case the quasi-monochromatic pulse is much longer than $\mathcal{E}(t)$, i.e. sufficiently narrow in frequency domain, the product $\mathcal{E}(t)\mathcal{E}_{QM}(t)$ can be approximated with $\mathcal{E}(t)\exp(-i\omega_1 t)$ and its spectral phase reads $\varphi(\omega-\omega_1)$, where $\varphi(\omega)$ is the spectral phase of the unknown field $\mathcal{E}(\omega)$ and ω_1 is the center frequency of the quasi-monochromatic field. Similarly with HOT SPIDER, a sequence of two measurements with different values ω_1 and ω_2 of the frequency of the quasi-monochromatic field yield by subtraction the quantity $\varphi(\omega-\omega_2)-\varphi(\omega-\omega_1) \approx (\omega_2-\omega_1)d\varphi/d\omega$.

One advantage of time-domain SPIDER is that it can be implemented in a remarkably compact experimental setup consisting of a pulse shaper, here a Dazzler™ device, directly followed by the two-photon detector. The Dazzler™ is an AOPDF pulse shaper which allows complete amplitude and phase control of the transmitted pulse [3,4]. Here the device is programmed in order to generate the sequence of three pulses shown in Fig. 1 and the data acquired on the two-photon detector are Fourier transformed and then treated numerically similarly with HOT SPIDER. Fig. 2 shows the spectral amplitude and phase retrieved for the pulses generated by a 1-kHz Ti:S amplifier after transmission through the Dazzler™, which was programmed in order to compensate for its own linear dispersion.

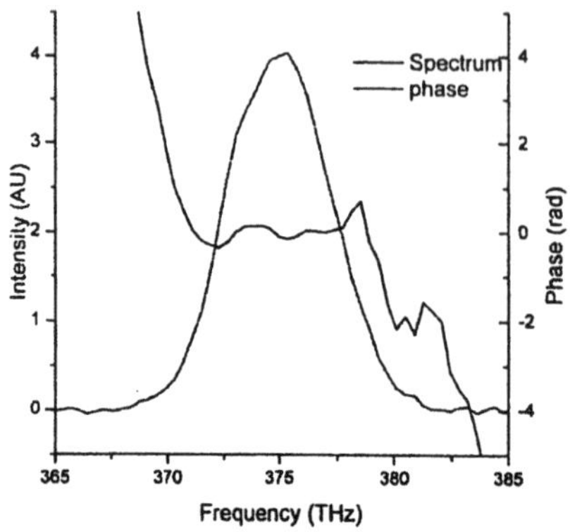

Fig. 2 : Spectral amplitude and phase retrieved from a time-domain SPIDER measurement.

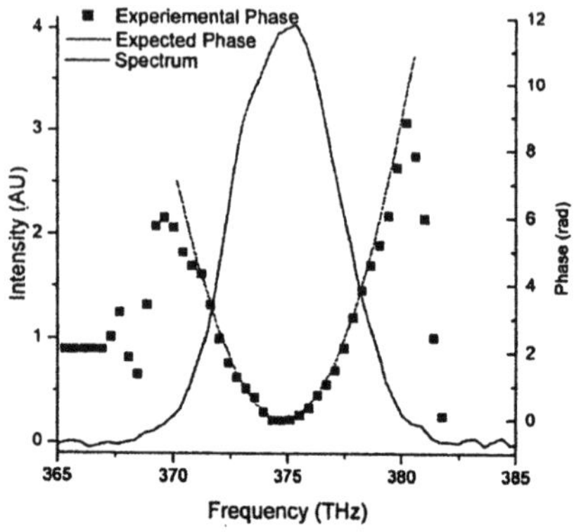

Fig.3. Difference in spectral phase between two time-domain SPIDER measurements differing by a known amount of second-order phase.

Fig.3 shows the phase difference between the phase shown on fig.2 and the phase measured after adding a known amount of second order spectral phase via the Dazzler™. The good agreement between the measured phase and the expected value demonstrates the validity of the technique.

To summarize we have demonstrated a new version of HOT SPIDER based on time-domain interferometry. This technique presents the general advantages of Fourier-transform spectroscopy and makes the use of a grating spectrometer unnecessary. Although we have implemented the method using a pulse shaper, it is also possible to use a modified Michelson interferometer with an additional beam providing a chirped pulse instead of the quasi-monochromatic pulse used here.

References

1 C. Iaconis and I.A. Walmsley, "Spectral phase interferometry for direct electric-field reconstruction of ultrashort optical pulses", Opt. Lett. **23**, 792 (1998).
2 C. Dorrer, P. Londero and I.A. Walmsley., "Homodyne detection in spectral phase interferometry for direct electric-field reconstruction", Opt. Lett. **26**, 1510 (2001).
3 F. Verluise, V. Laude, Z. Cheng, Ch. Spielmann and P. Tournois, "Amplitude and phase control of ultrashort pulses by use of an acousto-optic programmable dispersive filter: pulse compression and shaping," Opt. Lett. **25**, 575 (2000).
4 P.Tournois and D. Kaplan, "A simple theory of the acousto-optic programmable dispersive filter used for femtosecond laser pulse shaping" , XIII Conference on Ultrafast Phenomena, 2002 (same issue).

Amplitude and Phase measurements of femtosecond pulses shaped using spectral hole burning in free-base naphthalocyanine-doped films

S. Fraigne[1], J.-P. Galaup[1], J.-L. Le Gouet[1], L. Canioni[2], B. Bousquet[2], M. Joffre[3] and J.-P. Likforman[4]

[1]Laboratoire Aimé Cotton, CNRS UPR 3321, Université Paris-Sud, Bât. 505, 91405 Orsay cedex, France

[2]Centre de Physique Moléculaire Optique et Hertzienne, Université Bordeaux I, CNRS UMR 5798, 33405 Talence Cedex, France

[3]Laboratoire d'Optique et Biosciences, CNRS UMR 7645 - INSERM U451 – Ecole Polytechnique – ENSTA, Ecole Polytechnique 91128 Palaiseau Cedex, France

[4]Laboratoire d'Optique Appliquée, CNRS UMR 7639 - INSERM U451 – Ecole Polytechnique – ENSTA, Ecole Polytechnique 91128 Palaiseau Cedex, France
Present address: Institut de Physique et Chimie des Matériaux de Strasbourg, Groupe d'Optique Non-Linéaire et d'Optoélectronique, CNRS UMR 7504, Université Louis Pasteur, 23, rue du Loess, B.P. 20CR, 67037 Strasbourg Cedex
Email: jean-pierre.likforman@ipcms.u-strasbg.fr

Abstract. We performed pulse-shaping and time-reversal experiments using spectral holography based on persistent spectral hole-burning in free-base naphthalocyanine-doped films at 5K. We completely characterized the shaped pulses. The homogeneous linewidth was measured from a photon echo experiment.

Spatio-temporal holography using persistent spectral hole-burning in dye molecules embedded in plastic films is promising for applications such as femtosecond pulse shaping, data storage or phase conjugation. In addition to the common 2 spatial dimensions of classical holograms, each point in the sample plane can be used to store spectral informations such as the shape of a femtosecond pulse. On each point of the sample, spectral interferences between a reference pulse and an object pulse (for example chirped) are recorded via spectral hole-burning in the absorption band of the dye molecule. As early as 1983, photon echo and time-reversal experiments using persistent spectral hole-burning were performed in porphyrazine styrol solutions at 630 nm [1]. Then several developments, like the reconstruction of the polarization state of object pulses [2], time and space holography [3,4], image holograms using a single pair of femtosecond pulses [5] were demonstrated in different photosensitive media.

Few spectral holography experiments are done in the frequency domain of Titanium:Sapphire (Ti:S) lasers [5,6]. And, to our knowledge, none demonstrated

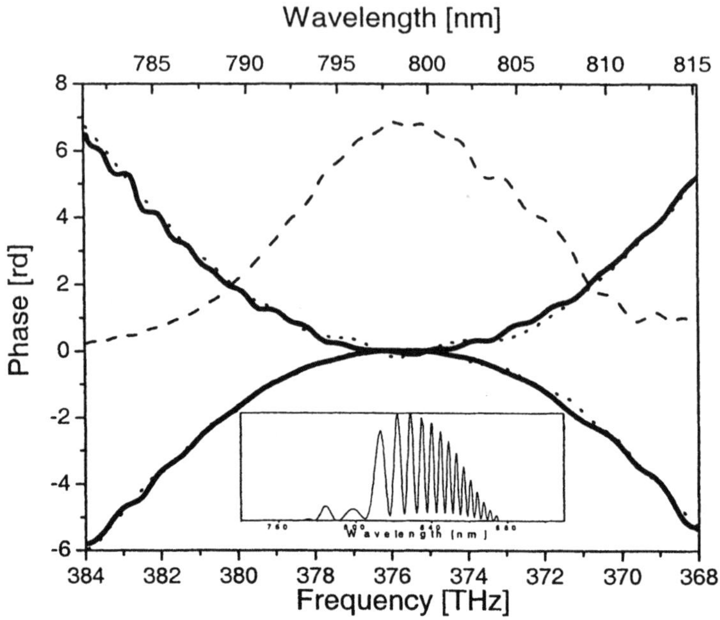

Fig. Solid lines: spectral phase of the diffracted pulse for two delays $+\tau$ and $-\tau$ between the writing pulses. Thin dotted lines : spectral phase difference between the 2 writing pulses for the two cases $+\tau$ and $-\tau$. We also show as a thin dashed line the diffracted pulse spectral amplitude. Inset: spectral interference fringes between the unchirped and chirped writing pulses at time delay equal 0 at 800nm. The chirp is formed by propagation through a SF58 glass plate of thickness 1.7 cm

the control of the phase of femtosecond pulses shaped using spectral holography based on persistent spectral hole-burning.

In this work we use a technique of spectroscopy adapted for measuring the amplitude and phase of photon echo signals [7] produced by diffraction of a fs pulse on a spectral hologram. We also improved the technique in terms of spectral resolution in order to measure photon echoes delayed by a few tens of picoseconds. Our study is focused on measuring the coherence time of the sample using a photon echo experiment in the photochemically accumulated regime and on demonstrating the pulse-shaping and time-reversal potentialities of our photo-sensitive material.

Spectral holograms are formed through persistent spectral hole burning using a sequence of 2 pump pulses separated by a time delay, in a collinear geometry. The sample is a free base naphtalocyanine embedded in polyvinylbutyral

(H_2NPc/PVB). The absorption peaks at 783 nm and is 20 nm large (FWHM). Our laser source is a 15-fs, 100MHz, Ti:S oscillator.

We performed a simple experiment that demonstrates the ability of H_2NPc/PVB for pulse shaping and time reversal. The experiment consists in introducing a chirp on one of the two pump pulses and engraving, in the absorption band, the spectral interferences between the chirped and unchirped pulses. Then the chirped (unchirped) pulse can be recovered by diffraction on the spectral hologram of the unchirped (chirped) pulse.

To change the sign of the spectral phase that we want to engrave in the sample, we simply change the time ordering between the two pulses, thus achieving time reversal. The amplitude and phase of the diffracted pulse are shown on the figure.

The next step for this kind of experiment is to write spectral holograms of arbitrary shape using a tunable cw source.

In order to be competitive with other pulse-shaping techniques it will be necessary to find materials that preserve a small homogeneous linewidth at higher temperature and present a large inhomogeneous bandwidth to shape 10-fs pulses

References

1 A. Rebane, R. K. Kaarli, P. M. Saari, A. Anijalg, K. Timpmann, Opt. Commun. **47**, 173

2 R. K. Kaarli, P. M. Saari, H. R. Sonajalg, Opt. Commun. **65**, 170 (1988).

3 P. Saari, R. Kaarli, A. Rebane, J. Opt. Soc. Am B **3**, 527 (1986)

4 A. Rebane, J. Aaviksoo, J. Kuhl, Appl. Phys. Lett. **54**, 93 (1989).

5 A. Rebane, M. Drobizhev, C; Sigel, Opt. Lett. **25**, 1633 (2000).

6 A. Rebane Opt. Commun. **147**, 429 (1998).

7 M. F. Emde, W. P. de Boeij, M. S. Pshenichnikov, D. A. Wiersma, Opt. Lett. **22**,1338 (1997)

Polarizations of Raman sidebands generated in impulsively stimulated rotational Raman scattering

Hiroyuki Kawano[1], Akira Suda[1], Kenichi Ishikawa[2], and Katsumi Midorikawa[1]

[1] Laser Technology Laboratory, RIKEN, 2-1 Hirosawa, Wako, Saitama, 351-0198 Japan
E-mail: kawano@riken.go.jp

[2] Department of Quantum Engineering & Systems Science School of Engineering, The University of Tokyo, 7-3-1 Hongo, Bunkyo-ku, Tokyo 113-8656, Japan

Abstract. Polarizations of high-order rotational Raman radiations are investigated via impulsively stimulated Raman scattering in hydrogen gas. Those polarizations are strongly dependent on the ellipticity of the polarization of a pump laser pulse.

1. Introduction

It is well-known that efficient stimulated Raman scattering is not easily carried out in the femtosecond regime because of the competitions against the other nonlinear optical phenomena, e.g., self-phase modulation or harmonic generation, etc. Recently, impulsively stimulated Raman scattering (ISRS) has been reported as a solution for this problem [1]. Stimulated rotational Raman radiation from hydrogen gas can become a powerful tool as a femtosecond multicolor probe laser source for spectroscopy or the other applications due to the small frequency separation of 587 cm^{-1} (v=0, J=1 to v=0, J=3) and broad spectral width of the sidebands . In addition, hydrogen gas is a transparent medium even in UV region, which is advantageous for ultrashort pulse generation. However, there have been few reports about the polarization of high-order rotational Raman sidebands via ISRS although its information is quite important for any applications. We, therefore, investigate the mechanism of the generation of rotational Raman radiation by measuring their polarizations and demonstrate the control and optimization for this technique.

2. Experimental Methods

We focused a pump pulse (785 nm, 70 fs, 200 μJ) and a probe pulse (393 nm, 60 fs, 5 μJ) to 2-bar hydrogen gas in a hollow fiber (126-μm inner diameter, 75-cm long) with a 570-fs time delay to generate rotational Raman sidebands. The polarization of the pump pulse was controlled by a quarter-wave plate from linear polarization (ellipticity = 0) to circular polarization (ellipticity = 1). The probe pulse was linearly polarized to obtain the largest number of Raman components [2]. The energy of the probe pulse was controlled to regulate the number of

Raman components. Polarizations of Raman sidebands were measured by use of a Fresnel rhomb, an analyzer, and a spectrometer.

3. Results and Discussion

Figure 1 shows the spectra of Raman sidebands from a probe pulse. When the pump pulse polarization is linear, no Raman components are observed. On the other hand, when the pump pulse polarization has some amount of ellipticity, multiple Raman sidebands immediately appear. Finally, when the pump pulse is circularly polarized, only 1st Stokes and 1st anti-Stokes lines can be seen and no other high-order lines.

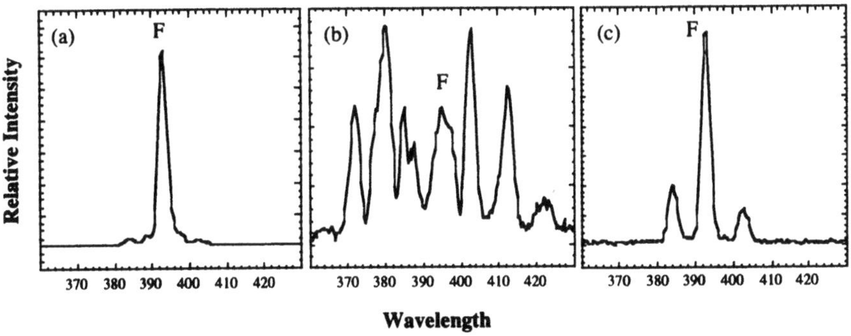

Fig. 1. Spectra of Raman sidebands from a probe pulse. The ellipticities of the pump pulse polarization is (a) 0.0, (b) 0.8, and (c) 1.0, respectively. Symbol F means a fundamental line i.e., the probe pulse.

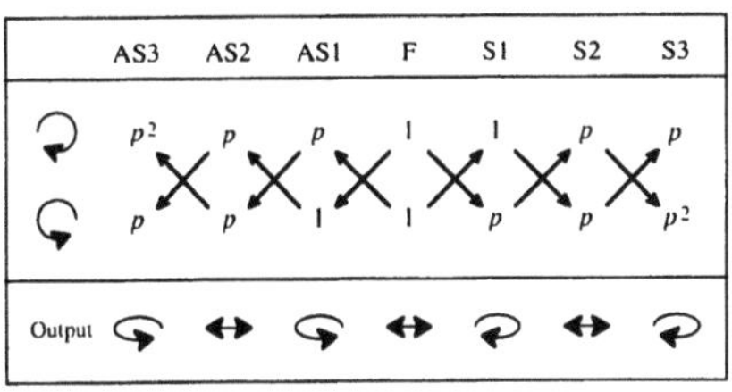

Fig. 2. Diagram of the synthesis of right- and left-circular polarizations with respect to rotational Raman components obtained by ISRS. Symbol p is the population rate of right- and left-rotated molecules driven by a pump pulse. ($0 \leq p \leq 1$)

These phenomena can be explained by the consideration as in Fig. 2. In ISRS, a probe pulse scatters Raman sidebands by the interaction with the Raman coherence *ab initio* induced by a pump pulse. When the pump pulse polarization is linear, it cannot induce rotational Raman coherence effectively due to parametric gain suppression, resulting in no Raman sideband generation from a probe pulse. When the pump pulse polarization has some amount of ellipticity, right- and left-circularly polarized Raman coherence is induced at the population rate of p. A linearly polarized probe pulse can be regarded as a synthesized light of right- and left-circularly polarized light. Each component interacts with proper molecular coherence in satisfying the angular momentum conservation law. In that case, the ellipticities of the polarizations of each-order Raman components obtained are also determined by the synthesis of right- and left-circular polarization components of which the amount rate is approximately estimated by p. Therefore, even order components should be linearly polarized, meanwhile odd order ones should be elliptically polarized as shown in Fig. 2. When the pump pulse polarization is circular, p is 0. Then, the energy flow from the probe pulse to the sidebands is limited up to 1st order Stokes and anti-Stokes radiations.

By changing the ellipticity of the pump pulse polarization, we measured the polarizations of rotational Raman sidebands from the probe pulse as shown in Fig. 3. As predicted in Fig. 2, when the pump pulse polarization is close to be circular, one can obviously see the significant difference between even and odd order components. On the other hand, when the pump pulse is almost (but not completely) linearly polarized, the Raman components are also almost linearly polarized. This is because p is quite close to 1.

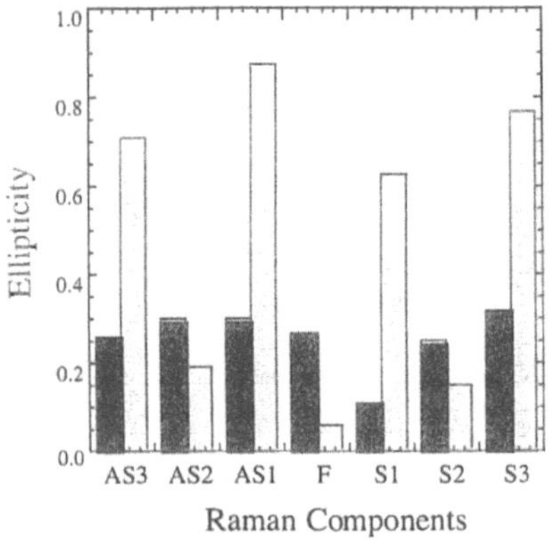

Fig. 3. Polarizations of the Raman components measured. The ellipticity of the pump pulse polarization is 0.8 (gray) and 0.003 (black), respectively.

4. Conclusions

We investigated the polarization of rotational Raman sidebands in ISRS and revealed its mechanism considering angular momentum conservation. Especially, we found the condition to achieve almost linearly polarized rotational Raman sidebands. It is quite useful to use them in many scientific and industrial applications.

Acknowledgements. We would like to thank Dr. Ushida and Dr. Wada for fruitful discussion.

References

1 M. Wittmann, A. Nazarkin, and G. Korn, Phys. Rev. Lett. **84**, 5508, 2000.
2 H. Kawano, A. Suda, and K. Midorikawa, Appl. Phys. Lett. **80**, 894 2002.

Nonlinear effects in microstructured fibers

Fiorenzo G. Omenetto[1], Anatoly Efimov[1], A. J. Taylor[1], J. C. Knight[2],

W. Wadsworth[2], P. St. J. Russell[2], J. Arriaga[3]

[1] Material Science and Technology Division, MST-10, Los Alamos National Laboratory, Los Alamos, NM 87545
E-mail: omenetto@lanl.gov
[2] Dept. of Physics, University of Bath, Bath UK, BA2 7AY, UK
[3] Institudo de Fisica (U.A.P), Apartado Postal J-48, 72570, Puebla, Mexico

Abstract. A variety of nonlinear effects are observed coupling low average power femtosecond pulses (up to 25 mW) in a short length (Z=20 cm) of microstructured fiber. We describe fundamental (N=1) and higher order (N=2) soliton formation, large self-frequency shift, visible harmonic generation and polarization dependent optical bistability.

1. Introduction

Control of nonlinear processes in optical fibers can provide the foundation for future all-optical signal processing applications that would impact a number of basic and applied fields. An ideal setting to study these phenomena is provided by microstructured optical fibers (or photonic crystal fibers –PCFs) [1]. The enhanced nonlinear effects in these structures have attracted considerable attention and have been the focus of intense research activity spanning a variety of applications [2]. We report here on a variety of effects which are observed experimentally by propagating 150 fs λ=1550 nm pulses in a short segment of microstructured fiber.

2. Results and Discussion

By operating at low average powers (up to 25 mW) and varying the input pulse polarization we can observe fundamental (N=1) and higher order (N=2) soliton formation in combination with a large self-frequency shift, visible harmonic generation and optical bistability in dependence of the input polarization state. Fundamental (N=1) and self-frequency shifted N=2 soliton formation is verified unambiguously by retrieving the temporal phase structure of the pulses after propagation, as opposed to indirect conjecture based on the combined spectral and temporal pulse features reported thus far in the literature.
For these measurements, we use a previously described experimental arrangement [3]. The pulses at λ=1550 nm are generated by an optical parametric oscillator (OPO) at a repetition rate of 80 MHz with average powers up to 80 mW and are coupled into a 25 cm segment of PC fiber. The latter is a microstructured fiber

with a silica core of 2.5 microns in diameter, suspended in air by a web of sub-micron silica strands with a cladding diameter of 90 microns.

It was previously noted [4] during studies of femtosecond pulse phase dynamics in conventional optical fibers, that the temporal phase function of the pulse is flat across the pulse when a soliton is formed, indicating the temporal coherence of the pulse. As the input power is increased, the pulse shortens, and the reconstruction reveals a flat temporal phase across the pulse for an average power of 6 mW at the fundamental wavelength. As the power is further increased, the phase varies and the pulse self-frequency shifts towards longer wavelengths. These shifts are particularly enhanced in microstructured and tapered fibers and large shifts have been reported in recent work [5]. A flat temporal phase feature appears again for pulse power through the fiber of 25 mW. The frequency shift in this short piece of PCF over this power range is in excess of 100 nm, and after the N=1 soliton is formed, the pulsewidths throughout the entire shifted wavelength range stay in the hundreds of femtoseconds range.

By rotating the input pulse polarization state, visible radiation can be generated [6]. An intensity transmission analysis as a function of input polarization reveals two principal polarization states α and β orthogonal to one another. When the input pulses are polarized along one of these two directions, one of two visible modes of different colors is detected at the output. These generated visible frequencies are higher-order modes exhibiting an eight-lobed structure and centered at λ_1=514 nm and λ_2=533 nm (fig. 2). It has been previously noted [6] that the observed components λ_1 and λ_2 bear a relation to the fundamental such that $\lambda_{FUND} \sim 3\lambda_1$ and $\lambda_{F,Shift} \sim 3\lambda_2$, (or $3\omega_{FUND} \sim \omega_1$ and $3\omega_{F,Shift} \sim \omega_2$ and where is the $\omega_{F,Shift}$ Raman self-frequency shifted portion of the fundamental). The mechanism for this process was described as temporal splitting of the fundamental pulse followed by Raman-self frequency shift of the split pulse, with subsequent third-harmonic frequency conversion of both pulses as the possible mechanism for the generation of the visible components.

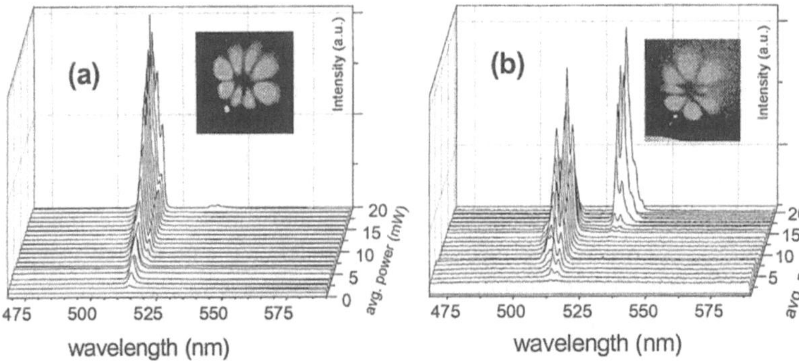

Fig. 1 - Detected spectra of the visible radiation at the output of the photonic crystal fiber as a function of power for both polarization states α and β. The component at 533 nm appears only in one polarization state. The inset shows the output mode structure of the visible radiation

A power dependent spectral analysis is carried out on the fundamental and on the visible harmonics for the two different polarizations. A comparison between the spectral features of the fundamental pulse polarized along the directions α and β show different spectral dynamics as a function of power but a fundamentally similar behavior: in both cases the spectrum shifts as expected toward longer wavelengths and the magnitude of the shift (~50 nm at P=25 mW) appears to be the same for both input polarization orientations (independent of the initial polarization). Since the SRS process is an intensity (i.e. pulsewidth) dependent process, this suggests that the dispersion properties are similar for both polarization states. The results for the visible components indicate a clear difference depending on the input polarization state. In the direction α (fig 1a) a spectral feature centered around λ=514 nm is detected for pulses of average power of 4 mW at 1550 nm. As the power of the fundamental is increased, more light is converted to the visible (with an estimated conversion efficiency of 0.2 % for a 25 mW pump) but only a feature at λ_1=514 nm is observed. The relationship of this component to the fundamental wavelength is consistent with the generation of third harmonic of the fundamental yet no third harmonic from the self-shifted component is observed. When the input pulse polarization is rotated along the direction β (fig. 1b), however, a sharp feature centered at λ_2=534 nm appears for pulses of average powers of 16 mW in addition to the λ_1 component that subsequently vanishes as the power if the fundamental is further increased. Whereas for both polarizations the fundamental shifts to 1600 nm (and beyond for higher input powers), only for the β polarization case the shifted fundamental is converted to third harmonic radiation. It must be noted that no frequency components that could be attributed to mixing processes are detected for the wavelength interval covered by the fundamental and its shifted components.
These results show that polarization selectivity combined with Raman self-frequency shifting and third harmonic generation provides an avenue to control the generation of specific visible harmonics. This approach can provide an avenue for all-fiber signal control and ultrafast optical switching based on nonlinear phenomena in microstructured fibers.

References

1. J. C. Knight et al. , Science, **282**, 1476 (1998), J.Ranka, R.Windeler, A.Stentz, Opt.Lett., **25**, 25 (2000).
2. Hecht, J., Laser Focus World **37**, 207 (2001).
3. F. G. Omenetto, J. W. Nicholson and A. J. Taylor, Opt. Lett. , **24**, 1780 (1999).
4. F. G. Omenetto, B. Luce, D. Yarotsky, and A. J. Taylor, Opt. Lett, **24**, 1392 (1999).
5. X. Liu, C. Xu, W. Knox, J. K. Chandalia, B. J. Eggelton, S. G. Kosinski and R. Windeler, Opt. Lett **26**, 358 (2001).
6. F. G. Omenetto, M. D. Moores, A. J. Taylor, J. C. Knight, W. Wadsworth, P. St. J. Russell, J. Arriaga, Opt. Lett. **26**, 1160 (2001).

Adaptive pulse shaping of amplified femtosecond pulse laser in phase and amplitude by directly referring to frequency resolved optical gating traces

Kimihisa Ohno, Takasumi Tanabe, and Fumihiko Kannari

Department of Electronics and Electrical Engineering, Faculty of Science and Technology, Keio University, 3-14-1 Hiyoshi, Kohoku-ku, Yokohama, Japan

Abstract. We experimentally demonstrated accurate pulse shaping for ultrashort pulses through a regenerative amplifier by adaptively control both phase and amplitude of spectral components of seed pulses with an acousto-optic programmable dispersion filter (AOPDF).

1. Introduction

Ultrashort laser pulses have been attracting increasingly broad interest in scientific applications such as measurement and control of ultrafast phenomena in atoms and molecules [1], and high-order harmonic generation [2]. Shaping of ultrashort laser pulses can provide additional functions for these applications. Since Weiner and co-workers demonstrated programmable pulse shaping by employing a liquid-crystal spatial light modulator (SLM) in a 4-f Fourier transform optical layout [3], this technique has been actively investigated, especially to compensate for the higher-order dispersion in chirped pulse amplifiers (CPA) so that a transform limited short pulse is obtained. Very recently, Verluise and co-workers experimentally demonstrated the novel pulse shaping with an AO programmable dispersion filter (AOPDF) [4], [5]. This device can shape both the phase and amplitude of ultrashort pulses by collinearly interacting an optical pulse with a programmed acoustic wave in a nonlinear crystal without a 4-f optical layout. They placed the AOPDF before a multi-pass amplifier and compensated for the nonlinear gain narrowing of CPA by suppressing the center part of spectrum of an input laser pulse. In principle, when a desired pulse is specified in both amplitude and phase, the explicit shaping mask is obtainable from the Fourier transform of the specified pulse and the initial pulse spectrum. When the pulse shaping is performed before amplification to preserve the optical power, however, the design of shaping mask cannot be obtained in such a straightforward manner. We have to take into account the complex transfer function of the CPA to design the spectral shaping masks. Precise description of the actual transfer functions including all the optical nonlinearities and the higher-order dispersion is not easy. For this case, we can achieve pulse shaping based on an optimization algorithm using closed-loop control by monitoring the complex profile of amplified pulses. In this paper, we propose and experimentally demonstrate a novel adaptive pulse shaping in phase and amplitude with an AOPDF by referring two-dimensional pattern of frequency resolved optical gating (FROG) measurement, which describes an entire profile of the amplitude and phase of ultrashort pulses, without any iterative calculations for pulse profile reconstruction.

2. Experiments and Results

The optical output $E_{out}(t)$ of the AOPDF is shown to be proportional to the convolution of the optical input $E_{in}(t)$ and of the scaled acoustic signal [4], [5]. In the rest of this paper, we take an example of producing a twin-pulse separated by a specified time. When simply combining two acoustic signals, which are programmed so that an optical pulse interacts with each acoustic wave at different position in the crystal, the optical pulse is shaped to two separated pulses. In order to cause the delay between the two acoustic waves, certain amount of first-order dispersion is added to the Fourier phase of one of the acoustic waves. Then, two acoustic waves are superposed in time domain. The acoustic wave designed in this manner corresponds to the spectral modulation in both amplitude and phase. In principle, with such an explicitly designed amplitude-and-phase filter, optical pulse shaping free from any ambiguity should be obtained when all the residual dispersion was pre-compensated. However, since a CPA exhibits nonlinear transfer function, such a straightforward shaping is not useful. Adaptive adjustment of the pulse shape is still necessary.

We shape the ultrashort pulses delivered through an amplifier to two pulses separated by 150 fs with the peak intensity ratio of 1 to 1. The pulse width is 42 fs (FWHM) for both the pulses. The coefficients of Fourier phase up to the seventh-order are adaptively optimized with SA algorithm by directly referring the FROG traces. The FROG traces of the target pulse and the optimized pulse are shown in Fig. 1. The temporal profiles reconstructed from each FROG traces are shown in Fig. 2. The output waveform is shaped to two pulses with the specified peak intensity ratio, pulse interval and pulse width. Some deviations in the peak intensity ratio from the target reveal that the small difference in the FROG trace results in larger difference in the reconstructed pulse shape.

3. Conclusion

The accurate pulse shaping in both phase and amplitude was experimentally demonstrated for ultrashort pulses delivered through the regenerative amplifier with the AOPDF located before the pulse stretcher by directly referring FROG traces. The AOPDF can modulate both phase and amplitude of ultrashort puls spectrum without a 4-f optical layout. Therefore, an explicit shaping function determined with the Fourier transform fo the specified output pulse can be realized. However, nonlinear transfer function of the CPA and optical passes due to SPM and gain narrowing prevents from the straightforward shaping using the explicit shaping mask. Adaptive pulse shaping can automatically compensate for the reisdual dsipersion and gain narrowing, and thus flexible shaping is obtainable in the temporal window defined by the AO crystal length.

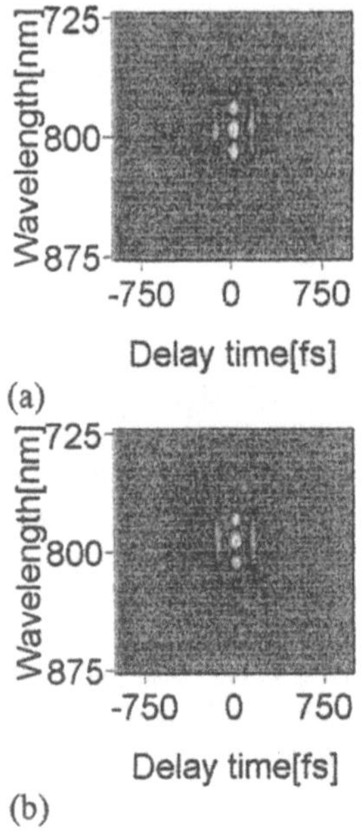

Fig. 1. FROG trace of (a) Target profile and (b) Optimized profile.

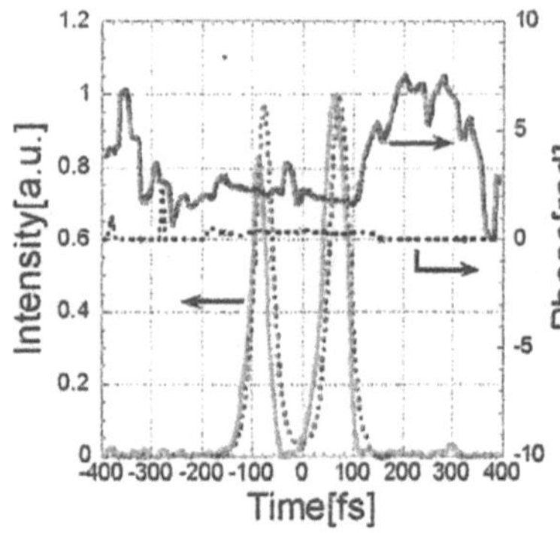

Fig. 2. Reconstructed temporal profile of target (dotted line) and optimized (solid line).

References

1 A. Anssion, T. Baumert, M. Bergt, T.Brixner, B. Kiefer, V. Swyfried, M Strehle, and G. Gerber, Science, 282, 919, 1998.
2 R. Bartles, S. Backus, E. Zeek, L. Misoguti, G. Vdovin, I. P. Christov, M. M. Murnane, and H. C. Kapteyn, Nature, 406, 164, 2000.
3 A. M. Weiner, Prog. Quant. Electr., 19, 161, 1995.
4 P. Tournois, Opt. Commun., 140, 245, 1997.
5 F. Verluise, V. Laude, Z. Cheng, Ch. Spielmann, and P. Tournois, Opt. Lett. 25, 575, 2000.

Joint Quantum Measurement Using Fourier-Transform Spectral Interferometry

M. Beck(1), C. Dorrer(2) and I.A. Walmsley(3)*

The Institute of Optics, University of Rochester, Rochester, New York 14627, USA
Permanent addresses:
(1) Dept. of Physics, Whitman College, Walla Walla, WA 99362, USA
(2) Bell Laboratories, Lucent Technologies, Holmdel, NJ 07733, USA
(3) Clarendon Laboratory, University of Oxford, Parks Rd., Oxford OX1 3PU, UK
*E-mail: walmsley@physics.ox.ac.uk

Abstract: We have measured the joint Q-function of a highly multimode field using unbalanced spectral heterodyne detection with a charge-coupled device array. We reconstruct the statistics of about 25 adjacent temporal modes each of about 100 fs duration.

1. Introduction

A complete description of the quantum state of a multimode field requires simultaneous measurement of the joint statistics for all modes. These yield information about the correlations between modes that cannot be obtained from single-mode measurements. The best joint measurements using balanced homodyne detection to date have determined some properties of two-mode fields.[1-3] It is difficult to extend the techniques used in these experiments to fields with many modes. No measurements of the full joint quantum state of even a two-mode system have been reported.

Theoretical approaches to multimode quantum state measurement using array detectors have usually adopted balanced detection to reduce the classical noise. Experimentally it is quite hard to properly register two detector arrays to subtract the LO noise. Here we show that a properly configured unbalanced array configuration still eliminates the classical noise of the local oscillator. We show that the method provides detection at the shot-noise limit (SNL), is sensitive to modes containing just a few photons and, in the case of temporal modes, should provide high time resolution. [4]

2. Homodyne detection

We use the technique of Fourier transform spectral interferometry to achieve the standard quantum limit in unbalanced array detection by encoding the relevant information about the signal pulse using a carrier modulation in the spectral domain.[5,6]The set up is shown schematically in Fig. 1. A short local oscillator pulse (LO) is mixed at a beamsplitter with the unknown quantum test pulse (S). S can be considered a coherent superposition of several replicas of the test pulse with appropriate amplitudes and phases. Both test and LO pulses are directed through a spectrometer (a low-loss prism in the diagram, a grating in the actual experiment), and are incident on a high quantum efficiency CCD array (Q.E. of 85% at 800 nm). The spectral intensity $\tilde{I}(\omega)$ is the sum of the LO pulse $\tilde{E}_{LO}$ and the signal pulse $\tilde{E}_S$. The temporal convolution $f(t) = E^*_{LO}(-t) \otimes E_S(t)$ between the LO and signal fields may be extracted from this interferogram. The second order classical LO noise is completely eliminated in our scheme by numerical filtering in the transform space.

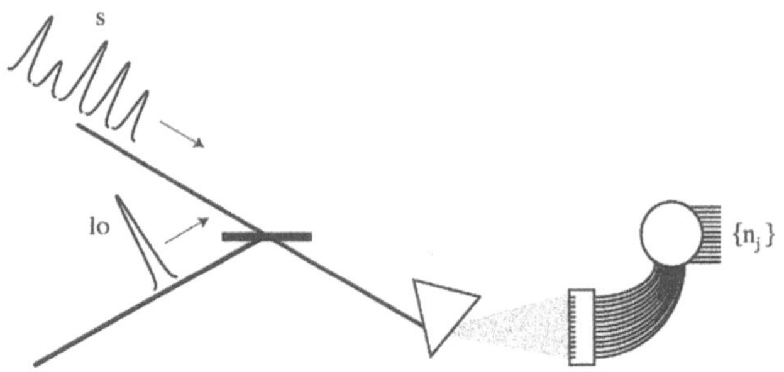

Fig. 1. Quantum Fourier-Transform Spectral Interferometry apparatus for ultrafast multimode quantum state measurement.

In our experiment the mode statistics are expressed in terms of the joint Q-function of the modes. The Q-function is a quantum mechanical phase-space, quasi-probability distribution that completely specifies the state. The operator that we measure in our experiments corresponds to the Fourier transform of $\hat{n}_j$, the number of photons incident on the j'th pixel,[7]

$$\hat{K}_p \equiv \sum_j \exp[-i2\pi pj/N]\hat{n}_j . \qquad (1)$$

Each run of the apparatus returns a set of $(N\text{-}M)/2$ complex numbers K_p; whose real and imaginary parts correspond to the realizations of the quadrature amplitudes of the signal modes (x_p, y_p). A large number of runs produces a probability distribution that approximates the Q-distribution for the multimode field quadratures $Q(x_p, y_p)$.[8]

3. Results

It is difficult to display these multivariable distributions graphically, so here we display correlations between the modes in terms of the joint distribution of the x-quadratures of the different modes. Figure 1 shows correlations between the x-quadratures of the modes in terms of joint distributions $Q'(x_p, x_{p'})$. Figure 1(a)

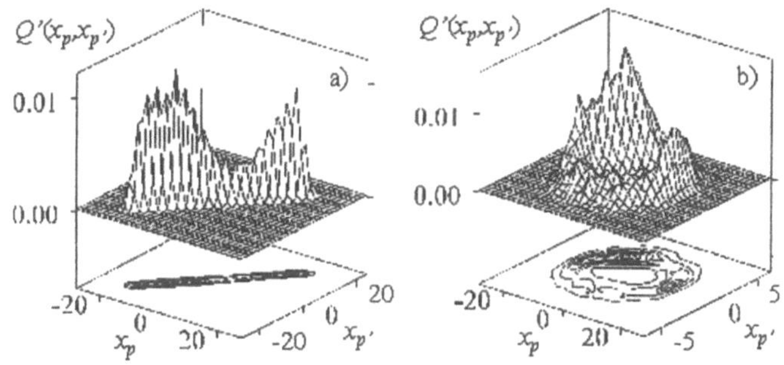

Figure 2. The measured joint distributions $Q'(x_p, x_{p'})$ for two different pairs of temporal modes of a chirped signal pulse. The mode delay times in a) are t=3.457 ps and t'=3.576 ps, while in b) they are t=3.457 ps and t'=3.785 ps.

shows correlations between two modes that are both near the peak of the pulse. This joint distribution indicates strong positive correlations between the x-quadratures, as expected for two modes whose phases are both random, but nearly the same. Fig. 2(b) shows correlations between two modes whose phase difference is nearly π/2, and this produces a joint distribution that is elliptical. Again, such a joint distribution is expected for two modes with random phases, but whose average phase difference is π/2.

We have demonstrated a new technique for performing joint quantum measurements of optical fields of over 500 modes with temporal resolution of several tens of femtoseconds. The measurements achieve the shot-noise limit, can detect modes containing only a few photons, and can in principle be extended to higher temporal resolution. We anticipate this method will be useful for understanding quantum effects in ultrafast nonlinear optics, such as squeezing in microstructure fibers, as well as in applications requiring large number of modes containing possibly nonclassical states, such as quantum information processing.

References

1 D.F. McAlister and M.G. Raymer, Phys. Rev. A **55**, R1609 (1997).
2 M. Vasilyev, S.-K. Choi, P. Kumar, and G. M. D'Ariano, Phys Rev. Lett. **84**, 2354 (2000).
3 A.G. White, D.F.V. James, P.H. Eberhard, and P.G. Kwiat, Phys. Rev. Lett. **83**, 3103, (1999).
4 M. Beck, C. Dorrer, and I. A. Walmsley, Phys. Rev. Lett. **87**, 253601 (2001).
5 L. Lepetit, G. Chériaux, and M. Joffre, J. Opt. Soc. Am. B **12**, 2467 (1995).
6 C. Dorrer, N. Belabas, J-P. Likforman and M. Joffre, J. Opt. Soc. Am. B **17**, 1795 (2000), and references therein.
7 C. Iaconis E. Mukamel and I. A. Walmsley, J. Opt. B, Quant. and Semicl. Opt. , **2**, 510 (2000).
8 N.G. Walker and J.E. Carrol, Opt. Quantum Electron. **18**, 355 (1986); N.G. Walker, J. Mod. Opt. **34**, 16 (1987).

Experimental and theoretical supercontinuum generation in a photonic crystal fiber by use of a-few-cycle optical pulse nonlinear propagation

Xiaojun Fang[1], Masakatsu Hirasawa[1], Robert S. Windeler[2], Ryuji Morita[1], Mikio Yamashita[1]

[1] Department of Applied Physics, Hokkaido University, and Core Research for Evolutional Science and Technology, Japan Science and Technology Corporation, Kita-13, Nishi-8, Kita-ku, Sapporo 060-8628, Japan.
Email: fang@eng.hokudai.ac.jp
[2] Bell Laboratories, Lucent Technologies, 700 Mountain Avenue, Murray Hill, New Jersey 070974

Abstract. The evolution of spectra and temporal profiles of 4.5-optical-cycle pulses propagating near zero dispersion wavelength (ZDW) in a photonic-crystal fiber (PCF) is clarified experimentally and theoretically by indicating that the self-steepening effect inducing the intrapulse four-wave mixing (FWM) is essential. This effect intensifies the spectral component shorter than the ZDW with the decay of higher-order solitons and consequently induces the intrapulse FWM. As a result, the anti-Stokes and Stokes components produced by the FWM enables us to generate a supercontinuum (SC) from 480 to 1020 nm.

1. Introduction

The mechanism in SC generation by PCFs is not yet well-understood because of the complicated interactions among many nonlinear optical processes such as self-phase modulation (SPM), stimulated Raman scattering (SRS), parametric FWM, soliton formation and self-steepening (SST) as well as the unusual dispersion profile. We clarify both experimentally and theoretically the mechanism in SC generation, especially for propagation of a-few-cycle pulses in a PCF. The employed theoretical equation is called the slowly-evolving-wave approximation (SEWA) [5]. The extended nonlinear envelope equation (NEE) [6], beyond the slowly-varying-envelope approximation (SVEA), includes the contribution of the exact delayed Raman response to nonlinear polarization, the more exact SST term and all orders of terms for effective linear dispersion of a PCF [1].

2. Experiment

With ZDW at 767 nm and negative group-delay dispersion (GDD) at its longer wavelength, high-delta PCFs used in our experiment consist of a 1.7-μm-diameter undoped silica core surrounded by an array of 1.3-μm-diameter air holes in a hexagonal close-packed arrangement [1]. Pulses of 12-fs (FWHW) with 800-mW average power centered 795 nm were generated from a Ti:sapphire laser at the repetition rate of 75 MHz (Femtolasers, M1). The corresponding spectrum ranged from 700 to 910 nm with 69-nm width (FWHM), as shown in (1) of Fig.1(a). Pulses were coupled into those PCFs with different lengths (4, 8 , 15, 61 mm) by a dispersion-free reflective objective of 36x magnification. Outputs from the fibers were collimated by another identical reflective objective. The spectra (Fig.1(a)) were measured by a calibrated spectrometer (Oceanoptics Inc., S200). From Fig.1(a) we find that the evolution of the intensity spectra during propagation has the following notable features. After a short propagation distance (4-mm), the spectrum broadens asymmetrically (625 to 915 nm) to form a shoulder on the blue side. With an increase in propagation distance (8-mm), the asymmetric spectrum continues to broaden (590 to 950 nm) with small peaks at the blue (625 nm) and red (915 nm) edges and the intensity-depleted region appears near 660 nm. After the 15-mm distance, with the further asymmetric broadening the peak at the blue edge (580 nm) grows in size and the red edge extends to 965 nm. In addition, the intensity-depleted region (650 nm) becomes more clear. Finally (the 61 mm distance), the spectrum broadens the most (480 to 1020 nm) with two peaks on the blue side (530 and 610 nm) and several peaks on the red side.

3. Theoretical Analysis and Discussion

The extended NEE for the pulse electric field is numerically solved by the split-step method [7]. The mechanism in SC generation for PCF propagation of a-few-cycle pulses with several-ten-kilowatt peak power (67 kW: the soliton number N=5.7) and a central wavelength (795 nm) slightly-longer than ZDW is found out as follows. At first (up to several millimeters), the combined effects of SPM and negative group-velocity dispersion play a main role (the nonlinear length of L_{NL}=0.2 mm and the dispersion lengths of the 2nd order L_D=18 mm and the 3rd order L_D'=21 mm). Most of pulse energy in the spectral band forms a higher-order soliton (N=5.7 and the ZDW soliton number $\tilde{N}$=4.7 [7]). Since the critical parameter of δ_3=0.24 [7,8] representing the relative effect of TOD to the GVD on soliton formation is considerably large, asymmetric multi-fold splitting of the temporal intensity profile occurs [7,8]. In addition, the SST effect influences gradually soliton propagation (the SST critical distance of L_s=2.3 mm) so that the spectrum is blue-shifted with asymmetric broadening ((2) of Fig. 1(a)), and the temporal intensity is steepened on the trailing side ((2) of Fig.1(b)). The combined effects of TOD and SST slow down higher-order solitions and hence the splitting pulse is delayed ((2) of Fig.1(b)). Around the 8-mm propagation distance ((3) of Fig.1(a)), the SST-induced short-wavelength component in the positive GDD region (<767 nm) efficiently causes the intrapulse FWM. This is because the intensified wavelength component sligthly-shorter than ZDW plays a role of the pump (ω_p) for the parametric FWM processes (for example, ω_p(766 nm)+ ω_p(766 nm)$\rightarrow$ ω_s(930 nm)+ ω_{as}(610 nm)), and phase-matches intrinsically with intrapulse Stokes (ω_s) and anti-Stokes (ω_{as}) components [2], as well as having the nearly group-velocity matching and the parametric gain. As a result, the anti-Stokes and Stokes wavelength components appear and the intensity-depleted region is yielded. The further propagation ((4) of Fig.1.a) extends their wavelength regions to try the phase matching. However, the degree of the wavelength-dependent-velocity matching worsens and the wavelength-dependent parametric gain is reduced [2,7]. The result of spectral broadening gives rise to the temporal pulse broadening to form the wings because of the large dispersion in their regions. In addition, the large TOD causes the oscillatory structure near the trailing edge of the pulse. Finally ((5) of Fig.1(a)), the intensified anti-Stokes and Stokes waves are spectrally-modulated due to the wavelength-dependent group-velocity matching and parametric gain. Consequently, the ultrabroad-band pulse from 480 to 1020 nm is generated. These findings suggest that the SST effect intensifying the spectral component shorter than ZDW with the decay of higher-order solitons and hence causing the intrapulse FWM is essential for the existence of the intensity-depleted spectral region. This spectral behavior is the most striking feature appearing in propagation of a-few-cycle pulses centered slightly-longer than ZDW in PCFs. The comparison between the calculated results with and without the SST term gives an evidence for this point. We also confirmed that neglect of the delayed Raman response (f_R=0) negatively affects the agreement with the experimental result for the propagation evolution of the whole spectral profile, especially for the spectral shift of the anti-Stokes component.

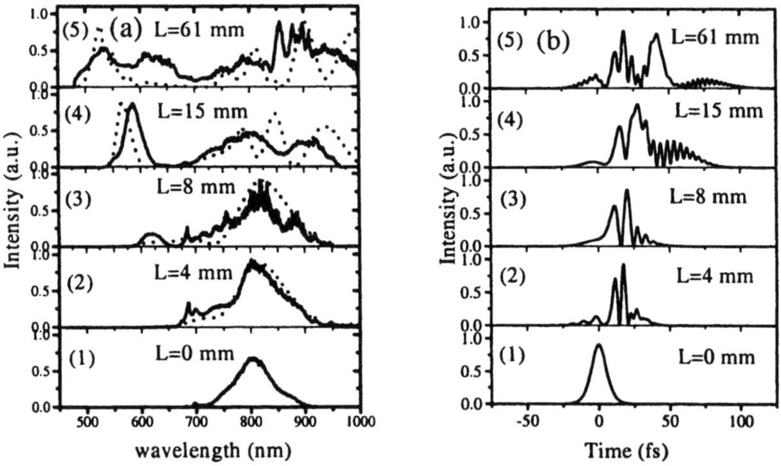

Fig. 1. (a) Experimental (solid lines) and theoretical (dashed lines) spectra of pulses in PCFs, (b) Numerically simulated temporal-intensity profiles.

4. Conclusions

The evolution process of nonlinear propagation of a-few-cycle pulses in PCF, has been well explained by a numerical analysis based on the extended NEE beyond the SVEA. The intensity-depleted region appearing in the spectrum is initially due to the SST effect that causes the blue-shifted spectral broadening with the decay of higher-order solitions and hence induces the intrapulse FWM with phase matching. Anti-Stokes and Stokes waves produced enable us to generate the SC.

References

1 J. Ranka, R. S.Windler, and A.J. Stentz, *Opt. Lett.*, vol. 25, pp. 25-27, 2000.
2 Philip St. J. Russell, *Opt. Lett.*, vol. 26, pp. 1356-1358, 2001.
3 A.V. Husakou, et al., *Phys. Rev. Lett.*, vol. 87, pp. 203901-1-203901-4, 2001.
4 K. J. Blow, et al., *IEEE J. Quantum Electron.*, vol. 25, pp. 2665-2673, 1989.
5 T. Brabec, and F. Krausz, *Phys. Rev. Lett.*, vol. 78, pp. 3282-3285, 1997.
6 N. Karasawa, et al., *IEEE J. Quant. Electron.*, vol.37, pp. 398-404, 2001.
7 Govind P. Agrawal, *Nonlinear Fiber Optics,3rd ed.*, Chaps. 2, 5 and 10.
8 P. K. A. Wai, et al., *Opt. Lett.*, vol. 7, pp. 464-466, 1986.

Photonic Wavepacket in Whispering-Gallery Modes

Hitoshi Kawashima and Toshifumi Hasama

National Advanced Industrial Science and Technology, Umezono 1-1-1, Ibaraki 305-8568, Japan
E-mail: kawashima-h@aist.go.jp

Abstract. We have generated a wavepacket in optical whispering-gallery modes of a dielectric sphere and observed its round trip motion along the periphery. The data is compared with the calculation based on a Mie scattering theory.

1. Introduction

A dielectric sphere supports optical eigenmodes. Some of them have a high quality factor and are termed whispering-gallery modes (WGM's) or equivalently morphology-dependent resonances [1]. Recent studies have addressed new advantages one can take of the WGM's [2,3]. In order to couple light to the WGM's, a prism coupling technique or its variation [2] has been used. With the effect of attenuated total reflection (ATR), one can identify wavelength of the WGM resonances. Frequency resolution of the measurement needs to be sufficiently high. This will be particularly the case if the focus of an investigation is on spectral fine structures that the WGM's with a high quality factor reveal [4,5]. Objective of the present study is to take an alternative approach in the time domain: one can acquire the characteristics of the eigenmodes by analyzing the impulse response. Some of recent studies [6,7] with the same aim have been done on the subnanosecond time scale. Below, we describe, to the best of our knowledge, the fastest WGM's transient and the first temporal analysis of the radiation emitted in the direction of total reflection in the ATR configuration.

2. Experimental Methods

As a dielectric sphere, we used a polystyrene sphere with index of 1.58 and of 41.7μm in diameter. The light source is a modelocked Ti:Sapphire laser. In a dispersion compensation optics, we limit its spectral bandwidth to 20 nm centered on 800 nm. The beam was divided with a half-mirror into two parts, one forwarded to the sample section, another serving as a reference pulse when analyzing the temporal profile of the pulse interacting with the sample. Along the sample arm, the beam is directed to a prism (SF10) coupler: the beam hits one face of the prism, then totally reflected at its base plane, and leaves from the other face. More than one sphere sit on the base plane. The light polarization lay in the scattering plane. By extending the scattering plane, we define the equatorial plane of the sphere. We made a preliminary calculation on a tangential component of the wavevector of a particular WGM resonance that would occur at 800 nm. The incidence angle to the prism was so determined that the tangential wavevector of the evanescent wave may become the calculated value. We translated the prism relative to the beam and confirmed that only one sphere was in the spot of internal total reflection across the interface. For temporal evaluation, we measured an intensity cross correlation with noncollinear second harmonic generation (SHG). In the absence of coupling to the sphere, the cross correlation had full width at half maximum of 76 fs.

3. Results and Discussion

Figure 1 shows the resultant SHG traces. The highest peak at the origin is assigned to the directly reflected pulse which spent no time in the sphere and was the first to come out of the prism. This is followed by a periodic revival of spikes. The rapidly oscillating features that go with the spikes suggest that the signal should result from some optical interference. From optical power dependence of the signal strength, the interference has turned out to be the one between the second harmonic field, $a_0 \cdot a_s$, contributed by both reflected and reference light, and the second harmonic field, $a_0 \cdot a_0$, of reference light, where the amplitude of reflected light is designated by a_s and that of reference light by a_0. The spikes, therefore, represent possible group delays held by the sample, in common with interferometric measurements with a low coherence source. They disappear upon placing the sphere out of the beam spot, and this evidences that they should arise from light coupling to the sphere. The time interval between the adjacent spikes is 680 fs. The periodic revival of the spikes depicts a round trip motion of a photonic wavepacket: it goes along the equator. At every passage through the contact point, a fraction of light goes back to the prism and to the reflected beam. Following argument is to give this interpretation further quantitative grounds.

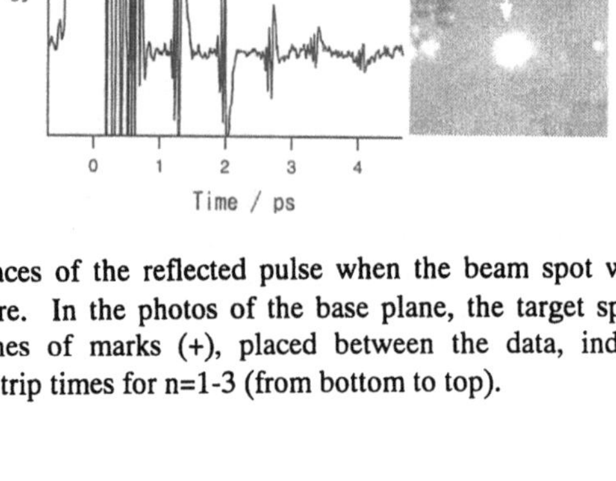

Fig. 1. SHG traces of the reflected pulse when the beam spot was off (upper) and on (lower) the sphere. In the photos of the base plane, the target sphere is pointed by the arrow. Three lines of marks (+), placed between the data, indicate multiples of the calculated round trip times for n=1-3 (from bottom to top).

The optical eigenmodes are labeled by three integers, n, l, and m, which designate the radial order, the angular and the azimuthal mode numbers, respectively. A high quality factor is generally expected for those modes which are labeled by $n \approx 1$, $l \gg 1$. The geometrical configuration of ATR fixes the azimuthal number m to $\pm l$ while the eigenfrequencies are degenerate with respect to it. Using an algorithm in Ref. [8], we calculated the resonance frequencies of the TM polarized modes with $n \approx 1$ and various l. For a series of resonances that belong to each n, the frequency spacing between l and $l+1$ resonances was almost constant across the spectrum of the optical pulse employed. This allows us to regard the reciprocal of the spacing as a round trip time, 682, 674, and 668 fs for $n = 1$-3, respectively. The diffractive nature of the WGM's lies in the fact that these values are shorter than the time for light to travel the optical path length equal to the circumference, 693 fs. In Fig. 1, the multiples of these round trip times are plotted for comparison with the data. The results of the calculation give a good fit to the data, although the damping is too fast to resolve the signal into respective mode orders. The agreement found here confirms the azimuthal localization of the wavepacket. This is analogous to the case of the wavepacket in atomic Rydberg states [9]. From the signal strength and its attenuation, it has been found that the rate of power transmission into the WGM's is about 1% and that the sphere as a cavity has the finesse of 3.6, which is typical to polymer spheres. It is to be noted that in Fig. 1, the sphere shines not only at the rim but as a whole body. In the present case, light has a broad spectrum, and it excites many eigenmodes including those of high order number n. The leakage of the high n modes occurs not only tangentially but radially as well, and thus causes our sphere to glitter as seen. In summary, we coupled light into a dielectric sphere and observed that a train of pulses followed the directly reflected pulse. A theoretical calculation shows that those delayed pulses arise from a round trip motion of a wavepacket composed of the WGM's.

References

1 L. Collot, V. Lefevre-Seguin, M. Brune, J.M. Raimond, and S. Harosche, Europhys. Lett. **23**, 327, 1993.

2 J.C. Knight, G. Cheung, F. Jacques, and T.A. Birks, Opt. Lett. **22**, 1129, 1997.

3 M. Cai, G. Hunziker, and K. Vahara, IEEE Photon. Technol. Lett. **11**, 686, 1999.

4 M.L. Gorodetsky and V.S. Ilchenko, J. Opt. Soc. Am. B **16**, 147, 1999.

5 S. Schiller and R.L. Byer, Opt. Lett. **16**, 1138, 1991.

6 J.C. Swindal, D.H. Leach, R.K. Chang, K. Young, Opt. Lett. **18**, 191, 1993.

7 R.W. Shaw, W.B. Whitten, M.D. Barnes, and J.M. Ramsey, Opt. Lett. **23**, 1301, 1998.

8 S. Schiller, Appl. Opt. **32**, 2181, 1993.

9 J.A. Yeazell and C.R. Stroud, Jr., Phys. Rev. **A35**, 2806, 1987.

Pulse dynamics in polarization-maintaining photonic crystal fibers

Arturo Ortigosa-Blanch, Jonathan C. Knight and Philip St.J. Russell

Optoelectronics Group, Department of Physics, University of Bath,
Bath BA2 7AY, United Kingdom
E-mail: a.ortigosa@bath.ac.uk

Abstract. The propagation of femtosecond pulses in the anomalous dispersion regime in a photonic crystal fiber is experimentally studied. Pulse breaking and soliton propagation is observed as a leading mechanism for supercontinuum generation.

Photonic crystal fiber (PCF) is a pure silica fiber in which waveguiding is due to the arrangement of a pattern of tiny air holes running down the fiber length. The massive nonlinearity and unusual group velocity dispersion characteristics that can be achieved in such fibers make them spectacular vehicles for nonlinear fiber optics. Anomalous group velocity dispersion at near infrared and visible wavelengths [1] and broadband supercontinuum generation [2] amply demonstrate the wide range of interesting effects which have been reported.

With the aim of understanding the physical processes governing ultrashort pulse propagation and supercontinuum generation in such fibers, we have performed a series of measurements using polarisation-maintaining PCF [3]. A scanning electron micrograph of the fibre is shown in Figure 1 (left-inset). The fiber birefringence is more than sufficient to remove bend-induced coupling between polarization modes and the two polarization eigenmodes show different linear characteristics, exhibiting anomalous dispersion at wavelengths longer than 781 nm and 806 nm respectively (Figure 1 left). A Ti:sapphire laser was used to launch pulses of different energies (76 MHz repetition rate, 200 fs FWHM pulse duration) and different pump wavelengths into the PCF. A half wave-plate was used to adjust the linear polarization state at the input so as to excite a single polarization mode. The output of the fiber was then analysed using an intensity autocorrelator and an optical spectrum analyzer.

We present results (Figure 2 (a)) for a 6 m piece of fiber and a pump wavelength of 850 nm. We first investigated the output pulse duration at relatively low input energies. At such energies (up to 6 W peak power) the pulse is substantially broadened by linear dispersion (output pulse length > 1 ps). The pulse then compresses down to 90 fs at a peak output power just below 66 W. We identify this value (corresponding to an input power of about 92 W) with the fundamental soliton energy, in accord with simple theory.

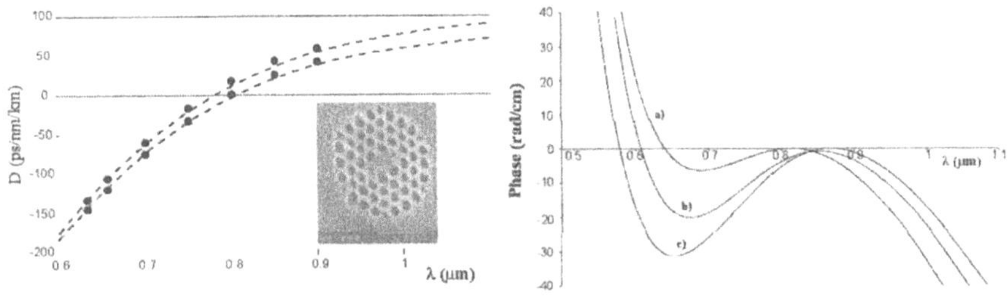

Fig. 1. (Left) Calculated (dotted line) and measured (points) dispersion for the eigenmodes of the polarization-preserving PCF shown in the inset, and (Right) phase mismatch between nonsolitonic radiation and solitons at (a) 817 nm, (b) 850 nm and (e) 870 nm.

As we increased the energy launched into the fiber, interesting features appeared in the recorded spectra. A pronounced pulse splitting and soliton shelf frequency shift (SSFS) was observed for input powers above the first soliton energy. A shift of up to 120 nm was observed for a peak output power of 203 W (corresponding to Figure 2 (a), output peak powers from 68 W to 203 W). As the power was increased further a second separate peak split off from the pump and also started shifting in frequency. At the same time, the original self-shifting soliton that appeared at a lower power kept shifting towards longer wavelengths (Figure 2 (a), 289 W output power). The generation of a second self-shifting peak is accompanied by the appearance of a blue-shifted peak at a wavelength around 600nm. As the input power was increased further, a third new spectral peak developed, being fully independent at around 618 W output power. Further increases in power lead to further self-frequency shifts in the existing peaks and the appearance of new peaks. These are accompanied by the appearance of new blue-shifted peaks as well. At yet higher pump energies (e.g. 6 kW peak power) all those peaks merge to give a continuous spectrum of peaks – a broadband supercontinuum. At these higher powers, pulse break up occurs after an increasingly short propagation length, so that just a few centimeters of fibre are required to observe supercontinuum generation [2]. At these higher powers, pulse breakup is surely accompanied by other nonlinear mechanisms, mainly modulation instability, and four-wave mixing which serve to diffuse the distinctive features observed at low energies. By considering the powers at which we were working, we can explain the appearance of this series of new spectral peaks in terms of the fission of higher- order solitons [4, 5]. Looking once more at Figure 2 (a), it is clear that, around the second soliton energy (289 W output power), we get two very well - defined peaks in addition to the pump. The autocorrelator confirms these as corresponding to discrete pulses. Furthermore, at the peak power corresponding to N = 3 we observe three spectral peaks (618 W output power) and four peaks (1.1 kW output power) for N = 4. Above the second soliton energy we also observe spectral components in the short wavelength part of the spectrum which we attribute to nonsolitonic radiation generated during soliton fission. Using the modal refractive indices from our

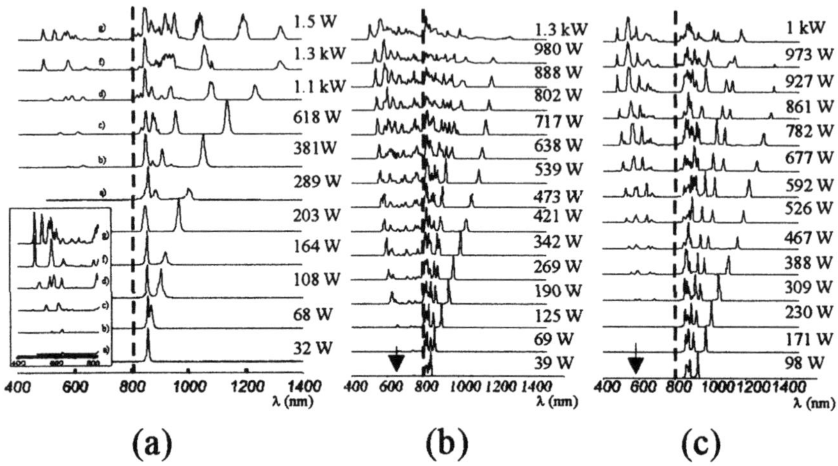

Fig. 2. Recorded output spectra for pump wavelengths of (a) 850 nm, (b) 817 nm and (c) 870 nm for different peak powers. The dotted line indicates the zero dispersion wavelength at 806 nm. (Inset in (a)) Detail of the spectra from 400 nm to 800 nm.

numerical model, we have confirmed that the blue-shifted nonsolitonic radiation appears at wavelengths which are phase-matched to the corresponding infra - red solitons. Phase matching curves for our fiber computed for different pump wavelengths are shown in Figure 1 (right). Figure 2 (b) and (c) shows the same mechanism when pumping at (b), 817 nm and (c), 870 nm a 16 m long piece of the same kind of fiber.

4. Conclusions

In conclusion, we have shown experimental evidence of the process that leads to supercontinuum generation in photonic crystal fibres: Soliton breaking and the associated phase-matched radiation at shorter wavelengths creates new frequency components that form the ultrabroad spectrum. The linear properties of the fiber, such as its effective index and GVD, play a very important role on the generation of the supercontinuum and allows one to tailor the spectra for different applications.

References

1 J. C. Knight *et al.*, IEEE Photon. Technol. Lett., **12,** (2000)
2 J.K. Ranka *et al.*, Opt.Lett., **25**, 25 (2000)
3 A. Ortigosa-Blanch *et al*, Opt. Lett **25**, 1325 (2000)
4 A.V. Husakou *et al.*, Phys. Rev. Lett., **87**, 203901 (2001)
5 J. Herrmann *et al*, Phys. Rev. Lett., **88**, 173901 (2002)

Generation of Ultrashort-Pulse Nondiffracting Beams and X-Waves with Thin-Film Axicons

R. Grunwald [1], V. Kebbel [2], U. Griebner [1], U. Neumann [1], A. Kummrow [1], E. T. J. Nibbering [1], M. Rini [1], M. Piché [3], G. Rousseau [3], and M. Fortin [3]

[1] Max-Born-Institut für Nichtlineare Optik und Kurzzeitspektroskopie, Max-Born-Str. 2A, D-12489 Berlin, Germany
E-mail: grunwald@mbi-berlin.de
[2] Bremer Inst. f. Angewandte Strahltechnik, Klagenfurter Str. 2, 28359 Bremen, Germany
[3] Université Laval, Faculté des sciences et de génie, G1K 7P4, 2624A Alexandre-Vachon, Quebec, Canada

Abstract. Ultrashort-pulse Bessel beams were generated from Ti:Sa lasers with transparent and reflecting thin-film axicons. By spatio-temporal autocorrelation, significance of optical X-wave formation was demonstrated for pulse duration down to 8 fs in accordance with simulations based on Rayleigh-Sommerfeld diffraction theory.

For spatio-temporal beam shaping of fs-lasers, structured optical elements of well-balanced refractive, diffractive and dispersive properties are of increasing interest [1]. Recently, the potential of broadband, low-dispersion thin-film components for the control of ultrashort-pulse light fields by specific interference effects was demonstrated [2]. The generation of nondiffracting beams of Bessel [3] or Bessel-like [4] intensity distributions with conical phase elements enables extremely homogeneous and extended focal zones. As it is known from surface acoustic [5] and THz-waves [6], non-dispersive wave phenomena with characteristic X-shape in space-time coordinates result from spectral interference of broadband conical beams [7]. X-wave detection was realized with Bessel beams shaped from the broadband radiation of a high-pressure Xe-lamp of 3 fs coherence time [8]. A first indirect indication of Bessel-X-pulses created from a short pulse laser by analysing dispersion was reported for 210 fs [9]. Reduced dispersion was interpreted to be caused by soliton-like behaviour. However, as it was found by theoretical calculations for typical Ti:Sa laser spectra [7], a distinct spatio-temporal X-structure should only be obtained for pulses < 20 fs. Initially, we studied Bessel-like beams at pulse durations down below 10 fs [10-11]. Here we report on recent progress in fs Bessel-like beam generation based on different types of reflective and refractive thin-film axicons. Experimental results are compared with simulations based on Rayleigh-Sommerfeld diffraction theory including space-time coupling (GVD, travel time differences). For the first time, to the best of our knowledge, reliable evidence is given for Bessel-X-pulse generation with optical few-cycle laser pulses down to a pulse duration of about 8 fs.

An amplified Ti:Sa laser system with Ar-filled hollow fiber and chirp-compensation generated pulses of 8...30 fs @ 790 nm, 1 kHz repetition rate and 150 µJ pulse energy [12]. The area of interest was imaged on a CCD-camera (1030 x 1020 pixels, zoom objective 1:1.2/12.5-75 mm, microscope M = 20x) and analysed with image processing software. For SHG a 100 µm BBO crystal was placed within the imaged plane. Two types of beam shapers have been studied: (a) Single reflective macroscopic axicons (diameter 1 cm, depth 4 μm, concave face made by Cu-vapor deposition using rotating masks, surface Au-coated), (b) hexagonal arrays of convex Gaussian-shaped refractive micro-axicons (SiO_2 on 2 mm thick quartz substrate, pitch 405 μm, sag heights 5.7 μm, made by vapor deposition in a planetary rotation system with shading masks. For details see [13]). The chirp caused by substrate was pre-compensated. For collinear autocorrelation, the beam was split symmetrically, and in case (a) expanded for suppression of effects caused by curvature of the central part (~ 280 μm) of axicon.

The analysis of time-integrated, spatially resolved fringe patterns generated with both types of axicons agrees well with results of simulation based on Rayleigh-Sommerfeld diffraction theory [2]. Bessel-like intensity profiles with large axial extension were found. The beam aspect ratio (diameter of central lobe / Rayleigh length) was $< 1 : 2\text{x}10^3$ for type-(a)- and $< 1 : 2\text{x}10^2$ for type-(b)-axicons corresponding to the average angles of incidence (0.05° and ~ 2°, respectively). Fig. 1 shows a set of linear cuts through the near-axis part of Bessel beam intensity distribution generated by type-(a)-axicon at axial distances z = 50 ... 120 cm. The profiles represent time-integrated interference patterns imaged with a microscope 20x objective. The laser pulse coherence time was determined independently by collinear and non-collinear 2nd order autocorrelation measurements to be 8.1 fs.

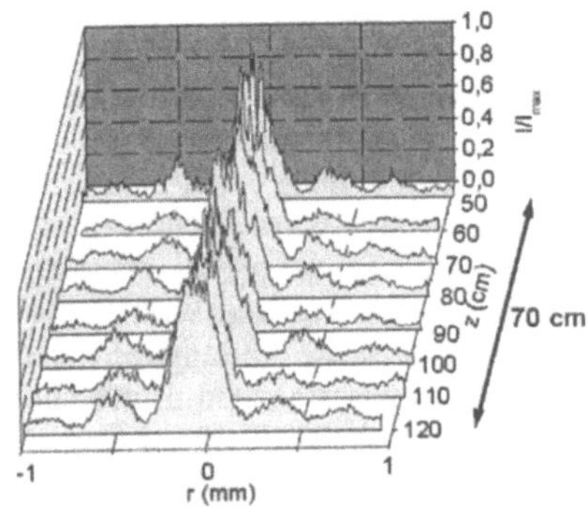

Fig. 1. Linear cuts through a Bessel beam intensity profile generated by reflecting type (a) axicon (time integrated, 20x objective, Ti:Sa laser @ 790 nm, 8.1 fs, z = beam direction).

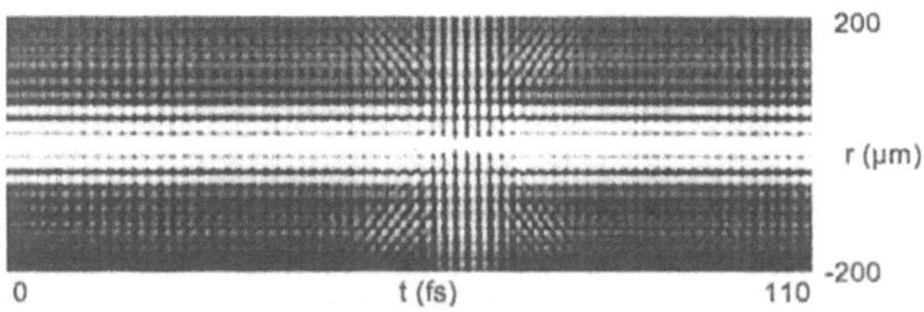

Fig. 2. Bessel-like X-pulse intensity profile generated with thin-film microaxicon type (b): 2nd order spatio-temporal autocorrelation (Ti:Sa laser @ 790 nm, 8.1 fs, step width 0.22 fs).

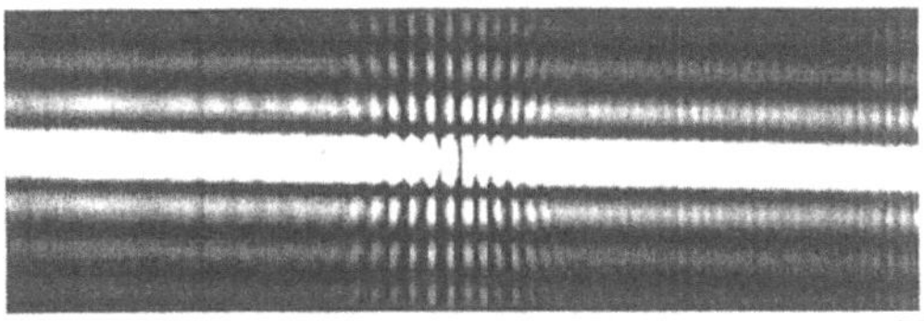

Fig. 3. Bessel-X-pulse intensity profile generated with thin-film axicon mirror type (a): 1st order spatio-temporal autocorrelation (data as Fig. 2, wave front tilt by incidence angle).

The results of 1st and 2nd order spatio-temporal autocorrelation measurements with transparent type-(b)-microaxicons (Fig. 2) and reflective axicon (Fig. 3) are in good accordance to patterns calculated on the basis of realistic spectral data. X-shaped subpatterns clearly can be recognized. The radially cut Bessel fringes (horizontal stripes) in Fig. 3 have been found to be slanting. It can be explained by the wave front tilt caused by the oblique angle of illumination of the axicon (11.5°). Substructures result from the non-ideal spectral distribution.

To conclude, ultrashort-pulse Bessel-like beams were generated with refractive thin-film microaxicon arrays as well as reflective axicons. In accordance to simulations, the formation of optical X-waves was demonstrated for sub-10-fs pulses. The propagation of few-cycle X-waves in dispersive media (where the formation of spatio-temporal solitons or "light bullets" might be possible [14]) and further nonlinear optical applications will be a subject of future investigations.

References

1 R. Piestun and D. A. B. Miller, Opt. Lett. **26**, 1373, 2001.

2 R. Grunwald, U. Griebner, T. Elsaesser, V. Kebbel, H.-J. Hartmann, and W. Jüptner, in *Fringe 2001,* Edited by W. Osten, W. Jüptner, Proc. 4th Int. Workshop on Automatic Processing of Fringe Patterns, Elsevier, Paris, 2001.

3 J. Durnin, J. J. Miceli, and J. H. Eberly, Phys. Rev. Lett. **58**, 1499, 1987.

4 R. M. Herman and T. A. Wiggins, J. Opt. Soc. Am. A **8**, 932, 1991.

5 J. Y. Lu and J. F. Greenleaf, IEEE Trans. Ultrason. Ferroelec. Freq. Cont. **39**, 19, 1992.

6 Z. Jiang and X.-C. Zhang, Opt. Express **5**, 243, 1999.

7 M. Piché, G. Rousseau, Ch. Varin and N. McCarthy, in Proc. SPIE **Vol. 3611**, 332, 1999.

8 P. Saari and K. Reivelt, Phys. Rev. Lett. **79**, 4135, 1997.

9 H. Sonajalg, M. Rätsep, and P. Saari, Opt. Lett. **22**, 310, 1997.

10 R. Grunwald, U. Griebner, F. Tschirschwitz, E. T. J. Nibbering, T. Elsaesser, V. Kebbel, H.-J. Hartmann, and W. Jüptner, Opt. Lett. **25**, 981, 2000.

11 R. Grunwald, U. Griebner, E. T. J. Nibbering, A. Kummrow, M. Rini, T. Elsaesser, V. Kebbel, H.-J. Hartmann, and W. Jueptner, JOSA **A 18**, 2923, 2001.

12 E. T. J. Nibbering, O. Dühr, and G. Korn, Opt. Lett. **22**, 1335, 1997.

13 R. Grunwald, H. Mischke, and W. Rehak, Appl. Opt. **38**, 4117, 1999.

14 F. Wise and P. di Trapani, Optics & Photonics News **13**, 29, Feb. 2002.

Part VI

THz Generation and Applications

Low-density optical pump – THz probe analysis of high-temperature superconductors

M. Brucherseifer, A. Meltzow-Altmeyer, P. Haring Bolivar, and H. Kurz

Institut für Halbleitertechnik, RWTH Aachen, Sommerfeldstr. 24, 52056 Aachen, Germany
E-mail: mbrucher@stevens-tech.edu

Abstract. Using time-resolved THz probing we analyze the dynamics of low-density optically excited, optimally doped $YBa_2Cu_3O_{7-\delta}$. A previously unknown additional relaxation component is found, tentatively indicating the separate relaxation of charge and spin excitations.

1. Introduction

Dynamic analysis of the relaxation dynamics of high-temperature superconductors (HTSC) after optical excitation offer a detailed insight into the superconductivity mechanism of these materials. Spectrally similar effects which would be superposed in stationary experiments can readily be distinguished in the time domain. In the past, most time-resolved experiments on HTSC have probed high energy states at photon energies around 1.0- 2.0 eV [1]. However, under these conditions, superconductivity related pairing dynamics cannot be distinguished from energy relaxation dynamics, as the probing energy is too far above the superconductivity gap. Recent experiments, probing pairing dynamics at lower energies e.g. MIR [2] or THz [3], have yielded interesting observations. However, these experiments were performed under very high optical excitation densities, which significantly modify the superconducting state [3]. Here we present a time-resolved optical pump-THz probe analysis with a drastically improved resolution better than $\Delta T/T_0=10^{-9}$, which allows to perform analysis at excitation densities 2 - 3 orders of magnitudes below previous approaches. Thus density dependent recombination dynamics in optimally doped $YBa_2Cu_3O_{7-\delta}$ probed at low energies are isolated for the first time. In contrast to a previously seen single relaxation component, two components are observed, tentatively indicating the separate relaxation of spin and charge excitations.

2. Experimental Methods

The experiments are performed with a time-domain THz transmission setup using a 76 MHz repetition rate 100 fs laser. Its great advantage is the possibility to directly measure the complex conductivity $\sigma = \sigma_1 - i\sigma_2$ of HTSC [4]. Based on a Two-Fluid-Model [5], one can simultaneously and separately analyze the behavior of quasiparticle x_n and superconducting condensate x_s densities. The samples used consist of optimally doped 20 nm $YBa_2Cu_3O_{7-\delta}$ films on $NdGaO_3$ substrates mounted in a cryostat. A stationary, temperature dependent evaluation of the complex conductivity delivers a critical temperature of $T_C = 88$ K confirming a good sample quality despite its thin thickness. For dynamical analysis a further optical laser pulse is focused, collinear to the THz beam, onto the sample with a 800 μm focus (= THz focus). Primary excitation densities in the range of $3 \cdot 10^{16}$ to $4.5 \cdot 10^{17} cm^{-3}$ are induced. The optical excitation breaks Cooper-pairs resulting in a decrease of x_s and increase of x_n, accordingly an increase of σ_1 and decrease of σ_2. The following relaxation process is dynamically analyzed by probing the THz transmission. A detailed numerical calculation, considering the optically altered conductivity due to the changed fractions of x_n and x_s, leads to expected THz transmission intensity modifications of maximally $\Delta T/T_0 \approx 10^{-8}$ which have to be resolved, compared to a free-space THz transmission T_0. To avoid long data acquistion measurements we do not analyze the full dependence of the THz transmission as function of optical and THz time delays but concentrate only on characteristic fixed delay points of the THz transient, and evaluate the dependence of the optical time delay. As indicated on the THz transmission transient in the inset of Fig. 1, we concentrate on the conductivity dynamics at the marked peak and zero crossing positions. Our numeric calculation reveals, that the signal increase measured at the peak of the THz-transient is dominated by the decrease of σ_2, whereas the signal at the marked zero crossing increases, depends equally on the σ_1 increase and the σ_2 decrease.

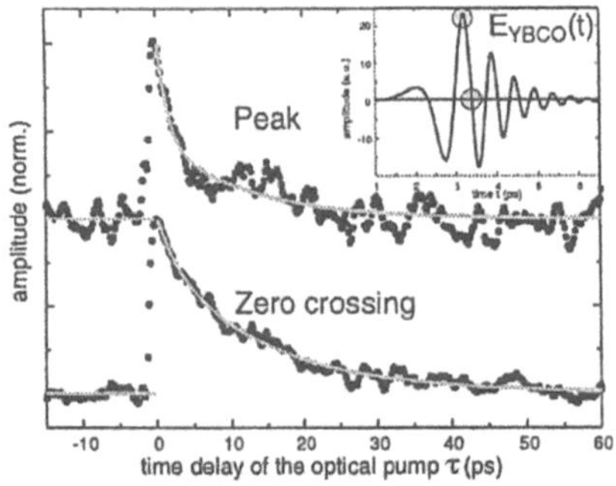

Fig 1. The inset shows a THz transient and the characteristic positions used for dynamical analysis.The main figure shows the respective relaxation signals after optical excitation.

3. Results and Discussion

Two exemplary relaxation transients taken at 10K are shown in Fig. 1. One can directly see different relaxation times of both transients indicating that at least two different components are necessary to explain the relaxation process. As the zero crossing signal has a better signal to noise ratio and contains both dynamics, we restrict in the following to such analysis and evaluate these transients via two component exponential fits with decay times τ_1 and τ_2. Fig. 2 shows the analysis of the density and temperature dependencies of both components. The left plot indicates a fast component τ_2 in the region of 1-3 ps that is in agreement with previous high-density observations [2,3]. This fast component exhibits a characteristic Rothwarf-Taylor density dependence. For the first time we observe an additional slow relaxation component τ_1 which decreases for increasing excitation density. We think that this additional slow component was not previously observed, as pre-

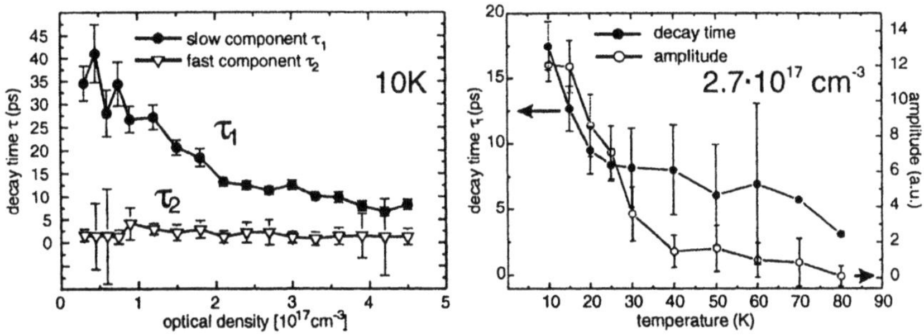

Fig2. Decay time density dependence of both components τ_2 and τ_2 at 10 K (left) and temperature dependence of amplitude and decay time of the slow component τ_1 (right).

vious experiments were performed at much higher excitation densities where both relaxation channels τ_1 and τ_2 become restricted by phonon dissipation. The temperature dependence of the amplitude and relaxation time of τ_1 are shown in the right part of Fig. 2. The decay time τ_1 decreases for higher temperatures. While the amplitude of the fast component τ_2 has a typical BCS type behavior with a sharp drop close to T_C (data not shown), the τ_1 component decays significantly already much below T_C.

4. Conclusions

The presence of two markedly different relaxation components proofs that the simple assumption of a single type of excitations is clearly not sufficient to explain the observed dynamics. We tentatively ascribe both components to the separate relaxation of charge and spin excitations (antiferromagnetic spin fluctuations), supporting the notion of independent charge and spin excitations within a collective state, the "quantum protectorate" [6]. The decrease of the respective relaxation times with increasing excitation density would then reflect the increasing probability for the coincidence of charge or spinon pairs, respectively. Further experimental and theoretical analysis will however be necessary to further support this conjecture and to assign the respective components.

References

1 S. G. Han, Z. V. Vardeny, K. S. Wong, O. G. Symko, Phys. Rev. Lett. **65**, 2708 (1990).
2 R. A. Kaindl, M. Woerner, T. Elsaesser, D. C. Smith, J. F. Ryan, G. A. Farnan, M. P. McCurry and D. G. Walmsley. Science **287**, 470 (2000).
3 R. D. Averitt, G. Rodriguez, A. I. Lobad, J. L. W. Siders, S. A. Trugman, and A. J. Taylor., Phys. Rev. Rapid Comm. **63**, 140502 (2001).
4 S. D. Brorson, , R. Buhleier, I. E. Trofimov, J. O. White, C. Ludwig, F. F. Balakirev, H.-U. Habermeier, and J. Kuhl. J. Opt. Soc. Am. B **13**, 1979 (1996).
5 A. J. Berlinsky, C. Kallin, G. Rose and A. C. Shi., Phys. Rev. B **48** (1993).
6 P. W. Anderson, Science **288**, 480 (2000).

Terahertz Wave Generation from Multi-Quantum Well with Transverse Electric Field

Isao Morohashi[1,3], Kazuhiro Komori[2,3], Hisashi Shimura[1], Takehiko Hidaka[1], Hirokazu Ikeda[4], Shinya Tujikura[4], Masami Yamamoto[4], and Yasunori Okamoto[4]

[1]Shonan Institute of Technology, 1-1-25 Tsujido-nishikaigan, Fujisawa, Kanagawa 251-8511, Japan
E-mail: i-morohashi@aist.go.jp
[2]National Institute of Advanced Industrial Science and Technology, Tsukuba-Central 2, 1-1-1 Umezono, Tsukuba, Ibaraki 305-8568, Japan
[3]CREST, Japan Science and Technology Corporation, 4-1-8 Honcho, Kawaguchi, Saitama 332-0012, Japan
[4]KUBOTA Corporation / Advanced Technology Laboratory, 1-1-1 Hama, Amagasaki, Hyougo 661-8567, Japan

Abstract. A multi-quantum-well terahertz wave generator with a transverse electric field was proposed and investigated. The terahertz radiation characteristics including excitation wavelength dependence was investigated and high-power terahertz generation, which was about 60-times higher than InAs references, was demonstrated.

1. Introduction

High power and high efficiency terahertz (THz) wave generators are important for the application to imaging and spectroscopy. THz generators using semiconductor nanostructure devices are promising, because of their high nonlinearity. Furthermore, the high radiation power is expected by applying the electric field parallel (transverse) to the MQW plane. In this transverse geometry, the high-efficiency THz radiation in the direction of the surface is expected. Such devices using the optical nonlinearity have already been reported for different applications such as photorefractive devices[1]. In this paper, we have proposed the MQW device with the transverse electric field as a THz wave generator, and report the preliminary THz radiation characteristics.

2. Experiments

A MQW sample (Fig. 1(a)) was consisted of sixty periods of 8nm-thick GaAs quantum wells separated by 10nm-thick $Al_{0.3}Ga_{0.7}As$ barrier layers, which is the same structure reported as photorefractive quantum well devices[1]. The MQW layer was grown on a (411)A semi-insulating GaAs substrate after a 100nm-thick AlAs layer was grown. The sample was fabricated by the molecular beam epitaxy (MBE) growth method at a temperature of 350°C. The substrate was removed by

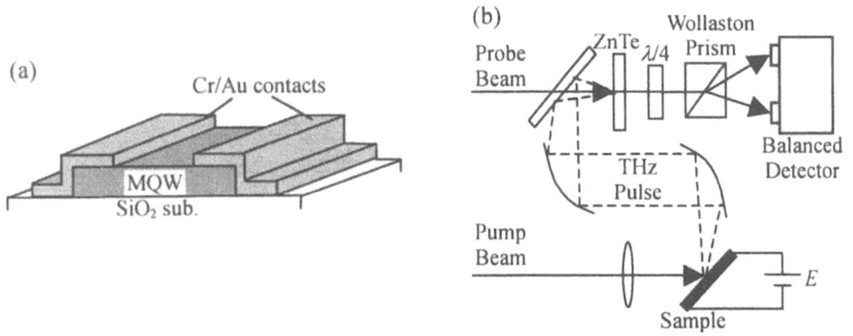

Fig. 1. (a) Sample structure and (b) the experimental setup using the EO sampling method

epitaxial lift-off, and the MQW layer was bonded on a glass (SiO_2) substrate by van der Waals (VDW) bonding. To apply an electric field, Cr/Au contacts were formed on the sample with 1mm gap.

THz wave was measured by the free-space electro-optic (EO) sampling method[2],[3] as shown in Fig.1(b). The MQW sample was illuminated by ultrashort optical pulses with a pulse duration of about 140fs at an angle of 45°. A radiated THz wave was collected with a pair of off-axis parabolic mirrors, and was incident on a ZnTe crystal together with a probe beam. The probe beam passed through the ZnTe crystal was analyzed by a Wollaston prism, and was detected by a balanced-photodetector.

3. Results and Discussions

Figure 2(a) shows detected THz waves as a function of the applied voltage, and Fig. 2(b) shows applied voltage dependence of the amplitude of the THz wave (THz amplitude). No signal was detected at E = 0V, and the THz amplitude was increased almost linearly with increasing the applied voltage. The power of the

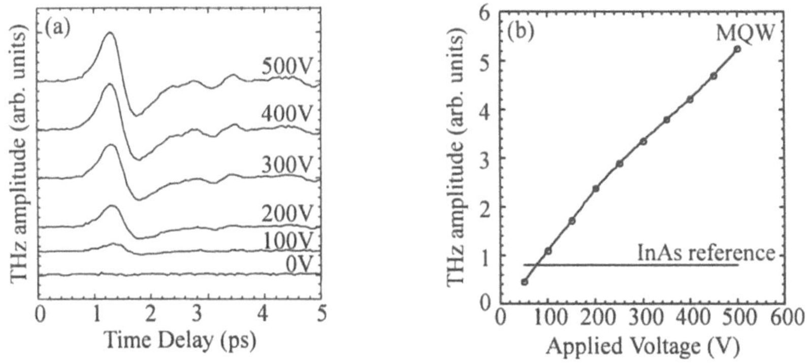

Fig. 2. (a) THz waves from the MQW sample and (b) Applied voltage dependence of the THz amplitude at P = 50 mW and room-temperature

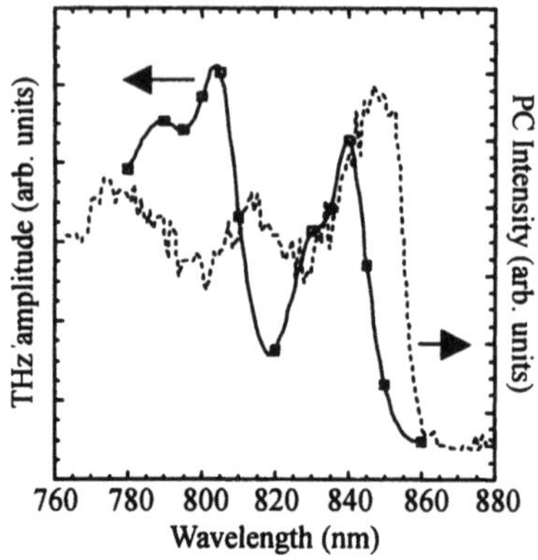

Fig. 3. Comparison between the THz amplitude and a PC spectrum at P = 60mW and room-temperature

THz wave at E = 500V was about 60 times (about 7.5-times in the THz amplitude) higher than that of a bulk InAs (reference sample) without an electric field (shown by dotted line in Fig. 2(b)). Figure 3 shows excitation wavelength dependence of the THz wave from the MQW sample. The solid line shows the THz amplitude, and the dashed line shows a photocurrent (PC) spectrum. The THz amplitude was at maximum at around the exciton absorption peak of the PC spectrum. Thus, observed THz wave was generated by the polarization of the photo-excited carriers around the exciton absorption wavelength. The shift between the THz amplitude peak and PC intensity peak may be originated from (1) screening effect or (2) an increase of sample temperature in the PC measurement.

4. Conclusions

We have demonstrated THz wave generation from a MQW device with a transverse electric field. THz waves were generated by ultrashort optical pulse excitation, and were detected by the free-space EO sampling method. The THz amplitude was almost linearly increased with increasing an applied voltage, and the THz amplitude was at maximum at around a PC spectrum peak. The power of the THz wave at E = 500 V was about 60-times higher than that of the bulk InAs sample. This device is expected as a high-power THz wave generation device.

References

1 Q. Wang, R. M. Brubaker, D. D. Nolte and M. R. Melloch, J. Opt. Soc. Am. B 9, 1626-1641 (1992)
2 Q. Wu and X. C. Zhang, Appl. Phys. Lett. 71, 1285-1286 (1997).
3 I. Morohashi, K. Komori, T. Hidaka, T. Sugaya, X. L. Wang, M. Ogura, T. Nakagawa and C. S. Son, Jpn. J. Appl. Phys. 40, 3012-3017 (2001)

Inversionless Amplification of Coherent THz Radiation

C. Janke[1], P. Haring Bolivar[1], A. Bartels[1], H. Künzel[2], and H. Kurz[1]

[1] Institut für Halbleitertechnik, RWTH Aachen, Sommerfeldstr. 24, D-52074 Aachen, Germany
E-mail: janke@iht-ii.rwth-aachen.de
[2] Heinrich-Hertz-Institut für Nachrichtentechnik, D-10587 Berlin, Germany

Abstract. A surface field THz emitter placed inside a high repetition-rate THz resonator is synchronously pumped by a femtosecond laser. The emitter efficiency is increased by more than one order of magnitude via amplification without inversion.

Time-resolved spectroscopy and imaging in the THz-region of the electromagnetic spectrum is an important and growing field of research with a wide range of applications. Most THz sources rely on the coherent photoexcitation of charge carrier distributions in semiconductor structures with femtosecond laser pulses. The radiation process is then given by the acceleration of the coherent charge carrier ensemble according to Faraday's law of induction. Up to now, a variety of broadband [1-3] or tunable [4-5] THz emitters has been developed for specific applications [6]. All these applications are, however, hampered by the lack of powerful and efficient THz sources. One of the main reasons restricting the efficiency of impulsively excited THz emitters is the fast loss of coherence. Typically, the coherence in the emitting system is destroyed by ultrafast scattering processes. This restricts the coherent motion of the photoexcited charge carrier ensemble and the coherent emission time is limited to a small fraction of the radiative lifetime [4], typically below 10^{-3}. Consequently, only a small portion of the energy available in the system is emitted coherently.

Here, we follow the approach of exposing the emitter to a coherent phasematched THz background field during the emission process [7] in order to radiate more energy *before* dephasing occurs and thereby amplify the amount of coherently emitted THz radiation. By placing the emitter inside a high repetition-rate THz resonator and synchronously optically exciting it with a mode-locked, high repetition-rate (1 GHz) Ti:sapphire femtosecond laser source [8], a net amplification of the THz emission power outside the resonator is achieved for the first time. The increase amounts to more than one order of magnitude.

A simple three-level model and a density-matrix approach [9] are used to study the amplification principle. The density matrix approach is expanded as described in [10] to take the THz background field and its influence on the intraband density matrix elements into account. Numerical simulations validate that the THz background field changes the intraband population densities, leading to emission or absorption depending on the relative phase of the background field. Hence, gain without population inversion can be achieved on the THz transition of the three-level system.

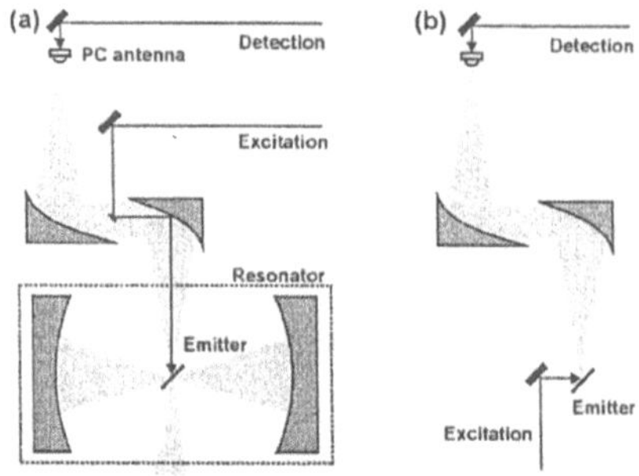

Fig. 1. Experimental setup. (a) The InGaAs emitter is placed between two spherical mirrors comprising the THz resonator. (b) Standard emission configuration without resonator.

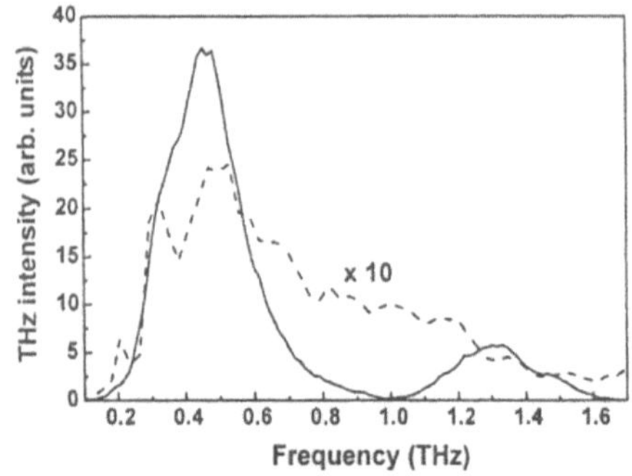

Fig. 2. Spectral intensity of amplified emission (solid line) and standard emission without resonator (dashed line) of the same emitter. The latter is magnified by a factor of ten to facilitate comparison.

This is in similar fashion to the intense activities regarding *amplification without inversion* and *lasing without inversion* [11,12]. The presented concept is therefore not only of practical but also of considerable fundamental interest. In addition, it is not limited to a specific type of THz emitter and should thus be transferable to any type of charge carrier related emitter system which can be coherently excited with ultrashort laser pulses.

The experimental setup is shown in Fig. 1. The pulse train from the Ti:sapphire laser is split into excitation and probe pulses. The excitation pulses are focused onto a thin film InGaAs surface field emitter placed between two spherical mirrors comprising the THz resonator. The generated THz pulses in specular reflection direction travel back and forth between the resonator mirrors and provide the background field for the emitter. The emitter is antireflection-coated to reduce intracavity losses. The residual reflectivity is exploited by using the emitter itself as output coupler, reflecting part of the radiation into the detection path consisting of two off-axis paraboloids and a photoconductive antenna as receiver. The antenna is gated by the optical probe pulses. The remaining part of the radiation is transmitted through the emitter and accumulates inside the resonator. However, due to the linear geometry of the resonator, half of the available radiation is coupled out opposite to the detection path and lost for the experiment.

To ensure optical pumping of the emitter synchronous to the impinging THz background field, the cavity lengths L (repetition rates, f_{rep}, respectively) of the exciting Ti:sapphire laser and THz resonator must be identical. This is achieved by carefully adjusting the distance between the spherical mirrors. The high repetition-rate of 1 GHz allows to build a compact THz resonator with a round trip length of only 30cm ($f_{rep}=L/c$, where c is the speed of light). The overall roundtrip phase change has to be an even integer of 2π to ensure phasematching of the background field and the subsequently generated THz pulse. Here, the Gouy effect, adding an extra phase shift of π for each focus passed through, has to be taken into account.

Time-resolved experiments are performed in the THz resonator configuration (Fig. 1a) and in standard emission configuration (Fig. 1b) to evaluate the amplification effect. The time-domain measurement data is converted into frequency space by Fourier transformation. The main experimental result is summarized in Fig. 2. The spectral intensity of the THz emission is amplified by roughly a factor of fifteen for frequencies between 0.3 and 0.6 THz (solid line) compared to the spectral intensity of the same emitter without resonator (dashed line). The dip around 1THz is caused by the transmission characteristics of the emitter, modulating both the intracavity transmission as well as the output coupling from the THz resonator.

In summary, by placing a surface field THz emitter inside a cavity we demonstrate for the first time a net amplification of the THz emission power. The increase amounts to more than one order of magnitude, although the results indicate that by adjusting the delicate balance between intracavity losses and output ratio more precisely, a further enhancement is feasible. Due to the inversionless nature of the amplification effect these results are of considerable fundamental and practical interest. Especially as this amplification effect is not limited to the specific type of THz emitter used, but should be applicable to any type of charge carrier related emitter system which can be coherently excited with ultrashort laser pulses.

References

[1] P.R. Smith, D.H. Auston, and M.C. Nuss, IEEE J. Quantum Electron. **QE-24**, 255 (1988).

[2] Ch. Fattinger and D. Grischkowsky, Appl. Phys. Lett. **54**, 490 (1989).

[3] X.-C. Zhang, B.B. Hu, J. T. Darrow, and D. H. Auston, Appl. Phys. Lett. **55**, 337 (1990).

[4] H.G. Roskos, M.C. Nuss, J. Shah, K. Leo, and D.A.B. Miller, Phys. Rev. Lett. **68**, 2216 (1992).

[5] C. Waschke, H. G. Roskos, R. Schwedler, K. Leo, H. Kurz, and K. Köhler, Phys. Rev. Lett. **70**, 3319 (1993).

[6] *For an overview see:* M.C. Nuss and J. Orenstein, in *Millimeter and Sub-Millimeter Waves*, ed. by G. Gruener (Springer Verlag, Heidelberg, 1998).

[7] R. Martini, F. Hilbk-Kortenbruck, P.Haring Bolivar, and H. Kurz , in *6th Int. Conf. On THz Electronics Proceedings THZ 98*, (IEEE, New York, 1998), pp. 242-245.

[8] A. Bartels, T. Dekorsy, and H. Kurz, Opt. Lett. **24**, 996-999 (1999).

[9] M.S.C. Luo, S.L. Chuang, P.C.M. Planken, I. Brener, H.G. Roskos, and M.C. Nuss, IEEE J. Quantum Electron. **QE-30**, 1478 (1994).

[10] T. Kuhn, in *Theory of Transport Properties of Semiconductor Nanostructures*, ed. by E. Schoell (Chapman and Hall, London, 1998).

[11] O. Kocharovskaya, Phys. Rep. **219**, 175 (1992).

[12] A. Belyanin, C. Bentley, F. Capasso, O. Kocharovskaya, and M.O. Scully, Phys. Rev. A **64**, 013814 (2001).

Transient Conductivity in Single-Crystal Al_2O_3 Probed by THz Time-Domain Spectroscopy

J. Shan[1], F. Wang[1], E. Knoesel[1,2], M. Bonn[3], M. Wolf[4], and T.F. Heinz[1]

[1] Departments of Physics and Electrical Engineering, Columbia University, 538 West 120 Street, New York, NY 10027
E-mail: tony.heinz@columbia.edu
[2] Current address: Department of Chemistry and Physics, Rowan University, 201Mulica Hill Rd., Glassboro, NJ 08028-1701
[3] Leiden Institute of Chemistry, P.O. Box 9502, 2300 Leiden, The Netherlands
[4] Freie Universität Berlin, Arnimallee 14, 14195 Berlin, Germany

Abstract: THz time-domain spectroscopy is applied to probe photo-induced charge transport in single-crystal Al_2O_3. The conductivity of the material and its temporal evolution yield the scattering rate, plasma frequency, and dynamics. The temperature dependence indicates that LO phonons play an important role in the scattering process.

1. Introduction

Terahertz time-domain spectroscopy (THz TDS) is a powerful probe of charge transport and ultrafast carrier dynamics in condensed matter. The method is particularly useful when used in conjunction with an optical pulse that generates charge carriers [1,2]. In this work, we report initial results of the application of this technique to investigate transient conductivity over a range of temperatures (77K – 300K) in single-crystal Al_2O_3 (sapphire). Electronic transport in sapphire and other insulators, unlike the case for semiconductors and metals, is very difficult to observe under equilibrium conditions [3]. It can, however, be readily studied by ultrafast pump-probe measurements. Aside from its intrinsic interest, characterization of charge transport in such dielectrics is essential for understanding electrical [4] and laser-induced [5] breakdown phenomena. While optical pump-probe techniques have been applied to study photo-induced carriers in sapphire [6], measurements of the THz response are particularly valuable because of their direct correlation with the transport properties of the mobile charge carriers.

2. Experimental Methods

In these measurements, the laser source was an amplified, mode-locked Ti:sapphire system, which provided pulses of approximately 100-fs duration at a wavelength of 810 nm. Charge carriers in sapphire (band gap of 8.7 eV) were created through a two-photon absorption process by femtosecond ultraviolet pulses at the third harmonic frequency. The pump-induced changes of the material were probed by THz electromagnetic pulses introduced at a defined delay time. The THz pulses were generated and detected, respectively, by optical rectification and electro-optic sampling of the femtosecond laser pulses in ZnTe crystals.

At each sample temperature, the electric-field waveform of THz pulses transmitted through the unexcited sapphire crystal and the pump-induced change in the THz waveform were recorded. The real and imaginary parts of the transient conductivity were then extracted [2].

3. Results and Discussion

Results at two representative temperatures, 295 K and 100 K, are illustrated in Figs. 1 and 2 (dots). We found that the photo-induced conductivity could be described approximately by a Drude model. From such a fit (lines), we obtained the plasma frequency and carrier scattering rate. At 295 K and 100 K, we found widely differing scattering rate of ~ 13.8 and 1.6 THz, respectively, corresponding to scattering times of 70 and 640 fs. The present approach allows us to separate the behavior of the scattering rate from other temperature-dependent effects (such as carrier capture in shallow traps) that complicate the interpretation of conductivity data [7]. The strong increase of the carrier scattering rate with temperature could be described roughly (above the lowest temperatures) by an activation energy of ~ 0.1 eV. This behavior suggests that LO phonons play a major role in the scattering process, as has been predicted theoretically [8].

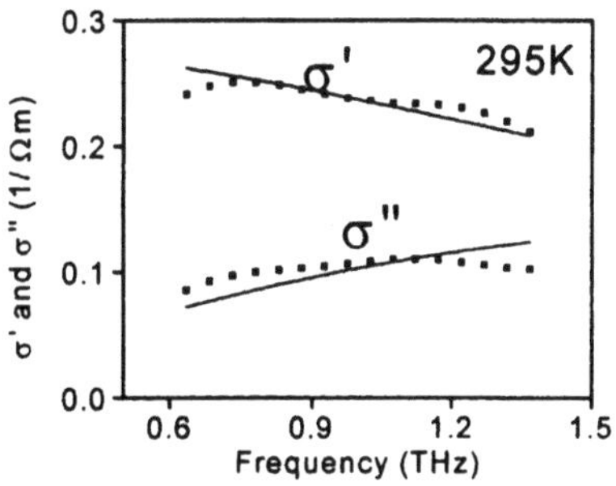

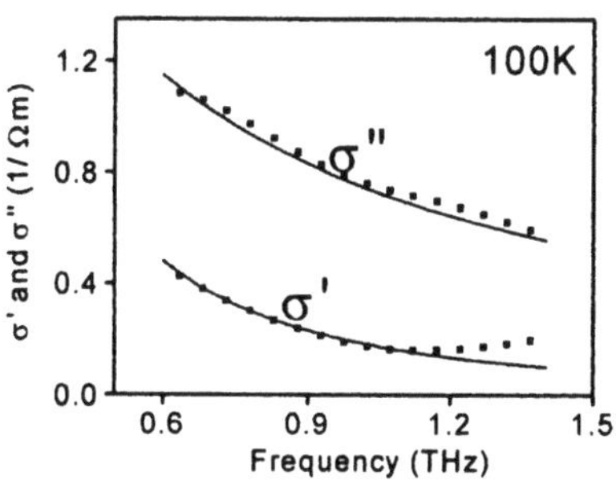

Figs. 1, 2. Frequency dependence of the change in real ($\Delta\sigma'$) and imaginary part ($\Delta\sigma''$) of the conductivity (dots) at 295 and 100 K. The UV pump fluence was 0.3 mJ/cm^2 at a pump-probe delay of 5 ps. The solid lines are from a fit to the Drude model.

By varying the delay time between the UV pump pulse and the THz probe pulse, we recorded the temporal evolution of the density of the photo-generated charge carriers (Fig. 3). We observed a decay time in the range of 10's of picosecond, which varied appreciably from sample to sample. The measured lifetimes showed no significant dependence on the pump fluence and relatively weak dependence on temperature.

The time scale for the decay of the induced THz response in our studies was found to be similar to that of an earlier measurement of optically induced THz response in sapphire reported by Haran *et al.* [9]. In this study somewhat different excitation conditions were used and the decay was attributed to phonon relaxation, rather than to the dynamics of charge carriers. To confirm our interpretation of the electronic nature of the response for our excitation conditions, we performed a

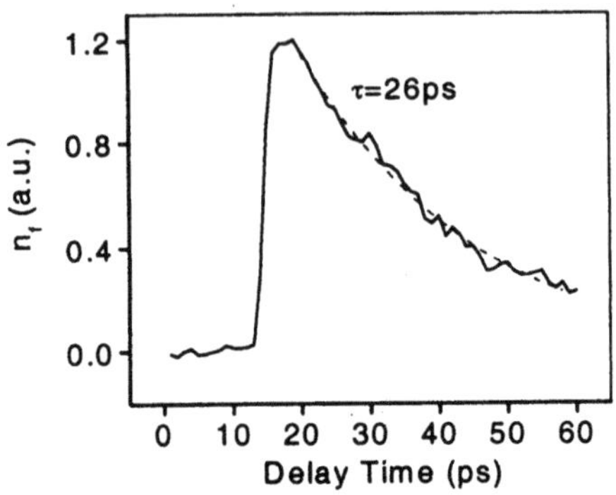

Fig. 3. Time-resolved measurements of the density of mobile carriers in sapphire for photo-excitation at a fluence of 25 J/m^2 by THz pump-probe spectroscopy. The fit is an exponential decay with a time constant of 26 ps.

complementary time-resolved measurements with near-infrared probe pulses. A similar decay was observed. This situation is expected for an electronic effect, but is difficult to explain as the consequence of an induced response of phonons. With respect to the measured decay rate of THz conductivity, the time scale of 10's of picosecond is too short to be attributed to the loss of charge carriers by radiative electron-hole recombination. Further, as noted above, the decay rate was found to vary from sample to sample. This behavior suggests the decrease in the density of mobile carriers results from localization at trap sites, such as impurities and defects.

Acknowledgment: This work was supported by the Nanoscale Science and Engineering Initiative of the NSF under award CHE-0117752 and by the US Air Force Office of Scientific Research.

References

1. For recent applications, see for example: M.C. Beard, G.M. Turner, and C.A. Schmuttenmaer, Phys. Rev. B **62**, 15764 (2000); R.D. Averitt et al. Phys. Rev. B **63**, 140502 (2001) and references therein.
2. E. Knoesel, M. Bonn, J. Shan, and T.F. Heinz, Phys. Rev. Lett. **86**, 340 (2001).
3. F.G. Will, H.G. deLorenzi, and K.H. Janora, J. Am. Ceram. Soc. **75**, 295 (1992).
4. J. J. O'Dwyer, *The Theory of Electrical Conduction and Breakdown in Solids and Dielectrics* (Clarendon, Oxford, 1973).
5. N. Bloembergen, IEEE J. Quantum Electron. **QE-10**, 375 (1974).
6. F. Queré et al., Appl. Phys. B **68**, 459 (1999); S. Guizard et al., Europhys. Lett. **29**, 401 (1995).
7. R.W. Klaffky, B.H. Rose, A.N. Goland, and G.J. Dienes, Phys. Rev. B **21**, 3610 (1980).
8. D.J. Howarth and E.H. Sondheimer, Proc. R. Soc. (London) **219**, 53 (1953); K.K. Thornber and R.P. Feynman, Phys. Rev. B **1**, 4099 (1970) and references therein.
9. G. Haran, W. Sun, K. Wynne, and R.M. Hochstrasser, Chem. Phys. Lett. **274**, 365 (1997).

T-ray Diffraction Tomography

Bradley Ferguson[1,2,3], Shaohong Wang[1], Doug Gray[2], Derek Abbott[3], and X.-C. Zhang[1]

[1] Department of Physics, Applied Physics and Astronomy, Rensselaer Polytechnic Institute, Troy, NY 12180, USA
Email: zhangxc@rpi.edu
[2] CRC for Sensor, Signal and Information Processing, Technology Park, Mawson Lakes Boulevard, Mawson Lakes, SA 5095, Australia
[3] Centre for Biomedical Engineering and Department of Electrical & Electronic Engineering, Adelaide University, SA 5005, Australia

Abstract. We demonstrate tomographic imaging using pulsed THz radiation. This diffraction tomography system provides a speed increase of an order of magnitude over previous T-ray computed tomography systems and allows diffraction effects to be observed and utilized during reconstruction.

1 Introduction

T-ray computed tomography (T-ray CT) is an important new tomographic imaging modality utilizing ultrafast laser sources to probe the optical properties of 3 dimensional (3D) structures in the far-infrared [1]. It promises to have important applications in non-destructive mail and packaging inspection, semiconductor testing and manufacturing quality control.

An ultrafast Ti:sapphire laser is used to generate broadband pulses of THz radiation. The T-ray CT system directly measures the transmitted amplitude and phase of the THz field at multiple projection angles and thus allows a wealth of information to be extracted from the target object including both its 3D structure and its frequency dependent far-infrared optical properties.

Using the previously described system[1] diffraction effects are neglected as the transmitted THz radiation is focussed and measured with a single detector. The target is scanned in x and y dimensions to determine the spatial properties of the target at each projection angle. This results in excessive image acquisition times. In this diffraction tomography system 2D electro-optic sampling is used together with modified reconstruction algorithms to dramatically improve the capabilities of T-ray tomography.

2 Diffraction Tomography

T-ray CT works well for targets with features that are large relative to the wavelength of the THz radiation (0.3 mm at 1 THz), however for more complex targets with fine structure the filtered back-projection algorithm is un-

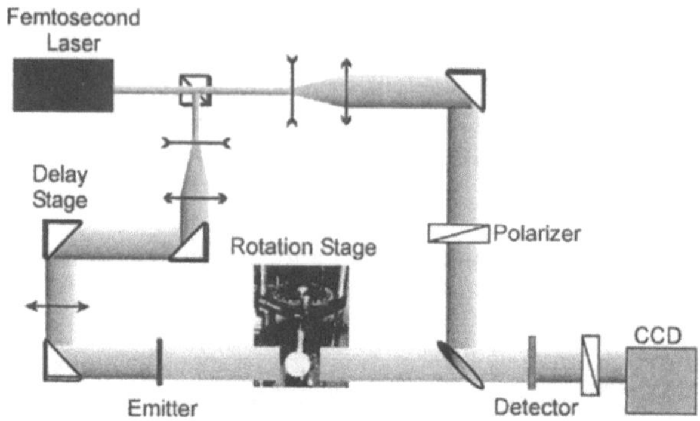

Fig. 1. T-ray diffraction tomography hardware schematic. Lenses are used to expand the pump and probe beams to a diameter of 2.5 cm. THz pulses are generated using a wide aperture photoconductive antenna and detected using 2D electro-optic sampling in a 4 mm thick ⟨110⟩ ZnTe detector crystal.

able to accurately reconstruct the target because diffraction effects dominate the measurements[2].

The T-ray diffraction tomography system is shown in Fig. 1. THz pulses are generated using a regeneratively amplified Ti:sapphire laser with an average output power of 700 mW, a pulse duration of 130 fs and a repetition rate of 1 kHz, incident on a biased photoconductive antenna. The pump and probe beams are expanded to a beam diameter of 2.5 cm using telescope lens systems. After transmission through the sample the 2D profile of the THz pulse is measured using a large 4 mm thick ZnTe electro-optic crystal. The crossed polarizers reduce the background CCD intensity [3]. A single motion stage is required to scan the THz temporal profile and the target is rotated to obtain an image at multiple projection angles. The full data acquisition period for a CCD exposure time of 50 ms per frame and a projection step size of 10° is approximately 8 minutes.

To test this system a simple polyethylene cylinder with a radius of 1 mm was imaged. Figure 2 shows the measured image for a given projection angle. The effects of diffraction are clearly visible. Reconstruction of this data using the filtered backprojection algorithm was not possible because of the diffraction effects. Authors have previously considered using the diffracted field to reconstruct the target [4]. We adopted a reconstruction based on the Born linearization of the wave equation called filtered backpropagation[5]. This approximation is only valid for small targets but was able to reconstruct this simple target as shown in Fig. 3. For more general targets iterative methods for solving the full non-linear wave equation must be employed[6].

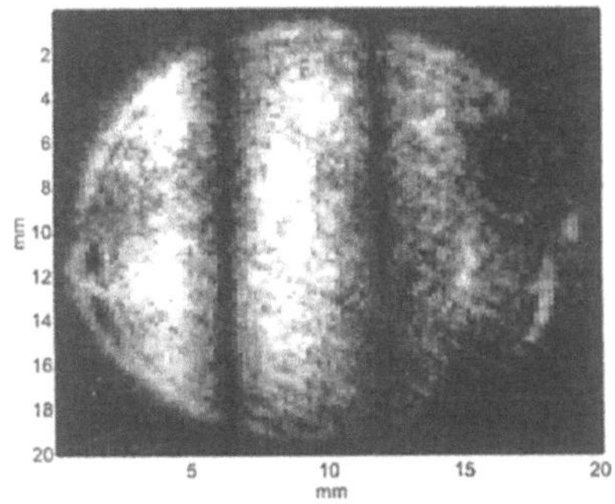

Fig. 2. THz image of a thin polyethylene cylinder for a single projection angle. The measured data was Fourier transformed and the amplitude at a frequency of 0.5 THz is plotted. The 2 cm diameter circular aperture of the detector is visible as are two vertical lines resulting from diffraction of the THz radiation.

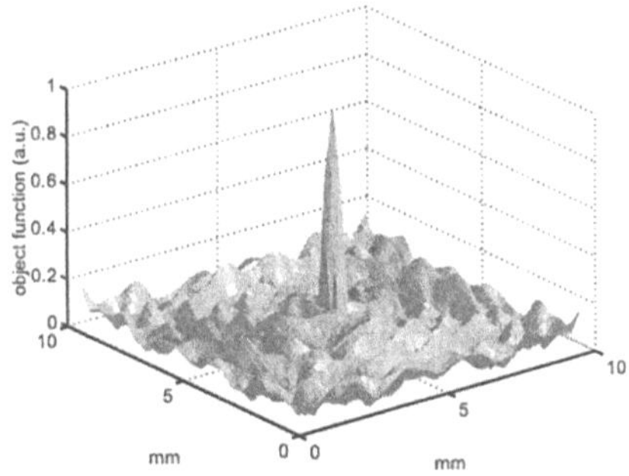

Fig. 3. Reconstructed cross-section of the polyethylene cylinder. The height shows the intensity of the reconstructed data and is related to the refractive index. The reconstruction was performed using the Born approximation and the Fourier Slice Theorem.

3 Conclusion

T-ray diffraction tomography extends the potential of terahertz time-domain spectroscopy to several new application areas. It is capable of reconstructing the 3D structure and frequency dependent far-infrared optical properties of an object, this can be performed relatively quickly by aid of 2D THz imaging with a CCD camera. Importantly the measurement of the target's diffraction pattern may allow complex targets to be reconstructed.

References

1. B. Ferguson, S. Wang, and X. Zhang, "T-ray computed tomography," in *2001 IEEE/LEOS Annual Meeting Conference Proceedings*, pp. PD1.7–PD1.8, IEEE, (San Diego) 2001.
2. B. Ferguson, S. Wang, D. Gray, D. Abbott, and X.-C. Zhang, *Optics Lett* (to be published) (2002)
3. Q. Wu, T. D. Hewitt, and X.-C. Zhang, *Appl. Phys. Lett.* **69**, 1026 (1996)
4. A. B. Ruffin, J. Decker, L. Sanchez-Palencia, L. Le Hors, J. F. Whitaker, T. B. Norris, and J. V. Rudd, *Optics Lett.* **26**, 681 (2001)
5. A. J. Devaney, *Ultrasonic Imaging* **4**, 336 (1982)
6. F. Natterer and F. Wübbeling, *Mathematical Methods in Image Reconstruction*, (Society of Industrial and Applied Mathematics, Philadelphia 2001)

Generation of tunable THz radiation by femtosecond tilted-pulse-front excitation

János Hebling[1], Gábor Almási[1], Ida Z. Kozma[1], and Jürgen Kuhl[2]

[1] Department of Experimental Physics, University of Pécs, H-7624 Pécs, Hungary
E-mail: Hebling@fizika.ttk.pte.hu
[2] MPI für Festkörperforschung, D-70569 Stuttgart, Germany
E-mail: j.kuhl@fkf.mpg.de

Abstract. We propose a generally applicable velocity matching method for efficient THz generation by optical rectification of ultrashort light pulses below the phonon frequency. It is based on tilting of the pulse front of the visible pump pulse. In a proof of principle experiment, tuning of the THz output in the 9.0 - 10.7 THz range is demonstrated.

1. Introduction

Optical rectification and electro-optic sampling are the most frequently used methods for the generation and time-resolved detection of ultrashort THz pulses, respectively [1]. Both effects are efficient only if the group velocity v_{vis}^{gr} of the ultrashort light pulse is matched to the phase velocity v_{THz}^{ph} of the generated THz radiation [2]:

$$v_{vis}^{gr} = v_{THz}^{ph} \tag{1}$$

Here, we suggest a general method to reach velocity matching below the phonon frequency and present the results of a proof of principle experiment. The proposed technique is applicable for velocity matched electro-optic sampling, too.

2. Experimental setup

The important feature of the novel excitation geometry proposed is illustrated in Fig. 1: the pulse front of the light beam is tilted with respect to the phase front, which is perpendicular to the propagation direction of the light beam. The THz radiation (in the nonlinear material the phonon-polariton) excited impulsively along the pulse front will propagate according to Huygens's principle perpendicularly to this front with the velocity v_{THz}^{ph}. In this situation, not the velocity of the light pulse, but its projection onto the propagation direction of the THz radiation has to be equal to v_{THz}^{ph}:

$$v_{vis}^{gr} \cdot \cos\gamma = v_{THz}^{ph} \tag{2}$$

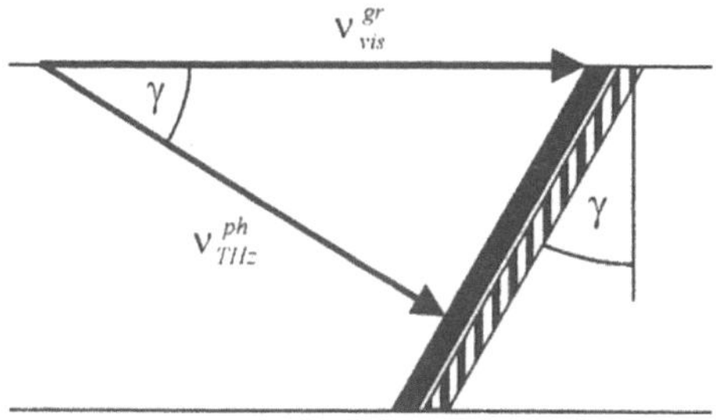

Fig. 1. Scheme of impulsive THz generation using a tilted pulse front. The ultrashort light pulse propagates to the right, while the THz radiation propagates perpendicularly to the generating pulse front, that is, under an angle γ with respect to the optical pulse.

According to Eq. (2), it is possible to change the matched velocity of the THz radiation by simply changing the tilt angle γ of the exciting light pulse front. Since v^{ph}_{THz} depends on the frequency of the excited THz radiation, this frequency can be tuned by changing γ.

In our experiment, 25-fs-long pulses from a Ti:sapphire laser at a repetition rate of 76 MHz were applied as pump pulses. Measuring the transmission of identical probe pulses with variable delay monitored the THz radiation generated by these pulses. The tilt of the pulse front of the pump and probe pulses was introduced by diffraction off a 600 line/mm reflection grating. The grating was imaged into the 130-μm-thick nonlinear material slab (GaP) by a reflecting microscope objective with a demagnification of about 3.

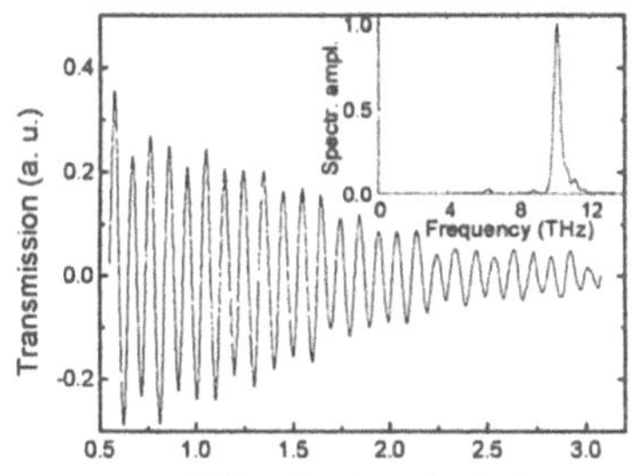

Fig. 2. Oscillation of the probe transmission caused by the generated phonon-polariton, and Fourier transform of the time-resolved signal (inset).

3. Results and Discussion

Fig. 2 shows the measured transmission versus the delay for γ = 39.3°. The beginning 0.5-ps-long part of the signal is omitted because it contains large amplitude oscillations caused by the optical Kerr effect. The signal observed for delay times larger than 0.5 ps shows strong oscillations caused by the excited phonon-polariton. The position of the corresponding 0.44 THz broad spectral peak at 10.16 THz in the Fourier transform (see inset) of the transmission trace is in perfect agreement with the 10.12 THz calculated from the polariton dispersion curve. We were able to continuously tune the polariton frequency from 9.0 to 10.7 THz by changing the demagnification.

Theoretical simulations of the polariton generation process show that the generated THz radiation should be tunable across the range from 0.7 - 0.95 times the phonon frequency (see Fig. 3a). On the other hand, these calculations predict a very broadband THz radiation if γ is adjusted for velocity matching at less than half of the phonon frequency (see Fig. 3b). For this case the oscillation in the time domain is strongly damped. Such THz pulses are well suited for measurements with high temporal resolution. The tunable narrow band THz pulses are expected to become useful for spectrally and temporally resolved THz imaging.

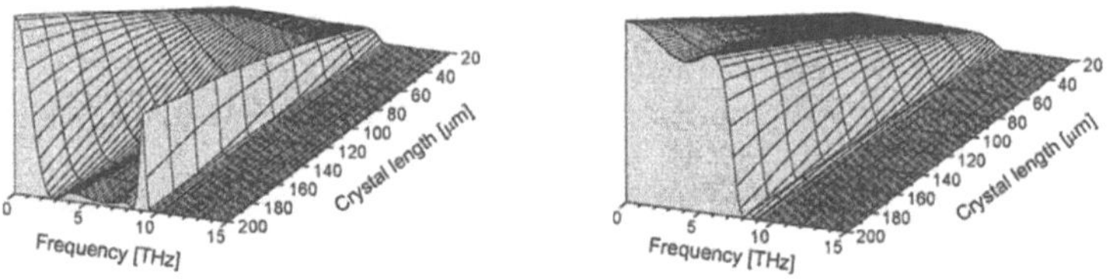

Fig. 3. Predicted spectra of the THz radiation vs. the length of the nonlinear crystal for (a) velocity matched, for 300 cm^{-1} and (b) 200 cm^{-1}, respectively.

In comparison to the usual Cherenkov-like THz excitation mechanism [3] the novel THz generation process using optical rectification with tilted-pulse-front excitation offers two important advantages: i) the THz power is simply scalable since the exciting beam area can be increased if higher laser pump powers are available, ii) the generated THz radiation is a plane wave which is easier to handle than the cone-shaped wave generated by the Cherenkov-like excitation.

Acknowledgements. This work was supported by the Hungarian National Research Fund under grant Nr.: T038372.

References

1 1. R. Huber, A. Brodschelm, F. Tauser, and A. Leitenstorfer, Appl. Phys. Lett. **76**, 3191, 2000.
2 A. Nahata, A.S. Weling, and T.F. Heinz, Appl. Phys. Lett. **69**, 2321, 1996.
3 D.H.Auston, K.P. Cheung, J.A. Valdmanis, and D.A. Kleinman, Phys. Rev. Lett. **53**, 1555, 1984.

Theory of THz-generation via optical rectification of fs-pulses in periodically poled nonlinear materials

Carsten Weiss[1], Garik Torosyan[1], Yuri Avetisyan[2], and René Beigang[1]

[1] Fachbereich Physik, Universität Kaiserslautern, Kaiserslautern, Germany
Tel.: +49 631 205 2180, Fax.: +49 631 3906
Email: beigang@physik.uni-kl.de
[2] Microwave engineering department, Yerevan State University, Yerevan, Armenia
Email: gerbarcr@ysu.am

Abstract: The generation of tunable THz-radiation via optical rectification of fs-pulses in periodically poled nonlinear materials is treated theoretically and compared with experimental results including generation in forward and backward direction, emission perpendicular to the surface and Cherenkov radiation.

1. Introduction

THz-generation via optical rectification of fs near infrared pulses has proven to be very efficient for the generation of tunable narrowband THz-pulses [1-5] which can easily be tuned by various methods. In particular, the emission of THz-radiation perpendicular to the propagation of the near infrared pump pulse allows for efficient generation as the influence of any strong absorption of THz-radiation in the nonlinear material can be reduced considerably.

2. Theoretical considerations

Assuming an ultrashort (pulse length $\Delta\tau$) bandwidth limited optical pulse with a Gaussian temporal and spatial profile polarized along the z-axis and propagating in direction of the x-axis (s. Fig. 1) which is rectified in a periodically poled nonlinear crystal the generated THz-pulse can be calculated. The source of the rectified field is the second order nonlinear polarization P_{nl}, which propagates along the x-axis and changes its sign after each domain with length $\Lambda/2$. The radiated field can be determined using the radiating antenna model which results in a far field electric field distribution in the frequency domain given by

$$E(\omega,R)=\frac{-\omega\mu_0 e^{-ikR}}{4\pi R}\cdot\int_V j(\omega,r)e^{ik(r-R)}dr \qquad (1)$$

The displacement current $j(\omega,r)$ is given by the time derivative of the nonlinear polarization. In the frequency domain the z-component of the displacement current can be written as

$$j_z(\omega,r)=i\omega P_z(\omega,r)\,. \qquad (2)$$

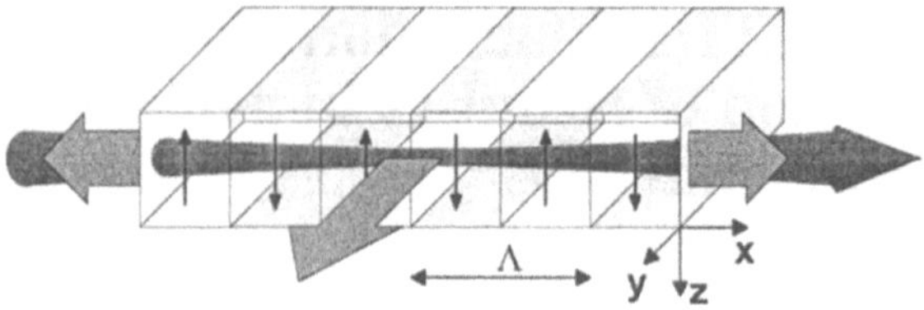

Fig. 1. Geometry of the periodically poled nonlinear crystal for optical rectification. THz radiation is emitted in forward and backward direction as well as perpendicular to the surface.

The spectral field amplitude caused by optical rectification of the ultrashort visible pulse is then given by

$$E_z(\omega,R) = const. \cdot \Delta\tau \cdot \frac{e^{-ikR}}{R} \cdot \omega^2 \cdot e^{-\frac{\Delta\tau^2\omega^2}{4}} \times \iiint d(x) \frac{w_0^2}{w^2(x)} \cdot e^{-2\frac{y^2+z^2}{w(x)}} \cdot e^{-i(k_x x - \omega \frac{x n_{IR}}{c} + k_y y + k_z z)} dxdydz \tag{3}$$

This equation is used to determine the temporal and spectral properties of the generated THz-pulses.

For THz-emission perpendicular to the direction of the pump beam a small spot size of the near infrared beam is required [4] in order to account for the conservation of momentum. The maximum spot size which still allows for emission of THz radiation perpendicular to the direction of pump beam propagation depends on the poling period and the refractive indices n_{THz} and n_{IR} at the generated THz- and the near infrared wavelengths, respectively. From our theoretical considerations we obtain

$$w_0^{\max} < \frac{\sqrt{2} \cdot \Lambda \cdot n_{IR}}{\pi \cdot n_{THz}}. \tag{4}$$

3. Comparison with experimental results

Our experimental set-up used to compare the theoretical simulations with experimental results consists of a conventional THz-setup (100-fs Ti:sapphire laser operating at 780 nm with a repetition rate of 82 MHz, average output power up to 1.5 W) with a fast silicon-on-sapphire (SOS) photoconductive switch as the THz detector. Our set-up allowed for the detection of THz radiation emitted in any arbitrary direction with respect to the direction of the pump beam including forward, backward and perpendicular emission. Periodically poled lithium niobate (PPLN) was used as nonlinear material.

A typical result is shown in Fig. 2 where experimentally determinded data are compared with theoretical simulations for THz-radiation emitted perpendicular to the direction of beam propagation. In the simulations of the temporal and spectral

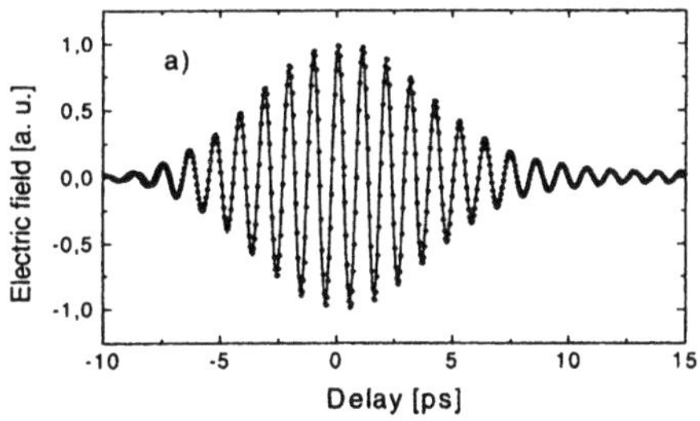

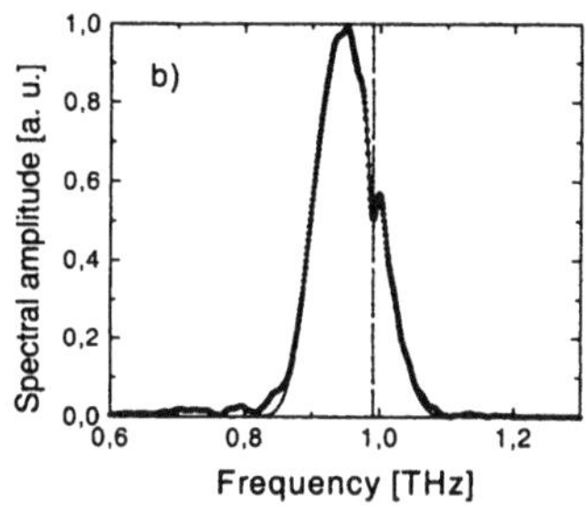

Fig. 2. Comparison between experimentally obtained data (dots) and theoretical simulations (solid lines) for a THz-pulse emitted perpendicular to the direction of beam propagation in PPLN: a) temporal profile and b) spectral profile. The dip in the spectral profile is caused by a water vapor absorption line.

wave form the spatial characteristics of our detection system as well as the spatial profile of the focused near infrared beam had to be taken into account. There is excellent agreement between theory and experiments.

4. Summary

In summary we have presented a theoretical analysis of THz-generation via optical rectification in periodically poled materials and compared the results with experimental data. The comparison includes emission in a direction perpendicular to the direction of beam propagation, emission in forward and backward direction and Cherenkov radiation. Starting from the nonlinear polarization induced in the nonlinear material and applying a simple radiating antenna model we have obtained excellent agreement both in the time and frequency domain.

References

1 Y. S. Lee, T. Meade, V. Perlin, H. Winful, T. B. Norris, and A. Galvanauskas, Appl. Phys. Lett. **76**, 2505, 2000.
2 Y. S. Lee, T. Meade, M. DeCamp, T. B. Norris, and A. Galvanauskas, Appl. Phys. Lett. **77**, 1244, 2000.
3 Y.-S. Lee and T. Meade and M. L. Naudeau and T. B. Norris and A. Galvanauskas, Appl. Phys. Lett. **77**, 2488, 2000.
4 C. Weiss, G. Torosyan, Y. Avetisyan and R. Beigang, Opt. Lett. **26**, 563 (2001)
5 C. Weiss, G. Torosyan, J.-P. Meyn, R. Wallenstein, Y. Avetisyan and R. Beigang, Optics Express **8**, 497, 2001.

THz generation by third-order non-linearities in air and air plasmas

Torsten Löffler, Susanne Eden, Mark Thomson, and Hartmut G. Roskos

Physikalisches Institut, Johann Wolfgang Goethe – Universität Frankfurt am Main, Robert-Mayer-Str. 2-4, D-60054 Frankfurt am Main, Germany
E-mail: t.loeffler@physik.uni-frankfurt.de

Abstract. Focusing of high-energy laser pulses overlapped with their second harmonic pulses into air leads to THz generation. We present a detailed study on the enlargement of this effect when a plasma is formed in the focus.

1. Introduction

Several processes lead to emission of a THz pulse upon focusing of a high-energy laser beam into air. If a plasma is generated, ponderomotive forces within the laser focus induce an ultrafast charge separation which leads to emission of a conical THz front [1,2]. Additional application of a dc electrical bias field to the air plasma significantly increases the THz emission intensity in the forward direction [3,4]. Recently, a third method to generate THz pulses in air was introduced [5]: An intense THz pulse was generated by the superposition of the fundamental and the second-harmonic (SH) pulses of a laser beam. The origin of the THz signal was attributed to four-wave rectification processes in the air. Unfortunately, the data of Ref. [5] only touched the threshold for photo-ionization of the air molecules leaving open the question whether the THz-generation efficiency is strongly changed when crossing the ionization threshold. In this contribution, we specifically explore this regime.

2. Experimental and Results

In our experiment, we employed the method of Ref. [5] and focused 150-fs, 775-nm laser pulses with an energy of up to 500 μJ through a 100-μm-thick SHG-phased-matched BBO crystal into air (at room temperature and 1 atm). The generated THz pulses were detected in forward direction (see details in Ref. [3,4]). The relative phase between the fundamental and the SH pulses was adjusted for maximum THz signal amplitude by rotating a thin quartz-glass plate placed in-between the BBO crystal and the focus. In contrast to Ref. [5], we focused the laser beams harder in order to achieve higher intensities in the focus and thus were able to cover the two intensity regimes with and without plasma generation.

As an experimental complication we found THz-pulse emission from the BBO crystal itself in addition to the THz-pulse generation in the focus of the laser beam. To distinguish between the two signal components, the visible light but not the THz signal was blocked between the BBO crystal and the focus in order to measure the pure BBO signal component. The pulse generated in the focus is then determined by subtracting the BBO signal component from the detected overall signal. In Fig. 1, the SH conversion efficiency (a) and the THz amplitude from the BBO crystal (b) are shown as a function of the rotation angle of the BBO crystal. The variation of the BBO-generated THz signal with the rotation angle and the linear dependency on the laser pulse energy (c) are in agreement with the results of model calculations assuming a $\chi^{(2)}$ optical rectification process based on the electro-optic tensor of BBO (see dashed line in Fig. 1(b)).

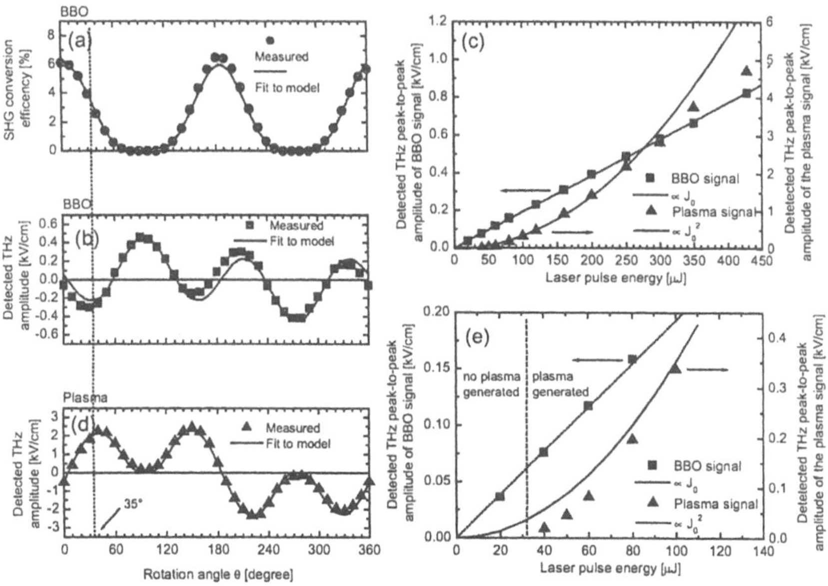

Fig. 1. SHG efficiency (a), detected THz amplitudes from the BBO crystal (b) and from the focus (d) as a function of the crystal rotation angle; (c) detected THz peak-to-peak amplitudes as a function of laser pulse energy; (e) zoom into the low-energy region of (c).

The amplitude of the THz signal emitted from the focus strongly varies with the rotation angle of the BBO crystal (see Fig. 1(d)). This is a consequence of the angle dependence of the polarization of the SH light generated in the type-I BBO crystal. At the angle setting for maximum conversion efficiency (0°, 180°, etc.), the SH light is polarized perpendicularly to the light at the fundamental wavelength. When the crystal is rotated, the polarization of the SH pulse changes gradually, until it is parallel to the polarization of the fundamental light just when the conversion efficiency drops to zero (see Fig. 1(a)). With respect to THz emission from the focus it is important to note that only the parallel components of the fundamental and the SH pulses will mix because the air, respectively air plasma, are isotropic media with the corresponding symmetry properties of the $\chi^{(3)}$ tensor representing the four-wave mixing process in the focus. This interplay of pulse intensities and polarization dependence results in a theoretical dependence of the THz emission process on the tuning angle of the BBO crystal as shown in Fig. 1(d) which matches the experimental data well. The model calculations explain why extrema of the THz amplitude are found at angles of 35°, 145°, 215°, and 325°.

Fig. 1 (c) displays the peak-to-peak amplitude of the THz signal from the focus as a function of the laser pulse energy. In a medium regime from 80 μJ up to 250 μJ, the signal varies quadratically with the laser pulse energy. The deviations at low and high energies will be addressed below. A quadratic variation of the THz amplitude is consistent with the $\chi^{(3)}$ model if one assumes a constant $\chi^{(3)}$ value and neglects both the depletion of the fundamental light during SHG and the saturation of the SHG efficiency. In this case, the amplitude of the THz field can be fitted with the following dependence on the energy J_ω of the fundamental laser pulse

$$E_{THz} \propto \chi^{(3)} \cdot E_\omega^2 \cdot E_{2\omega} \propto \chi^{(3)} \cdot J_\omega^2 \qquad \text{with: } E_\omega^2 \propto J_\omega \qquad \text{and: } E_{2\omega} \propto \chi_{BBO}^{(2)} \cdot J_\omega \,. \qquad (1)$$

Fig. 1 (e) displays the data of Fig. 1 (c) for low laser pulse energies on an expanded energy scale. In this representation it is obvious that the THz-pulse amplitudes for laser energies below 80 μJ are lower than expected from the $\chi^{(3)}$ fit discussed before. Interestingly, this deviation occurs in the regime where the plasma volume grows from zero (at a threshold pulse energy of about 30 μJ) to the size of the focal region (at about 80 μJ). This suggests that the quadratic fit to the data for energies above 80 μJ relates to the $\chi^{(3)}$ value of an air plasma and not to that of air itself. The deviation of the low-energy data from the fit hence leads to the interpretation that the THz mixing efficiency is reduced at lower laser pulse energies when no plasma is formed. A quantitative analysis not presented here indicates that the $\chi^{(3)}$ value of the air plasma is two orders of magnitude larger than the $\chi^{(3)}$ of air.

The deviation of the THz amplitude at high laser pulse energy from the quadratic fit is attributed to plasma-related effects such as phase mismatch of the waves and absorption.

3. Conclusions

In summary, we have shown that THz generation in the focus of high-intensity laser pulses becomes significantly more efficient when a plasma is generated.

4. Acknowledgement

This work was funded by the GSI (Gesellschaft für Schwerionenforschung).

References

1 H. Hamster, A. Sullivan, S. Gordon, W. White, and R. W. Falcone, in *Phys. Rev. Lett.*, Vol. 71, 2725, 1993.
2 H. Hamster, A. Sullivan, S. Gordon, and R. W. Falcone, in *Phys. Rev. E,* Vol. 49, 671, 1994.
3 T. Löffler, F. Jacob, and H. G. Roskos, in *Appl. Phys. Lett.,* Vol. 77, 453, 2000.
4 T. Löffler and H. G. Roskos, in *J. Appl. Phys.*, Vol. 91, 2611, 2002.
5 D. J. Cook and R. M. Hochstrasser, in *Opt. Lett.*, Vol. 25, 1210, 2000.

THz emission characteristics of dipole antennas fabricated on Multi-energy arsenic-ion-implanted GaAs and semi-insulating GaAs

Tze-An Liu [1], Masahiko Tani [2], Gong-Ru Lin [3], and Ci-Ling Pan [1]

[1] Institute of Electro-Optical Engineering, National Chiao Tung University, 1001, Ta-Hsueh Rd, Hsinchu, Taiwan 300, R.O.C.
E-mail: taliu.eo85g@nctu.edu.tw, clpan@cc.nctu.edu.tw
[2] Kansai Adv. Res. Ctr., Communications Res. Lab., Ministry of Posts and Telecommunications, 588-2 Iwaoka, Nishi-ku, Kobe, Hyogo 651-2401, Japan.
E-mail: tani@crl.go.jp
[3] Graduate Institute of Opto-Electronic Technology, National Taipei University of Technology, No.1, Sec. 3, Chung Hsiao East Rd, Taipei 10643, Taiwan.
E-mail: grlin180@ms15.hinet.net

Abstract. We compare the performance of THz dipole antennas fabricated on Multi-energy arsenic-ion-implanted GaAs (multi-GaAs:As^+) and Semi-insulating (S. I.) GaAs. High damage threshold biasing (> 60 kV/cm) and large saturating optical pumping power (> 30 mW) for multi-GaAs:As^+ based devices are reported.

1. Introduction

A large number of semiconductors have been investigated as broad band THz emitters and detectors for applications in THz sensing, imaging and spectroscopy [1,2]. In particular, one of the authors [3] have performed a comparative study of bow-tie and dipole antennas based on the material systems of semi-insulating (S. I.) GaAs and low-temperature molecular beam epitaxy (MBE) grown GaAs (LT-GaAs). Arsenic-ion-implanted GaAs or GaAs: As^+, exhibits structural, electrical, and ultrafast optoelectronic characteristics strikingly similar to those of LT-GaAs [4]. GaAs:As^+, could thus be an alternative to LT-GaAs for THz applications. In this work, we present a comparative study of THz dipole antennas fabricated on GaAs:As^+ and semi-insulating GaAs substrates. Pumping power and bias dependence of THz emission from the two class of emitters are reported.

2. Experimental Methods

The GaAs:As^+ antennas were prepared by first bombarding S. I. GaAs substrate with Arsenic ions at doses of 10^{16} ions/cm^3 and energies of 150keV and 200keV. This is followed by furnace annealing at 600^0C for 30minutes. The spacing of the dipole gap was 5μm. The photoconducting sampling technique with a 100-fs-class laser was used for detection [3].

3. Results and Discussions

The emitted THz waveforms from antennas fabricated on S. I. GaAs and GaAs: As^+ are shown side by side in Fig. 1. The uni-polar waveforms for the GaAs: As^+ device as opposed to the bipolar waveform for S. I. GaAs device is typical for devices using ion-implanted material as substrate [5]. Other features of GaAs: As^+ THz antennas include shorter THz radiation pulse width (about 275fs), as opposed to 310 fs for S. I. GaAs. The peak of the THz signal also arrives at an earlier time. Both features are tentatively attributed to the faster acceleration and earlier carrier transfer from the Γ to the L valley because of the lower carrier mobility of GaAs: As^+. This results in a rapid drop of the photocurrent. These features are reminiscent of THz radiation from dipole antenna fabricated on LT-GaAs [6]. The bias dependence of THz pulse width and peak shift from (a) S. I. GaAs and (b) multi-GaAs: As^+ fabricated dipole antennas are shown in Fig. 2.

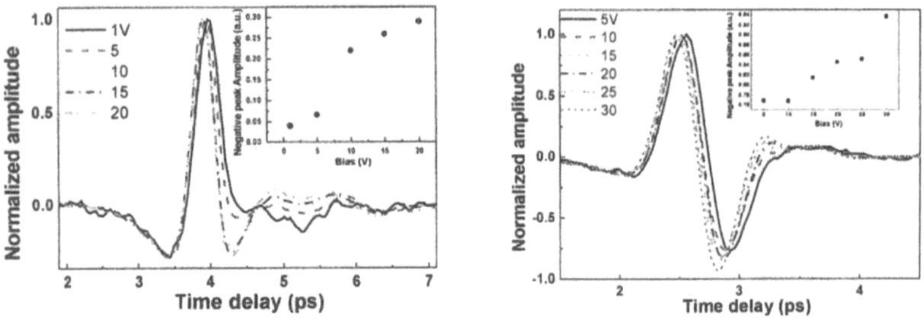

(a) (b)

Fig. 1. Bias dependence of THz radiation waveform from (a) S. I. GaAs and (b) multi-energy-implanted GaAs: As^+ fabricated dipole antenna. The pumping power was 10 mW. The Insets show the bias-dependence of the amplitudes of the negative peak after the main peak.

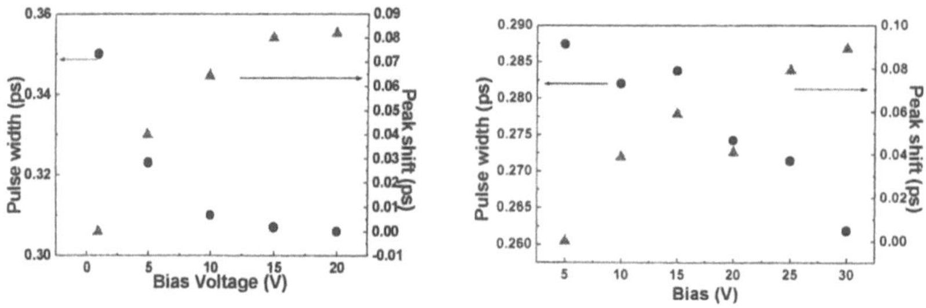

(a) (b)

Fig. 2. Bias dependence of THz pulse width and peak shift from (a) S. I. GaAs and (b) multi-GaAs: As^+ fabricated dipole antenna.

Figure 3 (a) and (b) illustrate the bias field and pumping power dependencies of the THz radiation amplitude from 5μm-gap-size dipole antenna fabricated on S. I. GaAs and GaAs: As^+. In Fig. 3(a), the pumping power and bias voltage were 5~20mW and 5V respectively. Clearly GaAs: As^+ exhibits high damage threshold voltage (>60kV/cm) in GaAs: As^+. This implies the possibility of even larger THz radiation amplitude than that from S. I. GaAs based devces. The saturation optical power (>30mW) of GaAs: As^+ based dipole antennas are also higher than S. I. GaAs based ones (~20mW). These attributes are reminiscent of trends observed in LT-GaAs [3].

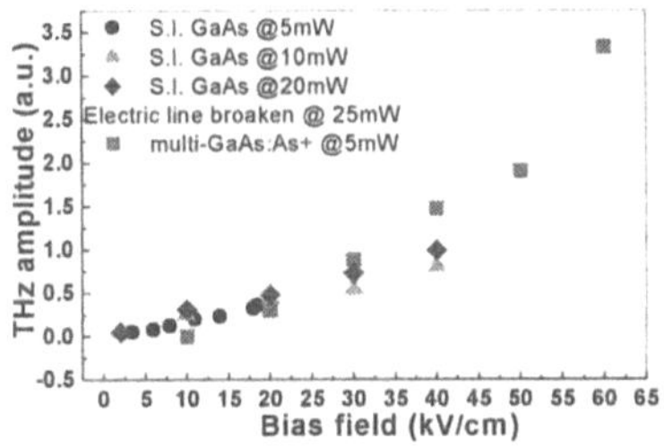

(a)

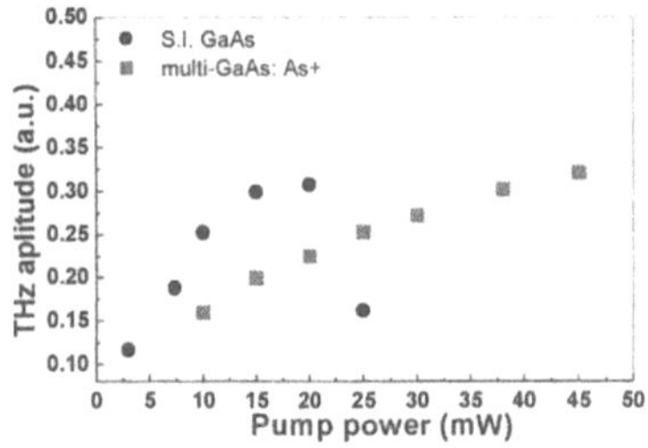

(b)

Fig. 3. (a) Bias field and (b) pumping power dependencies of the amplitudes of THz radiation waveforms for dipole antenna fabricated on S. I. GaAs and GaAs: As^+ with gap size of 5μm and pumped at pumping power of 5 to 20mW. The bias voltage for Fig. 3(b) is 5V

4. Conclusions

In summary, we show that the THz radiation waveforms of multi-energy-implanted GaAs: As^+ is uni-polar and exhibit a sharper negative peak, which reflects the ultrafast carrier lifetimes of GaAs:As^+ versus S. I. GaAs. Other potential advantages of Multi-energy-implanted GaAs: As^+ include high damage threshold biasing (> 60 kV/cm) and large saturating optical pumping power (> 30 mW).

References

1 X. -C. Zhang and D. H. Auston: J. Appl. Phys. Vol. 71, 326, 1992.
2 P. C. M. Planken, M. C. Nuss, W. H. Knox, D. A. B. Miller and K. W. Goossen: Appl. Phys. Lett. Vol. 61, 2009, 1992.
3 M. Tani, S. Matsuura, K. Sakai, and S. Nakashima: Appl Opts Vol. 36, 7853, 1997.
4 F. Ganikhanov, G. -R. Lin, and C. -L. Pan, Appl. Phys. Lett. Vol. 67, 3465, 1995.
5 G. R. Lin and C. L. Pan, Appl. Phys. B. Vol. 72, 151, 2001.
6 C. Ludwig and J. Kuhl: Appl. Phys. Lett. Vol. 69, 1194, 1996.

All-Optoelectronic CW THz Imaging for Tumor Recognition

Karsten J. Siebert[1], Torsten Löffler[1], Holger Quast[1], Rainer Leonhardt[1], Mark Thomson[1], Tobias Bauer[1], Stephanie Czasch[2], and Hartmut G. Roskos[1]

[1] Physikalisches Institut der J. W. Goethe-Universität, Robert-Mayer-Str. 2-4, D-60054 Frankfurt (M), Germany
[2] Institut für Veterinär-Pathologie, Universität Gießen, Frankfurter Str. 96, D-35392 Gießen, Germany

Abstract. We present an all-optoelectronic THz imaging system based on photomixing of two continuous laser beams using photoconductive antennas. Examples of images of pathological samples are shown and the capability to distinguish between different tissue types is demonstrated.

1. Introduction

Until recently optoelectronic THz imaging [1] has been restricted to pulsed systems providing data in the time domain which has to be Fourier transformed to obtain the frequency information. However the use of continuous-wave (cw) THz radiation generated by photomixing in combination with a coherent photoconductive detection scheme [2] promises higher spectral power in combination with a higher frequency resolution. Measuring in the frequency domain requires no Fourier transformation of the signals. As single-mode high-power diode lasers are becoming commercially available a strong reduction in size and cost is anticipated. Recently a coherent cw THz imaging system was demonstrated by us [3].

In this paper we describe the system and address what techniques can be applied to enhance the imaging contrast between different types of biological tissue.

2. Experimental Set-up

The experimental set-up of our cw THz imaging system is shown in Fig. 1. The two tunable optical cw wavelengths (centered around 780 nm) are generated by two unidirectional, single-mode ring cavities that share one Ti:Sapphire crystal as gain medium. The output at each wavelength is 160 mW. The two wavelengths are combined at a 50/50-beam-splitter to obtain the optical beat signal modulated at the difference frequency (tunable from 0 THz to 10 THz) of the two wavelengths.

The two modulated beams are then directed onto the LT-GaAs emitter antenna and detector.

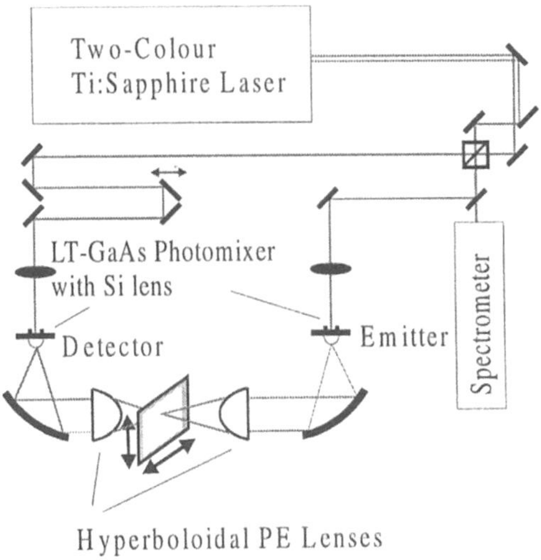

Fig. 1: Experimental set-up of the cw THz imaging system.

The cw THz radiation is guided with the help of off-axis paraboloidal mirrors and plano-hyperboloidal lenses (f = 20 mm) [3]. The signal detected by the lock-in amplifier is the current generated by the electric field of the THz radiation in the detector antenna which in turn is gated by the second THz-modulated optical beam. The system has a spatial resolution < 300 μm and the dynamic range (power) is 5000:1 at 1 THz. Images can be taken between 0.2 and 1.5 THz. The high-frequency cut-off is determined by the roll-off in the THz power generated by the photoconductive THz emitter.

3. Imaging Contrast

The biomedical samples studied were formalin-fixed, dehydrated and wax-mounted according to the standard histological procedure. An example is given in Figure 2 (photograph and THz transmission). The sample is a thin-cut piece of a dog's skin containing a mast-cell tumor. The THz image allows a distinction between the different types of tissue. The tumor can be identified. Imaging contrasts originate from the differences in absorbance and refractive-index of the various tissue types as well as from inhomogeneities which lead to scattering of the THz signal. Refractive-index related contrasts as well as scattering turn out to be very useful for imaging. Contrasts based on these mechanisms can be enhanced with optimized measurement techniques. Refractive-index changes can be exploited best in phase-change pictures [3] corrsponding to time-of-flight measurements in the case of THz pulses. Scattering contrast can be enhanced with the help of dark-field techniques [4].

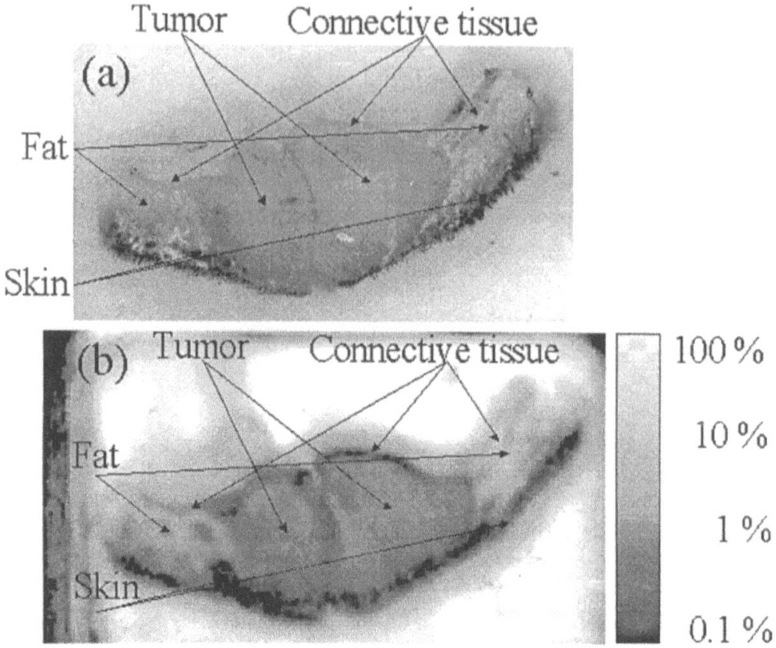

Fig. 2: a) Photograph of a wax-mounted thin-cut dog's skin tumor (mast cell tumor) embedded in healthy tissue; object size: 32 mm x 24 mm x 3mm; spacing between the centers of neighboring pixels: 0.25 mm. b) Logarithmic THz power transmission at 1 THz measured with the cw system.

Acknowledgements. This work is funded by DFG and by the TERAVISION project of the EU.

References

1 B. B. Hu and M. C. Nuss, Optics Lett. 20, 1716, 1995.

2 S. Verghese, K. A. McIntosh, S. Calawa, W. F. DiNatale, E. K. Duerr and K. A. Molvar, Appl. Phys. Lett. **73**, 3824, 1998.

3 K. J. Siebert, H. Quast, R. Leonhardt, T. Löffler M. Thomson, T. Bauer, H. G. Roskos and S. Czasch, Appl. Phys. Lett. **80**, 3003, 2002.

4 T. Löffler, T. Bauer, K. Siebert, H. G. Roskos, A. Fitzgerald and S. Czasch, Opt. Express **9**, 616, 2001.

Time-domain differentiation of terahertz signals

A. Filin, M. Stowe, and R. Kersting[1]

Department of Physics, Applied Physics, and Astronomy, Rensselaer Polytechnic Institute, Troy, NY 12180, USA
[1]E-mail: kerstr@rpi.edu

Abstract. We report on the transmission of terahertz signals through subwavelength grating structures. Time-resolved experiments show that the transmitted field is the first time-derivative of the incident wave. The experimental results are in accord with classical electromagnetic theory and may enable novel applications for ultrafast pulse shaping.

1. Introduction

Light transmission through metal structures is strongly reduced when the dimension of the apertures becomes smaller than the wavelength. Recent works on subwavelength structures have demonstrated an increased transmission mediated by surface plasmons [1]. In this work, we show that metallic subwavelength structures also significantly modify the shape of the pulse. The transmitted field is the first time-derivative of the incident pulse.

Terahertz time-domain spectroscopy (THz-TDS) is an ideal technique to investigate the transmission because of its subcycle time resolution. Our results show that the time-domain differentiation is scale invariant with respect to the wavelength and may permit novel applications for tailoring ultrafast light pulses.

2. Experimental Methods

Metallic transmission gratings were fabricated on semi-insulating silicon by electron beam evaporation. The 10 mm x 10 mm gratings have periods between 10 μm and 40 μm and a filling factor of 50%. The 200 nm gold films are significantly thicker than the skin depth, which is approximately 30 nm at 1 THz. We measured the transmission through the gratings by free-space THz-TDS, where time-resolution is achieved by electro-optic sampling [2].

3. Results and Discussion

The transmission through gratings of subwavelength period strongly depends on the orientation of the grating lines relative to the incident electric field of the THz pulse. Figure 1a) shows the case of perpendicular orientation. As expected, the THz pulses are perfectly transmitted through the grating. Both amplitude and phase of the incident light are preserved. In the following we will discuss exclusively the nontrivial case of parallel orientation. Figure 1b) shows that in this case the transmitted field amplitude is significantly reduced. Additionally, it appears as if the transmitted signal arrives earlier than the reference THz pulse, which was sent through an uncovered area on the same wafer. A detailed analysis shows that this effect is not due to superluminal propagation, but due to diffraction, which results in a time-domain differentiation of the incident light pulse.

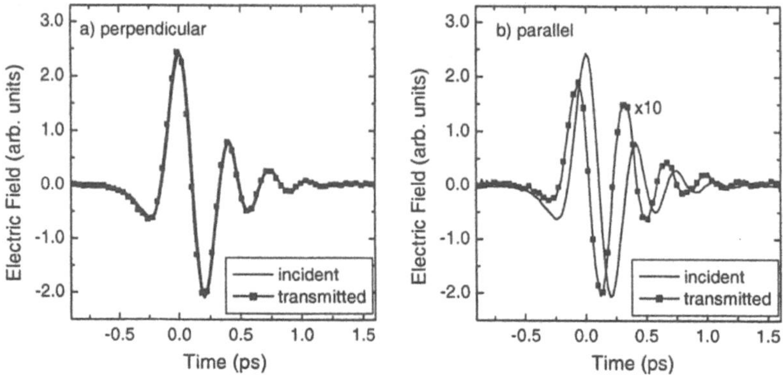

Fig. 1. THz signals transmitted through a metallic grating of 10 μm period compared to the incident pulse. a) Grating oriented perpendicular to the electric field of the incident THz pulse. b) Parallel orientation.

If the grating period is small compared to the wavelength of the incident pulse, only the zero-order diffraction signal is transmitted. We have shown that the transmission signal can be expanded in a series of time-derivatives [3]:

$$E^{tr}(t) = \frac{d \ln 2}{\pi c}\frac{\partial}{\partial t}E^{inc}(t) - \left(\frac{d \ln 2}{\pi c}\right)^2 \frac{\partial^2}{\partial t^2}E^{inc}(t) + \left(\frac{d \ln 2}{\pi c}\right)^3 \frac{\partial^3}{\partial t^3}E^{inc}(t) - \left(\frac{d \ln 2}{\pi c}\right)^4 \frac{\partial^4}{\partial t^4}E^{inc}(t) \pm \ldots \qquad (1)$$

where $E^{inc}(t)$ is the electric field of the incident pulse and d is the grating period. For frequency components, which fulfill $\omega d \ln 2/\pi c \ll 1$, the first term of the above equation is dominant and the transmission signal is the first time-derivative of the incident pulse.

Figure 2 shows a comparison between experimental results and calculation. The calculated transient reproduces the negative phase shift of the transmitted pulse. Furthermore, the transmission spectra follow the linear slope of the calculated data. The used 10 μm grating enables a precise time-domain differentiation between 0.75 and 3.5 THz. In general, the precision strongly depends on the relation between wavelength and grating period. According to eq. (1) an increase of the grating period would lead to contributions of higher order.

Measurements on gratings with increased period confirm the appearance of deviations from the first-order derivative. A relation of $d/\lambda = 0.02$ reduces higher order contributions and allows an analog differentiation with a precision of 97%. Further experiments demonstrated the successful application of the method to pseudo-arbitrary THz sequences.

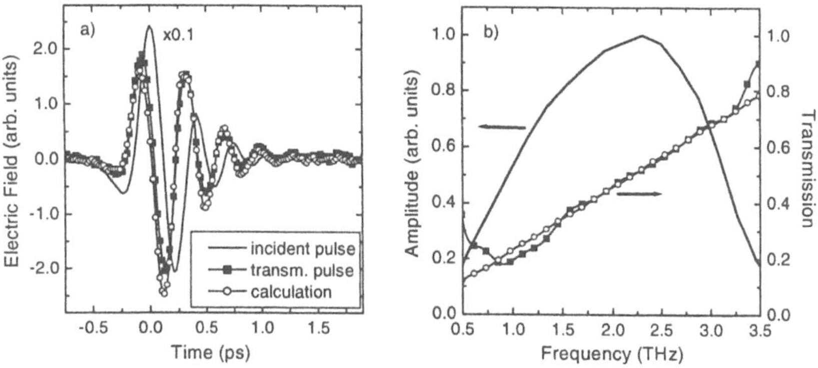

Fig. 2. a) Transients of the incident THz pulse (reduced by a factor of 0.1), transmission signal, and calculated transmission signal. The grating period is 10 μm. b) Amplitude spectra of the incident THz pulse and calculated and measured transmission amplitude.

4. Conclusions

Light transmission through metallic structures of subwavelength period leads to a differentiation of the incident field in time-domain. Our time-resolved THz experiments and calculations show that the differentiation results from diffraction. The phenomenon is not limited to few-cycle THz pulses. The above findings are scale-invariant and also hold when near-infrared laser pulses are transmitted through gratings of nanometer period. Such devices may permit novel applications for shaping of ultrafast laser pulses.

References

1 T. Ebbesen, H. Lezec, H. Ghaemi, T. Thio, and P. Wolff, *Nature* **391**, 667 (1998). L. Martin-Moreno, F. Garcia-Vidal, H. Lezec, K. Pellerin, T. Thio, J. Pendry, and T. Ebbesen, *Phys. Rev. Lett.* **86**, 1110 (2001). L. Salomon, F. Grillot, A. Zayats, and F. de Fornel, *Phys. Rev. Lett.* **86**, 1114 (2001).

2 Q. Wu and X. -C. Zhang, *Appl. Phys. Lett.* **67**, 3523 (1995).

3 A. Filin, M. Stowe, and R. Kersting, *Opt. Lett.* **26**, 2008 (2001).

Spectral amplitude and phase changes in diffraction of Terahertz pulses.

Oleg Mitrofanov, Mark Lee, L. N. Pfeiffer, K. W. West.

Bell Laboratories, Lucent Technologies, Murray Hill, NJ 07974 USA
E-mail: olegm@lucent.com

Abstract. Waveform characteristics of short Terahertz pulses, such as frequency chirp and carrier phase, induce complex transformations of the pulse in diffraction. We experimentally demonstrate that for a described class of Terahertz pulses the diffraction transformation reduces to the fundamental $\pi/2$-phase shift and spectral amplitude deformation.

Diffraction of Terahertz pulses usually results in complex changes of the pulse waveform. The deformation effects include changes of pulse duration [1], temporal advancement or delay [2], spectral blue shift, and generation of low-frequency components [3]. In near-field microscopy these effects occur when the pulse couples through a subwavelength aperture of the near-field probe [4]. Some of these effects are induced by characteristics of the incident pulse waveform such as carrier phase and frequency chirp. They can be avoided using a special preparation of the Terahertz pulses. The fundamental transformation due to diffraction includes only spectral filtering and a $\pi/2$-phase shift. The pulse envelope remains practically unchanged.

Terahertz pulses experience only the fundamental transformation in the near-field diffraction through a subwavelength aperture if all of the following conditions are satisfied: (1) the central wavelength is much longer than the aperture size $\lambda >> d$, (2) the pulse does not have a carrier frequency chirp, (3) the duration of the pulse is longer than the carrier oscillation period. Pulses that satisfy these conditions can be generated using photoconductive switches and will be discussed latter.

Spectral bandwidth of few-cycle pulses is extremely large, while the transmission coefficient of a small aperture is strongly wavelength dependent. The aperture size in the near-field microscopy is typically much smaller than the wavelength. In this regime the transmission coefficient of the aperture varies continuously with frequency according to a power law: $E/E_0 \propto \omega$ in the near-field zone ($E/E_0 \propto \omega^2$ in the far-field zone). If the pulse spectrum extends to wavelengths as short as twice the aperture size it excites the lowest aperture mode. The diffracted field exhibits resonant oscillations and the pulse loses its temporal confinement [1]. Fig. 1(a) shows waveforms of the incident and diffracted field in the case that the aperture is much smaller than the central wavelength. Note, that the aperture size does not affect the transformation because the transmission coefficient is a linear function of frequency for small apertures ($\lambda >> d$).

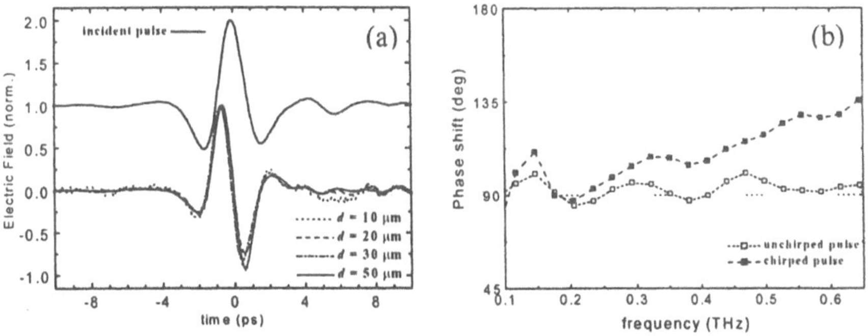

Fig. 1. (a) Normalized to its peak value waveforms of the incident and diffracted pulses. The electric field amplitude is 1.6, 5.6, 30, and 150 times smaller than that of the incident pulse for d = 50, 30, 20, and 10 μm respectively. (b) Phase shift due to diffraction on the aperture for negatively chirped and unchirped pulses (d = 30 μm).

The incident field induces edge currents on the aperture screen. E-field due to the induced currents equals the opposite of the incident field at the surface to yield zero total tangential E-field. The total field behind the aperture at distances comparable to the aperture size is $E_{inc}(t) - E_{inc}(t+\tau)$, where τ represents retardation of the induced field due to geometrical path difference. If $E_{inc}(t)$ is harmonic then the diffracted field amplitude is proportional to ω and the phase is $\pi/2$ shifted.

THz pulses generated by transient currents in photoconductive switches typically are negatively chirped. Phase shift due to diffraction of a pulse with varying carrier phase differs from $\pi/2$ [Fig. 1(b)]. In addition, the pulse intensity envelope of a chirped pulse is modified due to wavelength dependent transmission coefficient [2].

The phase shift of the carrier under the pulse intensity envelope results in the phase-amplitude modulation of the pulse spectrum. Spectral content of pulses with duration shorter than the oscillation period ($2\pi/\omega_0$) is not conserved under the phase shift transformation [3,5]. This effect results in enhancement of the low-frequency components of the diffracted pulses and depends on the carrier phase of the incident pulse. Fig. 2(a) shows the ratio of the aperture transmission coefficients measured using pulses with FWHM of the intensity envelope equal to 0.6 and 1.5 times the oscillation period.

Chirp of the THz pulses originates from the photocurrent dynamics in the photoconductive switch and it vanishes if duration of the optical excitation is longer than carrier lifetime. THz pulses with central frequency $\omega_0/2\pi$ = 0.27 THz and the intensity envelope FWHM of 2.4 ps are generated using stretched to 2 ps optical pulses out of Ti:Sapphire [Fig. 1(a)]. Longer pulses are produced by exciting the switch with two or three optical pulses delayed by ~ 2 ps. The inset of Fig. 2(b) shows a pulse generated by two 2 ps optical pulses. Diffraction transformation of this pulse consists only of the $\pi/2$ phase shift for all frequency components and spectral filtering proportional to ω. The transformation $E_{inc}(t) - E_{inc}(t+\tau)$ is also equivalent to the first time derivative of the pulse waveform.

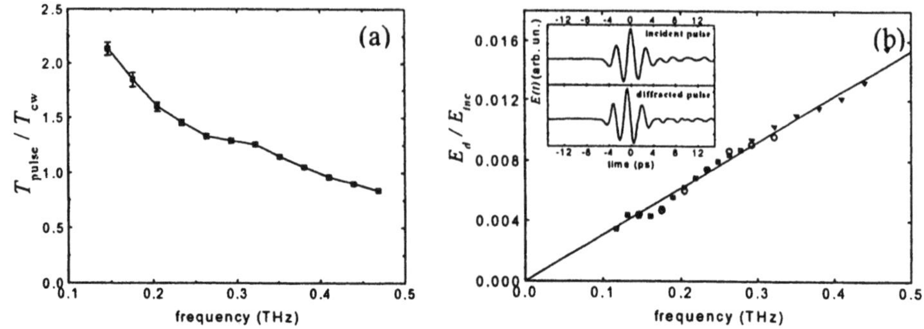

Fig. 2. (a) Ratio of the aperture transmission coefficients measured using a subcycle and 1.5-cycle (cw) pulses. (b) Filtering properties of a subwavelength aperture (d = 20 μm) characterized using 1.5-cycle pulses. Typical waveforms of the incident and diffracted pulses are shown in the inset.

Knowing deformation of the THz pulses due to subwavelength apertures used in near-field microscopy allows reconstruction of fields at an object. Relatively simple preparation of transform limited THz pulses removes chirp- and phase-shift induced deformations of the waveform and spectrum. In the limit of small apertures ($d \ll \lambda$) the deformation is independent of the aperture size.

References

1 O. Mitrofanov, M. Lee, J. W. P. Hsu, L. N. Pfeiffer, K. W. West, J. D. Wynn, J. F. Federici, in *Appl. Phys. Lett.*, 79, 907 (2001).
2 O. Mitrofanov, M. Lee, L. N. Pfeiffer, and K. W. West, in *Appl. Phys. Lett.*, 80, 1319 (2001).
3 O. Mitrofanov, L. N. Pfeiffer, and K. W. West, *to be published.*
4 O. Mitrofanov, M. Lee, J. W. P. Hsu, I. Brener, R. Harel, J. Federici, J. D. Wynn, L. N. Pfeiffer, and K. W. West, in *IEEE J-STQE*, 7(4) (2001).
5 T. Brabec and F. Krausz, in *Phys. Rev. Lett.* 78, 3282 (1997).

Resonant-cavity magnetic-field probe for mm-wave frequency-domain spatial field mapping

Ronald M. Reano[1], John F. Whitaker[1], and Linda P. B. Katehi[2]

[1] Electrical Engineering and Computer Science Dept., University of Michigan, Ann Arbor, 1301 Beal Ave., Ann Arbor, MI 48109-2122 USA
E-mail: reanorm@umich.edu
[2] School of Engineering, Purdue University, 1280 Engineering Administration, West Lafayette, IN 47907-1280 USA

Abstract. Magnetic-field distributions in the 55-75 GHz range obtained with magneto-optic cavity resonators and ultrashort probe pulses are described. Probe dimensions are tuned to achieve millimeter-wave-frequency resonance, resulting in a 10-dB sensitivity enhancement.

1. Introduction

Probes based on electro-optic and magneto-optic effects can provide a complete characterization of electromagnetic behavior. Time-domain magneto-optic sampling has been investigated by Freeman *et al.* to examine spin relaxation and magnetic resonance effects, by Elazzabi *et al.* to measure picosecond current pulses, and by Riordan *et al.* to sample transient THz beams [1-3]. In the frequency domain, Wakana *et al.* have examined low-frequency RF circuits using a magneto-optic probe with a CW laser [4]. The signal-to-noise ratio obtained from magneto-optic probes is typically far less than that obtained from electro-optic probes of similar physical dimensions [5]. In this paper, a frequency-domain technique for sensitivity enhancement is presented. By tuning the probe dimensions to form a resonant cavity at the frequency of interest, the magnetic-field inside of the probe becomes a magnified representation of the sampled field, thereby improving the resulting signal-to-noise ratio. A short pulse laser is utilized to achieve millimeter-wave operation, thus providing the means to perform internal node diagnostics on high-frequency circuits and radiating structures

2. Description of Probe and Optical Setup

The probe is a Faraday-effect-based polarization sensitive sensor consisting of a circular cylinder (length = 2.3 mm, radius = 1 mm) of terbium gallium garnet. Given the low frequency dielectric constant of 12.4, the length of the probe is approximately two wavelengths at 75 GHz. The free-space effective length of the probe is less than a half-wavelength at midband thereby allowing spatial field sampling within the Nyquist criteria without probe deconvolution.

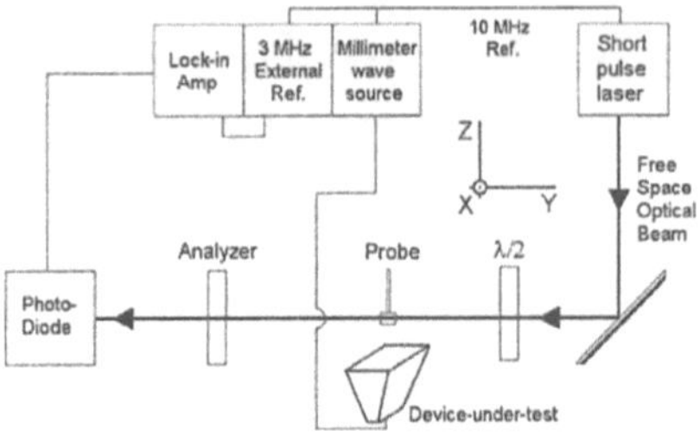

Fig. 1. Experimental setup for frequency domain spatial field mapping. The device-under-test (in this case a V-band rectangular horn antenna) is mounted on an xy translation stage.

The experimental arrangement is shown in Fig. 1. A Ti:Sapphire mode-locked laser tuned to 910 nm is used to generate a free-space sampling beam (80 MHz repetition rate, 80 fs pulse duration). The device-under-test (DUT) is fed via a millimeter-wave source configured for harmonic mixing in order to down-convert the sampled magnetic-fields to IF frequencies thereby allowing envelope detection with a MHz bandwidth photodiode [6].

3. Results and Discussion

The frequency response of the probe is shown in Fig. 2. The average millimeter wave power was 5 dBm over the frequency band. Two resonant peaks are observed at 60 GHz and 72 GHz. Off resonance, the signal-to-noise ratio is 10:1 whereas on resonance, this improves to 33:1 corresponding to a 10 dB improvement.

The resonant behavior of the magnetic-field inside the probe is confirmed with finite-element-method simulations. The simulation results are shown on the right graph of Fig. 2. In this simulation, as in the experiment, the longitudinal dimension of the probe is along the y-direction. The probe itself is centered at y = 5 mm. As shown, the field-solution for the magnetic-field at 60 GHz peaks at a value that is approximately three times larger than that at 56 GHz and 64 GHz.

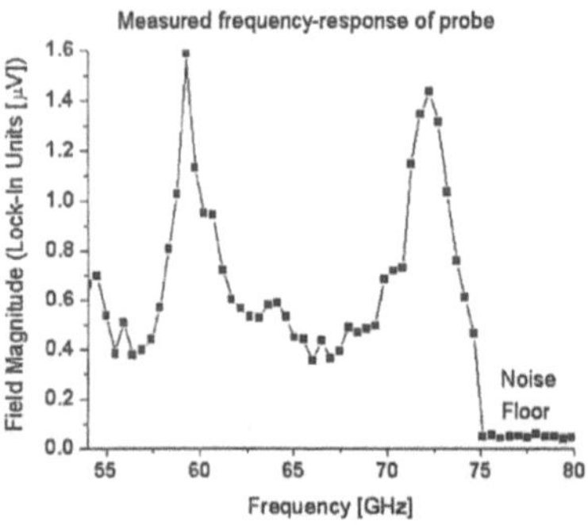

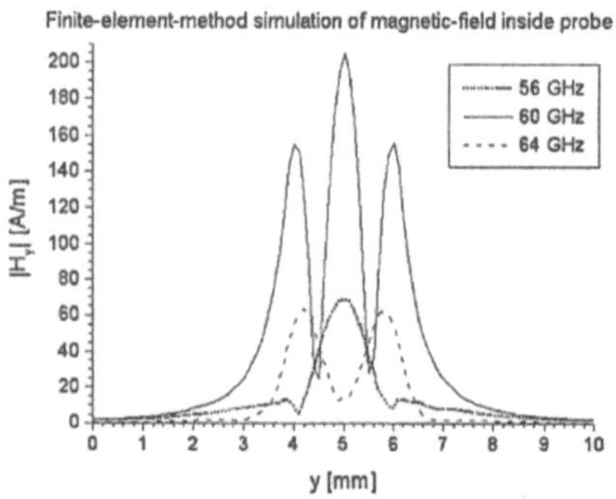

Fig. 2. Peaks in the frequency response of the probe due to the millimeter-wave field resonating in the probe. Simulations confirm the resonant behavior of the magnetic-field.

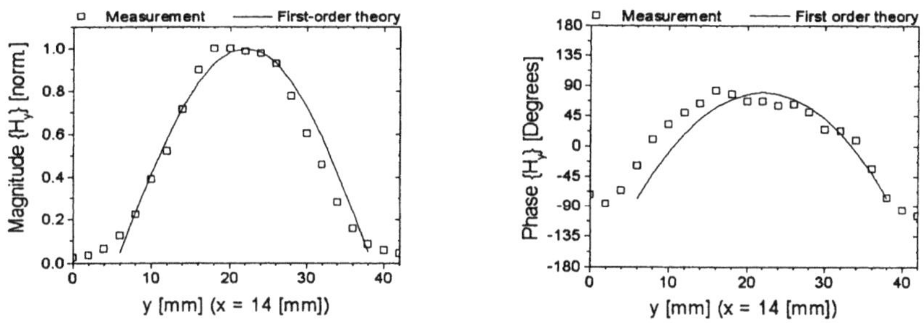

Fig. 3. Measured magnetic-field phasor as a function of position at the output of a horn antenna.

A two-dimensional field map of the horn antenna aperture was obtained at 60 GHz by raster scanning the DUT with an xy-translation stage. Line scans of these maps are shown in Fig. 3. The field was imaged in a plane approximately 1.5 mm above the antenna using a step size of 2 mm. The measurements are in good agreement with first-order aperture-antenna theory [7].

4. Conclusions

Magneto-optic probes that take advantage of millimeter-wave cavity resonances have been used to obtain spatial field maps at V-band with enhanced sensitivity. When used in conjunction with electro-optic sampling probes, the complementary probing techniques allow for a complete characterization of the electromagnetic field.

References

1. M. R. Freeman, "Picosecond pulsed-field probes of magnetic systems," Journal of Appl. Phys. **75,** 6194-6198 (1994).
2. Y. Elazzabi, M. R. Freeman, "Ultrafast magneto-optic sampling of picosecond current pulses," Appl. Phys. Lett. **68** 3546-8 (1996).
3. J. A. Riordan, F. G. Sun, Z. G. Lu, X.-C. Zhang, "Free-space transient magneto-optic sampling," Appl. Phys. Lett. **71** 1452-4 (1997).
4. S. Wakana, T. Ohara, M. Abe, E. Yamazaki, M. Kishi, M. Tsuchiya, "Fiber-edge electrooptic/magnetooptic probe for spectral-domain analysis of electromagnetic field," IEEE Trans. Microwave Theory Tech. **48** 2611-2616 (2000).
5. R. Reano, K. Yang, L. P. B. Katehi, J. F. Whitaker, "Simultaneous measurements of electric and thermal fields utilizing an electrooptic semiconductor probe," IEEE Trans. Microwave Theory Tech. **49** 2523-2531 (2001).
6. K. J. Weingarten, M. J. W. Rodwell, D. M. Bloom, "Picosecond optical sampling of GaAs integrated circuits", IEEE J. Quantum Electron. **24** 198-218 (1988).
7. W. L. Stutzman, G. A. Thiele, *Antenna Theory and Design, 2nd Ed.* (John Wiley and Sons, Inc., 1998).

Investigation of a THz Fresnel lens

E D. Walsby[1], S. Wang[2], B. Ferguson[2], J. Xu[2], T. Yuan[2], R. Blaikie[1], S.M. Durbin[1], D.R.S. Cumming[3] and X. –C. Zhang[2]

[1] Department of Electrical & Computer Engineering, University of Canterbury, Private Bag 4800, Christchurch, New Zealand.
[2] Department of Physics, Applied Physics and Astronomy, Rensselaer Polytechnic Institute, Troy, New York 12180, USA
[3] Department of Electronics and Electrical Engineering, University of Glasgow, Glasgow, G12 8QQ, UK

Abstract. Multiple level silicon Fresnel lenses have been investigated for use in the THz frequency range. Comparisons are made between a Fresnel lens and refractive silicon and polyethylene lenses at the focal point over a broad frequency range. The Fresnel lens showed an improved beam shape and a similar intensity to the refractive lenses.

1. Introduction

Due to their flexibility of design and usage, Fresnel lenses are very attractive for THz applications. For example, Fresnel lenses could be used for maneuvering THz wave fronts, fabricating THz diffraction optics or for making lens arrays for THz imaging applications [1]. Fresnel lenses are also lighter and more compact than conventional THz optics.

In broadband THz imaging and THz time domain spectroscopy applications, all the THz properties (amplitude, frequency and phase) are used to extract characteristic information from objects under study [2]. Therefore, it is crucial to obtain the time and space resolved THz field distribution for a Fresnel lens. Usually, the THz wave is detected on the focal plane of a lens or parabolic reflector hence studying the THz wave on the focal plane is especially important for THz applications. In this paper the focal plane properties of multi-level silicon Fresnel lenses fabricated using planar microfabrication techniques [3] are measured and compared with conventional refractive lenses.

2. Experimental Methods

Lenses were characterized using T-ray computed tomography, where both probe and pumping beams are expanded to 2.5 cm and collimated. Details have been published elsewhere [2]. The 2D THz image at the focal plane is transferred to the probe beam via a 4 mm thick ZnTe electro-optic crystal. This is carried by the probe beam and focused onto a CCD camera. By scanning the time delay between the THz and the probe beam, spatial and temporal THz distributions were obtained.

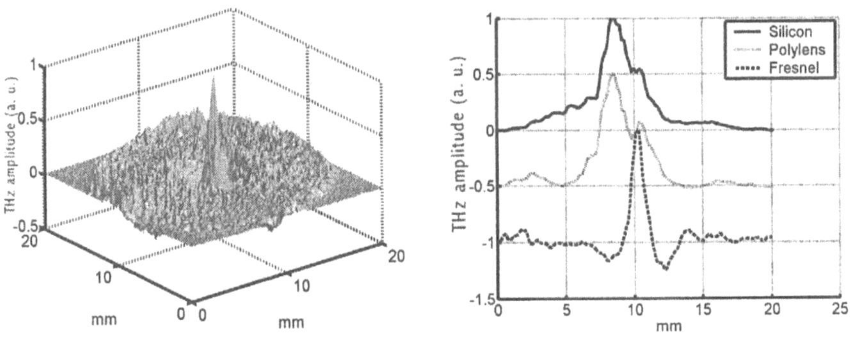

(a) (b)

Figure.1 (a) The THz 2D amplitude distribution on the focal plane of the Fresnel lens. (b) THz intensity distribution along a line across the focus center on the focal plane for all three lenses. (For clarity, the curves in (b) were vertically shifted).

In this experiment, a 4-level silicon Fresnel lens [3] (f = 2.5 cm, designed at 1 THz) was investigated. A polyethylene lens (f = 5 cm) and a biconvex silicon lens (f = 10 cm) were also considered for comparison. The Fresnel lens, polyethylene lens and silicon lens have the $F_{\#}$ of 0.83, 1 and 2.5 respectively. Fig. 1(a) shows the THz amplitude distribution on the focal plane of the Fresnel lens, and Fig. 1(b) provides the THz amplitude distribution along a line across the focal center on the focal plane for each lens.

On the focal plane of the Fresnel lens, a sharp THz focal point and a THz ring were formed. The focused THz distribution of the Fresnel lens is more symmetric compared to that of the silicon or polyethylene lenses. This means the aberration of the Fresnel lens is less than that of the silicon lens or polyethylene lens.

(a) (b)

Fig. 2 (a) THz waveforms and (b) Fourier spetcra of 3 lenses at their focal point. (For clarity, the curves in (a) were vertically shifted).

The THz waveform and Fourier spectrum of 3 lenses at the focal point were also obtained and are shown in Fig. 2. As can be seen, the Fresnel lens focused THz waves more efficiently in the frequency range of 0.5 to 1.1 THz. This result suggests that the Fresnel lens could be used as a focused THz band pass filter.

The focusing performance of the three lenses is compared at different THz frequencies and is shown in Fig. 3. The theoretical minimum focus spot is [4]

$$2w_0 = \frac{4}{\pi}\lambda F_{\#} \qquad (1)$$

where w_0 is the minimum focus radius, λ is the focus wavelength and $F_{\#}$ is the F-number of the lens. For convenience of comparison, we normalized the focal size by dividing the $F_{\#}$ number in Fig. 3.

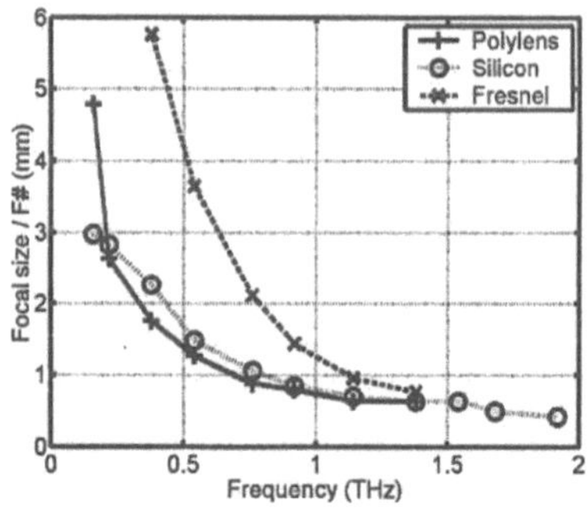

Fig. 3 Normalised spot size versus with THz frequency

The Fresnel lens shows good focusing performance at frequencies from 0.7 to 1.4 THz, and is comparable to the silicon lens at 1.4 THz.

3. Conclusions

We studied the THz temporal and spatial distribution on the focal plane of a 4-level Fresnel lens. The experimental results indicate that the Fresnel lens can effectively focus THz radiation around its design frequency. The Fresnel lens was observed to have a reduced aberration compared to the refractive lenses and provided stronger intensity in the focal spot in the 0.5 – 1.0 THz range.

References

1 S. Noach, A. Lewis, Y. Arieli, N. Eisenberg, Applied Optics, **35**, 3635, 1996.
2 B. Ferguson, S.Wang and X. -C. Zhang, "T-ray computed tomography", The 14th Annual Meeting of The IEEE Laser & Electro-Optics Society, PD 1.7, San Diego, CA November 2001.
3 E. D. Walsby, R. Cheung, R. J. Blaikie and D. R. S Cumming, Proc SPIE, **79**, 3879, 1999.
4 B. A. Saleh, M. C. Teich, in *Fundamentals of Photonics,* Wiley, New York, 1991.

Electro-optic detection of sub-wavelength THz spot sizes in the near-field of a metal tip

N.C.J. van der Valk, W. Th. Wenckebach, and P. C. M. Planken

Delft University of Technology; Faculty of Applied Sciences, Department of Applied Physics, Lorentzweg 1, 2628 CJ Delft, The Netherlands
email: N.C.J.vanderValk@tnw.tudelft.nl

Abstract. A THz pulse is incident on a metal tip. By *near-field* detection of *only* the THz electric-field components which are distorted by the tip, we measure THz spot sizes as small as 10 μm.

There is currently a strong interest to improve the spatial resolution in THz time-domain imaging for THz microscopy of single cells. To date, work on improving the resolution was based on the use of an aperture [1] or a dynamic aperture [2], to limit the THz spotsize. Apertures, however, block the longer wavelength components of the THz pulse and/or require microfabrication techniques to make them. Here, we present a new and simple method to obtain a sub-wavelength resolution in THz-TDS, in which we measure *only* the *distorted* electric-field component of a THz pulse, in the near field of a sharp metal tip. The setup is shown in Fig. 1.

A THz pulse, generated in a GaAs switch by a 15 fs, 800 nm, laser pulse, is incident on a metal tip which is held close to a (100) oriented GaP crystal. The metal tip distorts the THz electric field and creates a THz electric-field component

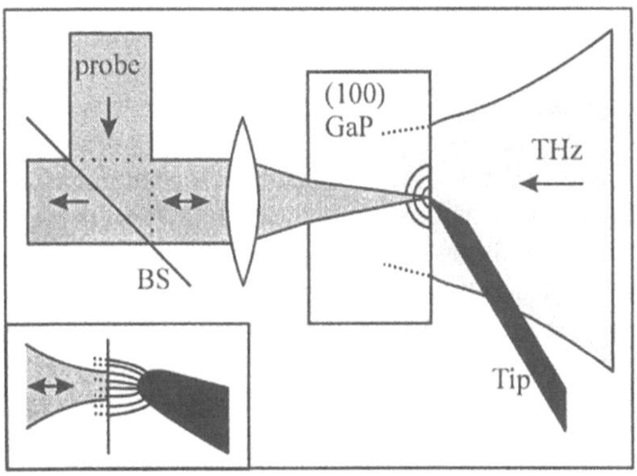

Fig. 1. Experimental setup. An 800 nm, 15 femtosecond, probe pulse is focused to about a 10 μm spot size in a (100) oriented GaP crystal. A THz pulse is incident on a sharp metal tip which locally creates a THz polarization component perpendicular to the crystal surface (inset). Through the electro-optic effect, the probe polarization is changed by this near-field THz polarization component. This is subsequently measured using a λ/4 plate, a Wollaston prism and a differential detector (not shown here).

perpendicular to the surface of the crystal (the "lightning rod effect"). This component is electro-optically detected in the GaP crystal by a tightly focused synchronized probe pulse. In a (100) oriented crystal only electric fields perpendicular to the surface generate an electro-optic effect, *making the detection process blind to the incident, parallel-polarized THz pulse.*

In Fig. 2, we plot the power spectrum of the THz pulse, calculated from the measured THz electric-field shown in the inset. For this measurement the tip was pressed against the crystal, directly above the probe focal spot. The spectrum peaks at 0.15 THz and contains frequency components up to 2.5 THz. The effect of the sharp tip is to bend and concentrate the THz electric field lines. This enhancement gives us a sufficiently large electro-optic signal to measure a 30 ps long THz electric field with a dynamic range of 50 in a total measurement time of 20 ms.

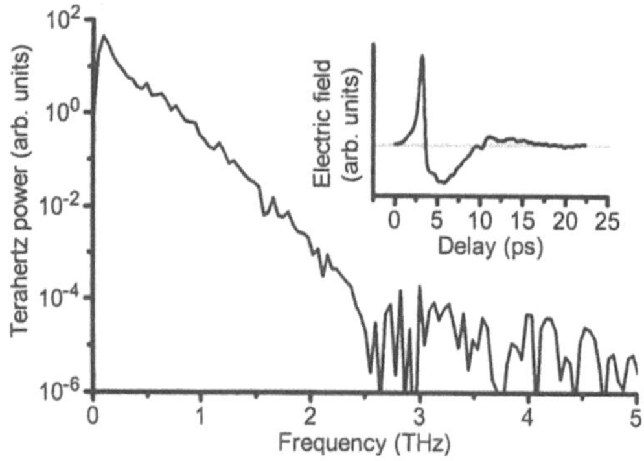

Fig. 2. THz power spectrum, calculated from the measurement of the THz electric-field component perpendicular to the crystal surface as a function of time, shown in the inset. The measurement was performed with the tip pressed against the crystal, directly above the probe focal spot.

To determine the spatial extent of the distorted electric field near the tip, we raster-scan the tip across the crystal surface. In Fig. 3, we plot the measured, integrated, THz intensity of the field component perpendicular to the crystal surface, as a function of tip position on the crystal. The plot corresponds to a 50x50 μm^2 area and clearly shows that the perpendicular component of the THz electric field is strongly localized near the tip. The plot also shows that there is no observable background from the incident, parallel polarized, THz pulse. The dashed circle indicates the FWHM of the THz intensity and gives a THz spot size of 10 μm, corresponding to λ/200 for our peak frequency of 0.15 THz, and λ/30 for a frequency of 1 THz. Presently, our smallest measured THz spot size is determined by the focal diameter of our probe beam. Thus, we expect that much smaller spot sizes, down to a few micrometers, can be obtained by tighter focusing of our probe beam.

Further measurements show that the THz signal decreases sharply to 30% of its maximum value, if the tip is moved only 6.5 μm away from the crystal surface. Also, we measure a strong increase of the THz spot size of 4 μm for every μm increase in the tip-crystal distance. Both measurement results are completely

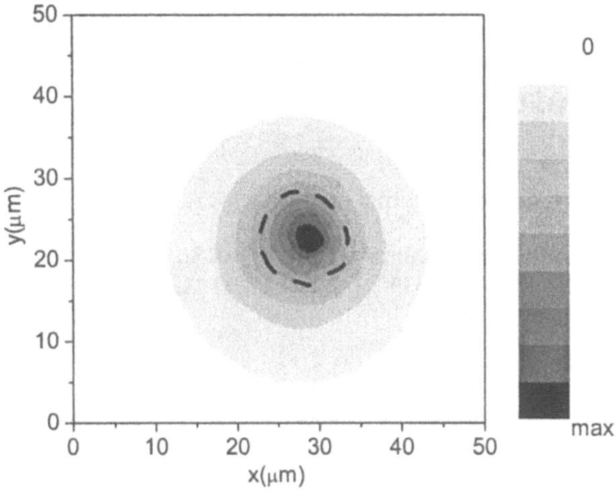

Fig. 3. THz intensity as a function of tip-position on the crystal. The dashed line is the FWHM of the THz spot and corresponds to a 10 μm diameter spotsize. Only the tip is scanned: both crystal and probe beam remain fixed.

consistent with the fact that we measure the near-field of the tip. With increasing tip-crystal distance the evanescent waves from the tip will decrease in amplitude at the detection area in the crystal. The near-field of the tip can be expected to be highly divergent, because of the small dimensions of the tip. This explains the strong increase in THz spot size.

The sharp decrease of the THz electric field when the distance to the crystal is increased only a few μm, is related to results obtained by Knoll et al. [3] at the much shorter wavelength of 10 μm. They found that the scattering cross-section of a tip close to a dielectric half-space is strongly increased due to an image charge induced in the dielectric medium. It is likely, that these results also apply in our case, and that an increase of the scattering cross-section is related to the strong enhancement of the local THz electric field when the tip approaches the crystal. We expect this effect to enhance our contrast in an imaging configuration, because the effect strongly depends on the dielectric constant of the medium that is close to the tip.

We emphasize that our technique to obtain a sub-wavelength resolution requires no microfabrication techniques. We expect that it can be widely applied to study small (biological) samples in the THz domain, by raster-scanning the crystal with the sample underneath the tip.

References

1 O. Mitrofanov et al., Appl. Phys. Lett. **77**, 3496 (2000).

2 Q. Chen and X. -C. Zhang, IEEE J. Select. Topics Quantum Electron. **7**, 608 (2001).

3 B. Knoll and F. Keilmann, Opt. Comm. **182**, 321 (2000).

Size dependent photoconductivity in CdSe nanoparticles as measured by time-resolved THz spectroscopy

Matthew C. Beard, Gordon M. Turner, and Charles A. Schmuttenmaer

Yale University, Department of Chemistry, 225 Prospect St., P.O. Box 208107, New Haven, CT 06520-8107 Tel: (203) 432-3979, Fax: (203) 432-6144,
Email: matt.beard@yale.

Abstract. The size-dependent transient photoconductivity in CdSe nanoparticles has been measured using time-resolved THz spectroscopy (TRTS). We find a size-dependent mobility that varies as r^3 where r is the radius of the NPs. The carrier dynamics display a biexponential decay, wherein both of the decay time constants increase linearly while the amplitude of the fastest time constant decreases linearly with increasing particle size. The measured frequency-dependent change in phase of the transmitted THz pulse falls into three distinct groupings as a function of size.

1. Introduction

The photoconductivity and charge transport properties of semiconductor nanoparticles (NPs) are critically important with regard to their use in electro-optic devices.[1] Measuring the photoconductivity in NPs, however, is a challenging problem because of the inherent difficulty in attaching wires to nanometer-sized objects. Furthermore, the carrier dynamics on a sup-picosecond to picosecond (ps) timescale play an important role in efficient charge separation and transport, but the low temporal resolution of traditional methods used for determining the photoconductivity precludes their use in studying the sub-ps dynamics. Time-resolved THz spectroscopy (TRTS), on the other hand, is a non-contact electrical probe capable of measuring the sub-ps to nanosecond (ns) photoconductivity,[2,3] and we report the TRTS results for CdSe NPs of various sizes.

Semiconductor NPs have sustained tremendous interest in the chemistry and physics communities because of dramatic size dependent effects.[4,5] The most well documented example is that the bandgap increases, and therefore the luminescence spectrum shifts to the blue, as the particle size decreases.[6,7] One of the goals of this work is to investigate the size dependent photoconductivity in semiconductor NPs. That is, how are the transport properties modified as the carriers are confined? At NP diameters smaller than the carrier mean free path, ~ 20 nm for CdSe, we expect a decrease in the carrier mobility. At NP sizes on the order of the Bohr radius, 4.9 nm for CdSe, an even more drastic reduction of on the mobility is expected.

2. Experimental Methods

The experiments consist of photogenerating electron – hole (*e-h*) pairs in a semiconductor NP sample with above bandgap light, and then probing them with low energy far-infrared (FIR) light, with frequencies of 0.1 to 3 THz, or 0.4 to 12 meV. TRTS experiments have been described in detail elsewhere.[2,3] The FIR transmission is directly related to the conductivity of the material.[2,3] Nearly monodisperse CdSe NPs with sizes ranging from 2.54 nm to 25 nm were prepared by the method developed by Peng et al.[8] and are dispersed into toluene for these studies.

3. Results and Discussion

Figure 1 (a) shows TRTS 1D pump scans for a variety of NP sizes. The data are well fit by a biexponential decay function convoluted with a Gaussian function. As the diameter increases, the amplitude of the fastest time constant decreases and becomes longer. The fast time constant represents e-h pairs that become trapped at the surface, and the time required for this process increases linearly with the diameter of the NPs.

We also measured the frequency-dependent response across the bandwidth of the THz pulse, which can be related to the complex conductivity of the NPs. Figure 1 (b) shows the change in phase $\Delta\phi$ of the raw collected data. The change in optical density ΔOD is not shown and is not size dependent. The $\Delta\phi$ show a distinct grouping as a function of NP diameter.

To extract the photoconductivity of the NPs, we model the sample as dielectric inclusions embedded in toluene. Bruggeman effective medium theory[8] is employed. The Drude model is used to represent the photoexcited NPs, and only those NPs that have been photoexcited contribute to the signal. It is found that the Drude model reproduces the observed photoconductivity in the NPs (smooth line). We find a size-dependent mobility that varies as r^3.

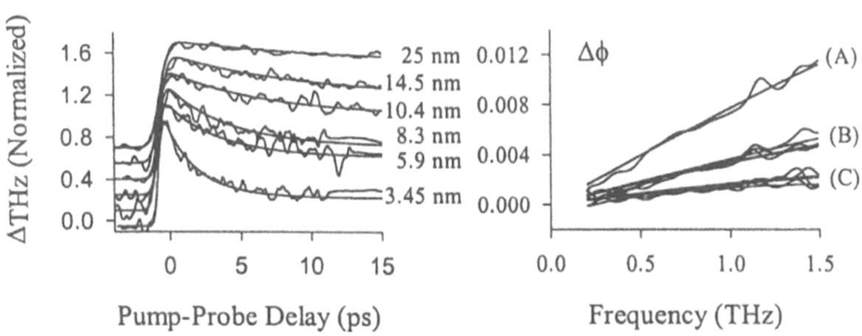

Fig. 1. (a) The size-dependent decay dynamics of the photoconductivity in CdSe NPs. (b) The change in phase ($\Delta\phi$) of the raw collected data for various CdSe NP sizes wit the best fit model (smooth lines). Group (A) are NP with 25 nm diameters, group (B) have 10.4 nm, 8.3 nm, and 5.9 nm diameters, and group (C) have 3.57 nm, 3.45 nm, an 2.5 nm diameters.

4. Conclusions

We have measured the size-dependent photoconductivity in CdSe NPs using time-resolved THz spectroscopy. TRTS provides dynamical information similar to other ultrafast optical spectroscopic techniques, yet differs in two fundamental ways. (1) TRTS measures the dynamics in the valence band for the smallest sizes, since the probe photon energies are too small to induce transitions in the conduction band. (2) TRTS measures the transport properties within the NPs in a non-contact fashion.

We observe that the photoexcited *e-h* pairs decay with a biexponential time-dependence and both time constants increase linearly with increasing NP radius. This suggests ballistic transport to the surface of the NPs. Finally, the frequency-dependent measurements display a distinct grouping as a function of NP size which results from a size-dependent mobility and a size-dependent absorption cross section.

1. Acknowledgments

The authors acknowledge the National Science Foundation and the Sloan Foundation for partial support of this research.

References

1. C. J. Wang, M. Shim, and P. Guyot-Sionnest, *Science* **291** (5512), 2390 (2001).
2. M. C. Beard, G. M. Turner, and C. A. Schmuttenmaer, *Phys. Rev. B* **62** (23), 15764 (2000).
3. M. C. Beard, G. M. Turner, and C. A. Schmuttenmaer, *J. Appl. Phys.* **90** (12), 5915 (2001).
4. A. N. Goldstein, C. M. Echer, and A. P. Alivisatos, *Science* **256** (5062), 1425 (1992).
5. S. H. Tolbert and A. P. Alivisatos, *Science* **265** (5170), 373 (1994).
6. C. D. Dushkin, S. Saita, K. Yoshie *et al.*, *Adv. Colloid Interface Sci.* **88** (1-2), 37 (2000).
7. L. E. Brus, *J. Chem. Phys.* **80** (9), 4403 (1984).
8. Z. A. Peng and X. G. Peng, *J. Am. Chem. Soc.* **123** (1), 183 (2001).

Integrated THz technologies for femtomol sensitivity label-free detection of DNA

Michael Nagel[1], Martin Brucherseifer[1], Peter Haring Bolivar[1], Heinrich Kurz[1], Anja Bosserhoff[2], and Reinhard Büttner[3]

[1] Institut für Halbleitertechnik, RWTH Aachen, Sommerfeldstr. 24, D-52074 Aachen, Germany
E-mail: nagel@iht-ii.rwth-aachen.de
[2] Institut für Pathologie, Universität Regensburg, Franz-Josef-Strauß-Allee 11 D-93053 Regensburg, Germany
[3] Institut für Pathologie, Universitätsklinikum Bonn, Sigmund-Freud-Strasse 25, D-53127 Bonn, Germany

Abstract. An integrated label-free approach for genetic analysis using time-domain femtosecond techniques for generating and electrooptically detecting THz signals is presented. Using planar THz resonators a femtomol sensitivity for DNA molecule detection is reached.

1. Introduction

One of the prominent challenges of biotechnology is the development of simple and flexible systems for the rapid analysis of genetic material. Due to the exceptional specificity of complementary DNA strands to hybridize, most approaches for identifying genetic sequences are based on detecting the hybridization of unknown DNA molecules (*target*) to known single-stranded oligo- or polynucleotide molecules (*probe* DNA). Fluorescent labeling of the target DNA and optical detection of the fluorophores is most commonly used for hybridization detection. Although fluorescent labeling has given rise to very efficient high-throughput diagnostic systems [1], a large interest for label-free detection schemes exists. Labeling possibly modifies the DNA strand conformation [2] and can therefore deteriorate the precision of gene detection. Additionally, fluorophore degradation, varying labeling yield and fluorescence efficiency site dependencies can deteriorate the quantificability of fluorescence based genetic diagnostic systems [3].

Theoretical calculations predict numerous intrinsic resonances in the THz frequency range associated with interbackbone excitations of DNA molecules (e.g. propeller-twist, hydrogen-bond breathing, base-roll and base-shift vibrational modes) [4,5], providing a unique potential of THz-probing-technologies for the label-free detection of the DNA binding state. Although investigations by Raman [6], Fourier-transform [7] or time-domain [8] THz techniques on hybridized DNA molecules have been performed in the past, only very few experiments have addressed binding state dependent analysis, e.g. via temperature dependent denaturation analysis [9]. Recently, the interplay between the binding state of DNA and its complex refractive index at THz frequencies has been investigated by us, present-

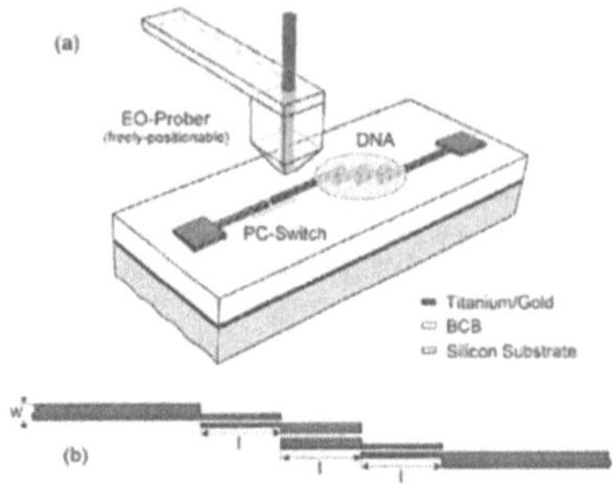

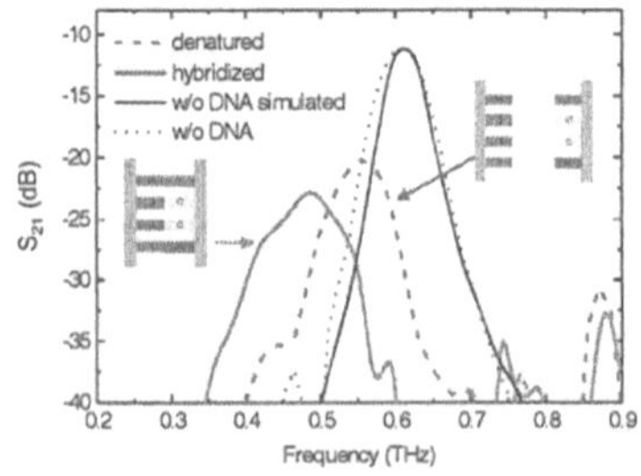

Fig. 1: Perspective schematic (not to scale) of the THz prober structure (a), top-view of the applied band-pass filter, w = 16 μm and l = 85 μm (b).

Fig. 2: Magnitude of the transmission S_{21} of the band-pass filters loaded with denatured or hybridized DNA. In comparison, the measured and simulated reference data of the unloaded filter.

ing a proof-of-principle [10]: With a free-space time-domain THz detection scheme a clear difference between the THz transmission spectra of hybridized (double stranded) DNA films and denatured (single stranded) counterparts of DNA molecules was observed, demonstrating a significantly higher refractive index and absorption for hybridized DNA. Here, we present an integrated guided THz wave approach for monitoring the binding state of DNA molecules, that reduces drastically the DNA material requirements and improves the sensitivity.

2. Experimental Methods

As indicated in Fig. 1 thin-film microstrip lines [11] are used to guide photoconductive generated THz signals in-plane through thin DNA films, thereby enhancing the interaction length of DNA with the THz waves and increasing the detection sensitivity in comparison to our former free-space approach. A further drastic sensitivity increase is obtained by integrating passive THz resonator structures on the waveguides, which amplify the measured DNA-specific differences of the THz signal transmission associated with the binding state dependent dielectric properties of DNA [10]. The idea of using resonator configurations for monitoring dielectric properties is well-known and frequently used at lower frequencies [12]. In our case a band-pass filter is used as a THz resonator. The propagating THz signals are electrooptically detected before and after the resonator to evaluate the THz transmission characteristics of the devices including the DNA films.

3. Results and Discussion

The experiments are performed on thin films of the DNA molecules (vector pcDNA3 with a size of 5400 bases) deposited from solution on the THz resonator structures. The time-domain experimental data are used to calculate transmission spectra. Fig. 2. presents such spectra. The thin continuous and dotted lines illustrate the theoretical and experimental transmission through the unloaded resonator structures, respectively, indicating an excellent agreement of resonator design and measurement with a center frequency of 610 GHz. The DNA loaded resonators show a marked shift to lower frequencies. The observed frequency shifts amount to 124 GHz (20%) and 55 GHz (9%) for the hybridized and denatured and material, respectively. The hybridized DNA clearly generates a stronger shift from the original center frequency. This behavior is in good agreement with free-space measurements [10] where the hybridized samples exhibited an increased refractive index and higher absorption in comparison to the denatured samples. This significantly different shift of resonance frequency upon the degree of hybridization represents the basis of THz probing as a label-free detection scheme. As our DNA spots exceed the THz resonator size by far, more DNA material is applied than necessary. The signal modifications stem from an extremely small DNA volume defined by the resonator size (265 x 50 μm^2) and the 40 - 80 nm DNA film thickness, which corresponds to 1.1 femtomol of probed DNA. If adequate pipetting and immobilization techniques are introduced, this femtomol sensitivity makes THz technologies comparably sensitive to commercial fluorescence-based systems [13].

4. Conclusions

Our analysis demonstrates that integrated THz technologies can efficiently probe the hybridization of nucleic acids. One inherent advantage of the applied planar approach is its scalability, leaving plenty of room for integrating two-dimensional arrays of THz transducers to analyze simultaneously a large number of genes. Further refinement of the technique is, however, required before this principle can be transferred to high-throughput genetic diagnostic systems. This work has been supported by the Deutsche Forschungsgemeinschaft (DFG).

References

1 M. Chee, et al., Science Vol. 274, 610, 1996.
2 H Ozaki and LW McLaughlin, Nucleic Acids Res. Vol. 20, 5205, 1992.
3 M.L. Larramendy, W. El-Rifai, and S. Knuutila, Cytometry Vol. 31, 174, 1998.
4 L. L. Van Zandt, and V. K. Saxena, Phys. Rev. A Vol. 39, 2672, 1989.
5 M. Bykhovskaia, et al., Theor. Chem. Acc Vol. 106, 22, 2001.
6 H. Urabe, and Y. Tominaga, Biopolymers Vol. 21, 2477, 1982.
7 J. W. Powell, et al., Phys. Rev. A Vol. 35, 3929, 1987.
8 A. G. Markelz, et al., Chem. Phys. Lett. Vol. 320, 42, 2000.
9 H. Urabe, and Y. Tominaga, J. Phys. Soc. Jpn. Vol. 50, 3543, 1981.
10 M.Brucherseifer, et al., Appl. Phys. Lett. Vol. 77, 4049, 2001.
11 M. Nagel, et al., Appl. Phys. Lett. Vol. 72, 1018, 1998.
12 M. D. Janezic, et al., IEEE Trans. on Microw. Th. and T. Vol. 47, 10, 2014, 1999.
13 R.J. Cho, et al. Proc. Nat. Acad. Sci. U.S.A. Vol. 95, 7, 3752, 1998.

Part VII

Dynamics at Interfaces and Solids

Ultrafast electron transfer and solvation dynamics in ice layers on Cu(111)

C. Gahl, U. Bovensiepen, C. Frischkorn, and M. Wolf

Fachbereich Physik, Freie Universität Berlin, 14195 Berlin-Dahlem, Germany
E-mail: wolf@physik.fu-berlin.de

Abstract. The dynamics of photoinjected electrons from copper into the conduction band of amorphous solid D_2O (H_2O) layers, the subsequent localization and energetic stabilization due to electron solvation in ice are investigated with time- and angle-resolved two-photon photoelectron spectroscopy.

The solvation dynamics of excess electrons in water have attracted special interest due to its fundamental importance for charge transfer and solvation processes in chemistry and biology and the apparent simplicity of the hydrated electron as a model solute [1-3]. Upon photoinjection in liquid water at room temperature an excess electron localizes and relaxes on a timescale of ~1 ps via several precursor states to the equilibrated solvated electron [3]. So far, most studies have focused on the solvation dynamics in the liquid phase, but very little is known about the initial stages of electron trapping in supercooled water and ice [4]. Photoinduced electron transfer into amorphous ice films leads to molecular rearrangements and crystallization [6]. Electrons in precursor states to solvated electrons in ice are known to enhance the dissociation of chlorfluorocarbons adsorbed on ice surfaces, a reaction which is of high relevance in the depletion of the ozone layer by released Cl atoms [5].

We have studied the dynamics of electron injection from a metal surface into ultrathin layers of amorphous ice and the subsequent localization, solvation and population decay using femtosecond time- and angle-resolved two-photon photoelectron (2PPE) spectroscopy. The inset in Fig. 1 illustrates the electron transfer, solvation and the 2PPE process in a schematic energy diagram: An UV pump pulse ($h\nu_1$=3.92 eV, 65 fs) excites electrons from the substrate into the conduction band (CB) of ice. The relaxation dynamics are monitored by a time-delayed probe pulse ($h\nu_2$=1.96 eV, 65 fs), which promotes the electron above the vacuum level, where its kinetic energy and momentum ($k_{\|}$) is analyzed [7]. D_2O (H_2O) is adsorbed on a Cu(111) single crystal surface in ultra-high vacuum at 100 K. Under these conditions, water is known to form an amorphous solid [8], while scanning tunneling microscopy indicates layer by layer growth [9]. The ice coverage Θ is determined in bilayers (BL) of bulk ice by the integrated intensity of thermal desorption spectra.

Time-resolved 2PPE spectra of amorphous ice on Cu(111) are depicted in Fig. 1. At early time delays the spectra exhibit two features: A pronounced peak e_S at

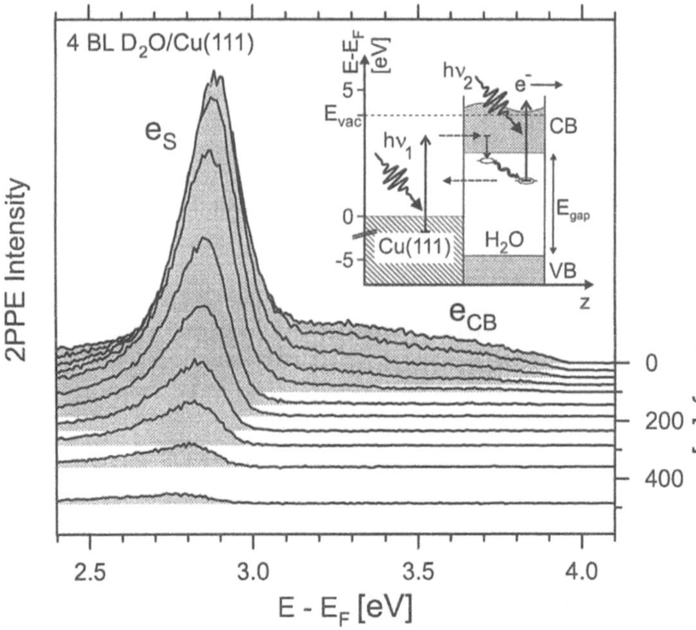

Fig. 1. Time-resolved 2PPE spectra of 4 BL D_2O/Cu(111) at normal emission; the inset represents the 2PPE process, the electronic structure of the ice layer on a Cu surface and the successive photoinjection, localization and solvation of electrons in the ice layer and their back transfer to the metal substrate.

~2.9 eV above the Fermi level (E_F) and a broad continuum e_{CB} extending up to 3.9 eV, which is attributed to photoinjected electrons in the conduction band of ice. Within the laser pulse duration the electrons of e_{CB} relax towards the band bottom or transfer back to the metal by coupling to electron-hole pairs in the metal [6]. The energetic position of e_S shifts with -270 meV/ps accompanied by a pronounced decrease in intensity. Angle-resolved measurements of the wave vector dispersion (not shown) unveil the transition from an extended to a localized state within 100 fs.

To understand the stabilization of e_S a sequence of elementary steps is considered: The CB intraband relaxation is followed by localization of the electron within 100 fs at favorable sites like free OH groups or a locally distorted environment. In analogy to charge transfer reactions, localization of the electronic wave function is achieved in a Marcus-type picture [10] by a change in the molecular configuration like the transition from V_{CB} to V_S along a configuration coordinate q_1 (Fig. 2). In addition, the electron is stabilized further on the timescale of 1 ps as evidenced by the time-resolved energetic shift of e_S (Fig. 1). This increase in binding energy is accomplished by electron solvation, i.e. screening of the excess electron through rearrangements of the water dipoles. The electron is stabilized on the potential V_S along a coordinate q_2, which is different from q_1 (Fig. 2). During all these steps,

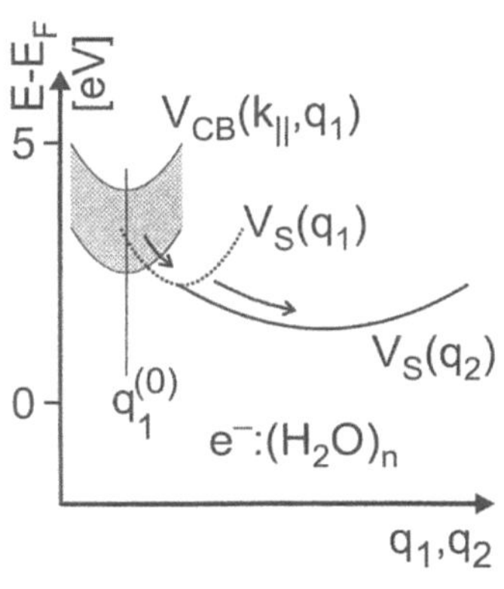

Fig. 2. Scheme of electron localization and solvation: Electron potentials of the conduction band V_{CB} and the solvated electron V_S along the configuration coordinates q_1, q_2; $q_1^{(0)}$ denotes the ground state configuration for the conduction band.

which follow the primary photoexcitation, back transfer to the metal substrate provides a more or less efficient relaxation channel. The crucial parameter is the wave function overlap of the respective initial and final states. Also electrons which formed a solvated electron complex $e^-:(H_2O)_n$ finally transfer back to the substrate as seen from the decay of the signal e_S (Fig. 1).

Coverage dependent experiments result in two distinct regimes of solvation dynamics for coverages $\Theta \geq 3$ BL and $\Theta \leq 2$ BL. At low Θ a four times faster peak shift of e_S is found at delays <300 fs compared to higher coverages. Considering the lower coordination of molecules at the surface, the fast solvation dynamics for $\Theta \leq 2$ BL are attributed to solvated electrons near the surface in the ice layer due to the higher mobility of the solvent at the surface with respect to bulk. For thicker layers the solvated electron resides in the bulk, because further increase of the coverage does not influence the dynamics.

Recently, it was suggested that the solvation dynamics are considerably slower in solid than in liquid water due to a proportionality to the dielectric relaxation time of the medium in the thermodynamic equilibrium [5]. In contrast, we have recorded timescales comparable to the solvation dynamics in the liquid phase [1-3] and small anion clusters [11]. Adsorption of water at surfaces, which serve as templates for the solvent, allows to modify the solvent structure and facilitates systematic studies of the relation between structure and solvation dynamics in low dimensional systems.

References

[1] T. W. Kee, D. H. Son, P. Kambhampati, and P. F. Barbara, J. Phys. Chem. A **105,** 8434 (2001)

[2] M. F. Emde, A. Baltuska, A. Kummrow et al., Phys. Rev. Lett. **80,** 4645-4648 (1998);
C-Y. Yang, K. F. Wong, M. S. Skaf, and P. J. Rossky, J. Chem. Phys. **114,** 3598-3611 (2001)

[3] R. Laenen, T. Roth, and A. Laubereau, Phys. Rev. Lett. **85,** 50-53 (2000).

[4] H. A. Gillis and T. I. Quickenden, Can. J. Chem. **79,** 80-93 (2001).

[5] Q. B. Lu and L. Sanche, Phys. Rev. B **63,** 153403 1-4 (2001).

[6] D. Chakarov and B. Kasemo, Phys. Rev. Lett. **81,** 5181-5184 (1998)

[7] E. Knoesel, A. Hotzel, and M. Wolf, Phys. Rev. B **57,** 12812-12824 (1998)

[8] R. S. Smith and B. D. Kay, Nature **398,** 788 (1999)

[9] K. Morgenstern and K. H. Rieder, private comm.

[10] N.-H. Ge, C. M. Wong, R. L. Lingle jr., J. D. McNeill, K. J. Gaffney, and C. B. Harris, Science **279,** 202 (1998)

[11] L. Lehr, M. T. Zanni, C. Frischkorn, R. Weinkauf, and D. M. Neumark, Science **284,** 635 (1999)

The Size of a Localized Electron at a Metal/Adsorbate Interface

Ilya Bezel[1], Kelly Gaffney[2], Sean Garrett-Roe[2], Simon H. Liu[2], Andrè D. Miller[2], Paul Szymanski[2], and Charles B. Harris[2]

[1] Miller Institute for Basic Research in Science, 2536 Channing Way #5190, Berkeley, CA 94720, USA
E-mail: ilya@xenon.cchem.berkeley.edu
[2] Department of Chemistry, University of California, Berkeley, CA 94720, USA and Chemical Sciences Division, E. O. Lawrence Berkeley National Laboratory, Berkeley, CA 94720, USA
E-mail: harris@socrates.Berkeley.edu

Abstract. The transient localized electron wavefunction shape at a metal-adsorbate interface has been directly measured in momentum space by using ~100 femtosecond time and angle resolved two-photon photoemission.

Electron localization in different systems has been a topic of study for over a half of a century; polaron formation in the solid state and electron solvation by liquids are just two examples. Structureless by itself but interacting strongly with the condensed phase environment, an electron is often utilized as a probe of its local microscopic structure. Despite that, there are few attempts to access the

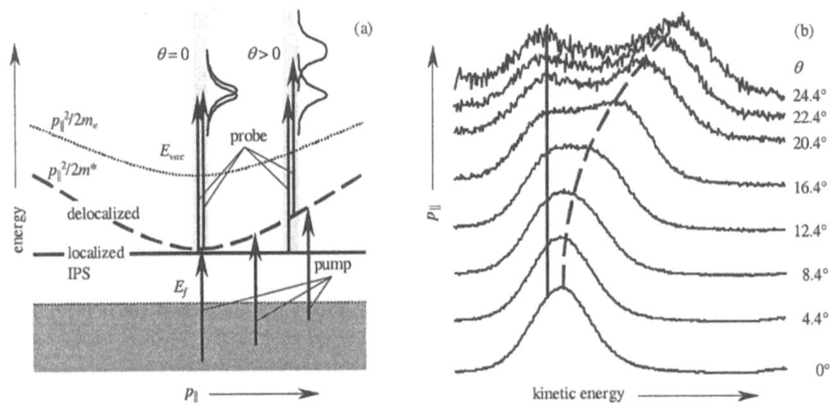

Fig. 1. (a) Schematic of TAR TPPE experiment. The pump pulse initially populates the delocalized band. Electrons which have different values of $p_\parallel$ are photoemitted at different angles θ relative to the surface normal. Thus, electron kinetic energy spectra taken different observation angle allow reconstruction of the dispersion curve. (b) Butyronitrile bilayer on Ag(111) experimental spectra taken at different θ (offset vertically by the value of $p_\parallel$). Delay between the pump and the probe pulses is 266 fs where both the localized and the delocalized states are populated. Solid and dashed lines show position of the localized and delocalized features, respectively.

shape of a localized electronic wave function directly from an experiment. Here, we report an elegant way of estimating the electron localization size parallel to a two-dimensional metal/adsorbate interface by using time and angle resolved two-photon photoemission (TAR TPPE).

In these experiments, the electronic kinetic energy spectrum is acquired as a function of the photoemission angle and delay between an ~80 fs pump and an ~80 fs probe pulses. The pump populates the interfacial electronic states and the probe photoionizes the electron from these states. For highly reflective substrates, electronic momentum parallel to the surface, $p_{\|}$, conserves during the probe step[1, 2] and thus TAR TPPE gives a unique opportunity to reconstruct the transient interfacial band structure and accesses information on the population of the intermediate electronic states (Fig. 1). Here, the intermediate interfacial electronic states are the image potential states (IPS) which reside a few Angstrom outside the surface covered with solvent adsorbate.[3] Being very close to the interface, the IPS electron is a sensitive probe of transient changes in the interfacial structure.[4]

In a number of systems, two distinct IPS are observed that are different by their spatial extent in the plane of the interface: the delocalized state, where the electron is free to move parallel to the surface and is characterized by a parabolic dispersion near the delocalized band origin (with the effective mass, m^*, approximately equal to the free electron mass, m_e), and the localized, self-trapped state, the energy of which is independent of the photoemission angle.[5] We argue that the photoionization signal amplitude at different observation angles for the delocalized and for the localized states reflects two qualitatively different phenomena. Namely, while the delocalized signal dependence on the observation angle is a measure of the delocalized band population, the localized signal conveys information regarding the localized state wavefunction shape.

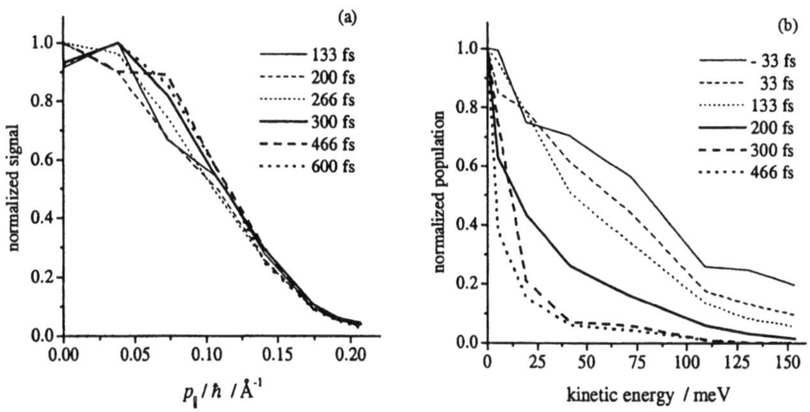

Fig. 2. Results for butyronitrile bilayer on Ag(111). (a) Localized signal amplitude as a function of the momentum parallel to the surface. Within the error of the fit, the shape of the wavefunction is time independent. (b) Delocalized signal as a function of the initial kinetic energy in the delocalized band (= $p_{\|}^2/2m^*$). Effective temperature drops with time.

According to the uncertainty principle, a spatially localized electronic wavefunction acquires a spread in the momentum space and the electrons detected off the surface normal reflect the probability amplitude of the localized electron to have the corresponding $p_{\parallel}$ before photoemission. An example of the delocalized and the localized signal behavior is shown in Fig. 2. Initially populated delocalized electronic band has high carrier temperature that drops gradually to the substrate temperature due to various relaxation processes. The localized state is dynamically populated due to electron transfer from the delocalized band. Fitting spectra taken at low observation angles where the lines overlap (Fig. 1b) to two peaks is made possible by fitting the entire dispersion simultaneously, where the positions and widths of the peaks at high angles dictate those at low angles. The width of the localized feature in the momentum space is time independent and corresponds to localization to ~ 12 Å FWHM. This is comparable to the size of a single overlayer molecule in accordance with small polaron formation theories in two dimensional systems, where the electron must collapse to a single lattice site. Energies of both states decrease as they are solvated with a time constant of ~200 fs, however this can be distinguished from the energy relaxation due to localization.

Acknowledgements. We would like to thank C. Wittig for valuable discussion. This work was supported by the Director, Office of Energy Research, Office of Basic Energy Sciences, Chemical Sciences Division of the U.S. Department of Energy, under Contract No. DE-AC03-76SF00098. The authors acknowledge NSF support for specialized equipment used in the experiments described herein.

References

1 N. V. Smith and S. D. Kevan, in *Angle-Resolved Photoemission: Theory and Current Applications*, Elsevier, Amsterdam, 1992.
2 H. Petek and S. Ogawa, in *Progress in Surface Science*, **56**, 239, 1997.
3 P. M. Echenique, J. B. Pendry, in *Progress in Surface Science*, **32**, 202, 1990.
4 S. H. Liu, et al., accepted to *Journal of Physical Chemistry B*.
5 N.-H. Ge, C. M. Wong, R. L. Lingle, Jr., J. D. McNeill, K. J. Gaffney, C. B. Harris, in *Science* **279**, 409, 1998.

Hot-Electron-Driven Charge Transfer Processes on Surfaces

Chi-Fong Lei[1], Michael Bauer[2], Kendall Read[1], Ra'anan Tobey[1], Margaret Murnane[1], and Henry Kapteyn[1]

[1] JILA, University of Colorado, Boulder, CO 80309-0440, USA
E-mail: clei@jila.colorado.edu

[2] Fachbereich Physik, University Kaiserslautern, 67663 Kaiserslautern, Germany

Abstract. We observe an ultrafast, hot-electron-driven, charge transfer process involving oxygen on platinum. Non-equilibrium electron dynamics are observed, where the hot electrons cool before the charge transfer process induces changes in atomic bonding on the surface.

Chemistry driven by intense ultrafast pulses (femtochemistry) is distinguished from conventional phonon-mediated chemical reactions in that the product yield can be greatly enhanced and more importantly, new chemical reaction pathways can be accessed [1]. In a surface-adsorbate system, non-equilibrium or "hot" electrons created by a femtosecond pulse can drive these reactions [2]. However, direct, time-resolved observation of hot-electron-driven reactions has proven to be difficult. Here, we present time-resolved observations of a hot-electron driven change of adsorbate molecules (O_2) on a metal surface (Pt) from one chemisorbed species to another. The process of time-resolved ultraviolet photoelectron spectroscopy makes it possible to observe both electronic energy relaxation, and structural changes in the surface/adsorbate, on their fundamental, femtosecond timescales.

These experiments use sub-10 femtosecond EUV light pulses generated using the process of high-harmonic generation to probe a platinum surface, after excitation by a 50 fs pump pulse. Photoelectrons from the surface are collected using a time-of-flight detector; details of the experimental setup can be found in Ref. 3. Depending on the coverage, molecular oxygen is known to exist in two chemisorbed states on Pt(111): superoxo (O_2^-) for high coverage and peroxo (O_2^{2-}) for low converage [4,5,6]. The spectra of the two species differ in the valence band region due to the different chemical binding. In particular, the spectrum of peroxo oxygen has a distinct peak ≈ 6eV below the Fermi edge, which is not present in the spectrum of superoxo oxygen [6].

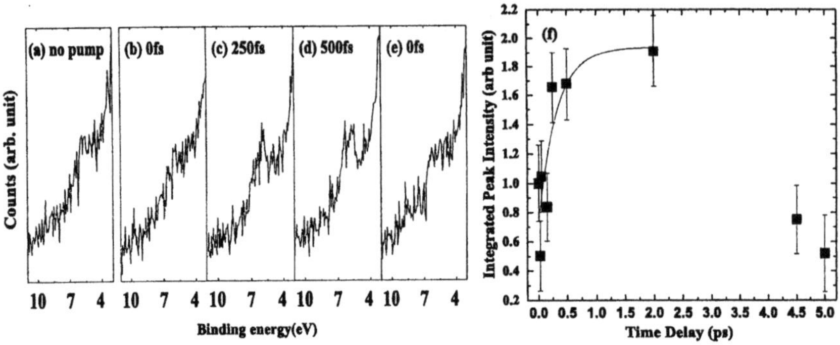

Fig. 1. Time-resolved extreme-ultraviolet valence-band photoemission spectra from a saturation layer of molecular oxygen adsorbed on Pt(111): (a) no pump beam; (b) with pump beam, zero delay between pump and probe; (c) 250 fs delay (c) 500 fs delay; (e) repeat of (b), taken immediately after (d); (f) Integrated amplitude of the characteristic 6eV peak feature as a function of pump-probe delay.

The results of a series of pump-probe experiments are shown in Fig 1. The sample was prepared such that superoxo oxygen initially dominates on the platinum surface. At 250-500fs delay between the pump and probe, the characteristic peroxo peak appears, indicating that the superoxo is changing to peroxo oxygen on a sub-ps timescale. The final trace of Fig. 1(e), at zero time delay, was taken at the end of the data run and shows that the excited peroxo oxygen relaxes back to superoxo oxygen. This indicates that the process is transient and reversible. Fig. 1(f) shows the integrated strength of the peroxo peak as a function of delay, indicating that the onset time of this process is 550 ± 140fs.

In more-recent data, we observed cooling of the hot electron distribution, and show that this time-scale is much faster than the time-scale for changes in the valence band spectrum. Figure 2 shows photoelectron spectra at the Fermi-edge of the sample for time delays of 0 and 250fs. Non-equilibrium, hot electrons are clearly seen at 0fs time delay, and subsequently cool after 250fs - well before the superoxo oxygen changes to peroxo. Since the charge transfer process involves an increase in bond length together with a possible rotation of the molecule on the surface, the existence of two different timecscales is reasonable. Thus, our interpretation of the data is that the initial hot electron distribution facilitates a charge-transfer to the oxygen molecule, which in turn changes its binding to the surface. There is an associated change in the valence-band electronic structure and photoelectron spectrum. After ~5ps (Fig. 1(f)), the electron transfers back to the surface and the molecule returns to it's original position.

We acknowledge support for this work from the National Science Foundation.

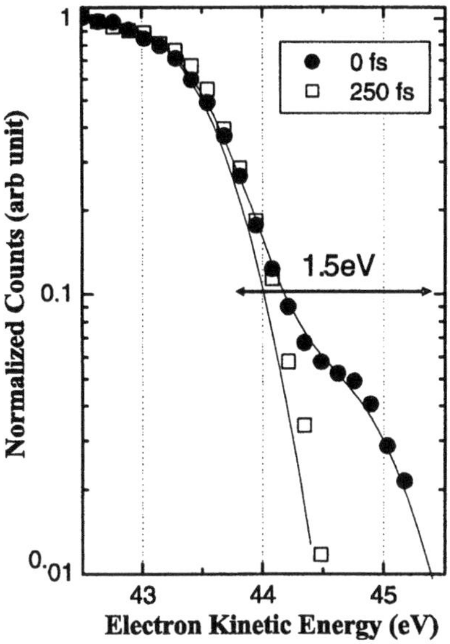

Fig. 2. Fermi edge of Pt at 0 fs (black circle) and 250 fs (hollow square) pump-probe delays. The electron distribution at time zero can be fit by a 77K Fermi-Dirac distribution, with 5% of the electrons excited by a pump photon energy (1.5eV). At 250fs delay, the non-equilibrium component of the distribution has cooled – before any change in the photoelectron spectrum is observed.

References

1 C. Hess, et al, "Femtosecond dynamic of chemical reactions at surfaces", Appl. Phys. A **71**, 477-483 (2000)

2 F. J. Kao, et al, "Femtosecond versus nanosecond surface photochemistry", Phys. Rev. Lett. **70**, 4098 (1993)

3 M. Bauer, et al, "Direct observation of surface chemistry using ultrafast soft-x-ray pulses", Phys. Rev. Lett. **87**, 25501 (2001); "Shooting an x-ray movie", Phys. Rev. Focus, July 2001; "X-rays light up chemical rections", PhysicsWeb, July 2001

4 A. Eichler and J. Hafner, "Molecular precursors in the dissociative adsorptoion of O_2 on Pt(111)" , Phys. Rev. Lett. **79**, 4481 (1997)

5 B. C. Stipe et al, "Atomic studies of O_2 dissociation on Pt(111) induced by photons, electrons and by heating", J. Chem. Phys. **107**, 6443 (1997)

6 C. Puglia, et al, "Physisorbed, chemisorbed and dissociated O_2 on Pt(111) studied by different core-level spectroscopy methods", Surf. Sci. **342**, 119 (1995).

Photoinduced electron transfer in dye/semiconductor systems on a sub-10-fs time scale

Robert Huber[1], Jacques E. Moser[2], Michael Grätzel[2] and Josef Wachtveitl[3]

[1] Lehrstuhl für BioMolekulare Optik, Oettingenstr. 67, Ludwig-Maximilians-Universität München, 80538 München, Germany
E-mail: robert.huber@physik.uni-muenchen.de
[2] Laboratory for Photonics & Interfaces, Department of Chemistry, EPFL, CH-1015 Lausanne, Switzerland
E-mail: je.moser@epfl.ch
[3] Institut für Physikalische und Theoretische Chemie, Marie-Curie-Str. 11, Goethe Universität Frankfurt, 60439 Frankfurt am Main, Germany
E-mail: wveitl@theochem.uni-frankfurt.de

Abstract. Pump-probe investigations of alizarin adsorbed on the surface of TiO_2 colloids reveal a 6 femtosecond electron transfer from the dye molecule to the semiconductor surface, allowing to identify the key factors for this ultrafast electron transfer.

1. Introduction

Sensitization of wide band gap semiconductors by organic dye molecules is a well established technique with widespread applications like the photographic process, detoxification and purification of water, and recently also solar energy to electricity conversion [1,2]. The physical interest in these photoresponsive mesoscopic systems arises from the special energetic situation, allowing ultrafast electron transfer from a system with discrete energy levels on the donor side, into a contiuum of energy levels on the acceptor side. This wide conduction band of empty electronic acceptor levels is the reason, why electron transfer from organic dye molecules to semiconductor-colloidal systems is among the fastest reported charge separation reactions. For systems with a moderate to high electronic coupling matrix element, the initial photoinduced electron injection into the conduction band of the semiconductor often escapes direct observation due to a lack of temporal resolution in the setup [3,4]. Here, we present for alizarin sensitized TiO_2 colloids a direct measurement of the electron transfer rate with an optimized NOPA-pump/SC-probe setup [5]. The finding that the initial charge separation reaction occurs on a sub-10-fs time scale is verified in three different, independent ways.

2. Results and Discussion

Figure 1 shows the transient absorbance changes for the alizarin/TiO_2 system after photoexcitation. We showed, that for times from 100fs up to many microseconds, the observed kinetic reflects the recombination reaction between the alizarin

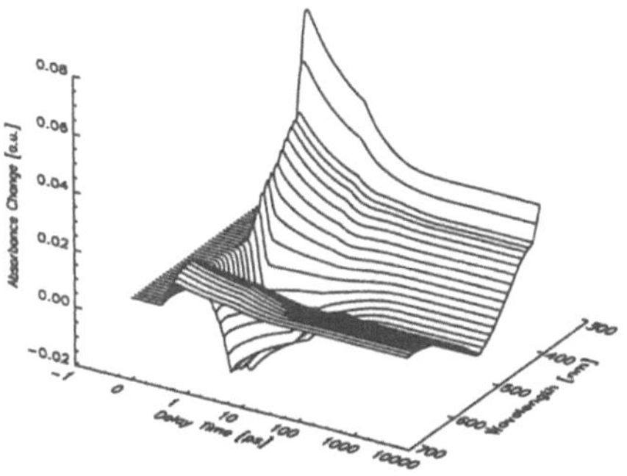

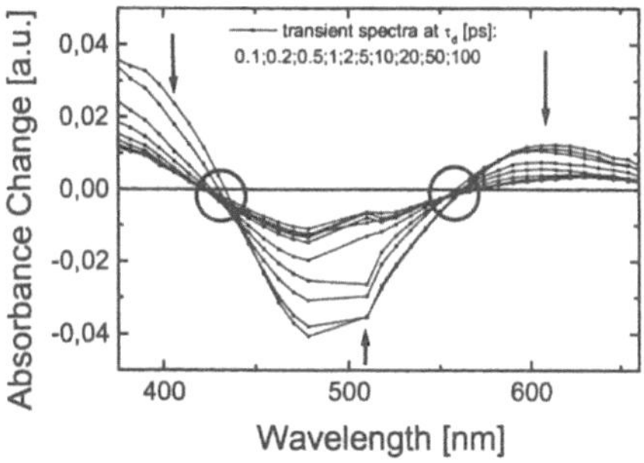

Fig. 1. (left): Transient absorbance changes for alizarin/TiO_2
Fig. 2. (right): Transient spectra at various delay times for alizarin/TiO_2

cation and the injected electron [3]. The transient spectra at various delay times in Figure 2 show two pronounced isosbestic points for this back reaction at 426nm and at 550nm. At these spectral positions, the slow, multiphasic back reaction does not contribute to the signal, a prerequisite for a reliable deconvolution of the ultrafast injection time. Therefore, a nonlinear fitting procedure of the 426nm transient was performed, where the signal is less disturbed by oscillatory contributions due to wave-packet motion than at 550nm.

Figure 3 depicts the best fit curve obtained with a time constant of 6.4fs, whereby the apparatus function was about 20fs (fwhm) cross-correlation time for this wavelength channel. The curves with decay times of 3fs and 9fs show systematic deviations from the experimental data and therefore represent an upper and lower limit for the initial time constant.

To include the spectral information provided by the multicolor experiments, a global fitting procedure over the whole investigated spectral range from 350nm to 650nm was performed [6]. Although there are spectral positions with dominant

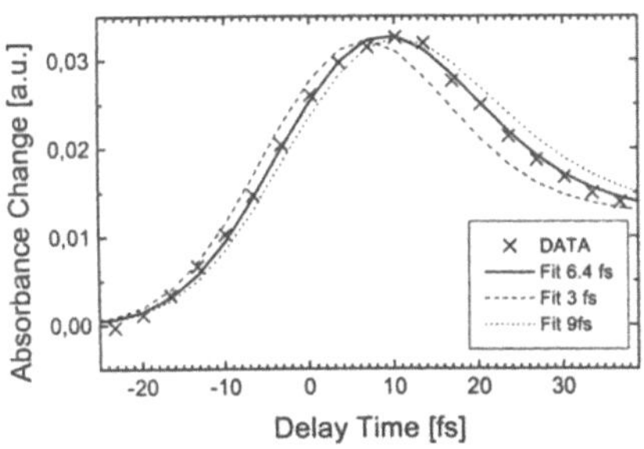

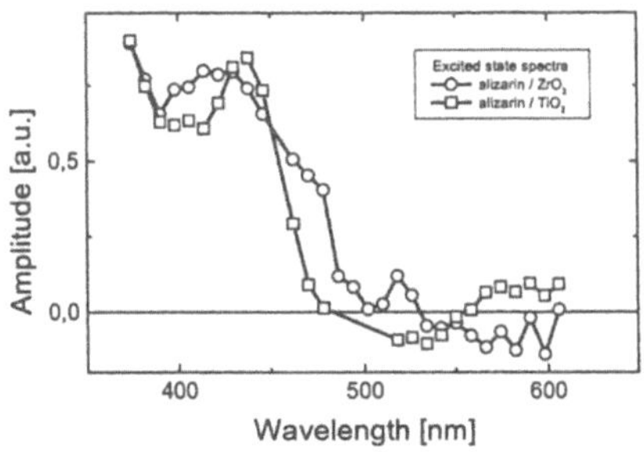

Fig. 3. (left): Transient data at λ_{pr}=426nm (isosbestic point);
Fig. 4. (right): Recalculation of excited state spectrum; comparison with ZrO_2

slow kinetics superposed on the ultrafast electron injection signal, the global fitting analysis avoids artificial contributions present at individual wavelengths. The result of the global fit is in very good quantitative agreement with the single wavelength fit for the λ_{pr}=426nm transient and results in a 6.5fs time constant. The spectral signature of the decay related spectrum for the 6.5fs time constant allows the unambiguous identification of the vanishing excited state absorption, the rising absorption of the cation and the injected electron. Thus the global fit yields a second independent way to determine the ultrafast kinetic component.

The third way is to recalculate the spectral characteristics of the excited state absorption from the transient measurements on TiO_2. Figure 4 shows the comparison between this deconvoluted spectrum and the excited state spectrum of alizarin coupled on ZrO_2. For the ZrO_2 system a significant population of the excited state is observed for a delay time of 1ps [3] and therefore the S* spectrum can be calculated easily. The two spectra are obviously in qualitative agreement providing further strong evidence for the reliability of the 6fs time constant.

3. Conclusions

These experiments demonstrate an ultrafast photoinduced charge separation reaction on a 6fs time scale in an unambiguous way. To the best of our knowledge this is the fastest electron transfer reaction ever directly observed by transient pump probe spectroscopy. It can be concluded, that even for dyes like alizarin, that exhibit a very strong electronic coupling, a direct charge transfer excitation into the conduction band of the semiconductor can be ruled out and a localized excitation of the dye with a subsequent electron injection into the TiO_2 colloid correctly describes the initial photoreaction.

References

1 B. O'Regan, J. Moser, M. Anderson, and M. Grätzel, in *J. Phys. Chem.*, Vol. 94, 8720-8726 (1990).
2 M. Grätzel, Nature 414, 338-344 (2001).
3 R. Huber, S. Spörlein, J. E. Moser, M. Grätzel, and J. Wachtveitl, in *J. Phys. Chem.*, Vol. 104, 8995-9003 (2000).
4 R. Huber, S. Spörlein, J. E. Moser, M. Grätzel, and J. Wachtveitl, in Ultrafast Phenomena XII, T. Elsaesser, S. Mukamel, M. M. Murnane, N. F. Scherer (ed.s), pp. 456-458, (Springer, Berlin, 2000).
5 R. Huber, H. Satzger, W. Zinth, and J. Wachtveitl, in *Opt. Commun.*, Vol. 194, 443-448 (2001).
6 R. Huber, J. E. Moser, M. Grätzel, and J. Wachtveitl, in *J. Phys. Chem.*, in press, (2002).

Carrier Relaxation Dynamics in Heavy Fermion Compounds

J. Demsar,[1] R.D. Averitt,[1] V.V. Kabanov,[2] L.A. Tracy,[1] S.A. Trugman,[1] J.L. Sarrao,[1] and A.J. Taylor[1]

[1] Los Alamos National Laboratory, Mail Stop K764, Los Alamos, NM 87545, USA
E-mail: jdemsar@lanl.gov
[2] "Jozef Stefan" Institute, Jamova 39, SI-1000, Ljubljana, Slovenia

Abstract. The first femtosecond carrier relaxation dynamics studies in heavy fermion compounds are presented. The carrier relaxation time shows a dramatic hundred-fold increase below the Kondo temperature revealing a dramatic sensitivity to the electronic density of states near the Fermi level.

1. Introduction

An important and unresolved issue in heavy fermion (HF) physics [1] is whether the electronic ground state is a many-body Kondo state that forms only below T_K, or is more simply described in terms of the thermal population of a narrow gap system with the gap arising from hybridization between the conduction electrons and localized f electrons (note that the gap is present at all temperatures in this case).

Femtosecond time-resolved optical spectroscopy is an excellent experimental alternative to conventional spectroscopic methods that probe the low energy electronic structure in strongly correlated electron systems [2]. In particular, it has been shown that carrier relaxation dynamics are very sensitive to changes in the low energy density of states (e.g. associated with the formation of a low energy gap or pseudogap) providing new insights into the low energy electronic structure in these materials.

In this report we present the first studies of carrier relaxation dynamics in heavy fermion systems by means of femtosecond time-resolved spectroscopy. Our results show that the carrier relaxation dynamics, below the Kondo temperature (T_K), are extremely sensitive to the low energy density of states (DOS) near the Fermi level to which localized f-moments contribute. Specifically, we have performed measurements of the photoinduced reflectivity dynamics as a function of temperature on the series of HF compounds $YbXCu_4$ (X=Ag, Cd, In) [3] in comparison to their non-magnetic counterparts $LuXCu_4$.

2. Experimental Methods

The experiments were performed on freshly polished flux-grown single crystals. We used a standard pump-probe set-up with a mode-locked Ti:Sapphire laser (30 fs pulses at 800 nm and 80MHz repetition rate) for both pump and probe pulse trains. The photoinduced (PI) change in reflectivity $\Delta R/R$ was measured using a photodiode and lock-in detection. The pump laser fluence was kept below 1 $\mu J/cm^2$ to minimize the steady-state heating effects and the pump/probe intensity ratio was ~30.

3. Results and Discussion

In Fig. 1a we show the PI reflectivity traces as a function of temperature on the HF system $YbAgCu_4$ (T_K ~ 100K, low temperature Sommerfeld coefficient γ ~ 210 mJ/mol K^2) [2]. The relaxation time is virtually temperature independent above ~ 150 K and shows a quasi-divergence (hundred-fold increase) as $T \rightarrow 0$ K. A similar divergence of the relaxation time below ~T_K is also also in the HF compounds $YbCdCu_4$ in $CeCoIn_5$ (not shown here) pointing to the universality of the observed behavior. On the other hand, the relaxation time of the non-HF compound $LuAgCu_4$ behaves similar to regular metals [4], showing a gradual decrease upon cooling – Fig. 1b. This indicates that the upturn in the relaxation time at ~ T_K and the subsequent divergence as $T \rightarrow 0$ K derives from the low energy electronic structure of these HF compounds.

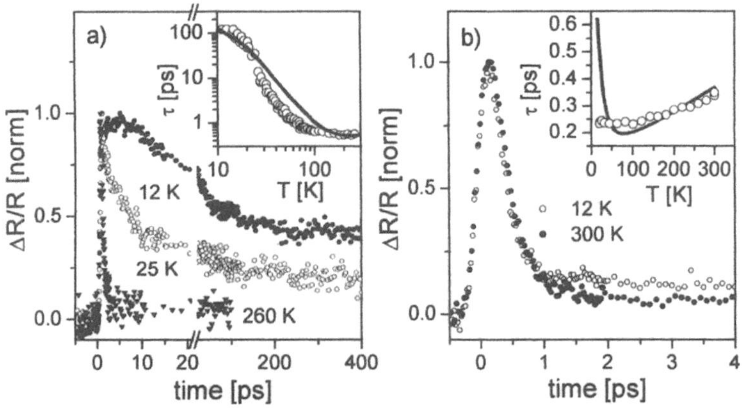

Fig. 1. Normalized photoinduced reflectivity of (a) $YbAgCu_4$ at various temperatures above and below T_K ~ 100K and (b) $LuAgCu_4$. Insets: the T-dependences of the relaxation times τ_R (open symbols) obtained by fitting $\Delta R/R$ with exponential decay, together with $\tau_{e\text{-}ph}$ obtained by numerical simulation (solid lines) using the two-temperature model – see text.

The two-temperature model (TTM) is commonly employed to interpret the ultrafast dynamics in metals [4]. It consists of two coupled heat equations

$$C_e\dot{T}_e = -H(T_e,T_L) + P(t) \quad ; \quad C_L\dot{T}_L = H(T_e,T_L) \qquad (1)$$

where C_e and C_L are the electron and lattice specific heats, P(t) is the absorbed power density and $H(T_e,T_L)$ is the energy transfer rate between electrons and phonons. In conventional metals, $C_L >> C_e$ holds down to very low temperatures

(~5 K), and the Fermi energy is much larger than the Debye temperature. In this case the TTM equations can be simplified giving an analytical expression for $\tau_{e\text{-}ph}$. For HF materials this is not the case, since at low temperatures C_e increases dramatically and becomes larger than C_L at temperatures on the order of 30 K. Therefore we must numerically solve Eqns. (1) using the experimental values for C_e and C_L [3].

In inset to Fig. 1(b) we plot (solid line) the calculated value of $\tau_{e\text{-}ph}$ for $LuAgCu_4$, showing good agreement with the measured relaxation time down to 40 K. At lower temperatures, similar to Ag and Au [4], the relaxation time does not show the expected $\tau_{e\text{-}ph} \propto T^{-3}$ increase, which has been attributed by several groups – see for example [4] – to nonthermal electrons.

The simulation of e-ph thermalization in HF $YbAgCu_4$ is more subtle. Namely, since $H(T_e,T_L)$ is also proportional to the DOS at E_F, it would, at first glance, appear to cancel out the effect of the increased electronic specific heat on $\tau_{e\text{-}ph}$. However, we have to consider the nature of states near E_F which, for HF compounds consist of f-electrons hybridized with the dispersive conduction band. In there was no hybridization, the f-electrons would not contribute to the e-ph relaxation because of the conservation of energy (i.e. typical phonon energies are much greater than the width of the f-level). For weak hybridization, the matrix element responsible for scattering of heavy electrons by phonons is proportional to hybridization strength. As a result, the contribution to the Eliashberg electron-phonon coupling function due to the hybridized electrons is small (i.e. the higher the DOS – the weaker the inelastic scattering of these states by phonons). Therefore, in the numerical simulation of e-ph thermalization in $YbAgCu_4$, we can, as a first approximation, use the same $H(T_e,T_L)$ function [4] as in the case of $LuAgCu_4$. The result of the simulation is plotted (solid line) in the inset to Fig. 1(a), showing surprisingly good agreement with the experimental data, giving further support to the analysis.

4. Conclusions

In conclusion, we have utilized ultrafast optical spectroscopy to study the carrier dynamics in the HF compounds $YbXCu_4$ and $CeCoIn_5$. We have observed a divergence in the relaxation time at low temperatures which is not seen in their non-magnetic analogs. The analysis of the relaxation dynamics in terms of the TTM suggests that the dramatic hundred-fold increase in the relaxation time at low temperatures is due to the anomalously large electronic specific heat below T_K.

References

1 L. Degiorgi, Rev. Mod. Phys. 71, 687, 1999 and references therein.
2 V. V. Kabanov et al., Phys. Rev. B 59, 1497, 1999, J. Demsar, D. Mihailovic, K. Biljakovic, Phys. Rev. Lett. 83, 800, 1999.
3 J. L. Sarrao et al., Phys. Rev. B 59, 6855, 1999.
4 R. H. M. Groeneveld et al., Phys. Rev. B 51, 11433, 1995, and references therein.

Ultrafast Electron Optics: Propagation Dynamics and Measurement of Femtosecond Electron Packets

Bradley J. Siwick, Jason R. Dwyer, Robert E. Jordan, R. J. Dwayne Miller

Departments of Physics and Chemistry, University of Toronto
80 St. George St., Toronto, ON, M5S 3H6
Canada
E-mail: dmiller@lphys.chem.utoronto.ca

Abstract. The propagation of femtosecond electron packets through a photoactivated electron gun is calculated and compared to experiment. It is shown that keeping propagation time below 0.5 ns is important in the design of photoelectron guns with temporal resolution below 500 fs. An all-optical electron pulse duration measurement scheme is also proposed. This method involves using the laser pulse ponderomotive potential to perform a cross-correlation with the electron packet.

1. Introduction

Ultrafast electron diffraction (UED), like ultrafast X-ray diffraction, holds the promise of providing a complete, global picture of structural dynamics on the femtosecond time-scale with atomic level detail. The main drawback of the electron technique is the difficulty associated with producing femtosecond pulses containing large numbers of electrons (10^4 or more). Space-charge effects (Coulomb repulsion) and the initial energy distribution of the electrons act to broaden the electron pulse as it propagates.

In this paper, the propagation dynamics of femtosecond electron packets produced by a photoactivated electron gun are discussed[1]. Discussion of this subject matter in the literature has been primarily limited to the context of streak camera technology[2,3]. A method for cross-correlating electron and laser pulses using the ponderomotive potential of the laser pulse is also proposed and discussed.

2. Modeling Electron Packet Propagation

The propagation dynamics of the photoelectron packets were computed using a Barnes/Hut tree algorithm[4] for the full N-electron equations of motion. The initial photoelectron energy distribution was assumed to have a width of 0.6 eV (Gaussian) in these calculations.

The results of the simulation were compared to the predictions of a mean-field model that approximates the electron pulse as a disk of charge to obtain the pulse length equation

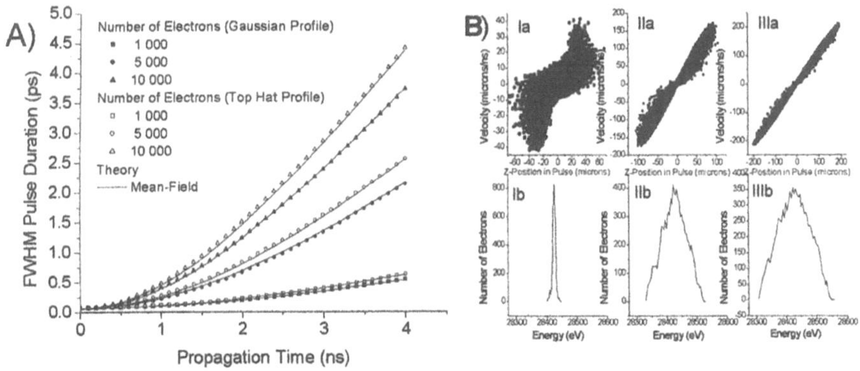

Fig.1 A) Electron pulse duration vs. propagation time for Gaussian and "top hat" beams, and its dependence on N (N = 1 000, 5 000 and 10 000, τ_o = 90 fs and r(0) = 250 μm). The results of the N-body simulation are presented along with the predictions of the mean-field theory. **B)** a) Velocity and b) Energy distribution of a 28.4 keV electron packet containing 10 000 electrons at 3 different times during propagation. I) Immediately after acceleration (95 kV/cm extraction field); II) After 0.5 ns free-space propagation; III) After 1 ns free-space propagation

$$\frac{d^2 l}{dt^2} = \frac{Ne^2}{m\varepsilon_o \pi r^2}\left[1 - \frac{l}{\sqrt{l^2 + 4r^2}}\right], \qquad (1)$$

where l is the pulse length, t the propagation time, N the number of electrons, r the beam radius, ε_o the permittivity of free space, e and m the electron charge and rest mass, respectively.

The results presented in Fig. 1A) indicate that if propagation time is kept below 0.5 ns, it should be possible to produce sub-picosecond electron pulses containing 10^4-10^5 electrons with a beam diameter of several hundred microns from photoactivated electron guns. The simulations agree well with measurements made in our laboratory. For a pulse containing 5 000 electrons, with r(0) = 75 μm and 4 mrad initial beam divergence followed by weak focusing, the pulse duration was 2.5 +/- 0.5 ps after 1.5 ns of propagation. Both the N-body simulation and mean-field theory predict 2.5 +/- 0.2 ps for these parameters
In addition to broadening the electron pulse to many times its original length, space charge effects also generate many eV of kinetic energy bandwidth in only a few nanoseconds (Fig. 1B).

3. All-Optical Electron Pulse Duration Measurement

In the non-relativistic case the ponderomotive force acting on an electron in a laser field takes the form[5]:

$$\vec{F}_{pond} = -\nabla U_{pond} \qquad (2)$$

$$U_{pond} = (e^2\lambda^2 I)/(2\pi m c^3), \qquad (3)$$

where I is the intensity , λ is the wavelength, c is the speed of light, and e and m are the electron charge and rest mass respectively

The effect of the ponderomotive potential on an electron pulse was simulated using Monte-Carlo techniques (non-interacting electrons). The results are shown in Fig. 2). If a 28.4 keV Gaussian femtosecond electron pulse is intersected at 90° with a laser pulse (1mJ, 400fs, 800nm, focused to 30μm), the pattern the electron beam forms on a detector 0.5m away is no longer Gaussian (Fig. 2, Inset). The height of the saddle point formed in the "far-field" beam pattern is sensitive to the delay between the laser and electron pulses and to the temporal duration of the two pulses.

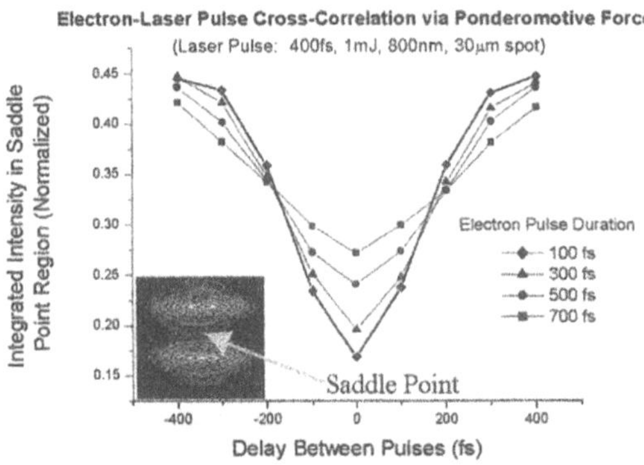

Fig. 2. Using the ponderomotive interaction to perform an electron-laser pulse cross-correlation. By measuring the electron beam current through a slit (30 μm wide in the above simulation) centered on the unperturbed beam maximum, one has a sensitive measure of the pulse overlap. This can also be used to determine T = 0 in a UED experiment.

Acknowledgements. The authors would like to thank John Dubinski, Dept. of Astronomy and Astrophysics, University of Toronto for providing the Barnes-Hut C-code used in the simulations. JRD acknowledges the support of a Walter C. Sumner fellowship.

References

1 B. J. Siwick et al., *J. Appl. Phys.* (Accepted. To be published Aug. 2002)
2 V. P. Degtyareva et al., *Opt. Eng.*, **37** (8), p. 2227 (1998)
3 K. Kinoshita et al., *Proc. SPIE* **981**, p. 62 (1988)
4 J. Barnes, P. Hut, *Nature*, **324**, p. 446 (1986)
5 V. S. Letokhov et al., *Adv. Chem. Phys*, **101** p. 185 (1997)

Ultrafast interfacial electron transfer from conjugated polymers to inorganic semiconductor nanoparticles

Neil A. Anderson, Xin Ai, Tianquan Lian

Department of Chemistry, Emory University, Atlanta, GA 30322
E-mail: tlian@emory.edu

Abstract. Ultrafast photoinduced electron transfer from conjugated polymers to inorganic semiconductor nanoporous thin films is studied using ultrafast IR spectroscopy. Composites for which charge transfer is energetically favorable exhibit subpicosecond electron injection. However, the rate of the charge transfer varies between different materials, likely due to differences in the conduction band composition and coupling between polymer and nanoparticle.

1. Introduction

There has been intense recent interest in composite materials of conjugated polymers with inorganic semiconductor nanoparticles, especially as photovoltaic materials for solar energy conversion [1,2]. This system differs from dye-sensitized nanoparticles, in that the polymer excitations are somewhat delocalized, excitations are mobile along the polymer chain, and the polymer-nanoparticle interface is not well defined. A clear picture of the mechanism and factors influencing the rate and efficiency of electron transfer from polymer to nanoparticles has not yet emerged. An understanding of interfacial ET dynamics will require systematic examination of various parameters of polymer, semiconductors and their interface.

Figure 1: Structures of (a) MEH-PPV (poly[2-methoxy-5-(2-e hyl-hexyloxy)-phenylene

vinylene)]) and (b) P3HT (poly(3-hexylthiophene)) conjugated polymers

In this work, we study photoinduced interfacial electron transfer in composites of the conjugated polymers MEH-PPV and P3HT (structures shown in Figure 1) with nanoporous thin films of SnO_2, TiO_2, and ZrO_2. Femtosecond IR spectroscopy is used to selectively probe the products of the charge transfer.

2. Experimental Methods

Colloidal SnO_2 and TiO_2 were synthesized according to p. 'blished procedures [3]. The solutions were cast onto sapphire substrates, dried in air, then baked. Films had good homogeneity and transparency. Average particle size was ~10 nm for SnO_2, and ~20 nm for TiO_2. Sensitized films were prepared by soaking the nanoporous films in a solution of polymer in chloroform, then washing with chloroform. Ultrafast experiments were carried out in a 400 nm pump-2000 cm^{-1} probe transient absorption scheme [3]. The pump beam selectively excites the polymer molecules. The pulsewidth at the sample was ~200 fs, determined by measuring the response from a silicon wafer. The film was constantly moved during data collection to avoid photobleaching effects.

3. Results and Discussion

The normalized kinetic traces obtained from MEH-PPV/SnO_2 and MEH-PPV/TiO_2 are shown in Figure 2a. The signals in these samples arise primarily from electrons injected into the inorganic semiconductor, although the charged defect on the polymer may have some absorption as well [3]. Electron transfer in MEH-PPV/SnO_2 is fit using an 850 fs exponential function. In MEH-PPV/TiO_2, it is instrument-limited (<100 fs). A noninjecting MEH-PPV/ZrO_2 film showed only a small signal attributed to interchain charge transfer [3].

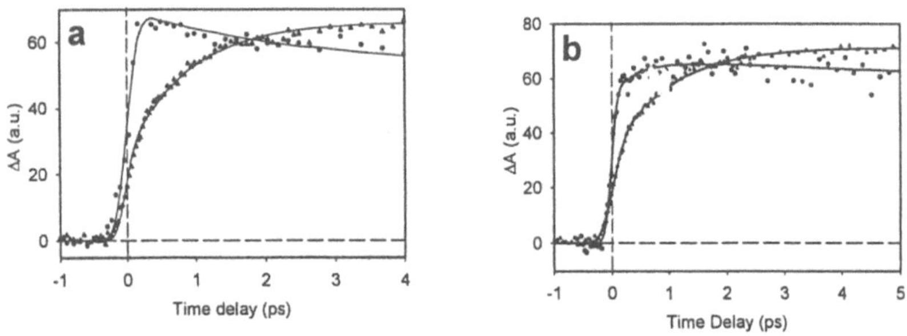

Figure 2: (a) Kinetic traces of MEH-PPV/TiO_2 (circles) and MEH-PPV/SnO_2 (triangles) probed at 2000 cm^{-1}. (b) Kinetic traces of MEH-PPV/SnO_2 (triangles) and P3HT/SnO_2 (circles).

The differences seen here are qualitatively consistent with the previous observation of slower electron injection in molecular dye-sensitized SnO_2 compared with dye-sensitized TiO_2 [4], although the difference is more dramatic with molecular dyes. This is notable as the SnO_2 conduction band is actually lower than TiO_2, making charge transfer more energetically favorable. However, the TiO_2 conduction band is primarily comprised of empty d orbitals, in contrast to SnO_2, which has a conduction band of empty s orbitals. This can affect the rate because the coupling between the polymer excited state and the conduction bands of the two semiconductors may be different due to the different accepting orbitals. Additionally, the conduction band composition results in a much lower effective mass for the SnO_2 conduction band electron, with a corresponding decrease in density of conduction band states, which can also slow injection.

Figure 2b shows a comparison of the injection dynamics for MEH-PPV and P3HT on SnO_2 film. Both polymers have similar excited state energies, so polymer energetics aren't expected to affect the observed rates. The P3HT shows substantially faster injection than MEH-PPV. It can be fit using a biphasic rise, with a large instrument-limited rise, and small 500 fs exponential component. P3HT/ZrO_2 showed no injection, and P3HT/TiO_2 showed instrument-limited injection.

The cause of the differences between the injection rates of the two polymers on SnO_2 is not yet clear. A contributing factor may be differences in the interaction between the polymer and nanoparticle. Sensitization of SnO_2 using P3HT required a shorter soaking time, while giving a higher sample absorbance than MEH-PPV. This may be due to a stronger Lewis acid-base interaction between the sulfur on the P3HT and the Sn of SnO_2. Furthermore, since the sulfur in P3HT is part of the conjugated polymer backbone, polymer excitations may be better coupled to the nanoparticle. Another possibility is that the shorter side chains of P3HT allow closer contact with the nanoparticle.

In summary, we have demonstrated ultrafast photoinduced electron transfer from conjugated polymer to inorganic semiconductor nanoparticles. The rate of transfer can be extremely fast, with differences that may arise due to differences in energetics, density of accepting states in the nanoparticle, coupling, and distance between the polymer backbone and nanoparticle.

Acknowledgements. This work is supported by the NSF CAREER award under grant 9733796, the ACS Petroleum Research Fund, and the Emory University Research Committee.

References

1 A.C. Arango, S.A. Carter, Appl. Phys. Lett. **74**, 1698, 1999.
2 S. Luzzati, M. Basso, M. Catellani, C.J. Brabec, D. Gebeyehu, N.S. Sariciftci, Thin Solid Films **403**, 52, 2002.
3 N.A. Anderson, E. Hao, X. Ai, G. Hastings, T. Lian, Chem. Phys. Lett. **347**, 304, 2001.
4 J.B. Asbury, E. Hao, Y.Q. Wang, H.N. Ghosh, T.Lian, J. Phys. Chem. B **105**, 4545, 2001.

Electronic coupling and coherence effects in ultrafast heterogeneous electron transfer

C. Zimmermann[1], F. Willig[1], S. Ramakrishna[1], R. Eichberger[1], R. Ernstorfer[1],N. Biswas[1], W. Storck[2], P. Persson[3]

[1] Hahn-Meitner-Institut, Dept. SE4, Glienicker Str. 100, D-14109 Berlin, Germany
E-mail: Zimmermann@hmi.de, Willig@hmi.de
[2] Fritz-Haber-Institut, Dept. Phys. Chem., Faradayweg 4-6, D-14195 Berlin, Germany
[3] Univ. of Uppsala, Dept. Quant. Chem., Box 518, S-75120 Uppsala, Sweden

Abstract. Electron injection from the excited state of the chromophore perylene was time-resolved for different anchor groups on the surface of anatase TiO_2. A modulation of the electron transfer rate was ascribed to the motion of a vibrational wavepacket.

Ultrafast electron injection into the conduction band of a semiconductor can be achieved via laser pulse excitation of a suitable chromophore that is located sufficiently close to the surface of the solid. It has been shown in model calculations [1,2] that generation of a vibrational wavepacket in the reactant state can lead to a modulation of the electron transfer rate. This effect can be observed as time dependent steps occurring in the occupancy of the reactant or product state of the light-driven heterogeneous electron transfer reaction. Such steps can only be observed if the density of electronic acceptor states is changing in the conduction band strongly over the energy range spanned by the Franck-Condon factors, i.e. the spectrum of the electron transfer reaction. To observe several steps electron transfer has to be sufficiently long compared to the vibrational period and sufficiently short against the lifetime of the vibrational wavepacket.

Thus, it is important to realize the desired electron transfer time for a given chromophore and find a suitable spacer-cum-anchor group that can provide the right electronic tunneling barrier at the surface of the semiconductor. The transfer time is controlled by the strength of the electronic coupling if the donor orbital of the chromophore is located high enough above the bottom of the conduction band and the density of states does not change too much over the energy range covered by the Franck-Condon factors [1,2]. The electronic coupling can be varied by changing the anchor-cum-spacer group that separates the edge of the chromophore from the surface atoms of the semiconductor. At present no simple recipe is apparent for predicting the electronic coupling or electron transfer times for a given set of spacer-cum-anchor groups. Model calculations show that the energy of the LUMO of the anchor-cum-spacer group is shifting significantly with respect to its electron affinity in vacuum when the anchor-cum-spacer group binds to the surface atoms of the semiconductor [3]. Theoretical calculations have also shown that there can be contributions of comparable magnitude to the electronic coupling through space and through saturated single C-C bonds of the spacer group [4]. It turns out, however, that electron injection experiments carried out with the chromophore perylene attached via different spacer-cum-anchor groups to anatase TiO_2 (mainly the (101) surface) are revealing meaningful qualitative trends in the measured transfer times. Earlier attempts by other groups to establish electron transfer times for different spacer-cum-anchor groups were performed with a Ruthenium chromophore. The latter shows a complicated behavior due to very fast intersystem crossing (ISC) within 30 fs that is occurring on the same time scale as electron injection from the excited singlet state [5]. At the same time electron transfer from the lower lying triplet state is slowed down since the latter's donor level is located close to the bottom of the conduction band [2]. In contrast, the perylene chromophore employed here offers several advantages. The donor level of the excited state is located high in the conduction band, ISC is sufficiently slow that it can be ignored for the electron injection process, and the absorption spectra of the ground, first excited singlet, and ionized states are all well separated in energy.

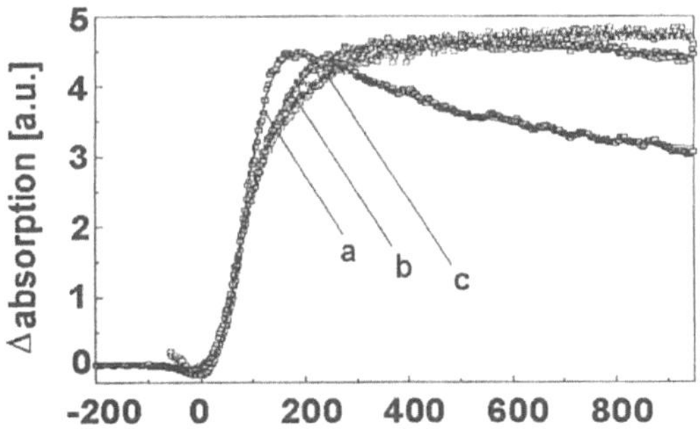

Fig. 1. Formation of the product state (ionized perylene) due to electron injection into anatase TiO_2, measured at 575nm with different spacer-cum-anchor groups (a): *carboxylic acid*; (b): *methyl-phosphonic acid*; (c): *methyl-(4-bicyclo[2.2.2]octane)-1-carboxylic acid.*

In Fig. 1 the three different transient absorption signals (λ_{probe} = 575nm) monitor the change in occupancy of the ionized state of perylene after laser pulse excitation. The transient absorption measurements were carried out with the samples mounted in an ultra-high-vacuum chamber. Dimerisation of the perylene chromophore was prevented at the surface of the semiconductor by two bulky tertiary-butyl side groups attached to the perylene chromophore. The different curves in Fig.1 correspond to different anchor-cum-spacer groups: *carboxylic acid* (a), *methyl-phosphonic acid* (b) and *methyl-(4-bicyclo[2.2.2]octane)-1-carboxylic acid* (c). Measurements with the phosphonate group instead of the carboxyl group as anchor gave essentially the same rise time as curve a. The rise of curve a in Fig.1 corresponds to a single exponential with time constant around 32fs. The decay of curve a shows that fast recombination has to be introduced into the fit if the injection time is short. The other clear case is curve c since the fit requires at least two time constants. Geometric models show that rotation around the $-CH_2-$ group in this spacer group can bring the perylene chromophore into two very different configurations, one where the distance between the rim of the chromophore and the atoms on the (101) anatase surface is short and the other where this distance is long. Thus, one expects two configurations, one giving rise to a short injection time and correspondingly short recombination time and the other to a much longer injection time a and correspondingly very long, i.e. negligible, recombination time for the given time window. The best fit to curve c applying this scenario gave a fast injection time of 45fs associated with a recombination time of 350fs and relative amplitude of 0.33 together with a slower injection time of 260fs associated with a long recombination time that can be ignored in the given time window. The same scenario should apply now to rotation around the $-CH_2-$ group in the other anchor-cum-spacer group corresponding to curve b. The rise of the latter curve allows for a good fit with a single exponential but this fit appears to be inconsistent in the light of the clearer case of curve c. Indeed, an equally good fit to curve b as with the single exponential rise can be obtained with relative amplitude 0.22 for one component with a rise time of 40fs associated with a recombination time of 622fs and a second component with a rise time of 84fs and a long recombination time in the picosecond time scale.

Hitherto, a completely rigid anchor-cum-spacer-group of suitable length has not yet been attached to the perylene chromophore such that only one configuration is allowed for the chromophore at the surface of anatase TiO_2. Such a compound would be best suited for investigating in the most promising time windows the theoretically expected [1,2] coherence effects that can show up e.g. as steps in the in the time-resolved electron transfer reaction. Thus, such investigations were carried out with the compound associated with the pseudo mono-exponential rise seen in curve b of Fig.1. Indeed, distinct steps can be seen in the rise of the electron transfer product state for this compound [6]. They can be ascribed to curve crossing effects of the vibrational wavepacket prepared in the excited donor state of the perylene chromophore. Such coherence effects yield improved insight into the reaction mechanism, e.g. identification of a vibrational mode that couples effectively to electron injection. The steps represent also a direct fingerprint of the reaction occurring from a hot vibrational state prior to redistribution of vibrational energy over lower energy vibrational modes in the chromophore. The described experimental result can be considered as an encouraging step towards realizing coherent control scenarios for at least certain reactions of rather large organic molecules.

References

1 S. Ramakrishna, F. Willig, V. May, Phys. Rev. B **62**, R16330, 2000.
2 S. Ramakrishna, F. Willig, V. May, J. Chem. Phys. **115**, 2743, 2001.
3 P. Persson, J. Chem. Phys. **112,** 3945, 2000.
4 C. Liang and M.D. Newton, J. Phys. Chem. **96**, 2855, 1992; C. Liang and M.D. Newton, J. Phys. Chem. **97**, 3199, 1993.
5 G. Benkö, J. Kallioinen, J.E.I. Korppi-Tommola, A.P. Yartsev, and V. Sundström, J. Am. Chem. Soc. **124**, No. 3, 489, 2002.
6 C. Zimmermann, F. Willig, S. Ramakrishna, B. Burfeindt, B. Pettinger, R. Eichberger, W. Storck, J. Phys. Chem. B **105**, 9245, 2001.

Photoconductivity of dye-sensitized titanium dioxide films measured by time-resolved THz spectroscopy (TRTS)

Gordon M. Turner, Matthew C. Beard, and Charles A. Schmuttenmaer

Chemistry Department, Yale University, P.O. Box 208107, New Haven, CT 06520-8107
E-mail:gordon.turner@yale.edu

Abstract. The transient photoconductivity of dye-sensitized nanocrystalline TiO_2 films displays non-Drude behavior. A model that describes backscattering of charge carriers describes the data well, and it is interpreted as scattering of mobile electrons within nanoparticles.

1. Introduction

Dye-sensitized solar cells (DSSCs) made from thin films of nanometer-sized titanium dioxide (TiO_2) particles have been thoroughly studied since their introduction by O'Regan and Grätzel in 1991 [1]. These DSSCs have been shown to reach solar-to-electric energy conversion efficiencies of greater than 10% [1,2]. The wide band gap TiO_2 is sensitized to visible photon absorption by chemisorption of a suitable organic dye. Photoexcited electrons in the dye molecules are then injected into the TiO_2 conduction band on a 50 fs time scale [3]. One of the best performing sensitizers reported thus far is Ru(4,4'-dicarboxy-2,2'-bipyridine)$_2$(NCS)$_2$ (Ruthenium 535) [2].

Recent UV pump / mid-IR probe studies have allowed researchers to obtain the first direct measurements of carrier dynamics in these systems [4,5]. The success of that method, as well as that reported in this paper, lies in its ability to detect carriers in the conduction band of the semiconductor directly, without being complicated by dye absorption at the probe wavelengths. Previous studies in the mid-IR have been limited to measuring the absorption only. TRTS measures the full complex permittivity in the FIR, and is therefore a more direct probe of conduction. The main advantage of TRTS is that it is a non-contact electrical probe capable of resolving the frequency-dependent conductivity on a sub-picosecond time scale.

2. Experimental Methods

Nanocrystalline TiO_2 films were prepared by spinning a few drops of a colloidal suspension of TiO_2 onto 1 mm thick fused silica substrates. The P25 TiO_2 from Degussa consists of ~25-nm diameter particles, and was prepared as described elsewhere [2]. Coumarin 343 (C343) was purchased from Aldrich, and Ru535 was purchased from Solaronix, Switzerland, and both were used as received. The spectrometer has been described in detail elsewhere [6], and is based on a regeneratively amplified Ti:Sapphire laser operating at a 1 kHz repetition rate, with 1mJ pulses at 800 nm. The types of experiments presented are 400 nm pump, THz (far-infrared: 3-100 cm^{-1}) probe.

3. Results and Discussion

Figure 1(a) is a comparison of 1D pump scans of Ru535-sensitized TiO_2 (Ru535/TiO_2) and C343-sensitized TiO_2 (C343/TiO_2) at early pump delay times. The dynamics for the two samples are nearly identical and, when longer scans are fit to a double exponential with a long time offset, only differ slightly in the decay time constant of the second exponential. Our full 2D experiments indicate an onset of photoconduction in 230 fs, which is limited by our detector response time [7]. This agrees well with the mid-IR studies that report the electrons are injected into the TiO_2 in under 100 fs [3,4]. Figure 1(b) illustrates the effect of dye sensitization, as well as the long-time offset in the carrier decay dynamics. We can bracket the long relaxation time component, τ_3, in photoexcited colloidal and single-crystalline TiO_2 to 100 ns < τ_3 < 200 μs.

Figure 2 shows the measured conductivity in a film of Ru535/TiO2, 3 ps after photoexcitation. It is immediately evident that the conductivity is non-Drude, as the real component does not peak at dc, and the imaginary component is actually negative. The data is fit to a classical modification to the Drude model [8]:

$$\hat{\sigma}(\omega) = \frac{\varepsilon_o \omega_p^2 \tau}{(1-i\omega\tau)}\left[1 + \frac{c}{(1-i\omega\tau)}\right] \tag{1}$$

where c is a term that describes the fraction of velocity retained after a collision. When $c = 0$, the scattering is isotropic and the standard Drude model is obtained. When $c = -1$, scattering is in the reverse direction, the real part of the conductivity is 0 at dc and the imaginary part of the conductivity is negative at low frequencies. Smith has applied this model to the optical conductivity of liquid Hg and Te, as well as the quasicrystal $Al_{63.5}Cu_{24.5}Fe_{12}$ (all systems that can be described as poor-conductors, or materials at the metal-insulator transition).

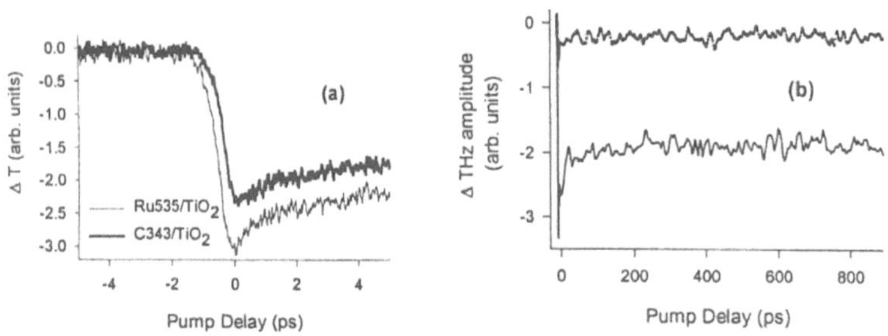

Fig. 1. (a) 1-D pump scans of Ru535/TiO_2 and C343/TiO_2 showing the average change in THz transmission when photoexcited at 400 nm (b) longer 1-D pump scans of unsensitized TiO_2 film (dots) and Ru535 sensitized TiO_2 film (solid line).

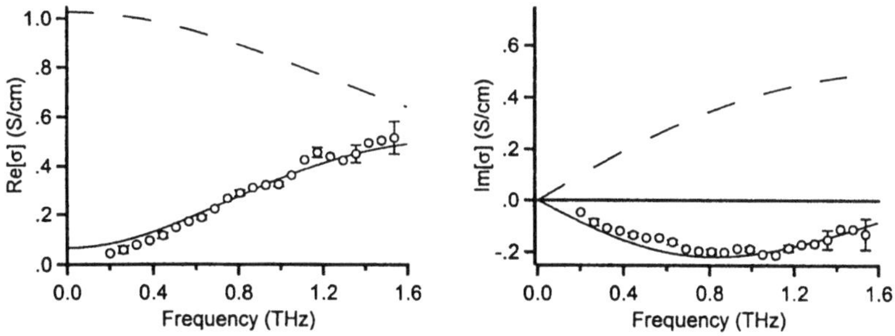

Fig. 2. Real and imaginary parts of the complex conductivity of Ru535/TiO_2, 3 ps after 400 nm photoexcitation (open circles). Solid lines are a fit of the Smith model to the data (ω_p = 12.3 x 10^{12} s^{-1} THz, τ = 76 fs, *c* = -0.93) and the dashed lines are the Drude model for the same parameters (*c* = 0).

Our data are fit by a *c* parameter of –0.93. We calculate the ac mobility, based on the scattering time, to be μ = 140 $cm^2V^{-1}s^{-1}$. According to the Smith model with a *c* parameter of –0.9, the dc mobility is a factor of 10 less than the ac mobility, or 14 $cm^2V^{-1}s^{-1}$.

4. Conclusions

Our TRTS measurements of the photoconductivity in nanocrystalline TiO_2 indicate that the photogenerated electrons are mobile charge carriers that display non-Drude behavior. The dc mobility of the electrons is decreased due to scattering off the particle walls or grain boundaries within the particles. This is an unprecedented level of detail about the conduction mechanism within a disordered nanometer-scale system, exploiting the non-contact sub-picosecond nature of the method.

Acknowledgements. Partial support from the Sloan Foundation and from the National Science Foundation CAREER award program (CHE-9703432) is gratefully acknowledged.

References

1. B. O'Regan and M. Gratzel, Nature **353**, 737 (1991).
2. M. K. Nazeeruddin, A. Kay, I. Rodicio *et al.*, J. Am. Chem. Soc. **115**, 6382 (1993).
3. R. J. Ellingson, J. B. Asbury, S. Ferrere *et al.*, J. Phys. Chem. B **102**, 6455 (1998).
4. J. B. Asbury, R. J. Ellingson, H. N. Ghosh *et al.*, J. Phys. Chem. B **103**, 3110 (1999).
5. T. A. Heimer and E. J. Heilweil, J. Phys. Chem. B **101**, 10990 (1997).
6. M. C. Beard, G. M. Turner, and C. A. Schmuttenmaer, Phys. Rev. B. **62**, 15764 (2000).
7. M. C. Beard, G. M. Turner, and C. A. Schmuttenmaer, J. Appl. Phys. **90**, 5915 (2001).
8. N. V. Smith, Phys. Rev. B **6415**, 5106 (2001).

Nonlinear Optical Probes of Vibrational Motion at Interfaces

T. Kikteva, D. Star, T. Johansson, and G.W. Leach

Department of Chemistry, Simon Fraser University, Burnaby, B.C. V5A 1S6 Canada.
E-mail: gleach@sfu.ca

Abstract. Femtosecond vibrational sum frequency generation allows for time-and frequency-domain vibrational spectroscopies and shows coherent nuclear motion (surface quantum beats) which dephases on a timescale dependent on surface inhomogeneity. A two-beam, equal pulse correlation technique has also been used to examine low frequency intramolecular motion of p-chloroaniline (PCA) at the fused silica/air interface. Time-resolved parametric signals are modulated by adsorbate vibrational motion driven through stimulated Raman excitation of low frequency lattice modes of crystalline PCA.

1. Introduction

The investigation of the vibrational motions of molecules has been a valuable source of information regarding molecular structure and bonding, the magnitude and nature of interactions between molecules, and the dynamics of energy transfer and redistribution. The introduction of the second-order nonlinear optical technique of sum frequency generation (SFG) [1-3] has provided a means of performing interface-specific vibrational spectroscopy and has been used to examine vibrational motion in a wide variety of condensed phase interfaces. One limitation of the sum frequency technique has been the difficulty in producing intense tunable infrared radiation over the entire spectral region of the adsorbate. Most studies to date have employed relatively high frequency infrared radiation and do not provide information regarding low frequency intra-and intermolecular motions. These motions are expected to be particularly sensitive to adsorbate interactions and environment, and to be active participants in energy transfer and relaxation processes. Here, we report the use of femtosecond vibrational sum frequency generation to perform time-and frequency-domain vibrational spectroscopies of adsorbates and a novel method for interrogating the low frequency Raman active motions of molecular adsorbates at the fused silica/air interface.

2. Experimental Methods

We employ a regeneratively amplified ti:sapphire laser system described in detail elsewhere [4]. Briefly, the amplified output consists of a 1 kHz train of 100 fs laser pulses with 1 mJ pulse energy, tunable about 800 nm. For the SFG experiments, tunable mid IR light is produced by mixing the signal and idler output from an optical parametric amplifier in a $AgGaS_2$ crystal. For the two-beam correlation measurements the 800 nm pulse train is split into two independent beams of equal intensity. Synchronization of the beams at the surface is achieved by incorporating a variable, computer-controlled optical delay line along one of the beam paths.

3. Results and Discussion

Fig. 1 shows the SFG vibrational spectrum of the CH region for a monolayer of the amphiphile stearic acid deposited by the Langmuir-Blodgett technique at the CaF_2/air interface (inset). The spectrum is obtained using a 100 fs mid IR pulse centered at 2900 cm^{-1} combined with a ~2 ps 800 nm upconverting pulse.

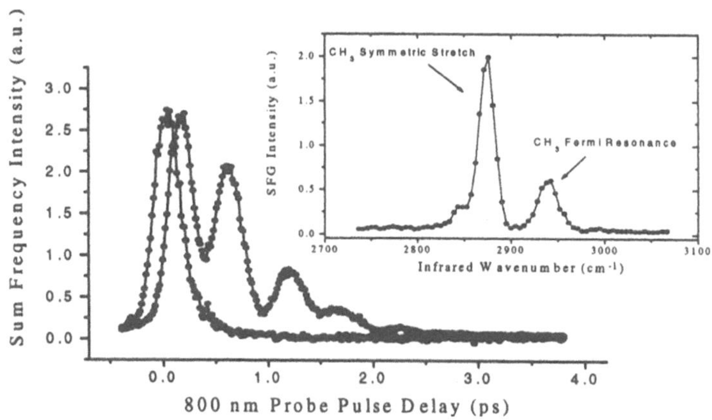

Fig 1. Femtosecond vibratonal sum frequency generation spectrum of a stearic acid monolayer (inset) and resulting quantum interference pattern

The spectrum shows the feasibility of this approach as well as the high degree of order within the monolayer. Fig 1 shows the results of a time-resolved pump/probe experiment. The coherent nuclear motion prepared by the 100 fs mid IR pulse is monitored through the time-resolved SFG response. The bandwidth of the infrared pulses is sufficient to span the observed vibrational resonances and excite these modes coherently. Coherent excitation prepares a nonstationary superposition state of the surface vibrational modes which evolves in time. We have followed the evolution of this vibrational wavepacket, again using sum frequency generation, but with a train of 100 fs, 800 nm pulses [4]. This pronounced intensity modulation is a direct result of the vibrational excitation and reflects the coherent motion of the vibrational modes excited. Thus we have directly measured the vibrational quantum beats of a molecular monolayer. Fig.2 shows the results of a new two-beam equal pulse correlation technique that probes low frequency vibrational motion of p-chloroaniline (PCA) at the fused silica/air interface through stimulated Raman excitation. The measured signal

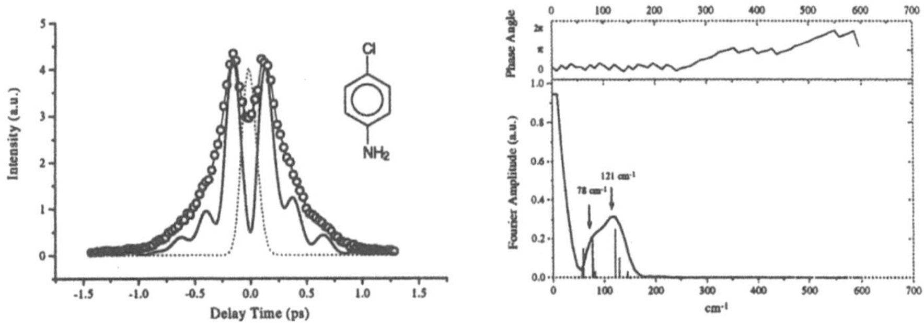

Fig 2. Two-beam correlation signal from p-chloroaniline (left). Fourier transform of correlation signal (right).

shows evidence of intensity modulation. Also indicated are the instrument response function centered at zero time, and the molecular response obtained by deconvolution of the signal and the instrument response. Removal of the instrument contribution from the signal gives a symmetric, oscillatory decay with a period of ~275 fs and 1/e decay time of approximately 250 fs. Fig. 2 also shows a frequency analysis of the oscillatory deconvoluted signal obtained by fast Fourier transform. The frequency spectrum shows substantial amplitude in the vicinity of 100 cm^{-1} consistent with the excitation of the low frequency lattice vibrations of crystalline PCA [5], the positions of which are also shown in Fig. 2. We have also studied ultrathin films of triphenylphosphine (TPP) and measured detection wavelength dependent in- and out-of-phase oscillations of the correlation signals associated with intramolecular C—P deformations.

4. Conclusions

Nonlinear optical methods have been used to examine coherent vibrational motion of adsorbates at the solid/air interface. Femtosecond vibrational sum frequency generation displays coherent nuclear motion which decays on a timescale of ~600 fs and is dependent on surface inhomogeneity. A two-beam equal pulse correlation technique has been used to examine low frequency intermolecular adsorbate motion driven through stimulated Raman excitation.

References

1 Y. R. Shen, *Annu. Rev. Phys. Chem.* **40,** 327 1989;Y. R. Shen, *Nature*, **337**, 519 1989.

2 G. L. Richmond, J. M. Robinson, V.L. Shannon, *Prog. Surf. Sci.*, **28**, 1 1988.

3 T. F. Heinz, *Nonlinear Surface Electromagnetic Phenomena,* Eds. H. Ponath and G Stegman, Amsterdam; Elsevier 1991. K. B. Eisenthal, *Annu. Rev. Phys. Chem.* **43**, 627 1992.

4 D. Star, T. Kikteva and G.W. Leach, *J. Chem. Phys.* **111**, 14 1999.

5 Righini and J. Califano, J. Mol. Spec. **43**, 272 1979.

Ultrafast dynamics of recombinative desorption of hydrogen from a Ru(001) surface

C. Frischkorn[1], M. Wolf[1], D. N. Denzler[2], C. Hess[2] and G. Ertl[2]

[1] Freie Universität Berlin, Fachbereich Physik, Arnimallee 14, 14195 Berlin, Germany
E-mail: christian.frischkorn@physik.fu-berlin.de
[2] Fritz-Haber-Institut d. Max-Planck-Gesellschaft, Faradayweg 4-6, 14195 Berlin, Germany

Abstract. Femtosecond-laser excitation induces recombinative desorption of molecular H_2 from atomic hydrogen on a ruthenium surface. The measured ultrafast timescale of this surface reaction reveals an excitation process mediated by hot substrate electrons. A coverage dependent isotope effect indicates a collective reaction mechanism whereby vibrationally excited H atoms modify the reactivity of coadsorbed species.

Introduction. The dynamics of energy transfer processes at surfaces are of fundamental importance for a microscopic understanding of chemical reaction at solid surfaces, e.g. in heterogeneous catalysis. In conventional chemistry with phonons and electrons always in thermal equilibrium, no detailed information can be obtained on the pathway of the energy flow between the substrate and the adsorbate. However, when a metal substrate is excited by an intense femtosecond-laser pulse, a transient non-equilibrium between the electronic and phononic temperatures is created, which lasts on the order of the electron-phonon coupling time (for ruthenium ~1 ps) and allows to distinguish electron-mediated from phonon-mediated surface reactions [1].

Experiment. The appropriate experimental approach is to measure the two-pulse correlation (2PC) of the respective reaction yield. To this end, the fundamental output of an amplified Ti:sapphire femtosecond-laser system is split in two equally intense portions and both beams are sent time-delayed onto the sample mounted in an ultrahigh vacuum chamber. The yield of the desorbed products is detected with a quadrupole mass spectrometer as a function of the pulse-pulse delay.

Results and discussion. Figure 1 shows such a 2PC of the desorption yield for the formation of molecular H_2 from atomic hydrogen on an H-saturated Ru(001) surface. The narrow full width at half maximum (FWHM) of ~1.1 ps unambiguously indicates that hot substrate electrons are the driving force behind this recombinative desorption [1,2]. The experimental data are well reproduced by a simulation (solid line) using the two-temperature model (electronic and phononic temperatures T_{el} and T_{ph}) [3]. Friction between the substrate and the adsorbate [4,5] couples energy from the substrate subsystems to the adsorbate vibrations and hence into the relevant reaction coordinate. The electronic friction model [5] used here to describe the 2PC data for hydrogen yields an ultrafast coupling time for the H_2 formation of only a few hundred femtoseconds.

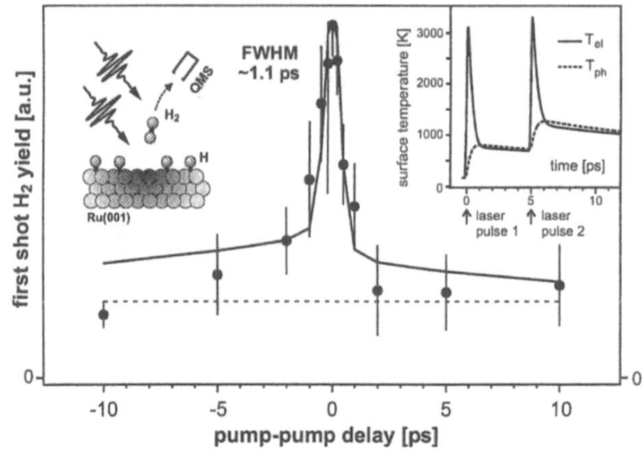

Fig. 1. Two-pulse correlation of the H_2 desorption yield showing a narrow FWHM of ~1.1 ps. The inset displays typical time profiles for the electronic and phononic temperature at an exemplary pulse-pulse delay of 5 ps.

In addition, a pronounced isotope effect in the reaction yield $Y(H_2)$:$Y(D_2)$ of 10:1 taken from isotopically pure saturation coverages corroborates the electronic origin of the desorption process.

Molecular hydrogen can also be formed "thermally" in this system, i.e. by conventionally heating the surface. Comparing desorption measurements occurring via either the thermally or the hot-electron driven pathway exhibit the uniqueness of the "femtochemical" reaction with clear evidence for a novel mechanism: The latter process involves a concerted action of more partners than only both reactants forming the respective hydrogen molecule. Experimentally, this is seen in significant changes of the isotope yield ratio of the desorbed molecular hydrogen if the mixture of the coadsorbed atomic H and D on the surface is varied.

For exemplary H:D mixtures, Table 1 summarizes the relative desorption yields of the three isotopic species obtained under thermal equilibrium and laser excitation conditions, respectively. Thermal desorption experiments (TD) with the mixed isotope HD result in the statistically expected relative isotope ratio H_2:HD:D_2 of 1:2:1. Additional data from mixtures ranging from ~15:85 up to ~85:15 H:D (prepared by precovering the clean surface with HD and subsequently saturating the surface with the respective pure isotope) further corroborate that second order rate equations hold in the case of "thermal" surface chemistry. According to chemical kinetics, the rate of a bimolecular reaction of the form $A + B \rightarrow AB$ scales with the concentrations of each of the reactants [A] and [B]. In contrast, clear deviations are found in the femtochemical case. As seen in the center column of Table 1 (italic numbers) and different from the TDS data, more D_2 with respect to H_2 (i.e. 1:3) is formed than one would expect supposing the isotope ratio of 1:10 (bold numbers) is independent of the reactants' concentration and is solely due to the mass difference.

Table 1. Relative yields of H_2, HD, and D_2 from Ru(001) by thermal heating under equilibrium conditions *versus* fs-laser excitation for differently mixed coverages.

	H : D ratio of the coverage in percent					
	100:0		50:50		0:100	
Yield	thermal	fs laser	thermal	fs laser	thermal	fs laser
H_2	1	**10**	1	*3*	-	-
HD	-	-	2	*4*	-	-
D_2	-	-	1	*1*	1	**1**

Summarizing, with femtosecond-laser excitation a higher H concentration [H] seems to enhance both the formation of H_2 *and* D_2 and thus second order chemical kinetics fail to describe this femtosecond-laser driven reaction.

Instead, we propose the following collective reaction mechanism: The lighter hydrogen H acts an antenna, when the surface is laser-excited with a femtosecond pulse. Due the electronic nature of the energy flow from the substrate to the adsorbate layer, atomic H in mixed layers is excited first, since its coupling time is by a factor of 2 shorter than that of atomic D. A subsequent change of the interaction between neighboring atoms, e.g. repulsion between the excited H and a "cold" D or changes in the potential energy landscape for the respective reaction partners, would then promote the recombination of the deuterated species. Theoretical calculations are still underway to more quantitatively describe the observed phenomena and might prove that our results have more general implications on surface chemistry.

Acknowledgements. Financial support by the Deutsche Forschungsgemeinschaft through Sfb 450 is gratefully acknowledged.

References

1 M. Bonn, S. Funk, Ch. Hess, D.N. Denzler, C.Stampfl, M. Scheffler, M.Wolf, and G. Ertl, Science **285**, 1342 (1999).
2 F. Budde, T.F. Heinz, M.M. Loy, J.A. Misewich, F. de Rougemont, and H. Zachrias, Phys. Rev. Lett. **66**, 3024 (1991).
3 M. Bonn, D.N. Denzler, S. Funk, M.Wolf, S.-S. Wellerhoff, and J. Hohlfeld, Phys. Rev. B **61**, 1101 (2000).
4 F. Budde, T.F. Heinz, A. Kalamarides, M.M. Loy, and J.A. Misewich, Surf. Sci. **283**, 143 (1993).
5 M. Brandbyge, P. Hedegard, T.F. Heinz, J.A. Misewich, and D.M. Newns, Phys. Rev. B **52**, 6042 (1995).

Chirp-controlled ground-state photodissociation of $Cr(CO)_6$

T. Witte, L. Windhorn, J. Yeston, M. Motzkus, D. Proch, and K.L. Kompa

Max-Planck-Institut für Quantenoptik, Hans-Kopfermannstr. 1,
D-85748 Garching, Germany, E-mail: tiw@mpq.mpg.de

Abstract. Femtosecond laser excitation of the T_{1u} CO-stretching manifold in ground state $Cr(CO)_6$ induces loss of CO. Enhancement and inhibition, respectively, of the dissociation are demonstrated by applying negative or positive chirp to the pulses.

1. Introduction

A prime goal of molecular control is the selective scission of chemical bonds by laser light. Attempts employing nanosecond IR laser pulses were hindered because the excitation time exceeded the timescale of intramolecular vibrational redistribution (IVR). With the advent of femtosecond infrared laser technology novel concepts have been developed. Chelkowski *et al.* [1] have proposed the use of properly chirped laser pulses to enhance vibrational excitation and dissociation. Assuming a sequential ladder climbing progression up the vibrational manifold in an anharmonic potential, negatively chirped IR laser pulses should ease excitation to higher levels, since the level spacing decreases with increasing energy. Although not yet realized for molecular dissociation, this concept has been applied to control vibrational population transfer in gas-phase NO [2] as well as solution-phase $W(CO)_6$ [3]. Recently, metal-carbonyls in the gas phase could be dissociated with intense unchirped mid-infrared fs laser pulses tuned to the C-O stretch resonance at 5μm (Fig. 1) [4]. We now investigate the role of chirp on the dissociation yield of $Cr(CO)_6$.

2. Experimental Setup

A glass cell of ~5cm^3 volume is filled with $Cr(CO)_6$ vapor to a pressure of ~40Pa. The MIR fs laser pulses (11μJ at 5μm, 1kHz) are generated by noncollinear difference frequencey mixing of the signal and idler pulses from a traveling-wave OPA. A typical pulse exhibits a spectral bandwidth of ~160cm^{-1} FWHM and a pulse width of ~110fs FWHM. This is equivalent to a time-bandwidth product of 0.53 (assuming Gaussian shape), indicating that the original pulses are close to the Fourier limit.

The macroscopic concentration of metal carbonyl is obtained from FTIR spectra [4] recorded after the sample has been exposed to the IR pulses over a period of

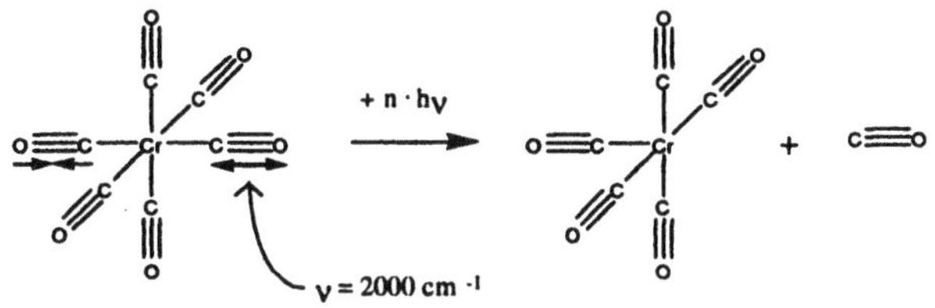

Fig. 1. Resonant excitation of the T_{1u} CO-stretching manifold with fs laser pulses leads to cleavage of CO [4].

several minutes. Each such spectrum thus reflects the accumulated dissociation yields of ~10^5 laser shots. Our diagnostics further comprise an autocorrelator to measure the pulse length, as well as a computer-controlled knife edge that is scanned through the focus to monitor the beam waist.

For the experiments the generation of both positively and negatively chirped laser pulses at 5μm central wavelength is a prerequisite. We have opted for a simple approach whereby chirp is induced by propagation through transmissive but dispersive material. The essential material property is the group velocity dispersion (GVD). For wavelengths around 5μm, lithium fluoride (LiF) shows a negative GVD and imposes a negative chirp, whereas germanium (Ge) exhibits a positive GVD and causes a positive chirp [5,6]. Thus, simply by transmitting fs laser pulses through substrates of LiF and Ge (or combinations thereof) prior to being focussed into the cell by means of a 5cm focal length mirror, we coarsely control the pulses' sign of chirp.

3. Results and Discussion

We performed three dissociation experiments, in which the MIR pulses were A) positively chirped after propagation through Ge, B) approximately unchirped after propagation through Ge+LiF, and C) negatively chirped after propagation through Ge+LiF plus additional LiF. Fig. 2 illustrates the experimental results. The bottom row displays the autocorrelation measurements. As can be seen in B), the chirps imnduced by Ge and LiF are opposite, since the AC width decreases with LiF added. the top row shows the respective change in relative absorbance and thus total concentration of $Cr(CO)_6$, as detected by the FTIR spectrum on a timescale of minutes. The arrow marks the time interval (8min) of exposure to the MIR light. The relative change after irradiation is -2% for positive chirp in A), -10% for unchirped pulses in B), and -28% for negative chirp in C). For each dissociation measurement, the focal beam waist was measured and shown to stay constant at a value of ~70μm. The Ge remained in the beam path since its high index of refraction causes a ~50% surface reflection loss of the pulse energy. The addition of LiF leads to further small losses, which, however, did not obscure the dissociation enhancement. We point out that, aside from the imposed chirp, all

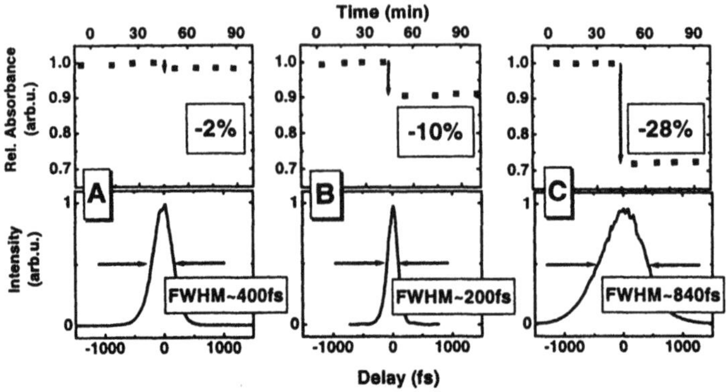

Fig. 2: Top row: Change of absorbance of $Cr(CO)_6$ after 8min irradiation with ~5μJ MIR laser pulses (arrow). Bottom row: Autocorrelation traces of MIR fs laser pulses: A) positive chirp with Ge, B) "unchirped" with Ge+LiF, and C) negative chirp with Ge+LiF+LiF. The dissociation yield is seen to be a steep function of chirp.

other experimental parameters remain constant. In particular, all diagnostics were conducted immediately before the molecular dissociation.

The chirp controlled dissociation of a molecule in the ground electronic state was demonstrated. In qualitative accordance with the picture of stepwise molecular excitation in an anharmonic potential, we find evidence for a ladder climbing mechanism in the excitation of the C-O stretch. The excitation occurs on a sub-picosecond time-scale, such that it outpaces the IVR processes. We conclude that negatively chirped MIR laser pulses optimize population transfer to levels above the dissociation threshold. Note that the excitation coordinate is not the reaction coordinate. After excitation the energy leaks out (with an as yet uncertain time constant) into the Cr-C reaction coordinate until dissociation occurs. Since it has been argued that excitation above dissociation threshold of a reaction coordinate would require excitation frequencies down to zero due to the anharmonicity, such an intermediate IVR step might be crucial for dissociation control.

Future experiments may implement a very recent concept for the programmable generation of shaped MIR fs laser pulses [7], in order to further elucidate the details of the dissociation mechanism and enhance its efficiency.

References

1. S. Chelkowski, A.D. Bandrauk, P.B. Corkum, Phys. Rev. Lett. **65**, 2355 (1990)D. Maas, D.I. Duncan, W.J. van der Zande, L.D. Noordam, Chem. Phys. Lett. **290**, 75 (1998)
2. V.D. Kleiman, S.M. Arrivo, J.S. Melinger, E. Heilweil, Chem. Phys. **233**, 207 (1998)
3. L. Windhorn, T. Witte, J. Yeston, D. Proch, M. Motzkus, K.L. Kompa, W. Fuss, Chem. Phys. Lett, *in press*
4. J.C. Diels, W. Rudolph, "Ultrashort Laser Pulse Phenomena", Academic Press, San Diego, 1996
5. L. Wolfe, "The Infrared Handbook", Environmental Research Institute of Michigan, 1989
6. T. Witte, D. Zeidler, D. Proch, K.L. Kompa, M. Motzkus, Opt.Lett., **27**, 131 (2002)

Part VIII

Coherent Effects in Semi- and Superconductors

Ultrafast coherent spectroscopy of a single quantum dot

Tobias Günther[1], Christoph Lienau[1], Thomas Elsaesser[1], Markus Glanemann[2], Vollrath Martin Axt[2], Tilmann Kuhn[2], Soheyla Eshlagi[3], and Andreas D. Wieck[3]

[1] Max-Born-Institut für Nichtlineare Optik und Kurzzeitspektroskopie, Max-Born-Str. 2A, D-12489 Berlin, Germany, E-mail: lienau@mbi-berlin.de
[2] Institut für Festkörpertheorie, Universität Münster, D-48149 Münster, Germany
[3] Lehrstuhl für Angewandte Festkörperphysik, Ruhr-Universität Bochum, Universitätstr. 150, D-44870 Bochum, Germany

Abstract. The nonlinear optical response of a single quantum dot is studied for the first time with femtosecond resolution. We find pronounced spectral oscillations in the differential reflectivity reflecting the perturbed free induction decay of the coherent polarization of a single exciton.

1. Introduction

Quantum dots currently attract much interest as model systems for artificial solid state atoms and as probes of the physics of wavefunction localization in disordered media [1]. The potential of using their optical excitations as bits for semiconductor-based implementations of quantum information processing is under intense investigation [2]. Although recently dephasing times of up to 600 ps have been demonstrated, such implementations are intimately linked to the ability to probe and control optical nonlinearities in single and coupled quantum dots on ultrafast time scales [3], a research field that remains to be explored.

In this paper, we describe and demonstrate a novel technique, combining near-field optics and femtosecond pump-probe spectroscopy, to analyze the nonlinear optical response of single quantum dots on ultrafast time scales. We resolve, in differential reflectivity spectra, transient oscillatory structures around the quantum dot exciton resonance. These oscillations persist up to negative time delays of more than 10 ps and reflect the perturbation of the free induction decay of the coherent QD polarization by transient manybody interactions.

2. Experimental Methods

We investigate excitons localized in interface quantum dots in a single 5.1 nm thick (100) GaAs QW layer buried at d = 120 nm below the surface. Differential probe reflectivity spectra $\Delta R(E_{det},\Delta t)/R_0 = \left[R(E_{det},\Delta t) - R_0(E_{det})\right]/R_0(E_{det})$ [$R_0(E_{det})$: steady state reflectivity at photon energy E_{det}] are recorded as a function of pump-probe delay Δt in a near-field pump-probe setup at 12 K with a combined spatial and temporal resolution of 200 nm and 100 fs (Fig. 1a). The 100 nW, 18meV broad probe laser is centered at 1.66 eV, around the QW absorption resonance. Photoluminescence (PL) from the sample and/or the reflected probe light is collected through the same fiber, spectrally dispersed and detected with a CCD camera. The pump-laser is centered at 1.675 eV to create about 5 electron-hole pairs per pulse in 2D continuum states.

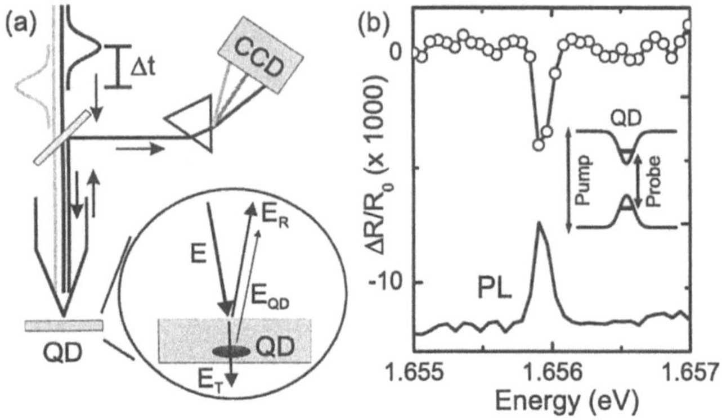

Fig. 1. (a) Schematic of the experimental setup. (b) Near-field PL spectrum of a single QD (solid line) and differential reflectivity $\Delta R/R_0$ at $\Delta t = 30$ ps. PL and ΔR are recorded with identical pump pulses centered at 1.675 eV, exciting electron-hole pairs in 2D continuum states. Inset: Schematic energy diagram.

3. Results and Discussion

For Δt>0, the $\Delta R/R_0$ spectra for above bandgap excitation at 1.675 eV show a few spectrally ultrasharp resonances, that are isoenergetic with single exciton emission peaks in simultaneously recorded PL spectra [Fig. 1(b)]. The reflectivity change ΔR displays a slow decay with time constants of 30 – 100 ps, depending on the size of the QD (Fig. 2a). Such time constants are determined by radiative recombination in the QDs.

The early time dynamics of $\Delta R(E_{QD})/R_0$ (Fig. 2b) show striking differences from those expected for an isolated two-level system. The signal displays an 8 ps rise at negative delay times that occurs much slower than 150 fs cross correlation of pump and probe pulses and is followed by a fast partial decay with a time constant of about 6 ps. The spectral characteristics of $\Delta R/R_0$ at Δt>0 and Δt<0 are markedly different. At Δt>0, the spectra show the expected bleaching of the dot resonance, while at Δt<0, (Fig. 2c), we find pronounced *spectrally symmetric* oscillations around the excitonic resonance.

As shown in Fig. 1(a). the reflected probe light, locally collected by the near-field tip, represents a superposition of the field $E_R(t)$ reflected from the sample surface, and the field $E_{QD}(t)$, radiated in back-direction from the probe-induced excitonic QD polarization $P_{QD}(t)=\int dt'\chi_{QD}(t)E_T(t-t')$. Here, $E_T(t)$ and $\chi_{QD}(t)$ denote the probe field interacting with the QD and the QD susceptibility, respectively. Excitation by the pump affects $\chi_{QD}(t)$ and – thus – results in a change of reflectivity. For $\Delta t < 0$, the pump perturbs the free induction decay of P_{QD} created by the probe (inset of Fig. 3), whereas for $\Delta t > 0$ the probe interacts with the system modified by the pump, resulting in a probe polarization different from the unexcited system. Then, the spectrally resolved differential reflectivity represents the spectral interferogram of $\tilde{E}_R$ and $\tilde{E}_{QD}$ ($\tilde{}$: Fourier transform):

$$\Delta R(\omega,\Delta t) \propto \text{Re}\left\{\tilde{E}_R^*(\omega)\left[\tilde{E}_{QD}(\omega,\Delta t)-\tilde{E}_{QD,0}(\omega)\right]\right\} \quad (1)$$

The line shape of ΔR evidently depends on the phase delay between E_R and E_{QD}. As a function of distance d it oscillates between absorptive and dispersive shape. In our measurements with d = 120 nm, this results in an absorptive (Lorentzian) shape at $\Delta t \gg 0$ and ΔR probes the imaginary part of the QD polarization, $\text{Im}(P_{QD})$.

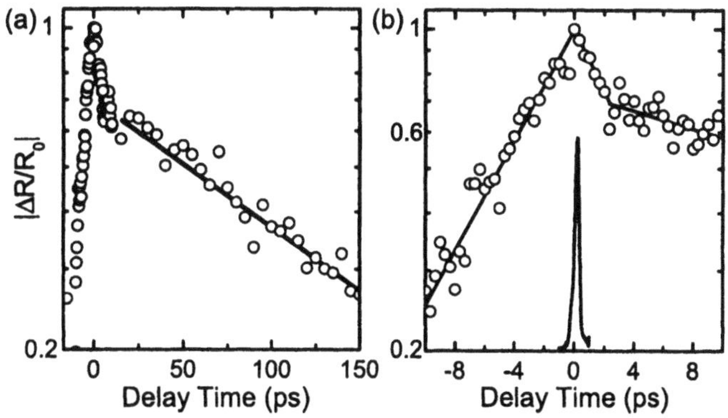

Fig. 2. (a) Time evolution of $\Delta R(E_{QD})/R_0$ for a single QD resonance at E_{QD} = 1.6598 eV (logarithmic ordinate scale). The decay is biexponential with a slow decay time 150 ps. (b) Early time dynamics of $\Delta R(E_{QD})/R_0$. A slow rise of $\Delta R/R_0$ is observed at negative delay times. Solid line around delay zero: 150 fs cross-correlation of pump and probe pulses.

In general, the free induction decay of $P_{QD}(t)$ can be perturbed by pump-induced many-body interactions changing the resonance energy, oscillator strength and/or dephasing rate of the QD polarization. Energy renormalizations give rise to spectrally asymmetric nonlinearities, as, e.g., for the dynamic Stark shift of exciton lines in quantum wells. Such energy shifts are absent in our data. Also, the spectrally integrated reflectivity change vanishes for each of the spectra recorded at negative time delays and – thus – changes of the oscillator strength are small. We conclude that the pump-induced damping of the QD polarization is the leading contribution to the QD nonlinearity at early times. Such excitation induced dephasing (EID) of the QD polarization arises from Coulomb scattering with the initial *nonequilibrium* carrier distribution in QW continuum states. The damping time of the QD polarization after the arrival of the pump pulse determines the energy range over which the reflectivity spectra display oscillations. In Fig. 3, spectral oscillations are observed in a window of about 0.5 meV around the

QD resonance, corresponding to a damping on a time scale of a few ps. The solid lines in Fig. 3 are calculated from a simple model in which the probe-induced QD polarization $P_{QD}(t)$ decays initially with an effective dephasing time T_2 = 15ps decreasing to T_{EID} = 3ps after the arrival of the pump laser (inset of Fig. 3). Evidently, this model is in quite good agreement with the experimental results. Due to the finite monochromator resolution T_2 = 15ps is a lower limit of the intrinsic QD dephasing time.

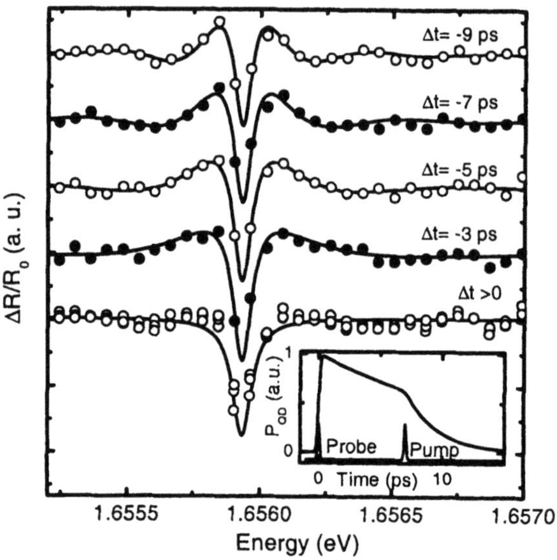

Fig. 3. Near-field $\Delta R/R_0$ spectra (circles) at different delay times Δt. The solid solid shows simulated spectra for the perturbed free induction decay of the coherent QD polarization assuming T_2 = 15 ps. Inset: Simulated dynamics of $P_{QD}(\Delta t)$.

To confirm our qualitative interpretation, we analyze the nonlinear optical QD response on the basis of the semiconductor Bloch equations (SBE) in mean-field approximation, including both disorder and Coulomb interaction. Nonlinear absorption spectra $\Delta A(\omega) \propto \mathrm{Im}\left[\tilde{E}_T^*(\omega)\tilde{P}(\omega)\right]$ are calculated, probing the imaginary part of $P_{QD}(\Delta t)$, as in the experiment. We chose a one-dimensional Gaussian-correlated two-band disorder potential with a correlation length of 10 nm and a disorder strength of 5 meV [Fig. 4(a), inset]. Excitation-induced dephasing is treated phenomenologically by a dephasing rate $\gamma = 1/T_2 + \gamma_1 n$ depending on the excitation density n. To reduce the numerical complexity, the relevant dephasing rates are scaled to T_2=2 ps and T_{EID}=0.4 ps. Pump and probe pulses are localized to 100 nm. The linear absorption spectra show resonances from single localized excitons at low energies. Transient, spectral symmetric oscillations govern the optical nonlinearity at negative delay times. The calculations clearly confirm the importance of EID contributions on the perturbed free induction decay and the minor role of coherent renormalizations of the QD resonance energy. Although

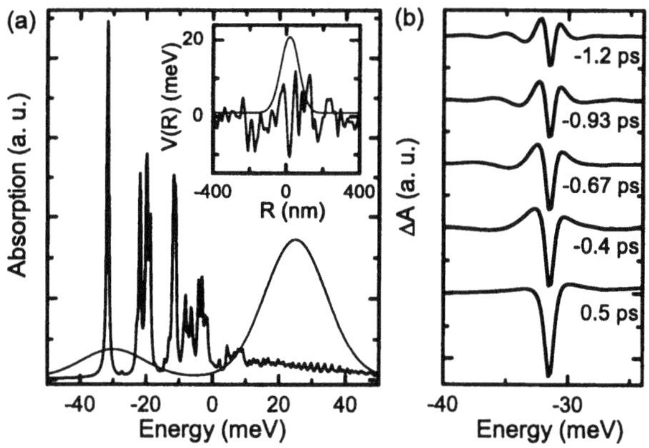

Fig. 4. (a) Spatially resolved linear absorption spectra of a QD ensemble calculated from the semiconductor Bloch equations. A Gaussian-correlated one-dimensional disorder potential is assumed (inset). Dashed lines: Spectra of pump (centered at 25 meV) and probe pulses (at -30 meV) with spot sizes of 100 nm. (b) Nonlinear absorption spectra for different pump-probe delays showing the perturbed free induction decay on the lowest resonance of Fig. (a).

single QDs resemble in many respects atomic systems, Coulomb interactions give rise to pronounced optical nonlinearities at negative delay times, which have been reported so far only for higher-dimensional systems

In conclusion, we have presented the first observation of the femtosecond excitonic nonlinearity of a single quantum dot. The free induction decay of the coherent excitonic polarization is shown to be perturbed by manybody interactions with continuum excitations on picosecond time scales. The exquisite sensitvity of the here demonstrated nano-optical pump-probe technique opens a new way to an ultrafast probing and manipulation of coherent quantum dot polarizations by external electromagnetic fields and/or material polarizations which is important for semiconductor-based implementation of quantum information processing.

Acknowledgements. We gratefully acknowledge support by the Deutsche Forschungsgemeinschaft (SFB296, SFB491), BMBF (01BM908/6), and the European Union through the EFRE and SQID programs.

References

1 F. Intonti, V. Emiliani, C. Lienau, T. Elsaesser, V. Savona, E. Runge, R. Zimmermann, R. Nötzel and K. H. Ploog, Phys. Rev. Lett. 87, 076801 (2001).

2 G. Chen, N. H. Bonadeo, D. G. Steel, D. Gammon, D. S. Katzer, D. Park and L. J. Sham, Science, 289, 1906 (2000)

3 E. Biolatti, R. C. Iotti, P. Zanardi, and F. Rossi, Phys. Rev. Lett. 85, 5647 (2000).

Ultrafast polarization dynamics induced by the "virtual excitation" of spin-polarized carriers in semiconductors

Arthur L. Smirl and Eric J. Gansen

Laboratory for Photonics & Quantum Electronics, 138 IATL, University of Iowa, Iowa City, Iowa 52242
E-mail: art-smirl@uiowa.edu

Abstract. The ultrafast changes in the polarization state of a weak probe pulse induced by the creation of spin-polarized virtual carriers in a multiple quantum well are measured, discussed in terms of saturation and exciton-exciton interactions, and used to explain the operation of an all-optical switch.

It has been known since the early 1970s that a spin-polarized carrier population can be produced in a semiconductor by the absorption of a circularly-polarized optical pulse with photon energy above the direct band gap. More recently, it has been shown [1-4] that by exciting near (but below) resonance a *virtual* carrier population can be produced. Consequently, by exciting with circularly polarized light, one would expect to produce a virtual spin-polarized carrier population that adiabatically follows the excitation pulse envelope. A great deal of information about the "instantaneous" nonlinearities associated with these virtual carriers has been obtained by performing conventional time-integrated differential transmission measurements for various combinations of input polarizations [1-4]; however, the polarization changes in the transmitted probe pulses have not been examined. Here, we combine the capabilities of differential spectroscopy with ellipsometric techniques to fully determine the changes in the polarization state of a weak probe induced by the creation of a spin-polarized population of virtual carriers as a function of time delay; and we demonstrate that this polarization information is of both fundamental and practical importance.

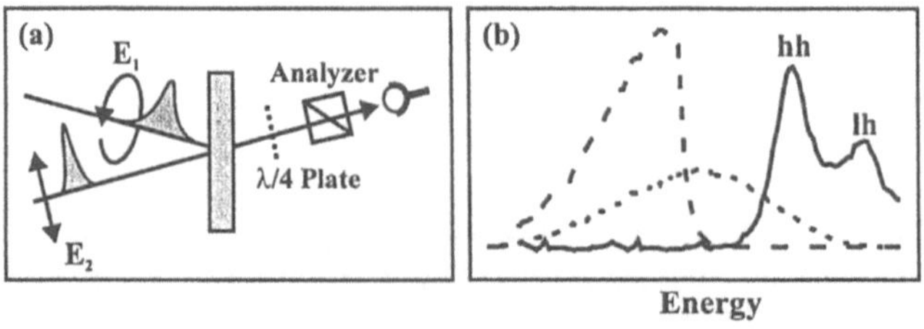

Fig. 1. (a) The experimental configuration for measuring the Stokes parameters of the time-averaged signal, and (b) the spectral tuning of the pump (dashed line) and probe (dotted line) pulses

Our measurements were performed using the geometry shown schematically in Fig. 1a. The 150 fs probe ($\mathbf{E}_2$) pulses (from a Ti:saphire laser) were linearly s-polarized and were tuned below the heavy hole (hh) exciton as shown in Fig. 1b. The pump ($\mathbf{E}_1$) pulses were right circularly polarized, were spectrally narrowed by a pulse shaper, and were tuned below the hh resonance by ~23 meV to avoid resonant excitation. The signal in the probe direction was measured as a function of analyzer orientation (both with and without the quarter-wave ($\lambda/4$) plate present) using synchronous detection techniques. In this way, we were able to measure all four Stokes parameters that determine the polarization state of the field emitted in the probe direction both with and without the pump present as a function of time delay. We emphasize that by measuring the vectorial nature of the probe field one obtains information about the phase of the right and left circularly polarized components of the probe emission that is not obtained by performing conventional differential transmission measurements. Initial measurements were performed at 100 K on a multiple quantum well (MQW) sample consisting of 40 periods of 10-nm-wide GaAs wells alternating with 10-nm-thick $Al_{0.3}Ga_{0.7}As$ barriers.

Typical results are shown in Fig. 2 for a pump intensity of ~445 MW/cm^2. A time-averaged change in ellipticity angle of ~0.8° and a time-averaged rotation of ~1.3° are observed when the pump and probe pulses are coincident (τ_{21}=0). Notice that these polarization changes adiabatically follow the pump pulse profile. In addition, an unpolarized component to the time-averaged probe polarization of ~1.7% is observed, indicating that the polarization state is changing rapidly on the time scale of the probe pulse width.

Our polarization results are at first surprising. For example, one can use a heuristic argument based on the assumption of a pure blue shift (which is expected to dominate at low fluences) to predict a simple rotation of the linearly polarized probe field (with no significant ellipticity). Similar arguments based on the assumption of a pure bleaching (expected to dominate at high fluences) can be

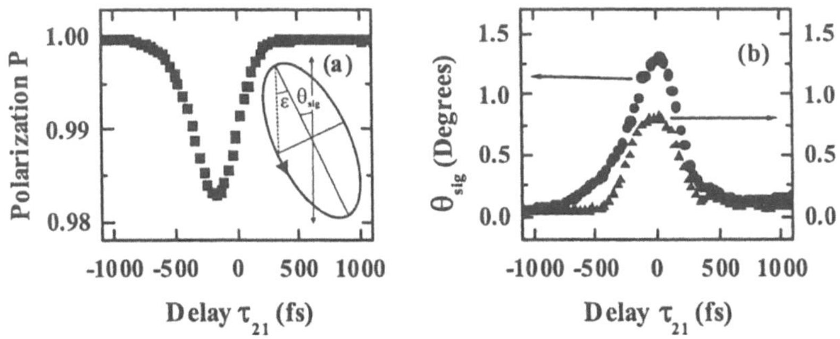

Fig. 2. The time-averaged (a) degree of polarization P, (b) ellipticity angle ε (triangles), and orientation θ_{sig} (circles) as a function of the time delay τ_{21} between the pump and probe pulses

used to predict an induced ellipticity in the probe field (with no rotation). But this is not what is observed. We observe changes in ellipticity and in orientation of similar magnitude. The polarization dynamics (for a full range of intensities) can be interpreted in terms of Pauli-blocking, first-order and higher-order Coulomb-induced exciton-exciton interactions and correlations, consistent with other recent studies of the optical Stark effect [1-4].

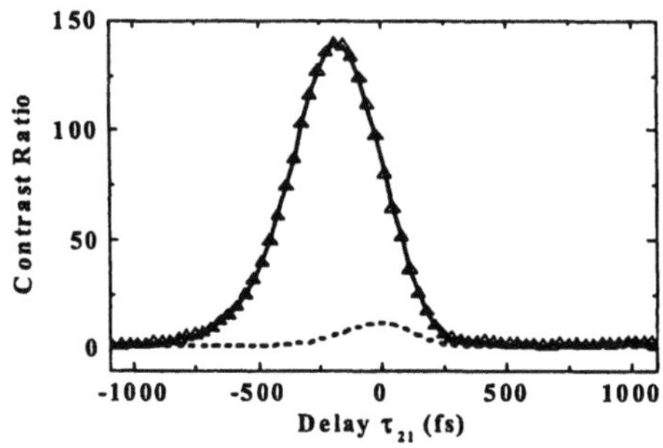

Fig. 3. The measured spin switch contrast ratio (open triangles) and the predicted contrast ratios calculated from polarization state data assuming completely polarized light (dashed line) and including the unpolarized component (solid line) as a function of time delay τ_{21}

To show the practical importance of this information, we use it to construct and explain the operation of an optically-addressed *coherent* polarization switch [4]. This switch is constructed by removing the quarter-wave plate in Fig. 1a and by orienting the analyzer to reject the probe in the absence of the pump. The switching performance is given by the open triangles in Fig. 3. As shown, by taking advantage of the nonlinearities associated with the *virtual spin-polarized* carrier population, this device exhibits a pulse-width-limited response and produces relatively large contrast ratios in thin samples. Moreover, the changes in the ellipticity and orientation alone will not account for the switching action, as shown by the dashed line in Fig. 3. In fact, the signal transmitted by the switch is dominated by the "unpolarized" component of the signal, as indicated by the solid line.

References

1 C. Sieh, T. Meier, F. Jahnke, A. Knorr, S. W. Koch, P. Brick, M. Hübner, C. Ell, J. Prineas, G. Khitrova, and H. M. Gibbs, Phys. Rev. Lett. **82**, 3112 (1999).

2 T. Meier , S. W. Koch, P. Brick, C. Ell, G. Khitrova, and H. M. Gibbs, Phys. Rev. B **62**, 4218 (2000).

3 S. W. Koch, C. Sieh, T. Meier, F. Jahnke, A. Knorr, P. Brick, M. Hübner, C. Ell, J. Prineas, G. Khitrova, and H. M. Gibbs, J. Lumin. **83-84**, 1 (1999).

4 E. J. Gansen, K. Jarasiunas, and A. L. Smirl, Appl. Phys. Lett. **80**, 971 (2002).

The Hall Current of Coherent Electron Wavepackets

T. Bauer[1], A. B. Hummel[1], J. S. Kolb[1], H. G. Roskos[1], Yu. A. Kosevich[2], and K. Köhler[3]

[1] Physikalisches Institut der Johann Wolfgang Goethe-Universität Frankfurt a. Main, Robert-Mayer-Str. 2-4, D-60054 Frankfurt a. Main, Germany
E-mail: a.b.hummel@physik.uni-frankfurt.de
[2] Max-Planck-Institut für Physik komplexer Systeme, D-01187 Dresden, Germany
[3] Fraunhofer-Institut für Angewandte Festkörperphysik, D-79108 Freiburg, Germany

Abstract. Terahertz-emission spectroscopy reveals two regimes of the Hall current of impulsively excited coherent charge-carrier wavepackets in a GaAs/AlGaAs superlattice as a function of the electric and magnetic bias fields.

1. Introduction

The Hall effect is one of the fundamental solid-state phenomena. Experiments on two-dimensional electron gasses at low temperatures have already shown that the wave nature of the charge carriers modifies the classical Hall effect leading to the integer and fractional quantum Hall effect, respectively. In this paper we report on the first observation of another wave phenomenon, the Coherent Hall Effect, which denotes the Hall effect of impulsively excited charge-carrier wavepackets in crossed electric and magnetic fields during the coherence time. We employ a semiconductor superlattice as a model for three-dimensional materials. Because of the ultrafast character of this effect we apply a femtosecond optoelectronic technique to investigate the transient current/voltage Hall response of the wavepackets [1].

A schematic representation of the experimental geometry is shown in Fig. 1a. The charge carriers, subject to an electric field $\boldsymbol{E}$ parallel and a magnetic field $\boldsymbol{B}$ perpendicular to the superlattice growth direction (x-direction), move perpendicularly to both $\boldsymbol{E}$ and $\boldsymbol{B}$ leading to a time-dependent Hall current along the y-direction [1-3]. This coherent, oscillatory Hall current manifests itself by emitting a THz radiation field in x-direction which is proportional to the second time derivative of the y-component of the electron/hole (intraband) polarization. For $B = 0$, the THz emission of pure Bloch oscillations (BO) can be observed in y-direction, while no signal is emitted in x-direction. The cosine-shaped dispersion relation of the superlattice minibands in the growth direction as a manifestation of the wave nature of the charge carriers is the reason for two regimes of motion depending on the ratio of E to B as shown in Fig. 1b. For large values of E (magneto-Bloch regime), the k-vector component k_x in growth direction increases continuously, while for dominating B (cyclotron-like regime), k_x oscillates around zero which

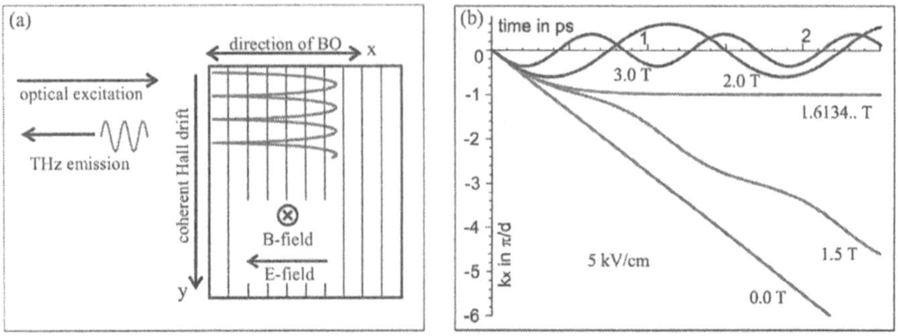

Fig. 1. (a) Experimental geometry. (b) The two regimes of motion of a wavepacket undergoing Bloch oscillations (BO) in a magnetic field B calculated in k-space for an electric field of 5 kV/cm. At low B (magneto-Bloch regime): oscillatory motion with k_x increasing monotonically (B = 1.5 T). At high B (cyclotron-like regime): motion restricted to miniband bottom, k_x oscillates around zero (B = 2 T and 3 T). Transition between the two regimes: all oscillations suppressed (B = 1.6134... T).

implies that the wavepacket motion is restricted to the bottom of the miniband. The transition between these two regimes occurs when the electric and magnetic forces exactly cancel each other. In this case all oscillations are suppressed after the equilibrium is reached.

2. Experimental Methods

We used a standard THz-emission setup to observe the coherent Hall current on a GaAs/$Al_{0.3}Ga_{0.7}As$ superlattice. The excitation beam impinged normally to the sample surface (along the growth direction). Therefore only coherent THz emission normal to the surface was detected to be selectively sensitive to magnetic-field-induced effects. The emitted THz radiation was separated from the optical excitation beam by a dicroic ITO-glass beamsplitter [4], focused onto an optically gated silicon-on-sapphire photoconductive antenna and detected in a time-resolved manner. The detection bandwidth is limited to 2.5 THz by the cryostat windows.

3. Results and Discussion

Fig. 2 shows measured THz-emission data which prove the existence of the two regimes. The oscillatory charge-carrier motion in the magneto-Bloch regime (upper three traces) is characterized by a decreasing frequency with increasing magnetic field. The frequency increases with the electric field (data not shown), similarly to pure BO. The coherent THz emission is strongly damped by the influence of the magnetic field (dephasing time: about 1 ps at low magnetic fields and smaller at higher fields). The relatively strong initial single-cycle oscillation is caused by the magnetic field and will not be considered further. In the cyclotron-like regime (lower four traces), the frequency increases linearly with increasing magnetic field. No influence of the electric field on the emission frequency is observed (data not shown). The oscillations are now long-lived with dephasing times of about 4 ps. The transition point is characterized by almost featureless THz signals (trace for $B = 2$ T) except for the initial single-cycle signal.

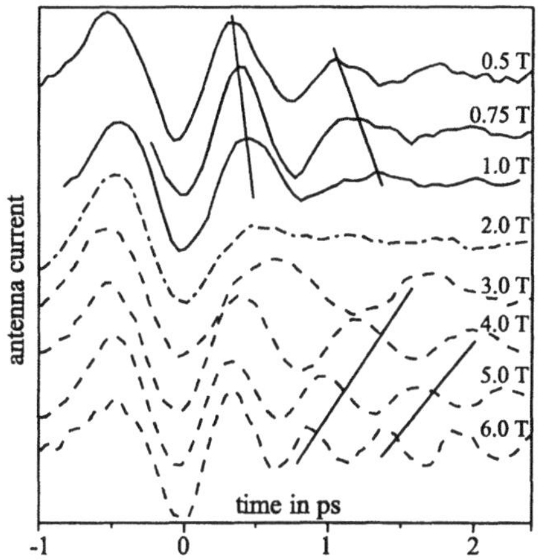

Fig. 2. Measured THz transients for an electric bias field of 6 kV/cm as a function of *B*.

It is surprising that the main features of the wavepacket behavior such as the existence and the character of two different regimes of motion and their dependence on the ratio of the electric to magnetic field can be explained within a semiclassical picture [1-3]. Coulomb effects apparently are not decisive for the global features but our data suggest that they influence the dephasing behavior as well as the *E* and *B* values where the transition between the two regimes occurs.

Acknowledgements. We thank G. Meinert, L. Banyai, and H. Haug for numerous discussions and K. Leo (TU Dresden) for bringing Yu. A. K. into contact with the group of JWG University. The work was funded by DFG (Schwerpunktprogramm "Quantenkohärenz in Halbleitern").

References

1 T. Bauer, J. Kolb, A. B. Hummel, H. G. Roskos, Yu. A. Kosevich, and K. Köhler, Phys. Rev. Lett. **88**, 086801 (2002).
2 Yu. A. Kosevich, Ann. Phys. **8**, SI-145-SI-148 (1999).
3 Yu. A. Kosevich, Phys. Rev. B **63**, 205131 (2001).
4 T. Bauer, J. S. Kolb, T. Löffler, E. Mohler, H. G. Roskos, and U. C. Pernisz, J. Appl. Phys. (submitted).

Ultrafast coherent electron transport in quantum cascade structures

Felix Eickemeyer[1], Klaus Reimann[1], Michael Woerner[1], Thomas Elsaesser[1], Stephano Barbieri[2], Carlo Sirtori[2], Gottfried Strasser[3], Thomas Müller[3], Rudolf Bratschitsch[3], and Karl Unterrainer[3]

[1] Max-Born-Institut für Nichtlineare Optik und Kurzzeitspektroskopie, 12489 Berlin, Germany
[2] Thales-CSF, Laboratoire Central de Recherches, 91404 Orsay, France
[3] Institut für Festkörperelektronik und Mikrostrukturzentrum, Technische Universität Wien, 1040 Wien, Austria

Abstract. The first mid-infrared pump-probe study of an electrically driven quantum cascade structure reveals an oscillatory gain recovery after femtosecond excitation, giving direct evidence for a coherent electron motion from the injector into the active region. Increased scattering at higher lattice temperatures reduces the amplitude of the coherent transport oscillation and leads to a concomitant increase of incoherent gain recovery processes.

The wave nature of electrons in solids gives rise to transport processes which are governed by the coherence properties of the electron wave function. So far, coherent transport has been studied mainly at low carrier concentrations where decoherence processes are comparably slow. Most optoelectronic devices work, however, at elevated carrier densities ($> 10^9$ cm^{-2}) where electron-electron and electron-phonon scattering are expected to result in a rapid loss of quantum coherence.

The recently invented quantum cascade laser (QCL) is an important prototype system for studying carrier transport at elevated densities. The QCL is an electrically driven unipolar semiconductor laser based on optical intersubband transitions in the mid-infrared. A QCL consists of a sequence of injector and active regions, as shown in the conduction band diagram in Fig. 1(a). The active part acts as a 3-level laser system with population inversion between level 3 and level 2 ($\tau_3 \approx 1.5$ ps $> \tau_2 \approx 0.1$ ps) whereas the injector serves as an electron reservoir, which is separated by a thick injection barrier (IB) from the active region. Contradictory pictures have been developed for electron transport from the injector into the active region, invoking either coherent resonant tunneling [1] or incoherent transport through the IB [2]. In the following, we present the first direct time-resolved study of transport in a QCL. The nonlinear depletion and recovery of electrically induced optical gain is studied in mid-infrared pump-probe experiments on three different GaAs/AlGaAs QCL structures without resonator. Strong gain oscillations originating from coherent electron motion between the injector and the active region give evidence for the coherent character of charge transport.

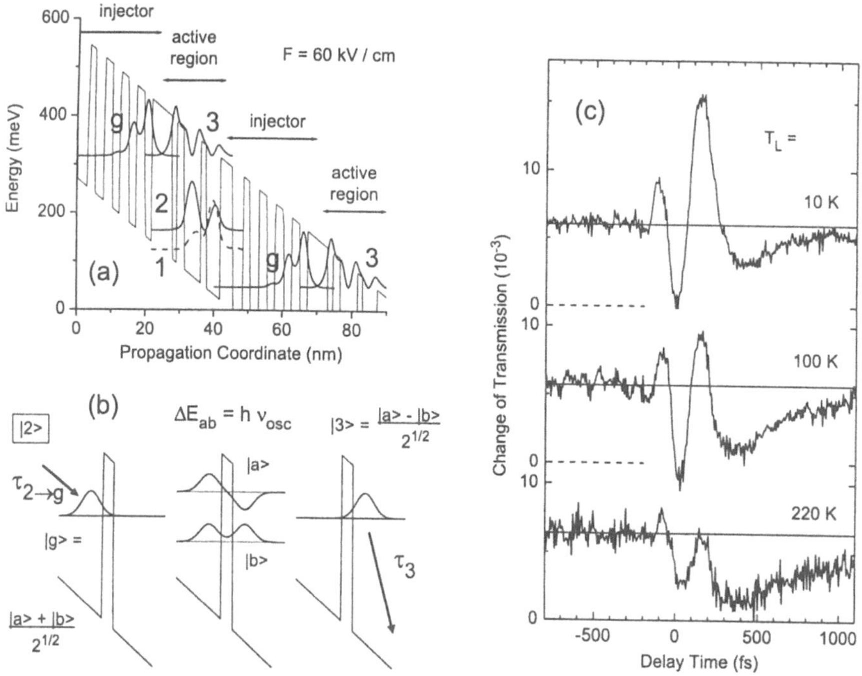

Fig. 1. (a) Conduction band diagram of a biased GaAs/AlGaAs QCL [3]. Probability densities $|\Psi(x)|^2$ are shown for $|g\rangle$ (ground state in the injector), $|3\rangle$ (upper laser state), $|2\rangle$ (lower laser state), and $|1\rangle$. (b) Coherent transport of electrons from $|g\rangle$ into $|3\rangle$ in the dynamic picture of resonant tunneling. $|g\rangle$ and $|3\rangle$ are coherent superpositions of the binding and anti-binding energy eigenstates $|b\rangle$ and $|a\rangle$, respectively. (c) Pump-probe transients obtained from this sample at a detection wavelength of $\lambda_{\text{det}} = 10$ μm for forward bias and different lattice temperatures T_L at a current density of 7 kA/cm^2. The current-induced optical transmission change $\Delta T/T_0(t_D) = T(j, t_D)/T(j = 0, t_D) - 1$ is plotted as a function of the time delay t_D between pump and probe pulses [$T(j, t_D)$: transmission for a current density j].

For the experiment we use 150-fs pump and probe pulses resonant to the $3 \leftrightarrow 2$ gain transition. The probe pulses are spectrally resolved (resolution: 0.15 μm) after interaction with the sample. We have investigated three different structures, all with 10 periods of injectors and active regions [4]. Sample A contains the same layer structure as the GaAs/$Al_{0.33}Ga_{0.67}$As QCL in [3] with a width of the injection barrier of $d_B = 6.2$ nm. Samples B and C have the same layer structure as the GaAs/$Al_{0.45}Ga_{0.55}$As QCL in [5] with $d_B = 4.0$ nm (sample B) and $d_B = 4.6$ nm (sample C).

In Fig. 1 (c), we show results obtained from sample A at a detection wavelength of 10 μm (center of the gain band). The measured transients represent the difference between a pump-probe curve with applied current and a pump-probe curve without current as a function of time delay between

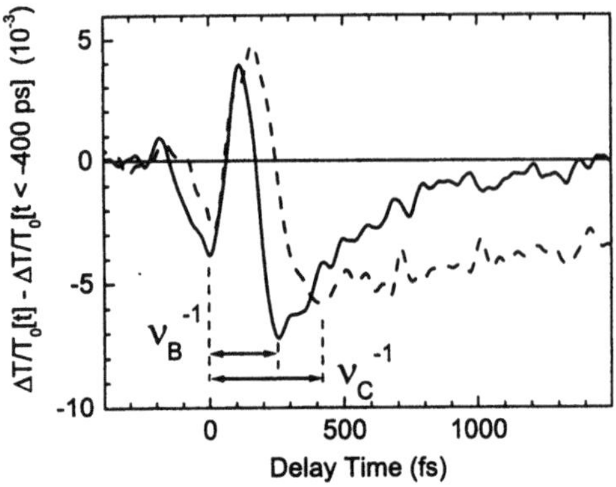

Fig. 2. Pump-probe transients of the gain saturation and recovery for sample B with a 4.0-nm injection barrier (solid line) and for sample C [5] with a 4.6-nm injection barrier (dashed line). For better comparison we have subtracted the constant signals at negative delay times $t_D < -400$ fs. The arrows indicate the corresponding oscillation periods ν_B^{-1} and ν_C^{-1}.

pump and probe pulses. The strong transmission increase before delay zero is due to current-induced gain [4]. Upon excitation the gain saturates and recovers subsequently in several hundred femtoseconds, displaying at early times pronounced gain oscillations. The oscillation frequency is determined by the height and width of the IB, as is evident from Fig. 2. The oscillation amplitude decreases with increasing lattice temperature [Fig. 1 (c)].

The gain oscillations demonstrate coherent transport between the injector and level 3 via resonant tunneling [see Fig. 1(b)]: Gain depletion by the femtosecond pump pulse creates a coherent superposition of wave functions $|a\rangle$ and $|b\rangle$, and, thus, motion of an electron wave packet from the injector through the IB into the upper laser state and back. The oscillation frequency is determined by the splitting ΔE_{ab} as is evident from the measurements on samples with different IB widths [cf. Fig. 2]. The strong damping of the oscillation is mainly due to carrier-carrier scattering in the injector. At elevated lattice temperatures carrier-carrier and carrier-phonon scattering become more efficient. As a result, both the oscillation amplitude [Fig. 1 (c)] and the lasing efficiency of the device [3] decrease. After the decay of the coherence between states $|a\rangle$ and $|b\rangle$ (for $t_D > 500$ fs), the optical gain returns to its quasi-stationary value by incoherent processes, which dominate at elevated temperatures.

In conclusion, our data clearly show that even at the high electron densities present in a quantum cascade laser the coherence properties of the electron wave function play a very important role for transport.

References

1. C. Sirtori et al.: IEEE J. Quant. Electron. **34**, 1722 (1998)
2. R. C. Iotti and F. Rossi: Phys. Rev. Lett. **87**, 146603 (2001)
3. C. Sirtori et al.: Appl. Phys. Lett. **73**, 3486 (1998)
4. F. Eickemeyer et al.: Appl. Phys. Lett. **76**, 3254 (2000)
5. H. Page et al.: Appl. Phys. Lett. **78**, 3529 (2001)

Role of the absolute optical phase in carrier-wave Rabi flopping with two-cycle pulses

O. D. Mücke[1], T. Tritschler[1], M. Wegener[1], U. Morgner[2], and F. X. Kärtner[2,3]

[1] Institut für Angewandte Physik,
Universität Karlsruhe (TH), D-76128 Karlsruhe, Germany
[2] Institut für Hochfrequenztechnik und Quantenelektronik,
Universität Karlsruhe (TH), D-76128 Karlsruhe, Germany
[3] Department of Electrical Engineering and Computer Science,
Research Laboratory of Electronics,
Massachusetts Institute of Technology, Cambridge, USA

Abstract. The influence of the absolute optical phase of few-cycle pulses on non-perturbative resonant extreme nonlinear optics in semiconductors should be measurable in GaAs samples and could in turn be used to determine the absolute optical phase of few-cycle pulses. To avoid distortions, sample design is an important issue.

The rapid development in few-cycle pulse generation [1] has now led us into a regime in which the phase between the rapidly oscillating light frequency and the electric field envelope has become a relevant quantity. This carrier-envelope phase, ϕ, sometimes also referred to as the *absolute optical phase*, can significantly influence the outcome of an experiment. Recently, several measurements of the change of ϕ between successive pulses derived from a mode-locked oscillator [2–6] as well as for amplified pulses [7–9] were reported.

The underlying basic idea [2,3] is to use the interference of the fundamental frequency of a laser pulse, which was spectrally broadened by self-phase modulation in a (photonic crystal) fiber of a few millimeter length, with the second harmonic generated with the help of a separate crystal. In Ref. [4] the same idea was used, except for the fact that the fundamental spectrum did already cover one octave, hence no need for additional broadening. Somewhat similar to this, Ref. [5] proposed to use the interference of the third harmonic, generated at a silicon wafer surface, with the second harmonic generated in a separate crystal. All the optical nonlinearities used in these and other [6] cases are off-resonant and within the perturbative regime. Furthermore, in most of these cases, the pulses have to propagate over a considerable distance within the apparatus and, hence, the difference between the phase velocity and the group velocity can change the absolute optical phase within the measurement set-up. Currently, this is the main obstacle in measuring ϕ.

A non-perturbative and, hence, distinctly different way to determine the absolute optical phase would be via x-ray generation in extreme nonlinear optics in atoms [1,10]. Furthermore, measurement of the absolute optical phase may benefit optical-frequency synthesis because it may create a simpler technique for determining/controlling the comb offset frequency [11].

In this contribution, we present numerical solutions of the coupled semiconductor Maxwell-Bloch equations. In particular, we neither employ the rotating wave approximation, nor the slowly varying envelope approximation, nor the concept of a background dielectric constant, nor the effective mass approximation and account for the important E_1 and E_2 resonances in GaAs. The modeling shows that, if the Rabi frequency becomes comparable to the light frequency, the different Rabi sidebands interfere around twice the laser center frequency, giving rise to a signal which strongly depends on the absolute optical phase.

Fig. 1(**a**) depicts the spectra of light emitted into the forward direction of a thin layer of GaAs for various values of the absolute optical phase ϕ. The inset schematically illustrates the meeting of different Rabi sidebands (also compare Fig. 1(**b**) and (**c**)). Furthermore, it becomes clear from the intensity dependence shown in Fig. 1(**b**) that it is *not* simply the interference of the tail of the laser spectrum itself with the third-harmonic signal – which would be similar to Refs. [2–6] – but rather the interference of different Rabi sidebands.

Fig. 2(**a**) shows the intensity spectra of light emitted into the forward direction versus carrier-envelope phase ϕ for a $L = 100$ nm thin GaAs film on a substrate with $\epsilon_s = (1.76)^2$ (e.g. sapphire). Note the dependence on ϕ with large visibility around $\omega/\omega_0 = 2.05 - 2.25$ and the period of π (rather than 2π) resulting from the inversion symmetry of the problem. (**b**) Same for $L =$

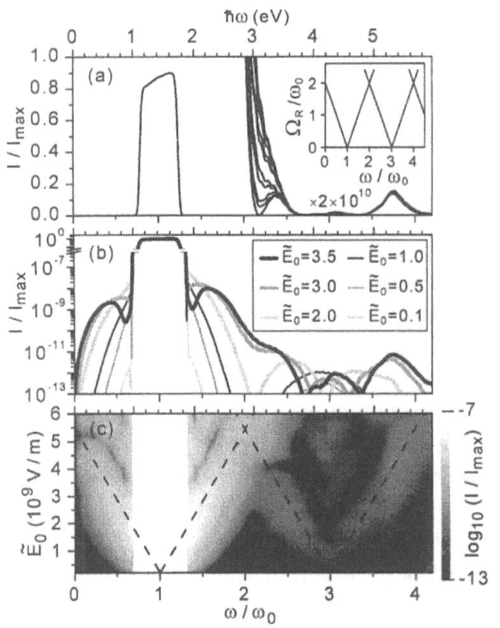

Fig. 1. (**a**) Signal intensity (normalized) emitted into the forward direction versus spectrometer frequency ω in units of the laser center frequency ω_0 for different values of ϕ. The GaAs film with $L = 20$ nm thickness on a substrate with $\epsilon_s = (1.76)^2$ has no barriers, but a front-side AR coating, $\tilde{E}_0 = 3.5 \times 10^9$ V/m. The inset illustrates the interference of peaks from the different pairs of Rabi sidebands as the Rabi frequency Ω_R increases. (**b**) As (**a**), but signal intensity (normalized) for fixed $\phi = 0$ and for different incident electric field amplitudes (in units of 10^9 V/m) as indicated. (**c**) As (**b**), but greyscale image. The dashed lines are guides to the eye.

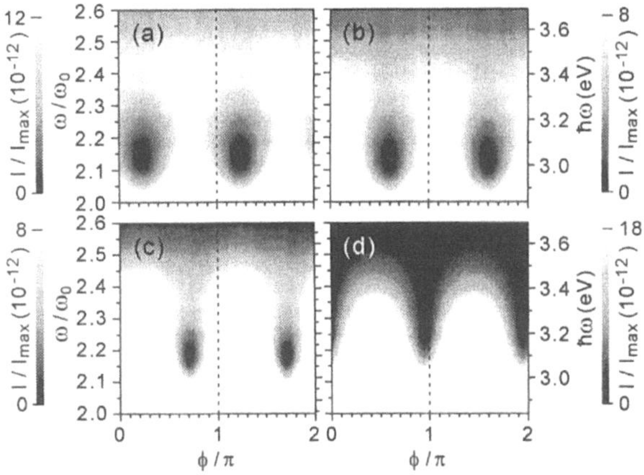

Fig. 2. Grey-scale image of the emitted intensity as a function of ω and ϕ for a thin GaAs film with thickness L on a substrate with dielectric constant ϵ_s. (**a**) $L = 100\,\text{nm}$, $\tilde{E}_0 = 3.5 \times 10^9\,\text{V/m}$, and $\epsilon_s = (1.76)^2$, (**b**) as (**a**) but $L = 20\,\text{nm}$, (**c**) as (**b**) but with an additional front-side AR coating (as in Fig. 1), (**d**) as (**c**) but for $\tilde{E}_0 = 4.0 \times 10^9\,\text{V/m}$.

20 nm, indicating that one already has some distortions in (**a**) due to the finite thickness of the sample as a result of different group and phase velocities. (**c**) As (**b**), but introducing a front-side $\lambda/4$-antireflection coating designed for the fundamental laser center frequency ω_0. Note that (**b**) and (**c**) are shifted with respect to each other horizontally, because the incident optical pulses, and thus also ϕ, are distorted as a result of multiple reflections. (**d**) As (**c**), but for a different incident electric field amplitude $\tilde{E}_0$. This variation also leads to a horizontal shift. This is interesting on the one hand because no such intensity dependence occurs in off-resonant perturbative nonlinear optics – pointing out the distinct difference between the two scenarios. It is disturbing on the other hand, because in order to determine the absolute optical phase one needs to calibrate the incident electric field amplitude via the measured splitting of the Rabi doublets. Finally, it is important to note that the electric field amplitudes and the absolute emitted signals shown here are comparable to the signals computed for samples corresponding to previous experiments [12] in which the nonlinear signals could easily be detected.

References

1. T. Brabec and F. Krausz: Rev. Mod. Phys. **72**, 545 (2000)
2. D. J. Jones et al.: Science **288**, 635 (2000)
3. A. Apolonski et al.: Phys. Rev. Lett. **85**, 740 (2000)
4. U. Morgner et al.: Phys. Rev. Lett. **86**, 5462 (2001)
5. M. Mehendale et al.: Opt. Lett. **25**, 1672 (2000)
6. H. R. Telle et al.: Appl. Phys. B **69**, 327 (1999)
7. A. V. Sokolov et al.: Phys. Rev. Lett. **87**, 033402 (2001)
8. G. G. Paulus et al.: Nature **414**, 182 (2001)
9. A. Baltuška et al.: Phys. Rev. Lett. **88**, 133901 (2002)
10. C. G. Durfee III et al.: Phys. Rev. Lett. **83**, 2187 (1999)
11. S. T. Cundiff et al.: Rev. Sci. Instrum. **72**, 3749 (2001)
12. O. D. Mücke et al.: Phys. Rev. Lett. **87**, 057401 (2001)

Femtosecond optically induced rectification, shift and injection currents in GaAs

N. Laman, D. Côté, J.E. Sipe and H.M. van Driel

Department of Physics, University of Toronto, Toronto, Canada
E-mail: vandriel@physics.utoronto.ca

Abstract. Second-order rectification and shift currents, and third order injection currents have been generated in GaAs using 150 fs pulses with photon energy below ($\hbar\omega$) and above ($2\hbar\omega$) the electronic bandgap and observed via the emitted THz radiation.

Introduction. Investigations of ultrafast electrical transients generated in semiconductors using optical pulses have been carried out since the early development of ultrashort pulse lasers. However, the underlying physics of the various mechanisms which can lead to current generation by optical pulses alone have not been well understood. Here we show how emitted THz radiation can be used to deduce the salient signatures of three physically distinct processes that are associated with all-optical current generation in GaAs. We refer to these as rectification, shift and injection currents.

The second order optical susceptibility, χ_2, represents the lowest order nonlinear optical response of a noncentrosymmetric semiconductor. In general, an optical pulse can induce [1] a DC polarization density via χ_2, through three independent contributions. The first represents the nonresonant (rectification) component and exists for all incident pulse frequencies: it leads to the production of a transient polarization ($P \propto I(t)$), where I is the envelope of the light intensity). A second component exists when the photon energy ($\hbar\omega$) is greater than the fundamental band gap and is associated with a shift of the center of charge within the unit cell of the semiconductor during valence-conduction band transitions; it leads to the production of a current ($J \propto \partial P/\partial t \propto I(t)$). The third component is associated with quantum interference between different absorption pathways and leads to current injection ($\partial J/\partial t \propto \partial^2 P/\partial t^2 \propto I(t)$). For symmetry reasons, the 2nd order current injection process does not exist in GaAs. For GaAs and other zincblende semiconductors the current injection process requires at least a χ_3 process, and indeed we have earlier shown how quantum interference between single and two photon absorption pathways can lead to current injection [2] in GaAs with the sign of the current controlled by the relative phases of the optical beams.

Experimental Methods. A parametric source was used to generate 150 fs pulses at 1.55 μm and 1.69 μm. The 1.55 μm beam (ω) was frequency doubled to generate a 0.775 μm beam (2ω). The relative phase of the beams was controlled by an interferometer before the beams were focused on the sample. The rectification current was generated with only the ω beam incident on the sample while the shift current had only the 2ω beam incident. For both these currents, the sample was a 100 μm thick (110) oriented SI-GaAs crystal. The THz radiation was collected and refocused by off-axis parabolic mirrors into a 500 μm ZnTe crystal along with a 0.845 μm beam (frequency doubled 1.69 μm) for electro-optic (EO) sampling. For the injection current, both pulses were simultaneously incident on a 1 μm thick (100) oriented SI-GaAs sample. The sample was directly mounted on a 120 μm GaP crystal, which was used for near-field EO sampling of the THz radiation.

Results and Discussion. Typical temporal THz traces for second-order rectification and shift currents [3] generated by pulses with peak intensities of 6 GW cm^{-2} and 500 MW cm^{-2} are shown in Figure 1. The long temporal traces are due to the imperfect phase matching of the probe pulse with the THz beam in the ZnTe detection crystal. However, the frequency response of the EO system for both rectification and shift current is rigorously identical and the phase shift is due to the difference in the generation of these currents. The shift current leads the rectification current by approximately π/2, which is the opposite of the expected phase. This is because the carrier density for the shift current is sufficiently high (~10^{18} cm^{-3}) for the plasma frequency ω_p = 30 ps^{-1} to be greater than the optical pulse bandwidth. A current reversal occurs during the optical pulse, resulting in an additional π phase shift.

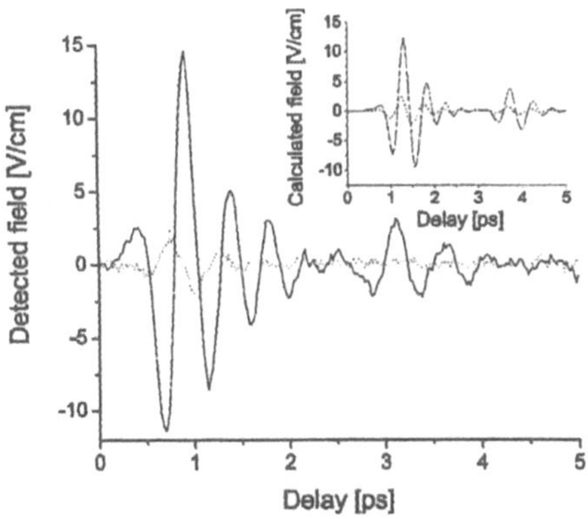

Fig. 1. Temporal traces of THz fields generated by rectification (solid line) and shift current (dotted line). The inset shows the calculated fields [3].

For the maximum optical intensities of 6 GW cm^{-2} and 500 MW cm^{-2}, the peak THz electric field in the ZnTe is 14 V cm^{-1} for the rectification current and 2 V cm^{-1} for the shift current. These translate into maximum rectification and shift current densities of 1 kA cm^{-2} and 60 kA cm^{-2}, when taking into account the interaction length of 95 μm for rectification current and 0.67 μm for the shift current. However, for the same low intensity (100 MW cm^{-2}) the shift current density is approximately 600 times higher than the rectification current density.

Figure 2. shows the near-field THz temporal traces for the third-order injection current at different carrier densities. Here, the EO signal in the GaP detector directly follows the current. A current reversal occurs at high carrier densities (10^{18} cm^{-3}) due to the plasma oscillations. This results in a decrease in the THz electric field for 2ω intensities above approximately 100 MW cm^{-2} and a maximum current density of 20 kA cm^{-2}. Analysis of the radiated THz temporal profile and its dependence on laser beam intensity yields a current relaxation rate of approximately 30 fs, which we attribute to carrier-carrier scattering.

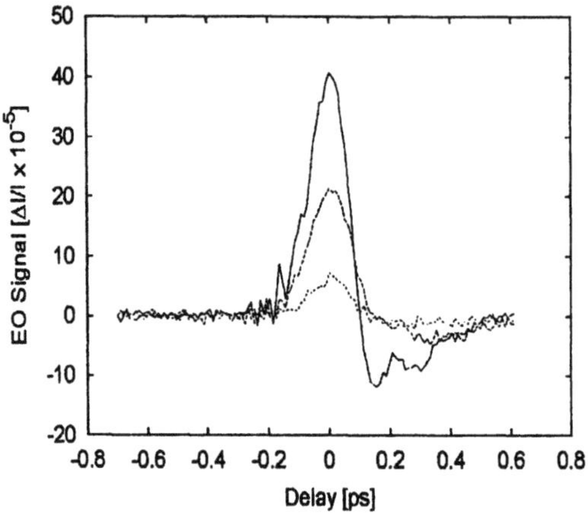

Fig. 2. Temporal traces of THz fields generated by injection current at carrier densities of 7×10^{17} cm^{-3} (solid curve), 8×10^{16} cm^{-3} (dashed curve) and 6×10^{15} cm^{-3} (dotted curve).

Conclusions. We have identified the rectification, shift and injection currents due to distinct nonlinear mixing processes in GaAs via the emitted THz radiation. Processes which involve real carriers (shift and injection currents) are strongly affected by plasma oscillations and carrier-carrier scattering.

Acknowledgements. We would like to thank R. Bhat, F. Nastos and O. Prepelita for useful discussions. We would also like to thank NSERC and PRO for financial support and A. Springthorpe at Nortel Networks for the thin GaAs sample.

References

1 J.E. Sipe and A.I. Shkrebtii, Physical Review B **61**, 5337, 2000.
2 D. Côté, J.M. Fraser, M. DeCamp, P.H. Bucksbaum and H.M. van Driel, Applied Physics Letters **75**, 3959, 1999.
3 D. Côté, N. Laman and H.M. van Driel, Applied Physics Letters **80**, 905, 2002.

Watching bare charges get dressed in an ultrabroadband THz experiment

Rupert Huber[1], Florian Tauser[1], Andreas Brodschelm[1], Max Bichler[2], Gerhard Abstreiter[2], and Alfred Leitenstorfer[1]

[1] Physik-Department E11, TU München, D-85747 Garching, Germany
E-mail: ruhub@ph.tum.de
[2] Walter-Schottky-Institut, TU München, D-85747 Garching, Germany

Abstract. In a photoexcited electron-hole plasma in GaAs, we test a key concept of many-body physics: the formation of dressed quasiparticles. A novel spectroscopic technique based on ultrabroadband THz electro-optics allows us to resolve the mid-infrared polarization response of the system with sub-cycle resolution of the electric field amplitude and phase. In this way, we directly observe the transient buildup of Coulomb screening on a femtosecond time scale.

Over the last few years, both generation and field-resolved detection of electromagnetic single-cycle transients have been extended into the mid-infrared spectral range using optical rectification and ultrabroadband electro-optic sampling [1-3]. In this way, a new class of ultrafast experiments has now become possible at frequencies often referred to as the "molecular fingerprint regime" - a region which is of central importance to condensed-matter physics, chemistry and biology. In the first non-linear femtosecond measurement exploiting these techniques we test an essential concept of many-body physics: the buildup of dressed quasiparticles [4].

A dense electron-hole plasma photoinjected in a semiconductor represents an ideal model system to study this phenomenon: Each charge carrier attracts a polarization cloud of opposite sign to form a dressed quasiparticle. These composite units interact via a dynamically screened potential. There has been an intense discussion on how to treat Coulomb interactions during ultrafast relaxation of a plasma under extreme nonequilibrium conditions. It has been presumed that immediately after ultrafast photogeneration of an electron hole plasma in a semiconductor no screening of the Coulomb interaction should exist. Quantum kinetic calculations predict that many-body correlations dressing up bare charges develop on an ultrashort time scale of the order of the inverse plasma frequency [5]. Here, we report the first direct experimental observation of this phenomenon.

We start with 10-fs laser pulses centered at 780 nm. A dense electron hole plasma (pair density $N = 2 \times 10^{18}$ cm^{-3}) is photogenerated in a 200-nm thin layer of GaAs via resonant interband excitation. After a femtosecond time delay t_D a single-cycle pulse (Fig. 1(a)) of a duration of 27 fs (FWHM of the intensity) and a central frequency of 28 THz probes the mid infrared polarization response of this nonthermal system in real time. The probe pulse contains frequency components between 1 THz and 90 THz. It is generated via phase matched optical rectification of a 10-fs laser pulse in a GaSe nonlinear crystal as thin as 30 µm. Any resonance in the sample will lead to a distortion of the wave form and a retarded tail of the

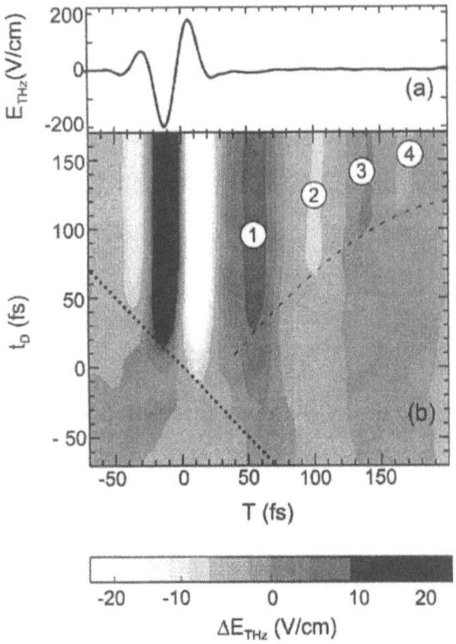

Figure 1: (a) Electric field amplitude E_{THz} of the single-cycle probe versus time T. (b) Electric field change ΔE_{THz} induced by 10-fs photoexcitation of 2×10^{18} cm^{-3} electron-hole pairs in GaAs versus delay times t_D and T. The diagonal dotted line denotes the position of the maximum of the 10-fs pump pulse. The dashed curve emphasizes the buildup of the retarded plasmon response.

mid-infrared probe transmitted through it. Ultrabroadband electro-optic detection is employed to fully monitor both amplitude and phase [1-3] of the probe electric field E_{THz} as a function of a second delay time T. Fig. 1(a) shows E_{THz} as transmitted through the unexcited GaAs sample. The THz electric field change ΔE_{THz} due to carrier excitation with the 10-fs pump is displayed in a gray scale map versus pump-probe delay t_D (vertical) and THz sampling delay T (horizontal) in Fig. 1(b). The polarization response of the carrier plasma follows instantaneously upon the single-cycle probe pulse at early delay times t_D between -20 fs and 20 fs. With increasing values of t_D the plasma develops a delayed oscillatory response after the THz probe burst: Starting with a first retarded half wave for t_D = 30 fs (label '1'), up to four delayed half cycles are observed for $t_D \geq 120$ fs (labels '1' to '4'). At pump-probe delays beyond t_D = 150 fs, the electric field change ΔE_{THz} versus T becomes stationary on a picosecond time scale determined by the carrier recombination time.

The two-time dependent data of Fig. 1 allow us to derive the dynamics of the complex dielectric function of the excited semiconductor at frequencies between 5 THz and 60 THz [4]. The imaginary part (see Fig. 2(a)) of the inverse dielectric function reflects the buildup of dissipation in the plasma for the limit of small momentum exchange q in an interaction process: 25 fs after carrier generation, a spectrally broad feature indicates an uncorrelated many-body ensemble with a quasi-instantaneous response. Within 100 fs a sharp maximum at the plasma frequency of $\omega_{pl}/2\pi = 14.4$ THz appears due to the transition to collective plasmon scattering. In the unexcited sample (lower solid line in Fig. 2(a)), $-\mathrm{Im}(1/\varepsilon_{q=0})$ vanishes exactly, except for a small peak at $\omega_{LO}/2\pi = 8.8$ THz. This feature is due to energy exchange with the crystal lattice via polar-optical interaction with LO phonons. The real part (see Fig. 2(b)) of the inverse dielectric function describes the buildup of Coulomb screening. The quantity $\mathrm{Re}(1/\varepsilon_{q=0})$ renormalizes the

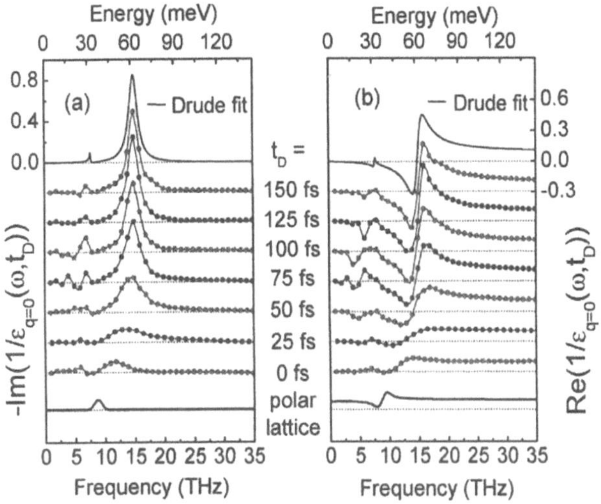

Figure 2: (a) Imaginary and (b) real part of the long-wavelength limit of the inverse dielectric function of GaAs versus frequency for different pump-probe delays t_D. The lower solid lines refer to unexcited GaAs. A Drude-fit of the data at t_D = 150 fs is represented by the upper solid line.

charge that is effective in a long-range Coulomb collision [4, 5]. For t_D = 25 fs, the dielectric function is flat above the plasma resonance ω_{pl} indicating still unscreened carrier-carrier collisions. The buildup of a screening charge around the bare particle and of a dressed interaction potential is completed after approximately 150 fs: A negative minimum develops below the plasmon energy due to overscreening by the forming polarization cloud. In contrast, a positive peak consistent with anti-screening emerges for an energy exchange slightly above the plasmon pole. Our findings are in agreement with recent quantum kinetic simulations [5]. As predicted, the approximate time scale for the dressing process is given by the inverse plasma frequency wich is 70 fs in our case.

In conclusion, our two-dimensional time-domain spectroscopy allows to resolve the field response of nonequilibrium systems in the mid and far infrared. The buildup of quantum correlations in a nonthermal electron-hole plasma in a semiconductor has been studied as a first example. In the same way, the dynamics of a variety of low-energy elementary excitations in important systems such as high-T_C superconductors, biological complexes and organic semiconductors have now become accessible with uncertainty-limited resolution.

References

1 Q. Wu and X.-C. Zhang, Appl. Phys. Lett. 71 (1997) 1285.

2 A. Leitenstorfer, S. Hunsche, J. Shah, M C. Nuss, W. H. Knox, Appl. Phys. Lett. 74 (1999) 1516.

3 R. Huber et al., Appl. Phys. Lett. 76 (2000) 3191. A. Brodschelm et al., Ultrafast Phenomena XII, Springer Series in Chemical Physics 66 (2000) 215.

4 R. Huber, F. Tauser, A. Brodschelm, M. Bichler, G. Abstreiter and A. Leitenstorfer, Nature 414 (2001) 286.

5 Q. T. Vu and H. Haug, Phys. Rev. B 62 (2000) 7179.

Observation of a hysteretic pseudogap behavior via coherent phonons in high temperature superconductors

T. Dekorsy[1], N. Georgiev[1], M. Helm[1], and O.V. Misochko[2]

[1] Institute of Ion Beam Physics and Materials Research, Forschungszentrum Rossendorf, Postfach 510119, D-01314 Dresden, Germany
E-mail: t.dekorsy@fz-rossendorf.de
[2] Institute of Solid State Physics, Russian Academy of Sciences, 142432 Chernogolovka, Moscow region, Russia

Abstract. We investigate ultrafast reflectivity changes of a near-optimally doped $YBa_2Cu_3O_{7-x}$ (123) superconductor. The amplitude of coherently excited phonons show strong changes as a function of temperature above the superconducting transition temperature. These changes give evidence for two crossover temperatures within the pseudogap regime. In addition, these crossover temperatures exhibit a clear hysteretic behavior as a function of temperature.

One of the intriguing subjects in the physics of high temperature superconductors (HTSC) is their behavior above the superconducting transition temperature T_c, i.e. in the normal state. Several well established non-time resolved techniques such as infrared, Raman and angle-resolved photoemisson spectroscopy revealed numerous anomalies in the normal state of these complex materials. These anomalies are attributed to a pseudogap state in a temperature range below a crossover temperature T* and above T_c. Although there exists no complete theory for the pseudogap regime, there is common agreement that the observed anomalies in the pseudogap range are indications that some superconducting properties persist above T_c [1]. Therefore more detailed investigation in this temperature range are highly desirable in order to reach a deeper understanding of the physics in HTSC.

Time-resolved ultrafast spectroscopy of the carrier and lattice dynamics has been shown to provide insight into the dynamics of HTSC, such as the optical break-up of Cooper pairs and their subsequent relaxation as well as the excitation of coherent phonons [2-4]. Recently, pseudogap dynamics have been investigated by ultrafast MIR spectroscopy and optical pump-THz probe spectroscopy in underdoped samples [5,6]. Here we report on a study of femtosecond relaxation and coherent phonon dynamics in a close to optimally doped $YBa_2Cu_3O_{7-x}$ sample by optical pump-probe spectroscopy.

The sample is a c-axis-oriented film of $YBa_2Cu_3O_{7-x}$ grown by off-axis laser deposition on <100> MgO. The film is 350 nm thick and has a superconducting transition at T_c = 88 K. Figure 1 shows the transient reflectivity changes for different temperatures obtained with 50 fs pulses at a wavelength of 780 nm from

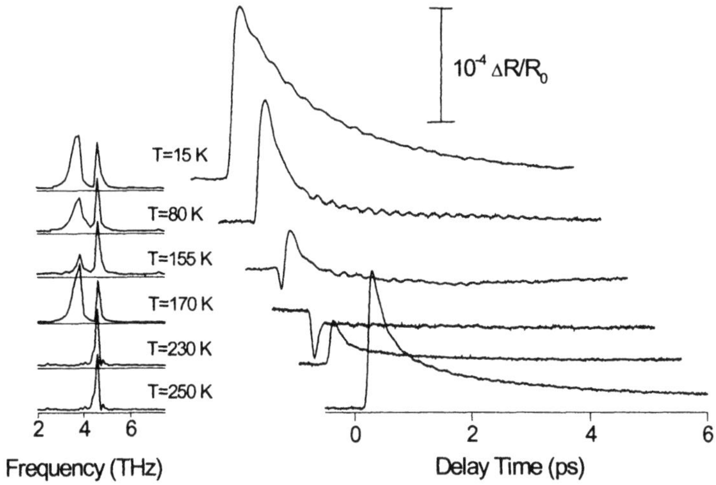

Fig. 1. Time resolved reflectivity change of $YBa_2Cu_3O_{7-x}$ for different temperatures. For clarity the time domain data are shifted vertically and horizontally. On the right the Fourier transforms are given which are shifted vertically only.

a Ti:Sapphire oscillator. The transients consist of an electronic background and a modulation due to coherently exited phonons. A strong change in both the phonon and relaxation dynamics is observed for temperatures below T_c. Here we will only focus on the phonon dynamics. For temperatures above T_c the dynamics is dominated by the A_g phonon generated by the Cu displacement (4.6 THz), while below T_c the A_g phonon of the Ba displacement (3.7 THz) becomes dominant in agreement with previous investigations [3,4]. However, at 170 K the phonon spectrum resembles remarkably the phonon spectrum in the superconducting phase with the amplitude of the Ba-mode being again increased in comparison to the amplitude of the Cu-mode.

Figure 2 shows the amplitude ratio of the Ba to Cu modes for smaller temperature steps. Besides the expected change of the phonon spectrum below T_c a very abrupt change in the phonon spectrum is observed at the crossover temperatures T_1* and T_2*. Most remarkably these crossover temperatures exhibit a strong hysteresis which is observed for the first time. It should be noted that changes in the relaxation dynamics appear abruptly at the same temperatures.

The excitation of the coherent Ba mode has been explained by the displacive excitation of coherent phonons (DECP) initiated by the break-up of Cooper pairs [3], although some other features point towards a Raman-like mechanism [4]. The

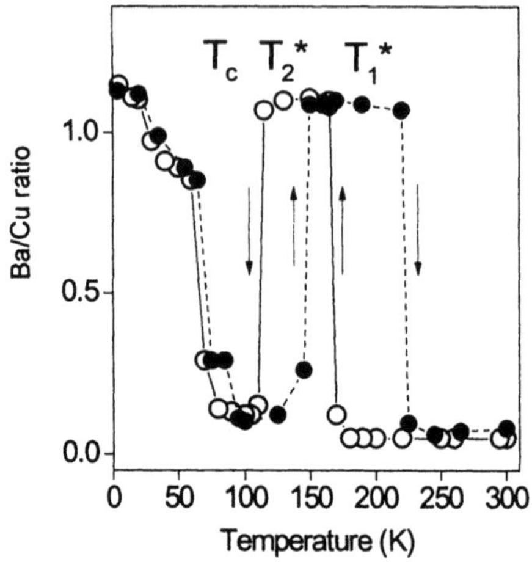

Fig. 2. Ratio of the Ba phonon amplitude to the Cu phonon amplitude versus temperature. The arrows indicate the direction of the temperature change.

re-appearance of the Ba mode between T_1* and T_2* could thus be explained by the break-up of localized, non-coherent pairs in the pseudogap regime. The regime between T_c and T_2* is presumably dominated by itinerant pairs. However, why the itinerant, non-coherent pairs are decoupled from the lattice remains unclear. The data presented here on the hysteretic behavior propose a new challenge for the theories describing the pseudogap state of which some predict a non-uniformity of the pseudogap phase [7].

References

1 For a review, see T. Timusk and B. Statt, Rep. Prog. Phys. **62**, 61 (1999).
2 D. H. Reitze, A. M. Weiner, A. Inam, and S. Etemad, Phys. Rev. B **46**, 14309 (1992).
3 W. Albrecht, T. Kruse, and H. Kurz, Phys. Rev. Lett. **69**, 1451 (1992).
4 O.V. Misochko, K. Kisoda, K. Sakai, S. Nakashima, Phys.Rev. B **61**, 4305 (2000).
5 R. A. Kaindl, M. Woerner, T. Elsaesser, D.C. Smith, J.F. Ryan, G.A. Farnan, M.P. McCurry, D.G. Walmsley, Science **287**, 470 (2000).
6 R. D. Averitt, G. Rodriguez, A. I. Lobad, J. L. W. Siders, S. A. Trugman, and A. J. Taylor, Phys. Rev. B **63**, 140502 (2001).
7 P. Devillard and J. Ranninger, Phys. Rev. Lett. **84**, 5200 (2000), and references therein.

Coulomb correlations, spectral line shapes, and size distributions of CdSe nanocrystals

Gregory D. Scholes, Margaret A. Hines, Anne Colonna, and Mayrose Salvador

Lash-Miller Chemical Laboratories, University of Toronto, 80 St. George Street, Toronto, Ontario M5S 3H6, Canada
E-mail: gscholes@chem.utoronto.ca

Abstract. We report 2-pulse and 3-pulse photon echo studies of CdSe nanocrystals at 293 K and 4.5 K. The data and simulations, which include the effects of many-particle correlations for the 2PE, show that the 3PE signal provides a precise measurement of the inhomogeneity in the exciton transition energy due to the size distribution of the quantum dots. The dephasing processes revealed by the 2PE are found to occur on at least two important time scales.

1. Introduction

We are exploring collective phenomena that emerge from the many-body interactions between subunits that comprise a complex assembly. Such an assembly can exhibit desirable characteristics, such as optical properties, that are unique to the assembly and quite different from those of the individual components. We can study such systems at a deep level by taking advantage of the hierarchy of electronic interactions present in certain molecular assemblies [1], so that electronic response is controlled by transient nonlinear response initiated by optical pulse sequences.

Here we report our studies of the origins of optical response in nanocrystalline semiconductor quantum dots of CdSe. In these systems we vary the size of the quantum dot, which changes the degree of quantum confinement, and we investigate the effects on the line shape and dephasing, the significance of Coulomb correlations in the nonlinear response, and the significance of the biexciton resonance.

2. Experimental Methods

Short visible laser pulses of 20 to 25 fs duration at the sample (intensity FWHM measured by autocorrelation in a 50 Hm thick BBO crystal and assuming a Gaussian pulse profile) were generated using a noncolinear optical parametric amplifier (NOPA) and precompensating for dispersion using a quartz prism pair. The NOPA was pumped by 200 HJ of the output of a Clark-MXR CPA-2001 amplified Ti: Sapphire laser system (1 kHz, 775 nm, ~140 fs pulses). The beam intensity used in the experiments was varied in the range 5 to 50 nJ per beam using a half-waveplate-polarizer combination. Time-integrated transient two-pulse photon echo (2PE) experiments were performed by splitting the laser beam into two parallel beams of equal intensity, which were focused into the sample using a silver-coated mirror with a radius of curvature of 0.5 m. 2PE signals in both $2\boldsymbol{k}_2 - \boldsymbol{k}_1$ and $2\boldsymbol{k}_1 - \boldsymbol{k}_2$ directions were detected simultaneously in order to determine accurately the temporal overlap $\tau = 0$ fs. The 3PEPS experiments were performed in an equilateral triangle beam geometry [2,3].

Highly fluorescent CdSe semiconductor nanocrystals were prepared according to the established method first reported by Murray, et al. [4]. Narrow size distributions were obtained by repeated size-selective precipitations using methanol and toluene. Samples were prepared by making thin films of CdSe in poly(butylmethacrylate) (PBMA) on quartz substrate using chloroform solvent. The films were prepared to have an optical density of ~0.4 at the excitation wavelength. The samples were held in a vacuum and cooled to 4.4 K in a continuous flow liquid He cryostat (Janis ST-100).

3. Results and Discussion

In Fig. 1 we show 3PEPS data for a 38 Å diameter CdSe sample. Intense 207 cm^{-1} beats, assigned to the CdSe LO phonon mode by Mittleman, et al. [5], are clearly evident in the data. A striking feature of the 3PEPS measurement is the non-zero asymptotic peak shift – a signature of inhomogeneous line broadening [3]. By simulating the data using Mukamel's response function formalism [2,6] we are able to extract the inhomogeneous broadening of the ensemble line shape that is attributed to the distribution of particle sizes in the sample. Assuming a Gaussian distribution of exciton transition energies, we obtain a standard deviation for the distribution of 280 cm^{-1} for this sample. We have demonstrated that the measured distribution broadens as we deliberately make the sample more polydisperse. The high initial peak shift at 4.5 K, compared to that at 293 K, is due to the much weaker coupling of the exciton transition to the fluctuating bath.

In Fig. 2 we show 2PE data recorded at 4.5 K for two different sized CdSe nanocrystal samples. At low temperature we observe a long decay component in the 2PE signal with a transverse relaxation time of ~40 ps. We attribute this to phase-destroying processes analogous to those observed in bulk CdSe, such as coupling to lattice modes. Coulomb interaction-mediated scattering processes appear to be

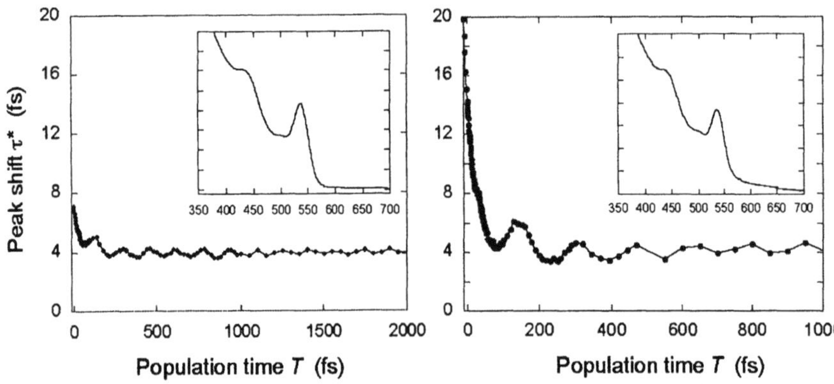

Fig. 1. 3PEPS data for 38 Å CdSe nanocrystals. Insets show the corresponding absorption spectra. Left: CdSe in toluene at 293 K. Right: CdSe in poly(butylmethacrylate) at 4.5 K

insignificant since we observe no long "decay" in the negative coherence time region. The intense third-order response at early positive coherence times is attributed to (1) detuning effects that represent the inhomogeneity of the nanocrystal ensemble, and (2) a rapid inertial dephasing process. The inertial solvation response is still relatively strongly coupled to the exciton transition at 4.5 K, as evinced from the 3PEPS data. Presumeably this reflects the large surface area-to volume ratio of the quantum dots, such that surface motions of the capping groups (TOP and TOPO in this case) can couple to the exciton transition. At short, negative coherence times the data exhibit a periodic modulation. The frequency of this modulation (more clearly seen in recent data) is equal to the energy difference between the first two electronic excited states of the quantum dot. We attribute this region in the signal to the biexciton coherence term in the 'average polarization model' for the full many-body problem [7].

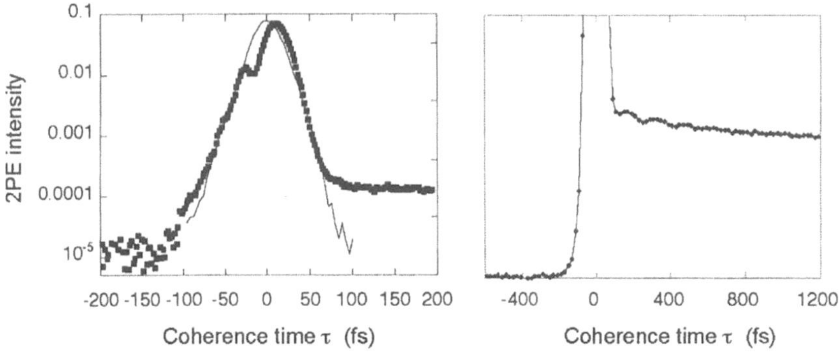

Fig. 1. Left: 2PE data (squares) for 39 Å diameter CdSe nanocrystals in PBMA at 4.5 K and laser pulse intensity autocorrelation (solid line). Right: 2PE data for 45 Å diameter CdSe nanocrystals in PBMA at 4.5 K, also plotted on a log scale

Acknowledgements. We are grateful for the assistance of Karyn Ang and Dr. Xiujuan Yang. The University of Toronto and NSERC are thanked for financial suppport.

References

1 V. M. Axt and A. Stahl, Z. Phys. B **93**, 195, 1994.
2 T. Joo, et al., J. Chem. Phys. **104**, 6089, 1996.
3 G. D. Scholes, et al., Phys. Rev. B **61**, 13670, 2000.
4 C. B. Murray, D. J. Norris, and M. G. Bawendi, J. Am. Chem. Soc. **115**, 8706, 1993.
5 D. M. Mittleman, et al., Phys. Rev. B **49**, 14435, 1994.
6 S. Mukamel, *Principles of* N*onlinear Optical Spectroscopy*, Oxford University Press, New York, 1996.
7 W. Schäfer, et al., Phys. Rev. B **53**, 16429, 1996.

Biexcitonic effects in coherently controlled six-wave-mixing experiments

H. G. Breunig, T. Voss, I. Rückmann, J. Gutowski

[1] Institut für Festkörperphysik, Universität Bremen, P.O.Box 330440, 28334 Bremen, Germany
E-mail: gbreunig@physik.uni-bremen.de

Abstract. The influence of biexcitonic and other polarization dependent contributions to optical coherent control of excitonic polarization is investigated using a ZnSe single quantum-well. Coherent control is achieved with a pair of phase-locked laser pulses. An additional pulse is used to generate a six-wave-mixing signal which oscillates as the delay between the phase-locked pulses is changed. This coherently controlled six-wave-mixing signal shows significantly different shapes for collinear (biexcitonic contributions enabled) and co-circular (biexcitonic contributions disabled) polarization states of the excitation pulses.

Coherently controlled four-wave-mixing (FWM) measurements have turned out to be a very sensitive tool to investigate the microscopic origin of the non-linear optical response of semiconductor quantum-well structures, in particular if excitons are excited [1-3]. The measured signal oscillates as a function of the phase difference of the two phase-locked pulses and is modulated additionally by contributions that oscillate with integer multiples of the excitonic resonance frequency. The influence of these harmonics can be measured in background free directions, i.e., the first (FWM) or second (six wave mixing, SWM) diffracted order in direction $2k_2$-k_1 or $3k_2$-$2k_1$, respectively. Their intensity is easily quantified by inspecting the corresponding Fourier transforms (FT). Here, we report on transient interferometric SWM measurements and the influence of biexcitons on the signal.

The sample consists of a 10 nm ZnSe single quantum well embedded in two 500 nm ZnSSe cladding layers and was pseudomorphically grown by molecular-beam epitaxy on a GaAs substrate. For the transmission experiments, the substrate was carefully removed by chemical etching. The experiments were performed at 4 K.

The mechanism of the interferometric SWM experiments is similar to that of the corresponding FWM measurements [2,3]. The phase relation of two (interferometric) pulses generated by an actively stabilized Michelson interferometer in direction k_1 can be tuned very accurately by controlling the length of one arm of the interferometer with a temporal resolution of 0.04 fs. If the time delay t_{int} between the interferometric pulses (which is always smaller than the dephasing time) is continuously changed a successive destructive and constructive interference of the induced coherent polarizations will occur. A polarization grating is formed with an additional pulse of wave vector k_2 delayed by t_{del} which is smaller than the dephasing time.

All pulses are generated by use of second harmonic pulses of a self-mode-locked titanium-doped sapphire laser with a pulse repetition rate of 83 MHz and a pulse width at half maximum of 120 fs. The laser pulses whose central wavelength was tuned to the biexcitonic resonance extended spectrally over the resonance of the heavy-hole (hh) exciton.

The $\chi^{(3)}$-signal that dominates the signal in the FWM direction $2k_2$-k_1 is generated already without action of a biexciton resonance. Biexcitonic contributions resulting from at least four-particle correlations should have a significant impact on the $\chi^{(5)}$-dominated signal emitted in $3k_2$-$2k_1$ since this is a $\chi^{(3)}$-free direction. All contributions emitted in direction $3k_2$-$2k_1$ are at least of fifth order in the optical field.

Figure 1 shows the time- and energy-integrated SWM signal obtained for two single collinearly (x,x) polarized incident pulses of directions k_1 and k_2 (dashed line). The polarization state of the pulses is controlled by two Pockels cells to avoid any mechanical adjustment. The signal exhibits strong exciton-biexciton beats for $t_{del} > 0$. For $t_{del} < 0$ a rapid increase of the signal without beating is observed which is in agreement with previous measurements [4]. The absence of beats for $t_{del} < 0$ has been explained by inhomogeneous broadening of the resonances in FWM experiments [5]. From the beating period a biexciton binding energy of (5.2 ± 0.3) meV is deduced. The beating is clearly absent in the co-circular ($\sigma^+\sigma^+$; Fig.1, solid line) configuration that does not allow the formation of biexcitons due to polarization selection rules [6].

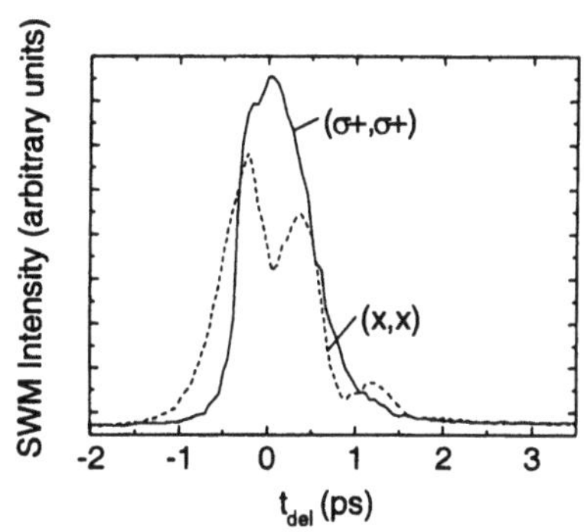

Fig. 1. SWM-signal for excitation with two co-circularly (solid) and collinearly (dashed) polarized pulses

The influence of biexcitons on the SWM signal is even stronger pronounced in interferometric measurements that enable the coherent control of the induced polarization. Fig. 2 shows interferometric SWM traces with differently polarized incident pulses. The delay of the k_2 pulse is fixed at $t_{del} = 0.6$ ps. In both cases, the signal oscillates with mainly the hh-exciton resonance frequency. In the collinear case an additional fine structure for small t_{int} occurs which is not observed in co-circular configuration. The polarization state of the incident pulses also strongly affects the envelope of the measured traces. A minimum in the envelope and smaller amplitudes are observed in the collinear case.

The corresponding FTs depicted in figure 3 reveal the nature of the fine structure. For collinear excitation contributions occur oscillating with up to five times the resonance frequency while in the co-circular case contributions of more than twice

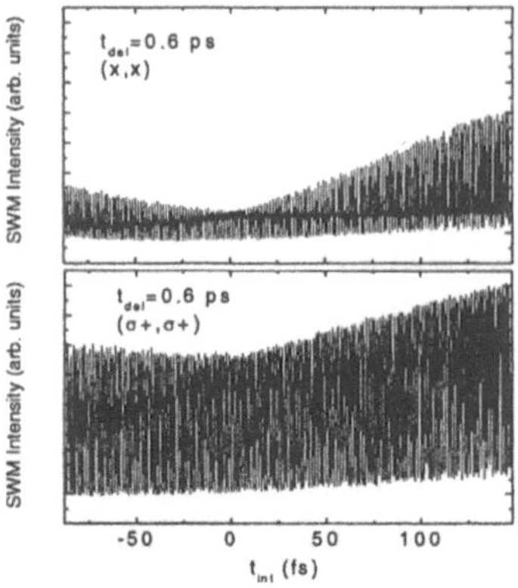

Fig. 2 Interferometric six-wave mixing-signal, for collinear (top) and co-circular (bottom) polarized incident pulses.

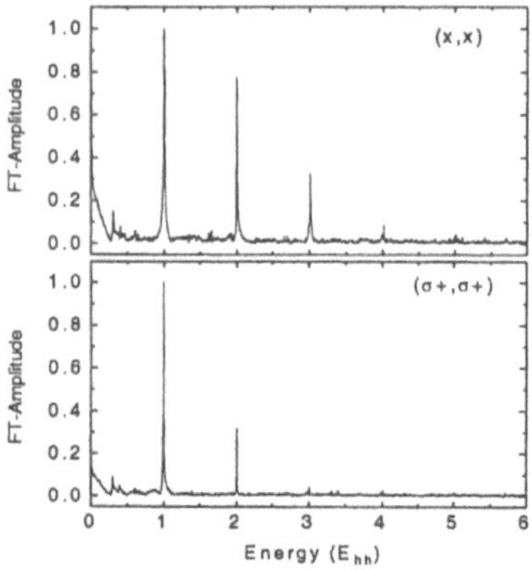

Fig. 3 Absolute values of the Fourier Transform with respect to t_{int} for collinear (top) and co-circular (bottom) polarized incident pulses.

the resonance frequency are absent or at least strongly suppressed. The lowest-order contribution of the non-linear polarization with respect to the optical field in the SWM configuration is $\chi^{(5)}$. A macroscopic model taking into account the field components emitted in $3k_2$-$2k_1$ shows that the interferometric signal due to $\chi^{(5)}$ oscillates with ω and 2ω only. Thus, apparently no higher-order contributions contribute to the $\sigma^+\sigma^+$ signal in Fig. 3 (bottom). The fact that in Fig. 3 (top) peaks occur at 3ω and 4ω indicates a strong influence of $\chi^{(7)}$ effects on the signal that are related to biexcitons. The 5ω peak reveals that even contribution of the order of $\chi^{(9)}$ are clearly visible in the experiment.

In conclusion, we have presented coherent control of the excitonic and biexcitonic polarization in a ZnSe quantum-well using a six-wave-mixing setup. By changing the polarization state of the incident beams we were able to separate the biexcitonic contribution in the SWM signal and to show that effects of at least seventh order in the optical field have to be taken into account in a microscopic model to theoretically explain our experimental results.

References

[1] M. U. Wehner, J. Hetzler, and M. Wegener, Phys. Rev. B 55, 4031 (1997).
[2] H. G. Breunig, T. Trüper, I. Rückmann, J. Gutowski and F. Jahnke, Phys. Stat. Sol. (b) Vol. 229, No. 2, p 621, (2002)
[3] H. G. Breunig, T. Trüper, I. Rückmann, J. Gutowski and F. Jahnke, article in press Physica B
[4] V. M. Axt, S.R. Bolton, U. Neukirch, L. J. Sham and D. S. Chemla, Phys. Rev. B 63, 115303, (2001)
[5] T. F. Albrecht, K. Bott, T. Meier, A. Schulze, S. T. Cundiff, J. Feldmann, W. Stolz, P. Thomas, S. W. Koch and E. O. Göbel, Phys. Rev. B 54, 4439, (1996)
[6] Y. Z. Hu, R. Binder and S. W. Koch, S. T. Cundiff, H. Wang and D. G. Steel, Phys. Rev. B 49, 14282, (1994)

Ultrafast investigation of exciton dissociation processes in polymeric semiconductors at high pump fluence

David M. Russell[1], R. Jane Essex[1], Richard H. Friend[1], Sepas Setayesh[2], Klaus Müllen[2], and Carlos Silva[1]

[1] Cavendish Laboratory, University of Cambridge, Madingley Road, Cambridge CB3 0HE, United Kingdom
E-mail: cs271@cam.ac.uk

[2] Max-Planck-Institut für Polymerforschung, Ackermannweg 10, D-55128 Mainz, Germany

Abstract. We report intensity-dependent femtosecond transient absorption measurements to explore the photophysics of a model step-ladder π-conjugated polymer. At low pump fluence, the lowest-lying singlet exciton ($1B_u$) is the only photoexcitation, but at high fluence triplet-exciton and polaron absorption features are evident in ultrafast timescales (within 100 fs). We find a two-step photogeneration mechanism for these long-lived species, with the $1B_u$ exciton as the intermediate. Detailed photophysical modelling reveals that the effective yield of sequentially excited species is 15%, with a branching ratio of triplet-excitons to polarons $\sim$ 5%.

1 Introduction

Polymeric semiconductor materials currently attract widespread international interest due to their intriguing new physics and promising applications in optoelectronics [1]. It is generally accepted that the lowest excited state in phenylene-based conjugated polymers is excitonic in nature [2]. We have reported previously that when thin-film conjugated polymers are subjected to femtosecond optical excitation, high yields of photoinduced charge generation result from two-step sequential resonant excitation [3]. Selection rules dictate that the high energy states accessed sequentially (with the odd-parity $1B_u$ luminescent exciton as an intermediate) have even-parity (A_g) symmetry. Lower-lying even-parity excited states have been classified into those that mediate charge transfer (which lie < 1.1 eV above $1B_u$) and those that undergo prompt internal conversion to $1B_u$ [4]. In this contribution, we present detailed photophysical modelling of femtosecond transient absorption data for step-ladder conjugated polymers. The effective yield of secondary photoexcitations is found to be $\sim$ 15%.

2 Experimental

The experimental apparatus is described in detail elsewhere [5], and consists of a dye-amplified Ti:sapphire laser at a repetition rate of 1 kHz. The pump photon energy was 3.18 eV, and the probe pulse consisted of single-filament white-light continuum generated in a sapphire window. Samples of poly-9,9-dioctylfluorene (F8) and poly-6,6′,12,12′-tetraocyl-2,8-indenfluorene (PIFTO) were spin-coated from anhydrouns p-xylene solution in a nitrogen-filled glove box to form 100-nm thick films on Spectrosil substrates. Films were kept under a dynamic vacuum of $< 10^{-5}$ mbar to prevent photooxidation.

3 Results and Analysis

Femtosecond transient absorption spectra for F8 and PIFTO are shown in Fig. 1. In both cases, probe-induced stimulated emission (SE) signal occurs in a spectral region covering the photoluminescence spectra at early time. Two photoinduced absorption features are observed at 1.5 (PA_1) and 2.1 eV (PA_2). Residual signal remains at long delays compared to the $1B_u$ lifetime, and these have been assigned to triplet-exciton and polaron absorption, respectively [6].

To investigate the photogeneration mechanism of these species, we investigated the pump fluence dependence of the $t = 0$ transient absorption signal, shown in Fig. 1(c). The SE, PA_1, and PA_2 signals display saturation fluences of 31.7, 96.6, 411 $\mu J/cm^2$. Therefore, a sequantial excitation mechanism is required to explain the instrument-limited generation of triplets and polarons at high fluence. Triplet generation by singlet fission of high-energy states has been proposed in ladder-type poly-p-phenylene [7]. From the measured

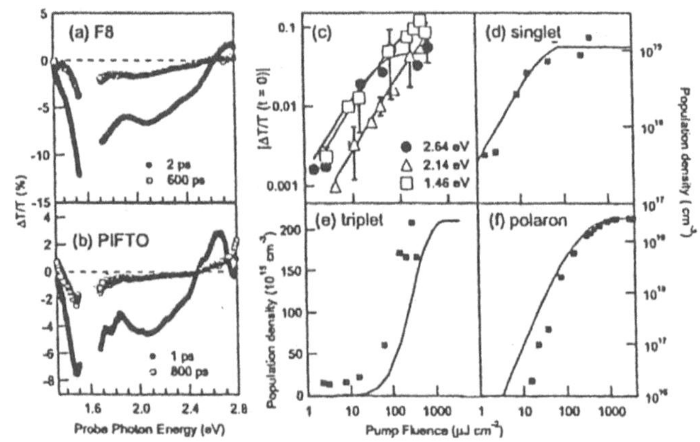

Fig. 1. Femtosecond transient absorption spectra for F8 (a) and PIFTO (b) at various pump–probe delays and moderately high pump fluence ($\sim 350\,\mu J\,cm^2$). (c) Pump fluence dependence of the initial transient signal at various photon energies. The solid lines are fits to $\Delta T/T \propto 1 - \exp(-\Phi/\Phi_{sat})$. Measured fluence dependence of population densities of singlets (d), triplets (e) and polarons (f) are also shown (filled squares), along with fits to the photophysical model (solid lines).

$\Delta T/T$, we extract the exciton population density with the Beer-Lambert law in the form $n_{ex} = (-\sigma_g d)^{-1} \ln(1 + \Delta T/T)$, with $\sigma_g = 4 \times 10^{-16}\,\mathrm{cm}^2$ [8]. Knowledge of this density, and singlet absorption cross sections at 1.5 and 2.1 eV (3.0×10^{-16} and $1.2 \times10^{-16}\,\mathrm{cm}^2$, respectively [8]) allows the extraction of triplet and polaron contributions to the PA, and these are plotted in parts (e) and (f). These data were modelled with the following.

$$\frac{dn_{ex}}{dt} = R(t)n_g - \frac{n_{ex}}{\tau_{ex}} - \beta n_{ex}^2 - \varphi R(t)n_{ex} \tag{1}$$

$$R(t) = \frac{\sigma_g f_{pu}}{\sqrt{2\pi}\gamma_{pu}} \exp[\frac{-t^2}{2\gamma_{pu}^2}] \tag{2}$$

$$\frac{dn_p}{dt} = \kappa\varphi R(t)n_{ex} + \frac{\beta}{2}n_{ex}^2 - \frac{n_p}{\tau_p} \tag{3}$$

$$\frac{dn_t}{dt} = (1-\kappa)\varphi R(t)n_{ex} + \frac{n_t}{\tau_t} \tag{4}$$

All parameters in this model are either known or independently measured [8], except φ (the effective photogeneration yield) and κ, which are varied here to fit the data in Fig. 1. The former affects the polaron and triplet saturation fluence, while the latter controls the branching ratio between the two. We find the best global fit with $\varphi = 0.15$ and $\kappa = 0.95$.

4 Conclusions

We conclude that at low excitation fluence, the $1B_u$ singlet exciton is the primary excitation, whereas at high pump fluence triplet excitons and polarons are produced by sequential excitation, first to $1B_u$, then to a high-energy state that dissociates efficiently ($\sim$ 15% effective yield).

References

1. R.H. Friend, R.W. Gymer, A.B. Holmes, J.H. Burroughes, R.M. Marks, C. Taliani, D.D.C. Bradley, D.A. dos Santos, J.L. Brédas, M. Lögdlund, and W.R. Salaneck: Nature **397**, 121 (1999)
2. J.L. Brédas, J. Cornil, and A.J. Heeger: Adv. Mater. **8**, 447 (1996)
3. C. Silva, A.S. Dhoot, D.M. Russell, M.A. Stevens, A.C. Arias, J.D. MacKenzie, N.C. Greenham, R.H. Friend, S. Setayesh, and K. Müllen: Phys. Rev. B **64**, 125211 (2001)
4. S.V. Frolov, Z. Bao, M. Wohlgenannt, and Z.V. Vardeny: Phys. Rev. Lett. **85**, 2196 (2000)
5. M.A. Stevens, C. Silva, D.M. Russell, and R.H. Friend: Phys. Rev. B **63**, 165213 (2001)
6. C. Silva, A.S. Dhoot, D.M. Russell, : J. Phys.: Condens. Matter (in press)
7. M. Wohlgennant, W. Graupner, G. Leising, and Z.V. Vardeny: Phys. Rev. B **60**, 5321 (1999)
8. D.M. Russell. PhD Thesis, University of Cambridge, Cambridge (2002)

Phase-Dependent Ultrafast Carrier and Lattice Relaxation: Amorphous, Crystalline, and Nanocrystalline Silicon

S.L. Dexheimer[1], C.P. Zhang,[1] J. Liu,[1] K.E. Myers,[1] J.E. Young,[1] Q. Wang,[2] and B.P. Nelson[2]

[1] Department of Physics, Washington State University, Pullman, WA 99164-2814, USA
E-mail: dexheimer@wsu.edu
[2] National Renewable Energy Laboratory, Golden, CO 80401, USA

Abstract. We investigate fast carrier and lattice dynamics in silicon-based semiconductor materials in which the degree of structural order has been systematically varied. We find carrier thermalization times of 150 fs in a-Si:H and 240 fs in c-Si, and report evidence for rapid lattice equilibration in a-Si:H. Dynamics in nc-Si:H are modeled in terms of distinct contributions from each of the component phases.

1. Introduction

The degree of order in a semiconductor lattice plays a fundamental role in determining the electronic properties of the material. In this work, we have carried out systematic studies of the fast carrier and lattice dynamics in amorphous, crystalline, and nanocrystalline silicon. The high time resolution of our measurements has allowed us to definitively establish the time scale for the fundamental carrier thermalization processes in a-Si:H, and comparison of the nature of the response in the amorphous and crystalline phases indicates that the degree of order has a dramatic impact on the associated lattice relaxation processes. The results provide a basis for understanding the fast carrier processes in heterogeneous nanocrystalline materials.

2. Experimental Methods

Recent advances in growth techniques for thin film hydrogenated amorphous silicon (a-Si:H) and nanocrystalline silicon (nc-Si:H), which consists of nanoscale crystallites of silicon embedded in an amorphous matrix, have allowed direct control of the degree of order [1]. We have carried out time-resolved measurements of the dynamics following photoexcitation in a series of a-Si:H and nc-Si:H thin films prepared by hot wire chemical vapor deposition (HWCVD) in which the degree of crystallinity has been controlled by varying the deposition conditions. For comparison, we have also directly measured the time-resolved differential transmittance of crystalline silicon (c-Si) using thinned Si (100) wafers prepared by boron implantation followed by etching to a thickness of ~ 350 nm.

Measurements of the time-resolved differential transmittance were made using pump pulses 35 fs in duration centered at a wavelength of 800 nm (1.55 eV) generated by an amplified Ti:S laser system operating at a repetition rate of 1 kHz. Probe pulses of variable wavelength were produced from a compressed femtosecond white-light continuum to give an overall temporal resolution of ~ 40 fs. Care was taken to minimize distortion of the measured pump-probe response by thin-film interference effects and by spatial nonuniformity of the photoexcited carrier density, and a separately detected reference beam was used to improve the sensitivity of the measurements.

3. Results and Discussion

In c-Si, excitation of the indirect optical transition at 1.55 eV generates a distribution of carriers confined in the lowest energy valleys in the conduction and valence bands. The measured response at probe wavelengths longer than the pump wavelength shows an increase in transmission that fits to a single exponential formation time of 240 fs. This response is consistent with relaxation of the carriers toward the band edge via optical phonon emission, establishing the time scale for the carrier-lattice interaction in crystalline silicon [2, 3].

In a-Si:H, optical excitation at 1.55 eV populates band tail states by one-photon absorption and extended states by two-photon absorption. The time-resolved response is dominated by induced absorbance from the photoexcited carrier population, which is enhanced by the lack of k-conservation in the disordered material. The high time resolution of our measurements has allowed us to clearly detect new fast components at short times [4] in addition to the well-established population decay via bimolecular recombination on picosecond time scales [5, 6]. Data were acquired at a series of pump fluences at probe wavelengths of 650 nm (1.91 eV), which lies above the nominal optical gap of ~ 1.7 eV for a-Si:H, and 940 nm (1.32), which lies in the band tail region below the nominal gap.

The response probed at 650 nm is fit to a model that includes a simple exponential decay of amplitude a and time constant τ, a population n_0 that undergoes bimolecular recombination with an associated rate constant k, and a component of amplitude c that decays slowly on the time scale of the measurements. The fluence dependence of the component amplitudes indicates the that dominant excitation process arises from two-photon absorption of the pump pulse, corresponding to an effective transition energy of 3.1 eV, well above the nominal optical gap of a-Si:H. The rapid initial decay of $\tau = 150$ fs reflects the fundamental carrier thermalization process in which the photoexcited carriers relax in energy via interaction with the lattice. The time scale of this important electronic process had remained in dispute in previous studies [5-7]. Interestingly, the time scale of the initial carrier cooling in a-Si:H is comparable, though somewhat faster than the 240 fs time scale that we observed for intraband relaxation in crystalline silicon.

The induced absorbance response detected at 940 nm is fit to a multi-component model that includes an exponential *rise* with formation time $\tau = 230$ fs in addition to the bimolecular and constant terms. We assign the rise time to heating of the lattice, which can be detected via a change in optical absorbance, since the band gap decreases with increasing temperature. Since the temperature dependence of the band gap is, in general, mediated by acoustic phonons, the associated change in optical absorbance reflects the time scale for redistribution of energy into these phonon modes. It is interesting to compare the time scales for phonon equilibration in the disordered vs. crystalline materials – this process appears to take place on a significantly more rapid time scale in amorphous silicon than the picosecond time scale typical of crystalline semiconductors, most likely as a result of increased coupling between the phonon modes.

The results on amorphous and crystalline silicon provide a basis for understanding the fast response in nanocrystalline silicon (nc-Si:H). A key issue in the physics of this heterogeneous material is the extent to which its optical and electronic properties can be understood simply as a combination of the properties of the separate phases, and in particular, the role of the grain-boundary regions. The composition of the material can be systematically varied by controlling the deposition conditions, and we have studied a series of nc-Si:H thin films of varying crystallite size and crystalline fraction. The pump-probe response detected at 940 nm is fit to a multi-component model that includes a fast exponential decay, bimolecular recombination, and a slowly varying (constant) component. These results, together with the known sample compositions, allow the photoexcited carrier dynamics in nc-Si:H to be understood in terms of the component phases of the material [2]. The initial fast exponential decay time $\tau = 240$ fs matches that detected in crystalline silicon, and the amplitude of the component scales with the crystalline fraction of the material, allowing this component to be assigned to intraband carrier relaxation within the silicon crystallites. In the materials studied, the size of the crystallites is larger than the ~ 5 nm quantum confinement threshold for c-Si, consistent with the observation of bulk-like properties associated with the crystalline fraction of the material. The bimolecular portion of the response is consistent with that observed in a-Si:H, indicating that this component is associated with the amorphous matrix. The constant *c* is expected to include contributions from the residual bleach and from lattice heating, but given its magnitude, it likely also includes contributions from long-lived states associated with the grain boundary regions in nc-Si:H.

Acknowledgements. This work is supported by the National Renewable Energy Laboratory and the National Science Foundation Division of Materials Research.

References

1 D. Han, *et al.*, J. Appl. Phys. 87, 1882 (2000).
2 K.E. Myers, Q. Wang, and S.L. Dexheimer, Phys. Rev. B, RC **64,** 161309 (2001).
3 K.E. Myers and S.L. Dexheimer, (in preparation).
4 S.L. Dexheimer, *et al.*, Mat. Res. Soc. Proc. **715,** A2.1 (2002), in press.
5 A. Esser, *et al.*, J. Appl. Phys. **73,** 1235 (1993).
6 P.M. Fauchet, *et al.*, J. Non-Cryst. Sol. **141,** 76 (1992).
7 J.A. Moon, *et al.*, Phys. Rev. B **50,** 10 608 (1994).

Photo-excitation of coherent acoustic phonons in InSb

M. D. Cummings and A. Y. Elezzabi

Ultrafast Photonics Laboratory, Department of Electrical and Computer Engineering, University of Alberta, Edmonton, AB, Canada, T6G 2V4
Tel: (780) 492-0756, Fax: (780) 492-1811
E-mail: mdc@ee.ualberta.ca

Abstract. We report on the first direct time-resolved measurement of coherent longitudinal acoustic phonon modes in highly photo-excited InSb. The impulsive nature of the excitation mechanism and an anisotropic softening of the InSb lattice are observed.

The increased availability of reliable ultrashort pulsed laser sources in the last decade has allowed for an intensification of the study of non-equilibrium coherent carrier and lattice dynamics in solid-state materials. Utilizing ultrashort laser pulses, direct observation of both coherent plasmon and phonon modes have been made by several groups [1-4]. While the study of coherent longitudinal optical (LO) phonons and their interactions with free carriers is well developed, the experimental investigation of coherent acoustic phonons has been predominantly carried out on semiconductor quantum structures. The aim of this investigation is to obtain a clearer understanding of lattice instabilities in semiconductors arising from a large photo-induced charge carrier concentration. These instabilities may occur when a high charge carrier density ($\geq 10^{21}$ cm^{-3}) is generated in the material, effecting a significant displacement of the semiconductor lattice, through changes in the deformation potential, thereby initiating coherent lattice oscillations [2,3,5].

Time-resolved pump-probe reflectivity measurements are performed, utilizing a Ti:Sapphire laser producing 10 fs pulses with a central wavelength of 810 nm. All measurements are performed at room temperature on an optically polished 500 μm thick sample of unintentionally n-doped (9×10^{13} cm^{-3}), high mobility, (111)-InSb. Differential and lock-in detection are utilized, thus, along with the use of multiple scan averaging, reflectivity changes as small as $\Delta R/R = 10^{-6}$ can be resolved.

As illustrated in Fig. 1, for a pump energy fluence of $F \approx 1.0$ mJ/cm^2 (corresponding to a carrier density of $n = 2.5\times10^{20}$ cm^{-3}), superimposed on a slowly varying electronic background is a strong modulation. It is important to note that the actual carrier density may be much higher than that due to photo-excitation alone as the photo-generated carriers may have an excess energy of up to 1.35 eV, approximately 7.5 times the band gap energy. Thus, impact ionization may enhance the charge carrier density. We interpret these oscillations as follows: above a critical energy fluence the charge carrier density, generated through photo-excitation and impact ionization, is large enough to slightly displace the lattice ions from their equilibrium positions. The resultant displacive motion of the lattice ions manifests itself as small amplitude coherent phonon oscillations as the lattice oscillates at or close to its normal vibrational modes.

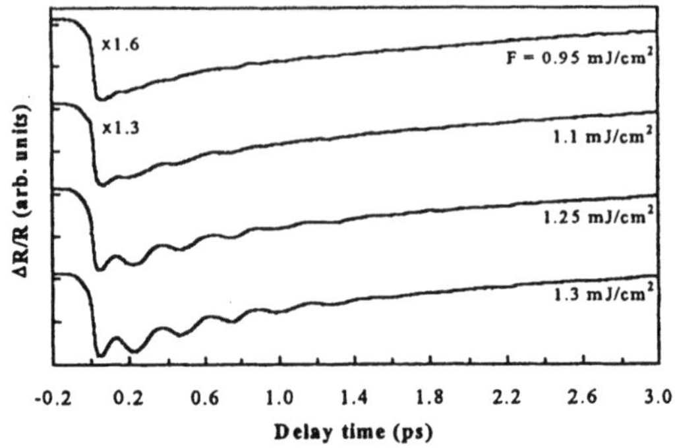

Fig. 1. Time-resolved reflectivity changes for InSb taken at four different excitation energy fluences.

The Fourier power spectrum of the reflectivity signals, taken for energy fluences above the threshold value of $F \approx 1.0$ mJ/cm^2, are shown in Fig. 2(a). Each curve clearly exhibits two frequency peaks. In order to accurately ascertain the peak frequency values and mode decay times, an exponentially decaying sinusoidal fitting function, for each mode, was Fourier transformed and compared with the data taken for $F = 1.1$ mJ/cm^2. The corresponding frequency values, $\nu =$ 3.75 THz and $\nu =$ 4.34 THz, identify these as the longitudinal acoustic (LA)-phonon L_1 and X_3 modes for InSb, respectively. Our experimental values are in excellent agreement with those determined via time independent means [6]. These coherent oscillations of the LA-phonons can be attributed to a purely impulsive electronic excitation of the lattice, where the (nearly) instantaneous atomic displacements result from the ultrafast generation of a high density of charge

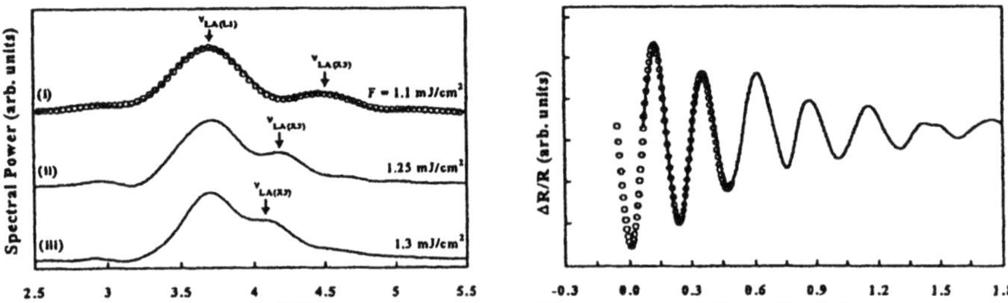

(a) (b)

Fig. 2(a). Fourier transforms of the lattice oscillation signals extracted from Fig. 1. The circles in (i) show the fit: $\sum_{i=1}^{2} A_i \exp(-t/\tau_i) \cos(2\pi\nu_i)$, where: $\nu_1 = 3.75$ THz, $\nu_2 = 4.34$ THz, $\tau_1 = 850$ fs, $\tau_2 = 550$ fs. **(b)** The extrapolated cosine-like fit to the data (circles) overlaid on the extracted lattice oscillations for F = 1.3 mJ/cm^2 (solid-line).

carriers. In order to verify this hypothesis, the first two oscillatory cycles of the reflectivity signal taken for a pump energy fluence of 1.3 mJ/cm^2 are fitted and extrapolated back to the zero time, as seen in Fig. 2(b). The amplitude of the oscillations follows a cosine-like behavior near time zero, supporting the conclusion that the excitation mechanism is of a displacive nature. A dramatic frequency red shift in $\nu_{LA(X3)}$ is observed as the pump energy fluence is increased beyond the oscillatory onset value of $F \approx 1.0$ mJ/cm^2, as shown in Fig. 2(a). The peak frequency decreases from $\nu_{LA(X3)} = 4.34$ THz to $\nu_{LA(X3)} = 4.04$ THz as the energy fluence is increased from $F = 1.1$ mJ/cm^2 to $F = 1.3$ mJ/cm^2. Contrary to $\nu_{LA(X3)}$, the peak frequency of the LA(L_1)-phonon mode is maintained at $\nu_{LA(L1)} =$ 3.75 THz over the same range of pump energy fluences. This differing behavior results from the anisotropic impulsive softening of the InSb lattice, as the lattice restoring forces along the (100) direction are weaker than the restoring forces along the (111) direction. The red shift of $\nu_{LA(X3)}$ can be explained by considering lattice instabilities due to transverse acoustic (TA) phonons and LO-phonon distortions. Using the model proposed by Stampfli et al. [5] the lattice instability onset time can be calculated. For InSb, the lattice instability will require, $\tau_{InSb} =$ 350 fs; thus, lattice instability occurs on a time scale shorter than the measured LA(X_3)-phonon decay time, $\tau_{LA(X3)} = 550$ fs (see Fig. 2(a) caption). This indicates that TA- and LO-phonon distortions are likely responsible for the softening of the LA(X_3)-phonon mode.

In conclusion, we have reported on the first direct time-resolved measurement of high frequency coherent LA-phonon oscillations in InSb. A cosine-like fit of the data, extrapolated back to time zero, indicates that the LA-phonon modes were excited through a displacive mechanism. At the lowest energy fluence for which oscillations were observed the measured frequencies are in excellent agreement with those obtained through time independent techniques. A frequency red shift in $\nu_{LA(X3)}$ as the energy fluence was increased is indicative of the initiation of impulsive lattice softening along the (100) direction.

References

1 G. C. Cho et al., "Subpicosecond time-resolved coherent-phonon oscillations in GaAs," Phys. Rev. Lett **65**, 764-766, 1990.

2 T. K. Cheng et al., "Mechanism for displacive excitation of coherent phonons in Sb, Bi, Te, and Ti_2O_3," Appl. Phys. Lett., **59**, 1923-1925, (1991).

3 S. Hunsche et al., "Impulsive Softening of Coherent Phonons in Tellurium," Phys. Rev. Lett., **75**, 1815-1818, (1995).

4 M. D. Cummings et al., "Femtosecond carrier dynamics in an asymmetrically excited GaAs photoconductive switch," Appl. Phys. Lett **78**, 3535-3537 (2001).

5 P. Stampfli et al., "Time dependence of the laser-induced femtosecond lattice instability of Si and GaAs: Role of longitudinal optical distortions," Phys. Rev. **B 49**, 7299-7305, (1994).

6 D. L. Price et al., "Lattice Dynamics of Grey Tin and Indium Antimonide," Phys. Rev. **B 3**, 1268-1279, (1971).

Ultrafast dynamics of coherent anharmonic phonons in bismuth

Muneaki Hase[1], Masahiro Kitajima[1], Shin-ichi Nakashima[2], and Kohji Mizoguchi[3]

[1] Materials Engineering Laboratory, National Institute for Materials Science, 1-2-1 Sengen, Tsukuba, Ibaraki 305-0047, Japan
E-mail: hase.muneaki@nims.go.jp
[2] Power Electronic Research Center, National Institute of Advanced Industrial Science and Technology, 1-1-1 Higashi, Tsukuba, Ibaraki 305-8561, Japan
[3] Department of Applied Physics, Osaka City University, 3-3-138 Sugimoto, Sumiyoshi-ku, Osaka 558-8585, Japan

Abstract. We report on the femtosecond dynamics of coherent anharmonic phonons in bismuth. The large amplitude coherent phonon generated by high-density photoexcitation exhibits a time-dependent phonon frequency, which is attributed to the anharmonicity of the lattice potential.

1. Introduction

In narrow band-gap materials, the photo-excitation of electrons from bonding into antibonding states involve with change in the equilibrium positions of the atoms; the atoms then oscillate around their new equilibrium positions. Thus photo-excitation of electrons leads to displacive excitation of coherent phonons (DECPs). [1] Under high-density excitation with several mJ/cm^2,[2-4] the amplitude of the coherent lattice displacement could increase up to several percent of the lattice constant. Under this condition, large amplitude coherent phonons would play an important role for structural modification. The purpose of this work is to examine the anharmonicity of the large amplitude coherent phonons; this is based on the classical mechanics[5] which indicates that the frequency of the anharmonic oscillator depends on the amplitude of the oscillation.

2. Experimental Methods

The bismuth polycrystalline films with 100 nm thickness were prepared by vacuum deposition on polished Si substrates at room temperature. Femtosecond seed pulses from a Ti:sapphire laser oscillator, operating at a wavelength of 800 nm with a pulse duration of ~70 fs, were amplified to a pulse energy of 500 μJ with a pulse duration of ~120 fs in a 1 kHz regenerative-amplifier system. The pump power density was reduced with a neutral density filter to below 7.6 mJ/cm^2 to prevent damaging the sample or causing laser ablation. The probe pulse energy was also reduced and fixed at 0.3 mJ/cm^2. The transient isotropic reflectivity change $\Delta R/R$ was recorded for pump power densities from 0.9 to 7.6 mJ/cm^2.

3. Results and Discussion

Figure 1(a) shows ΔR/R signal obtained at 7.6 mJ/cm^2. A large amplitude coherent A_{1g} phonon is observed.[1,6] The amplitude of the coherent lattice displacement estimated from the maximum photoexcited carrier density of 3.0×10^{21}cm^{-3} is ~1.3×10^{-2} nm.[1] This large displacement corresponds to 4.2 % of the interatomic distance of the nearest neighbors, which is below the critical displacement for the Lindeman melting criterion.[7] As shown in Fig. 1(a), the experimental datum cannot be fitted well with the harmonic oscillator, and there is a large deviation in the phase of the oscillation, which suggests that the mode frequencies are not constant but depend on the time. We performed a discrete wavelet transform (DWT)[8] in order to analyze the time-dependence of the phonon frequency. In Fig. 1(b) we show DWT spectra obtained for the two time delays of 0.3 ps and 3.0 ps as examples. The peak frequency of the A_{1g} mode is 2.52 THz at t = 0.3 ps and 2.78 THz at t = 3.0 ps. At t = 0.3 ps, the A_{1g} mode is asymmetrically broadened to the lower-frequency side, and the continuum, broad and structureless signal extending to zero frequency, appears.

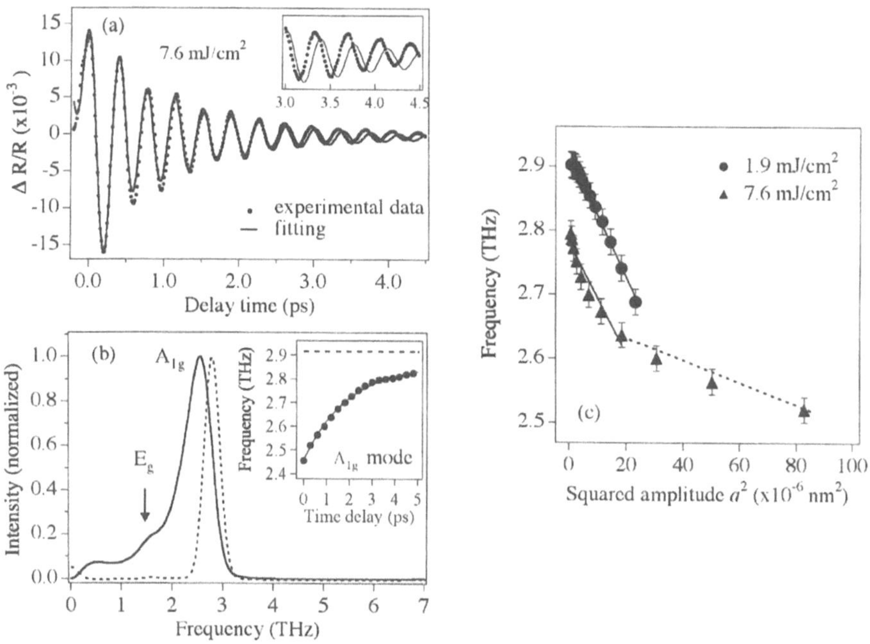

Fig. 1. (a) Transient reflectivity change observed in Bi film. (b) DWT spectra obtained at 7.6 mJ/cm^2 for the time delay of 0.3 ps (solid line) and 3.0 ps (dotted line). The E_g mode is broadened to lose its fine structure. The inset of (b) shows the variation of the A_{1g} mode frequency as a function of the time delay. The dashed line in the inset represents the ordinary frequency of 2.92 THz. (c) The variation of the frequency of the A_{1g} mode as a function of the squared amplitude of the coherent oscillation. The solid lines indicate the fitting of the data with Eq. (1). The dotted line for $a^2 > 20 \times 10^{-6}$ is an eye guide.

The appearance of the continuum implies a precursor of electronic melting under high-density photoexcitation. At t = 3.0 ps, only the A_{1g} mode with a nearly symmetric line shape is observed, and the continuum disappears. The A_{1g} mode frequency is initially down-shifted to 2.45 THz and changes toward the ordinary frequency of 2.92 THz (the inset of Fig. 1(b)).

In order to explain the transient frequency change, we consider an anharmonic potential. The frequency of the anharmonic oscillator is represented approximately by,[5]

$$\omega = \omega_0 + \gamma a^2 \left(\gamma \equiv -\frac{5\alpha^2}{12\omega_0^3} \right) \tag{1}$$

where a is the oscillation amplitude, and α is a coefficient of the third-order anharmonicity of the potential. Equation (1) implies that the phonon frequency shifts to a lower frequency side than the normal frequency of ω_0 as the amplitude increases. At 1.9 mJ/cm^2, the A_{1g} frequency decreases linearly with increasing the squared amplitude as shown in Fig. 1(c). This behavior obeys the relationship given by Eq. (1), accordingly verifying that the coherent A_{1g} phonon under the high-density excitation has anharmonic character.[6] At 7.6 mJ/cm^2 the A_{1g} frequency decreases almost linearly with the same slope as that at 1.9 mJ/cm^2 for smaller amplitudes than 20×10^{-6}nm^2. For the larger amplitudes, the A_{1g} frequency also shows linear decrease but with a more gradual slope. This suggests a dynamical change of the lattice potential due to the large lattice distortion, which could not be described by the rigid lattice potential.[5]

4. Conclusions

We have observed coherent anharmonic phonons in bismuth under high-density photoexcitation. The study of large amplitude coherent phonons will trigger further investigations of precursor phenomena of laser ablation and laser-induced phase transition as well as nonthermal melting.

References

1 H. J. Zeiger, *et al*., Phys. Rev. B **45**, 768, 1992.
2 T. K. Cheng, *et al*., Appl. Phys. Lett. **62**, 1901, 1993.
3 M. F. DeCamp, *et al*., Phys. Rev. B **64**, 092301, 2001.
4 S. Hunsche, et al., Phys. Rev. Lett. **75**, 1815, 1995.
5 L. D. Landau and E. M. Lifshitz, , in *Mechanics, Third Edition*, Vol. 1 of Course of Theoretical Physics, Butterworth-Heinenmann, Oxford, Chap. 5, 1976.
6 M. Hase, *et al*., Phys. Rev. Lett. **88**, 067401, 2002.
7 J. M. Ziman, in *Principle of the Theory of Solids*, 2nd edition, Cambridge University press, London, Chap. 2, 1972.
8 J. M. Combes, A. Grossmann, and Ph. Tchamitchian (Eds.), in *Wavelets: Time-Frequency Methods and Phase Space,* Second Edition, Springer-Verlag, Berlin Heidelberg, 1989.

Far-Infrared Carrier Dynamics in Superconducting MgB_2

R. D. Averitt[1], J. Demsar[1], D. W. Reagor[1], A. J. Taylor[1], W. N. Kang[2], H. J. Kim[2], E. M. Choi[2], S. I. Lee[2]

[1] Los Alamos National Laboratory, Mail Stop K764, Los Alamos, NM 87545,
E-mail: raveritt@lanl.gov
[2] Department of Physics, Pohang University of Science and Technology, Pohang 790-784, Korea

Abstract. We have performed optical-pump terahertz-probe measurements in transmission on MgB_2 thin films in the superconducting state. The initial optical perturbation of the superconducting condensate and subsequent pair recovery display a strong fluence dependence.

1. Introduction

The announcement in January 2001 of the discovery of bulk superconductivity in MgB_2 at $T_c \sim 39$ K [1] has generated a great deal of excitement since T_c is higher by nearly a factor of two in comparison to other previously known simple intermetallic superconductors (e.g. Nb_3Ge, $T_c \sim 23$ K). Far-infrared spectroscopy is a useful technique to characterize the low energy states in superconductors. Recently, the superconducting gap in MgB_2 has been studied using terahertz time-domain spectroscopy [2].

All-optical ultrafast time-resolved spectroscopy has also proven to be sensitive to the low energy states of correlated electron materials [3]. Combining ultrafast optical and far-infrared techniques to directly probe the low energy conductivity dynamics with picosecond resolution has allowed us to simultaneously measure the dynamics of quasiparticles and superconducting pairs in high-T_c superconductors [4], showing a similar temperature dependence of the relaxation time as in [3]. Time-resolved far-infrared studies of superconducting pair breaking have been performed on lead which is a conventional superconductor with $T_c = 7$ K [5]. However, given the low value of T_c in Pb, it is difficult to perform temperature and fluence dependent studies in the superconducting state. MgB_2, having a considerably higher T_c, offers the exciting possibility to look at the time-resolved far-infrared dynamics of a BCS-like superconductor in considerable detail. In the following we present our time-resolved studies of the pair breaking and recovery in MgB_2.

2. Experimental Methods

The MgB_2 films used in these studies had thicknesses of approximately 50 nm and 100 nm with $T_c \sim 34$ K [8]. The experiments were performed using 1-mJ 150-fs 1.5-eV pulses at 1 kHz from a commercial regeneratively amplified system. The

1.5 eV pulses were used to photoexcite the sample and generate and detect the terahertz (THz) pulses in ZnTe. Further details of the film growth and experiment are described elsewhere [4,6].

3. Results and Discussion

Figure 1(a) shows the THz pulses transmitted through the MgB_2 film at 35 K (above T_c) and at 7 K. The phase shift is the so-called kinetic inductance due to superconductor pairing (i.e. the conductivity due to the superconducting pairs is purely imaginary with a $1/\omega$ dependence that results in the observed phase shift in the time domain). The increase in amplitude below T_c is due to the increase in transmission above the superconducting gap compared to the normal state transmission. This is evident in Fig. 1(b) which shows the amplitude of the Fourier transform in the superconducting state divided by the amplitude in the normal state (35K). Below 30 K a gap opens up and increases in magnitude with decreasing temperature. The inset shows the gap Δ as a function of temperature as determined from data analysis using Mattis-Bardeen formulae [7] – the dashed line is a fit showing the BCS-like behavior of MgB_2. Fig. 1(c) and (d) show the temporal evolution of the complex conductivity at 7 K following photoexcitation with a fluence of ~1μJ/cm^2. Clearly, σ_{IM} (Fig. 1(c)), shows a decrease corresponding to Cooper pair-breaking. Correspondingly, there is an increase in the real conductivity σ_{RE} (Fig. 1(d)) due to an increase in the number of quasiparticles. Fig. 1(d) shows that the filling in of the superconducting gap takes several ps. At 300 ps (curves with open diamonds) the superconducting state has partially recovered. This is shown more clearly in Fig. 2(a) where σ_{IM}, and σ_{RE} (1

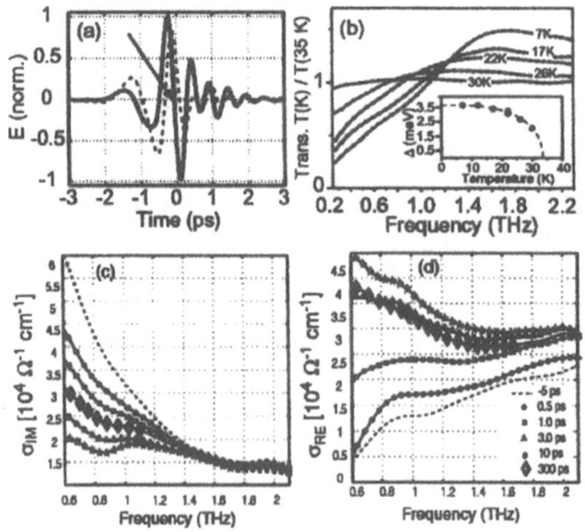

Fig1.(a)Transmitted E-field above (dashed line) and below (solid line) T_c. (b) Normalized transmission vs. frequency at various temperatures showing the gap opening – the inset show the gap versus temperature. Temporal evolution of (c) σ_{IM} and (d) σ_{RE} following photoexcitation.

THz) are plotted as a function of time. The superconducting pair recovery time as a function of temperature is shown in 2(b). The dynamics are much slower than in high-T_c materials which is consistent with the observation that the lifetime goes as $1/\Delta$ [3]. In addition, the temperature dependence is consistent with the gap model where the lifetime increases upon approaching T_c due to the anharmonic decay of 2Δ phonons (phonon bottleneck) [3], while at lower temperatures the lifetime increases due to a low density of quasiparticles (bi-particle recombination bottleneck) [8]. Fig. 2(c) shows the fluence dependence of the risetime that

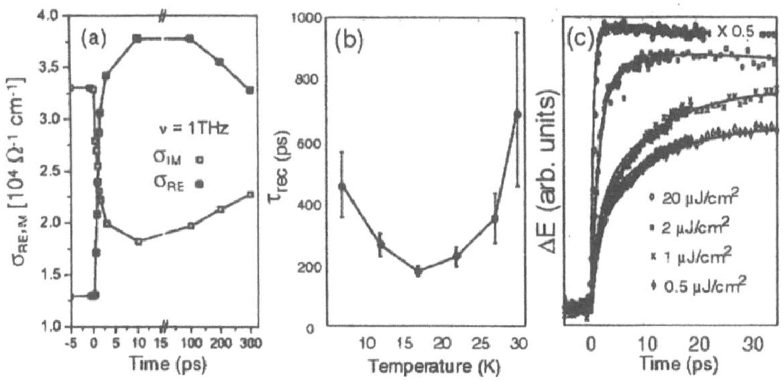

Fig. 2. (a) Temporal evolution of σ_{IM}, σ_{RE} at 1 THz. (b) Pair recovery time vs. temperature. (c) Fluence dependence of rise-time (i.e. pair-breaking).

corresponds to superconducting pair breaking (see Fig. 1(c) and (d)). The solid lines are fits with a two-exponential rise followed by a single exponential decay. The rise consists of a fast resolution limited component (~1 ps) and a slower (~15 ps) component. This slow risetime at lower fluences could result from phonon-mediated pair breaking following the initial fast quasiparticle avalanche pair-breaking process. This is consistent with the fact that at higher initial quasiparticle densities (e.g. 20 $\mu J/cm^2$) the risetime is dominated by the fast resolution limited response. In conclusion, we have performed the first time-resolved far-infrared studies on the new superconductor MgB_2.

References

[1] J. Nagamatsu, et al., Nature **410**, 63 (2001).
[2] R. A. Kaindl, et al., Phys. Rev. Lett. **88**, 027003 (2001).
[3] J. Demsar, et al., Phys. Rev. Lett. **83**, 800 (1999), V.V.Kabanov et al., Phys.Rev.B **59**, 1497 (1999).
[4] R. D. Averitt, et al., Phys. Rev. B **63**, 140502(R) (2001), R. D. Averitt, et al., Physica B **312-313**, 86 (2002).
[5] J. F. Federici, et al., Phys. Rev. B **46**, 11153 (1992).
[6] W. N. Kang, et al., Science **292**, 1521 (2001).
[7] D. C. Mattis, J. Bardeen, Phys. Rev. **111**, 412 (1958).
[8] J. Demsar, et al., Phys. Rev. B **63**, 54519 (2001).

Population dynamics in quantum structures

Thomas Müller[1], Wolfgang Parz[1], Rudolf Bratschitsch[1], Gottfried Strasser[1], and Karl Unterrainer[1]

[1] Technical University Vienna, Institute for Solid State Electronics, Floragasse 7, A-1040 Vienna, Austria
E-mail: thomas.mueller@tuwien.ac.at

Abstract. An interband pump/intersubband probe technique is applied to measure the dynamic intersubband absorption spectrum of an undoped GaAs/AlGaAs coupled quantum well excited by an ultrashort interband optical pulse.

1. Introduction

Ultrashort optical pulses with a large spectral width make it possible to measure the complete time-resolved absorption spectrum of a sample with a single pulse, offering simultaneously high resolution in both the time and frequency domains [1]. We have used ultrashort mid-infrared (MIR) pulses to measure the dynamic intersubband absorption spectrum of an asymmetric double quantum well excited by an ultrashort optical pulse.

In our experiment an interband pump pulse injects electrons into the first and second subband of the double quantum well structure (level-spacing E_{21} = 14 meV $< \hbar\omega_{LO}$). The time evolution of the decay of this coherent excitation into an incoherent electron population in the two subbands and their decay is monitored by probing the MIR absorption to a third (empty) subband [2]. The MIR probe pulse (10 μm central wavelength, 50 fs duration) is generated by phase-matched difference-frequency mixing in a 30 μm thin GaSe crystal [3,4]. The spectrum of the MIR pulse transmitted through the sample (T = 5 K) is obtained by measuring directly the electric field using a cross-correlation technique [5] and calculating the Fourier transform.

2. Intersubband Relaxation

Figure 1 (a) shows intersubband absorption spectra taken at different time-delays after the excitation by an interband pump pulse. The spectra clearly exhibit two absorption peaks: one around 112 meV, the other one 14 meV above this value. The low-energy peak is due to the (2-3) intersubband absorption, while the second peak is attributed to the (1-3) absorption. The amplitude of the first peak decreases with time-delay after excitation due to intersubband relaxation, while the amplitude of the second peak first rises slightly and subsequently it decreases due to carrier recombination.

Since the area under the peak (i-3) (i = 1,2) is directly proportional to the subband population $n_i(t)$, we are able to determine the population dynamics in the quantum well on the basis of the time-resolved absorption spectra. Figure 1 (b) shows the electron population of the first and second subband as a function of time delay after optical excitation (symbols). About 40% of the photo-excited carriers (1×10^{10} cm^{-2}) are injected into the second subband, while the remaining 60% are injected into the first subband at higher k-value. The population in the second subband (squares) shows an exponential decay. The carriers relaxing down from the second subband add to the population in the ground level. Subsequently, the population in the ground level drops due to carrier recombination. The lines in the inset of Figure 1 (b) are the results obtained from a simple rate equation model. By fitting the calculation to the experimental data we deduce an intersubband relaxation time of 14 ps. Taking only electron-electron scattering into account (which is the dominant intersubband scattering mechanism in our experiment), we calculated a relaxation time of 18 ps.

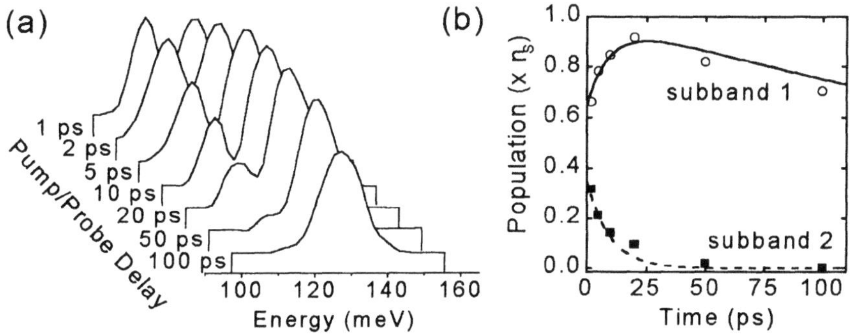

Fig. 1. (a) Time-resolved intersubband absorption spectra recorded at different delay times after interband excitation. (b) Population in the first and second subband (symbols: measurement, lines: model).

3. Intersubband Dephasing

The spectra above have been obtained by recording the response of the photoexcited quantum well to the MIR probe pulse at *fixed* pump/probe delays. Collecting time-resolved absorption spectra in this way requires that the response is short compared to the time-scale on which the absorption of the system changes. This requirement limits the time-resolution to a few picoseconds.

On the other hand it has been shown that in a setup with gated detection (electro-optic sampling or cross-correlation setup) *synchronously* scanning the gating pulse and pump delay *together* with fixed probe pulse delay results in a data set which is comprised of probe scans that have experienced a constant delay from the pump pulse [6,7]. In this case the time-resolution depends just on the duration of the pump pulse, and on the dispersion and thickness of the sample. We have applied this technique to record the time-resolved absorption spectra at delay times < 1 ps after excitation with a temporal resolution of about 150 fs.

Figure 2 shows cuts parallel to the pump-delay axis of the absorption data at probe energies of 115 and 130 meV, corresponding to the transitions (2-3) and (1-3), respectively. The (2-3) absorption shows an exponential decay, while the (1-3) absorption increases as discussed above. In addition both curves show a superimposed oscillation with a period equal to h/E_{21}. The oscillation in the absorption is caused by a quantum beat between levels 1 and 2 which is excited by the interband pump pulse. A density matrix calculation (lines in Figure 2) confirms the experimental results. Our experiment reveals the possibility to coherently control intersubband absorption. Furthermore we are able to determine intersubband relaxation and dephasing times in a single experiment.

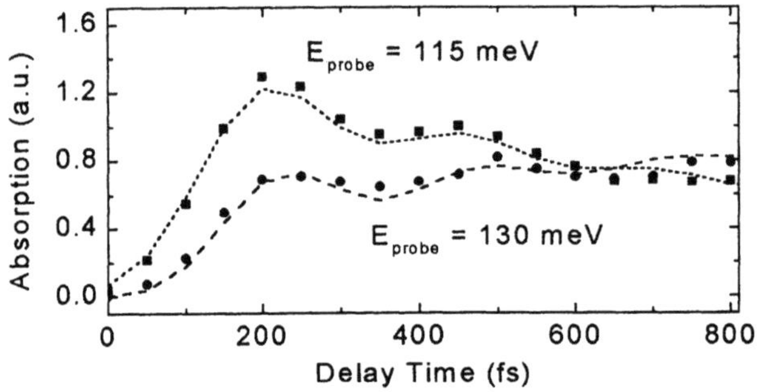

Fig. 2. Absorption at two different probe energies as a function of time delay after excitation (symbols: measurement, lines: density-matrix calculation).

Acknowledgements. We would like to acknowledge financial support by the Fond zur Förderung der Wissenschaftlichen Forschung (SFB-ADLIS).

References

1 R. Huber, F. Tauser, A. Brodschelm, M. Bichler, G. Abstreiter, and A. Leitenstorfer, Nature **414**, 286, 2001.

2 T. Müller, R. Bratschitsch, G. Strasser, and K. Unterrainer, Appl. Phys. Lett. **79**, 2755, 2001.

3 R. A. Kaindl, D. C. Smith, and T. Elsaesser, Opt. Lett. **23**, 861, 1998.

4 R. Huber, A. Brodschelm, F. Tauser, and A. Leitenstorfer, Appl. Phys. Lett. **76**, 3191, 2000.

5 J. N. Heyman, R. Kersting, and K. Unterrainer, Appl. Phys. Lett. **72**, 644, 1998.

6 J. T. Kindt and C. A. Schmuttenmaer, J. Chem. Phys. **110**, 8589, 1999.

7 M. C. Beard, G. M. Turner, and C. A. Schmuttenmaer, Phys. Rev. B **62**, 15764, 2000.

Quantum Interference between Real and Virtual Excitations

Y. H. Ahn[1], D. S. Kim[1], S. T. Cundiff[2], and J. M Shacklette[2]

[1] School of physics, Seoul National University, Seoul 151-742 Korea
E-mail: denny@phya.snu.ac.kr
[2] JILA, University of Colorado and National Institute of Standards and Technology, Boulder, CO 80309-0440

Abstract. Quantum interference between real and virtual polarizations is demonstrated in the time domain. By two-color pulse shaping, we simultaneously create virtual and real polarizations for excitions in GaAs quantum wells. Population modulation in time-domain results from the interference between the virtual and real amplitudes, and the modulation depth reveals the relative contributions of these two amplitudes. The fact that virtual and real amplitudes have a phase difference of 90-degree is demonstrated directly in time-domain.

1. Introduction

Off-resonant light excitations result in finite excited-state probability amplitude and polarization so long as the external perturbation persists. These *virtual* polarizations are responsible for the index of refraction and hence such well-known phenomena as the bending of visible light by a glass prism. At far off-resonance, the virtual polarization is inversely proportional to detuning, whereas the absorption is inversely proportional to the square of the detuning[1]. Thus, the virtual polarization survives, as a function of detuning, long after the absorption becomes practically zero. Here we report on the interplay between the virtual polarizations and the *real* (resonant) polarizations in semiconductors. Using the pump-probe (PP) and four-wave mixing (FWM) combined with pulse shaping technique, we precisely measure the virtual state amplitude normalized to that of the resonant excitation, and demonstrate in time domain that indeed, virtual polarizations are 90 degrees out of phase from the resonantly excited polarizations.

2. Experimental Results

Shown in Figure 1 is the experimental setup where by pulse shaping, we simultaneously excite the real and virtual polarizations (Fig. 1a), and measure their beating. To realize this idea, we performed femto/picosecond nondegenerate FWM [2] and PP in GaAs bulk and quantum wells (Fig. 1b). A two-color picosecond pump pulse from the pulse shaper excites both the resonant (ω_0) and off-resonant (ω_1) components simultaneously. Using the unshaped, twenty femtosecond probe

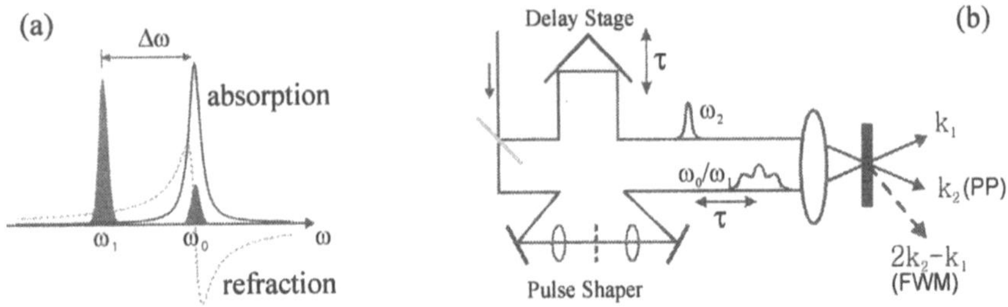

Fig. 1. (a) Schematic of experiments. One picosecond pulse excites below the exciton, and another collinear pulse excites the level resonantly. (b) Schematics of the setup.

pulse, which is much shorter than the picosecond two-color pump pulse, we observed oscillations in both polarization (TI-FWM) and population (PP) (Fig 2a).

The oscillations are essentially from the beating between the virtual and real polarizations. The depth of oscillations depends critically on the relative intensity of ω_0 and ω_1 components. Note that there is an approximately 90 degree phase difference between the cross-correlation (dotted lines), and PP and FWM (Fig. 2a).

These phase difference of 90 degrees can be explained simply by the fact that there exists a 90-degree phase difference between the real and the virtual amplitudes. The excited state population in our experiments, neglecting the phase relaxation and the envelope function, can be approximated by:

$$FWM, PP \approx \left| \frac{E_0 e^{i\omega_0 t}}{i\Gamma} + \frac{E_1 e^{i\omega_1 t}}{\Delta\omega} \right|^2 = \left(\frac{E_0}{\Gamma} \right)^2 + \left(\frac{E_1}{\Delta\omega} \right)^2 + \frac{2E_0 E_1}{\Gamma\,\Delta\omega} Sin\,\Delta\omega\, t \tag{1}$$

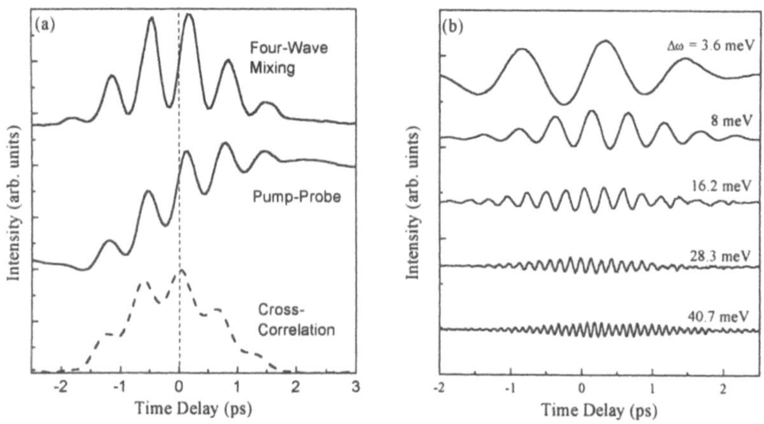

Fig. 2. (a) FWM, PP, CC data from up to bottom with the detuning Δω of 5.8 meV. (b) Oscillation parts in FWM signals as a function of detuning of the off resonant component of the pump pulse.

while the steady state pulse shape is given by:

$$CC \approx \left|E_0 e^{i\omega_0 t} + E_1 e^{i\omega_1 t}\right|^2 = E_0^{\ 2} + E_1^{\ 2} + 2E_0 E_1 Cos\, \Delta\omega t \quad (\Delta\omega = \omega_0 - \omega_1) \tag{2}$$

In the case of "virtual-virtual" beating where ω_0 component is slightly off-resonant we can see the oscillation that is nearly in phase with CC. This is the first direct demonstration of the off-resonant and resonant polarizations being out of phase by 90 degrees.

We also note that the beating period goes precisely as $2\pi/\Delta\omega$ as shown in Fig. 2b. Our experiments show in addition that the virtual polarization normalized to the resonant polarization decreases with detuning as $1/\Delta\omega$ while maintaining its phase. So in such a way we can see how the virtual probability amplitude falls off.

For comparison, we have numerically solved the Optical Bloch equations (OBEs) for a two level system using a shaped first pulse to reproduce the experiments. The calculations are performed by expanding the optical fields and OBEs in a spatial Fourier series and numerically integrating the resulting coupled differential equations. The results are shown in Fig. 3. The calculated FWM signal and excited state population (to simulate pump-probe) show the same 90-degree phase with respect to the cross-correlation as occurs in the experiment.

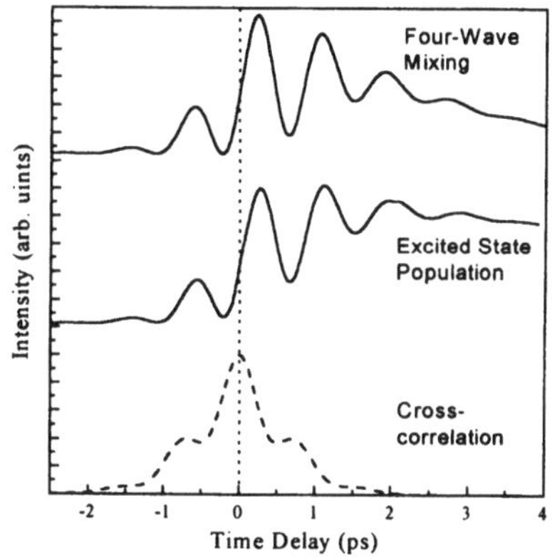

Fig. 3. Numerical calculations for a 4.8 meV.

4. Conclusions

We have demonstrated the first controlled virtual-real quantum interference and shown that there does exist a 90 degree phase difference between the real and virtual polarizations and that virtual polarization decreases with the detuning. Our experiments agree well with theory. Experiments in an atomic vapor are in progress as well as those for above-band-gap excitations in semiconductors.

References

1. See, M. Sargent, M. O. Scully, and W. E. Lamb, *Laser physics* (Addison-Wesley, USA, 1974).
2. S. T. Cundiff et al., "Optical Coherence in Semiconductors: Strong Emission Mediated by Nondegenerate Interactions", Phys. Rev. Lett. **77**, 1107 (1996).

Coherent spin-lattice interaction in a ferromagnetic metallic film

Jean-Yves Bigot, Luca Guidoni, Valérie Halté

Laboratoire IPCMS-GONLO, 23 Rue du Loess 67037 Strasbourg, France
e-mail: Bigot@ipcms.u-strasbg.fr

Abstract. A coherent transfer of energy between the spin and lattice waves in a ferromagnetic film is reported. The associated oscillating magnetization dynamics, with a period of 6.8 ps, is observed with time resolved polarimetric measurements.

A few years ago, it has been demonstrated that it is possible to demagnetize ferromagnetic metallic thin films using femtosecond optical pulses [1]. The magnetization dynamics of Ni and CoPt metallic films was then investigated with time resolved Kerr measurements as a function of applied magnetic field and laser intensity when the materials undergoes its ferro-paramagnetic phase transition. These observations have been confirmed with a variety of femtosecond techniques like for example the second order nonlinear Kerr effect [2] or spin resolved two photon photoemission [3]. More recently, we have shown that, under a weak perturbation of the magnetization with pulses of 20 fs duration, the initial ultrafast spin dynamics follows the thermalization of the electronic populations to a Fermi-Dirac distribution [4]. In this paper, we report the first observation of a coherent exchange of energy between the spins and the lattice while both populations are cooling down.

We have investigated the time dependent magneto-optical response of magnetized CoPt thin films using a pump-probe polarimetric technique. The samples are epitaxially grown on a (0001) oriented sapphire substrate. The 21.5 nm film is made of consecutive one Co and three Pt monolayers. The resulting spontaneous magnetization is perpendicular to the plane of the film. The rotation and ellipticity of a linearly polarized probe is analyzed, for the two opposite remnant magnetization directions ±M, in the following way. The probe intensity $I_{\pm M}$ is measured as a function of the analyzer angle α, which is set perpendicular to the incident probe polarization. For small angles α one can obtain:

$$I_{\pm M} \approx I_0\left[\alpha^2 - 2\alpha(\alpha_0 + \theta_0 \pm \theta_M) + (\alpha_0 + \theta_0)^2 + \varepsilon_0^2 + \theta_M^2 + \varepsilon_M^2 \pm 2(\alpha_0\theta_M + \theta_0\theta_M + \varepsilon_0\varepsilon_M)\right] \quad (1)$$

where Θ_M (resp. ε_M) is the magneto-optical rotation (resp. ellipticity), Θ_0(resp. ε_0) is a static residual rotation (resp. ellipticity) induced by the experimental apparatus and α_0 an arbitrary angular offset. The same expression holds both for the Kerr (ε_{MK}, θ_{MK}) and Faraday (ε_{MF}, θ_{MF}) configurations. The effect induced by the pump pulse is taken into account via the time dependent rotation and ellipticity $\theta_M(t)$ and $\varepsilon_M(t)$. In Figure 1a) we have represented the parabola $I_{\pm MF}(\alpha,t)$ for the pump-probe delays t = –600 fs and t = 700 fs. The negative delay then corresponds to the static magneto-optical response since no permanent thermal effect occurs owing to the low repetition rate of the laser system (5 kHz amplified Ti:S; pump: 100 fs, 800 nm; probe : chirp compensated continuum 500-700 nm; $I_{pump}/I_{probe} > 100$). The shift of the parabola contains the full information about the magnetization. The rotation can be directly inferred from equation 1 noticing that: $(I_{+M}-I_{-M})/4I_0 = -\Theta_M\alpha + \Theta_M(\Theta_0 + \alpha_0) + \varepsilon_0\varepsilon_M$. This quantity is displayed in Figure 1b) for several delays. For increasing delays, the slope first increases and then decreases ($t > 700$ fs). The corresponding rotation $\Theta_{MF}(t)$ is represented in Figure 2a). A striking feature of Figure 1b) is that all lines cross at the same point. An immediate consequence is that the ellipticity and rotations have the same relative temporal dependency, i.e. $\Theta_{MF}(t)/\varepsilon_{MF}(t) = C^{te}$. Therefore any of these two quantities represents the magnetization dynamics. This confirms that the time dependent magneto-optical response is dominated by the spins (via the magnetization) and not by the electronic charges, in contrast to recent claims [5].

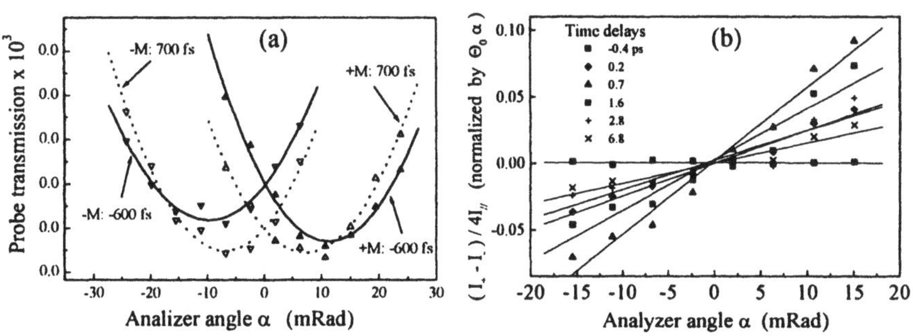

Figure 1. a) Probe intensity $I_{\pm M}$ as a function of the polarimetric angle α for two pump-probe delays: -600 fs (fit in full line) and +700 fs (fit in dotted line)
b) $(I_{+M}-I_{-M})/4I_0 = -\Theta_M\alpha + \Theta_M(\Theta_0 + \alpha_0) + \varepsilon_0\varepsilon_M$ obtained for different pump-probe delays. The magneto-optical rotation $\Theta_M(t)$ can be extracted from the slope of the lines

The maximum relative variation $\Delta\theta/\theta = \Delta\varepsilon/\varepsilon$ in figure 2a) is of 33%. For such conditions of pump excitation the partial demagnetization process is reversible. During the recovery of the magnetization well contrasted oscillations occur on the magneto-optical signal as seen in Figure 2b). The period of these oscillations is ~6.8 ps. It shows that a long lived coherent process affects the spins dynamics. These oscillations also manifest in the differential transmission and reflectivity, although much less contrasted and rapidly damped, as shown in the inset of Figure 2b). The origin of this coherent process may be due to different mechanisms. A first one would involve a resonant contribution of the lattice waves (acoustic phonons), coupled both to the electrons and the spins. Another possibility is a resonant contribution of the spin waves coupled to the lattice, which would also affect the electrons. The present observations may have important consequences in the context of the ultrafast control of magneto-optical devices.

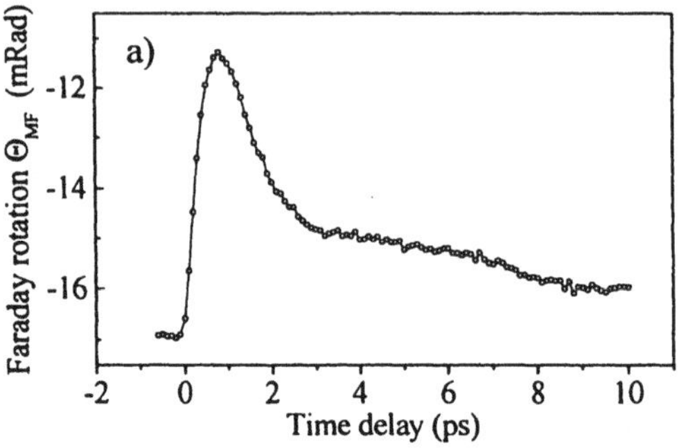

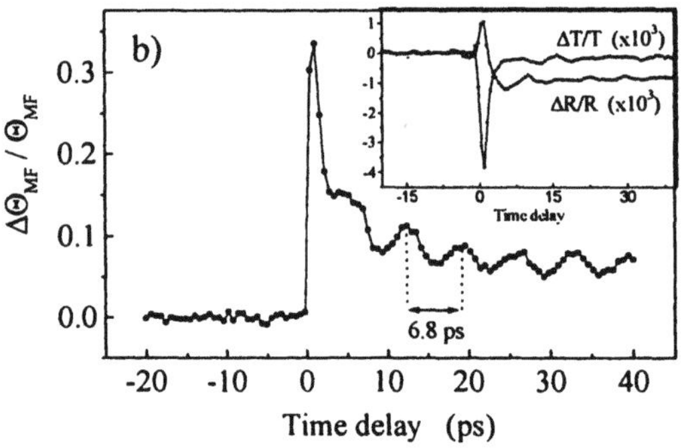

Figure 2. a) Time evolution of the polarimetric angle $\Theta_{MF}(t)$ during the first 10 ps
b) Oscillatory behavior of the magnetization during the magnetization recovery
Inset: differential transmission and reflectivity signals

We thank J. ARABSKI, F. HUBER and M. ALBRECHT for their technical support.

References

1 E. Beaurepaire, J.-C. Merle, A. Daunois, and J.-Y. Bigot, *Phys. Rev. Lett.* **76**, 4250 (1996); E. Beaurepaire, M. Maret, V. Halté, J.-C. Merle, A. Daunois, and J.-Y. Bigot, *Phys. Rev. B* **58**, 12134 (1998).
2 J. Hohlfeld, E. Matthias, R. Knorren, K. H. Bennemann, *Phys. Rev. Lett.* **78**, 4861 (1997).
3 A. Scholl, L. Baumgarten, R. Jacquemin, W. Eberhardt, *Phys. Rev. Lett.* **79**, 5146 (1997).
4 L. Guidoni, E. Beaurepaire, and J.-Y. Bigot, *Phys. Rev. Lett.* **89**, 17401 (2002)
5 B. Koopmans, M. van Kampen, J. T. Kohlhepp and W. J.M. de Jonge *Phys. Rev. Lett.* **85**, 844 (2000)

Direct Visualization of the Gouy Phase Shift

Nikolay S. Stoyanov, T. Feurer, David W. Ward, and Keith A. Nelson

Dept. of Chemistry, Massachusetts Institute of Technology, Cambridge, MA 02139, USA
E-mail: kanelson@mit.edu

Abstract. Using an excitation pulse in the form of a ring we generate a phonon-polariton wavepacket that propagates inward toward a focal point at the center. This results in a sharp increase in the polariton intensity. It also permits direct visualization and quantitative measurements of the Gouy phase shift and of anisotropy in polariton propagation.

The Gouy phase plays an important role whenever focusing of a wave to the diffraction limit is considered, as in resonators, optical tweezers, or phase-sensitive nonlinear interactions of light with matter in the focus of a high intensity beam. With the advent of single cycle and even sub-cycle THz pulses, interest in the Gouy phase was revived because it is responsible for dramatic changes of the amplitude of those pulses as they pass through the focus. To date, the Gouy phase has only been indirectly observed. The first realization was performed by Gouy himself using white light interferometry [1]. Later experiments used microwaves to infer the existence of the Gouy phase [2,3]. More recent experiments with single cycle THz pulses have confirmed the predicted amplitude changes caused by the Gouy phase for cylindrical [4] and spherical [5] single cycle pulses.

We report direct observation of the Gouy phase shift through real-space imaging of a phonon-polariton wavepacket passing through a focus [6]. The curved wavefront is generated through spatial semicircular excitation and is therefore aberration-free and the phase front is parallel to the pulse front. The spatiotemporal evolution recorded by a femtosecond imaging pulse at different times following the excitation allows direct visualization of the phase fronts as the wavepacket moves through the focal region, thereby revealing the phase anomaly through direct observation. The experiments were performed using a home-built femtosecond Ti:sapphire laser system with a repetition rate of 1 kHz, pulse width of 50 fs, and energy of 0.7 mJ. The laser output was split into a pump and a variably delayed imaging probe pulse. The pump pulse was expanded, sent through an axicon-lens combination that yielded a ring-like focal intensity distribution, and imaged onto an x-cut 5 mm thick $LiTaO_3$ single crystal. The excitation ring diameter at the sample was 2.3 mm with a thickness of 0.20 mm. The pulse energy at the sample was adjusted to 60 μJ, leading to an excitation intensity of 83 GW/cm^2. For semicircular excitation, half of the ring was blocked. A phonon-polariton mode represents an eigenmode of an ionic crystal which has both an electromagnetic and a polar lattice vibrational contribution, which has a frequency in the THz region. In $LiTaO_3$, which is a uniaxial crystal, phonon-polaritons may be generated coherently by femtosecond laser pulses via impulsive stimulated Raman scattering (ISRS) with the optical and the phonon-polariton polarizations aligned parallel to the crystal axis [7]. The semicircular excitation pattern remains well collimated throughout the sample, so the geometry is essentially two-dimensional, i.e. the phonon-polariton propagation component of interest is parallel to the crystal surface and perpendicular to both the excitation and the imaging beam. The coherent phonon-polariton response modulates the refractive index through the electro-optic effect, and a variably delayed wide probe beam illuminating essentially the whole crystal is used to image the resulting phase modulation using the Talbot effect [8].

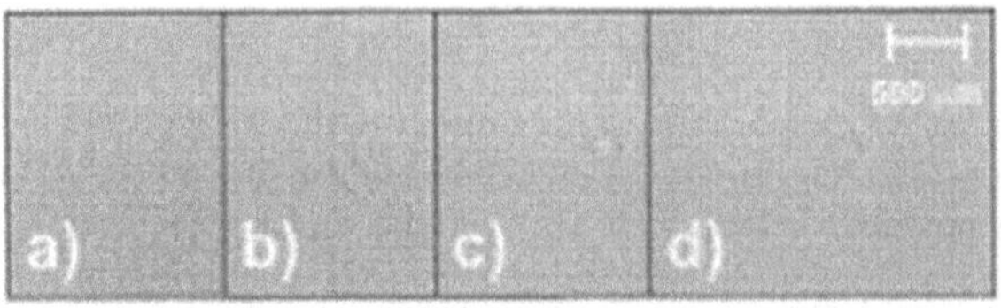

Figure 1. Semicircular phonon-polariton wavepacket propagating from left to right at different probe time delays. The images are recorded at a) 12.6 ps, b) 18.9 ps, c) 25.6 ps, and d) 32.3 ps after the excitation pulse.

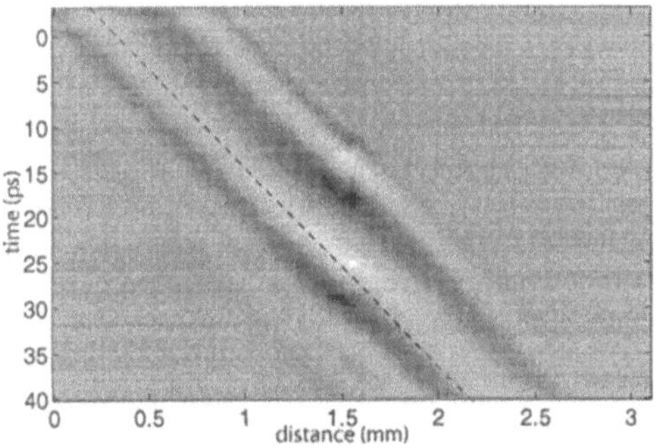

Figure2. The time axis corresponds to a particular probe delay and each horizontal line is constructed by averaging around a 36 deg angle with respect to the propagating axis. The dashed line which initially follows a peak falls between a peak and a trough after the focus indicating a phase shift of $\pi/2$.

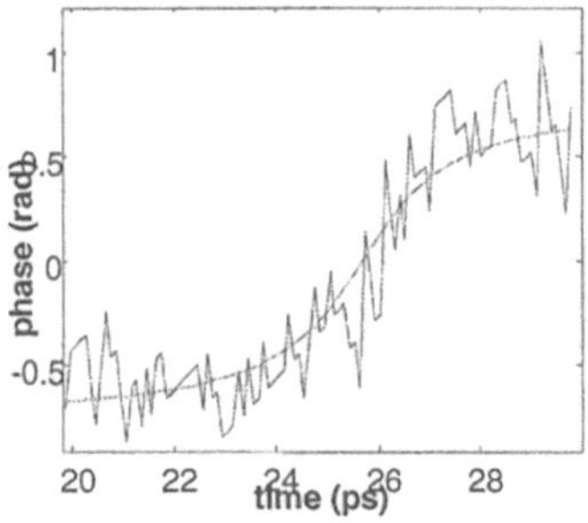

Figure 3. Plot of the experimental and calculated phase change versus distance as the polariton ripple travels through the focus indicating the $\pi/2$ phase shift.

Fig. 1 shows a sequence of snapshots taken at different delay times between the pump and the imaging pulse. The phase and the pulse front curvature are solely determined by the spatial profile of the excitation laser light at the sample, and are almost perfectly circular. The initial phonon-polariton amplitude, however, is maximal when the generated wavefront is parallel to and minimal when it is perpendicular to the excitation polarization direction. The central wavelength of the phonon-polariton response is about 165 μm and the beam waist at the focus is approximately 55 μm. The phonon-polariton passes through the focus about 25.6 ps after excitation.

Extracting the phonon-polariton amplitude along the propagation axis from each image and stacking the resulting horizontal lines vertically yields the spatiotemporal amplitude distribution shown in fig. 2. As the wavepacket passes through the focus, the peaks and valleys of the phase accelerate, propagate with superluminal phase velocity, decelerate again, and finally continue to propagate with the normal phase velocity. If we trace a single phase peak and subtract the phase k_{0z}, the phase acquired due to free space propagation, any anomalous phase contribution should become obvious as shown in fig.3. Clearly, there is an additional anomalous phase acquired as the wavepacket passes through the focal region. This is the Gouy phase. Most of the additional phase is occurring within one or two Rayleigh ranges. The wavepacket launched by the excitation pulse has cylindrical symmetry and is focused in one dimension only, parallel to the optical axis of the crystal. The expected Gouy phase is $\pi/2$ [9], in excellent agreement with experiment. In physical terms, the Gouy phase causes the propagation constant within the focal region to decrease and the wavelength to increase. Consequently, the phase fronts of the cylindrical wavepacket advance by a quarter of a wavelength as it passes through the focus.

In conclusion, we have shown that through specific excitation geometry it is possible to generate an aberration-free phonon-polariton wavepacket with semicircular phase fronts. The pump-probe imaging scheme allows observation of the phase fronts as they propagate through the focal region and therefore direct visualization of the Gouy phase.

Acknowledgements. This work was supported in part by National Science Foundation Grant No. CHE97-13388. T. Feurer acknowledges financial support by the Max Kade Foundation.

References

1 C.R. Gouy, *Acad. Sci. Paris* **110**, 1251 (1890)
2 M.P. Bachynski, G. Bekefi, *J. Opt. Soc. Am.* **47**, 428 (1957)
3 G.W. Farnell, *Can. J. Phys.* **36**, 935 (1958)
4 R.W. McGowan, R.A. Cheville, D. Grischkowsky,*Appl. Phys. Lett.* **76**, 670 (2000)
5 A.B. Ruffin, J.V. Rudd, J.F. Whitaker, S. Feng, H.G. Winful, *Phys. Rev. Lett.* **83**, 3410 (1999)
6 T. Feurer, N.S. Stoyanov, D.W. Ward, K.A. Nelson, *Phys. Rev. Lett.* **88**, 257402 (2002)
7 T.P. Dougherty, G.P. Wiederrecht, and K.A. Nelson, *Opt. Soc. Am. B* **9**, 2179 (1992)
8 H. F. Talbot, *Philos. Mag.* **9**, 401 (1836)
9 S. Feng, H.G. Winful, *Opt. Lett.* **26**, 485 (2001)

Part IX

Reaction and Solvation Dynamics in Liquids

Exploring sub-picosecond dynamics in the far-infrared with THz spectroscopy

Charles A. Schmuttenmaer, Matthew C. Beard, and Gordon M. Turner

Yale University, Department of Chemistry, 225 Prospect St., P. O. Box 208107, New Haven, CT 06520-8107, USA
Email: charles.schmuttenmaer@yale.edu

Abstract. Time-resolved THz spectroscopy has been used to characterize the time-dependent photoconductivity in a variety of semiconductors, ranging from bulk GaAs to colloidal sintered TiO_2. It has also been used to directly measure sub-picosecond intramolecular electron transfer.

1. Introduction

Terahertz spectroscopy emerged about 13 years ago with the demonstration that nearly single cycle pulses of far-infrared (FIR) radiation could be generated, sent through free space, and subsequently detected in the time-domain.[1,2] Since then, THz spectroscopy has become a very active area with studies ranging from condensed matter physics to gas-phase spectroscopy to biomedical imaging. *Time-resolved* THz spectroscopy (TRTS), on the other hand, is much younger. One of the most unique aspects of THz spectroscopy is that the pulses are of sub-picosecond duration, and thus it is a powerful method for characterizing the FIR optical properties of a material, *i.e.*, absorption coefficient and refractive index, on a ultrafast timescales.

We have spent considerable effort over the last few years characterizing and understanding the method of time-resolved THz spectroscopy (TRTS),[3] and have now begun to apply it to novel systems such as low-temperature grown GaAs,[4] semiconductor quantum dots, and sintered colloidal TiO_2. We have also measured the electromagnetic (EM) pulse emitted when a collection of dipolar molecules are partially oriented and coherently excited.[5].

2. Experimental Methods

The THz system has been described in detail previously.[3] It is based on a regeneratively amplified Ti:Sappire laser (1 kHz repetition rate, 100 fs pulse duration, 1 mJ/pulse, 800 nm wavelength). The spectrometer consists of a optical pump pulse at the fundamental of the Ti:Sapphire laser (800 nm), or 400 nm if it is frequency doubled, or anywhere throughout the visible spectrum (at the cost of reduced pulse energies) when employing an optical parametric amplifier. The THz probe pulse has bandwidth from 0.1 to 3 THz (3 to 100 cm^{-1}).

3. Results and Discussion

3.1 Low-temperature grown GaAs

Low-temperature grown GaAs (LT-GaAs) is a material with extremely fast carrier trapping rates, and is used in high speed applications. It is impossible to characterize the mobility of LT-GaAs using conventional techniques such as a four point probe because the material remains conductive for only ~2 to 3 ps after photoexcitation. Optical studies established the timescale for the electron dynamics, but can not determine the mobility of the carriers prior to trapping.

LT-GaAs is grown and annealed in an excess As flux. During the annealing phase, the As atoms form clusters which then act as deep traps. The properties of LT-GaAs are strongly influenced by the anneal time and temperature. Long anneal times and/or high temperatures lead to larger As clusters which are more widely separated. Because the material is grown in an excess As flux, there are initially many As antisites and interstitials. Annealing helps remove these, but shallow defects will remain. Thus, the material is not as good a conductor as regular GaAs, even before the carriers are trapped. The mobility in regular GaAs is ~8500 $cm^2 V^{-1} s^{-1}$.

Figure 1 displays the band structure of LT-GaAs and the time-dependent mobility of the electrons after photoexcitation at 400 nm. Data can only be obtained for ~3 ps after photoexcitation because the carriers are trapped with a time constant of ~ 1.1 ps. It is seen that when exciting with 400 nm light, the mobility increases from 400 $cm^2 V^{-1} s^{-1}$ to 1200 $cm^2 V^{-1} s^{-1}$ as the carriers scatter back and relax in the Γ valley. When exciting close to the bottom of the Γ valley with 800 light, the mobility remains constant at 3000 $cm^2 V^{-1} s^{-1}$ (not shown). This level of detail simply could not be attained with any other experimental technique.

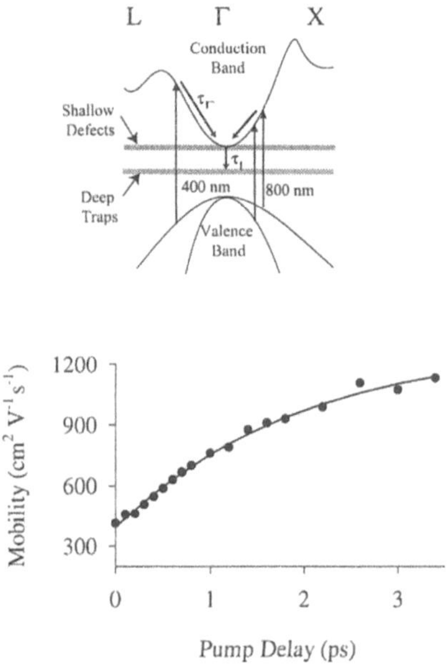

Figure 1. Top: Band structure of LT-GaAs. It is assumed that it is identical to that for regular GaAs, except that it has shallow defects due to interstitials and lattice dislocations, as well as deep traps from the As clusters. Bottom: Time-dependent mobility of the carriers in LT-GaAs after photoexcitation at 400 nm.

3.2 Colloidal TiO_2

Colloidal sintered TiO_2 forms the heart of the so-called "Grätzel cell", a dye-sensitized solar cell that has attracted attention since 1991.[6] Most of the ultrafast spectroscopy

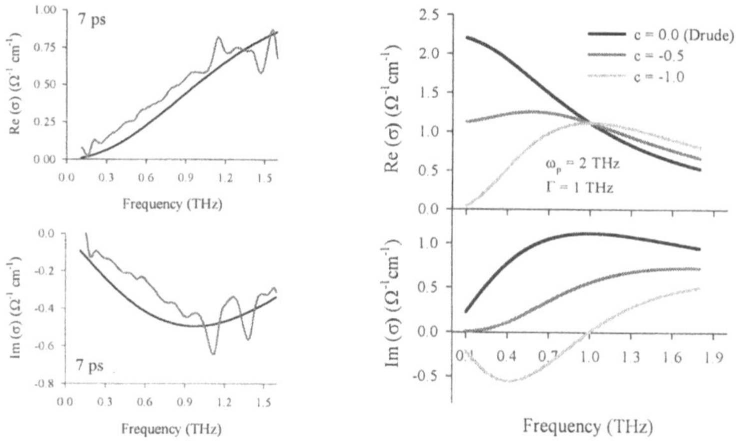

Figure 2. Left: Real and imaginary conductivity of dye-sensitized colloidal TiO_2 7 ps after photoexcitation. Smooth curves are fit, noisier curves are experimental data. Right: Dependence of generalized Drude model on "persistence of velocity" parameter.[9]

on this system has focused on the dye molecule. That is, a sharp reduction in fluorescence quantum yield and lifetime when the molecule is adsorbed to TiO_2 compared to when it is in solution. This behavior is attributed to rapid electron injection upon photoexcitation. We have characterized the photoinjected electron dynamics following excitation by monitoring the transient THz absorption. The only other closely related studies were carried out in the mid-IR region of the spectrum.[7,8] While those studies provide some measure of carrier dynamics, they must *assume* Drude-like behavior of the electrons in that they attribute the transient IR absorption to free carrier absorption by the electrons. This assumption need not be made in our studies because we measure the frequency-dependent complex conductivity, and the FIR region of the spectrum is very sensitive to scattering rates.

One of the most significant results of this work is that the carriers behave in a highly non-Drude manner in colloidal, sintered TiO_2. Figure 2 displays the real and imaginary components of the frequency-dependent conductivity. The analogous properties for a Drude conductor, as well as that based on the modification to the Drude model developed by Smith[9] are also shown in the figure.

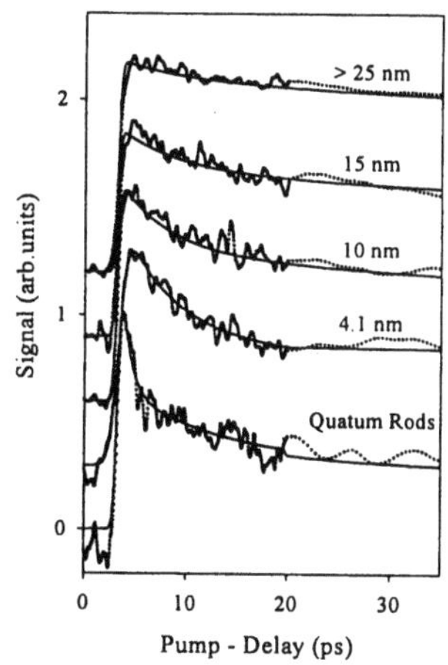

Figure 3. 1D pump delay scans of CdSe quantum dots of various sizes. Smooth curves are triple-exponential fits.

3.3 Quantum Dots

Semiconductor quantum dots (QDs) have received considerable attention over the last 2 decades. One of the most important types of studies involving quantum dots is the determination of how a particular property varies as a function of particle size. The optical band-gap has been the most probed property of semiconductor QDs, however melting temperature, lattice constant, among other things have all been determined for one or more materials. The basic goal of our work is to determine the size-dependence of the "photoconductivity".

Certainly, in large enough particles, the photoconductivity will be identical to that in bulk, and in very small particles it will be nonexistent because there is not enough room for the electrons or holes to move. There should still be THz absorption by small QDs upon photoexcitation, but that is due to the fact that the electron-hole pair is highly polarizable rather than due to transport. In any case, the ability to characterize the time-resolved, frequency-dependent photoconductivity without attaching wires to the sample will allow us to characterize the size-dependent properties of semiconductor QDs.

Figure 3 shows pump-delay scans of several sizes of CdSe QDs. It is seen that the response is strongly size dependent. The fast component in the multiexponential decay is much more pronounced in the smaller particles, which have higher surface to volume ratios.

3.4 Rapid intramolecular electron transfer

When a molecule undergoes intramolecular charge transfer, some fraction of an electron's charge moves from one portion of the molecule to another, and therefore must undergo acceleration. Thus, this process radiates an electromagnetic (EM) pulse. If the charge transfer occurs on a ps timescale, then the radiated pulse will fall in the THz region of the spectrum. If there is a random, isotropic distribution of photoexcited molecules, then the pulses emitted by each of them will interfere destructively and no net signal will be measured. Also, if they are excited continuously, then different molecules will radiate at different times, again leading to no net signal. However, if the molecules are oriented prior to photoex-

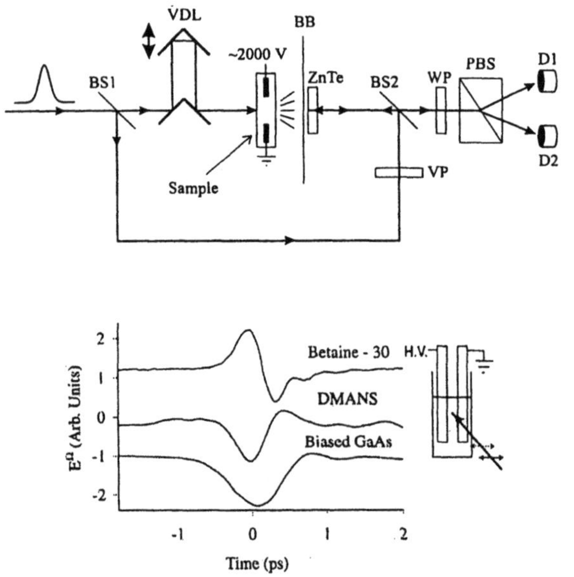

Figure 4. Top: Experimental set up used in rapid ET studies. Electrically biased solution is photoexcited to generate THz pulse, which is read out using free space electro-optic sampling in ZnTe crystal. Bottom: Comparison of pulse generated from Betaine-30 and DMANS (biased GaAs is used for reference). The polarity of the pulse is determined by the direction of charge transfer, and its shape is determined by the forward and back ET rate constants.

citation, and are excited coherently with a short laser pulse, then the individual emissions will add together constructively resulting in a signal that can be measured.[5]

The apparatus is similar to those described in Ref. [3], and is shown in Figure 4. One portion of the visible pulse is used to photoexcite the sample, and the other portion is used to gate the detector. The direction of charge transfer determines the polarity of the pulse, and since we detect the amplitude, not the intensity of the THz pulse, we determine the direction of charge transfer unambiguously. Furthermore, the dynamics of forward and back electron transfer affect the shape of the waveform emitted from the sample as long as they are on a timescale of 0.1 to 10 ps.

4. Conclusions

We will continue to pursue *time-resolved* THz spectroscopy because this aspect of THz spectroscopy is truly unique in that there simply does not exist any other way to generate subpicosecond FIR pulses. We have shown that it is possible to characterize transient photoconductivity in bulk semiconductors, nanocrystalline TiO_2, and CdSe quantum dots. Furthermore, charge transfer reactions that occur on a ps to sub-ps timescale can be studied in a very direct manner by detecting the EM pulse that accompanies the charge transfer event.

Acknowledgements. Partial support from the Sloan Foundation and from the National Science Foundation CAREER award program (CHE-9703432) is gratefully acknowledged.

References

1. C. Fattinger and D. Grischkowsky, Appl. Phys. Lett. **54,** 490 (1989).
2. P. R. Smith, D. H. Auston, and M. C. Nuss, IEEE J. Quantum Electron. **24,** 255 (1988).
3. M. C. Beard, G. M. Turner, and C. A. Schmuttenmaer, Phys. Rev. B **62,** 15764 (2000).
4. M. C. Beard, G. M. Turner, and C. A. Schmuttenmaer, J. Appl. Phys. **90,** 5915 (2001).
5. M. C. Beard, G. M. Turner, and C. A. Schmuttenmaer, J. Phys. Chem. A **106,** 878 (2002).
6. B. O'Regan and M. Gratzel, Nature **353,** 737 (1991).
7. H. N. Ghosh, J. B. Asbury, and T. Q. Lian, J. Phys. Chem. B **102,** 6482 (1998).
8. T. A. Heimer and E. J. Heilweil, J. Phys. Chem. B **101,** 10990 (1997).
9. N. V. Smith, Phys. Rev. B **6415,** 5106 (2001).

THz Emission from Charge-Transfer Reactions in Molecules Aligned in Solutions and Crystals

John J. Carey, David Jones, Steven P. Jamison and Klaas Wynne

Department of Physics, University of Strathclyde, Glasgow, Scotland, UK, Email: john.carey@strath.ac.uk, URL: http://dutch.phys.strath.ac.uk/FRC/

Abstract: A detailed experimental and theoretical study of THz-pulse generation by photo-induced charge transfer in solution and crystals will be presented. In addition, a novel class of molecular crystals will be introduced with optical rectification efficiency superior to ZnTe.

The generation, detection, and application of sub-picosecond pulses in the tera-hertz (THz) region of the electromagnetic spectrum have become an area of intense research activity in the past few years. We have performed new comprehensive experimental and theoretical studies of THz generation by intra-molecular charge transfer (CT) in solution and molecular crystals. Some of the molecular crystals studied exhibit optical-rectification efficiencies and bandwidths exceeding that of the commonly used ZnTe generators. A new analytical model based on the linear optical properties of the generator, optical system, and detector allows for full prediction of THz pulse shapes in all our experiments.

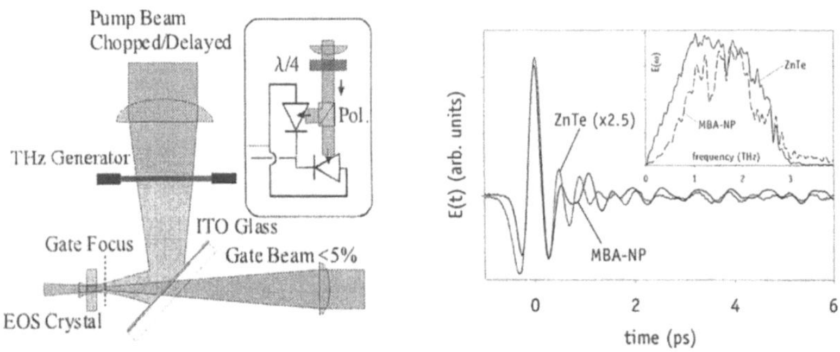

Fig. 1. (left) Layout of the set-up used in the experiments. (right) THz pulses produced by optical rectification in MBANP (0.2 mm) and ZnTe (0.5 mm) in the time and frequency domain.

Recent studies on semiconductors [1] and ultrafast CT in solution [2] have shown that the detailed measurement of THz pulse shapes can be used in principle to extract the ultrafast dynamics of microscopic currents. We have studied THz emission by photo-induced CT in Reichardt's dye in a variety of solvents resulting in CT on time scales ranging from 500 fs to 5 ps. In addition, THz emission due to photo-induced CT in molecular crystals of (-)-2-(α-methylbenzylamino)-5-nitropyridine (MBANP) and analogues have been studied in a preliminary attempt to measure a three-dimensional path of electron transfer. A new analytical model has been developed incorporating the microscopic dynamics of the photo-induced current, phase matching in the generating medium, (near field) propagation, and detection of electromagnetic waves by electro-optic sampling (EOS). This model can be used to obtain a thorough understanding of the potential of the technique for measuring ultrafast CT in various systems of interest.

The configuration for detection and generation in all experiments is shown in Fig. 1. An 800 or 400-nm, 40-fs, 0.5-mJ p-polarised pump beam generating the THz light is focused (f = 10 cm) through the sample. The emitted THz beam will then focus in the same location as the pump beam during alignment. The THz beam reflects off an ITO-coated dichroic beamsplitter and focuses in a ZnTe detection crystal. The THz field was detected by standard EOS [3] and it can be shown that in this arrangement the THz pulse is measured in the far field.

To study THz emission from CT in liquids it is essential to align the molecules with a high voltage DC bias. We have developed a novel pulsed-voltage supply that can supply up to 40-kV 100-ns pulses. This allows the application of high bias fields without arcing or corona discharge. In crystalline form, the molecules are static and require no alignment. To study THz emission due to CT in MBANP, the crystal was excited at 400 nm. The MBANP molecules absorb all of the blue light and the polarisation change caused by the subsequent CT results in a strong THz signal (Fig. 2).

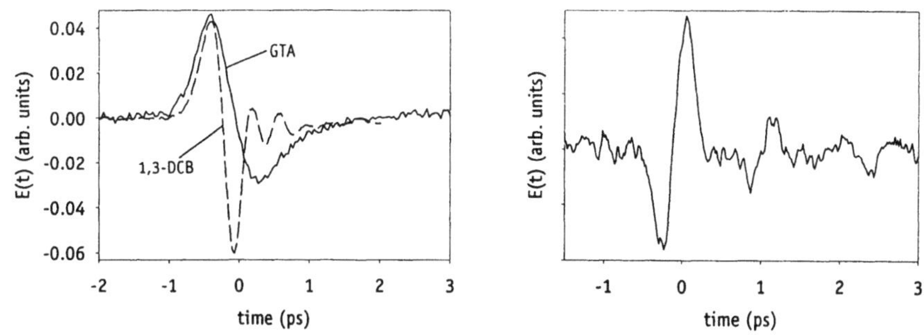

Fig. 2. (left) THz pulses generated by photo-induced CT in Reichardt's dye in 1,3-dichlorobenzene and glycerol triacetate. (right) *Idem*, in an MBANP crystal excited at 400 nm.

The inorganic crystal ZnTe is typically used for THz generation if broad bandwidth pulses are essential but high power is not required. Organic molecular crystals have been proposed as THz generators and have been reported to generate stronger fields than ZnTe [4]. However, all the current organic THz generators fail to match the quality of ZnTe for THz applications. Optical rectification of 800-nm femtosecond pulses in the organic crystal MBANP and analogues generates THz pulses that are stronger and more broadband than those from ZnTe (Fig. 1). MBANP has the added advantage of being easy to grow [5] and could potentially be grown in large sizes for high-power applications. In our experiments, 200-μm long <001>-cut MBANP crystals were used and compared to 500-μm long <110>-cut ZnTe crystals. We estimate the field strength of THz pulses generated in MBANP to be five times that from ZnTe. The spectrum of MBANP has a slight dip at 1.3 THz but otherwise has more bandwidth than ZnTe.

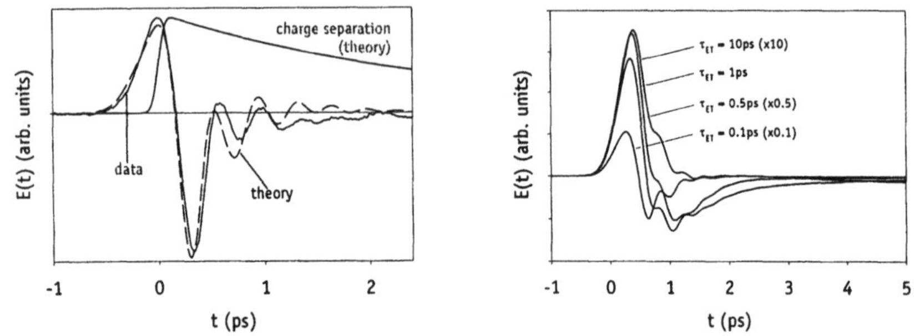

Fig. 3. (left) Comparison between experimental THz signal generated by direct CT in Reichardt's dye in chloroform and theory (80-fs pulses at 800 nm, 2.8-ps electron-transfer time, 0.5-mm ZnTe EOS crystal). (right) Theoretical simulation of the THz signal produced by indirect CT in chloroform solution for various rates (80 fs, 0.5-mm ZnTe EOS crystal, 1-mm sample).

An analytical model has been developed to describe the generation, propagation, and detection of THz pulses by a change in microscopic polarisation. This means it can be applied to all current THz-pulse generation media when the appropriate response function is used. The basic premise is of an intensity envelope (the femtosecond pump pulse) travelling through the THz generator creating a nonlinear polarisation along the way. The equation includes the dielectric response of the medium (obtained from THz spectra), dispersion, and group/phase-mismatch of the exciting and generating fields as well as propagation to the far field. Analytical equations are applied in the frequency domain and a final (numerical) Fourier transform provides one with a predicted THz pulse shape. Appropriate response functions for generation by CT in solutions and crystals as well as for photoconductive antennas and optical rectification have been derived. This model can be used to understand and analyse the data as well as predict signal shapes that may be measured in future experiments (Fig. 3).

Acknowledgements. Financial support from the Royal Society, the Petroleum Research Fund, and EPSRC are gratefully acknowledged.

References

[1] A. Leitenstorfer, S. Hunsche, J. Shah, M. C. Nuss, and W. H. Knox, Phys. Rev. B **61**, 16642 (2000).

[2] M. C. Beard, G. M. Turner, and C. A. Schmuttenmaer, J. Phys. Chem. A **106**, 878 (2002).

[3] J. J. Carey, J. Zawadzka, D. A. Jaroszynski, and K. Wynne, Phys. Rev. Lett. **84**, 1431 (2000).

[4] P. Y. Han, M. Tani, F. Pan, and X.-C. Zhang, Opt. Lett. **25**, 675 (2000).

[5] R. T. Bailey, G. Bourhill, F. R. Cruickshank, D. Pugh, J. N. Sherwood, G. S. Simpson, and S. Wilkie, Mol. Cryst. Liq. Cryst. **231**, 223 (1993).

Directly measuring the coupling of solvent intermolecular modes in solvation

Sungnam Park, Jeongho Kim, and Norbert F. Scherer

Department of Chemistry, University of Chicago, Chicago, IL 60637
E-mail: nfschere@uchicago.edu

Abstract. Resonant pump off-resonant probe OKE spectroscopy was performed to study the coupling of solvent molecular modes to photoexcitation of the solute. The intermolecular solvent modes of a range of polar and polarizable solvents including $CHCl_3$ and CH_3CN are coupled in distinct to the electronic excitation of Coumarin 153.

1. Introduction

Polarization response spectroscopy (PORS), a resonant pump off-resonant probe optical Kerr effect (two-color OKE) spectroscopy, was employed to investigate the coupling of solvent molecular modes to photoexcitation of the solute. In this spectroscopy, the pump field is electronically resonant with the solute and off-resonant with solvent and the probe field is electronically off-resonant with both solute and solvent. The pump field (i) excites the solute and (ii) induces solvent molecular motions. The induced solvent motions are coherently driven as a result of their coupling to the new electronic structure resulting from photoexcitation of the solute. The modes observed in OKE experiments of symmetric top (non-hydrogen bonded) solvents can be classified by the symmetry of the motion into diffusive reorientational, interaction-induced, librational, and intramolecular oscillatory motions. Diffusive reorientation is an individual molecular orientational motion occurring on a picosecond time scale. Interaction-induced is translational motion arising from molecular density fluctuations. Libration is a hindered molecular rotation in the intermolecular potential.[1] These solvent modes, which are Raman-active, can couple in a mode-specific manner to the photoexcitation of the solute. Related experiments have been reported that have concentrated on determining all tensor elements of $\chi^{(3)}$[2, 3] as opposed to establishing the molecular couplings as in the present case.

2. Experimental Methods

Femtosecond pulses centered at 800 nm and of ~ 25 fs duration were generated from a home-built cavity-dumped Ti:Sapphire oscillator.[4] The 400 nm second harmonic pump pulse was generated with a 200 μm thick BBO crystal. The FWHM of the cross-correlation obtained and fitted to a Gaussian temporal profile was ~ 85 fs including group velocity mismatch. Coumarin 153 (C153) was used as a solute molecule because its solvation dynamics have been characterized in several organic solvents[5] and does not exhibit a transient absorption of the 800 nm probe pulse. The chemical perturbation results from a large change in dipole moment from ~ 6 D to ~14 D upon photoexcitation.[6]

3. Results and Discussion

Two-color OKE measurements were performed with different concentrations of C153 in $CHCl_3$ and CH_3CN (and other solvents) as shown in Fig.1. The negative long decay component due to the reorientational motion of C153 was fit with an exponential function and subtracted to obtain "reduced" PORS transients. The reduced PORS signals measured in solutions of varying solute concentrations were normalized by taking into account the pump intensity attenuation by C153 absorption. Polarizability susceptibilities shown in the center panels of Fig. 1 are the imaginary part of the Fourier transformation of reduced PORS transients divided by Fourier transformation of the cross-correlation function. The low frequency region (0 ~ 250 cm^{-1}) contains information on diffusive reorientational, interaction-induced, and librational motions.

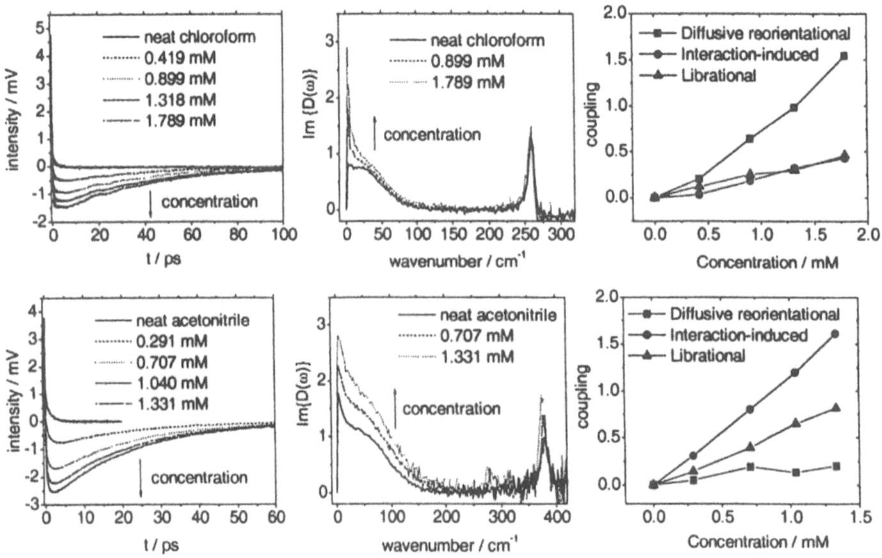

Fig. 1. Two-color OKE signals of C153 in $CHCl_3$ or CH_3CN (left panels), polarizability susceptibilities (center panels) and coupling (($Area_{solution}$-$Area_{solvent}$)/$Area_{solvent}$) (right panels) are plotted in $CHCl_3$ (upper panels) and CH_3CN (lower panels).

The amplitudes in the low frequency region increase with solute concentration reflecting enhancement of the solvent motions. The enhancement in $CHCl_3$ and CH_3CN is quite different indicating that the solvent modes in each solvent are coupled to the photoexcitation of C153 in distinct ways. A peak near 0 wavenumbers grows greatly in $CHCl_3$ with solute concentration while in CH_3CN the overall shape looks the same but the amplitude and width of the low frequency region increase. The peaks in higher frequency region in Fig.1 correspond to intramolecular vibrational modes of $CHCl_3$ (260 cm^{-1}) and CH_3CN (380 cm^{-1}) observed in the measurement. There appears to be no significant change in the $CHCl_3$ vibrational mode but that of CH_3CN may be increasing.

More detailed information on which solvent motions are strongly coupled to the photoexcitation of the solute is obtained by lineshape analysis. The low frequency region is assumed to be described by terms that represent diffusive reorientational, interaction-induced, and librational motions.[7] All parameters are determined by nonlinear least square fitting. Areas under the fitted curves are calculated and coupling plotted against the concentration as shown in Fig. 1. In $CHCl_3$, the diffusive reorientational motion is strongly coupled to the changed electronic structure of C153 in S_1 while in CH_3CN, the interaction-induced motion is strongly coupled showing that solvent specificity of the interaction.

PORS is complementary to time-resolved fluorescence Stokes shift and photon echo spectroscopies. The PORS approach does not depend on chromophore fluorescence[5] and will be applied to photodissociation with non-emissive reaction products. It is better suited to examine strongly interactive solute-solvent systems than echo methods. [8]

References

1 D. McMorrow and W. T. Lotshaw, J. Phys. Chem. 95, 10395-10406 (1991).
2 G. D. Goodno, V. Astinov and R. J. D. Miller, J. Phys. Chem. B 103, 603-607 (1999).
3 P. Vohringer and N. F. Scherer, J. Phys. Chem. 99, 2684-2695 (1995).
4 Y.-H. Liau, A. N. Unterreiner, D. C. Arnett and N. F. Scherer, Applied Optics, 38, 7386-7391 (1999).
5 M. L. Horng, J. A. Gardecki and M. Maroncelli, J. Phys. Chem. 99, 17311-17337 (1995).
6 F. Cichos, R. Brown and P. A. Bopp, J. Chem. Phys. 114, 6834-6842 (2001).
7 Y. J. Chang and E. W. J. Castner, J. Phys. Chem. 100, 3330-3343 (1996).
8 G. R. Fleming and M. Cho, Annu. Rev. Phys. Chem. 47, 109-134 (1996)

Low-frequency, condensed phase solvation dynamics using a novel two-dimensional, resonant-pump, third-order Raman-probe technique in the time domain

David F. Underwood and David A. Blank[1]

Department of Chemistry, University of Minnesota
207 Pleasant ST SE, Minneapolis, MN 55455

For the first time, we present data from a novel femtosecond spectroscopic technique which probes the molecular polarizability and the low frequency response (0.1 cm^{-1} to 500 cm^{-1}) of the solvent, resulting from the electronically-resonant excitation of the solute. Conceptually, the information is obtained in the framework of a pump-probe experiment; however, while the pump pulse is the same as a traditional pump-probe (a single electronically resonant pulse), the probe is a non-resonant third-order Raman (TOR) sweep in the time domain (pulses are 1 kHz, 24-30 fs (intensity, FWHM)). We label this new technique (RAPTORS, ResonAnt Pump Third Order Raman (probe) Spectroscopy).

A 400 nm pump pulse is generated by frequency doubling in a BBO crystal and compressed to 28 fs using a standard prism pair geometry. Pulse diagnostics are accomplished by frequency-resolved optical gating in a third-order transient grating setup. The electronically resonant pulse excites a chromophore in solution (in this study 5×10^{-4} M 9,10-diphenylanthracene (9,10-DPA) in chloroform), which initiates a change in the electric field immediately adjacent to the surrounding solvent. This is manifested as a change in the molecular polarizability of the effected molecules, which is immediately probed by TOR spectroscopy. The initiating resonant pulse is swept in time (t) relative to the two time-coincident non-resonant pulses. For a given resonant pulse delay, the third TOR pulse is swept in time, τ, which yields the dynamics of the solvent response to the changing electric field of the chromophore. The non-resonant 800 nm beams are set in the standard 'box' geometry using a beamsplitter and diffractive optic. This ensures phase stability for the first two time-coincident TOR probe pulses, and allows intrinsically stabilized heterodyne detection [1] for the retrieval of the third-order field (discussed below).

Data are collected by iteratively stepping the excitation pulse delay (t) and sweeping the TOR signal (τ) in whatever range is suitable for a given solvent, resulting in a two-dimensional plot as depicted in Figure 1. Since the probe step is carried out in the time domain, all the Raman active motions within the bandwidth of our femtosecond laser pulses, currently ca. 500 cm^{-1}, are measured simultaneously and extracted directly from the data via Fourier transformation. The signals in Figure 1 are initially positive for t = 0-300 fs with a peak at ca. 170 fs, consistent with the time scale of the inertial solvation component, and represent the greater polarizability provided by the inertial reaction of the solvent molecules in response to the field of the excited solute. Following the inertial response, the solvent molecules relax into an equilibrium configuration in the field of the excited solute molecules and the response becomes negative, demonstrating that these molecules are now less polarizable than the bulk.

The RAPTORS experiment detects a cross-term between the third order field due to resonant excitation $\Delta E^{(3)}(\tau, t)$, and the one-dimensional third order response of the sample, $E^{(3)}(\tau)$. We make a separate heterodyne detected measurement of $E^{(3)}(\tau)$ and can divide this out of the RAPTORS response to leave only the desired change $\Delta E^{(3)}(\tau, t)$ in the two-dimensional data matrix. Thus, $\Delta E^{(3)}(\tau, t)$ holds the complete response of the system to the perturbation caused by the initial resonant pulse.

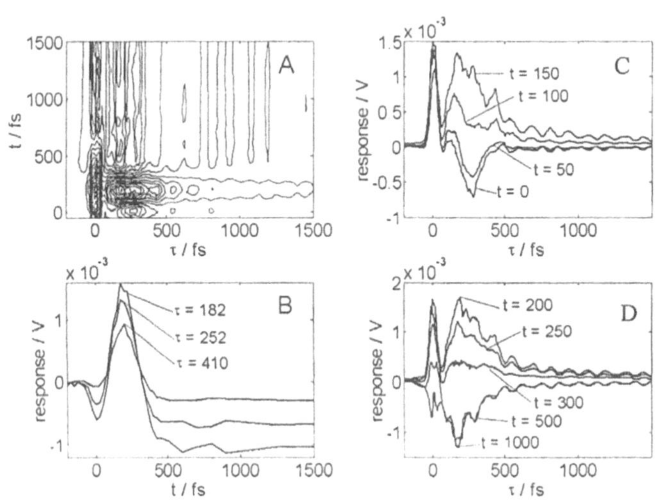

Fig. 1. Resonant-pump, nonresonant-probe spectra of 9,10-diphenylanthracene in chloroform with TOR polarizations selected to enhance the nuclear response. (A) Contour plot of the 2D spectrum. (B) Slices along the pump-probe delay, t, for the sweep value τ. (C and D) Slices along the TOR delay, τ, for various pump-probe delays, t.

References

1. M. Khalil, N. Demirdven, O. Golonzka, C. J. Fecko, and A. Tokmakoff: J. Phys. Chem. A, **104**,*24* 5711 (2000)

Optical Heterodyne Detected Transient Grating (OHD-TG) Studies on the Reactive and Non-reactive Resonant Systems

Qing-Hua Xu, Ying-Zhong Ma and Graham R. Fleming

Department of Chemistry, University of California, Berkeley and Physical Biosciences Division, Lawrence Berkeley National Laboratory, Berkeley, CA 94720
E-mail:grfleming@lbl.gov

Abstract. OHD-TG measurements were performed to separate the real and imaginary components of the $P^{(3)}$. They are found to be similar for a non-reactive system while different a the reactive system. A simple model is proposed.

Four-wave mixing spectroscopic methods such as the transient grating (TG) method have been widely applied to the study of photophysical and photochemical processes in the condensed phase. Conventional TG measurements are usually performed in a homodyne manner and the signal is proportional to the modulus squared of the third-order polarization $P^{(3)}$. The dynamic time constants were usually extracted by doubling the fitting results [1], which assumes that the real and imaginary components follow the same dynamics. In this work, we carried out heterodyne detection of the TG signals with a simplified diffractive optics method to measure the real and imaginary components of $P^{(3)}$ separately. Our results show that the above assumption is true in a simple two-level system, while very different real and imaginary responses were observed for a multi-level system.

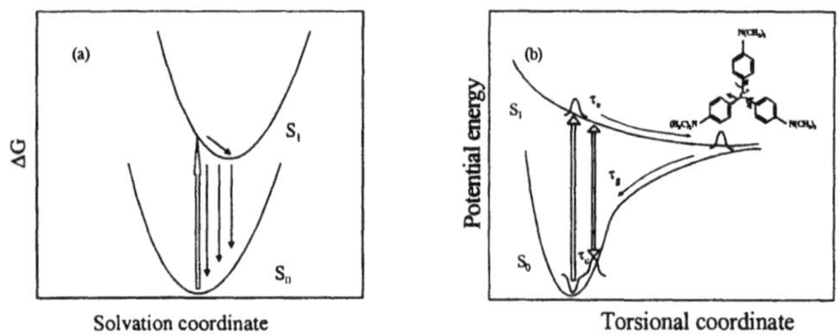

Fig.1 Schematic energy surface for non-reactive and reactive systems. The molecular structure of crystal violet is shown in the inset of (b).

The arrangement for the OHD-TG with diffractive optics has been described in detail previously [2]. Two different systems have been studied. One is a laser dye, rhodamine 640 (R640), dissolved in dimethylsulfoxide (DMSO), representing a

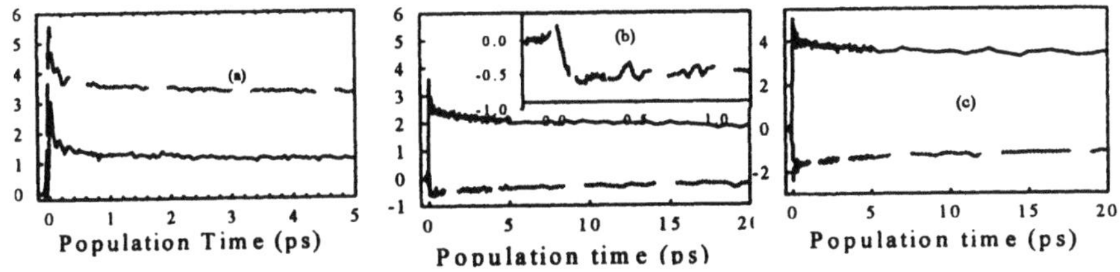

Fig.2 OHD-TG measurements of the real (dashed lines) and imaginary (solid lines) components for R640/DMSO at 570 (a), 607 (b) and 615 nm (c). The inset of (b) shows the short time behavior at 607 nm.

simple two-level system modulated by the solvent fluctuation (Figure 1a). The other is crystal violet (CV) in ethanol, where the nonradiative relaxation will quickly bring the excited state population down to the ground state potential energy surface (Figure 1b). OHD-TG measurements on these two systems are shown in Figure 2 and Figure 3 respectively.

OHD-TG in R640/DMSO was carried out at three different wavelengths, 570, 607 and 615 nm. It was found that the imaginary response is positive for all detection wavelengths, while the real component is strongly dependent on the wavelength and flips sign from positive to negative as the wavelength is shifted from the blue to the red edge of the absorption spectrum. Such a wavelength dependence of the real component can be understood considering the Kramers-Kronig relation. However, except for the difference in the sign of the magnitude and the initial decay behavior (the initial decay is contaminated by interference with the solvent non-resonant signal) [3], if the sign of the negative signal is reversed, the profiles of the real and imaginary components can be overlapped quite well. The identical decay behavior for the two responses confirms that they result from the same solvation dynamics, electronic and vibrational relaxations.

The situation is different for CV/ethanol. It is found that not only the amplitudes, but also the decay profiles of the real and imaginary components are strongly dependent on the detection wavelength. A simple model is proposed to describe the wavelength dependent differences in the two responses. By considering the hot ground state contribution, the total response can be represented by [4]

$$P_{tot}^{(3)} \approx P_e^0 e^{-T/\tau_e} + P_g^0 e^{-T/\tau_c} - P_{g'e}^0 (e^{-T/\tau_c} - e^{-T/\tau_g}) \qquad (1)$$

where P_e^0, P_g^0 and $P_{g'e}^0$ represent the contributions of the excited state population, ground state bleaching and hot ground state absorption; τ_e, τ_g and τ_c represent the lifetime of the stimulated emission, ground state recovery and hot ground state cooling rate, respectively. By choosing the proper amplitudes of P_e^0, P_g^0 and $P_{g'e}^0$ according to the spectrum, we can reproduce the qualitative behavior with fixed values of τ_e, τ_g and τ_c (Figure 3c and 3d). The striking differences between the decay profiles and wavelength dependencies of the real and imaginary components predominantly originate from the different wavelength dependence of

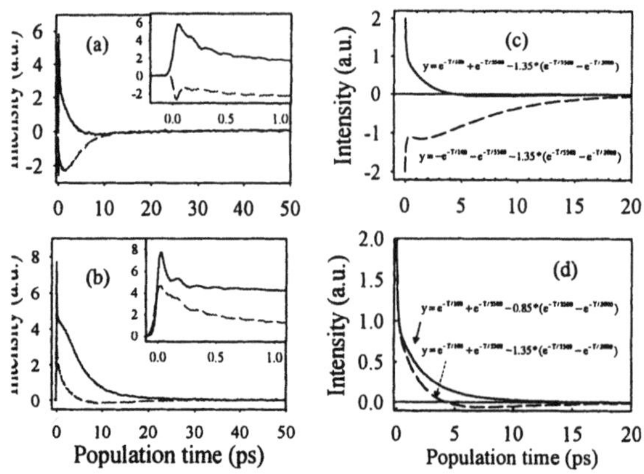

Fig.3 OHD-TG measurements of the real (dashed line) and imaginary (solid line) components for CV/ethanol at 593 nm (a) and 620 nm (b) and the corresponding model simulation results (c) and (d). The insets of (a) and (b) show the short time behavior. The formulas used for the simulation are also included in (c) and (d).

the amplitudes of the various contributions, rather than changes in the time scales and the observation of new contributions at particular wavelength.

In summary, the OHD-TG measurements in non-reactive and reactive systems demonstrate that the real and imaginary components of $P^{(3)}$ may not be identical in a multi-level system. Such a difference will complicate the analysis of the homodyne detection measurements, where the signal is a sum of squared real and imaginary components and contains the cross terms of two different responses. Care should be taken in the characterization of homodyne measurements on complex systems.

Acknowledgments. This work is supported by a grant from NSF.

References

1. T. Joo, Y.W. Jia, J.-Y. Yu, M. J. Lang, and G. R. Fleming, *J. Chem. Phys.* **104**, 6089-6108 (1996).
2. Q.-H. Xu, Y.-Z. Ma, and G. R. Fleming, *Chem. Phys. Lett.* **338**, 254-262 (2001)
3. Q.-H. Xu, Y.-Z. Ma, I. Stiopkin, and G. R. Fleming, *J. Chem. Phys.* **116**, 9333-9340 (2002)
4. Q.-H. Xu, Y.-Z. Ma, and G. R. Fleming, *J. Phy. Chem. A.* submitted (2002).

Polar and non-polar solvation in the femtosecond evolution of 2D Fourier transform spectra

Anchi Yu, John D. Hybl, Darcie A. Farrow, and David M. Jonas

Department of Chemistry and Biochemistry, University of Colorado, Boulder, CO 80309-0215 USA
E-mail: David.Jonas@colorado.edu

Abstract. A combination of magic angle pump-probe and femtosecond 2D Fourier transform spectra is used to probe polar and non-polar solvation in two structurally related molecules with and without a dipole moment change upon electronic excitation.

It is well known that solvent fluctuations can control charge transfer reactions in polar solvents by forcing the charges to wait for a favorable solvent configuration. Because intramolecular vibrations and polar solvation cause frequency shifts within 100 fs, recent studies have used femtosecond four-wave mixing for time resolution [1,2]. In femtosecond spectroscopy, high time resolution usually implies low frequency resolution, necessitating a compromise between time and frequency resolution. Recently, the ability to measure femtosecond optical fields has allowed "two-dimensional" experiments with high excitation and signal frequency resolution. Since the frequency resolution comes from Fourier transformation of the signal field with respect to time intervals between pulses, the frequency resolution limit is decoupled from the time resolution limit dictated by the pulse duration [3]. In such 2D FT experiments, the time and frequency resolution are set by the sample itself, allowing a nearly undistorted view of the solvation process.

Electronic excitation usually produces a change in molecular size which leads to both an intramolecular vibrational response and a non-polar solvent response. In polar solvents, it is expected that the slowest polar (dielectric relaxation) and non-polar responses (shear relaxation) will have similar timescales because they are both controlled by viscosity. Recent work indicates that the fastest polar solvation processes, thought to consist of small rotations of nearby solvent molecules, are strongly coupled to translation through solute size changes [4]. Although polar and non-polar solvation are strongly coupled, we have isolated the changes in the solvation process caused by a change in solute charge distribution by comparing 2D FT spectra and magic angle pump-probe transients for two structurally related cyanine dyes. The linear absorption and emission spectra of the dye HDITCP exhibit weak non-polar shifts dependent only on the solvent polarizability. The addition of three polar functional groups to the HDITCP molecule makes IR144, whose absorption and fluorescence spectra have a strong dependence on orientational solvent polarity. Magic angle pump-probe transients recorded using 30 fs pulses show that on a picosecond timescale, *every* non-polar response has a corresponding polar counterpart and the polar response *always* has a larger amplitude and a slower time constant. The larger amplitude reflects the greater energetic importance of polar solvation. The slower timescale may reflect longer range electrostatic forces probing solvation at greater distances from the solute. The intramolecular vibrational quantum beats are nearly identical for the two molecules, indicating that the structural changes upon electronic excitation are extremely similar. Non-polar solvent responses should be similarly identical.

2D FT spectra are used to characterize the fast solvent response. A three pulse sequence is used to generate 2D Fourier transform spectra. Near transform limited 22 fs pulses (time-bandwidth product 0.43 by SHG FROG) from a cavity dumped titanium sapphire laser are split into 4 pulses which are delayed by translation stages (100 nm steps) and propagate in parallel on the corners of a square. The first pulse sets electronic dipoles oscillating and the second pulse enhances or suppresses these oscillations depending on the dipole frequency, delay τ, and position in the sample. After a relaxation time T, the third pulse stimulates a background free signal that is phase modulated by the dipole oscillation during τ. For Fourier transformation, delays between excitation pulses are measured to ±0.04 fs by spectral interferometry for each signal. To separate the real (absorptive) and imaginary (refractive) parts of the 2D spectra, the time delay and phase shift of the signal field relative to the excitation pulses are separately determined by high precision spectral interferometry using a tracer beam to correct for propagation between sample and detector. The complete signal field $E_s(t;\ \tau, T)$ is detected at the sample for a set of τ. The set is double Fourier transformed with respect to τ and t to make a 2D spectrum with frequency axes ω_τ (excitation) and ω_t (detection). The 2D FT spectra are absolutely phased by comparing the projection over ω_τ to the spectrally resolved pump-probe signal [3].

2D spectra are calculated with a multi-mode Brownian oscillator model for which a few percent of the picosecond relaxation has been empirically included in the same manner as the population lifetime. This might reflect a departure from the relation between vibrational dephasing and vibrational population relaxation rates built into the Brownian oscillator model. The model assumes that the lineshape function for IR144 is the same as that for HDITCP plus polar solvation. The T=0 2D spectrum is dominated by intramolecular vibrations, but the rapid evolution of the 2D spectra over the first 100 fs is quite sensitive to the solvent response. Because sharp features in the time domain transform into broad low amplitude features in the frequency domain, we have found it useful to compare experiment and model in both domains. Projections of the 2D data have higher signal to noise ratios and allow sharp comparisons between theory and experiment. Figure 1 shows time domain "echo slices" through the signal fields $E_s(t=\tau, T)$ at the time $t=\tau$ when a photon echo would be emitted for large inhomogeneous broadening. (Signal is improved by Fourier transforming a filtered projection of the 2D spectrum to get the slice.) The slight changes due to polar solvation are significant.

Despite sufficient time resolution, the spectral density of polar solvation recovered from the model does not show methanol OH libration near 600 cm^{-1}, but peaks around 95 cm^{-1}, slightly lower than the lowest frequency rotation about the *c*-inertial axis perpendicular to the CO bond in pure methanol. This frequency reduction might reflect either the local influence of the solute or the role of quasi-bound solvent translational motions.

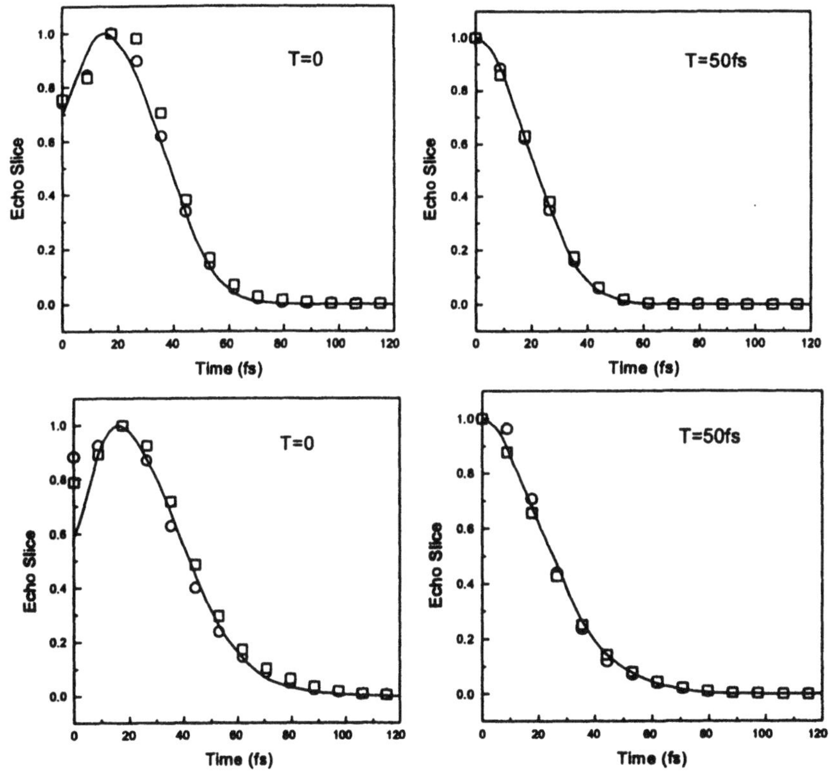

Fig. 1. Echo slices through the signal fields $E_s(t=\tau, T)$ computed from 2D spectra of IR144 (top panels) and HDITCP (bottom panels) in methanol at relaxation times T=0 and T= 50 fs. Coherence is rapidly destroyed by non-polar solvation and intramolecular vibrations in both molecules, but polar solvation destroys coherence more rapidly in IR144.

References

[1] W. P. de Boeij, M. S. Pshenichnikov, and D. A. Wiersma, "Heterodyne-detected stimulated photon echo: applications to optical dynamics in solution," Chem. Phys. **233**, 287-309 (1998)

[2] L.D. Book and N. F. Scherer, "Wavelength-resolved stimulated photon echoes: Direct observation of ultrafast intramolecular vibrational contributions to electronic dephasing," J. Chem. Phys. **111**, 792-795 (1999).

[3] J. D. Hybl, A. Albrecht Ferro, and D. M. Jonas, "Two-Dimensional Fourier-Transform Electronic Spectroscopy," J. Chem. Phys. **115**, 6606-6622 (2001).

[4] E. R. Barthel, I. B. Martini, and B. J. Schwartz, "How does the solvent control electron transfer? Experimental and theoretical studies of the simplest charge transfer reaction," J. Phys. Chem. B **105**, 12330-12241 (2001

Photodissociation of ICN in polar solvents: evidence of long lived rotational excitation in real liquids

A. C. Germaine and S. E. Bradforth

Department of Chemistry, University of Southern California, Los Angeles, CA 90089
E-mail: agermain@usc.edu and bradfort@usc.edu

Abstract. The ICN Ã band photodissociation in polar solvents is studied using ultrafast pump-probe spectroscopy. The transient anisotropy reveals that highly rotationally excited CN photoproducts survive in methanol, ethanol, and water for several picoseconds.

Introduction: Photodissociation dynamics in a liquid environment are of considerable complexity and interest. In liquids, effects such as caging and collisions with surrounding solvent molecules that cause dissipation of electronic, translational, rotational and vibrational energy to the surroundings must be considered in addition to the often complex intrinsic dissociation dynamics manifested in the gas phase. The Ã band photodissociation of ICN is an excellent candidate to study the effects of the solvent on dissociation. This system has been studied extensively in the gas phase, both experimentally and theoretically. Three electronic states comprise this band and two product channels, ICN*→ I+CN and I*+CN, are observed [1]. In the I + CN channel, the CN is produced highly rotationally excited, N~50, also making the system an excellent choice for studies of relaxation of highly rotationally excited molecules in the liquid phase.

Time resolved anisotropy measurements are a powerful tool to study rotational excitation and relaxation after photodissociation in a solvent environment. For example, in recent experiments on the photodissociation of HgI_2 and I_3^- in ethanol, the pump-probe polarization anisotropy results indicate an initial super-thermal rotational distribution of the diatomic fragments after dissociation followed by rapid thermalization. This is evidenced by a very fast Gaussian component in the anisotropy decay with correlation times shorter than a calculated thermal free-rotor time, in addition to the longer, ~10 ps, rotational diffusion time constants for HgI and I_2^- [2,3]. In contrast, the experiments in this paper present evidence for surprisingly long-lived rotational excitation and unperturbed free-rotor type motion of the CN product for ~100 rotational periods.

Experimental Methods: Pump-probe spectroscopy is used to study the ICN photodissociation dynamics in the polar solvents methanol, ethanol, H_2O and D_2O. An amplified Ti:Sapphire and OPA laser system with appropriate frequency doubling provides tunable UV pump and probe. The ICN Ã band is excited at two wavelengths, 224 nm and 255 nm. Due to the considerable blue shift of the Ã band in polar solvents this corresponds to ~260 nm (band center) and ~300 nm in the gas phase absorption. The reaction dynamics are probed via the CN B←X transition at 385 nm. Time resolution is optimized with prism pairs in both the

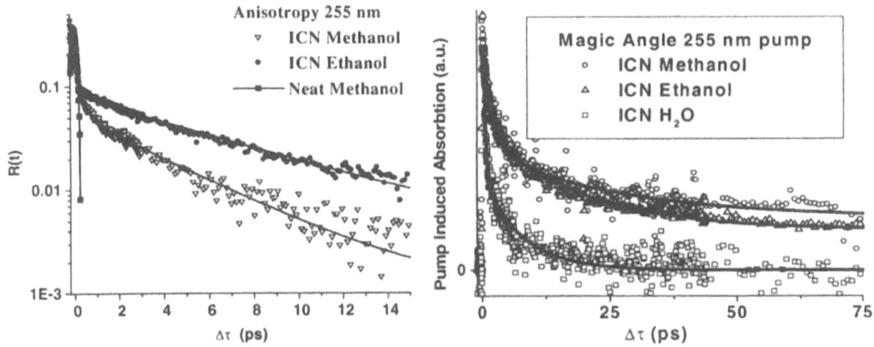

Figure 1 (Left) Transient anisotropy of CN fragment in methanol (lower trace) and ethanol (upper trace) with a 255 nm pump; neat solvent signal vanishes by 0.2 ps; (Right) Isotropic pump-probe signal of ICN in MeOH, EtOH and H_2O at 255 nm.

pump and probe beam paths to compensate for the considerable dispersion in the UV. Experimental time resolution is ~140 fs at 255 nm and ~180 fs at 225 nm as determined from the FWHM of the solvent coherent response. The sample is a ~200 μm free flowing jet of ~0.1 M ICN solution. High signal-to-noise pump-probe anisotropy measurements are achieved through the simultaneous detection of the parallel and perpendicular signals. The probe polarization is set 45° relative to the pump polarization and a Wollaston polarizing prism is used after the sample to separate the parallel and perpendicular components of the probe.

Results and Discussion: *Fig. 1(left)* shows the pump-probe anisotropy data for ICN in ethanol and methanol at 255 nm. The early signal is dominated by a strong solvent coherent spike. Thus, it is difficult to separate out the early portion of the anisotropy signal due to ICN. After ~200 fs the solvent signal has reached zero and the anisotropy is entirely due to the CN photofragment. These data sets show relaxation on two time scales from an initial anisotropy of 0.1. The short time constant is 400 fs in methanol and 500 fs in ethanol. These are approximately the timescales expected for rotational diffusion of thermalized CN and the characteristic time in ethanol is slightly longer due to the higher viscosity. The longer time constants at 255 nm are 3.7 ps and 6.7 ps in methanol and ethanol respectively. This several picosecond relaxation process is much longer than would be expected for rotational diffusion of a thermalized CN photofragment. We note that CN^- in room temperature water is believed to drag considerable solvent in its tumbling and has a characteristic reorientation time of 0.8 ps[4]. One possible assignment is this feature is due to long-lived rotationally hot CN fragments. In the gas phase, a large fraction of the CN product is released with the majority of the excess kinetic energy going into rotation, yielding a rotational distribution centered near N=50 (~ 0.6 eV). The highly rotationally excited CN requires ~1000 gas phase collisions to dissipate the rotational kinetic energy [5].

After photodissociation of a linear or bent triatomic in a collision-free regime, due to conservation of angular momentum, the long time anisotropy is not expected to be zero but 0.1 for parallel pump and probe transition dipoles or –0.05 for perpendicular transition dipoles [6]. The same will hold in a liquid over the period where collisions are ineffective at interrupting the rotational motion. It is thus reasonable to believe the surprisingly long several picosecond anisotropy decay may be attributed to the highly torqued CN photofragments requiring a large number of collisions with the solvent before slowing enough for their orientations to be completely randomized. We are making the analogy with a classical spinning top – it is hard to push the top off its axis of rotation. A similar result holds for water: the long time constant is 3.7 ps. This implies that water is no more effective at interrupting the CN rotations than the alcohols. With a 224 nm pump, we see similar decay times in methanol and ethanol with a lower initial anisotropy of 0.07. Additionally, preliminary data in weakly interacting cyclohexane suggest an order of magnitude longer lived anisotropy for the CN product.

Fig. 1(right) shows the isotropic signals at 255 nm for ICN in three polar solvents at a common pump wavelength. There are two timescales for consumption of the CN photoproduct. The short timescale ranges from 0.55 ps in water to 3 ps in methanol and is assigned to geminate caging recombination. The timescales are similar, but shorter than a previous experiment in chloroform [7]. Our results suggest faster cage recombination and a smaller fraction escaping for solvents with more rigid cages due to hydrogen bonding. The longer time process ranges from 7.7 ps in water to 27 ps in methanol. In previous experiments on chloroform, a second ~70 ps time constant was observed and assigned to abstraction of a H or Cl atom from the solvent by the CN [7,8]. We have not yet confirmed that this is the case in these polar solvents. The CN may also escape the cage and undergo a secondary recombination with its geminate partner. A kinetic isotope effect k_D/k_H~2.2 is observed for the long time constant in D_2O. Another point of interest is that no CN fragments survive to timescales longer than we can measure (400 ps) in H_2O and D_2O while in ethanol and methanol 20-25% survive. Finally, we note that at 224 nm caging recombination occurs more rapidly; however, the longer time constant remains the same in all solvents.

References

1. Y. Amatatsu, S. Yabushita, K. Morokuma, *J. Chem. Phys.* **100**, 4894-4909 (1994) and references within.
2. H. Bürsing and P. Vöhringer *Phys. Chem. Chem. Phys.* **2**, 73-82 (2000).
3. T. Kühne and P. Vöhringer, *J. Phys. Chem.* **102**, 4177-4185 (1998).
4. P. Hamm, M. Lim, R.M. Hochstrasser, *J. Chem. Phys.* **107**, 10523-10531 (1997).
5. S. Hay, S. Shokoohi, S. Callister, C. Wittig, *Chem. Phys. Lett.* **118**, 6-11 (1985).
6. M. MacPherson, J. Simons, R. Zare, *Mol. Phys.* **38**, 2049-55 (1979).
7. C. Wan, M. Gupta, A.H. Zewail, *Chem. Phys. Lett.* **256**, 279-287 (1996).
8. D. Rafferty, E. Gooding, A. Romanovsky, R.M. Hochstrasser, *J. Phys. Chem.* **101**, 8572-8579 (1994).

Dynamics of water in ionic solvation shells

Michel F. Kropman, Han-Kwang Nienhuys, and Huib J. Bakker

FOM-Institute for Atomic and Molecular Physics, Kruislaan 407, 1098 SJ Amsterdam, The Netherlands

In the last decade it has become possible to investigate the molecular dynamics of liquid water with femtosecond mid-infrared nonlinear spectroscopy [1–5]. However, up to now there have been no spectroscopic studies on the dynamics of water molecules in the solvation shells of dissolved molecules and ions. Such an investigation is of great interest, since the interactions between solvating water molecules and dissolved molecules and ions are essential for chemical and biochemical reactions occuring in aqueous media.

We performed femtosecond mid-infrared pump-probe experiments on the O–H stretch vibration of the HDO molecule in different aqueous salt solutions. These solutions consist of a low concentration of HDO (0.1 M) in D_2O and a high concentration (1, 3 or 6 M) of a particular salt (NaCl, NaBr, NaI, KF). The mid-infrared pulses required to pump and probe the O–H stretch vibration are generated via parametric generation and amplification in BBO and KTP crystals. They are tunable between 2.7 and 4 μm, have a typical pulse energy of 20 μJ, and a pulse duration of approximately 200 femtoseconds.

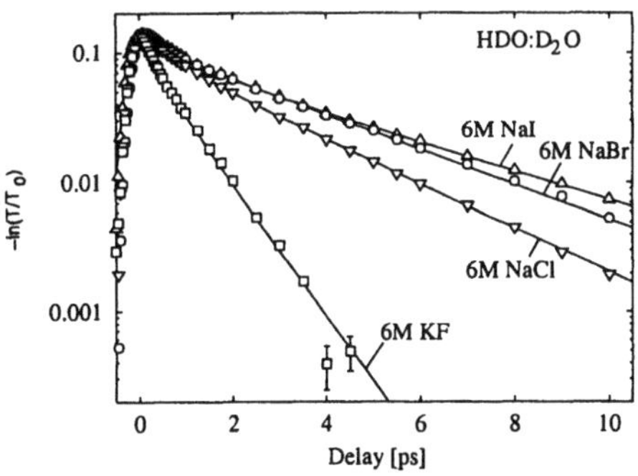

Fig. 1. Isotropic transmission change of the O–H stretch vibration of HDO in solutions of D_2O and 6 M of a particular salt as a function of delay, using a central frequency of 3450 cm^{-1} for the pump and 3200 cm^{-1} for the probe.

In the experiment, an intense pump pulse generates an anisotropic distribution of vibrationally excited O–H groups. The vibrational and orientational dynamics of the water molecules are probed using weaker delayed probing pulses that are polarized parallel and perpendicular to the polarization of the pump pulse. The vibrational relaxation is monitored by plotting the isotropic transmission change $\ln(T(\tau)/T_0)_{\parallel} + 2\ln(T(\tau)/T_0)_{\perp}$ as a function of delay τ, with T the transmission of the probe and T_0 the transmission of the probe in absence of the pump. The orientational relaxation is monitored by plotting the anisotropy parameter $[\ln(T(\tau)/T_0)_\parallel - \ln(T(\tau)/T_0)_\perp]/[\ln(T(\tau)/T_0)_\parallel + 2\ln(T(\tau)/T_0)_\perp]$ as a function of delay.

In Fig. 1, the isotropic transmission change is presented as a function of delay for several salt solutions. For a solution containing KF, the decay is single-exponential with a time constant of about 800 fs [6], identical to the neat liquid. For the solutions containing NaCl, NaBr or NaI, the relaxation becomes strongly non-exponential and contains a fast component and a slow component of which the amplitude is observed to correlate with the concentration of the dissolved anion. This observation can be understood from the fact that the O–H stretch frequency and its relaxation rate are strongly determined by the hydrogen-bond interaction. One of the O–H groups of a water molecule in the solvation shell of an anion X^- will form an $O\text{–}H\cdots X^-$ hydrogen bond, that is different from an $O\text{–}H\cdots O$ hydrogen bond in bulk water and in a cationic solvation shell. The time constant of the slow component increases with the anion's mass: 2.6±0.1 ps for Cl^-, 3.4±0.1 ps for Br^-, and 3.9±0.1 ps for I^-(6 M solutions). A great advantage of the large difference in vibrational lifetime is that after a few picoseconds only HDO molecules remain excited that are hydrogen bonded to an anion. This enables a very selective study of the dynamics of the aqueous anionic solvation shell.

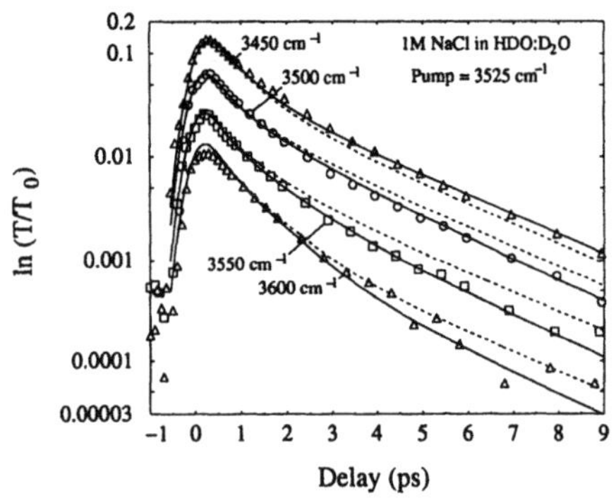

Fig. 2. Isotropic transmission change in a solution of 1 M NaCl in $HDO{:}D_2O$ at different probing frequencies, revealing spectral diffusion on a 10 ps timescale. The dashed lines are calculated assuming infinitely slow spectral diffusion.

In a second experiment [7,9] we measured the rate at which the frequency of an excited O–H group changes, resulting from variation of the $O\text{–}H\cdots X^-$ hydrogen-bond length (X = Cl, Br, or I). In Fig. 2, it can be seen that the isotropic transmission change decays at different rates for different probing frequencies. The difference in decay rates results from a spectral diffusion process, that produces an 'outflow' of molecules that vibrate at a frequency near the pump frequency, and an 'inflow' at frequencies away from the pump. We modeled the data using two Brownian oscillators, one for the water-bound O–H groups ($O\text{–}H\cdots O$) and one for the anion-bound ($O\text{–}H\cdots X^-$). The parameters for the $O\text{–}H\cdots X^-$ oscillator can be determined with high accuracy since after a few picoseconds the $O\text{–}H\cdots O$ oscillators have all relaxed, and no longer affect the signal. We obtained values for the O–H stretch frequency (hydrogen-bond length) correlation time of several anions. The values are 12±3 ps (Cl^-), 25±5 ps (Br^-, in agreement with Ref.[8]), and 18±5 ps (I^-), which is much longer than the value of 500 fs found for the neat liquid [4]. Thus, the solvation shell is a stable, well-ordered structure compared to bulk liquid water.

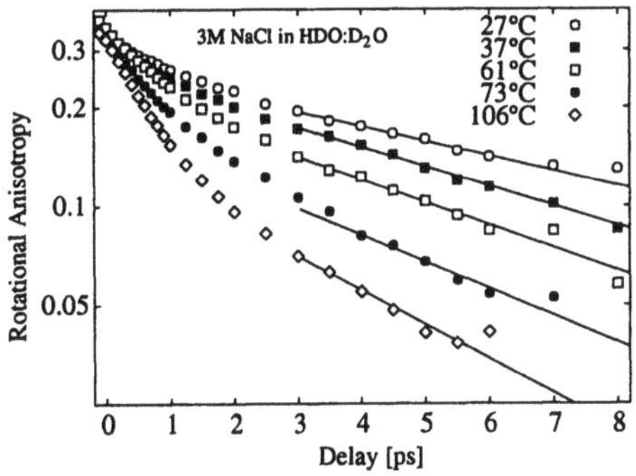

Fig. 3. Anisotropy of the O–H stretch vibration of HDO in a solution of 3 M NaCl in D_2O as a function of delay using a central frequency of 3450 cm^{-1} for the pump and of 3200 cm^{-1} for the probe.

In Fig. 3, the rotational anisotropy is shown as a function of delay for a solution of 3 M NaCl in HDO:D_2O at several temperatures [10]. The anisotropy decays with a time constant of 8±0.5 ps at room temperature, and of 3±0.5 ps at 106 °C, which is much slower than in neat liquid water (about 2.6 ps at room temperature [6]). The decay of the anisotropy is however faster than the hydrogen-bond motion, which means that it stems from a rotation of the complete solvation structure, not of the individual solvent-shell molecules.

The solvation shell of Br^- reorients somewhat slower than that of Cl^-, while the solvation shell of I^- is faster, possibly implying a smaller solvation shell of the larger I^-.

References

1. S. Woutersen, U. Emmerichs, and H.J. Bakker: Science **278**, 658 (1997)
2. G.M. Gale, G. Gallot, F. Hache, N. Lascoux, S. Bratos, and J.-C. Leicknam: Phys. Rev. Lett. **82**, 1086 (1999)
3. J. C. Deàk, S. T. Rhea, L. K. Iwaki, and D. D. Dlott: J. Phys. Chem. A **104**, 4866 (2000)
4. S. Woutersen and H. J. Bakker: Phys. Rev. Lett. **83**, 2077 (1999)
5. J. Stenger, D. Madsen, P. Hamm, E. T. J. Nibbering, and T. Elsaesser: Phys. Rev. Lett. **87**, 7401 (2001)
6. H.-K. Nienhuys, S. Woutersen, R. A. van Santen, and H. J. Bakker: J. Chem. Phys. **111**, 1494 (1999)
7. M. F. Kropman and H. J. Bakker: Science **291**, 2118 (2001)
8. S. Raugei and M. L. Klein: J. Chem. Phys. **116**, 196 (2002)
9. M. F. Kropman and H. J. Bakker, J. Chem. Phys. **115**, 8942 (2001)
10. M. F. Kropman, H.-K. Nienhuys, and H. J. Bakker: Phys. Rev. Lett. **88**, 77601 (2002)

Non-Equilibrium Charge Recombination Dynamics of Excited Donor-Acceptor Complexes

Olivier Nicolet and Eric Vauthey

Dpt. of Physical Chemistry of the University of Geneva, CH-1211 Genève 4, Switzerland
E-mail: eric.vauthey@chiphy.unige.ch

Abstract. The effect of free energy, solvent and temperature on the charge recombination dynamics of excited donor-acceptor complexes has been investigated. The results are better explained with a model where charge recombination can occur before the excited complex has reached thermal equilibrium.

1. Introduction

Contrarily to the prediction of the semi-classical Marcus theory, the rate constant of charge recombination (CR) of excited donor-acceptor complexes (DACs) does not exhibit a bell-shaped free energy dependence, but rather shows a linear relationship when plotted in a logarithmic scale [1]. Moreover, this behavior continues down to very low driving forces, and the so-called normal region, i.e. the decrease of the rate constant with decreasing exergonicity, is not observed. There has been several hypothesis to account for this discrepancy, including variation of other parameters than free energy when going from one DAC to another and non-equilibrium dynamic effects [2], but there is still no conclusive explanation.

We have investigated the effect of free energy, solvent and temperature on the dynamics of CR in excited DACs composed of pyromellitic dianhydride (PMDA) as electron acceptor and a series of methoxy-benzenes as electron donors [3]. Previous investigations have shown that these donors can be used as a homogenous series for investigating the influence of the driving force on CR dynamics [4].

2. Experimental

The CR dynamics of excited DACs was investigated using the multiplex transient grating (TG) technique [5]. The DACs were excited in their charge transfer band using two time coincident 100 fs pulses at 400 nm crossed with an angle of 1°. The resulting gratings were probed using white light pulses generated by self-phase modulation. As discussed in detail elsewhere [4,5], the TG spectrum is very similar to a transient absorption spectrum, the advantage of the former technique being its superior sensitivity.

3. Results and Discussion

The CR dynamics was investigated by monitoring the decay of the excited DAC population. Figure 1A, shows the TG spectrum obtained after excitation of a DAC at 400 nm. The 670 nm band is due to PMDA$^{\cdot-}$, the excited state of a DAC being essentially a contact ion pair. Apart from the decrease of intensity, a shift and a narrowing of the PMDA band can be observed. This spectral dynamics is essentially biphasic with a fast component (around 1 ps), which can be ascribed to solvation, and a slower one, which depends strongly on the solvent viscosity and polarity. This component is attributed to small structural changes of the excited complex, involving diffusional reorientation and also a variation of the interionic distance.

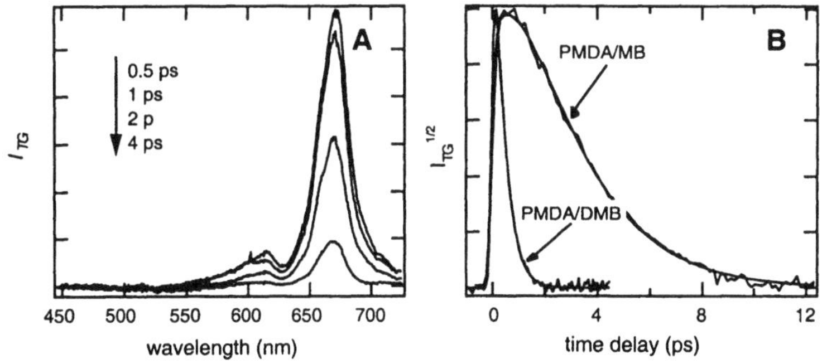

Fig. 1. A) TG spectra recorded at different time delays after excitation of PMDA/methoxybenzene in acetonitrile. B) Time profiles of the square root of the TG intensity measured with PMDA/methoxybenzene and PMDA/1,4-dimethoxybenzene

In acetonitrile, the decay time of the excited state population decreases from 3.5 ps with methoxybenzene to 240 fs with tetramethoxybenzene (see Figure 1B). Figure 2A shows that the observed relationship between the logarithm of the CR rate constant (k_{CR}) and the free energy is not linear and depends markedly on the dynamic properties the solvent, contrarily to previous reports [6]. However, in the lower exergonicity regime, the CR dynamics is faster than diffusive solvation. Directly after excitation, the solvent configuration around the excited state is essentially the same as the equilibrium configuration of the neutral ground state. The excited DACs are thus formed far from equilibrium and the very large CR rate constants measured here indicate that CR occurs before the excited population reaches its equilibrium. Theoretically, such a situation can be analyzed in terms of a hybrid model developed by Barbara and coworkers [7] to account for the temperature and solvent dependence of ET dynamics in betaines and in mixed valence compounds. However the prediction of this model for the free energy dependence of the ET dynamics has not been explored.

Figure 2B shows the free energy dependence of k_{CR} in acetonitrile (ACN) and valeronitrile (VaCN) calculated using the hybrid model. For comparison, the free energy dependence assuming instantaneous solvent relaxation is also reported. In our analysis, the solvent relaxation has been considered to occur via inertial and diffusive motion and the only difference between both solvents was assumed to be the diffusional solvation time (500 fs in ACN and 4.7 ps in VaCN). This figure shows that both the observed free energy and solvent dependence of k_{CR} can be qualitatively well reproduced. This is also true for the temperature dependence. If the contribution of inertial solvation is not included in the analysis, unrealistically large values of the electronic coupling matrix element have to be used to reproduce the observed rate constants.

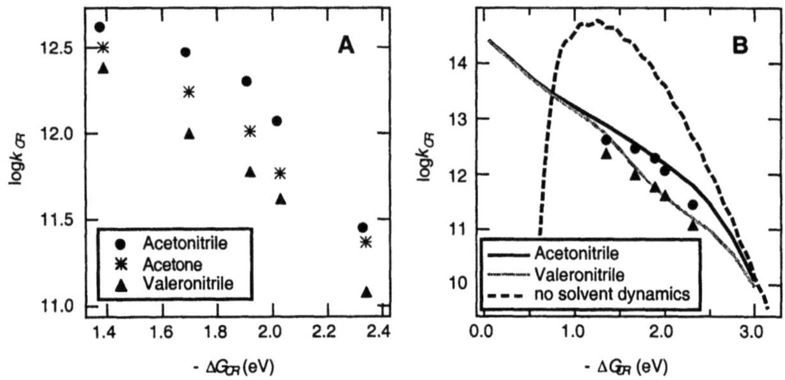

Fig. 2. A) Free energy and solvent dependence of CR in PMDA/methoxybenzenes DACs; B) Free energy dependence of CR calculated using the hybrid model for non-equilibrium electron transfer in acetonitrile and valeronitrile with the measured data

Finally, this figure also shows that the hybrid model does not predict a 'normal region'. This is due to the fact that CR occurs very efficiently before the equilibrium configuration of the excited population is reached, as already suggested theoretically [2].

This is to our knowledge the first experimental investigation of free energy dependence of electron transfer dynamics in the non-equilibrium regime.

References

1 T. Asahi and N. Mataga, J. Phys. Chem. **93**, 6575, 1989.
2 M. Tachiya and S. Murata, J. Am. Chem. Soc. **116**, 2434, 1994.
3 O. Nicolet and E. Vauthey, J. Phys. Chem. A **106**, in prin t.
4 E. Vauthey, J. Phys. Chem., J. Phys. Chem. A **105**, 340, 2001.
5 C. Högemann, M. Pauchard and E. Vauthey, Rev. Sci. Instrum. **67**, 3449 1996.
6 T. Asahi, M. Ohkohchi and N. Mataga, J. Phys. Chem. **97**, 13152, 1993.
7 G. C. Walker, E. Akesson, A. E. Johnson, N. E. Levinger and P. F. Barbara, J. Phys. Chem. **96**, 3728, 1992.

Femtosecond Fluorescence Lifetimes Determined by an Intrinsic Raman Gate

Qingrui An, Wolfgang Zinth, and Peter Gilch

Sektion Physik, Ludwig-Maximilians-Universität München, Oettingenstr. 67, D-80538 München, Germany

Abstract. A technique to record fluorescence lifetimes in the femtosecond regime which relies on the inverse Raman effect is presented. This technique allows to use a conventional pump-probe set-up for the lifetime measurement.

The standard methods for the determination of fluorescence lifetimes in the femtosecond regime are based on optical gates. Currently, two non-linear effects serve as optical gates; (i) sum frequency generation in a non-linear crystal (fluorescence up-conversion) [1] and (ii) the optical Kerr effect (see e.g. [2]). Both techniques require a gating set-up spatially separated from the fluorescing sample. Here a novel method [3] is presented which uses a pump-probe set-up to record fluorescence decay profiles. The method is founded on the inverse Raman effect [4]: Two light fields (frequencies ν_{probe} and ν_{fl}, $\nu_{probe} < \nu_{fl}$) propagate within a material with a Raman active vibration at the frequency ν_{Raman}. If the frequencies of the light fields fullfil the condition $\nu_{fl} - \nu_{probe} = \nu_{Raman}$, then photons with the frequency ν_{fl} are annihilated and photons with the frequency ν_{probe} are created. For pulsed light fields in addition the two pulses must overlap in time. This constitutes a gating mechanism which is utilized here. The light field with the frequency ν_{fl} is generated by excitation of the fluorophore with a pump laser pulse and a probe laser pulse supplies the other field (ν_{probe}). Since the fluorescence emissions show normally a broad spectrum, the resonance condition for the inverse Raman effect can be easily fulfilled. The Raman active mode involved may either be a mode of the fluorophore itself or of its solvent.

The effect is demonstrated using a solution of triiodide I_3^- in ethanol. After optical excitation with 407 nm laser pulses I_3^- emits a very weak and broad emission that peaks at 480 nm but extends to $\sim$ 800 nm [5]. When these 407 nm pulses overlap in space and time with 815 nm probe pulses a dip in the emission spectrum is observed (see insert in Figure 1). This dip occurs at a wavelength of 660 nm which corresponds to a shift of 2900 cm^{-1} with respect to the probe pulse. The spectrum of the dip is identical with the stationary Raman spectrum of the ethanol solvent. When the pump probe delay is increased, this signal reduction decreases (see Figure 2). For I_3^- in ethanol the time dependence of the signal is too close to the experimental response function to decide whether the effect depends on the fluorescence

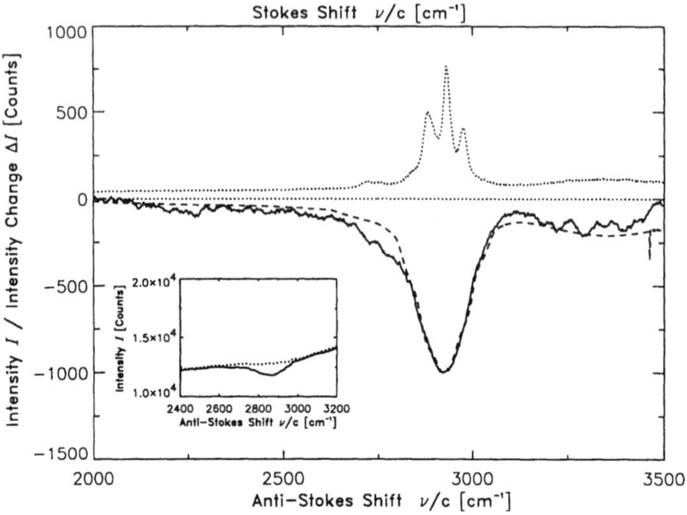

Fig. 1. Section of the fluorescence spectrum of I_3^- around 660 nm. The insert shows this spectrum in absence (dotted line) and presence (solid line) of the 815 nm probe pulse. Taking the difference of the two (solid line) yields the spectral signature of the inverse Raman effect. This spectral signature is identical with Raman spectrum of the ethanol solvent (dotted lines) provided that the resolution of both is the same (the dashed line symbolizes the Raman spectrum with a numerically reduced resolution).

lifetime. For the two other samples (para-nitroaniline (pNA) and para-N,N-dimethylnitroaniline (pDNA) in n-butanol) one finds exponentially decaying signals which can be nicely fitted using a single exponential convoluted with the instrumental response function. The time constants derived compare well with values reported in the literature (see Table 1). Since the maximal signal reduction is proportional to the ratio of the probe pulse duration and the fluorescence lifetime, the relative amplitude of the effect decreases with increasing fluorescence lifetime. In conclusion, it is demonstrated that the

Table 1. Fluorescence lifetimes determined via the inverse Raman effect

Sample	Lifetime determined here	Lifetime reported	Relative Amplitude of the Effect
I_3^- in ethanol	~ 100 fs	~ 200 fs [6]	6 %
pNA in butanol	400 fs	580 fs [7]	2 %
pDNA in butanol	1000 fs	1100 fs [7]	1 %

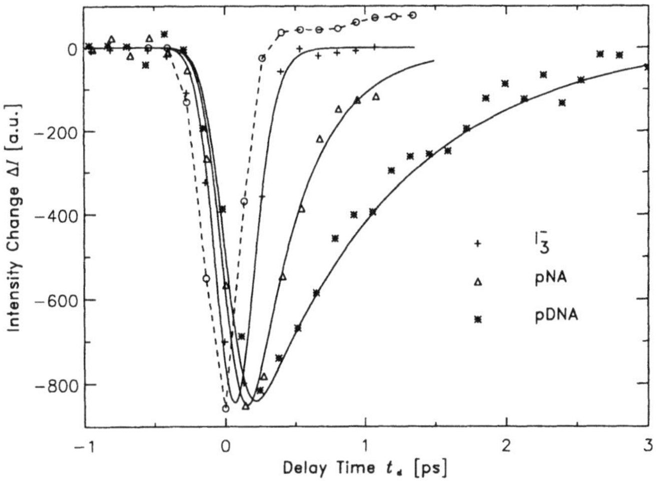

Fig. 2. Temporal dependence of the inverse Raman effect at a frequency shift of 2900 cm^{-1}. The dashed line symbolizes the experimental response function. The solid lines depict fit results as described in the text.

inverse Raman effect allows to record femtosecond fluorescence decay profiles with a pump-probe set-up. For probe pulses of the order of 100 fs fluorescence lifetimes can be measured in the interesting window between 100 fs and some picoseconds.

References

1. D. Block, J. Shah, A.C. Gossard: Solid State Commun. **59**, 527 (1986)
2. J. Takeda, K. Nakajima, S. Kurita, S. Tomimoto, S. Saito, T. Suemoto: Phys. Rev. A **62**, 10083 (2000)
3. Q. An, W. Zinth, P. Gilch: Opt. Comm. **202**, 209 (2002)
4. W.J. Jones, B.P. Stoicheff, Phys. Rev. Lett.: **15**, 657 (1964)
5. P. Gilch, I. Hartl, Q. An, W. Zinth: J. Phys. Chem. A **106**, 1647 (2002)
6. T. Kühne, P. Vöhringer: J. Chem. Phys. **105**, 10788 (1996)
7. S.A. Kovalenko, S. Schanz, H. Hennig, N.P. Ernsting: J. Chem. Phys. **115** 3256 (2001).

How Fast is the Ultrafast Electron Transfer in Solute-Solvent Systems?

P. Gilch[1], E. Baigar[1], W. Zinth[1], M. Stöckl[2], P. Härter[2], T. von Feilitzsch[3], and M.E. Michel-Beyerle[3]

[1] Sektion Physik, Ludwig-Maximilians-Universität München, Oettingenstr. 67, D-80538 München, Germany
[2] Anorganisch-chemisches Institut, Technische Universität München, Lichtenbergstr. 4, D-85747 Garching, Germany
[3] Institut für Physikalische und Theoretische Chemie, Technische Universität München, Lichtenbergstr. 4, D-85747 Garching, Germany

Abstract. Comparison of the ultrafast electron transfer kinetics of a solute-solvent system and its intramolecular analogue shows that the large solvent to solute rates are intrinsic and not due to accumulation of reaction probabilities.

One fascinating application of femtosecond technology is the determination of the lowest timescale for chemical reactions. Because of their practical and theoretical importance electron transfer (ET) reactions have been continuously in the focus of such studies. The fastest ET rate constants in molecular donor/acceptors (D/A) systems have been observed for solute-solvent type ET systems. For instance, after photoexcitation oxazine 1 accepts an electron from the N,N-dimethylaniline solvent within less than 50 fs as revealed by transient absorption [1] and up-conversion [2] measurements. Similar values have been observed for ferrocenes as electron donors [3]. Since in these systems the acceptor can be surrounded by as many as 20 donors [4], the observed large rates could be due to the accumulation of reaction probabilities instead of an instrinstically fast ET rate. There are quantumchemical calculations on such systems [4,5] which indicate that among the $\sim$ 20 donors there are only a few (1–3) which participate effectively in the ET. However, experimental evidence for this conclusion is missing. By determination of the ET kinetics of an intramolecular analogue of an ultrafast solute-solvent system this evidence is given here.

The intramolecular D/A system $FcNb^+$ used consists of a ferrocene (Fc) donor moiety and nile blue Nb^+ acceptor part connected by a methylene chain (for the structure see Figure 1). The lowest energy transition of the Nb^+ moiety was photo-excited by a 7 fs laser pulse tuned to 635 nm. The response of the sample was probed by second broadband pulse, the absorption of which was spectrally resolved. The cross-correlation time was determined to be 15 fs (FWHM). The spectrum of the probe pulse covered the ground state absorption and the stimulated emission region of Nb^+.

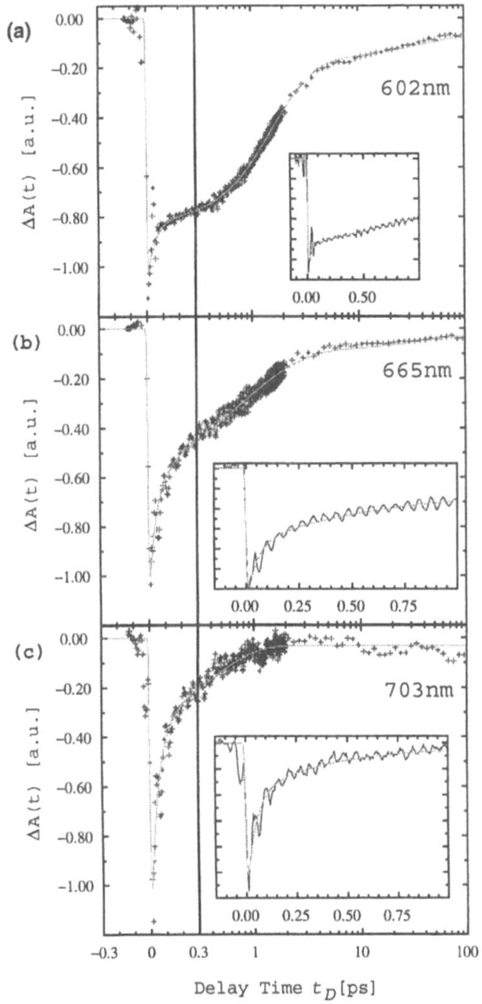

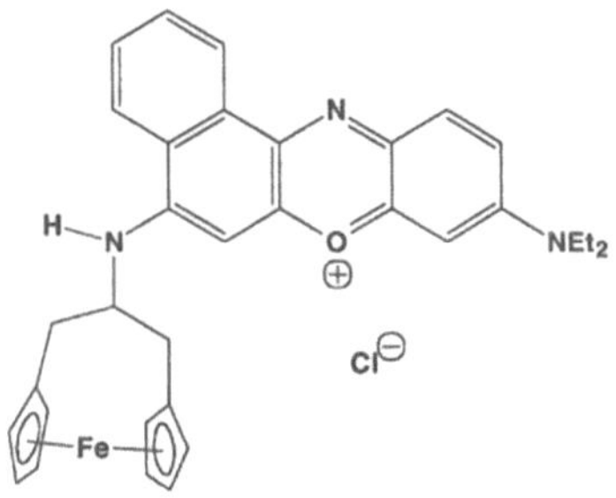

FcNb$^+$

Fig. 1. Transient absorption signals of $FcNb^+$ after excitation of the $S_0 \rightarrow S_1$ transition of the Nb^+ moiety. Experimental values are represented by crosses, the results of a fit (see text) by lines. The apparent scattering of the experimental values at early times is not due to noise but to vibrational wavepacket motion (see insets). The signals have been normalized, the peak values for each wavelength are given below. **(a)** Transient absorption (peak value of -15 mOD) at 602 nm, the signal is dominated by ground state recovery. **(b)** Signal at 665 nm (-50 mOD), contributions of the decay of the stimulated emission and the ground state recovery are detected. **(c)** Signal at 703 nm (-12 mOD), only the stimulated emission contributes to the signal.

Representative time traces shown in Figure 1 exhibit two features: oscillation and multi-exponential kinetics. The oscillations are caused by vibrational wavepacket motions in the ground and the excited state of the Nb^+ moiety and are not discussed here. A global analysis of the multi-exponential kinetics (for details see ref. [6]) yielded two dominant contributions, a 90 fs and a 1.4 ps component with decay associated spectra which can clearly be identified as stimulated emission decay and ground state recovery, respectively. In addition, minor components of 400 fs (stimulated emission decay) and 40 ps (ground state recovery) result from the fit. The 90 fs major component is readily assigned to the decay of photo-excited state quenched by ET and the 1.4 ps contribution is related to backward electron transfer and the recovery of the ground state. The weak components might stem from minority conformations of $FcNb^+$ with weaker electronic coupling. The assigment of the observed ultrafast processes as ET reactions is strongly supported by magnetic field dependencies of the ground state recovery (data not shown), which demonstrate that a radical pair, i.e. the product of the ET process, is indeed formed.

In an intramolecular D/A system the number of donors involved in the ET is definitly one. Therefore the observed 90 fs time is an intrinsic ET time. This value is very close to the $\sim$ 50 fs reported for the fastest solute-solvent systems and even faster than the $\sim$ 200 fs time [3] seen in a bi-molecular system containing ferrocene. This identity of magnitudes gives strong evidence that in the solute-solvent systems the observed rate constants are intrinsic ET rate constants and not the sum of (rather small) rate constants of all the possible (up to 20) D/A pairs. This result is in line with theoretical calculations [4,5] which pin the number of donors with a large electronic coupling to the acceptor and, therefore, a large ET constant down to 1–3.

References

1. S. Engleitner, M. Seel, W. Zinth: J. Phys. Chem. A **103**, 3013 (1999)
2. I.V. Rubtsov, H. Shirota, K. Yoshihara: J. Phys. Chem. A **103**, 1801 (1999).
3. P. Gilch, F. Pöllinger-Dammer, C. Musewald, M.E. Michel-Beyerle, U.E. Steiner: Science **281**, 982 (1998)
4. P.O.J. Scherer: J. Phys. Chem. A **104**, 6301 (2000)
5. Jr. E.W. Castner, D. Kennedy, R.J. Cave: J. Phys. Chem. A **104**, 2869 (2000)
6. E. Baigar, P. Gilch, W. Zinth, M. Stöckl, P. Härter, T. von Feilitzsch, M.E. Michel-Beyerle: Chem. Phys. Lett. **352** 176 (2002)

Photoisomerization of small molecules in solution: Studies of ICN(aq) and NO_3^-(aq)

Dorte Madsen, Jane Larsen, Jens Aage Poulsen, Jan Thøgersen, and Søren R. Keiding

Department of Chemistry, Aarhus University, Langelandsgade 140, DK-8000 Århus C, Denmark
E-mail: keiding@chem.au.dk

Abstract. Photoisomerization of aqueous iodine cyanide and aqueous nitrate is studied by transient absorption spectroscopy. The experiments show that constitutional isomerization and subsequent vibrational relaxation in aqueous solution may occur in only 1 ps. The photolysis of ICN(aq) and NO_3^-(aq) furthermore represent striking examples of how solvents may impede or assist isomerization.

1. Introduction

Photoisomerization plays a major role in chemistry and biochemistry with one of the most prominent reaction being the cis-trans isomerization of rhodopsin, which initiates vision. Of particular interest are isomerizations in which fragments emanating from photodissociation of a molecule in solution are contained by the surrounding solvent and forced to recombine in a geometry different from the original (constitutional isomerization). Detailed investigation of these processes requires solvent-solute systems with few degrees of freedom. Several photolysis experiments on small molecules in solution have been performed over the years, but rarely do these reactions lead to formation of species different from the parent molecule. Exceptions are photo-isomerization of iodine cyanide in chloroform [1] and diodemethane in acetonitrile and cyclohexane [2,3].

Here we present far-UV femtosecond transient absorption studies of the photoisomerization of ICN and NO_3^- in water. Common to both molecules is a large energy difference between the stable and metastable isomer and a significant barrier separating the two geometries [4,5]. Dissociation of ICN leaves the CN fragment spinning rapidly and simulations suggest that solvent induced recombination with the iodine atom may result in the formation of INC [6]. The INC isomer is initially vibrationally excited and simulations predict several interconversions between the two geometries as the molecule relaxes to the ground state of one of the two isomers. The isomerization of the nitrate ion to form peroxynitrate, $ONOO^-$, may proceed differently. Thus, the pyramidalization of excited state nitrate brings the oxygen atoms closer together enabling the O-O bond formation of $ONOO^-$.

2. Experimental Methods

The photoisomerization is studied by transient absorption spectroscopy. The pump pulses needed for dissociating ICN (λ=266 nm) and NO_3^- (λ=200 nm) are generated by frequency tripling or quadrupling the output of a 1 kHz amplified Ti:sapphire laser delivering 750 μJ, 90 fs pulses centered at 800 nm. The pump pulse is sent through a scanning delay line and overlaps the probe in the sample. The probe pulse is generated by an OPA covering the spectral range from 460 nm to 2.5μm. Shorter wavelengths are obtained by either frequency doubling the OPA pulses or sum frequency mixing them with harmonics of the Ti:sapphire laser. The iodine cyanide measurements are done using 0.11 M aqueous ICN solution in a 1.8 mm flow cell equipped with CaF_2 windows, while the nitrate measurements are done in a jet consisting of 0.014 M of aqueous NO_3^- solution. The temporal resolution of the measurements on ICN(q) and NO_3^-(aq) is 0.5 ps and 250 fs, respectively.

3. Results and Discussion

The transient absorption spectrum induced by exciting NO_3^-(aq) at 200 nm is depicted in Figure 1.

The induced bleaching observed below 230 nm is caused by the removal of groundstate NO_3^- molecules. The absorption partially recovers in 3ps. Simultaneously the wide absorption feature initially extending out to 360 nm is blueshifting towards the nitrate absorption maximum at 200 nm. This spectral behavior is a clear indication of fast geminate recombination and subsequent vibrational relaxation of NO_3^-. Concurrently, $ONOO^-$ is observed as the constant

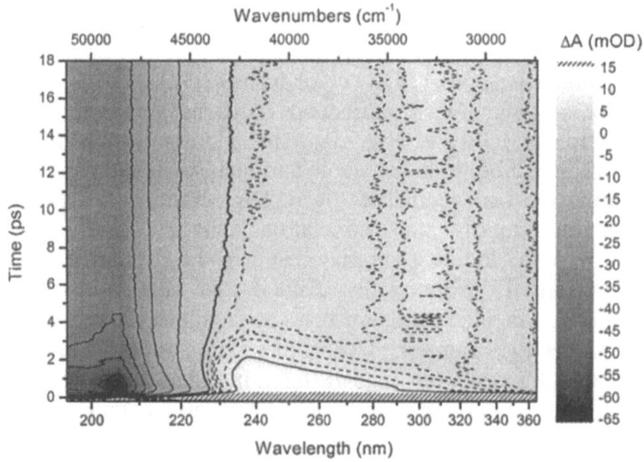

Fig. 1. Contour plot of the transient absorption measured from 194 nm to 400 nm when aqueous NO_3^- is photolyzed at 200 nm

induced absorption from 250 nm to 400 nm. Thus, $ONOO^-$ is formed in 1 ps and the lack of spectral evolution indicates that crossing to the peroxynitrate isomer occurs close to its equilibrium geometry or, alternatively, that the timescale for vibrational relaxation of $ONOO^-$(aq) exceeds that of recombination.

Transient absorption measurements of the iodine cyanide photolysis (not shown) show that approximately half of the excited molecules recombine to form ICN, while about 10 % isomerize to INC within 0.5 ps (instrument limited response time). The remaining iodine and CN fragments escape the solvent cage. The INC has only little vibrational excitation after 1 ps implying that most of the excess excitation energy (3 eV) remaining after the formation of INC is dissipated to the solvent in less than a picosecond. This relaxation time is in good agreement with theoretical estimates based on the golden rule approximation and with MD simulations assuming relaxation in low-frequency IN stretching and INC bending modes. The fast recombination and relaxation, however, makes it impossible to determine if INC is formed by recombination or due to early interconversion during vibrational relaxation of ICN.

Diffusive recombination involving fragments that have escaped the solvent cage further reduces the I and CN quantum yield. However, this recombination process exclusively produces ICN. The lack of diffusive formation of INC is ascribed to steric hindrance by water molecules hydrogen bonding to the nitrogen atom of the solvated CN radical.

4. Conclusions

The experimental data on aqueous ICN and NO_3^- show that constitutional isomerization and subsequent vibrational relaxation occur in 1 ps. This is significantly faster than the isomerization of CH_2I_2 in cyclohexane and acetonitrile and than ICN in chloroform, thus demonstrating the efficient energy dissipation in water.

Acknowledgements. The authors thank Tina D. Poulsen and Svend Knak Jensen for comments and calculations.

References

1 C. Wan, M. Gupta, and A. H. Zewail. Chem. Phys. Lett. **256**, 279, 1996
2 A. N. Tarnovsky, J. –L. Alverez, A. P. Yartsev, V. Sundström, E. Åkesson. Chem. Phys. Lett. **312**, 121, 1999.
3 W. M. Kwok, C. Ma, A. W. Parker, D. A. Phillips, M. Towrie, P. Matousek, D. L. Phillips. J. Chem. Phys. **113**, 7471, 2000.
4 U. Samuni, S. Kahana, R. Fraenkel, Y. Haas, D. Danovich, and S. Shaik. Chem. Phys. Lett. **225**, 391, 1994.
5 J. D. Ray. Journ of Inorg. Nucl. Chem, **24**, 1159, 1962.
6 I Benjamin. J. Chem. Phys. **103**, 2459, 1995.

Ultrafast Intersystem Crossing in a Transition Metal Complex

Eric A. Juban and James K. McCusker

Department of Chemistry, Michigan State University, East Lansing, MI 48824, USA
E-mail: mccusker@cem.msu.edu

Abstract. Femtosecond electronic absorption spectroscopy of $Cr(acac)_3$ in solution allows direct observation of the $^4T_2 \rightarrow {}^2E$ intersystem crossing event, yielding a time constant of ~1 ps. Pump and probe energies were varied to differentiate electronic surface crossing from vibrational cooling.

1. Introduction

Intersystem crossing is an important aspect of virtually all photophysical and photochemical processes involving transition metals. One of the most extensively studied transition metal ions is chromium(III) [1, 2]. A diagram of this ion's electronic structure, depicting d-electrons under octahedral symmetry, is illustrated in Figure 1.

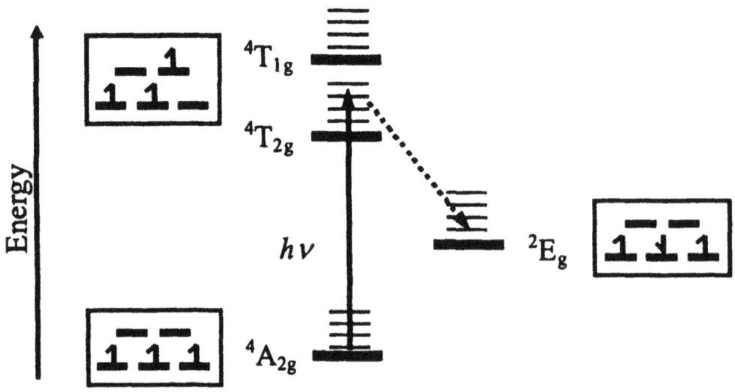

Fig. 1. Schematic representation of the low-lying electronic structure of Cr(III). Solid and dotted lines represent radiative and nonradiative processes, respectively.

The photophysics implied by this figure are typical of many simple transition metal complexes; that is, the initial, spin-allowed transition does not correspond to direct population of the lowest-lying excited state. The subsequent formation of these lower-lying excited states is our primary interest.

Using Cr(III) as an example, three principal processes may occur in the course of excited-state evolution: 1) vibrational relaxation within the ^4T states, 2) intersystem crossing to the ^2E manifold, and 3) vibrational relaxation within the ^2E

state. Conventional wisdom holds that these processes are well-separated in time: vibrational relaxation within the ^4T states will be essentially complete before the spin-forbidden transition to the ^2E occurs. However, we have found that for complexes of ruthenium(II) [3] and iron(II) [4] this mechanistic picture is incorrect. In the case of Fe(II), for example, intersystem crossing has been shown to be significantly faster than vibrational cooling [4]. In the present study we have used femtosecond time-resolved electronic absorption spectroscopy to examine such ultrafast dynamics in Cr(III).

2. Results and Discussion

The molecule we have studied is $Cr(acac)_3$; the acetylacetonate ligand is shown in Figure 2, along with the absorption and low-temperature emission spectra of the compound:

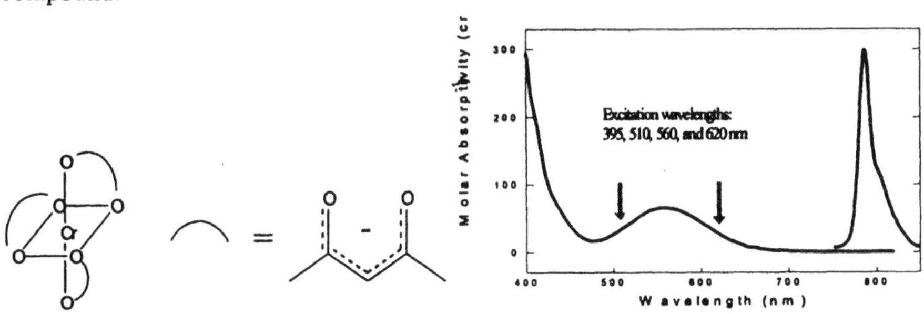

Fig. 2. Geometric structure, room-temperature electronic absorption spectrum (in acetonitrile, left), and low-temperature photoluminescence spectrum (in 4:1 ethanol/ methanol at 100 K, right) of $Cr(acac)_3$

Typical data on $Cr(acac)_3$ are shown in Figure 3 [5]. In all cases there is no further evolution after the initial, single exponential process until the onset of ground state recovery (~700 ps).

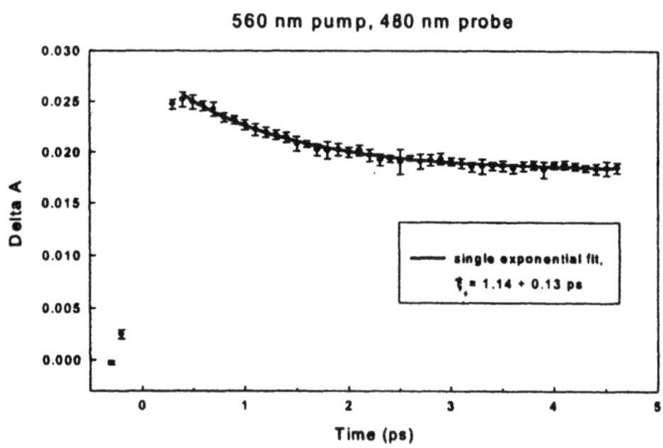

Fig. 3. Ultrafast transient absorption data of $Cr(acac)_3$ in acetonitrile solution at 480 nm, following ~100 fs excitation at 560 nm

In an effort to distinguish vibrational relaxation processes from intersystem crossing, multiple probe and pump wavelengths were examined. No clear trend was identified as a function of probe wavelength at a given excitation wavelength; however, preliminary data reveal a distinct trend as a function of excitation wavelength (Table 1):

Table 1. Kinetics for initial excited-state evolution in $Cr(acac)_3$, as a function of excitation energy. All data correspond to a probe wavelength of 480 nm.

Excitation Wavelength (nm)	520	530	560	580	610	620
τ_{obs} (ps)	1.53 +/- 0.13	1.32 +/- 0.14	1.23 +/- 0.11	1.21 +/- 0.30	1.08 +/- 0.13	0.91 +/- 0.15

4. Conclusions

Vibrational cooling is typically hallmarked by an increase in the relaxation rate as the probe wavelength is increased [6]. The relaxation rate should decrease, however, as the pump wavelength is increased. Our data are inconsistent with this model, suggesting that the kinetic process being monitored is not vibrational cooling. The ~1 ps time constant is therefore assigned as the $^4T_2 \rightarrow {}^2E$ conversion. We are currently working to extrapolate an absorption spectrum for the $^4T_2$ state from the initial absorption amplitudes in order to further delineate the dynamics of this system.

Acknowledgements. We gratefully acknowledge the U.S. National Science Foundation for financial support of this research.

References

1 L.S.Forster, in *Chem. Rev.*, Vol. 90, 331-353, 1990.
2 A.D. Kirk, in *Chem. Rev.*, Vol. 99, 1607-1640, 1999.
3 A.T. Yeh, C.V. Shank, and J.K. McCusker, in *Science*, Vol. 289, 935-938, 2000, and references therein.
4 J.E. Monat and J.K. McCusker, in *J. Am. Chem. Soc.*, Vol. 122, 4092-4097, 2000.
5 Details about the femtosecond spectrometer and data acquisition protocols can be found elsewhere. See: N.H. Damrauer and J.K. McCusker, *J. Phys. Chem. A*, Vol. 403, 8660, 1999.
6 D.A.V. Kliner, J.C. Alfano, and P.F. Barbara, in *J. Chem. Phys.*, Vol. 98, 5375-5389, 1993, and references therein.

Time-Resolved Resonance Raman Studies of Halooxide Photoisomerization Kinetics

Sophia C. Hayes, Catherine C. Cooksey, and Philip J. Reid

Department of Chemistry, Box 351700, University of Washington, Seattle WA 98195, USA
E-mail: preid@chem.washignton.edu

Abstract. The condensed-phase photoisomerization dynamics of chlorine dioxide and dichlorine monoxide are investigated using femtosecond pump-probe and time-resolved resonance Raman. These studies demonstrate that photoisomerization occurs on the tens-of-picoseconds timescale for both compounds.

1. Introduction

The photochemistry of halooxides has been of interest due to the participation of these compounds in a variety of stratospheric photochemical processes.[1,2] The most studied member of this class of molecules is chlorine dioxide (OClO). Of particular interest with respect to atmospheric chemistry is the ability of OClO to photochemically produce atomic chlorine (Cl).[1] We are interested in determining the mechanism of Cl formation following OClO photoexcitation, and in particular the role of an intermediate species, ClOO, in this process. Photochemical studies of OClO isolated in low-temperature matrixes provided the first spectroscopic evidence for ClOO formation, and only recently have time-resolved studies been performed which demonstrate the formation of this intermediate in solution.[2,3] Although ClOO production has been observed, the mechanism by which this isomer is formed remains unclear. To explore the mechanism of photoisomer formation in halooxides, we have performed time-resolved resonance Raman and femtosecond pump-probe studies of OClO and dichlorine monoxide (ClOCl).

2. Experimental Methods

The amplified Ti:Sapphire based spectrometer employed in these studies has been described in detail elsewhere.[2] The time-resolved resonance Raman studies employed pump and probe beams at 390 and 260 nm, respectively, obtained from the second and third harmonic output of a Ti:Sapphire amplifier. The instrument response as measured by the optical Kerr effect in water was 1.2 ± 0.1 ps. In pump-probe studies of ClOCl, photoexcitation at 260 nm was provided by the third harmonic of the amplifier output. Probe wavelengths spanning the UV were generated using the second or third harmonic of the amplifier output, or using the fourth harmonic of the signal field from an optical parametric amplifier (TOPAS). The instrument response as measured by the optical Kerr effect in water was 250 ± 50 fsec.

3. Results and Discussion

Time-resolved resonance Raman (TRRR) studies of OClO dissolved in water and trichlorofluoromethane (freon-11) were performed. Freon-11 was chosen for study since it is an aprotic solvent in which that absence of self-association via hydrogen bonding provides for a weaker solvent cage and a corresponding reduction in geminate recombination relative to water.[2] Representative spectra are presented in Figure 1, where the production of ClOO is evidenced by the scattered intensity at ~1442 cm^{-1}.

Kinetic analysis of these data reveals that the appearance and decay kinetics of ClOO are solvent dependent. Specifically, aqueous ClOO appears and decays with time constants of 27.9 ± 4.5 ps and 398 ± 50 ps, respectively, whereas in freon-11 these timeconstants become 172 ± 30 ps and 864 ± 200 ps. The quantum yield for ClOO production was found to be solvent independent; however, the geminate-recombination quantum yield is dramatically reduced in freon-11.[2] This observation suggests that ClOO is not produced by geminate recombination, but is instead formed via photoisomerization.

To determine if photoisomerization is representative of halooxide photoreactivity, we have investigated another halooxide, ClOCl. Little is known about the solution-

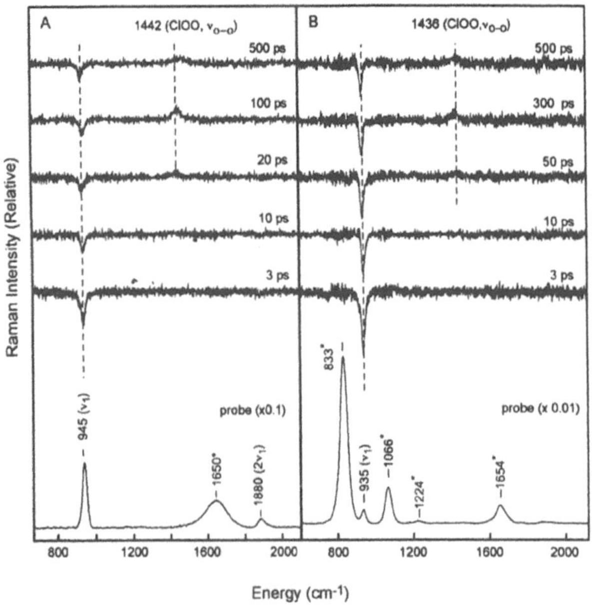

Fig. 1. Time-resolved resonance Raman Stokes difference spectra of OClO dissolved in **(A)** water and **(B)** freon-11. The temporal delay between the pump and the probe is indicated. Transitions marked with an asterisk in the probe-only spectra are due to the solvent.

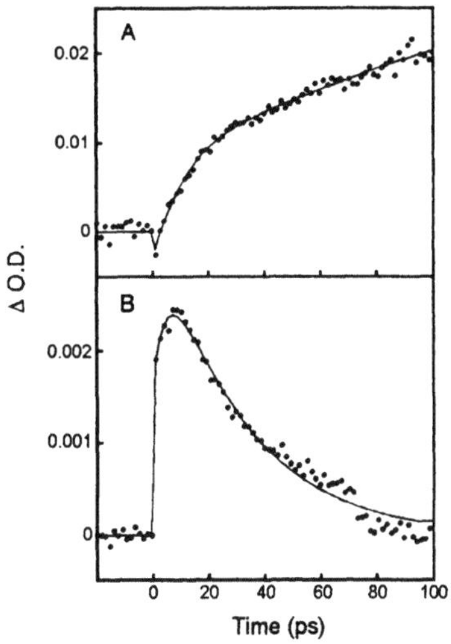

Figure 2. Femtosecond pump-probe studies of ClOCl dissolved in perfluorohexane. (**A**) Pump and probe wavelengths of 260 nm. The increase in optical density that appears with a time constant of ~30 ps is consistent with ClClO production. (**B**) Pump and probe wavelengths of 260 nm and 400 nm, respectively. A rapid increase in optical density is observed followed by decay with a time constant of ~30 ps.

phase reactivity of this compound; however, resonance Raman studies of ClOCl dissolved in carbon tetrachloride performed with 5-ns temporal resolution have observed ClClO formation.[4] Femtosecond pump-probe results are presented in Figure 2. Following photoexcitation, a significant increase in optical density is observed at 260 nm consistent with ClClO formation. The combined studies of OClO and ClOCl photochemistry presented here demonstrate that photoisomerization is a general feature of halooxide photoreactivity in condensed environments.

Acknowledgements. This work was supported by the NSF and by the Petroleum Research Fund administered by the American Chemical Society. PJR is an Alfred P. Sloan Fellow and is a Cottrell Fellow of the Research Corporation.

References

1. V. Vaida and J. D. Simon, Science 268, 1443, 1995.
2. S. C. Hayes, C. L. Thomsen and P. J. Reid, J. Chem. Phys., 115, 11228, 2001.
3. C. L. Thomsen, P. J. Reid, and S. R. Keiding, J. Am. Chem. Soc. 122, 12795, 2000.
4. A. P. Esposito, K. Rousslang, and P. J. Reid, J. Photochem. and Photobio. A, 129, 9, 1999.

Femtosecond Fluorescence Studies of Two-Dimensional Dynamics of Bridged and Unbridged Michler's Ketones

Emile van Veldhoven, Hong Zhang, and Max Glasbeek

Laboratory for Physical Chemistry, University of Amsterdam,
Nieuwe Achtergracht 129, 1018 WS Amsterdam, The Netherlands
E-mail: glasbeek@science.uva.nl

Abstract. Femtosecond fluorescence upconversion experiments are reported for [4,4'-bis(*N*,*N*-dimethylamino)-benzophenone (Michler's ketone) and its bridged derivative compound 3,6-bis(dimethylamino)-10,10-dimethylanthrone (BMK). For Michler's ketone a (sub)picosecond dynamic Stokes shift and a simultaneous decay of the integrated emission intensity are found. It is concluded that the excited-state dynamics of this molecule is determined by phenyl-group twisting *and* by solvation dynamics. In the case of BMK, twisting is absent and the excited-state dynamics is determined solely by (fast) initial solvation followed by population decay to the ground state.

1. Introduction

For di- and triphenylmethane dye molecules in liquid solution it is well-known that the fluorescence quantum yield varies drastically with the viscosity of the solvent. Torsional motions of the phenyl rings of the dye molecules are believed to control the fluorescence quantum yield, i.e., if the solvent viscosity permits phenyl-group twisting the fluorescence is weakened. However, recent theoretical studies of photoexcited TICT states in liquid solution, have emphasized the importance of a two-dimensional reaction coordinate approach [1]. In the latter, the dynamics in the excited state is determined not only by the reaction coordinate (e.g., twisting), but also by the solvation coordinate. In this paper, we present results of femtosecond fluorescence upconversion experiments of two diphenyl methane dyes (Figure 1), Michler's ketone (MK) and its bridged analog 3,6-bis(dimethylamino)-10,10-dimethylanthrone (BMK). For the unbridged MK in alcoholic solution very fast fluorescence decay kinetics of only a few picoseconds is reported, whereas its bridged counterpart BMK has a lifetime at least one order of magnitude longer. The molecules are identified as prototype examples of the slow and fast solvation limits in the model of Kim and Hynes.

2. Experimental

Details of the fluorescence upconversion setup (time response ~ 150 fs) and the time-correlated single-photon counting picosecond fluorescence set-up (time response ~16 ps) were as in [2]. All measurements were performed at room temperature. Pulsed photo-excitation was near 360 nm, fluorescence upconversion detection was under magic angle conditions. For the measurement of fluorescence transients in the time span from 15 ps up to 5 ns, a second laser system, outfitted with a time-correlated single-photon-counting (TCSPC) detection system, was used [2].

3. Results and Discussion

Fluorescence upconversion transients were measured for MK and BMK in the protic solvents methanol, ethanol and decanol, and in the aprotic solvents acetonitrile and DMF. For MK, in alcoholic solution, the upconversion transient shows a fast decay (e.g., in ethanol ~ 2 ps) when detection is at the blue side (λ<440 nm) of the broad fluorescence band; detection at the red side (λ>550 nm) yields a rise with similar time constants. Each transient could be fitted with a multi-exponential function convoluted with the system response function. Typical transients and fittings for MK and BMK in ethanol are shown in Fig. 1.

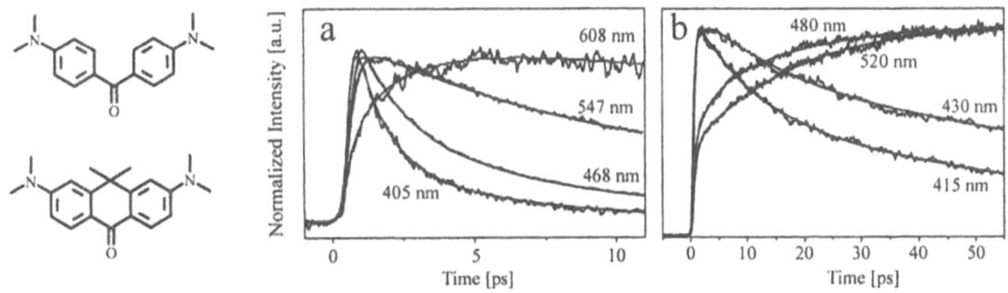

Fig. 1. Structure of MK (upper left) and BMK (lower left). Fluorescence upconversion transients for MK (a) and BMK (b) in ethanol. Detection wavelengths are indicated. Drawn lines are best fits to convoluted multiexponential functions.

The temporal behavior of the emission spectra of MK and BMK in the protic and aprotic solvents was obtained by means of the spectral reconstruction method. For MK, in alcoholic solution, a salient feature is a fast decay of the integrated emission intensity concomitant with the dynamic Stokes shift of the emission band maximum (Fig. 2). In the aprotic solvents acetonitrile and DMF, on the other hand, the dynamic Stokes shift is on a subpicosecond time scale and not accompanied by a drop in the integrated intensity. Moreover, in addition to the fast initial Stokes shift, the fluorescence of MK displays a long decay component (~ 800 ps) that is absent for MK in a protic solvent. The excited-state dynamics of MK in alcoholic solution can be understood applying the same model as given previously for the analog compound auramine [2]. (Auramine has a molecular structure similar to that of MK except that the C=O group should be replaced by a =HNH cation). In this model excited-state relaxation, induced by twisting motions, is along a (quasi) barrierless adiabatic potential energy surface. During this relaxation, the nature of the electronic wave function varies from radiative to nonradiative.

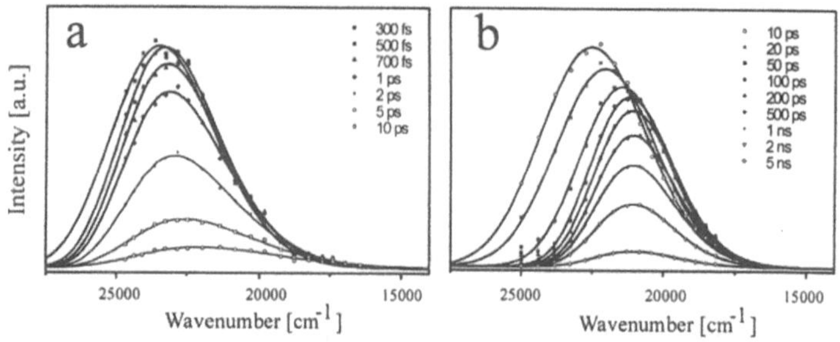

Fig. 2. Time dependence of reconstructed emission spectra of MK (a) BMK (b) in ethanol

The role of twisting is further illustrated by comparing the kinetic data of MK and BMK. In BMK, bridging of the phenyl groups blocks twisting of the latter and, as can be inferred from the solvent dependence of the dynamic Stokes shift, solvation prevails. The solvation times in alcoholic solution (e.g., in ethanol ~ 16 ps) are slower than the characteristic times for the dynamic Stokes shift and the intensity decay of MK (in ethanol ~ 2 ps). Clearly, twisting dynamics is crucial in determining the short-time excited-state dynamics of MK.

The data for MK in the aprotic solvents DMF and acetonitrile are characteristic of fast solvation dynamics only. Now the dynamic Stokes is not accompanied by a drop in the integrated emission intensity. Furthermore, the time dependence of the spectral response function follows the typical solvation times of the liquids [3]. As was discussed recently by Kim and Hynes for TICT states in liquid solution, the excited-state dynamics in general is determined by twisting as well as solvation [1]. The ratio of the twisting and solvation rate constants determines which of the processes dominates. For MK in alcoholic solution we find that twisting prevails (slow solvation limit) while for MK in the aprotic solvents DMF and acetonitrile solvation is rate determining (fast solvation limit).

4. Conclusions

The excited-state dynamics of MK dynamics is found to vary dramatically depending on the typical solvation times known for the solvents used. The dynamics is controlled by twisting (alcoholic solution) and by solvation (DMF, acetonitrile). The data demonstrate that for MK a two-dimensional reaction coordinate model is applicable in the fast and slow limits.

References

1. H.J. Kim and J.T. Hynes, J. Photochem. Photobiol. A: Chem. **105,** 337, 1997.
2. M.J. van der Meer, H. Zhang, and M. Glasbeek, J. Chem. Phys., **112,** 2878, 2000.
3. L. Reynolds, J.A. Gardecki, S.J.V. Frankland, M.L. Horng, and M. Maroncelli, J. Phys. Chem. **100**, 10337, 1996.

Ultrafast Pump-Probe and Resonance Raman Spectroscopic Study of Excited State Electronic Coupling in Two Related Hexanuclear Charge Transfer Compounds

Carolyn J. Mordas[1], Brian W. Pfennig[1], Elmar Schreiber[2], and Andrew B. Bocarsly[1]

[1] Department of Chemistry, Frick Chemical Laboratory, Princeton University, Princeton NJ 08544, USA
E-mail: bocarsly@princeton.edu
[2] Hochschule Bremen, University of Applied Sciences, Neustadtswall 130, 28199 Bremen, Germany
E-mail: eschreib@hs-bremen.de

Abstract. Photoinduced multi-electron transfer is reported in two hexanuclear intervalant charge transfer complexes of Pt(IV) and Fe(III). Femtosecond pump-probe spectroscopy, resonance Raman and electrochemical data suggest the degree of excited state electronic coupling is tunable by changing the electronic nature of the bridging ligand.

1. Introduction

We have previously reported the synthesis [1] and charge transfer dynamics of $[(NC)_5Fe^{II}\text{-}CN\text{-}Pt^{IV}(NH_3)_4\text{-}NC\text{-}Fe^{II}(CN)_5]^{4-}$ (hereafter referred to as Fe(II)-Pt(IV)-Fe(II) or **1**), which is known to undergo a photochemically induced net two-electron transfer when irradiated with light in the visible region. The resulting one-electron [unstable] Fe(II)-Pt(III)-Fe(III) excited state then undergoes a thermal electron transfer to yield two equivalents of Fe(III) and one of Pt(II). Single color pump-probe experiments performed on **1** determined the excited state lifetime of the metal-to-metal charge transfer (MMCT) to be 1920 ± 85 fs [2]. Recent work has involved the synthesis of two related compounds whose structures are given below.

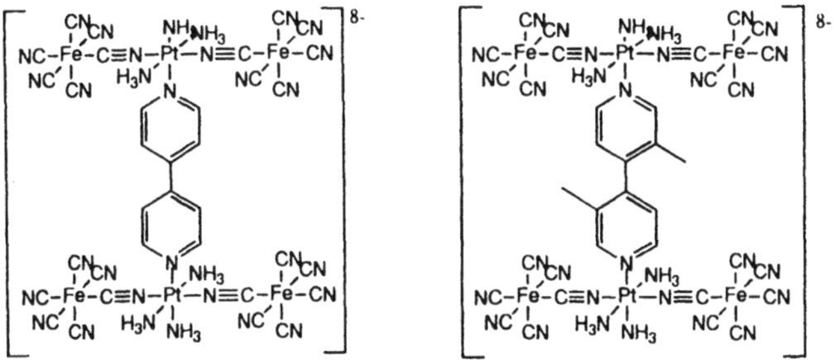

Figure 1. Structures of Compounds 2 and 3

It is expected that in the case of 2, the two platinum centers are in electronic communication with each other within the excited state. Upon irradiation into the Fe(II) → Pt(IV) intervalent charge transfer band, the excited electron becomes delocalized along the Pt-L-Pt bridge. However, in 3, electronic communication between the platinum atoms is not expected since the pyridyl rings in the bridging ligand are non-coplanar due to steric hindrance from the methyl groups.

2. Experimental Methods

The kinetics and dynamics of the electron transfer and the excited state were studied with resonance Raman and ultrafast pump-probe spectroscopy. Resonance Raman spectra were obtained by irradiating into the MMCT band of both compounds at 488 nm.

Single color pump-probe experiments were performed on 2 and 3 to determine the MMCT excited state lifetime. Pulses at 400 nm (200 uJ, 100 fs pulsewidth, 1 kHz repetition rate) were generated by frequency doubling the output of a regeneratively amplified Ti:Sapphire (Spectra-Physics Tsunami) beam. The beam was attenuated and split into a pump and probe beam (10:1 power ratio). The pump beam was modulated at 100 Hz through the use of a chopper and probe intensity was detected using a lock-in amplifier. Samples of 2 and 3 in water were flowed through a quartz cuvette with a 1 cm pathlength. Data were collected at 100 fs intervals by signal averaging 20 readings. A schematic of the pump-probe setup is given in Figure 2.

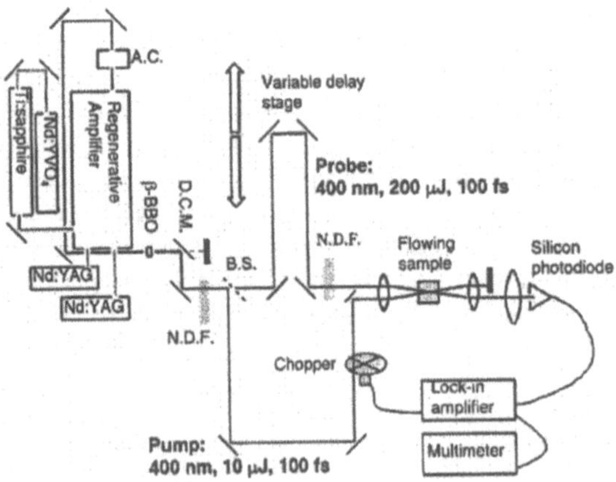

Figure 2. 400 nm pump-probe schematic

3. Results and Discussion

The resonantly enhanced Raman spectra of **2** and **3** are nearly identical, except for the presence of three peaks at 1234, 1291 and 1604 cm^{-1} corresponding to ring modes of the bridging ligand in **2**.[3] This yields evidence that the dipyridyl ligand in **2** is associated with the charge transfer transition, while the dimethyl dipyridyl ligand in **3** is not. The transient absorption data are well fit by a single exponential, yielding an excited state lifetime of 3240 ± 60 fs for **2** and 1780 ± 60 fs for **3**. These values are consistent with delocalization along the conjugated dipyridyl ligand in complex **2**. In the case of **3**, the excited state lifetime is comparable to that of **1**, since there is no delocalization of the transferred electron in the excited state.

4. Conclusions

The degree of delocalization of the electron in the excited state in the intervalent charge transfer systems studied here is tunable by changing the ligand bridging the platinum centers. Resonance Raman is a useful method in determining the extent of delocalization in these systems because it yields information about ligand modes associated with the electron transfer. Ultrafast pump-probe spectroscopy gives additional information about the lifetime of the excited state since the conjugation of the bridging ligand leads to delocalization and increased stability of the Pt^{III} excited state, lengthening the lifetime.

Acknowledgements. The authors would like to thank A. McCloskey for contributions to the syntheses of **2** and **3**. BWP would like to thank the Donors of The Petroleum Research Fund, administered by the American Chemical Society, for their generous support of this research. CJM and ABB acknowledge the National Science Foundation for funding under grant number CHE-0079169. The authors are also grateful to the State of New Jersey for funding the Center for Ultrafast Laser Applications at Princeton University.

References

1 M. Zhou, B.W. Pfennig, J. Steiger, D. Van Engen, and A. B. Bocarsly, *Inorg.Chem*, **29**, 2456-2460 (1990).
2 D.F. Watson, H.S. Tan, E. Schreiber, and A.B. Bocarsly, in preparation.
3 B.W. Pfennig, C.J. Mordas, A.McCloskey, J.V.Lockard, P.M.Salmon, J.L.Cohen, D.F. Watson, and A.B.Bocarsly *Inorg. Chem.*, accepted.

Ring Opening Reaction Dynamics of a Photochromic Diarylethene Derivative

Sangdeok Shim and Taiha Joo

Division of Molecular and Life Sciences, Department of Chemistry, Pohang University of Science and Technology, Pohang, 790-784, Korea
E-mail: thjoo@postech.ac.kr

Abstract. Ring opening reaction dynamics of a photochromic dye in solution has been studied by transient absorption and time-resolved fluorescence of the closed form. Reaction rate is determined accurately and a wave packet motion in the excited state is observed.

1. Introduction

Diarylethene derivatives have received much attention as a photochromic material for optoelectronic devices such as optical memories and optical switches [1]. Upon UV irradiation, the open isomer of 2,3-Bis(2-methylbenzo[b]-thiophen-3-yl)hexafluorocyclopentene (o-BTF6, see Fig. 1) undergoes ring closure reaction, whereas the closed isomer (c-BTF6) undergoes ring opening reaction by visible light. Thermal stability and resistance to fatigue as well as prompt conversion rates are the key elements for successful applications. Most studies on the dynamics focus on ring opening reaction, which has added complexity due to conformational diversity of the open form. In addition transient absorption (TA) has been used exclusively. In this work, we have investigated the ring opening dynamics of c-BTF6 by TA and time-resolved fluorescence (TRF) of the closed form to observe molecular reaction dynamics in the excited state.

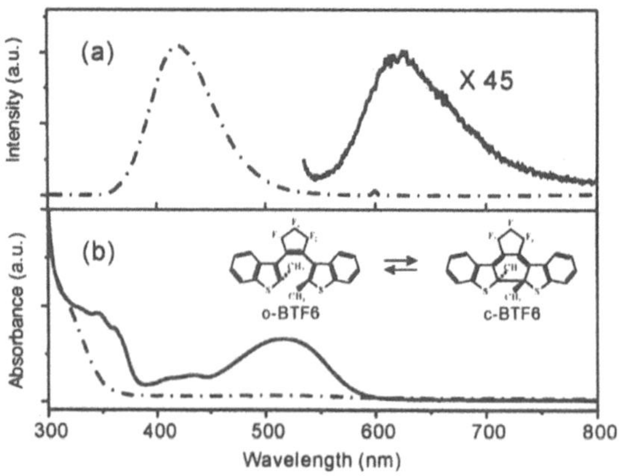

Fig. 1. Emission (a) and absorption (b) spectra of the open (dash dot) and closed (solid) forms of BTF6 in n-hexane.

2. Experimental Methods

c-BTF6 in n-hexane solution was prepared from o-BTF6 by shining UV light. The excitation light at 530 nm for both TA and TRF measurements was generated by a home-built Ti:sapphire amplifier and optical parametric amplifier system generating tunable 50 fs pulses in the visible at 5 kHz. TRF was measured by fluorescence up-conversion technique. Time resolutions of the TA and TRF measurements were 50 fs and 150 fs, respectively.

3. Results and Discussion

Fig. 1 shows stationary absorption and emission spectra of the open and closed forms of BTF6 in n-hexane. TA spectra measured by continuum probe pulse display several bands over the entire visible region originating from ground state bleach (GB), excited state stimulated emission (ESE), and excited state absorption (ESA). Fig. 2 shows TA time profiles near the absorption maximum (530 nm) and at 710 nm, where the signal is due to the excited state absorption exclusively. The signal at 530 nm is negative at early times, which indicates that it is dominated by ESA from S_1. The signal then rises by two time constants 4.7 ps and 22 ps to a constant positive value (GB) indicating that the ring opening reaction has occurred. If the 22 ps transient is due to the relaxation to the ground state as assigned in other studies [2,3], the signal should converge to zero. The TA signal at 710 nm is negative initially and rises to zero by exactly the same time profile observed for the 530 nm transient.

Fig. 2 also shows time resolved spontaneous fluorescence of the c-BTF6. The TRF signal decays by a double exponential with time constants 3.0 ps and 21 ps. Since the ring opening reaction quantum yield of this particular molecule is close to 1, together with the TA results, the 21 ps process can be safely assigned to the ring opening reaction.

Two possibilities may be considered for the origin of the 3 ps component: solvation process or a conformational change to a precursor state in the S_1 surface.

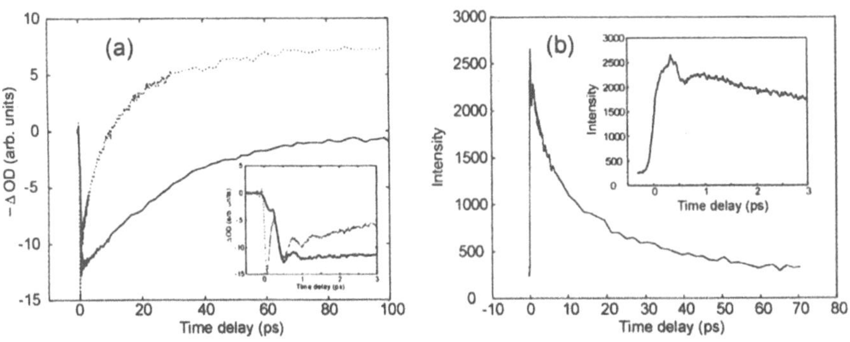

Fig. 2. (a) Transient absorption of c-BTF6 in n-hexane at 530 nm (dotted line) and at 710 nm (solid). (b) Time resolved fluorescence measured at the emission maximum of the c-BTF6.

Solvation process was ruled out from the detection wavelength dependence: The TRF time profile does not change when the detection wavelength is varied over the entire emission band. It is therefore concluded that the 3.0 ps component is not due to solvation but due to a change of the conformation to possibly a precursor state, which may show lower transition dipole moment between S_1 and S_0.

The TA signals at 530 nm and at 710 nm and the TRF transient show a single oscillation of 66 cm^{-1}, although the pump pulse is capable of creating wave packets with much higher frequencies. This shows that the S_0-S_1 transition is coupled almost exclusively with this 66 cm^{-1} mode. Quantum mechanical calculation [4] shows large displacement between the S_0 and S_1 potential surfaces along the reaction coordinate, the C-C bond where the bond break occurs. Therefore, the 66 cm^{-1} vibrational mode is likely to be a vibrational mode involved with the reaction coordinate. In addition, the TRF profile does not change as the excitation wavelength is varied from 500 to 530 nm. These observations suggest that the Franck-Condon region is a bound state with rather high barrier to the open form and the reaction proceeds via curve crossing to the ground state instead of the barrier crossing in the excited state.

4. Conclusions

It is shown through time-resolved fluorescence and transient absorption that the ring opening reaction of the diarylethene to the open form (or to a precursor state) proceeds somewhat slowly compared to the previously believed a few ps time scale. In addition, it is not likely that the reaction proceeds through barrier crossing in the excited state. It is likely that the reaction undergoes by curve crossing to the ground surface.

References

1 M. Irie, Chem. Rev. **100**, 1685, 2000.
2 J. Ern, A. T. Bens, A. Bock, H.-D. Martin, and C. Kryschi, J. Lumin. **76/77**, 90, 1998.
3 J. Ern, A. T. Bens, H. D. Martin, S. Mukamel, S. Tretiak, K. Tsyganenko, K. Kuldova, H. P. Trommsdorff, and C. Kryschi, J. Phys. Chem. A **105**, 1741, 2001.
4 H.-G. Cho and B.-S. Cheong, Bull. Kor. Chem. Soc. **19**, 308, 1997.

Solvent Control of Electron Transfer Dynamics

Erik R. Barthel, Ignacio B. Martini, Ernö Keszei, and Benjamin J. Schwartz

UCLA Department of Chemistry and Biochemistry, Los Angeles, CA 90095-1569 USA
Tel. (310) 206 4113; Fax (310) 206 4038
E-mail: schwartz@chem.ucla.edu

Abstract. We investigate the charge-transfer-to-solvent (CTTS) dynamics of sodide in several solvents. Since this CTTS system has only electronic degrees of freedom, the dynamics provide a window into how solvent motions control electron transfer.

It is of great interest to understand how motions of solvent molecules control the dynamics of electron transfer (ET). Here we tackle this problem using the charge-transfer-to-solvent (CTTS) transition of sodide (Na^-) [1]. Absorption of a visible photon by Na^- produces a localized excited state bound only by the polarization of the solvent, from which solvent motions promote electron detachment. Since the CTTS band of Na^- is similar in a variety of ethers and amines [2], we can use ultrafast spectroscopy to determine how the local solvent structure controls the dynamics of CTTS detachment. We find that there is a sub-ps delayed detachment of the electron from the CTTS excited state (Na^{-*}) followed by back ET (geminate recombination) on a time scale of a few ps. For ether solvents, the back ET times correlate with solvent polarity, with more polar ethers having shorter back ET times. We also find that this trend does not hold in non-ether solvents, and there appears to be no correlation of the forward ET (detachment) time with solvent polarity. This points to the importance of local solvent structure in determining the CTTS detachment dynamics.

Figure 1A shows the absorption spectrum of the Na^- CTTS transition in tetrahydrofuran (THF), tetrahydropyran (THP), diethyl ether (DEE), and hexamethylphosphoramide (HMPA); Figure 1B shows the chemical structures of each of these solvents. The spectrum of Na^- is similar not only in the three ethers but also the more polar HMPA. The ~2 μm absorption of the solvated electron in HMPA is also similar to that in ethers. This is a result of the diffuse nature of the positive dipole of HMPA and the difficulty for large HMPA molecules to form cavities around negative particles with radii smaller than 4-5 Å [3], leading

Figure 1. A: Absorption spectrum of Na^- in different solvents; B: Chemical structures of the different solvents; C: Absorption band of Na^{-*} as determined by fitting to model in [1].

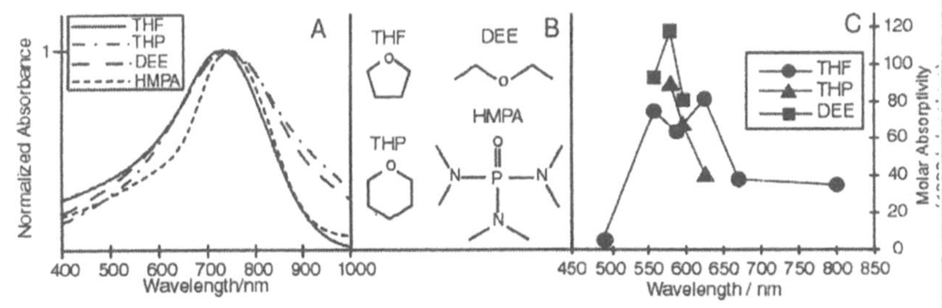

to a similar cavity size as for electrons in ethers. These factors also explain why the spectrum of Na^- in HMPA is so similar to that in the ethers.

Figure 1C shows the visible transient absorption spectrum of the sodide CTTS excited state, Na^{-*}, as determined by fitting pump-probe data to a detailed kinetic model presented previously [1]. This species has its absorption maximum near 590 nm, the location of the sodium D line. The similarity of the spectrum in all solvents investigated supports our assignment of this absorption to a Na atom core that does not interact strongly with the surrounding solvent.[1]

Figure 2 shows pump-probe data for Na^- excited at ~500 nm and probed at both ~580 nm and at ~2 μm. The solid curves are fits to the data using our kinetic model [1]. Probing at 580 nm monitors both Na^{-*} and the ground state bleach of Na^-, while probing at 2 μm monitors solvated electrons. In the three ether solvents, the 580-nm transient absorption is instrument-limited, and then decays as the electron detaches from Na^{-*} and the solvent rushes in to destroy Na^{-*} and create a solvated Na atom. Although the ejection mechanism is qualitatively similar, it is clear that the time for the forward ET, as measured by the decay of the Na^{-*} absorption, is quite different in the three ethers.

This conclusion is also supported by the matching rise times for the solvated electron's absorption at ~2 μm, which also reflect the e^- detachment time. The back ET (geminate recombination) time scale is also different for each solvent. As shown in Table 1, the back ET times increase with decreasing solvent polarity for the ethers [4]. However, this trend does not include the highly polar solvent HMPA, in which the back ET time is considerably slower.

The trend in back ET times with solvent polarity, and the lack of such a trend for the forward ET times, suggests different mechanisms for the two ET processes. The initial photoexcitation of Na^- produces a *p*-like state that must detach and localize into an *s*-like solvated electron. We believe that translational motions of the solvent molecules into the *p*-like angular node promote this detachment. In the fastest solvent, THF, inward translations of the nearly planar ring can be assisted by translation-rotation coupling, as seen previously in MD simulations [5]. In the next-fastest solvent, DEE, we believe that free rotation about the C-O bond decreases the ability (relative to THF) of the first-shell molecules of this solvent to undergo inward translation. Finally, for slow THP, it is likely that packing of the solvent into chair-like structures makes it difficult for the first-shell solvent molecules to translate into the angular node, resulting in

Figure 2. Pump-probe data for Na^- in different solvents. Pump and probe wavelengths are indicated in the legends. Solid curves are part of global fits to the data using the kinetic model described in [1].

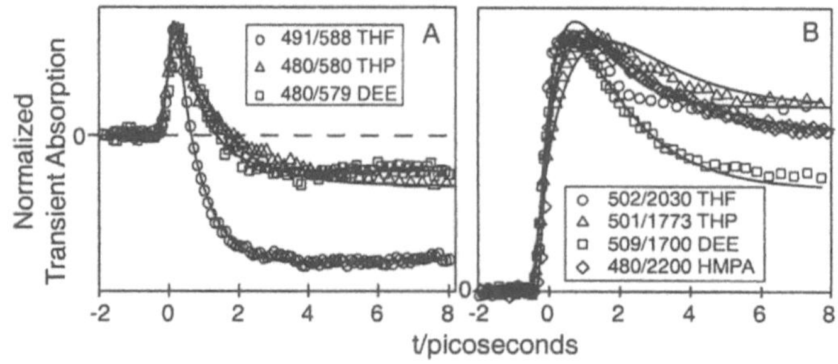

Table 1. ET times for Na^- in solvents with different dielectric properties; the dielectric values are taken from [4]. The ET times have an uncertainty of ±0.05 ps.

solvent	abbreviation	$\varepsilon/\varepsilon_0$	μ/D	ET Times (ps) forward	reverse
diethyl ether	DEE	4.42	1.15	0.82	1.53
tetrahydropyran	THP	5.68	1.63	1.26	1.38
tetrahydrofuran	THF	7.47	1.75	0.70	0.78
hexamethylphosphoramide	HMPA	29.00	5.55	0.41	2.51

a slower response. This type of packing effect is also seen in the pronounced liquid structure for nearest neighbors in the related solvent, cyclohexane [6].

In contrast to detachment, the fast back ET is due to the recombination of contact pairs [1] in which the sodium atom is in the electron's first solvation shell. The recombination progresses via a nonadiabatic transition, the rate of which depends on the overlap of the electron's wave function with its geminate Na atom partner. This is why the macroscopic dielectric properties of the solvent become important in stabilizing the contact pair: the less polar the solvent, the better solvated the neutral sodium atom and the less favorable the back ET.

The large differences in the CTTS dynamics in very polar HMPA likely are due to a different molecular solvent structure. The bulky nature of this solvent creates large cavities even before electron detachment [3]. Thus, the forward detachment can proceed rapidly since little solvent translational motion is needed, and the unusually high stability of the newly-formed solvated electron cavity could then result in the observed slow back ET time

Overall, the results show that it is not sufficient to consider only the dielectric properties of a solvent when considering ET dynamics. The local molecular structure of a solvent can have direct effects on the motions that control the dynamics of ET. For the Na^- CTTS reaction in weakly polar ether solvents, the forward ET time is dictated primarily by the local solvent structure, whereas the back ET depends mostly on solvent polarity. For the bulky polar solvent HMPA, the slow back ET reaction likely is due to the tendency for the solvent to form unusually stable cavities for the solvated electron.

References

[1] E. R. Barthel, I. B. Martini, B. J. Schwartz, "How Does the Solvent Control ET? Experimental and Theoretical Studies of the Simplest Charge Transfer Reaction," *J. Phys. Chem. B* **105**, 12330-41 (2001).

[2] M. T. Lok, F. J. Tehan, J. L. Dye, "Spectra of Na^-, K^-, and e^-solv in Amines and Ethers," *J. Phys. Chem.* **76**, 2975-81 (1972).

[3] E. A. Shaede, L. M. Dorfman, G. J. Flynn, D. C. Walker, "Spectrum, Kinetics, and Radiation Chemical Yield of Solvated Electrons in Hexamethylphosphoric Triamide," *Can. J. Chem.* **51**, 3905-3913 (1973).

[4] C. Laurence, P. Nicolet, M. T. Dalati, J. -L. M. Abboud, R. Notario, "The Empirical Treatment of Solvent-Solute Interactions: 15 Years of π*," *J. Phys. Chem.* **98**, 5807-5816 (1994).

[5] V. Tran, B. J. Schwartz, "Role of Nonpolar Forces in Aqueous Solvation: Computer Simulation Study of Solvation Dynamics in Water Following Changes in Solute Size, Shape, and Charge," *J. Phys. Chem. B* **103**, 5570-80 (1999).

[6] J. G. Harris, F. H. Stillinger, "Isomerization and Inherent Structure in Liquids. A Molecular Dynamics Study of Liquid Cyclohexane," *J. Chem. Phys.* **95**, 5953-65 (1991).

Reexamining CTTS dynamics of Na^- – Probing the excited state and electron using multipulse pump-probe.

Roseanne J. Sension*, Zhaohui Wang, Ofir Shoshana, Bixue Hou, and Sanford Ruhman

Department of Physical Chemistry, and the Farkas Center for Light Induced Processes, the Hebrew University, Jerusalem 91904, Israel. E-mail: sandy@fh.huji.ac.il
*on sabbatical leave from: Dept. of Chemistry and the FOCUS Center, University of Michigan, Ann Arbor, MI 48109-1055

Abstract: Probing electrons ejected from photorexcited $Na^-_{(THF)}$ with 30 fsec 2 and 3 pulse experiments suggests that the CTTS excited state decays in less than 200 fsec. An intermediate earlier assigned as $Na^{-*}$is reassigned to a "dry" Na^0.

Recent ultrafast investigation of photoinduced Charge Transfer To Solvent (CTTS) from atomic anions is aimed at understanding how solvent motions control curve crossing and relaxation leading to the thermal products. In particular, lack of any internal structure in the ion leaves the solvation coordinate alone to moderate the ejection dynamics. Accordingly, the precise chronology of events leading to the charge transfer is of central interest. In this spirit study of the CTTS dynamics of Na^- in the cyclic ether THF has been initiated. Schwartz and coworkers have bridged the gap between reactants to the known nanosecond products in THF solution, using ~150 fsec time resolution [1]. Excitation of the broad and intense CTTS band (λ_{max}=730 nm) was observed to generate solvated electrons (λ_{max} =2 μm), and neutral sodium (λ_{max}=900 nm), with a rapid component of recombination whose amplitude depends heavily on the photon energy. A single additional intermediate, with a narrow 600 nm absorption, was detected and assigned to the CTTS excited state, decaying exponentially in 700 fsec to generate the products.

Our initial effort, which concentrated on probing from 500-950 nm following excitation at a number of frequencies demonstrated that the CTTS absorption band is inhomogeneously broadened, with sub-bands polarized in definite directions in the solvent [2]. This was presumably due to symmetry breaking of the ion environment by solvation. Also, a rapidly relaxing excess bleach to the red of the excitation frequency was observed and suggested to correctly reflect the excited state decay period. This was however difficult to determine with confidence without directly observing the rise of the ejected electrons. Here this system is reexamined with similar time resolution, probing in the range 1.1-1.6 μm. At the blue edge of this range the extinction of solvated neutral sodium far exceeds that of the free electron, while at the other end the opposite is true, providing an opportunity to compare the rise of both, even though we are not accessing the peak absorbance of the electron.

Experiments were conducted with a multipass amplified Ti;sapphire laser system built in our lab, producing 28 fsec pulses containing 600 μJ of energy centered at 790 nm. A commercial NOPA provided tunability, with a booster stage of amplification to increase the output in the IR. This consisted of a 1mm type 2 BBO crystal pumped by ~120 μJ of the amplifier fundamental, and served both to shorten and boost the IR energy ten fold to ~30 μJ. Excitation pulses were derived by splitting off portions of the fundamental or by generating 1.1 μm in the NOPA/booster setup. Probes were generated by filtering segments of white continua generated either with 790 or 1100 nm pumping. Generating stable probe pulses at the red edge of the probing range proved challenging, presumably due to the change in the sign of sapphire GVD near 1.5 μm. White light beyond 1.4 μm was accordingly generated in 3 mm of SF11 glass, which is free of this problem in the said range. The SF11 continuum generated with 1.1 μm pump was very energetic and stable in the IR, extending only marginally into the red portion of the visible. Na^- in THF was prepared as described before. Most data were obtained with OD 2 solutions. The three pulse experiment was conducted in OD 0.8 solution, to limit the absorbance of the sample at 1.1 μm.

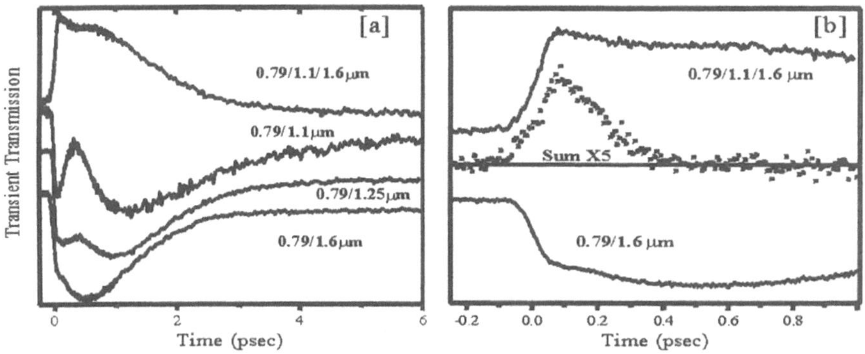

The results are portrayed in the figure. Panel a shows 6 psec scans at 1.1, 1.25, and 1.6 μm following sodide excitation at 790 nm. In all cases the excitation leads to a net reduction in transmission, since the parent ions absorption is very low throughout the range. The first striking observation is that despite delayed increases in absorption, a very substantial instantaneous absorption rise is observable in all. At 1.1 μm, this component is superceded by a very rapid decay in absorption, giving way to a delayed absorption rise peaking at ~1.3 psec. At 1.6μm, the instantaneous rise is almost directly continued by a slower rise, peaking near 600 fsec, and later decaying to a constant background. Finally, probing at the intermediate wavelength exhibits separate delayed rises, which follow the initial instantaneous jump in absorption.

The instantaneous absorption rise, and its rapid decay at 1.1 μm must indicate that a hitherto unobserved short lived intermediate exists, absorbing throughout the NIR. We tentatively assign this feature to the CTTS excited state. The very distinct peaking times of absorbance at 1.1 μm, and 1.6 μm demonstrates that solvated neutral sodium, dominating delayed absorption at 1.1μm, and the electron, dominating absorption at 1.6 μm, rise separately, indicating that a chemical stage separates them. This is significant, since they were previously assumed to appear together in the wake of photoejection.

In order to investigate the nature of the seemingly distinct components of the 1.1μm absorption, a three pulse experiment was conducted. When excited at ~800 nm, nearly 90% of the sodide is known to recombine within picoseconds. Assuming that the fast initial feature is the CTTS excited state, adding an additional 1100nm quantum would be equivalent to initial excitation with a 460 nm photon –known to lead to negligible ultrafast recombination. In contrast the neutral solvated sodium should not be effected at all by 1.1μm excitation. Accordingly the following 3 pulse experiment was devised: at time zero a 790nm pulse initiates the sodide excitation, with a 1.6 μm probe kept at a constant delay of 33 psec, following the residual absorption of the ejected electrons. An intense and chopped 1.1 μm excitation pulse is scanned from before to after the CTTS excitation, and its effect on the electron recombination is recorded. The results are depicted at the top of panel a, and all of panel b. It shows clearly that 1.1μm excitation during the short lifetime of the excited state reduces rapid recombination and disappearance of electron absorption. The later portion of the signal does not follow the 1.1 absorption scan, resembling instead the appearance and decay dynamics of the electron (probed at 1.6 μm). In fact, panel b demonstrates how the three pulse signal can be separated into a superposition of the two. Obviously then, the nascent electrons still absorb significantly at 1100 nm.

The dynamic scenario arising from these results is as follows. Excitation in the CTTS band produces a very short lived excited state, which preserves the photon energy and decays by electron ejection in less than 200 fsec. The fragment formed along with an electron is not the known neutral, but a "dry" sodium atom which quite naturally absorbs narrowly very near where it would in vacuum. The transition to the solvated neutral takes time, and explains the kinetic distinctness of the neutral sodium band and the electron band. The delayed buildup of 1.6 μm absorbance could represent electron solvation, but due to the seeming overlap in the electron and CTTS excited state bands, this question will require further study. Finally, the new assignment of the excited state means that the process of charge transfer is at least 4 times faster than earlier interpreted. The implications of this on the mechanisms of the transfer must be scrutinized further.

Acknowledgements. This work was supported by the Israel-US Binational Science Foundation. The Farkas Center is supported by the Minerva Gesellschaft, GmbH, Munich, Germany. RJS was supported by NSF 0078972 and 9727948 and a Lady Davis Visiting Faculty Fellowship.

References

1. E. R. Barthel, I. B. Martini, and B. J. Schwartz, *J. Phys. Chem,* ***B*****, 105** 12230 (2001).

2. Z. Wang, O. Shoshana, and S. Ruhman, *Ultrafast Phenomena XII*, 624-626, Springer Verlag, Heidelberg 2001.

Ultrafast site-specific mid-infrared spectroscopy of excited-state intramolecular proton transfer

M. Rini, A. Kummrow, J. Dreyer, E. T. J. Nibbering, and T. Elsaesser

Max Born Institut für Nichtlineare Optik und Kurzzeitspektroskopie,
Max Born Strasse 2A, D-12489 Berlin, Germany
E-mail: nibberin@mbi-berlin.de

Abstract. We present an ultrafast mid-infrared study of excited-state intramolecular proton transfer in 2-(2'-hydroxyphenyl)benzothiazole. Monitoring the C=O stretching mode of the product state reveals the reaction time, anharmonic coupling features and vibrational relaxation time scales.

Photoinduced excited-state proton transfer represents an elementary structural change that frequently occurs on ultrafast times. In the femtosecond time domain, such reactions have mainly been studied by measuring the build-up of fluorescence or stimulated emission from the reaction product [1,2]. Though the rise of emission has been resolved, the rise times depend on the spectral position within the emission band and - thus - an interpretation in terms of proton transfer dynamics may be difficult. In contrast, with ultrafast vibrational spectroscopy one can monitor changes in the vibrational spectra of relevant functional groups, in this way providing direct site-specific access to molecular structure [3]. Here, we apply femtosecond mid-infrared spectroscopy for the first time to study the intramolecular excited state proton transfer in 2-(2'-hydroxyphenyl)benzothiazole (HBT), transforming the enol into the keto tautomer of HBT. We monitor the dynamical characteristics of the C=O carbonyl stretching mode of the product keto state. The onset of the vibrational band of this mode indicates the appearance of the product state. In addition, the shape of the band is potentially sensitive to vibrational energy redistribution after proton transfer, either when the C=O stretching mode is vibrationally excited, or by anharmonic coupling to other intramolecular modes that are themselves transiently excited

HBT dissolved in C_2Cl_4 is excited by a 40 fs pulse in the range between 315 and 350 nm [4] and the resulting change of vibrational spectra is probed with 100 fs pulses tunable in the mid-infrared [5]. In Fig. 1 we present transient infrared spectra, i.e. the change of vibrational absorption recorded between 1410 and 1590 cm^{-1} together with the infrared spectrum of the enol ground state. The bands present in the ground state show up in the transient spectra as bleached signals. A prominent new band is formed at 1530 cm^{-1}, representing the C=O stretching band of the keto species. Strength and spectral position of this band agree with earlier picosecond measurements [6]. The new C=O band builds up with a delay of 30-50 fs (Fig. 2a), representing the formation time of keto-HBT. Based on the expected value for the anharmonic shift for this type of vibration we conclude from the lack of change in strength and width of the C=O band after its formation that the new band is formed in its v=0 vibrational ground state. On the other hand, the center frequency of the C=O band blue-shifts by about 6 cm^{-1} on a picosecond time

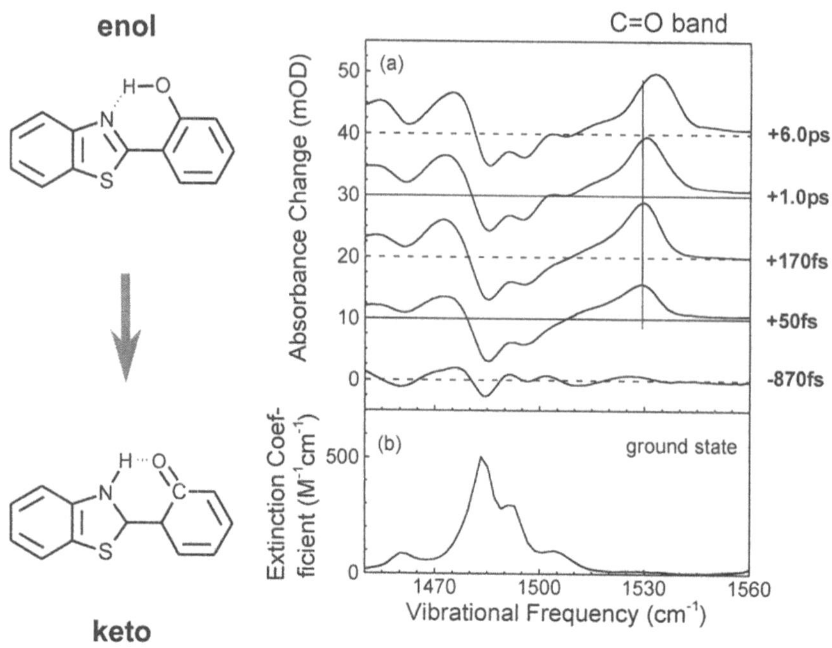

Fig. 1. Proton transfer reaction scheme of HBT (left panel); transient (a) and steady-state (b) infrared absorption spectra (right panel). Transient spectra are corrected for the frequency-independent solvent contribution.

scale, an effect attributed to IVR and vibrational cooling. We ascribe these findings to be the most likely consequence of non-thermal distributions, generated during the ESIPT reaction, for other modes that are anharmonically coupled to the C=O stretching mode.

Our data provide new insight into the anharmonic couplings between high-frequency modes involving proton motion and low-frequency modes. The C=O absorption measured as a function of delay time (see Fig. 2b/c) displays oscillatory features that are due to the anharmonic coupling of low-frequency modes at about 50 and 120 cm^{-1}. The latter vibration represents a Raman-active in-plane deformation mode that is impulsively excited upon electronic excitation [7]. The coherence of this mode is also detectable in optical studies of the product keto state of HBT [2]. On the other hand, the 50 cm^{-1} oscillation that we ascribe to coherence in an infrared-active out-of-plane deformation mode, can neither be impulsively excited by the pump pulse nor be detected in optical pump-probe studies of electronic spectra due to selection rules. Excitation of this mode is thus driven by nonradiative processes after preparation of the excited enol-state. From the observations that a) the 50 cm^{-1} mode is only observable at excitation

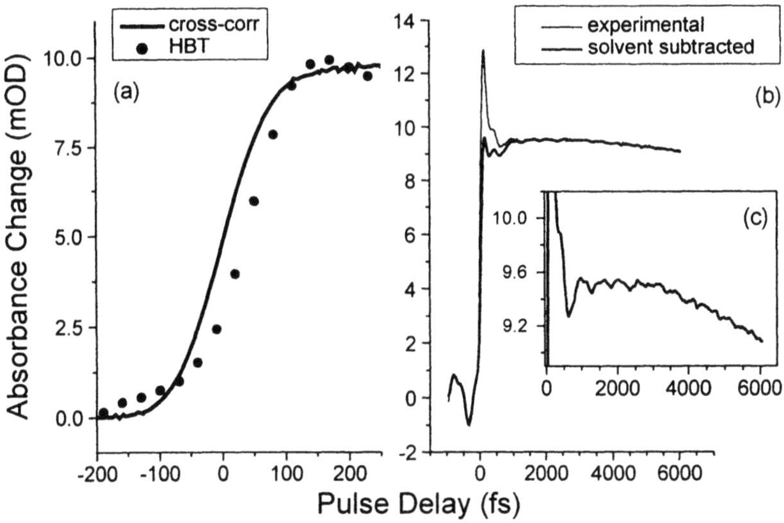

Fig. 2. Transient infrared absorbance change recorded at 1530 cm^{-1} of HBT excited at 335 nm. The data show a solvent contribution around zero delay, as well as a coherent modulation due to anharmonic coupling to low-frequency modes.

wavelengths of 315 and 335 nm, but not at 350 nm where the origin transition is located, whereas b) the reaction time appears to be independent of the excitation energy, we are led to the tentative explanation that the coherent excitation of the infrared-active 50 cm^{-1} out-of-plane deformation mode is driven by an IVR mechanism in the enol excited state.

References

1 F. Laermer, T. Elsaesser, and W. Kaiser, Chem. Phys. Lett. **148,** 119, 1988.
2 S. Lochbrunner, A. J. Wurzer, and E. Riedle, J. Chem. Phys. **112**, 10699, 2000.
3 C. Chudoba, E. T. J. Nibbering, and T. Elsaesser, Phys. Rev. Lett. **81**, 3010, 1998.
4 A. Kummrow, M. Wittmann, F. Tschirschwitz, G. Korn, and E. T. J. Nibbering, Appl. Phys. B **71**, 885, 2000.
5 M. Rini, A. Kummrow, J. Dreyer, E. T. J. Nibbering, and T. Elsaesser, Faraday Discussions, 2002 (in press).
6 T. Elsaesser, and W. Kaiser, Chem. Phys. Lett. **128**, 231, 1986.
7 M. Pfeiffer, K. Lenz, A. Lau, T. Elsaesser, and T. Steinke, J. Raman Spectrosc. **28**, 61, 1997.

Ultrafast Relaxation Dynamics of the Solvated Electron in Liquid Ammonia

Andreas-N. Unterreiner[1], Jörg Lindner[2], and Peter Vöhringer[2]

[1] Institute for Physical Chemistry, University of Karlsruhe, Kaiserstraße 12, 76128 Karlsruhe, Germany, E-mail: Andreas.Unterreiner@chemie.uni-karlsruhe.de
[2] Biomolecular and Chemical Dynamics Group, Max-Planck-Institute for biophysical Chemistry, Am Faßberg 11, 37077 Göttingen, Germany, E-mail pvoehri@gwdg.de

Abstract. We present what is to our knowledge the first femtosecond pump-probe studies on the solvated electron in liquid ammonia. Ultrafast near-infrared excitation of the electrons can be understood quantitatively in terms of a simple temperature jump which red-shifts within our time-resolution the electronic resonance by 4800 cm^{-1} into the mid-infrared spectral region. A subsequent blue-shift of the time-dependent absorption spectrum indicates the dynamics of electron cooling and occurs on a time scale of 200 fs. Both pump pulse-induced temperature jump and cooling occur under shape stability of the electron resonance.

1. Introduction

Solvated electrons are key intermediates in radiation chemistry [1]. Therefore, their optical response to impulsive preparation and/or excitation are of central importance. However, despite being first discovered as a metastable species in liquid ammonia solutions, the relaxation pathways of electrons have only been studied in polar, protic liquids like water [2] or alcohols. Here, we present the first femtosecond pump-probe experiments on the solvated electrons in liquid ammonia at a temperature of 223 K.

2. Results and Discussion

Solvated electrons in liquid ammonia solution with a concentration of $(2 \pm 1)\times10^{-4}$ mol/l were prepared by dissolving sodium metal in thoroughly dried liquid ammonia at a temperature of 223 K. Pump and probe pulses were derived from two independently tunable optical parametric amplifiers driven by 70-fs, 800-nm pulses from a Ti:Sapphire oscillator/regenerative amplifier system running at 1 kHz. The linear absorption spectrum of the sample is displayed in Fig. 1a. The electron resonance is peaked at 1500 nm and has a width of 3300 cm^{-1} (FWHM). The electrons were excited with 1 μJ, 100 fs-pulses centered at 1280 nm. The probe pulses with durations of 100 fs were tunable from 1200 to 2400 nm and had energies of 150 nJ. Representative transients with detection on the red and the blue edge of the stationary electron resonance are displayed in Fig. 1b and c.

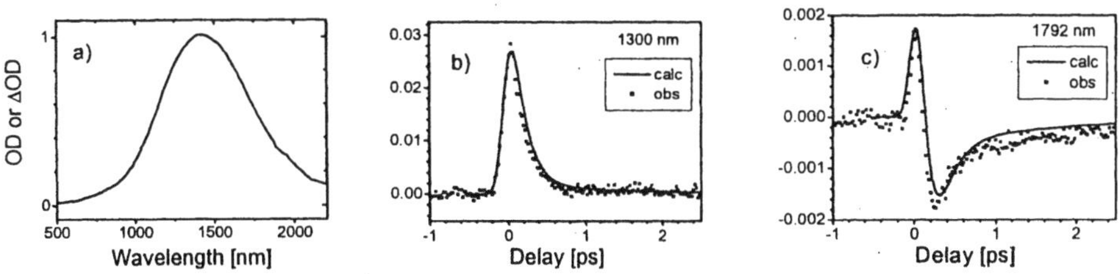

Fig. 1. Probe wavelength-dependent transients for the solvated electron in liquid ammonia at 223 K (symbols). The solid curves represent model simulations based on the cooling model as explained in the text.

Qualitatively, on the high-frequency side of the linear absorption spectrum, the pump-probe transient is dominated by a strong bleaching signal which recovers on a time scale of well below 1 ps. Simultaneously, on the low-frequency side, an immediate bleach is turned into a transient absorption within the instrument response function.

We discuss these results in terms of a simple temperature-jump scenario which has proven sufficient to fully reproduce our recent femtosecond pump-probe studies on the relaxation dynamics of aqueous electrons. At time $t = 0$, the solvated electron is photo-excited. Internal conversion back to the ground state occurs instantaneously within our time resolution. Furthermore, the excess energy delivered by the 1280-nm excitation photon is instantaneously converted into thermal energy resulting in the prompt formation of a canonical ensemble of hot electrons characterized by a temperature, $T(t = 0)$. Spectroscopically, the electrons appear as "canonically hot" because the excess energy is equi-partitioned among those nuclear degrees of freedom of the solvent that can couple to the electron transition dipole moment. As a result, the electron absorption spectrum at $t = 0$ is strongly red-shifted with respect to the stationary spectrum at $t = -\infty$ but its spectral shape remains unperturbed. This is because the linear absorption spectrum of the solvated electron in liquid ammonia is known to continuously shift to lower frequencies with increasing temperature without changing its spectral shape and width. This thermal shift is given by the well-known temperature coefficient, $\alpha = (d\nu_{max}/dT)$ of -18 cm^{-1}/K where ν_{max} is the resonance central frequency. We emphasize that this often forgotten finding also holds for solvated electrons in liquid water [2]. Subsequent to effective pump-induced heating (and red-shifting), transport of excess energy into the remaining solvent modes results in cooling and manifests itself as a time-dependent blue-shift of the electron absorption spectrum. Provided that at each instant in time, the remaining excess energy is equilibrated among the solvent degrees of freedom relevant to line broadening of the electron resonance, an assignment of a time-dependent temperature is warranted. Furthermore, since for all delays including $t = 0$, the spectral shape of the resonance remains stable, this temperature can be read off directly from the spectral position of the time-dependent absorption spectrum.

To simulate the optical response of the solvated electron, the temperature-dependent spectral shape, $\varepsilon(\nu, T)$, of the canonical absorption spectrum is represented by the following functional form

$$\varepsilon(\nu,T)=\begin{cases}\dfrac{\gamma^2}{(\nu-\nu_{max}(T))^2+\gamma^2} & \text{for } \nu>\nu_{max}(T)\\[2ex] \exp\left[-4\ln(2)\dfrac{(\nu-\nu_{max}(T))^2}{\sigma^2}\right] & \text{for } \nu\le\nu_{max}(T)\end{cases}$$

where $\nu_{max}(T) = \nu_{max}(223\text{ K}) + \alpha\,(T - 223\text{ K})$. The high-frequency side is described by a Lorentzian line-shape whereas a Gaussian is used for the low-frequency side [1]. The temperature is assumed to relax double exponentially according to $T(t) = \Delta T_0\,((1\text{-}f)\exp(-t/\tau_1) + f\exp(-t/\tau_2)) + 223\text{ K}$ where ΔT_0 corresponds to the pump-induced heating at time $t = 0$ and τ_1 and τ_2 are the relaxation times for the cooling process. Model simulations are reproduced in Fig. 1b-c as solid curves. Using relaxation times of $\tau_1 = (170 \pm 50)$ fs and $\tau_2 = 1.5$ ps, $f = 0.043$ and an initial temperature jump, ΔT_0, of 267 K corresponding to a pump-induced red-shift of 4800 cm^{-1}, an excellent agreement between simulation and experiment is obtained.

3. Conclusions

In conclusion, we have measured for the first time the relaxation dynamics of the solvated electron in liquid ammonia. The near-infrared response of the electron can be simulated quantitatively using a temperature jump model in which femtosecond excitation leads to a quasi-instantaneous formation of thermally hot electrons. During the entire process of thermalization, the time-evolving absorption spectrum of the electrons is identical to a canonical spectrum characterized by a temperature, T(t). This is because for all time delays, the electronic resonance shifts only spectrally while its shape remains constant. Non-adiabatic transitions which bring the electron back to its ground state occur within our time resolution. Furthermore, no evidence for any precursor states such as the "wet" electron is found. Our data indicate, that during the entire process of thermalization, the electron is always in thermal equilibrium with the nuclear degrees of freedom of the solvent bath which can couple to the electronic transition dipole moment. The observed blue shift is brought about by energy flow from this specific set of solvent modes to other bath degrees of freedom which are not of relevance for electronic line broadening.

References

1 T.R. Tuttle and S. Golden, *J. Phys. Chem.* **95,** 5725, 1991.

2 A. Hertwig, H. Hippler, A.N. Unterreiner, and P. Vöhringer, *Ber. Bunsenges. Phys. Chem.* **102,** 805, 1998; A. Hertwig, H. Hippler, and A.N. Unterreiner, *Phys. Chem. Chem. Phys.* **1,** 5633, 1999.

Mechanisms for Photodetachment in Water

J.A. Kloepfer, V.H. Vilchiz, V.A. Lenchenkov, and S.E. Bradforth

Dept. of Chemistry, University of Southern California, Los Angeles, CA 90089, U.S.A
E-mail: bradfort@usc.edu

Abstract. The effect of excitation energy and solvent deuterium substitution is explored for photodetachment of aqueous iodide. The results are explained in part by a new *ab initio* characterization of the vertically excited state.

Threshold electron ejection processes, such as photodetachment via optical charge-transfer-to-solvent (CTTS) excitation, provide very simple models for understanding solvent-controlled electron transfer (ET). In previous work, we have demonstrated that ejection from the prototypical CTTS system, the aqueous iodide anion, is to short range into a caged pair [1]. The electron subsequently either diffuses out of the cage, an activated process due to the attractive interaction between the nascent electron and the polarizable iodine neutral, or undergoes reverse electron transfer from the cage to reform I^- [2]. In the current experiments, our goal was to gain additional insight into the molecular mechanism of ejection by exploring the effect of a large increase in the excitation energy and by isotopic substitution of the solvent. The results are rather unexpected and an ab initio characterization of the vertically excited bulk CTTS wavefunction turns out to be particularly useful in rationalizing the dynamics of the electron.

Aqueous iodide ions are excited at the red-most edge (250 nm) or at the band center (225 nm) of their CTTS absorption. 250 nm pulses of ~ 50 fs are generated by doubling the signal output of a 390 nm pumped OPA and 225 nm pulses (~ 150 fs) are produced by sum frequency mixing of the OPA signal with residual 390 nm. The ejected electrons once trapped are detectable by their absorption across the visible and near IR. Mapping the spectral evolution of this absorption at several probe wavelengths reports on the subsequent solvent response to the trapped electron [3]. Results are shown here for a 700 and 1000 nm probe selected by an interference filter from a variably delayed white light continuum. Fig. 1 shows the delayed appearance of the electron at earliest times (longer than instrument response) and a tens-of-picoseconds population decay due to geminate recombination at two different excitation energies. The fraction remaining has escaped iodine and any slow population decay after ~200 ps is due to secondary recombination [2]. The temporal signature of the geminate recombination directly relates the location of the newly formed electron with respect to its geminate partner (and parent) [1]. It is observed that there is no significant difference in the traces at any phase of the ejection and recombination dynamics. This suggests that ejection proceeds essentially unaltered despite the large increase in available energy that one might have predicted would be available for longer range, one would see a slower population decay phase and a higher escape fraction, as is

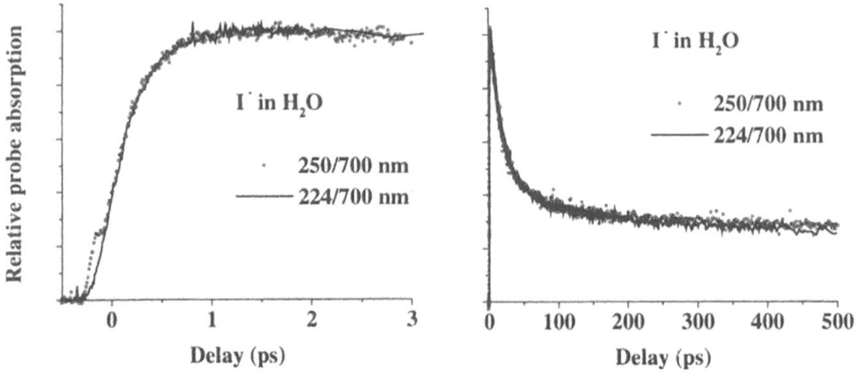

Fig. 1. Femtosecond transient absorption of aqueous I^- at 250 and 224 nm (5.0 and 5.5 eV).

observed in the detachment of Na^- [4] and in the multiphoton ionization of liquid water above 9 eV [5].

Figure 2 demonstrates the effect on the ejection mechanism on changing the solvent from H_2O to D_2O. It is observed that neither the appearance, as judged by the rising edge of the 1000 nm data, or the spectral evolution is significantly altered on isotopic substitution. One might expect that the dominant response of the water molecules in the first shell would be inertial libration of the adjacent solvent (giving a large deuterium effect) in response to the change in local charge density as observed for dye solvation dynamics in water. It appears that *translation* of the immediate solvent (which gives a small deuterium effect due only to the change in overall mass) dominates [6] and facilitates the budding of the electron into an adjacent solvent cavity. Interestingly, there is a large effect on the geminate recombination on solvent deuteration. This cannot be explained simply by the higher viscosity of D_2O. The fits shown in Fig. 2 are numerical diffusion simulations [2] for the encounter pair which include a reaction probability for non-adiabatic electron transfer if the electron and iodine are in contact. It is found that the effective rate constant for return ET must be a factor of 0.55 slower in D_2O to fit the data.

To consider these results it is useful to compare with benchmark mixed quantum classical molecular dynamics simulations for halide CTTS detachment [7]. These suggest that electron separation is solvent driven, but that the CTTS band is made up of both $p \rightarrow s$ and $p \rightarrow d$ electron promotions with the detachment timescale controlled by transitions between different angular momentum waves. Thus, tuning the pump over the absorption band should vary the angular momentum composition of the initially prepared state again predicting different ejection dynamics [7]. We have performed our own *ab initio* calculations in an attempt to get an improved description for the vertically-prepared wavefunction [8]. These calculations provide an explanation for the results in Fig. 1. Inclusion of all iodine valence electrons reveals that promotions into *d* character orbitals are at much higher energy and do not contribute to the CTTS band. Instead, exciting at varying photon energies selects from different non-degenerate occupied iodide

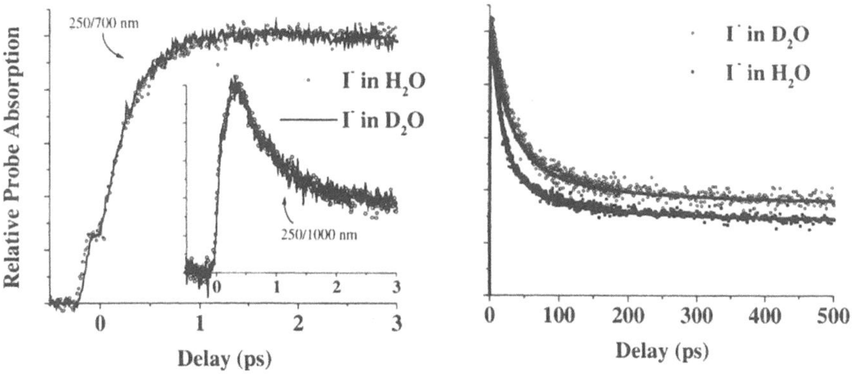

Fig. 2. The effect of isotopic substitution on the detachment dynamics. (Right panel) Lines represent numerical diffusion simulations for geminate recombination/escape, in the presence of a charge-polarizable atom potential, and include non-adiabatic reverse ET.

p-orbitals and more or less solvated ground state configurations for promotion to a single acceptor orbital which is primarily of *s* symmetry. Therefore, changing the excitation wavelength has rather little effect on the character of the optically prepared state in terms of its shape or kinetic energy. The pronounced difference in the excitation energy dependence in the geminate recombination in I^- versus Na^- [4] can be explained in terms of the reversal of the occupied and unoccupied orbital symmetries between the two systems. The calculation also shows that there is a large expansion in electron density and the CTTS electron occupies void space primarily within the first solvent shell. The pre-existing asymmetry of the local water network defines an initial direction for charge translation and this budding is presumably continued by displacement of the adjacent solvent.

References

1. J. A. Kloepfer, V. H. Vilchiz, V. A. Lenchenkov, A. C. Germaine, and S. E. Bradforth, *J. Chem. Phys.* **113**, 6288 (2000).
2. J. A. Kloepfer, V. H. Vilchiz, V. A. Lenchenkov, X. Chen, and S. E. Bradforth, *J. Chem. Phys.* in press (2002).
3. V. H. Vilchiz, J. A. Kloepfer, A. C. Germaine, V. A. Lenchenkov, and S. E. Bradforth, *J. Phys. Chem. A* **105**, 1711 (2001).
4. I. Martini, E. Barthel, and B. J. Schwartz, *J. Chem. Phys.* **113**, 11245 (2000).
5. R. A. Crowell and D. M. Bartels, *J. Phys. Chem.* **100**, 17940 (1996).
6. V. Tran and B. Schwartz, *J. Phys. Chem. B* **103**, 5570 (1999).
7. W.-S. Sheu and P. J. Rossky, *J. Phys. Chem.* **100**, 1295 (1996).
8. S. E. Bradforth and P. Jungwirth, *J. Phys. Chem. A* **106**, 1286 (2002).

Part X

Control and Dynamics of Condensed Phase Processes

Feedback control of photosynthetic light harvesting

W. Wohlleben[1], J. L. Herek[2,3], R. J. Cogdell[4], D. Zeidler[1], K.-L. Kompa[1], V. Sundström[2] and M. Motzkus[1]

[1] Max-Planck-Institut für Quantentoptik, 85748 Garching, Germany
E-mail: mcm@mpq.mpg.de
[2] Chemical Physics, Lund University, 22100 Lund, Sweden
[3] Department of Biophysics, Vrije Universiteit, 1081 HV Amsterdam, The Netherlands
[4] Division of Biochemistry and Molecular Biology, IBLS, University of Glasgow, G12 8QQ, UK

Abstract. We present coherent control of biological function. Feedback optimisation of the excitation pulse shape exploits molecular quantum interferences to steer the energy flow in the light-harvesting complex LH2 from *Rhodopseudomonas acidophila*.

1. Introduction

Quantum control of photochemistry exploits molecular interferences by adjusting the phases of multiple quantum pathways to different product channels via adapted excitation light. Experiments implement a learning loop in which molecular feedback iteratively refines the excitation pulse until an optimum solution directs the photoreaction towards a desired channel [1]. So far, feedback control has been absent from the world of biology. Furthermore, only few examples have proven explicitly the role of molecular coherence. Here we report feedback-optimised coherent control of biological function, namely regulation of the energy flow pathways in the 125 kDa light-harvesting antenna complex LH2 from *Rhodopseudomonas acidophila*, a photosynthetic purple bacterium.

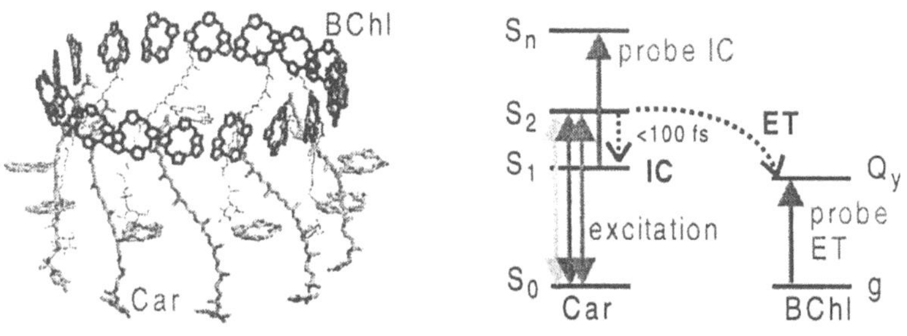

Fig. 1. The LH2 antenna complex of *Rps. acidophila*. Following excitation of the Car molecules to the S_2 state energy flows into two discrete channels of energy flow: intramolecular internal conversion and intermolecular energy transfer

The function of a photosynthetic antenna complex is to harvest sunlight and, through a series of energy transfer steps both within and between complexes, to make the energy available to the reaction centre, where it is used to fuel the primary dark reactions of photosynthesis [2,3]. The peripheral antenna LH2 (see Fig. 1) contains two pigment species, carotenoid (Car) and bacteriochlorophyll-a (BChl). BChls absorb in the near-IR, while Car pigments absorb at higher energy, in the region 450-540 nm. In this particular species, only 50% of the light harvested by Car is transported to the BChl acceptor states; the remaining energy is dissipated via intramolecular internal conversion processes [4]. Here we explore the competition between intra- and intermolecular channels with shaped excitation light.

2. Experimental Methods

In order to see if specially designed excitations can influence the fate of excitation energy, we construct a learning loop consisting of a pulse shaper, the LH2 sample and an evolutionary algorithm. We excite the Car to the S_2 state with shaped fs pulses from a non-collinear optical parametric amplifier (centered at 525 nm) and monitor the ensuing energy flow into internal conversion (IC) to the Car S_1 state and energy transfer (ET) to the final BChl acceptor state (Fig. 1). Specifically, we probe the transient absorption kinetics of Car (S_1-S_n transition, 580 nm) and BChl (Q_y bleach/stimulated emission, 880 nm). These spectrally separated, discrete signals of the IC and ET channels provide feedback for the evolutionary algorithm that optimises the excitation pulse shape for a given target.

3. Results and Discussion

We begin with conventional "blind" optimisations, allowing each frequency component of the pulse spectrum to take phases from 0 to 2π and attenuation from

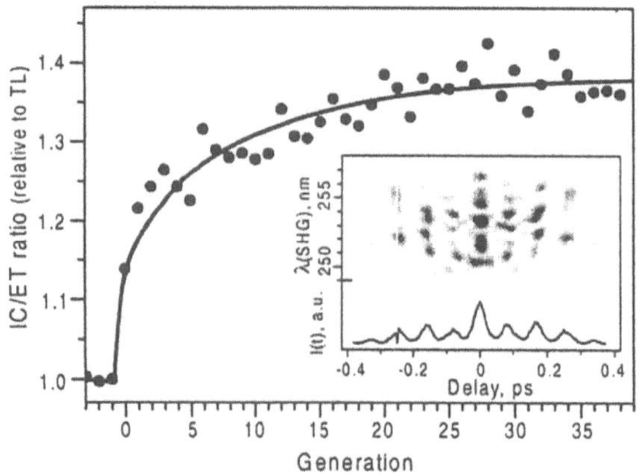

Fig. 2 Convergence curve for the optimisation of IC/ET with 64 parameters. The inset shows the optimal pulse FROG trace (λ(SHG)) and autocorrelation (I(t))

0 to 100% (a total of 64 parameters for phase and amplitude modulation). The target for the evolutionary algorithm is to maximise the ratio of the IC/ET signals. The optimal pulse is 30% lower in energy and yet the total excitation rate is higher. The optimisation achieves a ~35% enhancement of the feedback signal (see Fig. 2 and inset). We carefully checked that the spot overlap on the sample did not move due to any spatio-temporal shaping side effects. Additionally we recorded the full transient absorption spectrum (by dispersing the probe light through a monochromator), and verified that the shaped pulse affects the amplitudes of the Car and BChl bands uniformly.

As the optimisations described above consistently found solutions based on multi-pulse excitations, we now restrict the pulse shape to equispaced phase coherent subpulse sequences: in the Fourier plane of the 4f zero-dispersion compressor, the typical shaping setup, a sinusoidal spectral phase $\Phi(\lambda)$ is added as $\Phi(\lambda)=\pi*\sin(a*\lambda+b)$. By varying parameters a and b, the algorithm adjusts the temporal spacing of the subpulse comb and the optical carrier phase pattern across the comb, respectively. The target is again to optimise for maximum IC/ET. The algorithm achieves again convergence on all three parameters.

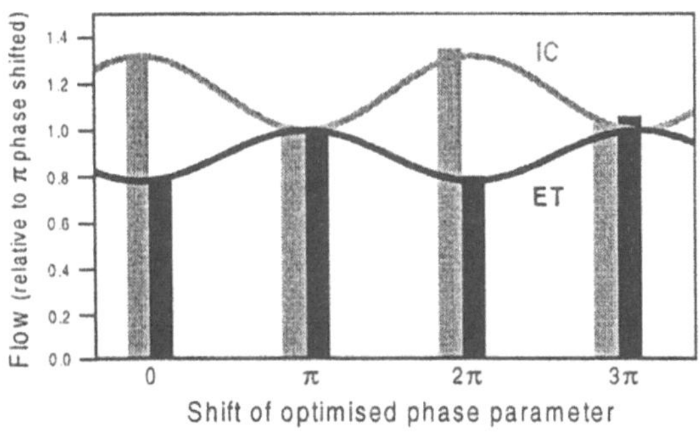

Fig. 3. Evidence of coherent control

We now shift the carrier phases of the subpulse comb by adding π to the optimised value for the phase pattern parameter b. Of course, the absolute phase fluctuates and is of not importance here. However, the relative phase of subpulses that are diffracted (in time) by the shaper from a single ncOPA pulse, is under control *via* parameter b. We stress that changing this phase parameter has no bearing on the pulse envelope, energy or spectrum (as confirmed by careful pulse characterisation). Mechanisms that do not involve coherence in molecular excitations will not be sensitive to such changes. However, the kinetics from the phase shifted subpulse comb are markedly different from the optimised solution.

Notably, this phenomenon is robust against attenuation of the shaped excitation pulse, except for saturation effects at energies approaching 250 nJ. To quantify exclusively the coherent effects on the energy flow-partitioning ratio, we attenuate to 25 nJ, far below saturation and annihilation onset and compare optimised excitation shapes to phase shifted. As shown in Fig. 3, the IC/ET ratio oscillates by ~30%. This observation is the *experimentum crucis* of coherent control: The optical carrier phase under an optimised envelope is indeed decisive for the control effect [5].

Given the temporal spacing of the subpulses, this interaction most likely does not occur in the initially excited S_2 manifold, but either in the S_1 or S_0 states, involving low frequency bending modes of the Car.

4. Conclusions

Complex matter waves in the internal degrees of freedom can hence be used to control the flow of excitation energy in LH2. The optimised pulse shapes feature repetitive light-matter interactions, creating coherent wave packets that interfere in a manner that enhances Car IC. We note that in none of our experiments could we find a pulse shape that would enhance the biological function of light harvesting, i.e. result in less IC losses and more ET.

We present to our knowledge the first study of optical biological function with feedback optimised coherent control techniques. By investigating optimisation goals and the specific pulse shapes, one will gain insight into how nature has designed biological systems for optimal performance.

Acknowledgements. We thank A. Gardiner for help with sample preparation, T. Hornung for stimulating discussions and M. Mathys for help with experiments.

References

1 H. Rabitz, R. deVivie-Riedle, M. Motzkus and K.-L. Kompa "Whither the future of controlling quantum phenomena?" Science **288**, 824-828 (2000) and references therein.

2 T. Pullerits and V. Sundström "Photosynthetic light-harvesting pigment-protein complexes: Toward understanding how and why" Acc. Chem. Research **29**, 381-389 (1996).

3 W. Zinth, T. Arlt and J. Wachtveitl "The primary processes of bacterial photosynthesis - ultrafast reactions for the optimum use of light energy" Ber. Bunsenges. Phys. Chem. **100**, 1962-1966 (1996).

4 A. Macpherson, J. B. Arellano, N. J. Fraser, R. J. Cogdell and T. Gillbro "Efficient energy transfer from the carotenoid S_2 state in a photosynthetic light-harvesting complex" Biophysical Journal **80**, 923-930 (2001).

5 J. L. Herek, W. Wohlleben, R. J. Cogdell, D. Zeidler and M. Motzkus "Quantum control of energy flow in light harvesting", Nature, in press (2002)

Adaptive Femtosecond Quantum Control in the Liquid Phase

T. Brixner, N. H. Damrauer, P. Niklaus, and G. Gerber

Physikalisches Institut, Universität Würzburg, Am Hubland, 97074 Würzburg, Germany
E-mail: gerber@physik.uni-wuerzburg.de

Abstract. Despite the failure of single-parameter approaches (wavelength, intensity, or linear chirp control), adaptive femtosecond pulse shaping discovers complex laser fields for chemically selective molecular excitation in the liquid phase.

Coherent light sources can be used to actively manipulate the outcome of light/matter interactions by exploiting time and frequency domain interference phenomena. This has led to the development of the field of quantum control [1,2] and one of its most exciting new advances: adaptive femtosecond pulse shaping. In this multiparameter control methodology, the quantum system itself guides an automated search (within a 128-parameter search space) for coherent fields best suited for achieving a control task. Thus the method is ideal for complex control experiments as first shown for gas-phase photodissociation processes [3]. To date, all demonstrations of this technique on molecular systems have focused on controlling the outcome of unimolecular phenomena, and little attention has been paid to selectively controlling mixtures of different molecules. The experiments discussed here consider simultaneous but selective multiphoton excitation of two distinct electronically and structurally complex dye molecules in solution [4]. Single-parameter quantum control schemes (variation of wavelength, intensity, and linear chirp) and many-parameter adaptive control are investigated by use of a femtosecond pulse shaper (Fig. 1) capable of producing complex shapes by spectral phase modulation.

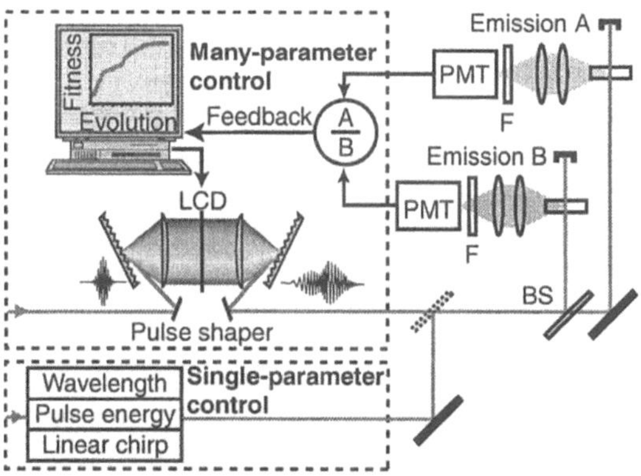

Fig. 1. Experimental setup for adaptive control of liquid-phase photoexcitation

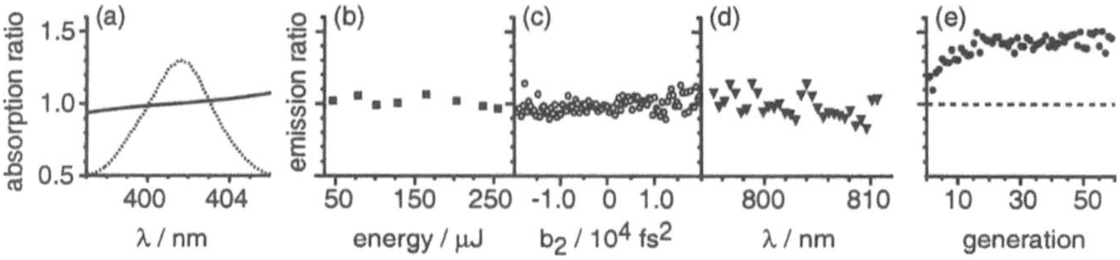

Fig. 2. Control of molecular excitation. (a) The DCM/$[Ru(dpb)_3]^{2+}$ ratio of the linear absorption profiles (solid line) is shown with the second-order power spectrum (dotted line) of a bandwidth-limited laser pulse. The DCM/$[Ru(dpb)_3]^{2+}$ emission ratio is shown (b) for varying pulse energies, (c) for varying second-order spectral phase, (d) for scanning a symmetric and rectangular transmission window of 5 nm width over the laser spectrum, and (e) for many-parameter phase-shaping as a function of generation number within the evolutionary algorithm

These laser pulses irradiate methanol solutions of DCM (4-dicyanomethylene-2-methyl-6-*p*-dimethylaminostyryl-4H-pyran) and $[Ru(dpb)_3](PF_6)_2$, where dpb=4,4'-diphenyl-2,2'-bipyridine. Following two-photon absorption and excited-state dynamics, charge-separated configurations are reached in both molecular species. Spontaneous emission yields are then collected at approximate right angles, spectrally filtered, detected by photomultiplier tubes (PMT), and temporally integrated by boxcar averagers. The emission ratio of molecule A versus molecule B is used as feedback signal in the evolutionary learning loop as it reflects the excitation ratio between the two molecules.

An obvious starting place for selective excitation of molecules using light is to adequately choose the corresponding excitation wavelength. However, if the absorption spectra are very similar (as is often the case with large molecules under solution conditions), one-photon excitation does not select a single species. The power spectrum of $E^2(t)$ for a bandwidth-limited laser pulse (Fig. 2a, dotted line) indicates which transition frequencies are possible in two-photon processes with our laser system. The DCM/$[Ru(dpb)_3]^{2+}$ linear absorption ratio within this spectral range (Fig. 2a, solid line) does not vary significantly, and therefore selective excitation of one molecule with respect to the other is not possible by varying the corresponding (one-photon) excitation wavelength.

In Fig. 2b, the DCM/$[Ru(dpb)_3]^{2+}$ emission ratio is plotted as a function of pulse energy. It is seen that despite the variation of laser power, the excitation ratio between the two molecules remains constant. This reflects the fact that two-photon absorption is responsible for the emissive yield in both molecules, and the second-order power dependencies cancel each other when considering emission ratios. Varying linear chirp is another possibility which has been exploited successfully in previous experiments on quantum control. However, under the conditions here the DCM/$[Ru(dpb)_3]^{2+}$ emission ratio does not change as a function of linear chirp (Fig. 2c). Introduction of second-order spectral phase increases the laser pulse duration, reduces the temporal peak intensity, and therefore reduces the transition probability for the induced nonresonant two-photon excitation processes, but in the same way for both molecules. Still another possibility of selective control in two-photon absorption would be given if the two-photon absorption cross sections were different for the two molecules. This was investigated by scanning a rectangular transmission window over the laser spectrum while recording the emission signals (Fig. 2d). Within the statistical uncertainties, the DCM/$[Ru(dpb)_3]^{2+}$ emission ratio does not change and selective excitation is not possible. All these results seem to indicate that the excitation processes for each molecule in the case of spectrally broad absorption bands are only trivially second-order intensity-dependent and therefore cannot be separately controlled.

However, this assumption is wrong, and selective excitation can be achieved if many-parameter adaptive quantum control is employed. Using the DCM/$[Ru(dpb)_3]^{2+}$ emission ratio as a feedback signal, the evolutionary algorithm iteratively improves the signal by approximately 50 % (Fig. 2e). This demonstrates that one molecule (DCM) can be selectively excited with respect to the other molecule ($[Ru(dpb)_3]^{2+}$). It should be emphasized that control is possible in the presence of complex solvent/solute interactions.

The failure of single-parameter control schemes is helpful for the interpretation of relevant control mechanisms. Since the initial excitation step is not controllable, nontrivial control of excited-state wavepacket dynamics must be involved which goes beyond the initial excitation step. This in turn shows that molecular phase coherences must exist which are not destroyed by interaction with the surrounding solvent on the time scale of the optimized pulse shape (100 fs to 1 ps). In summary, chemically selective molecular excitation in the liquid phase was achieved by nontrivial many-parameter adaptive quantum control. The presence of long-lasting phase coherences opens the door to a variety of control schemes in which coherent superpositions of molecular states are required. The adaptive technique can be used in all cases where chemically selective excitation is desired provided the molecular responses are not identical. Applications are envisioned in chemical analysis where specific molecular properties could be made visible, or in bimolecular reaction control where specific educt molecules could selectively be prepared in reactive states ("activated") for purposes of chemical synthesis. Another application could be found in multiphoton fluorescence microscopy. Here, it would be possible to obtain emission signals preferentially from a specific fluorophore, and thus spatial (microscopic) resolution could be combined with "chemical resolution".

Acknowledgements. We gratefully acknowledge financial support from the European Coherent Control Network (COCOMO): HPRN-CT-1999-00129, the German-Israeli Cooperation in Ultrafast Laser Technologies (GILCULT): FKZ-13N7966, and the "Fonds der chemischen Industrie".

References

1 S. A. Rice, and M. Zhao, *Optical Control of Molecular Dynamics*, Wiley, New York (2000).
2 T. Brixner, N. H. Damrauer, and G. Gerber, Adv. At. Mol. Opt. Phys. **46**, 1 (2001).
3 A. Assion, T. Baumert, M. Bergt, T. Brixner, B. Kiefer, S. Seyfried, M. Strehle, and G. Gerber, Science **282**, 919 (1998).
4 T. Brixner, N. H. Damrauer, P. Niklaus, and G. Gerber, Nature **414**, 57 (2001).

Real-time dynamics of a breather in *trans*-polyacetylene revealed by 4.7-fs visible pulses

T. Kobayashi[1], S. Adachi[1], and V. M. Kobryanskii[2]

[1]Department of Physics, Graduate School of Science, University of Tokyo, Hongo 7-3-1, Bunkyo, Tokyo 113-0033, Japan. e-mail: kobayashi@phys.s.u-tokyo.ac.jp
[2]Institute of Chemical Physics, Russian Academy of Science, Kosygin Street 4, Moscow 117977, Russia

Abstract: The formation and the relaxation dynamics of a *breather*, a dynamic bound-state of a soliton pair, with a 44-fs period and a 50-fs decay-time in *trans*-polyacetylene was revealed by a time-frequency analysis of pump-probe signals.

1. Introduction

Conjugate polymers are of great interest because of their possible applications to light-emitting diodes and optical memories. Among them polyacetylene is considered to be most basic because it is the simplest conjugated polymer and has been attracting scientists' interest because of its peculiar features of conductivity and spectroscopic properties. Especially it has nonlinear excitations of solitons in a *trans* form, because of the degeneracy between $(-C{=}C-)_n$ (A-phase) and $(=C-C=)_n$ (B-phase). It was concluded by analysis using the Su-Schrieffer-Heeger Hamiltonian[1] that a photogenerated electron-hole pair evolves into a soliton-antisoliton pair within the 100-fs following the photoexcitation because of one-dimensionality of the system. It has been also theoretically predicted[2] that within this initial time the soliton pair is spatially localized, and it forms a dynamic bound state called a breather. In spite of many theoretical predictions, no clear experimental evidence of the existence of the breather has emerged so far probably because of the insufficient time-resolution of the experiments.

2. Experimental results

All the pump-probe signals at several probe wavelengths as shown in Fig. 1(a) exhibit periodically oscillating features(Fig, 1(b)) which are due to the modulation of electric transition probability induced by molecular vibrations. The oscillating components are extracted from the pump-probe kinetics and power spectra of the Fourier transform that were computed over the whole pump-probe delay range are shown in Fig. 1(c). They are composed of two peaks located at 1080 and 1470 cm^{-1}. The lower and the higher frequencies are close to those observed by stationary resonant Raman scattering, which are identified as the stretching modes of single and double bonds of carbon atoms, respectively[3]. In order to examine the time-evolution of the amplitude of each vibrational mode, we performed a spectrogram analysis using the Gaussian window function with a HWHM Δt = 96 fs. In Fig. 1(d), t_1 and t_2 correspond to two distinct peaks in Fig 1(c). In addition to these two modes, several broad bands (t_1', t_1'', t_2' and t_2'') appear in the spectrogram. Their broad bandwidths are due to their short time and partly due to inhomogeneity.

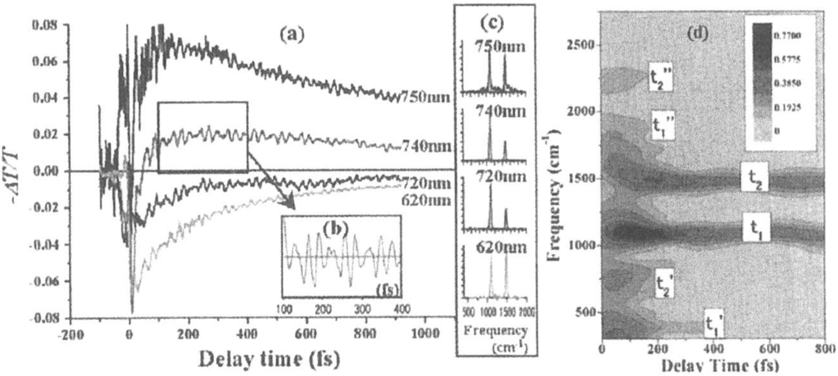

Fig. 1 (a) Probe delay-time dependence of pump-probe signals. (b) Periodically oscillating components of the pump-probe signals. (c) Power spectra of Fourier transform of the oscillating components of pump-probe signals integrated over the whole range of pump-probe measured delays. (d) Spectrogram of periodically oscillating components exhibits two-dimensionally (gate-delay time vs. frequency).

3. Discussion

It is theoretically predicted that the breather formation is accompanied by the collective oscillation of the bond-length, which can be detected in the pump-probe experiment as modulation of the instantaneous vibrational frequencies. This phenomenon is originated from the mode coupling between the carbon-carbon stretching modes and the breather mode. Based on these theoretical predictions, listed above, the time trace(Fig. 2(a)) and the frequency distribution of the vibrational frequencies (Fig. 2(b)) of the simulation signal with a modulation period of 44 fs and a modulation duration-time of 50 fs are shown. Fourier power spectrum(Fig. 2(c)) and the spectrogram (Fig. 2(d)) calculated using the simulation signal reproduce all the characteristic features of the Fourier power spectra and the spectrogram computed from the experimental pump-probe traces. In the Fourier power spectra, modulation cannot be detected because of its short duration time. In the analysis of the spectrogram, any modulation with the characteristic period shorter than the time-width of the window function (HWHM of 96 fs) does not appear clearly, which results in the absence of fast modulation in Fig. 2(d). The evidence of the modulation appears in the spectrogram indirectly in the shape of satellite-bands t_1', t_1'', t_2' and t_2''on both sides of the main vibrational modes t_1 and t_2, respectively. This result suggests that the unidentified short-lifetime modes visible in Fig. 1(d) can be explained by the modulation of the vibration modes caused by the mode coupling. The period of the modulation is determined to be 44 ± 3 fs from the frequency separation between the main mode and the satellite-bands, which on the average is 760 cm^{-1}. This value is consistent with the theoretical prediction of 33-50 fs [1,2,4].

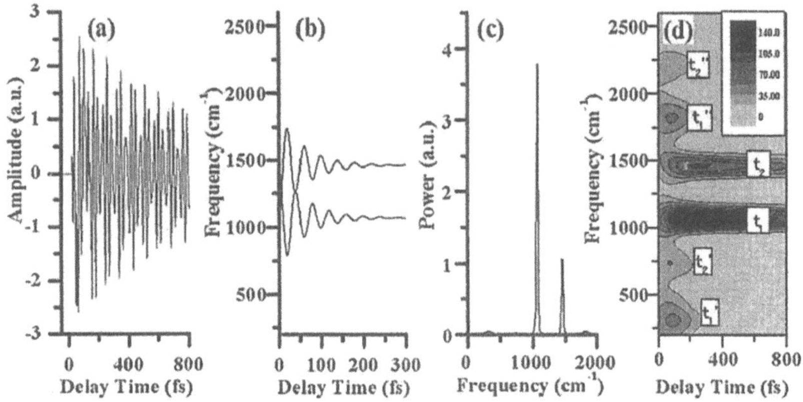

Fig. 2 Simulation based on the theoretical predictions. (a) Time trace. (b) Frequency distribution. Instantaneous frequencies corresponding to the vibrational modes t_1 and t_2 are modulated with the modulation period of 44 fs and the modulation duration-time of 50 fs. (c) Fourier power spectrum and (d) spectrogram are calculated using the simulation signal.

In conclusion, using 4.7-fs pulses we could for the first time directly observe the formation and the decay of the breather mode in *trans*-polyacetylene, which was predicted more than 15 years ago but has not been able to be experimentally detected.

References

1. W. P. Su and J. R. Schrieffer, Natl. Acad. Sci. USA Proc. **77,** 5626-5629 (1980).
2. A. R. Bishop, D. K. Campbell, P. S. Lomdahl, B. Horovitz and S. R. Phillpot, Phys. Rev. Lett. **52,** 671-674 (1984).
3. V. M. Kobryanskii, SPIE Proceedings **3790**, 227–234 (1999).
4. M. Sasai, and H. Fukutome, *Prog. Theor. Phys.* **79,** 61 (1988)

Control of an electron transfer reaction using a sequence of femtosecond laser pulses

Ignacio B. Martini; Erik R. Barthel; Benjamin J. Schwartz

Department of Chemistry, University of California, Los Angeles; Los Angeles, CA, 90095
tel: 310-206-4113, fax: 310-206-4038; Schwartz@chem.ucla.edu

Abstract: A sequence of femtosecond laser pulses is used to control the electron transfer dynamics of sodide in solution. The electron recombination can be either enhanced or suppressed depending on the electron re-excitation delay time.

We report the optical control of electrons during a Charge-Transfer-to-Solvent (CTTS) reaction. The picosecond and sub-picosecond CTTS dynamics of sodide (Na^-) in tetrahydrofuran (THF) has been extensively studied.[1-4] The basic picture of this photoinduced electron transfer reaction is summarized in Figure 1.[1] Upon photoexcitation, the excess electron is promoted to a "*p*-like" state, and subsequent molecular solvent motions induce the delayed detachment and localization of the electron near the Na^0 partner. At low excitation energies the electron is ejected to the same solvent cavity that contains the sodium atom, into an *immediate* contact-pair. These species share significant wave function overlap that leads to direct nonadiabatic recombination in ~1 ps. In contrast, higher excitation energies produce *solvent-separated* contact pairs that reside one or more solvent shells away from the sodium atom. Negligible wave function overlap prevents rapid recombination, which happens on the hundreds-of-picoseconds timescale. The relative number of electrons ejected into immediate and solvent-separated pairs can be smoothly adjusted by tuning the excitation energy across the visible CTTS absorption band.[1] Finally, *free* electrons are also created by high-energy excitation and possibly by dissociation of solvent-separated contact pairs. The free electrons have lifetimes longer than nanoseconds.

By taking advantage of the vastly different recombination rates of the two types of contact pair electrons and the great separation of the absorption spectra of the

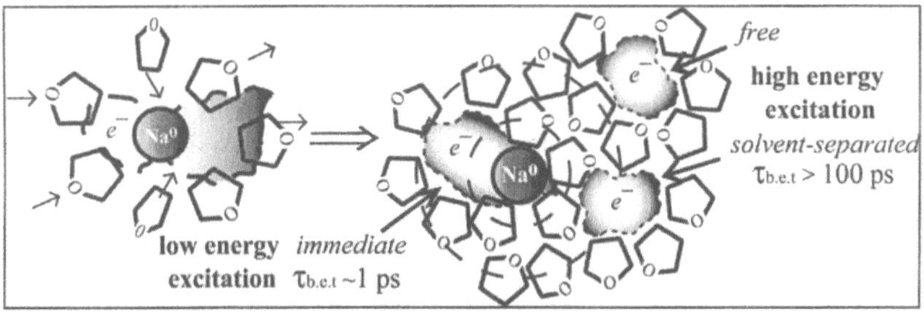

Fig. 1. Schematic representation of the different kinds of sodium-solvated electron contact pairs created in the sodide CTTS reaction.

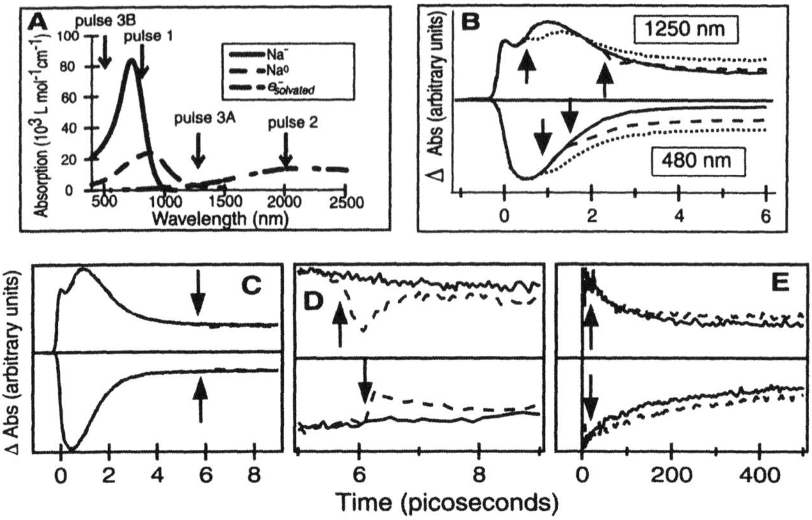

Fig. 2. **A** Absorption spectrum of all the species involved in the CTTS reaction of sodide in THF. **B-E** The solid lines represent the regular pump-probe data of the CTTS reaction of sodide in THF after 780 nm excitation. The positive signals correspond to the transient absorption of the electron probing at 1250 nm, while the negative transient bleach is probed at 480 nm. The dashed lines correspond to the three-pulse experiments, with the 2000 nm light interacting with the sample at different delay times indicated by the arrows. **B** is the three-pulse experiment performed on the immediate contact pairs, **C** is the three-pulse experiment performed on the solvent-separated contact pairs, **D** is a magnification of **C**, **E** is the long time dynamics of **C**.

electron and the sodide parent (Figure 2A), we designed a three-pulse experiment to modify the population dynamics of the recombining electrons.[2,3] A 780-nm pulse is used to excite Na^- and create a distribution of solvated electrons with most in immediate contact pairs. A second pulse at 2000 nm then excites predominantly electrons in immediate contact pairs if it arrives at the sample within the first 3 ps of the initial pump. At times later than 3 ps, all the immediate pairs have recombined, so the re-excitation pulse acts exclusively on electrons in solvent-separated contact pairs and free electrons. A third pulse is used to probe the resulting dynamics by monitoring the Na^- bleach recovery at 480 nm or the electron absorption at 1250 nm. This type of electron re-pump experiment is similar to that recently performed by Barbara and co-workers.[5,6]

The solid curves in Figures 2B-E show the results of regular two-pulse pump-probe experiments over different time scales. The upper panels reflect the 1250-nm e^- absorption dynamics, while the lower panels show the 480-nm Na^- bleaching dynamics. The CTTS photodetachment is represented by the slow rise (~700 fs) of the electron absorption signal (positive signals, the pulse-width limited component is explained elsewhere [3]). The negative bleach signal, on the other hand, appears instantaneously, and both traces decay at short times (<3 ps) due to the nonadiabatic geminate recombination of the electrons in immediate contact pairs (Figs 2B&C). The hundreds-of-ps decay presented in Fig 2E corresponds to the recombination of electrons in solvent-separated contact pairs.

The dashed and dotted curves in Fig. 2B-E present the effect of the 2000-nm electron re-excitation pulse on the recombination reaction. Re-excitation of immediate contact pairs (Fig 2B) is achieved by applying the second pulse within the first 3 ps after the initial 780 nm pulse, while solvent-separated contact pairs are excited at delay times longer than 4 ps (Figs. 2C,D&E).

A larger signal offset is observed for the three-pulse experiment performed on immediate contact pairs, as well as at long times for the solvent-separated contact pairs (Figs 2B&E). This indicates that the effect of the second pulse is to produce more electrons and less Na^- at long times than would have been created with just the initial pump pulse. This result is explained as a partial suppression of the geminate recombination of both kinds of contact pairs. On the other hand, the instantaneous effect of the 2000-nm light on solvent-separated contact pairs is a decrease in the size of the signal offset (Figs 2C&D), which is assigned to enhanced recombination. Thus, the immediate effect of the second pulse is to decrease the number of electrons relative to that produced by the initial pump.

The microscopic dynamics of the three-pulse experiment can be explained with the help of Figure 1. Upon excitation, the solvated electron is promoted to a quasi-continuum state, characterized by a large delocalization of the electronic wave function.[2,3] The excited immediate contact pair electron will now access regions of the solvent further away than its original solvent cavity. The electron can now re-localize, effectively transforming into a solvent-separated contact pair or a free electron, thus shutting off the rapid immediate nonadiabatic geminate recombination. Likewise, the original solvent-separated contact pairs can convert into either free electrons, shutting off the long time recombination, or into immediate contact pairs, with the latter resulting in enhancement of the geminate recombination. A detailed analysis of the number of solvent-separated electrons that recombine upon excitation suggests that recombination must occur both from the excited and ground states of the solvated electron.[3]

To the best of our knowledge, this is the first time a charge transfer reaction reaction has been photo-controlled during the process of electron transfer.

References

[1] E. R. Barthel, I. B. Martini, B. J. Schwartz, "How does the solvent control electron transfer? Experimental and Theoretical Studies of the simplest charge transfer reaction" J. Phys. Chem. B. **105,** 12330-41 (2001).

[2] I. B. Martini, E. R. Barthel, B. J. Schwartz, "Optical control of electrons during electron transfer," Science **293,** 462-465 (2001).

[3] I. B. Martini, E. R. Barthel, B. J. Schwartz, "Manipulating the production and recombination of electrons during electron transfer: Femtosecond control of the charge-transfer-to-solvent (CTTS) dynamics of the sodium anion (Na^-)," J. Am. Chem. Soc., in press.

[4] Z. Wang, O. Shoshana, S. Ruhman, "Probing CTTS dynamics of rhe Na^- in THF: Novel relaxation and polarization effects" in Ultrafast Phenomena XII., Vol. 66 of OSA Proceedings Series (Optical Society of America, Washington, D. C., 2000), pp. 624-26.

[5] D. H. Son, P. Kambhampati, T. W. Kee, P. F. Barbara, "Delocalizing electrons in water with light" J. Phys. Chem. A **105**, 8269-72 (2001).

[6] D. H. Son, P. Kambhampati, T. W. Kee, P. F. Barbara, "One-photon UV detrapping of the hydrated electron" Chem. Phys. Lett. **342**, 571-77 (2001).

Watching and controlling intramolecular energy flow in solution: Effects of selective vibrational excitation and chemical substitution

Jens Aßmann,[1] Ales Charvat,[1,2] Rebekka von Benten,[1] and Bernd Abel [1,2*]

[1] Institut für Physikalische Chemie, Universität Göttingen, Tammannstraße 6, D-37077 Göttingen, Germany
[2] MPI für biophysikalische Chemie, Am Faßberg, D-37077 Göttingen, Germany
[*] E-mail: babel@gwdg.de

Abstract. Femtosecond transient absorption spectroscopy has been used to excite CH-overtones in small organic molecules in solution and to monitor intramolecular as well as intermolecular energy flow for the case of different chemical substitutions, functional groups, and structure.

1. Introduction

Recently, we have measured intramolecular vibrational energy redistribution (IVR) and intermolecular vibrational energy transfer (VET) in solution for several molecules [1-3]. One of the major motivations for our studies has been to understand the relationship between the molecules' chemical structure and the rates of VET and particularly the rates of IVR. Another aspect was the question how long a vibrational excitation (wave packet) and vibrational energy is located in specific degrees of freedom. Progress in an understanding of these trends may help to design molecules in which IVR rates and reactivity in solution can be controlled by (temporary) chemical modification and/or selective vibrational excitation.

2. Experimental Methods

In recent articles [1-3] we have given a detailed description of the ideas, concepts and hardware of our experimental approach. Briefly, for the selective excitation and probing of molecules in solution in our experiments we employ femtosecond (fs) IR-pump and UV-probe transient absorption spectroscopy with a time resolution of about 100 fs [1,2]. A special feature of the experiment is that near IR-laser pulses are used to selectively excite molecules in overtones of CH-stretch vibrations and excited state populations are monitored by transient UV-absorption spectroscopy [4]. Because the initially excited "state" is a non-stationary state of the molecular Hamiltonian it evolves in time and vibrational energy flows from the initially excited CH-bond to other degrees of freedom including the Franck-Condon (FC) active modes (for the electronic transition) which were not excited initially. Resonance Raman experiments help to unambiguously identify such modes. Their increasing population enhances the absorption of a time delayed probe pulse tuned to the long wavelength wing of the electronic absorption spectrum. At the same time vibrational energy transfer to the solvent takes place and in turn decreases the absorption. Transient absorption traces are converted into energy profiles via calibration against high temperature absorption spectra [1,2].

3. Results and Discussion

In a first set of experiments we focused on the effect of chemical substitution on the mechanisms and time scales of IVR and VET for two differently substituted hydrocarbons in solution (CCl_4). Near infrared fs-laser pulses tuned to the corresponding spectral absorption features near 1.7 μm selectively excited CF_3CH_2I and CH_3CH_2I in the overtone region of the CH-stretch vibration in solution.

The time delayed rise and subsequent slower decay of the transient absorption in the UV spectral range at 335 nm and 340 nm, induced by the IR-pump, monitored IVR and VET for both molecules on a picosecond (ps) timescale. In order to extract rate constants for IVR and VET we adopted the model introduced in Ref. [1]. Fig. 1 shows fits of the model to the experimental data and τ_{IVR} and τ_{VET} are given in Table 1. In other experiments we excited CH_2I_2 and allyl iodide

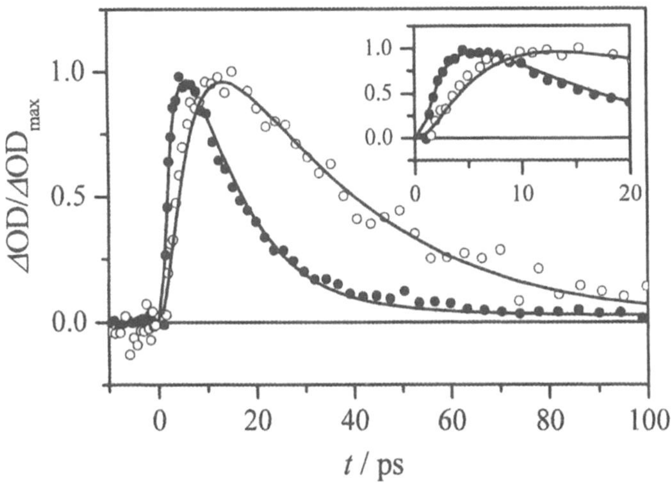

Fig. 1. Normalized transient absorption time profiles for CH_3CH_2I (•) and CF_3CH_2I (o) in CCl_4 for 1715 nm (•) and 1695 nm (o) excitation. λ_{pr}=335 nm and 340 nm. Also shown is a fit using the model described in Ref. [1].

in order to elucidate the dependence of IVR on the size and structure of the molecules. Finally, we studied the impact of other functional groups like OH which increase intermolecular interaction significantly. All molecules (in CCl_4) were excited in the two quanta region of the CH-stretch vibration (υ_{CH}=2) and in all cases the FC-active C-I mode was probed. Corresponding relaxation constants are given in Table 1. The obtained results provide firm evidence for our conclusion that hierarchical IVR in these molecules "survives" in solution which may be governed by specific low-order intramolecular interactions and which can be rationalized with a simple low-order coupling model [3]. If we assume that the rate of IVR of the molecules depends on the number of low(est)-order interactions to next neighbors in quantum number space the qualitative trends in Table 1 can be easily explained. While the timescales of VET for both molecules are in the range observed for other molecules the effect of chemical substitution on the rates is somewhat surprising.

Table 1. Relaxation coefficients τ_{IVR} and τ_{VET} for different molecules excited in υ_{CH}=2 in CCl_4 as determined from the simple model described in Ref. [1] and number of low-order interactions of the CH-stretch overtone (υ_{CH}=2, ±100 cm^{-1}) with states in the first tiers [3].

molecule	τ_{IVR} [ps]	τ_{VET} [ps]	number of 3rd order resonances	number of 4th order resonances	number of 5th order resonances
CH_2I_2	10±1	63±8	1	4	19
C_3H_5I	2±1	17±2	5	87	258
CH_3CH_2I	1.5±1	20±2	7	45	100
CF_3CH_2I	3.5±1	50±9	1	50	298
$HOCH_2CH_2I$	1.2±0.5	17±3	5	73	263

4. Conclusions

In the present study it has been found that the timescales of IVR depend upon the specific excitation as well as chemical substitution and structure, eventually opening new avenues for a (chemical) control of IVR and reactivity of vibrationally excited molecules in solution.

Acknowledgements. Financial support by the DFG and FCI and helpful discussion with Dr. D. Schwarzer are gratefully acknowledged.

References

1 A. Charvat, J. Aßmann, D. Schwarzer, and B. Abel, *J. Phys. Chem. A* 105, 5071-5080 (2001).
2 A. Charvat, J. Aßmann, B. Abel , D. Schwarzer, K. Henning, K. Luther, J. Troe, *Phys. Chem. Chem. Phys.*, 3, 2230-2240 (2001).
3 J. Aßmann, A. Charvat, D. Schwarzer, C. Kappel, K. Luther, and B. Abel, *J. Phys. Chem. A.*, (June 2002).
4 D. Bingemann, A. King, F. F. Crim, *J. Chem. Phys.* 113, 5018 (2000).

Ultrafast Dynamics of Molecular Wires based on Multiporphyrin Linear Arrays

[1]Hyun Sun Cho, [1]Dongho Kim, [2]Akihiko Tsuda and [2]Atsuhiro Osuka

[1]Center for Ultrafast Optical Characteristics Control and Department of Chemistry, Yonsei University, Seoul 120-749, Korea
[2]Department of Chemistry, Kyoto University, Kyoto 606-8502, Japan

1. Introduction

A major source of inspiration for the design and synthesis of optical nanostructures comes from the light-harvesting antenna complexes of natural photosynthetic systems. The antenna complexes are comprised of a large number of pigments that are arranged in a rigid three-dimensional matrix. The natural antenna complexes absorb light and funnel the resultant energy to the reaction centers by means of excited-state energy transfer processes. The versatile optical (absorption and emission), redox, and photochemical properties of the porphyrins make them ideally suited as components of nanostructures. Continuing efforts to realize the mimicry of solar energy harvesting complexes have enabled the design and synthesis of various types of covalently linked porphyrin arrays with the goal of applying these arrays to molecular photonic devices and artificial light-harvesting array systems. The recent success in elaborating various porphyrin architectures using several types of linkers *via meso* position attachment has brought up the issues on the electronic coupling, of which the extent is determined by the interconnection length and relative orientation between the adjacent porphyrin moieties.

2. Results and Discussion

Recently, it became possible to connect the porphyrin molecules directly up to one hundred twenty eight Zn(II)porphyrin units with controlling the number of porphyrin moieties in arrays.[1] The direct *meso-meso* connection with ~8.4 Å center-to-center distance results in exceptionally strong electronic coupling between the adjacent porphyrin units to exhibit a large exciton coupling energy of approximately 2,100 cm^{-1} derived from the plots of the exciton splitting in Soret transitions (Fig. 1). But the Q-bands reamin at nearly the same position as the number of porphyrin units increases, suggesting that the electronic communication between the adjacent porphyrins is almost prohibited due to the orthogonal arrangement of the neighboring porphyrin moieties. The fluorescence rise and decay dynamics of these porphyrin arrays were investigated by femtosecond fluorescence up-conversion and time-correlated single photon counting methods, respectively, which exhibits a gradual decrease in their time constants as the

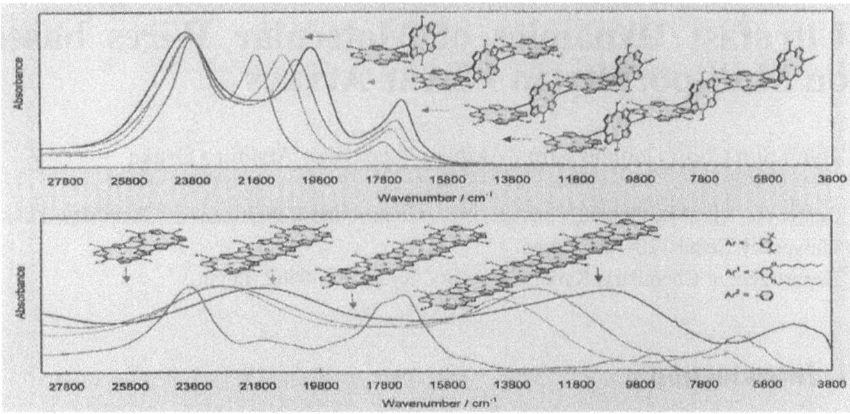

Fig. 1. The absorption spectra of the directly-linked orthogonal porphyrin arrays (upper) and the fused porphyrin arrays in toluene.

number of porphyrin units increases. Specifically, the fluorescence lifetime of 2.54 ns and the $S_2 \rightarrow S_1$ internal conversion process of 1.6 ps of Zn(II)porphyrin monomer were reduced to 1.9 ns and 0.7 ps, respectively, in the dimer.[1] Based on the natural radiative times as a function of the number of porphyrin units in the arrays, the radiative coherent length of the S_1 states of the porphyrin arrays was estimated to be 5-6 porphyrin units, indicative of a large coherent length due to ample interporphyrin interactions.

The population dynamics of the S_1 states which was measured by the fluorescence rise upon excitation of the S_2 states reveals faster internal conversion rates from the initially prepared S_2 states as the number of porphyrin units increases in the arrays mainly because the exciton split Soret transitions provide a ladder-type deactivation channel in returning to the S_1 states of the arrays. These porphyrin arrays would be promising molecular photonic wires due to a large excitonic interaction arising from a short interporphyrin distance with a maintenance of individual porphyrin characteristics.

For the possible future application of molecular electric wires, fused porphyrin arrays up to twelve porphyrin units by connecting two additional β positions from the adjacent porphyrin moieties in the directly linked Zn(II)porphyrins were successfully synthesized.[2] These porphyrin arrays exhibit even much stronger exciton coupling between Soret bands along with a systematic red-shift of the Q bands to IR region (up to a few μm) due to much enhanced π-electron delocalization throughout the entire fused porphyrin arrays (Fig. 1). The femtosecond UV/visible/IR pump/probe experiments were carried out to investigate the electronic relaxation process of each state of the fused porphyrin arrays (Fig. 2). The energy relaxation dynamics of the lowest excited states of the fused porphyrin arrays was much accelerated as the number of attached porphyrin units increases due to a smaller HOMO and LUMO gap. For instance, the lowest excited S_1 state of the fused porphyrin dimer decay to the ground state with a time constant of 4.5 ps, which is much faster than that of the orthogonal dimer (1.9 ns) (Fig. 3). In parallel with the faster S_1 state dynamics with an increase of the porphyrin units in the fused porphyrin arrays, the internal conversion rates from

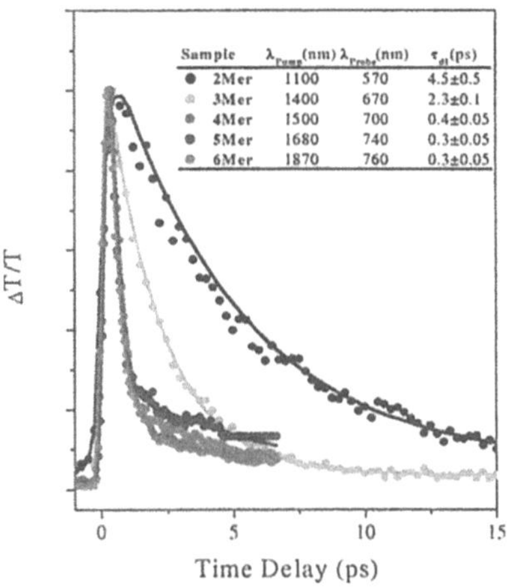

Fig. 2. The decay profiles of the bleaching signals of the fused porphyrin arrays upon photoexcitation of the lowet energy Q-bands.

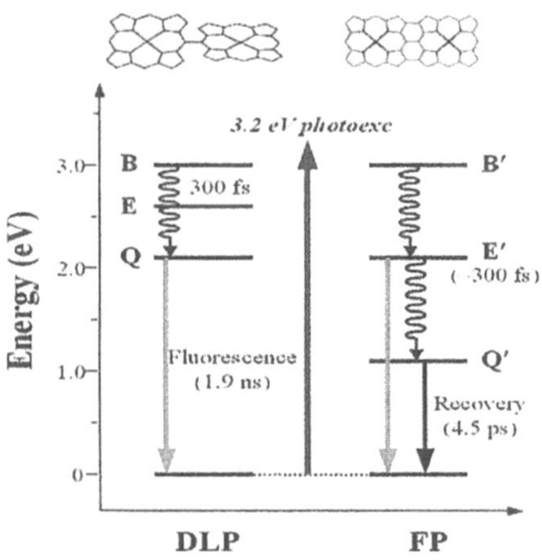

Fig. 3. The relaxation scheme for the directly linked orthogonal dimer and the fused porphyrin dimer.

the initially pumped S_2 states to the lowest energy S_1 states of the fused porphyrin arrays also exhibit a systematic increase presumably due to the participation of the many electronics states existing between the initially prepared S_2 and the S_1 states as seen in the broader absorption spectra in the entire wavelength region. The relatively unperturbed Q-states of the orthogonal porphyrin arrays do not affect much the energy relaxation dynamics of the lowest excited states of the porphyrin arrays as compared with that of the monomer. On the other hand, the continuous lowering of the lowest excited states upon elongation of the fused porphyrin arrays arising from extensive π-electron delocalization throughout the flat fused porphyrin array plane enhances their energy relaxation rates to the ground states. At the same time, the absorption bands of the fused porphyrin arrays becomes gradually broader, indicating the increased density of states between the excited states of the fused porphyrin arrays. These fused arrays provide a good prospect for molecular electric wires due to much reduced band gap energy between HOMO and LUMO levels. Thus, the combination of the reduced energy gap and the increased intermediate energy levels between the excited states is likely to facilitate the electronic deactivation processes to the lower excited or ground state as the fused porphyrin arrays become longer. Since the lowest electronic transitions in the fused porphyrin arrays occur closer to the vibrational excitations, these molecules can be good candidates for the investigation of vibronic coupling in large organic molecules. In addition, the unique photophysical properties of the fused porphyrin arrays deserve much attention for the future applications as IR sensors and nonlinear optical materials.

References

1. Kim, Y. H.; Jeong, D. H.; Kim. D.; Jeoung, S. C.; Cho, H. S.; Kim, S. K.; Aratani, N.; Osuka, A. *J. Am. Chem. Soc.* **2001**, *123*, 76.
2. Tsuda, A; Osuka, A. *Science* **2001**, *293*, 79.

Femtosecond Energy Concentration in Nanosystems Coherently Controlled by Excitation Phase

Mark I. Stockman[1], Sergey V. Faleev[1,2], and David J. Bergman[3]

[1] Department of Physics and Astronomy, Georgia State University, Atlanta, Georgia 30303, USA
E-mail: mstockman@gsu.edu
[2] Present address: Sandia National Laboratories, Livermore, CA 94551, USA
E-mail: gs01sfx@hotmail.com
[3] School of Physics and Astronomy, Raymond and Beverly Sackler Faculty of Exact Sciences, Tel Aviv University, Tel Aviv, 69978 Israel
E-mail: bergman@post.tau.ac.il

Abstract. We predict and quantitatively evaluate the unique possibility of concentrating the energy of an ultrafast excitation of a nanosystem in a small part of the whole system by means of coherent control (phase modulation of the exciting ultrashort pulse) [1]. Such concentration is due to dynamic properties [2] of surface plasmons and leads to local fields enhanced by orders of magnitude. This effect exists for both engineered and random nanosystems. For linear processes, the coherent control affects the temporal distribution of the local fields, while the integrated energy of the local fields is not affected. For nonlinear processes, both the temporal distribution and integral energy are coherently controllable. We also discuss possible applications.

1. Introduction

The problem of ultrafast optical excitation at a given part of a nanosystem arises in many fundamental and applied fields, including but not limited to ultrafast probing with nanometer spatial resolution, site-selective photochemistry, and computations at the nanoscale. In particular, for the nanoscale computations, such ultrafast concentration of energy on nanoscale is desired to energize and control nano-circuits on time scale where the bandwidth of metal wires is insufficient.

We propose to use phase modulation of an exciting femtosecond pulse as a functional degree of freedom to coherently control spatial distribution of the excitation energy on the nanoscale. This possibility exists due to the fact that polar excitations (surface plasmons) in inhomogeneous nanosystems tend to be localized with their frequency correlated with position. In that case, changing the phase of the excitation will cause the exciting field to take energy from excitations in some parts of the system and to add energy to the surface plasmon oscillations of other parts as desired. We have carried out quantitative computations to confirm this heuristic idea.

2. Theory and Numerical Results

The theory of the predicted effect is based on the Green's function method in the spectral representation [2,3]. This Green's function is numerically calculated using the eigenmode-expansion solution of the Dirichlet-Neumann (3+1)-dimensional boundary problem in real space and frequency domain [3]. As a numerical method, such an approach is exceptionally stable and reliable, and it is sufficiently efficient.

The results of the simulations [1] for a V-shaped planar nanostructure made of silver (8 nm thick) subjected to a phase-modulated 25-fs exciting light pulse (the carrier frequency $\hbar\omega_0 = 0.8$ eV) with a negative chirp are illustrated in Fig. 1. This example shows that, as the pulse progresses, the maximum of the local fields and, correspondingly, the energy of the excitation on the nanoscale move from the opening of the nanostructure [see Fig. 1(a)] toward the apex where they strongly concentrate for a significant time interval on the femtosecond scale from 80 to 118 fs [see Fig.1 (b) and (c)]. The maximum local field at the apex exceeds the exciting field by almost two orders of magnitude, which means that the *local optical intensity is enhanced by four orders of magnitude.* Note that here this energy concentration is due to the phase modulation of the exciting pulse and not to randomness of the underlying nanostructure as in the "Ninth Wave" effect [4]. Such energy concentration may be used to energize or control the nanostructures for computations on the nanoscale, for directed nano-modification, to excite nonlinear photoprocesses localized at the nanoscale for ultrafast optical probing, and in other fields.

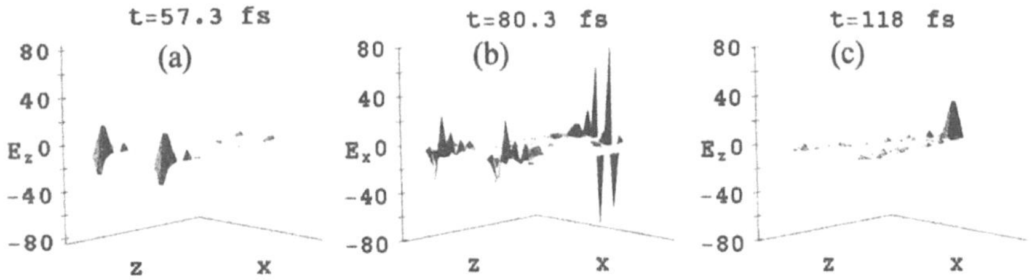

Fig. 1. Spatial distributions of local electric fields (in the units of the exciting pulse amplitude) at the nanostructure for moments of time indicated. The duration of the exciting pulse is 50 fs, and the phase modulation parameter $\alpha = -0.3$ [1].

In the absence of the phase modulation of the exciting pulse, such energy concentration is completely absent. This phase modulation affects the excitation due to two factors: *(i)* The modulated pulse has wider bandwidth and *(ii)* Nontrivial effect of the coherent control, i.e., dependence of the excitation kinetics on the specific time dependence of the light-wave phase. To distinguish between these two mechanisms, we have carried out the computations for the excitation pulse of the exactly the same shape and spectral composition, but the opposite (positive) chirp. Note that such a pulse is time conjugated with respect to the original pulse with the negative chirp.

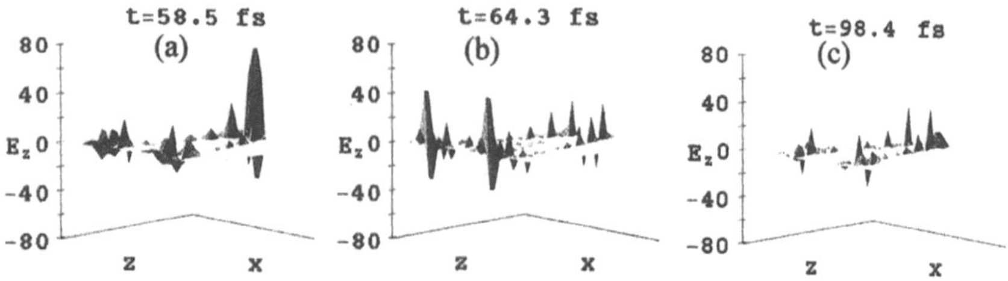

Fig. 2. Same as in Fig.1, but for the positive chirp, with the phase modulation parameter $\alpha = 0.3$ [1].

The results for this time-conjugated pulse are shown in Fig. 2. In that case, the kinetics is radically different from Fig. 1. The energy concentration does not occur for any significant interval of time. Rather, this excitation energy is transferred forth and back between the opening and the tip of the V-shape [Fig. 2 (a) and (b)]. This shows that not only the spectral composition of the excitation pulse but *the nontrivial effects of the coherent control are of primary importance.*

We have also examined the two-photon fluorescence, which is a second-order nonlinear photoprocess. In this case, the average intensity of the emitted light is efficiently coherently controllable, which may be the simplest approach to observe this effect.

3. Conclusions

We have shown by a stable and reliable numerical procedure that by using the appropriate phase modulation of the exciting femtosecond pulse, it is possible to coherently control the spatio-temporal dynamics of ultrafast excitation of nanosystems.

Acknowledgements. This work was supported by the Chemical Sciences, Biosciences and Geosciences Division of the Office of Basic Energy Sciences, Office of Science, U.S. Department of Energy. Partial support for the work of DJB was provided by grants from the US-Israel Binational Science Foundation and the Israel Science Foundation.

References

1 M. I. Stockman, S. V. Faleev, and D .J. Bergman, in *Phys. Rev. Lett.*, 88, 067402, 2002.
2 M. I. Stockman, S. V. Faleev, and D. J. Bergman, in *Phys. Rev. Lett.* 87, 167401, 2001.
3 D. J. Bergman and D. Stroud, in *Solid State Phys.*, 46, 147, 1992.
4 M. I. Stockman, in *Phys. Rev. Lett.*, 84, 1011, 2000.

Ultrafast laser-induced structural changes in anisotropic solids

K. Sokolowski-Tinten[1], V.V. Temnov[1], P. Zhou[1], D. von der Linde[1], S. Ashitkov[2], M.B. Agranat[2], and V.E. Fortov[2]

[1] Institute for Laser and Plasma Physics, University of Essen, 45117 Essen, Germany
E-mail: kst@ilp.physik.uni-essen.de
[2] Institute of High Energy Density, Joint Institute of High Temperatures, Russian Academy of Sciences, 13/19 Izhorskaya St., 127412 Moscow, Russia

Abstract. Time-resolved polarization sensitive microscopy has been used to study laser-induced structural changes in graphite and tellurium. In both materials the transition from an optically anisotropic to an isotropic state, indicating the loss of crystalline order, is possible on a subpicosecond time scale.

1. Introduction

Ultrafast light-induced phase transitions have been studied in a number of covalently bonded materials and with many different techniques, including time-resolved measurements of the reflectivity [1] and of the reflected second harmonic [2]. More recently, the direct observation of disordering of solids during melting has become possible through the use of ultrafast time-resolved X-ray diffraction [3]. While simple reflectivity measurements provide information about the changes of the electronic properties, which can not be uniquely related to transient structural changes, femtosecond time resolved X-ray diffraction has been applied, up to now, to a few selected materials (Ge, InSb) only.

We present here the results of a femtosecond time-resolved study of fast structural changes in laser-excited graphite and tellurium exploiting the optical anisotropy (birefringence) of these materials.

2. Experimental Methods

If a linearly polarized electromagnetic wave is reflected from the surface of an anisotropic medium, reflectivity depends on the orientation of the polarization vector relative to the crystallographic axes. Moreover, the state of polarization changes upon reflection, even at normal incidence. In uniaxial crystals this effect is strongest for reflection from surfaces containing the anisotropy axis, the situation realized in our experiments. Supposed that the angle between the incident polarization and the anisotropy axis is not 0^0 (extraordinary polarization) or 90^0 (ordinary polarization) a small fraction (a few times 10^{-3}) of an incident linearly polarized wave will be transformed to the orthogonal polarization component after reflection. However, the orthogonal component should disappear and ordinary and extraordinary reflectivity should become equal, if the crystal is converted to an isotropic medium for example as a result of a phase transition [4]. Polarization sensitive ultrafast time resolved optical microscopy (λ = 400 nm) at normal incidence has been used to follow the transient changes of the anisotropic optical properties of single crystalline graphite and tellurium after excitation with single 100 fs laser pulses at 800nm.

3. Results and Discussion

Results obtained on graphite as a function of pump-probe time delay for a fixed pump fluence (0.5 J/cm^2), and as a function of fluence for a fixed time delay (1 ps) are shown in the left and the right part of Fig. 1, respectively. R_o (black curve) and R_e (blue curve) denote the reflectivity for ordinary and extraordinary polarization, respectively. R_{cr} (red curve) represents the transformed orthogonal component (*ansiotropic reflectivity*), mentioned above, for an incident polarization angle of 45^0 with respect to the anisotropy axis.

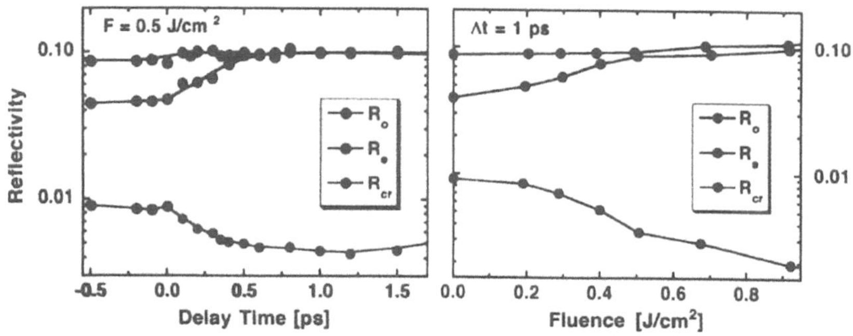

Fig. 1. Ordinary reflectivity R_o (black), extraordinary reflectivity R_e (blue), and *anisotropic reflectivity* R_{cr} (red) of graphite as a function of time delay for a pump fluence of 0.5 J/cm^2 (left), and as a function of pump fluence for a fixed delay time of 1 ps (right)

R_o and R_e increase and become nearly equal, both, as a function of time and with increasing pump fluence, while over the same time and fluence ranges the *anisotropic reflectivity* R_{cr} drops significantly. This behavior demonstrates that upon strong optical excitation the material changes from the initial anisotropic state to an isotropic state [5]. If these changes are interpreted as a structural transition, the decay of optical anisotropy in less than one picosecond for pump fluences above 0.4 J/cm^2 (below the threshold for optical breakdown and plasma formation!) indicates a *non-thermal* loss of crystalline order, faster than the usual electron-lattice equilibration time.

In tellurium we observed a similar behavior, namely a rapid loss of anisotropy (R_o and R_e becoming equal, R_{cr} decreasing). However, tellurium exhibits a pronounced fluence dependence, as demonstrated by Fig. 2. It shows the *anisotropic reflectivity* R_{cr} as a function of time delay for different pump fluences.

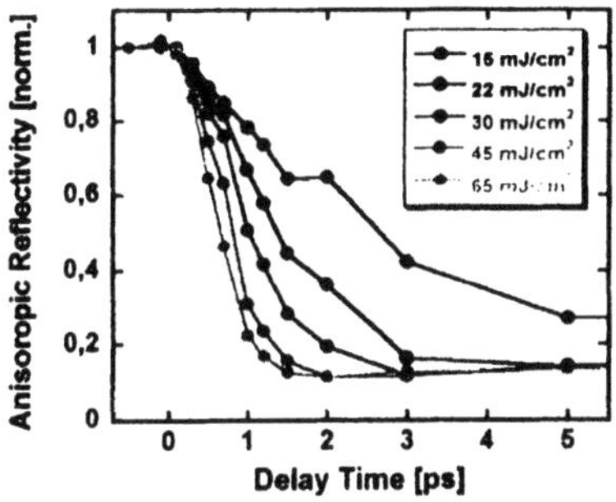

Fig. 2: Anisotropic reflectivity R_{cr} of tellurium as a function of pump-probe time delay for different pump fluences above the melting threshold (approx. 10-12 mJ/cm^2).

For the highest fluences the signal decay, indication of the loss of anisotropy, occurs again on a sub-ps time-scale. For lower fluences the loss of anisotropy slows down considerably resulting in decay times of a few picoseconds.

The data in Fig. 2 seem to indicate that different disordering mechanisms are operative in tellurium. While the sub-picosecond decay at higher fluences can be again attributed to a *non-thermal* pathway, the slower dynamics observed at low fluences seems to be compatible with a *homogenous thermal* melting scenario [6].

References

1 C.V. Shank, R. Yen, and C. Hirlimann, Phys. Rev. Lett. **50**, 454 (1983); K. Sokolowski-Tinten, J. Bialkowski, and D. von der Linde, Phys. Rev. B **51**, 14186 (1995); L. Huang, J. P. Callan, E. N. Glezer, and E. Mazur, Phys. Rev. Lett. 80, 185 (1998); D.H. Reitze, H. Ahn, M.C. Downer, Phys. Rev. B **45**, 2677 (1992).

2 C.V. Shank, R. Yen, and C. Hirlimann, Phys. Rev. Lett. **51**, 900 (1983); H.W.K. Tom, G.D. Aumiller, and C.H. Brito-Cruz, Phys. Rev. Lett. **60**, 1438 (1988).

3 C.W. Siders, A. Cavalleri, K. Sokolowski-Tinten, Cs. Tóth, T. Guo, M. Kammler, M. Horn-von-Hoegen, K.R. Wilson, D. von der Linde, and C.P.J. Barty, Science **286**, 1340 (1999); A. Rousse, C. Rischel, S. Fourmaux, I. Uschmann, S. Sebban,.G. Grillon, Ph. Balcou, E. Förster, J.P. Geindre, P. Audebert, J.C. Gauthier, and D. Hulin, Nature **410**, 65 (2001); K. Sokolowski-Tinten, C. Blome, C. Dietrich, A. Tarasevitch, M. Horn-von-Hoegen, D. von der Linde, A. Cavalleri, and M. Kammler, Phys. Rev. Lett. **87**, 225701 (2001).

4 M.B. Agranat, S.I. Anisimov, and V.E. Fortov, JETP **88**, 1184 (1998); M.B. Agranat, S.I. Ashitkov, V.E. Fortov, Appl. Phys. A **69**, 637 (1999).

5 S.I. Ashitkov, M.B. Agranat, P.S. Kondratenko, S.I. Anisimov, V.E. Fortov, V.V. Temnov, K. Sokolowski-Tinten, P. Zhou, and D. von der Linde, JETP Letters **75**, 87 (2002).

6 B. Rethfeld, K. Sokolowski-Tinten, D. von der Linde, and S.I. Anisimov Phys. Rev. B **65**, 92103 (2002).

Induced Optical Tunnelling

Charles Hirlimann and Benjamin Thomas

Institut de Physique et Chimie des Matériaux de Strasbourg, Groupe d'Optique Non Linéaire et d'Optoélectronique, UMR 7504 CNRS-ULP, 23, rue du Lœss BP 20 CR, 67037 Strasbourg Cedex, France.
E-mail: Charles.Hirlimann@ipcms.u-strasbg.fr

Abstract. Following a theoretical study performed by François Bardou and Dominique Boosé, this work shows the first experimental evidence for induced tunnelling of a particle when it endures a momentum transfer during its reflection on a barrier.

1. Theoretical grounds

François Bardou and Dominique Boosé recently published [1] their discovery of a new induced tunnel effect, they named "quantum evaporation". In this work they showed that, for a one-dimensional electron reflecting on an energy barrier, the stationary character of its eigen-functions could be broken under special conditions.

If a small momentum is transferred to the electron at the time it reflects on the barrier, its wave-function is projected on the complete set of its stationary eigen-functions. A small momentum means that the energy transfer that accompanies the momentum transfer is negligible compared to the barrier height or at least that it does not allow for a jump of the particle above the barrier. The tunnelling probability of the electron is strongly enhanced by the momentum perturbation because the weight of the eigen-states with energies above the barrier becomes preponderant in the calculation of the tunnelling probability.

This probability enhancement is independent on the sign of the momentum transfer: the electron can be pushed toward or pull out of the barrier; it only depends on the momentum amplitude in a quadratic way. The authors have showed that the momentum transfer is equivalent to a kinetic translation of the barrier.

2. Experimental Method

The original prediction by Bardou and Boosé has been made for a one-dimensional electronic system, while we performed an optical transposition of the effect. The tunnelling barrier, in our case, is a Newton's optical barrier [2]between two silica UV grade 90° prisms, in a slightly frustrated total internal reflection geometry. The thin air layer that separates the prisms joined along their hypotenuse faces is crossed through (via the tunnel coupling) by 80 fs, 620 nm, probe light pulses generated by an amplified CPM dye laser. This train of pulses simulates a train of particles reflecting on a barrier. The probe intensity on the barrier ranges from 6.5

to 90 MW/cm^2, the energy per pulse being 1 to 13 pJ. A 10 times weaker pump pulse simulates the momentum transfer via the index of refraction change (Kerr effect) induced by the pump in the silica that varies the length of the optical path for the probe, mimicking a kinetic move of the barrier. On the leading edge of the probe intensity envelope, the optical path lengthens while it comes back to normal on the trailing edge because the non-linear index of refraction of silica is positive. This way we could test an important signature of the effect: its independence upon the sign of the momentum transfer. Recording the probe intensity variation versus the time delay between the space-superimposed probe and pump eliminates the integrated index change induced in silica by the probe. This allows testing the necessary coincidence of the reflection and perturbation as being another signature.

3. The Bardou-Boosé effect

In this experiment we have therefore looked for a signal increase of the probe slightly transmitted (≈1%) through the optical tunnel when it is coincident with the pump.

At high intensity (> 10 GW/cm^2, Figure 1) the Kerr effect produces a beam defocusing and a spectral broadening of the probe pulses. Maximising the light collection aperture has eliminated the decrease in intensity induced by defocusing on the detector. The barrier height has been controlled by having a probe incident angle larger than the total internal reflection angle in such a way that the maximum spectral broadening could not generate photons jumping above the barrier and therefore increasing the transmitted signal. This large incidence angle is also unfavourable to the non-linear Goos-Hänchen [3] effect. This last effect is known to take place under high-intensity experimental condition not respected here.

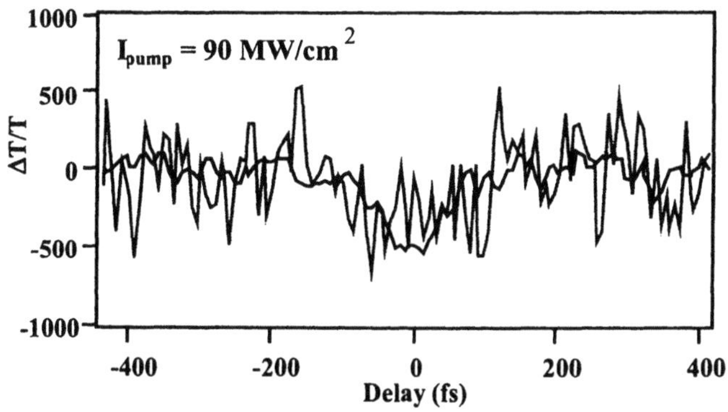

Fig. 1. At "high" intensity only quadratic residual absorption can be observed.

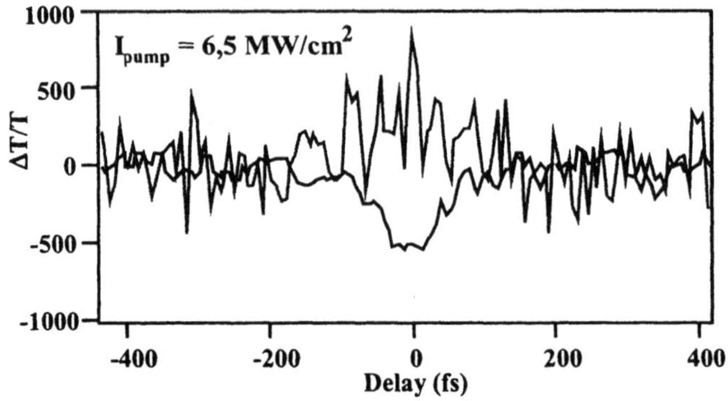

Fig. 2. At "low" intensity an enhancement of the transmitted intensity is observed during the superimposition of the two pulses. The result is compared to the smoothed residual absorption.

At low intensity (in the range of a few tens of MW/cm^2, Figure 2), when the probe and the pump are time coincident, crossed three photon absorption in silica takes place which is dominated by the strong probe two photon absorption associated to the extraneous simultaneous absorption of a pump photon. This three-photon absorption measured on the probe beam exhibits a quadratic dependence versus intensity that diminishes the transmitted intensity. In this regime the quantum-evaporation effect is also quadratic with opposite sign and therefore competes with the three-photon absorption process.

The numerical simulations of reference [1] show that at sufficiently low intensity the induced tunnelling probability decreases less rapidly than a quadratic effect. We experimentally observed that, down at 60 MW/cm^2, no more differential signal was observed in the limit of the 4 ‰ sensitivity limit of our experimental set-up. A statistically significant positive signal appeared for weaker intensities down to 6.5 MW/cm^2. The envelope of this signal follows the convolution function of the pulses as directly measured from the three-photon absorption at slightly higher intensities. This fact directly demonstrates: i) the instantaneous character of the measured effect ii) the absence of the effect when the perturbation is not applied during the reflection time iii) its independence against the sign of the barrier perturbation. These behaviours are all signatures of the Bardou-Boosé effect.

References

1 D. Boosé and F. Bardou ; Europhys. Lett. 53 (2001) 1-7
2 Sir Isaac Newton in "Optique" (Christian Bourgois, Paris, 1989) p. 185.
3 O. Émile, T. Galstyan, A. Le Floch, F. Bretenaker, Phys. Rev. Lett. 75 (1995) 1511-1513.

Wavepacket Interferometry for Energy Transfer: Ultrafast 2D Electronic Spectroscopy

Dmitri Kilin[1] and Jeffrey A. Cina[1,2]

[1] Oregon Center for Optics and Department of Chemistry, University of Oregon, Eugene, OR 97403, USA

[2] E-mail: cina@oregon.uoregon.edu

Abstract. We present theoretical studies of nonlinear wavepacket interferometry in a model energy-transfer complex. The calculated two-dimensional time-domain electronic spectra—quadralinear in four phase-controlled ultrashort pulses—exhibit amplitude-level information on the vibrational motion accompanying energy transfer.

Advances in controlling [1] or monitoring [2] with sub-optical-period accuracy the inter-pulse delays within sequences of ultrashort laser pulses have led to the development of electronic interference spectroscopies in which the signal is *quadralinear* in external fields bearing precisely specified optical phase relationships to one another. [3,4] These techniques generate coherent signals that are directly proportional to the overlap between multi-pulse target and reference nuclear wavepackets (in a pure-state description [5]) or to a multi-pulse density matrix increment (in a mixed-state treatment [6]), and approach their ideal realization with pulses short enough to freeze nuclear motion on the fastest relevant timescales.

To assess nonlinear wavepacket interferometry (WPI) as a method for gaining amplitude-level information on the vibrational dynamics accompanying electronic energy transfer, we are analyzing calculated signals from an idealized model system consisting of two chromophores with energy-transfer coupling J, each with an internal molecular vibration of frequency ω whose equilibrium position is displaced by d upon electronic excitation. [7] To simplify the treatment and allow each molecule to be separately addressed, we assign them perpendicular lab-fixed electronic transition moments of (equal) magnitude μ.

A particularly interesting WPI signal from this system yields the four-pulse interference contribution to population of the acceptor-excited electronic state (1') following resonant interaction with a pair of phase-locked pulse-pairs $(A_{acc}B_{don})(C_{don}D_{don})$ with the polarizations indicated. This signal contribution—isolable from competing pathways by phase cycling and the combination of measurements with various excitation and emission polarizations—directly determines the complex overlap between the reference $\alpha_{1'}(q_{don}, q_{acc})$ and energy-transfer target $\xi_{1'}(q_{don}, q_{acc})$ wavepackets whose phase-space trajectories are schematically illustrated in Fig. 1.

The WPI signal, like the target $\xi_{1'}$ itself, is *first order* in J. Fig. 2 shows the real part of the complex overlap $\langle\alpha_{1'}|\xi_{1'}\rangle$ isolated from a calculated interferogram for a case of energy transfer that is downhill by twice the Franck-Condon energy (here $E_{FC} = 3.69\omega$).

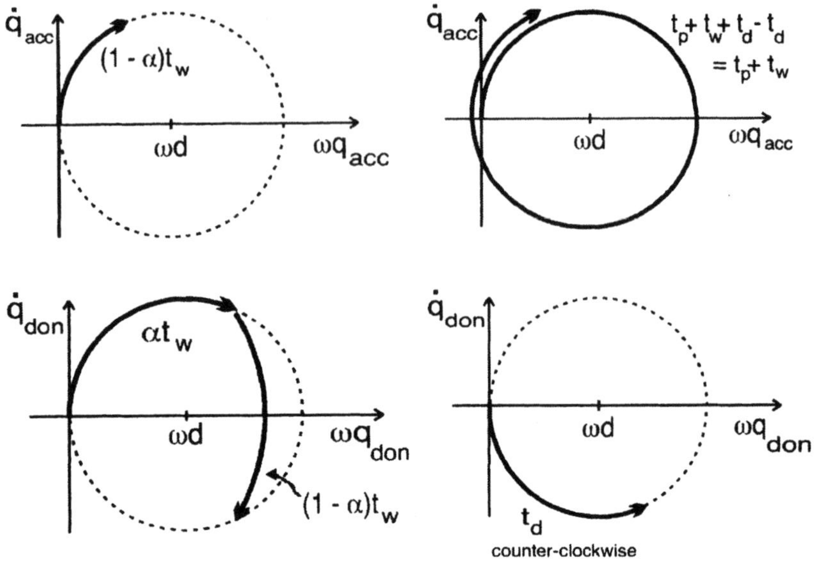

Fig. 1. Schematic target (left) and reference (right) phase-space trajectories for vibrational modes of the donor and acceptor molecules. t_p (t_d) is the *AB* (*CD*) intrapulse-pair delay; t_w is the interpulse-pair delay between *B* and *C*. Transfer of the target trajectory from the donor-excited surface to the acceptor-excited surface is assumed to occur αt_w after Franck-Condon excitation by the *B* pulse

As suggested by the semiclassical description of Fig. 1, the temporal location of the signal peaks (at $t_p = 0.68(2\pi/\omega)$ and $t_d = 1.29(2\pi/\omega)$ in Fig. 2 for example) indicate when during t_w the energy-transfer surface-crossing event occurred. We have obtained a semi-analytic expression relating the fringe structure of the individual peaks to the dynamics and energetics of target and reference wavepacket motion whose predictions are in accord with the oscillations seen in Fig. 2. [7]

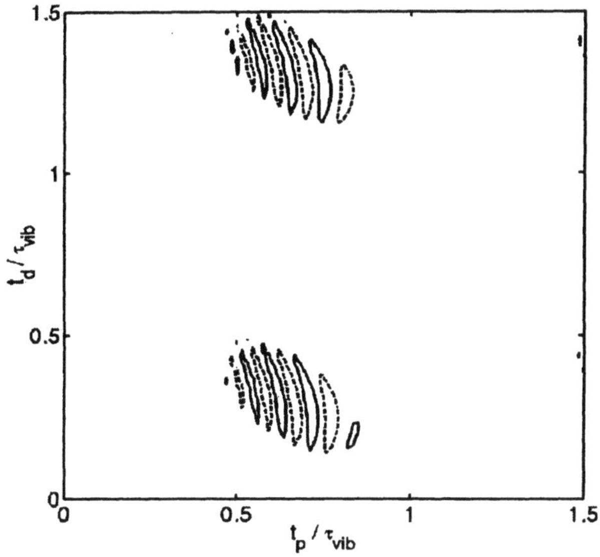

Fig. 2 Real part of target-reference overlap isolated from calculated WPI signal for the model system with the acceptor site energy downhill from that of the donor. The pulses are abrupt on the vibrational timescale, the locking frequency for both pulse-pairs matches the vertical excitation frequency for the donor, and t_w is fixed at half a vibrational period. [8,9]

References

1 N. F. Scherer et al. J. Chem. Phys. **95**, 1487 (1991); **96**, 4180 (1992).
2 A.W. Albrecht, J.D. Hybl, S.M. Gallagher Faeder, D. M. Jonas, J. Chem. Phys. **111**, 10934 (1999); Erratum, J. Chem. Phys. **115**, 5691 (2001).
3 W.P. de Boeij, M.S. Pshenichnikov, D.A. Wiersma, Chem. Phys. **233**, 287 (1998).
4 J.D. Hybl, A. Albrecht Ferro, D.M. Jonas, J. Chem. Phys. **115**, 6606 (2001).
5 J. A. Cina, J. Chem. Phys. **113**, 9488 (2000).
6 M. Cho, N.F. Scherer, G.R. Fleming, and S. Mukamel, J. Chem. Phys. **96**, 5618 (1992).
7 J.A. Cina and D. Kilin, in preparation for J. Chem. Phys. (2002).
8 Our paper in the technical abstracts for this conference showed a similar plot of the overlap in the case of equal donor and acceptor site energies. We now believe that the fringe structure in that figure is incorrect; Ref. [7] will present the corrected overlap.
9 This work is supported by the US/NSF.

Liquid/Glass Solvent Dynamics: from 300 to 3 K

Kees Lazonder, Maxim S. Pshenichnikov and Douwe A. Wiersma

Ultrafast Laser and Spectroscopy Laboratory, Materials Science Centre
University of Groningen, Nijenborgh 4, 9747 AG Groningen, The Netherlands
phone: +31-50-3634324, fax: +31-50-3634111, e-mail: M.S.Pshenichnikov@chem.rug.nl

Abstract. Temperature-dependent echo-peak shift and heterodyned-echo experiments in glass-forming liquids are discussed. Freezing out a large part of the bath fluctuations by varying the temperature from 300 K to 3 K allows for testing current solvation models.

1. Introduction

The effects of solvent molecules on reactants are essential to the outcome of chemical reactions in the liquid phase. Ultrafast spectroscopies, and especially photon echo spectroscopy, have most efficiently been used to study solvent dynamics because the resonance frequency of dissolved chromophores is extremely sensitive to its surrounding. By charting the time-resolved correlation function of optical transitions upon ultrashort excitation, detailed insight into the dynamics of the solvent can be obtained.

The multimode Brownian oscillator (MBO) model [1], which is successfully used to describe the solvent dynamics at room temperatures in the liquid phase, has also been tested at lower temperatures [2-4]. Previous studies on temperature dependent solvation dynamics have been conducted in polymer hosts that form an amorphous solid already at room temperature. The sample was cooled down to 30 K, where the MBO model should still be readily applicable. Experimentally, no major deviations from theoretical predictions were reported.

In this Contribution, we describe echo-peak shift (EPS) and heterodyne-detected photon echo (HDPE) experiments in liquids that form glasses upon cooling below 150 K. At this point diffusive solvent motion is frozen out. By cooling further down to 3 K, all other relaxation channels are blocked as well. At this temperature regime, the tunneling two-level system (TLS) model [5-6] is expected to provide a more adequate description than phonon-based models.

2. Experimental

As a probe for the liquid/glass dynamics, a DTTCI dye molecule diluted in an ethanol/propanediol mixture was used. Two echo signals, induced by three 20-fs excitation pulses from a cavity-dumped Ti:Sapphire oscillator, are detected at the conjugated phase-matched $-\boldsymbol{k}_1 + \boldsymbol{k}_2 + \boldsymbol{k}_3$ and $\boldsymbol{k}_1 - \boldsymbol{k}_2 + \boldsymbol{k}_3$ directions. In the HDPE experiment, the echo field is heterodyned by a fourth pulse, the local oscillator [8,9].

3. Results and discussion

In the EPS experiment [7], the offset of the maximum of the time-integrated echo signal from zero delay is measured as a function of the waiting time t_{23} between the second and the third pulses (Fig.1a). As the temperature drops, the dynamics become visibly slower.

The decreased slope of the EPS function at times >100 fs clearly indicates that the inertial/diffusional modes are being frozen out. Figure 1b plots the detailed temperature dependent data at two waiting times that correspond to the first vibrational recurrence (~210 fs) and almost full relaxation at room temperature (100 ps). Both plots can be approximated by linear functions with similar slopes within the experimental uncertainty. The offset between the two lines originates from the vibrational relaxation of the 150-cm^{-1} intramolecular mode. Therefore, as the temperature decreases, the system becomes more inhomogeneous as predicted by the MBO model (solid lines). Note that the glass transition that occurs at T~150 K, is not apparent in the EPS plot.

In the HDPE experiment, scanning the delay between the last two pulses (t_{34}) for a set of fixed timings of the first two (t_{12}), time resolved echo-peak profiles are obtained (Fig.2a; T=100 K). For extracting information regarding the homogeneous *vs.* inhomogeneous contributions to the line broadening, the position of the echo maxima has been proven to be extremely instructive [7]. For instance, in the framework of Bloch model, the echo maximum peaks at

$$t_{34} = t_{12} - T'^2_2 / T_2, \quad (1)$$

where T'^2_2 and T_2 are inhomogeneous and homogeneous phase relaxation times. If the line broadening is perfectly static, $t_{34} = t_{12}$ at all times that $t_{12} > 0$. At the other extreme, a totally homogeneous system produces profiles that always peak at $t_{34} = 0$. In more sophisticated models of solvent dynamics [1], a relation similar to Eq.1 also holds although the interpretation of the relaxation times becomes quite different [7-8]. Figure 2 shows the temperature dependence of the maximum of several slices of echo profiles at different temperatures. It is immediately clear that the system grows to be more static at lower temperatures. From 300 K to the glass

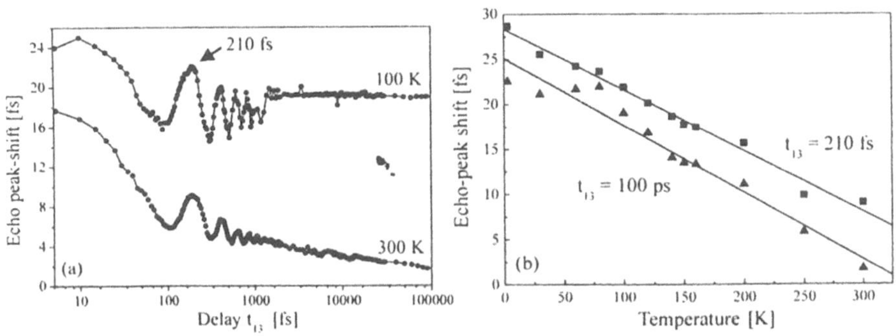

Fig. 1. (a) EPS traces of DTTCI in a propanediol/ethanol mixture at two different temperatures. (b) The temperature dependence of the EPS for two population times, the lines are linear fits.

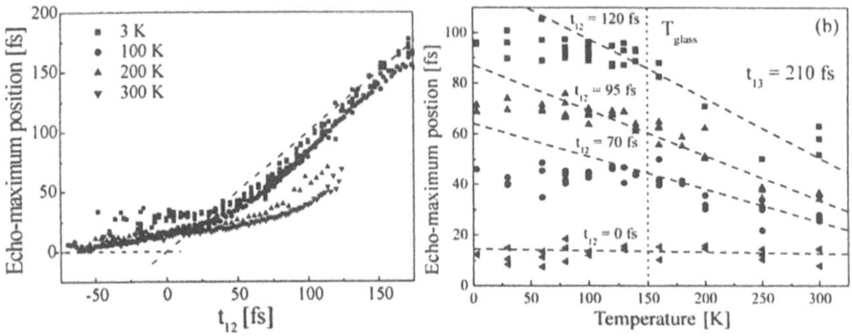

Fig. 2. (a) The maxima of the time-resolved 3 pulse echo profiles of DTTCI in a propanediol-ethanol mixture at several temperatures. (b) The temperature dependence of these maxima for several coherence times. The dashed lines are fits using the MBO-model. The glass temperature is indicated by the vertical dotted line. From Fig.1 one can conclude that the plot of the time-resolved echo maximum position obeys Eq.1 reasonably well. The estimated dephasing time at 100 K is $T_2 \cong 60 fs$.

temperature at 150 K, the dynamics slow down in a linear fashion, which is consistent with the EPS data (Fig.2b). However, deceleration of the solvent dynamics seemingly comes to a halt around the glass temperature. Most probably, at these temperatures the spin-boson formalism is not applicable any longer, and a different model should be applied. Currently, we attempt to establish the connection between the MBO and TLS models. The latter has been proven quite successful in describing the low-temperature dynamics in glasses [5-6].

References

1 S. Mukamel, *Principles of Nonlinear Optical Spectroscopy* (Oxford University Press, New York, 1995).
2 Y. Nagasawa, S. A. Passino, T. Joo and G. R. Fleming, J. Phys. Chem., **106**, 4840, 1997.
3 Y. Nagasawa, J. Y. Young and G. R. Fleming, J. Phys. Chem., **109**, 6175, 1998.
4 C.J. Bardeen, G. Cerullo, and C.V. Shank, Chem. Phys. Lett., **280**, 127, 1997.
5 W. A. Phillips, *Amourphous Solids. Low-temperature Properties* (Springer-Verlag, Berlin, 1981).
6 K. Lazonder, K Duppen and D. A. Wiersma, J. Phys. Chem. B, **104**, 6468, 2000.
7 W.P. de Boeij, M.S. Pshenichnikov, and D.A. Wiersma, J. Phys. Chem., **100**, 11806, 1996.
8 W.P. de Boeij, M.S. Pshenichnikov, and D.A. Wiersma, Chem. Phys., 233, 278, 1998.
9 S.M. Gallagher, A.W. Albrecht, T.D. Hybl, B.L. Landin, B. Rajaram, D.M. Jonas, JOSA B 15, 2338, 1998.

Ultrafast Dephasing of Photoexcited Polarons in Primary Doped Polyaniline

J. Kim, A. N. Unterreiner, S. Park, and N. F. Scherer

Department of Chemistry, University of Chicago, Chicago, IL 60637, USA
E-mail: nfschere@uchicago.edu

Abstract. Ultrashort pulse pump-probe and transient anisotropy methods are used to study the dynamics of primary doped conductive polyaniline in solution. Photoexcited carrier dynamics are identified on timescales of sub-100 fs up to >100 ps timescales.

1. Introduction

Polyaniline (PANI) is a conducting polymer with properties of fundamental interest and technological importance, e.g. electrical conductivity. Depending on the degree of oxidation, its conductivity changes dramatically; acidification of the half-oxidized form (emeraldine base) increases the conductivity by ten orders of magnitude – a process termed primary doping [1]. The primary doped (emeraldine salt) form of PANI shows a broad absorption centered at 806 nm attributed to the formation of polarons. While decay dynamics of the photoexcited polaron have been studied with ultrafast pump-probe measurements on thin films [2], the non-exponential (modeled as bi-exponential) decay could not be distinguished as originating from only intra-chain or mixed intra- and inter-chain dynamics. Measurements on isolated polymer chains are required for comparison. This paper addresses the latter ambiguity through ultrafast pump-probe investigations of the primary doped conductive form of PANI.

2. Experimental Methods

Pump-probe and transient anisotropy measurements of primary doped PANI were performed around the center of the polaron absorption band with sub-20 fs ultrafast pulses centered at 800 nm generated from a home-built cavity-dumped Ti:Sapphire oscillator [3]. Primary doped PANI solution was prepared by dissolving emeraldine base powder in chloroform containing d,l-camphorsulfonic acid.

3. Results and Discussion

The pump-probe signal of PANI is shown in Figure 1 on a long timescale, up to 100 ps, and a short timescale (inset). The whole decay up to 100 ps was fitted with a 4th order exponential decay with two longest time constants of 117.0 and 9.36 ps in a manner described elsewhere [4]. A very fast decay time constant of 34 fs is obtained by convoluting the instrument response function with a tri-exponential

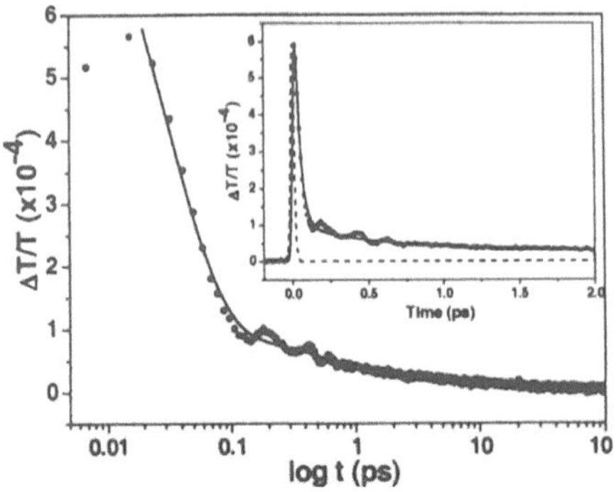

Fig. 1. Pump-probe differential transmission signal (points) and 4th order exponential fit (solid line). Inset: short timescale data (points) with tri-exponential fit (solid line) and instrument response function (dashed line).

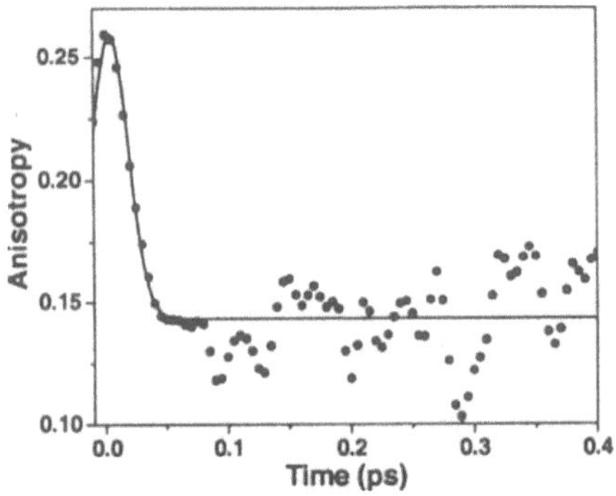

Fig. 2. Transient anisotropy (points) and a 30 fs FWHM Gaussian fit (solid line).

decay while fixing the long component at 9.36 ps, as determined above, and resolving a 560 fs intermediate component. The decay is accompanied by oscillations with components of ~160 and ~210 cm^{-1}, which are in good agreement with ground state modes in a recent Raman and infrared study [5]. Figure 2 shows the transient anisotropy, r(t), calculated from the pump-probe transient responses obtained with parallel and perpendicular relative polarizations. The initial value of 0.26 decays within 40 fs to 0.14 and remains at this value for at least 0.4 ps. The initial anisotropy decay is best fit by a 30 fs FWHM Gaussian.

To establish a mechanism for the relaxation dynamics of the photoexcited PANI system, the relaxation dynamics are modeled with three-level kinetic schemes involving ground state, excited state, and an intermediate state, with inhomogeneity considered for the excited and intermediate states. Kinetic equations are solved for four possible schemes and the pump-probe signal is constructed in the range of 0.2 to 100 ps, considering ground state bleaching and stimulated emission contributions. The initial dephasing is not considered in the modeling so the first 200 fs is not included in the analysis. As shown in Figure 3, the scheme providing the statistically best fit to the pump-probe data involves inhomogeneity in both excited and intermediate states, which suggests that inclusion of significant heterogeneity is necessary to adequately model the data.

From the results, photoexcited carrier dynamics of primary doped PANI have been identified on two distinct time scales. The very fast component, with a 34 fs exponential decay is attributed to the relaxation of the initially delocalized polarons within individual polymer chains. This is supported by the transient anisotropy signal exhibiting a very fast Gaussian decay with a characteristic time of 30 fs. The extremely rapid anisotropy decay results from dephasing of the initially delocalized polaron excitation due to diagonal disorder (i.e. variations in the monomer energies that may be associated with structural disorder in the polymer chain) and vibronic coupling, in analogy with such dynamics observed in light harvesting complexes of photosynthetic bacteria [6]. The longer timescale

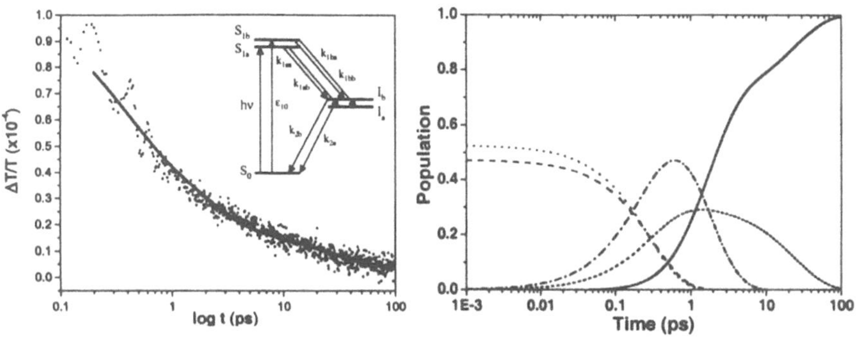

Fig. 3. (Left) Simulated signal (line) from a kinetic scheme (inset) providing the best fit to the pump-probe data (points). (Right) Population evolution of states involved in the kinetic scheme; S_0 (solid), S_{1a} (long-dashed), S_{1b} (dotted), I_a (dash-dotted), I_b (short-dashed).

dynamics, two or more exponential decays of >500 fs to >100 ps time constants, reflect the decay from the initially excited Franck-Condon region and the slow recovery of the ground state via a long-lived intermediate state. When polyaniline is photoexcited, it will relax to a geometrically distorted "twisted" structure by way of torsional motions between the adjacent rings [5]. It was reported by Martinez and coworkers that the photoexcited state of ethylene rapidly evolves to a twisted geometry that nonradiatively quenches to the ground state through the conical intersection of the excited and ground state with charge-separated character [7]. The decay must occur by a nonradiative channel as opposed to charge recombination in electroluminescent conducting polymers [8] since the quantum yield for radiative emission from PANI is very small. We propose that the relaxation in photoexcited PANI occurs via a "twisted" intermediate state with charge-separated character as in the isomerization of ethylenic species including retinal and PYP, giving rise to the slow ground state recovery over a broad range of timescales.

References

1. A. G. MacDiarmid and A. J. Epstein, Synth. Met. **69,** 85, 1995.
2. M. J. Feldstein, W. Zheng, Y. H. Liau, S. Rane, D. C. Arnett and N. F. Scherer, to be submitted to J. Phys. Chem., 2002.
3. Y. H. Liau, A. N. Unterreiner, D. C. Arnett and N. F. Scherer, Appl. Opt. **38,** 7386, 1999.
4. J. Kim, A. N. Unterreiner, S. Rane, S. Park, J. Jureller, L. Book, Y. H. Liau and N. F. Scherer, submitted to J. Phys. Chem., 2002.
5. P. Colomban, S. Folch and A. Gruger, Macromolecules **32,** 3080, 1999.
6. L. D. Book, A. E. Ostafin, N. Ponomarenko, J. R. Norris and N. F. Scherer, J. Phys. Chem. B **104,** 8295, 2000.
7. M. Ben-Nun and T. J. Martinez, Chem. Phys. Lett. **298,** 57, 1998.
8. A. Sakamoto, O. Nakamura, G. Yoshimoto and M. Tasumi, J. Phys. Chem. A **104,** 4198, 2000.

Ultrafast thermalization dynamics of water nanodroplets by infrared pump-probe spectroscopy

Gerhard Seifert, Toralf Patzlaff, and Heinrich Graener

Martin-Luther-University Halle, Physics Institute, Hoher Weg 8, D-06099 Halle, Germany
E-mail: g.seifert@physik.uni-halle.de

Abstract. The ultrafast thermal relaxation of reversed micelles was investigated by time-resolved infrared pump-probe spectroscopy. Modeling this picosecond cooling process in terms of heat conduction provides a new, selective way to determine the water nanodroplet radii.

1. Introduction

Water can be dispersed in a nonpolar solvent in the form of nanometer-sized spherical droplets by help of appropriate amphiphiles. In a distinct range of composition the water in these solutions is surrounded by a monolayer of surfactant molecules, and the whole sphere is usually denoted as reverse micelle. These microemulsions have raised considerable interest in the last years, not least because of their attractivity for a variety of technical applications like chemical catalysis, drug delivery or nanocluster synthesis. A great number of experimental and theoretical investigations revealed information about, e.g., the essential role of the surfactant layer for structure and phase behavior, or dealt with questions like solvation dynamics both in the water pool and close to the surfactant layer. While the structure and dynamics of water in the reverse micelles are strongly modified directly at the interface [1,2], the basic properties of bulk water reemerge within a few molecular layers [1,3]. So in appropriate ternary mixtures with droplet sizes of at least several nanometers most of the water in such a microemulsion is found in its bulk configuration.

Recent femtosecond infrared laser experiments revealed that vibrational relaxation in the OH stretching region and the subsequent energy thermalization in neat water happen on a subpicosecond time scale [4-7]. So time-resolved infrared spectroscopy with an appropriate pulse length can be used to selectively heat the water inside reverse micelles within a few picoseconds, and trace the cooling process by monitoring the temperature dependent changes of the OH stretching band, as we demonstrated recently [8]. Here we report a series of such experiments on octane-AOT-water ternary mixtures (where AOT denotes sodium di-2-ethylhexyl sulfosuccinate). The composition of the mixture is characterized by the molar water-to-AOT ratio $\mu = [H_2O] / [AOT]$ (also referenced as w_0) and water mass fraction $\alpha = m_{water} / m_{total}$. While μ determines the size of the reversed micelles (often a simple linear relation assuming a monodisperse size distribution can be used [9]), α is correlated with the micelle concentration at given μ. In this study we used samples with $\alpha = 0.005$ and $10 < \mu < 55$, i.e. rather large reverse micelles at low concentration.

2. Results and Discussion

The experiments were performed in an IR double resonance spectrometer, which provides separately tunable pump and probe pulses of 2.5 ps temporal and 8-10 cm^{-1} spectral width (full width at half maximum). The intense pump pulses of typically 25 μJ energy are tuned to the water OH stretching band, creating an ultrafast temperature rise of the water droplets via vibrational excitation and the subsequent subpicosecond relaxation processes [4-7]. The much weaker probe pulses (energy < 1 μJ) monitor the corresponding transient transmission changes, which are directly proportional to the water temperature in certain frequency ranges (e.g. 3150-3350 cm^{-1}). Fig. 1 gives some examples for time-resolved results obtained on water-AOT-octane microemulsions of different μ; note the logarithmic time scale after the axis break. The non-exponential decay of the droplet temperature apparently depends strongly on μ, i.e. on the micelle size, in particular the cooling is faster for smaller micelles.

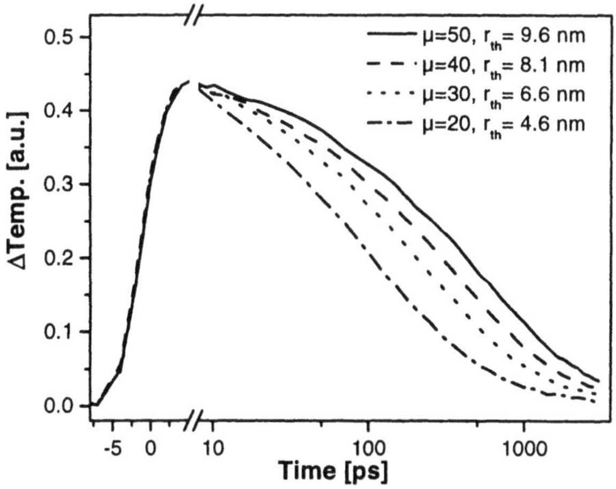

Fig. 1. Measured temperature changes after excitation at 3300 cm^{-1}, and probed via the ν(OH) band at frequency position 3200cm^{-1} for different water-to-AOT ratios μ. The values of r_{th} given denote the corresponding thermodynamic water droplet radii as obtained fitting the data on the basis of heat conduction in a two component system.

This general behavior can well be explained applying the macroscopic heat equation to the spherically symmetric problem of an isolated, initially hot, sphere. Numerical simulations on this basis provide excellent fits of the experimental data, even if a simplified two component system neglecting the AOT interface is used [10]. Using equilibrium values for the thermal diffusivities χ of water and octane the only free parameter in the simulations is the 'thermodynamic' droplet radius r_{th}, the best fit values of which are given in the legend of Fig. 1. While the data given in Fig. 1 refer to a pump fluence of approximatley 0.1 J/cm^2, some novel data measured at lower pump intensities indicate that the real radius of the aqueous core of the reverse AOT micelles is slightly smaller than the values given in Fig. 1. Thus our experiments can, in the low intensity limit (i.e., at small initial ΔT), be interpreted as a new, very reliable method to determine selectively the size of the water droplets in microemulsions. Furthermore the results give evidence that the amphiphilic AOT interface allows for a very efficient transfer of heat from water into the nonpolar solvent. On the other hand the high intensity range of our data (large initial $\Delta T > 20$ K) clearly shows that the micelles are transiently (for tens to hundreds of picoseconds) in a thermodynamic non-equilibrium state, where, e.g., they remain stable in spite of transient temperatures far above their equilibrium stability range. In particular this last finding will be an important aspect of future work in this field, looking for example for the laser induced control of chemical reactions inside the water pools of microemulsions by creating temperatures above a certain threshold for well-defined sub-nanosecond time intervals.

References

1 J. Faeder, B.M. Ladanyi, J. Phys. Chem. B **104**, 1033 (2000).

2 D.S. Venables, K. Huang, C.A. Schmuttenmaer, J. Phys. Chem. B **105**, 9132 (2001).

3 C. Borsarelli, S.E. Braslavsky, J. Phys. Chem. B **101**, 6036 (1997).

4 S. Woutersen, U. Emmerichs, H.J. Bakker, Science **278**, 658 (1997); A.J. Lock, S. Woutersen, H.J. Bakker, J. Phys. Chem. A **105**, 1238 (2001).

5 R. Laenen, C. Rauscher, A. Laubereau, Phys. Rev. Lett. **80**, 2622 (1998).

6 G.M. Gale *et al.*, Phys. Rev. Lett. **82**, 1068 (1999).

7 J. Deak et al., J. Phys. Chem. A **104**, 4866 (2000).

8 T. Patzlaff, M. Janich, G. Seifert, H. Graener, Chem. Phys. **261**, 381 (2000).

9 *Structure and reactivity in reverse micelles*, M.P. Pileni (ed.), Elsevier, Amsterdam, 1989

10 G. Seifert, T. Patzlaff, H. Graener, Phys. Rev. Lett. **88**, 147402 (2002).

Spectroscopy of macrocyclic poly (9,9-dimethyl-2-vinylfluorene) as an artificial light harvester

J. M. Johnson, X. Chen, A.C. Germaine, X. Zhang, K. Alberty, T. E. Hogen-Esch and S. E. Bradforth

Dept. of Chemistry, University of Southern California, Los Angeles, CA 90089, U.S.A.
E-mail: jerainnj@usc.edu

Abstract. The photophysical properties of macrocyclic poly (9,9-dimethyl-2-vinylfluorene) have been characterized by steady state fluorescence and ultrafast spectroscopy. Ultrafast transient absorption anisotropy studies show that energy migration occurs on a picosecond time scale.

Three-dimensional crystal structures for the light harvesting antenna complexes of photosynthetic bacteria have recently been resolved [1]. This has led to a more profound understanding of how nature harvests sunlight in an efficient fashion. The antenna complex (LH2) is highly symmetric and the deployment of chromophores is cyclic. It is therefore of interest whether this architecture is important for the very rapid and efficient energy transfer - excitation hopping times at room temperature of ~100 fs have been reported [2]. In our investigations, the ring-like structure of LH2 is mimicked in novel two-dimensional macrocyclic polymers, which have recently been synthesized [3]. Such poly-vinyl aromatic polymers can be prepared with 10 – 50 chromophore groups (here the chromophore is 9,9 dimethylfluorene) that are pendent to the polymer backbone. In this report, the steady state and time-resolved spectroscopic properties of the macrocyclic polymers are compared with their linear analogs.

A 1kHz Ti:S regenerative amplifier producing 100 fs pulses centered at 800 nm pumps an optical parametric amplifier (OPA). Sum-frequency generation of the OPA signal with residual 800 nm pump followed by frequency doubling is used to produce pump pulses at a wavelength of 300 nm. The variably delayed probe beam is generated from a separate portion of the 800 nm fundamental. The pump-probe instrument response function is ~170 fs as determined by difference frequency cross-correlation measurements. Both the pump and probe pulses are focused into a 0.5 mm path length CaF_2 flow cell containing poly(9,9-dimethyl-2-vinylfluorene), linear or cyclic, dissolved in cyclohexane. The probe is polarized at 45° with respect to the pump in order that parallel and perpendicular transient absorption components may be simultaneously detected. This is achieved by separating the parallel and perpendicular probe fields by a Wollaston prism after the sample and employing two detection photodiodes, which are in turn connected to lock-in amplifiers referenced to a mechanical chopper in the pump path. The two time-resolved signals obtained are then used to calculate the anisotropy, which is fit with a bi-exponential decay function. Absorption and steady-state fluorescence measurements for fluorene, linear and cyclic poly (9,9-dimethyl-2-vinylfluorene) with different degrees of polymerization (DP_n) have also been recorded. All experiments were performed at room temperature.

The absorption spectra of both the linear and cyclic polymers have very similar characteristics. As seen in *Figure 1a*, the spectra for both systems overlay and, other than a small blue shift, are in fact virtually identical to the chromophore 9,9-dimethylfluorene free in solution. This indicates that in the ground state there is minimal electronic interaction between adjacent chromophore molecules as expected for a non-conjugated polymer. The fluorescence spectra for both polymers, however, reveal an enhancement in emission in the cyclic polymers opposed to their linear analogs (*Fig. 1b*).

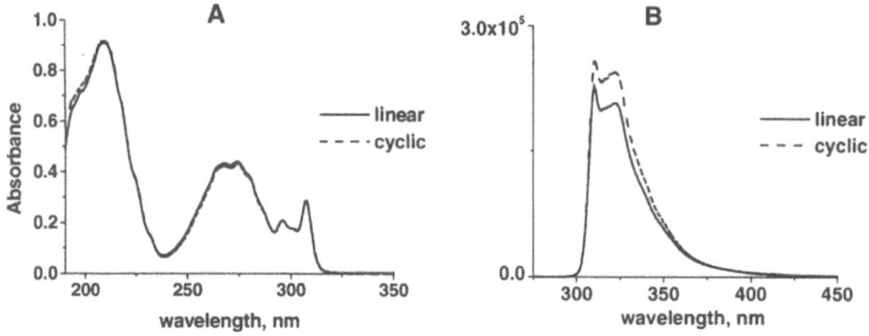

Fig. 1. Absorption and fluorescence spectra of macrocyclic polymer and its linear analog

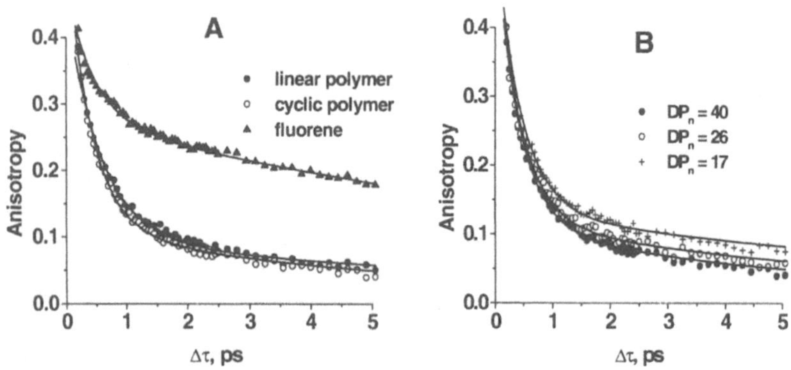

Fig. 2. (a) Anisotropy decay for fluorine monomer and linear and cyclic polymers; (b) as a function of DP_n in macrocyclic polymers.

Figure 2 shows the femtosecond transient absorption anisotropy data for isolated fluorene and the linear and macrocyclic forms of poly (9,9-dimethyl-2-vinylfluorene) in cyclohexane. The dependence of the anisotropy decay with size of the macrocyclic polymer (expressed as its degree of polymerization, DP_n) is also shown. For fluorene, which represents the free chromophore in solution, an initial anisotropy of 0.4 is observed, indicating that the excited state absorption transition dipole is parallel to the pumped ground state transition dipole. The anisotropy decay is characterized in terms of 2 time constants, 460 fs and 11 ps. The longer time constant corresponds to the rotational reorientation time of the free chromophore, and this value is consistent with other aromatic compounds with similar structure (*e.g.*, anthracene, 8 ps) [4]. At times longer than 20 ps, the residual anisotropy for fluorene is 0.01, which implies that the distribution of the transition dipole moment is almost completely random.

With the chromophore pendent to the polymer backbone, we no longer expect to observe rotational reorientation by tumbling on the reported timescales. The anisotropy decay is substantially altered for the polymers, however the short decay component recovered from the fits, is essentially the same, ~420 fs for both linear and cyclic polymer. We speculate that the early anisotropy decay phase for the polymer system is associated with librations of pendent chromophores on the polymer backbone[5]. The values for the longer decay component differ slightly between the linear (τ=7.2 ps) and cyclic (τ=5.0 ps) form of the polymer. As rotational diffusion of the entire polymer takes place on much longer timescales, the decay of the anisotropy must be interpreted in terms of energy transfer between adjacent chromophore groups along the polymer backbone. When the size of the macrocycle is varied, the observed trend is that the rate of energy transfer decreases with a decrease in the degree of polymerization. This is possibly due to increasing strain in the smaller rings. The strain may cause steric repulsion between pendent chromophores in smaller rings, which will in turn make the inter-chromophore distance larger resulting in a slower rate of energy transfer. We note that there is a residual anisotropy observed that increases with decreasing ring size. This observation suggests that the transition dipoles in the smaller rings are closer to coplanar than in larger ones. The rates for energy transfer implied by our results agree well with predictions based on Forster's Theory. However, the extent of depolarization in a single hop of the excitation depends intimately on the angle between adjacent chromophore molecules. Therefore, to thoroughly extract information from the datasets on the rate and dispersion of energy migration in these chromophore architectures, energy transfer simulations based on an ensemble of polymer structures are required and such calculations are in progress.

References

1. G. McDermott, S. M. Prince, A. A. Freer, A. M. Hawthornthwaite-Lawless, M. Z. Papiz, R. J. Cogdell, N. W. Isaacs, *Nature* **374**, 517 (1995).
2. R. Jimenez, S. N. Dikshit, S. E. Bradforth, G.R. Fleming, *J. Phys. Chem.* **100**, 6825 (1996).
3. Y. Gan, D. Dong and T. E. Hogen-Esch, *J. Am. Chem. Soc.* **122**, 2130 (2000).
4. G. S. Jas, Y. Wang, S. W. Pauls, C. K. Johnson and K. Kuczera, *J. Chem. Phys.* **107**, 8800 (1997).
5. A. T. Sengupta, M. D. Fayer, *J. Phys. Chem.* **96**, 8619 (1992).

Control of Ground State Dynamics in Polymers with Shaped Pulses

D. Zeidler[1], S. Frey[1], W. Wohlleben[1], M. Motzkus[1], F. Busch[2], T. Chen[2], W. Kiefer[2], and A. Materny[3]

[1] MPI für Quantenoptik, Hans-Kopfermann-Str. 1, D-85748 Garching, Germany, Tel.: +49 89 32905-216, mcm@mpq.mpg.de
[2] Institut für Physikalische Chemie, Universität Würzburg, Am Hubland, D-97074 Würzburg, Germany
[3] School of Engineering and Science, International University Bremen P.O.Box 750 561, D-28725 Bremen, Germany

Abstract. Coherent control of the ground state vibrational dynamics via a Raman process is demonstrated in polydiacetylene. The relative amplitudes and phases of the normal modes could be set.

Mode-selective chemistry involves deposition of energy in the vibrational motion of a certain bond (local mode) by creating a suitable superposition of normal modes. Thereby, the dynamics can be steered along the desired reaction coordinate. If the relative phases and amplitudes of these modes are chosen suitably upon excitation with a broadband femtosecond laser pulse, the preparation of local modes should be possible.

We demonstrate selective excitation of ground state vibrational modes in a polymer, polydiacetylene (PDA), by a stimulated Raman process. The process is optimized in a self-learning loop [1] using the anti-Stokes Raman signal of a four-wave-mixing (FWM) process as feedback. The samples used in the experiments were FBS diacetylenes embedded in the monomer single crystal cooled to approximately 15 K [2]. Its Raman-active normal modes $\tilde{\nu}_1 - \tilde{\nu}_4$ are composed mainly of stretching and bending motions of the backbone bonds and were all excited by the pump and Stokes laser pulses of the fs–CARS experiment (Fig. 1).

The left part of Fig. 2 shows the wavelength-resolved transient of the anti–Stokes signal with nearly Fourier limited pump, Stokes, and probe pulses. With the laser wavelengths as given in the caption of Fig. 1, the recorded anti-Stokes wavelengths show that all modes, $\tilde{\nu}_1 - \tilde{\nu}_4$, have been excited.

The spectral bands of the wavelength-resolved CARS signal which correspond to the different normal modes of PDA served as the feedback signal for the optimization algorithm. The optimization algorithm, based on an evolutionary strategy [4], searched for the optimal settings of the LC mask that enhance the desired mode. It has been shown that the distribution of energy in the different modes depends strongly on chirp and delay of the Stokes pulse [5]. In a first feedback-controlled experiment using 100 fs pulses

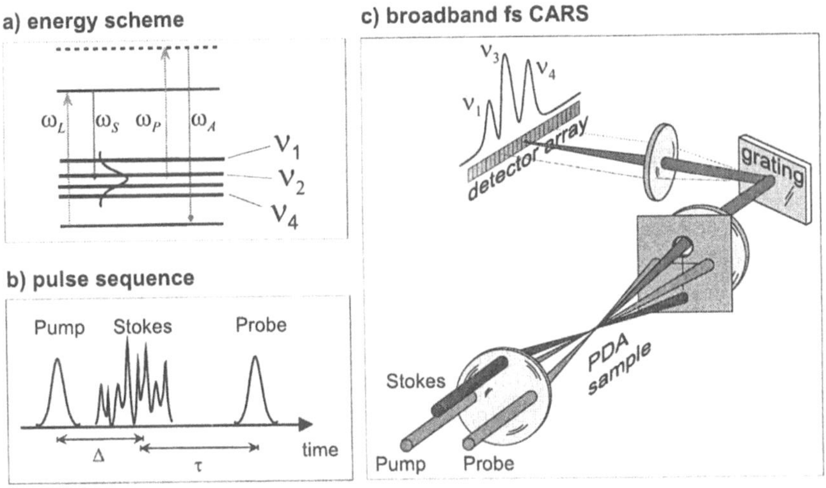

Fig. 1. Experiment. (a) Energy diagram of the CARS scheme. A commercial Ti:Sapphire fs laser system pumps two non-collinear optical parametric amplifiers. Their outputs are tuned to 605 nm and 654 nm (FWHM $\approx$ 35fs). The 605 nm pulses are split into pump and probe pulses. The 654 nm pulses are routed through a 4f pulse shaping setup [3] and serve as Stokes beam. The sequence of the three incoming pulses with shaped Stokes pulse is shown in subfigure (b). All pulses ($\approx$ 100 nJ) are arranged in a folded BoxCARS configuration (c) and focused (f = 60 cm mirror) into the sample. The CARS signal is collimated, spectrally dispersed and detected with a CCD line camera. (A lens is shown instead of the mirror for graphical reasons.)

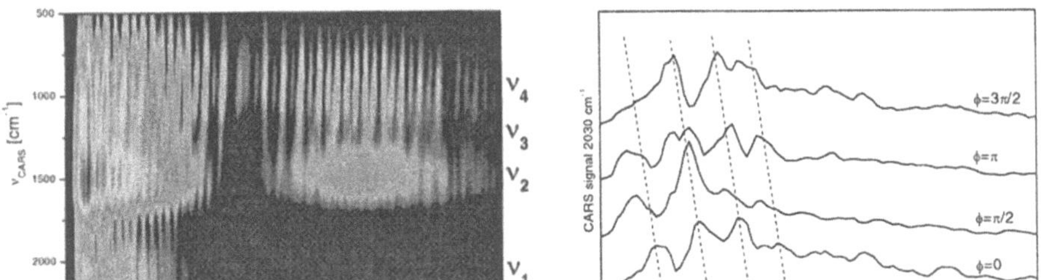

Fig. 2. Left: Temporally and spectrally resolved CARS transient on polydiacetylene with an unshaped Stokes pulse. With a constant time delay of Δ between pump and Stokes pulse, the time delay τ between Stokes and probe pulses was varied. Right: Cuts through the transients at a Stokes wavelength corresponding to the $\tilde{\nu}_1$ mode for phase steps 0, $\pi/2$, π, and $3\pi/2$ between the wavelengths of the Stokes spectrum corresponding to the $\tilde{\nu}_1$ and $\tilde{\nu}_{22'}$ modes.

we thus restricted the algorithm to a search for polynomial phase functions, without changing the spectrum of the pulses. With the feedback-controlled maximization of the ratio of the mode strengths, a highly selective excitation of the normal modes has been demonstrated [6]. This could be verified with the 35 fs pulses, although a more general, free parameterization with more independent parameters than in the previous "chirp approach" proved to be necessary. In the 100 fs experiments, the retrieved phase functions after each optimization reveal that the frequency components responsible for excitation of the other two modes are shifted beyond the Franck-Condon window which connects the excitonic state and the electronic ground state, thereby minimizing their contributions. In addition, the shape of the phase function might be interpreted in terms of symmetric and antisymmetric phase functions around the corresponding Raman resonances which suppress or enhance this two-photon transition [7]. The first hypothesis has been cross-checked in the 35 fs experiments by a combined phase- and amplitude parameterization which allowed the algorithm to adjust the frequencies and their timing of the Stokes pulses by a polynomial phase and Gaussian amplitude representation.

For the focusing of a wave packet in order to produce a local mode, it is essential to be able to control the phases of the normal modes. An additional experiment was performed to verify that the phase of the oscillations changes with the phase of the controlling Stokes pulse. Here, a phase step was introduced onto the spectrum of the Stokes pulse between the pixels corresponding to the Stokes wavelength of the $\tilde{\nu}_1$ and $\tilde{\nu}_{22'}$ mode. The dependence of the temporal evolution of the $\tilde{\nu}_1$ mode on the magnitude of the phase step is shown in the right part of Fig. 2, indicating that the relative phase of the modes can be influenced.

References

1. R.S. Judson and H. Rabitz, Phys. Rev. Lett. 68 (1992) 1500.
2. A. Materny, and W. Kiefer, Macromolecules 25 (1992) 5074.
3. A.M. Weiner, Rev. Sci. Instrum. 71 (2000) 1929.
4. D. Zeidler, S. Frey, K.-L. Kompa, and M. Motzkus, Phys. Rev. A 64 (2001) 023420.
5. T. Chen, A. Vierheilig, P. Waltner, M. Heid, W. Kiefer, and A. Materny, Chem. Phys. Lett. 326 (2000) 375.
6. D. Zeidler, S. Frey, W. Wohlleben, M. Motzkus, F. Busch, T. Chen, W. Kiefer, A. Materny, *J. Chem. Phys.*, in print
7. D. Meshulach, Y. Silberberg, Phys. Rev. A 60 (1999) 1287

The role of nanoaggregate domains in the exciton dynamics of poly(p-phenylene vinylene)

C. J. Bardeen, S. H. Lim, T. G. Bjorklund, and K. M. Gaab

Department of Chemistry, University of Illinois, 600 S. Mathews Ave., Urbana, IL 61801, USA
E-mail: bardeen@uiuc.edu

Abstract. Transient absorption and emission spectroscopy are used to determine the role of nanocrystalline aggregates in the photophysics of the conjugated polymer PPV.

1. Introduction

The photophysics of conjugated polymers have been intensely studied in the hope that they will provide insight into the electronic structure and dynamics of these materials. Conjugated polymers are often considered as an inhomogeneous collection of weakly-coupled chain segments, among which energy can be transferred via a Forster mechanism [1]. In this model, aggregation leads primarily to nonradiative traps for the luminescent Frenkel excitons. We have made a detailed study of the photodynamics of the prototypical conjugated polymer poly(p-phenylene vinylene) (PPV) and have found that the effects of aggregation are more complex, and not as detrimental as previously thought [2].

2. Results and Discussion

In the disordered environment of a polymer film, there is a mix of species, from amorphous regions of isolated molecules to aggregates consisting of hundreds of molecules. Both powder x-ray diffraction and transmission electron microscopy confirm the existence of these different domains in PPV.

The prevailing opinion is that these domains lead to interchain exciton species that act as weakly emissive trapping states, e.g. excimers. We see evidence for such states in the nanosecond dynamics of PPV, where a redshifted, broadened emission appears at very long delays. We also observe changes in the short-time emission spectrum that are consistent with delocalization of the singlet exciton at low temperature. Figure 1 shows the transient emission spectrum of PPV at 15 K. After the initial excitation pulse, the emission spectrum shifts and changes shape. The apparent decrease of the electron-phonon coupling with delay (as seen by the change in relative intensities of the 0-0 and 0-1 vibronic peaks) is consistent with exciton relaxation and delocalization [3]. This change in the spectrum is also observed in the steady-state fluorescence of PPV as the temperature is lowered.

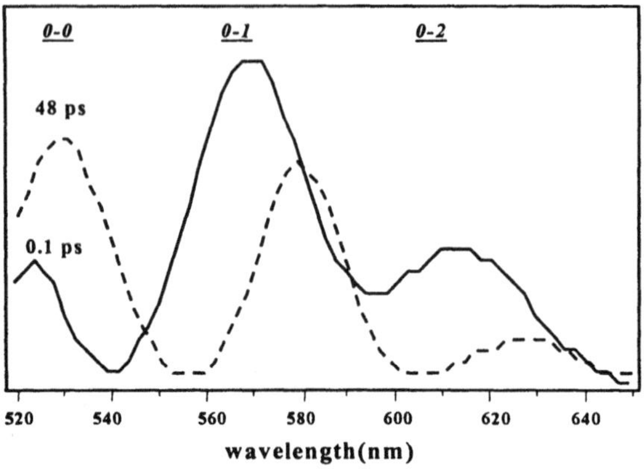

Fig. 1. Transient emission spectrum of PPV at 20 K. 100 fs after excitation (solid), the emission spectrum is characteristic of a high energy, localized exciton. After 48 ps (dashed), the exciton relaxes and its effective wavefunction delocalizes, leading to a redshift and growth of the 0-0 vibronic peak at the expense of the 0-1 and 0-2 transitions.

The question of whether this delocalization occurs over a single polymer chain or over an aggregate of chains is addressed by several different types of experiments. Using calibrated transient absorption and bleaching experiments, we find no evidence for an enhanced absorption cross-section. These measurements are complicated by exciton-exciton annihilation processes which turn on at fluences of less than 10 $\mu J/cm^2$. The fact that the radiative lifetime does not increase as temperature decreases argues against the existence of superradiance in this system.

However, polarization anisotropy experiments indicate that the chromophore actually extends over several chains rather than a single segment [4]. At 20 K, the steady-state fluorescence spectrum has a wavelength-dependent anisotropy consistent with delocalization over multiple chromophores, where the 0-0 peak exhibits lower anisotropy than the other vibronic transitions. The transient emission shows that this wavelength dependence grows in on a timescale of roughly a picosecond. At room temperature, no spectral dependence of the anisotropy is observed. The temperature dependent luminescence can be understood in terms of a transition from a delocalized exciton at low temperatures, similar to what is observed in molecular crystals, to a localized or "self-trapped" exciton at higher temperatures, which can undergo thermally activated diffusive hopping and quenching [5].

3. Conclusions

In many ways, the spectroscopy of conjugated polymers represents a mixture of the strong intermolecular coupling observed in molecular crystals and the weak,

Forster coupling seen between isolated chromophores. The luminescence properties depend sensitively on molecular-level interactions, and other types of chemically modified PPV's can display intermediate behavior. But PPV is a case that demonstrates that aggregation need not necessarily lead to quenching and decreased fluorescence.

References

1 R. Kersting, B. Mollay, M. Rusch, J. Wenisch, G. Leising, and H. F. Kauffmann, J. Chem. Phys. **106**, 2850, 1997.

2 P. K. H. Ho, J. S. Kim, N. Tessler, and R. H. Friend, J. Chem. Phys. **115**, 2709, 2001.

3 S. H. Lim, T. G. Bjorklund, and C. J. Bardeen, Chem. Phys. Lett. **342**, 555, 2001.

4 F. C. Spano, J. Chem. Phys. **114**, 5376, 2001.

5 T. G. Bjorklund, S. H. Lim, and C. J. Bardeen, J. Phys. Chem. B **105**, 11970, 2001.

Electronic Excitation Oscillations in PPV – Femtosecond Luminescence Interferometry

A. Tortschanoff[1], F. Milota[1], J. Sperling[1], V. Szöcs[2], and H. F. Kauffmann[1]

[1] Institut für Physikalische Chemie der Universität Wien, Währingerstr. 42, A-1090 Wien, Austria
E-mail: kauffm@ap.univie.ac.at
[2] Institute of Chemistry, Comenius University, Mlynska solina CH 2, SK-842 15 Bratislava, Slovakia

Abstract. Oscillatory relaxation of the conjugated polymer poly-(p-phenylenevinylene) is probed by fs-luminescence interferometry. Intersite polarization interference may account for the fast beating in the presence of strong damping and inhomogeneous site dephasing.

The measurement of electronic coherences in the complex multi-chromophoric ensemble of a conjugated polymer is a formidable task and only beginning to be exploited, nowadays. A very useful picture describes the PPV as a conjugated chain where static disorder-breaks [1] disrupt the backbone chain into an ensemble of segmental sites [1,2] with fluctuating self-energies and intersegmental electronic coupling energies. Accordingly, the excited state of the polymer system splits into a density-of-states (DOS) of localized S_1-levels [2,3] with excitations, predominantly, site-oriented for weakly coupled segments. In this concept, coherent, oscillating kernels, at early femtoseconds and on short spatial scales, are subject to dephasing with the many-body array of the DOS thereby translating into uncorrelated populations and subsequent incoherent, energy-dispersive excitation funneling on later fs-time-scales and longer migrative pathways [4]. One expects, therefore, the Hamilton dynamics of the coherent transients to be measured under extreme temporal boundaries and very particular excitation/probing conditions, only. The energetic width of the DOS (σ) in PPV, the frequency-bandwidth of the fs-pulse ($\delta\omega$) and their spectral overlap are central issues that require a delicate balance in making experimental coherences feasible.

In the absence of substantial nuclear degrees-of-freedom, as indicated by the rigid chromophore-backbone of PPV, fs-pulse-excitation into the DOS of the 0-0 S_1–S_0 transition will, naturally, promote an electronic wavepacket (WP). For excitation near the center of the DOS, the sudden polarization of the electronic WP is a pseudo-continuous superposition of oscillations carrying a multitude of optical cycle frequencies, beatings from site polarization interferences, due to site-to-site coupling, and dampings brought about, first of all, by inhomogeneous dephasing. In this limit case, we find purely dissipative interferograms of a few fs duration, only, where the electronic coherences - due to multi-chromophore interferences - are damped out, quasi-instantaneously, the overdamped behavior of the inhomogeneous multi-line relaxation thus completely masking electronic coupling and delocalization phenomena. However, for PPV-segmental sites in the low-energy regime of the DOS-bottom states, the site-density is lower and hence intersite-coupling less pronounced. In the energetic region below the mobilisation

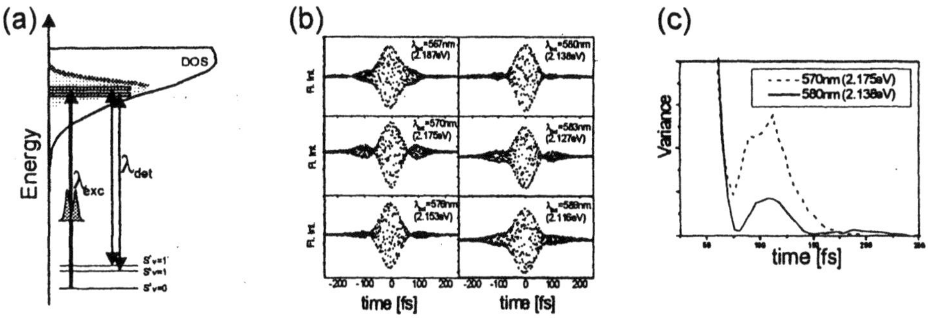

Fig. 1. (a) Scheme indicating the ground state levels involved and the DOS with the states excited by the pulse indicated together with the subensembles that are detected at λ_{det}. (b) interferometric scans and (c) COIN scans for different detection energies λ_{det} as indicated (excitation always at λ_{det}=540nm (2.296eV)). (see text for details)

threshold, where intersite energy transfer is absent, it should be possible to probe the individual site dephasing.

With this goal we performed fs-interferometry on PPV with excitation tuned to the low energy edge of the $S_1 \leftarrow S_0$ transition and with spectrally narrow detection which allows to circumvent the broad inhomogeneous width of the DOS. As outlined in Fig. 1(a) excitation with two collinear time delayed pulses excites a subensemble within the low energy region of the DOS where the formed population grating contains the signatures of the dephasing in the sample which is caused by structural and vibration relaxation. While the pure dephasing is expected to occur on timescales comparable and slower than the pulsewidth, in bulk PPV it is overhelmed by the inhomogeneous broadening, that hides all coherent signals. Using a detector with narrow spectral bandwidth, will selectively detect a subensemble given by the width of the detector, and the observed decay will be limited not by the broad inhomogeneous manifold, but by the coherence lifetime of the subensemble. Further we can choose the sites that are monitored by tuning the detection energy, which offers a convenient way to study in much detail effects from detuning of the laser central frequency from the transition energy.

The well known low temperature flourescence spectrum of PPV features a typical 180meV vibronic progression associated with a phenyl-stretching mode. The peak corresponding to the $S_0^{v=1} \leftarrow S_1^{v=0}$ transition shows a pronounced shoulder on the high energy side resulting from a strong Raman-active mode assigned to phenyl C-H bending [5]. The energy difference between these two vibronic levels of 400 cm^{-1} and the situation is outlined in the level scheme of Fig.1(a). In the experiment the fluorescence of the 0-1 transition was detected. Thus we monitor two subensembles of the DOS which are situated at a constant energy (λ_{det}) above the v=1, v=1' vibrational groundstate levels. Since there is no energy transfer among different sites on the red edge of the DOS, the signal should show a typical polarization beating which is damped with the coherence dephasing time T2. We further measured the corresponding COIN (Coherence Observation of Interference Noise, COIN [6]), traces, where the optical cycle frequency of the electronic transition is averaged. COIN is, therefore, a very stable technique where the information, in particular, about interfering cohererences with frequencies, lower in comparison with ω_0, is of preserved, and aliasing effects can be excluded. Details concerning the measurement and analysis of COIN profiles are given elsewhere [7].

Experiments were performed at 1.4K with 70 fs excitation pulses from a Coherent RegA-OPA, 250 KHz configuration with a carrier wavelength of λ_{exc}=540 nm. Fig. 1(b) shows experimental results from the linear inerferometric scans for different detection energies (spectral width ~2nm (0.0075eV)), where the total fluorescence is plotted as a function of the delay time between the pulses. Note that the relevant information is contained in the amlitude modulation, which can well be recognized even without a phase stable interferometer and is also confirmed by the COIN scans (Fig. 1(c)). A strong dependence on the detection wavelength can be observed over a finite region above and below of which we only recover the interferogram corresponding to the pulse-autocorrelation. The sensibility of the signal on the dection window is well understood and the principal features are recovered in numerical calculations[8]. The shape of the patterns is a complex convolution of the spectral pulse-shape, the width and energetic position of the detection window and the intrinsic sample properties and dephasing characteristic of the absorbing sites, as well as the fluorescence spectrum. The interesting patterns that show up in our experiments are a result from the very special situation of the finite pulsewidth and fast dephasing on a similar timescale combined with energy selective detection. The beatings vanish as the spectral window of detection gets broader.

To summarize, we conjecture that we observed a particular polarization interference effect brought about by the coherent, in-phase superposition of optical free induction decays from spatially separated, sparse-level sites. Very recent, somewhere related experiments on spatial interferences of resonant Rayleigh scattering signals from discrete PPV-tails states [9] seem to confirm the concept of temporal coherence. Our excitation tuning experiments allow to study the delicate balance of coherent/incoherent relaxation dynamics triggered by inhomogeneous dephasing, in more detail. Detailes studies are, currently, in work.

Acknowledgements. This work was done within the framework of the Austrian Special Research Project F016 (ADLIS) and the FWF project P12566-PHY. J.S. is funded by the Austrian Academy of Sciences.

References

1 G. Rossi, R. R. Chance, and R. Silbey, J. Chem. Phys. **90**, 7594, 1989; S. N. Yaliraki and R. Silbey, J. Chem. Phys. **104**, 1245, 1996
2 U. Rauscher, *et al.*, Phys. Rev. B. **42**, 9830, 1990
3 H. Bässler and B. Schweitzer, Acc. Chem. Res. **32**, 173, 1999
4 R. Kersting, *et al.*, Phys. Rev. Lett. **70**, 3820, 1993
5 L. Orion, *et al.*, Phys. Rev. B **57**, 7050, 1998
6 O. Kinrot, I. Sh. Averbukh, and Y. Prior, Phys. Rev. Lett. **75**. 3822 (1995)
7 A. Tortschanoff, *et al.*, J. Chem. Phys. **110**, 4493, 1999
8 F. Milota, J. Sperling, A. Tortschanoff, and H. F. Kauffmann, to be published
9 S. P. Kennedy, N. Garro, and R. T. Phillips, Phys. Rev. Lett. **86**, 4148, 2001

Vibrational energy relaxation in H_2O

A.J. Lock[1] and H. J. Bakker[1]

[1] FOM Institute for Atomic and Molecular Physics, Kruislaan 407, 1098 SJ Amsterdam, The Netherlands
E-mail: a.lock@amolf.nl

Abstract. We have determined the lifetime of the OH-stretch vibration in pure liquid water as a function of temperature using femtosecond mid-infrared pump-probe spectroscopy. The lifetime T_1 increases from 260 ± 18 fs at T = 298 K to 320 ± 18 fs at T = 358 K. The increase in lifetime with temperature can quantitatively be explained from the decrease in spectral overlap of the OH-stretch vibration and the overtone of the H-O-H bending mode.

1. Introduction

Water is the most commonly used solvent for chemical reactions. Water molecules are believed to play an important role in the reaction dynamics by rearranging around the reactants and by lowering the energy barrier for the reaction. In addition, water molecules can also accept energy via their molecular vibrations. Hence, a detailed study of the energy dynamics of the molecular vibrations of the liquid can contribute to our understanding of the role of water in reactions. In the past few years, extensive studies were performed on a dilute solution of HDO molecules in D_2O [1-5], for which it was found that the lifetime of the OH-stretch vibration T_1 was 740 ± 20 fs [4]. Not much is known yet on the energy dynamics of "real" liquid water, i.e. H_2O. Deàk et al. found that after excitation of the OH-stretch band of H_2O, population is transferred to the bending mode [3]. However, the time resolution of their experiment was not sufficient to determine the time constant of the relaxation. Here we present the first measurement of the timescale of the energy relaxation of H_2O. We measured this time constant as a function of temperature. From the temperature dependence we conclude that the bending mode forms the main accepting mode, but that the hydrogen bond strongly influences the relaxation behavior [6].

2. Results and Discussion

We measured the lifetime of the OH-stretch vibration of H_2O with fs mid-infrared pump-probe saturation spectroscopy. The pulses have a spectral width of 80 cm^{-1}, a time duration of about 220 fs (FWHM) and an energy of about 20 μJ. A typical result of a pump-probe experiment on liquid water is shown in Figure 1. The pump frequency is in the middle of the OH-stretch absorption band, the probe frequency is shifted to the red in order to probe the $\nu_{OH} = 1 \rightarrow 2$ transition. Around $t = 0$ an induced absorption is observed that decays with the lifetime of the OH-stretch vibration. The signal, however, does not decay back to zero, but to a somewhat higher final level of transmission, which results from the fact that the absorption band has shifted somewhat to the blue after the equilibration of the excess energy over all degrees of freedom. This second process is characterized by a so-called equilibration time τ_{eq} which has a value of 550 ± 50 fs [7].

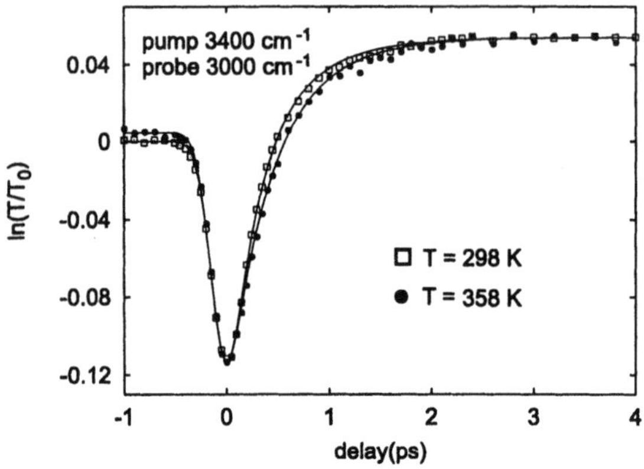

Figure 1: The transmission change as a function of delay time between pump and probe pulse for two different temperatures. The pump frequency was fixed at 3400 cm^{-1}, the probe frequency at 3000 cm^{-1}.

The measured lifetime of the OH-stretch vibration is astonishingly short in comparison with that of HDO dissolved in D_2O. Instead of a value of about 740 ± 20 fs, we here find a value of about 260 ± 18 fs at room temperature. This lifetime increases to 320 ± 18 fs at $T = 358$ K. This result is surprising, because in case the OH-stretch vibration relaxes to the first overtone of the bending mode, as proposed by Deàk et al. [3], one would expect that the lifetime T_1 would decrease with temperature, according to the multi-phonon theory of vibrational relaxation [8,9]. The key to solve this contradiction is to study the temperature dependence of the expression for the transition probability, resulting from Fermi's Golden Rule (FGR), from the OH-stretch to the overtone of the bending mode [10]. The lifetime of the OH-stretch vibration is inversely proportional to this transition probability P_{IVR}:

$$P_{IVR} = 2\pi / \hbar \int \left| \left\langle 0_\nu 2_\delta \left| R_\nu R_\delta^{\,2} \right| 1_\nu 0_\delta \right\rangle \right|^2 \delta(E_\nu - E_{2\delta}) \rho(E_\nu = E_{2\delta}) dE$$

where R_ν and R_δ are the transition dipole operators for the stretch (ν) and bending mode (δ), respectively. The integration has to be carried out for all energies E for which the donor (ν) and acceptor mode (2δ) have the same energy. Hence, this integral represents the overlap between the spectral bands of the stretch mode and the overtone of the bending mode.

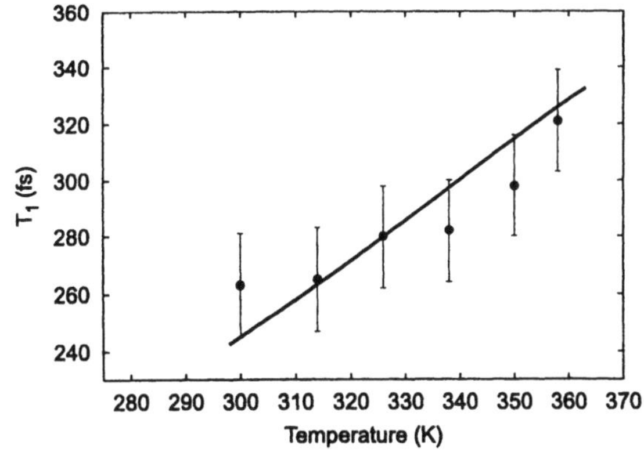

Figure 2: The lifetime T_1 of the OH-stretch vibration as a function of temperature is indicated with closed circles. The solid line represents the calculated trend for the FGR expression (see text).

In Figure 2, the experimental observed temperature dependence of the lifetime T_1 is compared to the results of the above calculation. As one can see, the FGR expression describes the experimentally observed trend very well. The observed increase of T_1 with temperature can be explained as follows. The OH-stretch band undergoes both a blue shift as well as a decrease of the absorption cross section with increasing temperature. The reason for these changes lies in the fact that the average hydrogen bond length increases. The bending mode shows an opposite temperature behavior, i.e. an increase in temperature leads to a small red shift. The net effect is a decrease of the spectral overlap of the stretch mode and the overtone of the bending mode as a function of temperature. Hence, although practically all of the energy of the excited OH-stretch of the H_2O molecule is transferred to the bending mode, the hydrogen bond still plays a decisive role in determining the rate of this energy transfer.

Acknowledgements. The research described in this contribution is part of the research program of the Foundation for Fundamental research on Matter (FOM) and was made possible by financial support of the Netherlands Organization for Scientific Research (NWO).

References

1 G. M. Gale, G. Gallot, F. Hache, N. Lascoux, S. Bratos, and J.-Cl. Leicknam, Phys. Rev. Lett. **82**, 1068 (1999)
2 R. Laenen, C. Rauscher, and A. Laubereau, Phys. Rev. Lett. **80**, 2622 (1998)
3 J. C. Deàk, S. T. Rhea, L. K. Iwaki, and D. D. Dlott, J. Phys. Chem. A **104**, 4866 (2000)
4 S. Woutersen, U. Emmerichs, H. K. Nienhuys, and H. J. Bakker, Phys. Rev. Lett. **81**, 1106 (1998)
5 H. K. Nienhuys, S. Woutersen, R. A. van Santen, and H. J. Bakker, J. Chem. Phys. **111**, 1494 (1999)
6 A. J. Lock and H. J. Bakker, *Accepted for publication in J. Chem. Phys.*
7 A. J. Lock, S. Woutersen, and H. J. Bakker, J. Phys. Chem. A **105**, 1238 (2001)
8 A. Nitzan, and J. Jortner, Mol. Phys. **25**, 713 (1973)
9 H Lin, J. Chem. Phys. **65**, 1053 (1976)
10 H. J. Bakker, J. Chem. Phys, **98**, 8496 (1993)

Two-Color Three Pulse Photon Echo Peak Shift Spectroscopy

Ritesh Agarwal, Bradley S. Prall, Abbas H. Rizvi, Mino Yang, and Graham R. Fleming

Department of Chemistry, University of California Berkeley and Physical Biosciences Division, Lawrence Berkeley National Laboratory, Berkeley, CA 94720 USA
E-mail: grfleming@lbl.gov

Abstract. Two-color three pulse photon echo spectroscopy has been used to probe correlation in two different regions of the electronic spectrum of a chromophore in the condensed phase.

Ultrafast resonant four-wave mixing spectroscopy has seen steady development since the pioneering work of Yajima and coworkers and Weiner and Ippen. Recently the photon echo peak shift (3PEPS) has been demonstrated and applied to a number of systems [1]. In this paper we describe a variant of the 3PEPS experiment in which the third of the three input pulses is centered at a different frequency from the first two pulses, i.e., two-color 3PEPS (2C3PEPS). The experiment was first proposed by Yang and Fleming [2] as a probe for electronic mixing in molecular complexes. In our initial work, however, we simply select both colors to be within the bandwidth of the broad absorption band of dye molecules in solution. The results obtained are significantly more interesting and informative than might be expected from such a simple variation of the 3PEPS experiment. Beyond population dynamics which are simultaneously available from the two color transient grating (2CTG), the 2C3PEPS experiment is highly sensitive to the degree of correlation of transition frequency between prepared and probed states and raises the prospect of many novel studies of dynamics.

We define a new quantity to characterize 2C3PEPS experiments, the difference peak shift, $\Delta\tau*$, obtained by taking the difference between peak shifts obtained by two different delay scanning procedures. This is because the two-color experiment cannot be treated in the impulsive limit and particular care must be paid to the influence of finite pulse duration, especially during the region of pulse overlap for pulses of different colors. Interestingly the difference peak shift seems to strongly reduce the effects of convolution and of vibronic interference in the dephasing dynamics. The two scanning procedures correspond (I) to the standard scan for 3PEPS and (II) to a nonrephasing or free-induction decay sequence. If the two peak shifts are identical the difference peak shift ($\Delta\tau^*$) is zero and this implies the absence of rephasing capability (memory). The pulse convolution effects are partially removed because the interaction sequences, in which the odd colored pulse interacts last, dominate all other interaction sequences for time coincident pulses. This induces a positive shift in the peakshift for both scanning methods in 2C3PEPS during the region of pulse overlap which is normally only found in the type I scan for one color 3PEPS.

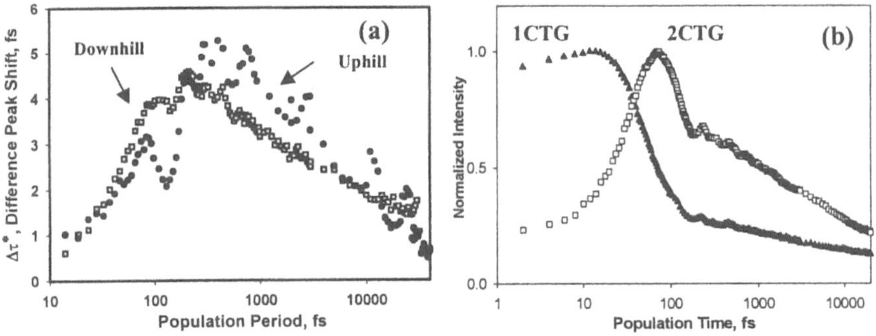

Fig. 1. IR144 in Methanol (a) The difference peak shift for downhill (750-750-800 nm) and uphill (800-800-750 nm) case. (b) One color (750-750-750 nm) and two color (750-750-800 nm) transeint gratings show population dynamics.

Fig. 1(a) shows the difference peak shift for IR 144 in methanol with the first two pulses centered at 750nm and the third at 800nm (downhill case). The pulses were derived by filtering from a single 15fs pulse and had durations of 42-45fs. The initial difference peak shift starts close to zero, rises to a maximum and then decays on a much longer timescale back to zero. The implication is that the initial rephasing capability (not the peak shift, which is finite in both type I and type II scans) is zero and increases before decaying. Comparison of 1(a) and 1(b) shows that the rise in the difference peak shift is not just tracking motion of population from the pump to probe regions. Numerical modeling shows that the rise in the initial peak shift corresponds to the inertial component of the solvation dynamics, which has been very difficult to characterize by single color experiments. Modeling suggests that the value of the inertial component timescale is insensitive to the vibronic structure of the transition, unlike standard echo measurements. We obtain a timescale of 220 fs for IR 144 in methanol. The subsequent decay corresponds to the diffusive (irreversible) portion of the solvation dynamics.

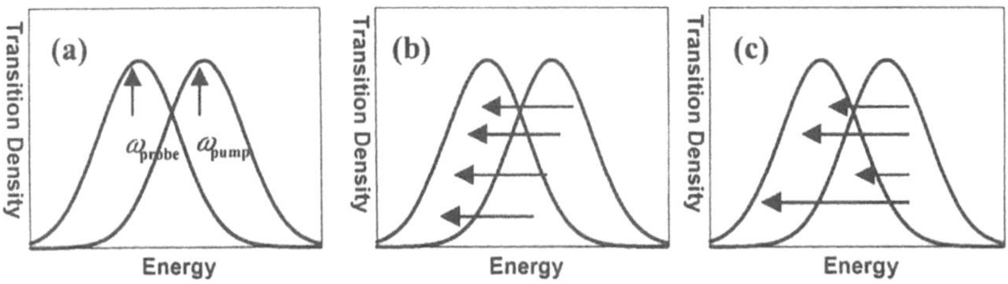

Fig.2. Time–dependent spectral shifts. Panel (a) shows the spectral distribution at the pump and probe wavelengths. Panel (b) corresponds to the "rigid" case, where no memory is lost, i.e. no scrambling of transition frequencies occurs during the shift. Panel (c) depicts another extreme scenario where all the memory is lost, i.e. complete scrambling takes place during the shift

In Fig. 2 we show our thoughts on the microscopic origin of the rise in the difference peak shift. After excitation the spectrum of the various members of the ensemble shifts to lower frequency via the solvation. Two extreme scenarios can be considered: (1) the shift is 'rigid' in the sense that the relative position with respect to the mean frequency is conserved or (2) the shift randomizes the position within the band. This suggests that a conditional probability approach is appropriate to describe the fraction of the ensemble that conserves its relative transition energy. We take the following form for the correlation function describing the response at the probe frequency given that the excitation was at ω_{pump}.

$$\langle \delta\varepsilon_{pr}(t)\delta\varepsilon_{pu}\rangle = \frac{P_{\text{inhom}}(t)}{P_{\text{inhom}}(t)+P_{\text{hom}}(t)}\langle \delta\varepsilon_{pr}(t)\delta\varepsilon_{pu}\rangle_{\text{inhom}} + \frac{P_{\text{hom}}(t)}{P_{\text{inhom}}(t)+P_{\text{hom}}(t)}\langle \delta\varepsilon_{pr}(t)\delta\varepsilon_{pu}\rangle_{\text{hom}} \quad (1)$$

The first term is a correlation function in which the fluctuation dynamics of the inhomogeneous distribution of transition energies is described by a conditional probability weighted by the relative contribution of the fluctuation. Except at very short times we expect that $P_{inhom}(t)$ to dominate $P_{hom}(t)$ allowing the rephasing capability of the inhomogenous distribution to dominate the correlation function. At short times the contribution of the homogenous distribution is expected to dominate the overall correlation function, but for most cases it is believed to have a low rephasing capability leading to a near zero correlation function. Eq. (1) seems to provide an adequate physical picture of the rise and decay of the difference peak shift, when the form for the conditional probability $P(\omega_2, t|\omega_1)$ derived by several authors [3] is inserted into the correlation function [4].

The 2C3PEPS experiment may be useful for a variety of studies. The requirement of a change in the transition energy while probing the memory of the transition energy allows 2C3PEPS to provide information not readily available from other experiments. For example, incoherent energy transfer in an inhomogeneous ensemble leads to partial correlation of donors and acceptors as the energy transfer proceeds [5]. This correlation appears as a rise in the two-color difference peak shift. Such studies could be attractive for characterizing both diagonal and off-diagonal disorder in antenna systems.

Acknowledgements. This work was supported by a grant from the National Science Foundation (NSF), and in part by the Director, Office of Science, Office of Basic Energy Sciences, Chemical Sciences Division, of the U.S. Department of Energy under Contract No. DE-AC03-76SF00098.

References

1 G.R. Fleming, S. A. Passino, and Y. Nagasawa, Phil. Trans. Roy. Soc. A **356**, 389, 1998.
2 M. Yang and G. R. Fleming, J. Chem. Phys. **110**, 2983, 1999.
3 R. F. Loring, Y. J. Yan, and S. Mukamel, J. Chem. Phys. **87**, 5840, 1987.
4 R. Agarwal, B. S. Prall, A.H. Rizvi, M. Yang, and G. R. Fleming, J. Chem. Phys. **116**, 6243, 2002.
5 M. Yang and G. R. Fleming, J. Chem. Phys. **113**, 2823, 2000.

Probing solvation dynamics at aqueous lipid-membrane interfaces with 3-pulse photon-echo peak-shifts

Helge Bürsing, Santi Kundu, and Peter Vöhringer

Biomolecular and Chemical Dynamics Group, Max-Planck-Institute for biophysical Chemistry, Am Faßberg 11, 37077 Göttingen, Germany, e-mail pvoehri@gwdg.de

Abstract: Three-pulse photon-echo peak-shift (3PEPS) experiments were performed to explore structural relaxations relevant to solvation of a cyanine dye non-covalently anchored to aqueous interfaces of unilamellar phospholipid vesicles. The influence of the lipid polymorphism is studied through the temperature-dependence of the 3PEPS decay. As opposed to the main order/disorder phase transition connected with packing phenomena in the hydrophobic membrane core, the rippled-to-gel pre-transition strongly affects the solvation response on time scales below 2 ps. This finding indicates that the pre-transition is connected with structural modifications of the hydration shell surrounding the polar head groups and is consistent with a dehydration of the interface.

1. Introduction

Solvation dynamics at aqueous phospholipid membrane (PLM) interfaces are of considerable importance for charge and mass transport across biological boundaries mediated either by diffusion or by integral membrane-bound proteins. We have recently been able to study the dynamics of solvation of a cyanine dye that was non-covalently anchored to the hydrophobic core of PLMs by means of long-chain alkyl residues attached to its optical chromophore [1]. The latter itself has a very high specificity to the aqueous PLM interface as evidenced by high-resolution confocal microscopy [2]. Here, we present 3PEPS studies to elucidate in more detail the temperature dependence of biointerface-specific solvation. PLMs exhibit a rich polymorphism that can be traced back primarily to ordering phenomena within the hydrophobic core [3]. The present study is our first attempt at exploring possible interrelations between the complex thermodynamic phase behavior of PLMs and short-time structural relaxations occurring at their interface to the aqueous surroundings.

2. Results and Discussion

Unilamellar vesicles of dimyristoyl (DMPC) and dipalmitoyl (DPPC) phosphatidylcholine in water were prepared and labeled with the chromophore, 1,1'-dioctadecyl-3,3,3',3'-tetramethylindodicarbocyanine chloride (C18-DiD) as described in Ref. [1]. To characterize the thermodynamic phase behavior of the PLMs, differential scanning calorimetry was carried out. The specific heat data are

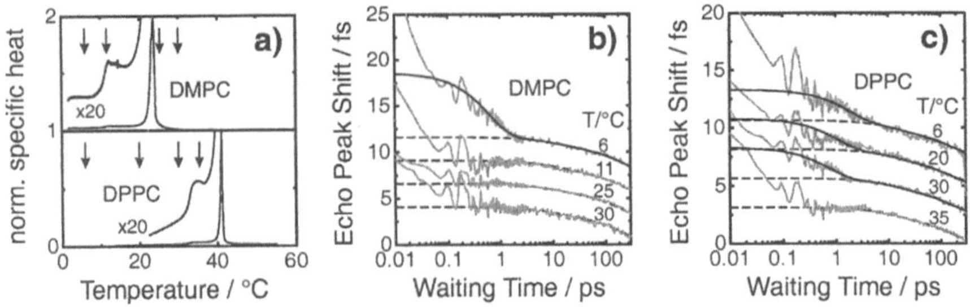

Fig. 1. Differential specific heat for DMPC and DPPC labeled with C18-DiD (a) and temperature-dependent 3PEPS for C18-DiD anchored to DMPC (b) and DPPC (c). Solid (dashed) lines represent fits to a triple- (double-) exponential decay.

displayed in Fig. 1a. For both lipids, the major peak indicates the gel-to-fluid phase transition at which the degree of ordering of the hydrocarbon tails in the membrane core drastically changes (chain melting, 23°C for DPMC and 41°C for DPPC) [3]. Both systems exhibit a rather broad pre-transition located roughly 12°C (for DMPC) or 9°C (for DPPC) below the main transition, i.e. fully in the gel phase [3]. This transition is indicative for the PLMs to form mesoscopic ripple structures as evidenced by electron microscopy. The phase intermediate to the two phase transitions is therefore denoted rippled gel [3].

For both lipids, 3PEPS experiments were conducted in each of the three phases as indicated in Figure 1a by vertical arrows. The experiments were carried out with a home-built non-collinearly phase-matched optical parametric amplifier whose output was characterized by pulse durations of 20 fs at a center wavelength of 650 nm. The temperature dependent 3PEPS decays for C18-DiD in DMPC are shown in Fig. 1b. In the linear response limit, the 3PEPS is capable of tracking the decay of the solvation correlation function relevant to the optical chromophore except on time scales comparable to the pulse durations where free induction decay-like effects (FID) give rise to extraneous chromophore-specific contributions [4]. Hence, we will only discuss the 3PEPS data for time scales longer than 80 fs (i.e. 4 times the pulse duration). We will also discard chromophore vibrational wave packet contributions giving rise to periodic modulations of the 3PEPS data on time scales below 1 ps.

Above the pre-transition, the temperature dependence of the 3PEPS data is very weak and the main chain-melting transition at 24°C does not affect the 3PEPS decay significantly (see Fig. 1b). For these temperatures, a fit to a double-exponential decay can reproduce the data satisfactorily. Time constants of 150 ps and 12 ps with amplitudes of 3.2 and 0.9 fs can be used independently of the temperature. We have previously assigned these two components to rotational diffusion of the label and to rotational diffusion of water molecules in the PLM interface region. Such a preliminary assignment was based on a qualitative agreement of the time constants with those obtained by other experimental techniques such as dielectric dispersion and transient anisotropy data [1]. Most remarkably, for temperatures below the pre-transition, the 3PEPS decay contains an additional ultrafast component, which is clearly absent in the rippled gel and the fluid phase.

To adequately represent the 6°C-data, a third exponential is needed whose time constant is 600 fs with an amplitude of 7 fs.

Fig. 1c displays complementary 3PEPS data for C18-DiD anchored to DPPC vesicles. Again, for temperatures above the pre-transition, a double-exponential fit is sufficient to satisfactorily reproduce the data. The time constants are 160 and 12 ps with amplitudes of 2.3 and 0.85 fs. Again, below the pre-transition, a third ultrafast component with a time constant of 700 fs becomes evident. Upon further cooling, this component remains visible and its time constant increases (1.0 ps at 20°C and 1.7°ps at 6°C). Most importantly, it should be emphasized that similar to DMPC, this third component is detected only in the gel phase but neither in the rippled nor the fluid phase. In fact, this correlation between 3PEPS and specific heat observed for two distinctly different PLMs provides convincing evidence for the sub-picosecond solvation response as being truly gel phase-specific.

Having established its gel phase-specific nature, the microscopic origin of the ultrafast sub-2 ps component remains to be clarified. Calorimetric studies on multilamellar vesicles of phosphatidylcholines as a function of the water content indicate that the rippled gel-to-gel pre-transition involves primarily a dehydration of the PLM interface with its highly polar lipid head groups [4]. Such an effect will also be experienced by the cyanine chromophore. It is quite possible that upon dehydration, the optical chromophore is slightly expulsed from interface region while still being fully anchored with its alkyl tails to the nonpolar membrane core. As a result, above the pre-transition, the chromophore is mostly exposed to the lipid head groups and strongly hydrogen-bonded interstitial water. Below the pre-transition, dehydration of the head groups results in the formation of a water layer above the lipid membrane interface whose properties are reminiscent of bulk water. If the chromophore is able to penetrate into this bulk-like water layer, the solvation response is likely to become similar to aqueous bulk solvation. In fact, we have previously shown that the room temperature solvation response of C18-DiD at DMPC/water interfaces differs from aqueous bulk solvation mostly for waiting times below 2 ps [1]. It is on such short time scales where bulk water responds with restricted intermolecular translational modes typical for its random hydrogen-bonded network. It is obvious that these bulk liquid modes do not exist in the rippled-gel or the fluid phase since the lipid head groups are strongly coordinated to interstitial water thereby fully disrupting the bulk water network of hydrogen bonds. Finally, the high interface sensitivity of the cyanine label naturally explains why the solvation response is totally insensitive to the main chain-melting transition where the head-group hydration is complete.

References

1 H. Bürsing, D. Ouw, S. Kundu, and P. Vöhringer, Phys. Chem. Chem. Phys. **3,** 2378, 2001.

2 P. Schwille, J. Korlach, and W. W. Webb, Cytometry **36,** 176, 1999

3 G. Cevc, Biochem. Biophys. Acta **1062,** 59, 1991

4 W. P. de Boeij , M. S. Pshenichnikov, and D. A. Wiersma, Ann. Rev. Phys. Chem. **49,** 99, 1998

Charge carrier recombination in poly(9,9-dioctylfluorene) (PFO) studied by electric field-assisted femtosecond spectroscopy

T. Virgili[1], G. Cerullo[1], C. Gadermaier[1],L. Lüer[1], G. Lanzani[1], S. De Silvestri[1], D.D.C. Bradley[2]
[1] Istituto Nazionale per la Fisica della Materia, IFN-CNR Dipartimento di Fisica, Politecnico di Milano, Italy E-mail: tersilla.virgili@polimi.it
[2] Centre for Electronic Materials and Devices, The Blackett Laboratory, Prince Consort Road, Imperial College SW7 2BW London

Abstract. Femtosecond electric-field assisted pump-probe measurements were performed on organic light-emitting devices based on poly(9,9-dioctylfluorene). The dynamics of polarons generation, due to field induced singlet dissociation, and recombination into triplet and singlet states are investigated. The efficiency of polarons recombination into singlet states was found to be around 70%. Carriers mobility at the microscopic scale was found to be $\mu(t)=1\text{-}10^{-2}$ cm^2/Vs indicating the presence of a dispersive and hot transport.

1. Introduction

Electroluminescence in organic light-emitting diodes (OLEDs) arises from recombination of the injected positive and negative charges, to form singlet excitons, and their subsequent radiative decay [1]. The quantum efficiency η of this process was thought to have an upper limit of 25%, given by the statistics of spin multiplicity. Recent experimental and theoretical work [2-4], however, showed that higher values of η are possible. Recently, we introduced electric field assisted femtosecond pump-probe spectroscopy [5, 6]. In this work we employ it to study the charge generation and recombination dynamics in poly(9,9-dioctylfluorene) (PFO) OLEDs.

2. Experimental Methods

The sample was excited at 390 nm and conventional pump-probe experiments were carried out with an excitation energy density of 0.4 mJ/cm^2. During the experiment the sample was kept in vacuum. The sample used is a single layer device, consisting of an indium-tin-oxide (ITO) layer acting as anode, a 100 nm film of PFO and a thermally evaporated Aluminum layer acting as cathode. In this experiment the pump and the probe beam pass the glass, ITO, and polymer to be reflected by the aluminum. The reflected probe beam is then focused on a photodiode. With the device reverse biased (F=1.7MV/cm) we measure the total dynamics in presence of the electric field $(\Delta T/T)_F$. By modulation of the applied voltage, we collect only field-induced changes in the $\Delta T/T$ spectra, which we call Δ^2T/T. Following our nomenclature, Δ^2T/T =

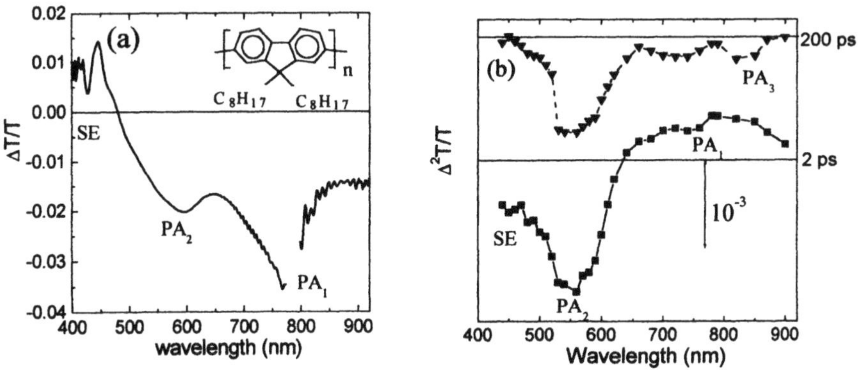

Fig 1: (a) ΔT/T spectrum of PFO (the chemical formula is given in the inset) 2 ps after photoexcitation with 150 fs pulses at 390 nm; (b) Δ^2T/T spectra, for an applied field of 1.5×10^6 V/cm, at 2 ps and at 200 ps.

$(\Delta T/T)_F$-ΔT/T. The spectral features of the species enhanced by the field keep the same sign as in ordinary ΔT/T, while those of the species quenched change sign.

3. Results and Discussion

Fig. 1(a) shows the ΔT/T spectrum of PFO 2 ps after photoexcitation; we observe a stimulated emission (SE) band peaking at 450 nm, and two photoinduced absorption (PA) bands, PA_1 peaking at ≈ 800 nm and PA_2 at 580 nm. In agreement with previous work [7], we assign PA_1 to singlet exciton and PA_2 to charge carrier absorption. The corresponding Δ^2T/T spectrum at 2 ps pump-probe delay is shown in Fig. 1(b): we observe a quenching of SE and PA_1 and an increase of PA_2, which correspond to a field-assisted dissociation of singlet excitons into charge carriers. The Δ^2T/T spectrum at 200 ps delay changes dramatically: at this time, the singlet features have vanished because the population has mostly recombined. In addition we observe a partial decay of the field-generated charge population and the growing in of the new PA_3 band. Based on previous work [8], we assign PA_3 to triplet excitons generated by the non-geminate charge carrier recombination.

Fig 2 shows kinetics of the singlet decay (680nm) (A), the field-induced polaron decay (580 nm) (B), the field-induced singlet decay (C). The corresponding fits are shown by solid and dashed lines. The field-induced decay at 680 nm is the superposition of two different processes: the singlet decay as in figure 2(A)and the singlet generation from polarons recombination. Three fits are shown in the figure using different values for $\eta(S_1)$ (efficiency of polarons recombination into singlets). From the fits we can obtain a value for η between 0.6 and 0.8. This large value can be attributed to strong intermolecular coupling, the electronic states being coherent superpositions of localized states [4].

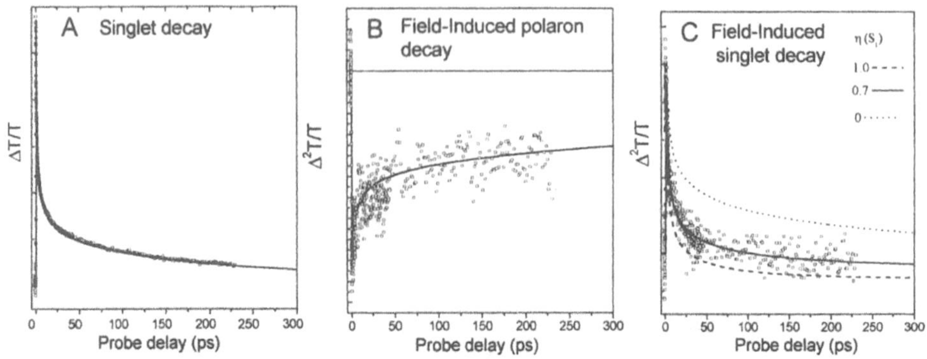

Fig. 2: Δ^2T/T dynamics at different wavelengths; solid and dashed lines are fits.

Δ^2T/T kinetics also allows estimates for the microscopic mobility of the charge carries. We calculated a value of $\mu(t) = 1 - 10^{-2}$ cm^2/Vs attributed to a dispersive and hot transport.

4. Conclusion

Field-assisted pump-probe spectroscopy enables us to generate charge carriers in an OLED and then selectively follow their temporal evolution, under conditions very similar to those of device operation. This technique allows to extract information that is crucial both to the understanding of the physics of organic semiconductors and to the optimization of OLED efficiency.

References

[1] R.H. Friend et al., Nature **397**, 121 (1999).
[2] Y. Cao, I.D. Parker, G. Yu, C. Zhang, and A.J. Heeger, Nature **397**, 414 (1999).
[3] M. Wohlgenannt, K. Tandin, S. Mazumdar, S, Ramasesha, and Z.V. Vardeny, Nature **409**, 494 (2001).
[4] Z. Shuai, D. Beljonne, R.J. Silbey, and J.L. Bredas, Phys. Rev. Lett. **84**, 131 (2000).
[5] W. Graupner, G. Cerullo, G. Lanzani, M. Nisoli, E.J.W. List, G. Leising, S. De Silvestri,, Phys. Rev. Lett. **81**, 3259 (1998).
[6] C. Zenz, G. Cerullo, G. Lanzani, W. Graupner, F. Meghdadi, G. Leising, S. De Silvestri, Phys. Rev. B **59**, 14336 (1999).
[7] B. Kraabel et al. Phys. Rev. B **61**, 8501 (2000)
[8] A.J. Cadby et al. Phys. Rev. B **62**, 15604 (2001).

Coherent Control Over Collective Polariton Excitations: The Dawn of Polaritonics

T. Feurer[1], Joshua C. Vaughan[1], Nikolay S. Stoyanov[1], and Keith A. Nelson[1]

Department of Chemistry, MIT, Cambridge, MA 02139, USA
e-mail: kanelson@mit.edu

Abstract. Automated spatiotemporal coherent control over lattice waves propagating at light-like speeds offers prospects of novel THz spectroscopy and ultrafast signal processing. We demonstrate programmable phonon-polariton steering, focusing, and amplification. In addition, passive control by laser-machined waveguide structures, dispersive elements, and complex integrated devices is shown.

1 Introduction

Coherent optical control efforts to date have been aimed almost exclusively at manipulation of excitations that are localized in space. Control over coherently propagating modes offers several important new prospects. First, it opens up the use of spatial as well as temporal parameters as control elements in the design of optimized excitation waveforms. This enables new types of control in which interferences among coherent responses that originate from different regions of a sample may be used to selectively enhance or suppress the net response at a specified location and/or time. Coherent amplification of responses is possible through constructive interferences, allowing large coherent amplitudes to be achieved with modest excitation light intensities. Collective responses may be optically excited in one region of a heterogeneous sample and directed toward other regions including those at which the responses could not be driven directly by the excitation light due to selection rules, opacity, or other restrictions. In this manner the normally inaccessible regions may be studied or subjected to otherwise unattainable coherent control. In addition, propagating modes may possibly be used as coherent signals that can be launched at specified locations or "addresses" and received, manipulated further, or read out at other specified addresses, all at ultrahigh bandwidths.

2 Results and Discussion

Achieving control over propagating coherent responses places new demands on both the excitation and probing processes. In particular, it must be possible to generate excitation light fields that are specified functions of both time and position, and it must also be possible to monitor the time and position-dependent responses. Programmable spatiotemporal femtosecond pulse shaping, through which a single light beam with a single femtosecond pulse may be transformed into many spatially separated light beams, each of which has a specified time-dependent waveform, has been developed [1] and very recently improved [2] to achieve the high fidelity necessary for coherent control applications. Separately, spatiotemporal imaging of propagating, coherent lattice modes has been reported [3]. The combination of these two capabilities enables automated spatiotemporal coherent control over lattice waves that propagate at light-like speeds. Preliminary attempts at such control [4] yielded only barely visible responses because of limitations in the quality of the spatiotemporally shaped excitation waveforms.

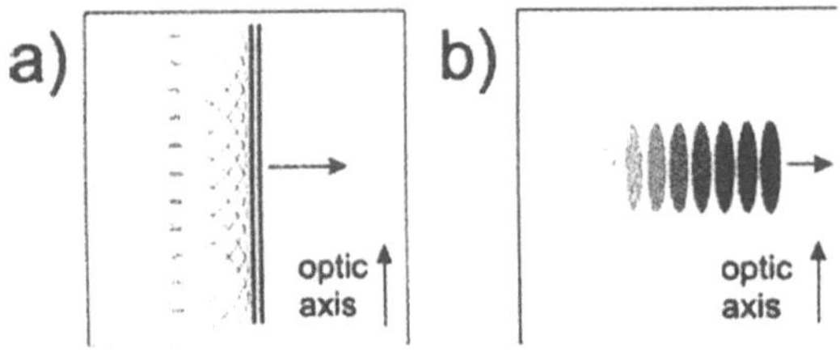

Fig. 1. (a) Arrangement for polariton steering and focusing experiments. (b) Polariton amplification geometry. The polarization of the excitation beam is parallel to the optic axis, and the propagation direction of the phonon-polariton is perpendicular to the optic axis.

Here we demonstrate versatile, programmable manipulation of coherent phonon-polaritons in crystalline solids. Coherent responses from roughly twenty separate excitation regions are generated through the use of spatiotemporal femtosecond pulse shaping. Depending on the excitation geometry, their interferences permit polariton steering and focusing [see figure 1a)], or amplification [see figure 1b)], or other control objectives to be achieved. The capabilities demonstrated here should be applicable as well to other types of mixed electromagnetic waves including exciton-polaritons and plasmon-polaritons.

Figure 2 shows results of experiments conducted in $LiTaO_3$ crystals with the geometries in fig. 1a) (LH and center columns) and fig. 1b) (RH column), using 800 nm spatiotemporally shaped excitation waveforms and variably delayed, 400 nm probe pulses. The images of the polariton responses were projected onto a CCD [3]. The LH column shows the result of a linear temporal sweep in which the uppermost excitation spot was irradiated first, then the spot below it, and so on. Different linear sweeps were used to steer polariton responses along selected directions. The center column shows the result of a parabolic temporal sweep in which the first pulses arrived at the top and bottom spots, the next pulses arrived at the adjacent spots, and so on, with

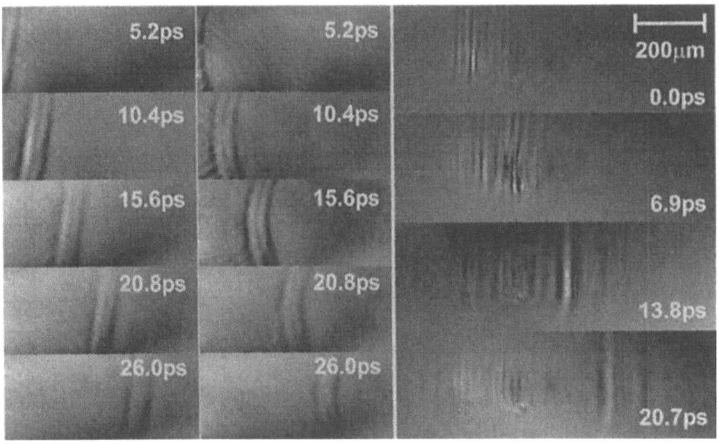

Fig. 2. Polariton responses to a series of about 20 spots, generated through spatiotemporal femtosecond pulse shaping. (LH column) Tilted polariton wavefronts resulting from pulses arriving at excitation spots separated along the vertical dimension as in fig. 1a) at progressively later times, with a linear relation between temporal delay and spatial location of the spot. (center column) Polariton focusing specified through a parabolic relation between temporal delay and spatial location of the spots along the vertical dimension. (RH column) The excitation spots are horizontally arrayed as in fig. 1b). Excitation pulses arrive at the LH spot first and sweep toward the right, resulting in polariton amplification.

the last pulses arriving at the middle spots. Polariton focusing with various focal lengths was conducted by varying the curvature of the parabolic temporal sweep. A combination of linear and parabolic delays was used for both focusing and steering, i.e. to direct focused polaritons to selected locations. The RH column shows polariton amplification, achieved with a sequence of pulses moving from left to right with temporal and spatial delays fixed to match the polariton group velocity. The phonon-polariton response propagating to the right is amplified substantially, while the unamplified responses propagating toward the left are only barely visible.

Besides active coherent control we demonstrate passive control over phonon-polariton propagation by laser-machining integrated waveguide structures on a $LiNbO_3$ platform. Because the wavelengths of phonon-polaritons is typically on the order of a few ten to a few hundred microns, femtosecond laser-machining offers a versatile tool to fabricate waveguides and more complex integrated waveguide structures.

Figure 3 shows a series of snapshots of a phonon-polariton being focused into a laser-machined waveguide structure. The waveguide is 0.2 mm wide and 1.7 mm long. The walls of the waveguide appear blurry, because the CCD camera was moved out of the image plane to achieve high contrast phase imaging. The white bars have been added by image processing and indicate the location and width of the waveguide walls. Immediately after entering

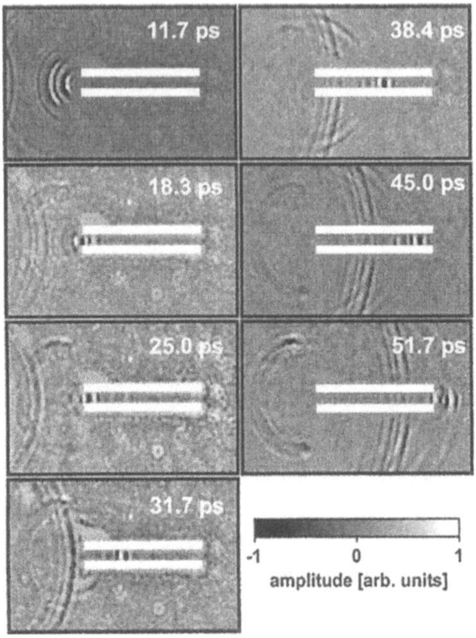

Fig. 3. Phonon-polariton propagation in a laser-machined waveguide structure.

the waveguide the phonon-polariton wavefront becomes perfectly plane and remains so until the phonon-polariton exits the waveguide. The last image shows the diverging phonon-polariton as it comes out of the waveguide.

More complex structures, for example an integrated Mach-Zehnder interferometer, have been fabricated and their functionality has been shown.

Femtosecond laser-machining may also be used to fabricate integrated THz refractive and dispersive elements, such as beamsplitters, lenses, prisms, or gratings. Figure 4 shows a plane phonon-polariton wavepacket intercepting a laser-machined grating structure. The grating consists of 10 'grating lines', each of which is 0.20 mm wide and the distance between two neighboring lines is 0.15 mm. The delay time between the pump and the imaging pulse is indicated. The generated line-like phonon-polariton response travel from the left to the right and intercepts the grating at normal incidence. Part of the wavepacket is reflected and part is transmitted, therefore, the grating may either be used in reflection mode or transmission mode. As time evolves the single, almost point-like sources originating from each 'grating line' start to interfere and to develop wavefronts that correspond to zero, first, and second order etc. diffraction in the far field.

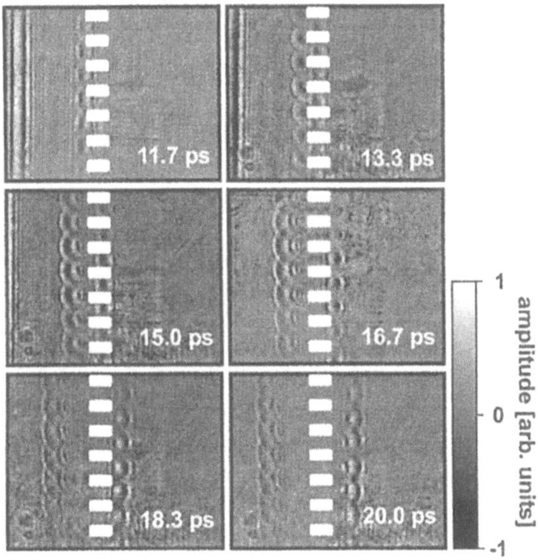

Fig. 4. Phonon-polariton amplitude for different pump probe delays. The wavepacket travels from the left to the right and intercepts the grating at normal incidence. The images show the diffraction pattern of the grating in reflection and transmission. The grating structure has been enhanced (white rectangles) by image processing.

3 Conclusions

The capabilities demonstrated here illustrate versatile, active and passive control over propagating responses that can be used for linear and nonlinear THz spectroscopy and for multiplexed solid-state THz signal generation and processing. A wide range of fundamental and practical applications of polaritonics is anticipated.

Acknowledgements. This work was supported in part by National Science Foundation grant no. CHE97-13388 and MRSEC program award no. DMR98-08941, and Army Research Office grant no. DAAD19-99-1-0034. T. Feurer acknowledges financial support by the Max Kade Foundation.

References

1. R.M. Koehl, T. Hattori, K.A. Nelson, Opt. Commun. **157**, 57 (1998)
2. T. Feurer, J.C. Vaughan, R.M. Koehl, K.A. Nelson, Opt. Lett. **27**, 652 (2002)
3. R.M. Koehl, S. Adachi, K.A. Nelson, J. Phys. Chem. A **103**, 10260 (1999)
4. R.M. Koehl, K.A. Nelson, J. Chem. Phys. B **114**, 1443 (2001)

Coherent valence fluctuation in photo-induced neutral-ionic transition of an one-dimensional charge transfer complex

S. Iwai[1], S. Tanaka[2], K. Fujinuma[2], H. Kishida[2], H. Okamoto[1,2], Y. Tokura[1,3]

[1]*Correlated Electron Research Center (CERC), National Institute of Advanced Industrial Science and Technology (AIST), Tsukuba 305-0046, Japan, tel:+81-298-61-4757, fax: +81-298-61-2663, e-mail: s.iwai@aist.go.jp*
[2]*Department of Advanced Materials Science University of Tokyo, Tokyo, 113-0033, Japan*
[3]*Department of Applied Physics, University of Tokyo, Tokyo, 113-0033, Japan*

Abstract. Initial dynamics of the photo-induced neutral(N)-ionic(I) transition of Tetrathiafulvalene-p-chloranil was investigated by visible and mid-infrared pump-probe spectroscopy. Charge transfer excitation leads to the ultrfast generation of a one-dimensional ionic (D^+A^-) string in N phase or neutral (D^0A^0) string in I phase within 200 fs. A coherent oscillation of a degree of charge transfer (ρ) coupled with a intra-chain lattice phonon was also observed .

1. Introduction

Recently, many efforts have been made to find materials showing photo-induced phase transition (PIPT), since giantic responses in optical and/or other electronic properties are stimulated by a weak light irradiation in the PIPT [1]. The essence of PIPT is that a macroscopic change in electronic state (or equivalently a phase transition) is triggered by a local photo-excitation in a few sites.

Tetrathiafulvalene-p-chloranil (TTF-CA) is a one-dimensional (1-D) mixed stacked charge transfer (CT) crystal which shows a temperature-induced neutral (N) to ionic (I) phase transition at T_{NI}=81 K. The crystal lies in the N phase at the temperature higher than T_{NI}. By lowing temperature, the degree of CT (ρ) between TTF (donor, D) and CA (acceptor, A) changes abruptly from 0.3 to 0.7 at T_{NI} and the I phase is formed [2]. The DA stacks in the I phase undergo the dimeric lattice distortions due to the spin-Peierls mechanism. The ferroelectric ground state is stabilized in the I phase since the dimeric molecular displacements are three-dimensionally ordered. Since the discovery of the N-I transition by the light irradiation at the temperature near T_{NI}, TTF-CA has been one of the prototypical materials showing PIPT [3].

Very recently, we have investigated the initial processes of photo-induced I-N transition by using sub-picosecond pump-probe reflection spectroscopy [4]. We found that 1-D microscopic N domains or D^0A^0 strings are photogenerated within 1 ps as a precursor state for a macroscopic PIPT to the N phase. However, the primary dynamics of the photo-induced change in the degree of CT (ρ) and that of the dimeric lattice distortions have not been clear. In this study, we focus on the sub-ps ~ several ps dynamics of N-I transition involving the valence fluctuations and the lattice vibrations.

2. Experimental Methods

We have carried out the pump-probe experiments whose probing window is 0.1-2.5 eV to observe both the photo-induced change in ρ and the dimeric lattice distortion at the same time. The intramolecular reflection band of TTF (~2.25 eV in I phase) is quite sensitive to the change in ρ [5], while the CT band (0.3-1.3 eV) is strongly affected by the dimeric lattice distortion [6]. Photo-excitations were made at the peak energy of the CT band, which corresponds to an excitation of a neutral DA pair (D^0A^0) in the I stacks or an excitation of an ionic DA pair (D^+A^-) in the N stacks.

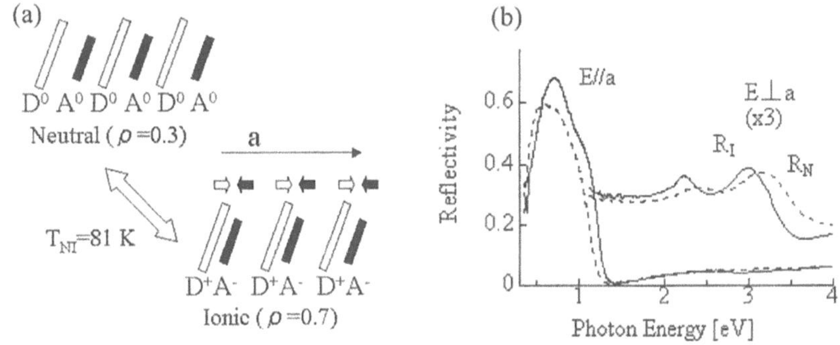

Fig.1 (a) Schematic illustration of the ionic (I) and the neutral (N) phases in TTF-CA. (b) Polarized reflectivity spectra for E//a and E⊥a of TTF-CA on a (100) surface at 77K (the solid lines: R_I) and at 90 K(the broken lines: R_N)

3. Results and Discussion

Figure 2 shows the time evolutions of photo-induced reflectivity changes reflecting the change in ρ due to the I-to-N (a, 77K) and N-to-I (b, 90 K) transition. At 77K, we observe the initial rise within 200 fs which is followed by the further increase of the signal up to 30 ps. The fast rise and the subsequent growth of the signal in Fig.1(a) are attributable to generation of microscopic D^0A^0 strings and their multiplication process to macroscopic N domains, respectively [4]. At 90 K, the initial reflectivity change also occurs very fast, that is attributable to the formation of the D^+A^- string. This initial reflectivity change decays in several ps, indicating that the photogenerated D^+A^- string is not stable in the N stacks.

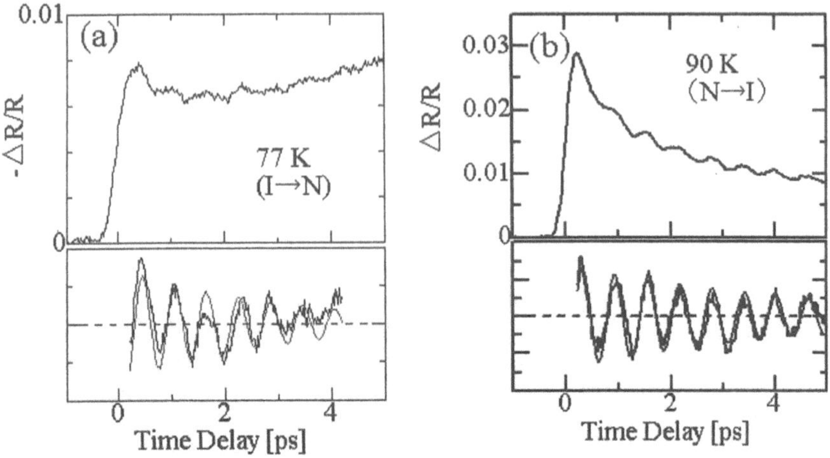

Fig. 2 Time evolutions of the reflectivity change observed at 2.25 eV (a: 77K, b: 90 K). Pumping photon energy is 0.65 eV. The polarizations of the pump and the probe pulse are parallel and perpendicular to the 1-D chain axis. The lower panels show the oscillating component of the reflectivity changes.

On the other hand, the dimeric molecular distortion occurs in the time scale of 1 ps (not shown) at 90 K, which is deduced from the photo-induced reflectivity change observed at the CT band region. This shows that the undimerized D^+A^- strings are initially produced (< 200 fs) before the dimeric lattice distortion occurs at ~1 ps.

As seen in Fig.2 (a, b), the instantaneous rise of the signal is followed by the oscillating structure showing the temporal modulation in ρ. The oscillations with the periods of 0.60 ps (77 K) and 0.61 ps (90 K) are considered to be caused by the coupling between the change of ρ and the coherent lattice vibration. The frequencies of the corresponding phonons are 56 cm^{-1} (77 K) and 55 cm^{-1} (90 K), respectively. The intra-chain optical phonon, whose frequency is several tenths cm^{-1} , possibly couples with the charge transfer process and drives the coherent oscillation in ρ.

References

1 K. Nasu, *Relaxations of Excited states and Photo-Induced Structural Phase transitions* (Springer,1997).
2 J. B. Torrance et al., Phys, Rev. Lett. **47**, 1747 (1981).
3 S. Koshihara, Y. Takahashi, H. Sakai , J. Phys. Chem. B,**103**, 2592 (1999).
S. Koshihara et al., Phys. Rev. **B**, 42, 6853 (1990).
4 S. Iwai et al. Phys. Rev. Lett. 88, 057402 (2001).
5 Y. Tokura et al., Solid State Commun. 43,757(1982).
6 H. Okamoto, Y. Tokura, T. Koda, Phys. Rev. B36, 3858 (1987).

Part XI

Multidimensional Spectroscopy and Dynamics

Diffractive optics based heterodyne detected six-wave mixing: "Dutch Cross" fifth-order Raman

K. J. Kubarych, C. J. Milne, S. Lin, and R. J. D. Miller

Departments of Chemistry and Physics, 60 St George Street, University of Toronto, Toronto, Ontario M5S 1A7 Ph: 416-978-1528/Fax: 416-978-2537 E-mail: *dmiller@lphys.chem.utoronto.ca*

Abstract: A novel polarization configuration predicted to yield enhanced contrast of the direct fifth-order Raman signal is demonstrated in liquid carbon disulfide and shows a rapid decoherence in the liquid dynamics.

Fifth-order Raman spectroscopy (six-wave mixing) is the lowest order electronically non-resonant method suitable for investigating the details of low-frequency motions in liquids.[1] The fifth-order response is capable of revealing homogeneous and inhomogeneous contributions to the vibrational line shape while offering insight into the anharmonicity of the underlying intermolecular potential energy landscape. First proposed by Tanimura and Mukamel[1], the fifth-order Raman processes involves an initial pulse pair that create a coherence between two vibrational levels, followed by a second pulse pair (after a time delay, τ_2) that transfer the system to a second coherence. After a time delay, τ_4, a probe pulse interacts with the net induced polarization in the material generating a signal. Signal along the time diagonal, $\tau_2 = \tau_4$, indicates echo-like rephasing in analogy to echoes in spin, electronic, and infrared transitions.

Recently, several groups have reported experimental results where considerable effort has been made to reduce unwanted cascaded third-order processes that can obscure the desired fifth-order observable.[2-4] While a consensus has yet to be reached, it is becoming clear that the rephasing ability of CS_2 is limited as the most recent experimental results to-date indicate that the response decays faster than the free-induction decay (FID). This rapid decay in the rephased coherence is in contrast to even homogeneous dephasing limits that typify echo response functions. For the broad spectral density of states characterizing liquids, additional decay mechanisms, such as mode-mixing and polarization quenching[5] offer a possible explanation. Additionally, since the generation of an echo requires a one-quantum coherence followed by transfer to a two-quantum coherence, differences in the dynamics of the two excited vibrational states will also affect the response.[2]

The principal concern in interpreting fifth-order Raman data is an assessment of the degree to which cascaded third-order processes contribute to the measured signal. Jansen *et al.* have performed finite-field molecular dynamics (FFMD) simulations of the fifth-order Raman response of liquid CS_2 in order to obtain directly comparable signal amplitudes for the direct fifth-order and cascaded third-order responses.[6] Their work included a proposal for a unique polarization configuration capable of reducing the cascade contribution by a factor of ~10^4 compared to the fully-polarized tensor element based on the FFMD results. The suppression can be shown to arise from the near perfect cancellation of the different cascaded third-order intermediate terms. This polarization configuration, which we will denote the "Dutch Cross", has been implemented experimentally and is presented here.

In our previously measured results for the fifth-order Raman response of CS_2, a major signal component was found to be a prolonged ridge feature along the τ_4 axis (probe delay). The generation of this signal occurs when the four pump pulses are coincident at the sample, and is thus most susceptible to two-photon resonance enhancement that may obscure the true non-resonant fifth-order response. In order to address this issue, a novel three-color interaction scheme was implemented where the second pulse pair is at shorter wavelength (530 nm) than the first, 400 nm pair. The third color is generated by noncollinear optical parametric amplification (NOPA), producing a broadband pulse that is capable of rephasing the entire low-frequency spectrum while reducing any potential resonant absorptive response.

The Dutch Cross polarization configuration is implemented using the same two-color diffractive optic (DO) apparatus that has been reported previously.[2] With the 70 fs, 800 nm output of a 1 kHz Ti:Sapphire regenerative amplifier one 800 nm beam and two 400 nm beams are generated. A diffractive optic generates the necessary beam geometry and a reference beam that is collinear with the resulting signal beam and the two beams are detected on a photodiode. One of the pump beams is chopped and a lock-in amplifier detects the photodiode signal. In order to adjust the phase of the reference beam relative to the signal, a thin microscope cover slip mounted on a rotation stage can be tilted to impart a small path length difference.

The polarization configuration proposed by Jansen *et al.* is as follows: the first pump pair is polarized at –60° to the vertical, the second pump pair is vertically polarized and the probe and signal are polarized +60° to the vertical.[6]

The three-color modification utilizes the residual 800 nm beam that remains after generation of the 400 nm beam. This IR beam, is used to create the 400 nm pump (30μJ) and white light seed required for a NOPA based on BBO. The NOPA output is compressed with 12 bounces on chirped mirrors. The visible

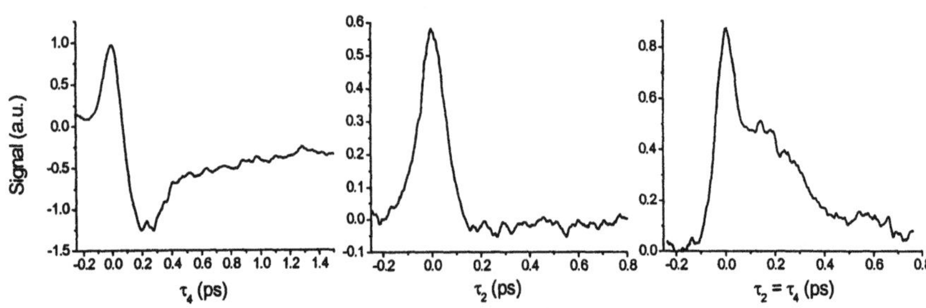

Fig. 1: Dutch Cross fifth-order response of CS_2. Slices are along τ_4, τ_2, and the diagonal (from left to right).

beam (λ=530 nm, $\Delta\lambda$=25 nm), is delivered to the diffractive optic with all-reflective optics in order to limit dispersion.

Fig. 1 shows the heterodyne detected fifth-order Raman data obtained with the Dutch Cross configuration. The data are presented as three slices through the two-dimensional space. Typically a comparison is made with the cascaded third-order simulation to assign the phase.[2] However, all the data is distinct from the cascaded simulations, and the largest signal component in the phase scans shows a pronounced ridge along τ_4 as observed previously.[2] This finding is consistent with the assumption that the Dutch Cross sufficiently decreases the cascade contribution significantly and is used to assign the relative phase. The τ_4 scan has a fast rise at the origin followed by a negative feature that peaks at 250 fs followed by a slow decay after 400 fs. Along the τ_2 axis, the signal decays much more rapidly than the signal along the τ_4 axis, but it is not pulsewidth limited. The diagonal slice shows a very fast decay (electronic hyperpolarizability) followed by a broad peak at around 100 fs ultimately leading to a slower decaying feature. The pronounced rephasing signature along the time diagonal is in excellent agreement with the FFMD simulations of Jansen *et al.*[6] as well as the general trend found by Saito and Ohmine that nuclear rephasing occurs at ~100 fs with a rapid decay to zero by 500 fs.[5]

Fig. 2 shows the fully polarized, $R^{(5)}_{111111}$, data collected using the three-color pulse sequence.. The data agree with our previous reports of the fifth-order Raman response of CS2, with a persistent ridge along the diagonal with very rapid dephasing along the diagonal. This result confirms that only a minimal signal component along the τ_4 axis is due to resonant contributions.

1. Y. Tanimura and S. Mukamel, J. Chem. Phys. **99**, 9496 (1993).
2. V. Astinov, K. J. Kubarych, C. J. Milne, and R. J. D. Miller, Opt. Lett. **25**, 853 (2000); *ibid*, Chem. Phys. Lett. **327**, 334 (2000); K. J. Kubarych, C. J. Milne, S. Lin, V. Astinov, and R. J. D. Miller, J. Chem. Phys. **116**, 2016 (2002).
3. D. Blank, L. J. Kaufman, and G. R. Fleming, J. Chem. Phys. **113**, 771 (2000); L. J. Kaufman, D. A. Blank, and G.R. Fleming, J. Chem. Phys. **114**, 2321 (2001).
4. O. Golonzka, N. Demirdoven, M. Khalil, and A. Tokmakoff, J. Chem. Phys. **113**, 9893 (2000).
5. S. Saito and I. Ohmine, J. Chem. Phys. **108**, 240 (1998); *ibid*, **88**, 207401 (2001).
6. T. I. C. Jansen, J. G. Snijders, and K. Duppen, J. Chem. Phys. **114**, 10910 (2001).

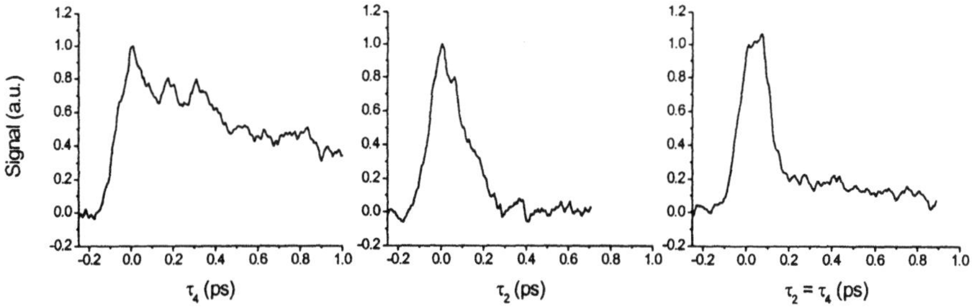

Fig. 2: Fifth-order response of CS_2 using the three-color pulse sequence. Slices are along τ_4, τ_2, and the diagonal (from left to right).

Heterodyne detected fifth-order nonresonant Raman spectroscopy of CS_2: Evidence for anharmonic coupling

Laura J. Kaufman[1], Shinji Saito[2], Larry D. Ziegler[3], Iwao Ohmine[2], and Graham R. Fleming[1]

[1] Department of Chemistry, University of California, Berkeley and Physical Biosciences Division, Lawrence Berkeley National Laboratory, Berkeley, CA 94720
E-mail: laura@zepto.cchem.berkeley.edu
[2] Department of Chemistry and The Photonics Center, Boston University, Boston, MA 02215
[3] Department of Chemistry, Faculty of Science, Nagoya University, Nagoya, Japan

Abstract. We present actively phase-locked heterodyne detected fifth-order data. Nodal lines are found in several tensor elements, and these are also present in molecular dynamics simulations of the signal. The nodes are a result of anharmonic coupling between modes.

1. Introduction

Tanimura and Mukamel first proposed nonresonant fifth-order spectroscopy [1], and it was soon realized that the experiment could reveal both couplings within and between inter- and intramolecular motions as well as the anharmonic potential of a liquid. Because early fifth-order experiments were plagued by contamination from interfering third-order cascade signals, the development of the technique was slowed [2]. Substantial challenges were also encountered in the various theoretical approaches to the problem.

Recently, several advances have been made on both the theoretical and experimental fronts. Normal mode theory has been extended to allow for anharmonic modes, and this has been shown to account for many of the qualitative features of the liquid Xe fifth-order signal [3]. Further, molecular dynamics simulations of the fifth-order signal of liquid CS_2 have been completed, and they reveal features not seen in previous simulations [4]. On the experimental side, two groups have presented heterodyne-detected results on CS_2 that are believed to be free of cascade contamination [5]. While certain aspects of these results are in agreement with each other, there are still clear discrepancies that need resolution.

3. Results and Discussion

Here, we present several tensor elements of actively phase-stabilized heterodyne detected nonresonant fifth-order signal of CS_2 collected in a geometry chosen to reduce the presence of third-order cascades. This geometry has reduced the phase-matching factors for the third-order cascades by at least 10^6 over the original geometries used to measure the fifth-order signal, which were shown to allow cascade signals to dominate [2]. We also are confirming the absence of significant cascade and cascade-fifth-order cross-term contamination by employing a geometry that improves on the ratio of phase-matching factors of the fifth-order:cascade signals by an additional factor of ~ 4.

The heterodyne measurements that were obtained confirm many of the observations (shape and magnitude) made via homodyne measurements taken in this same geometry [6]. However, one key element appears in the heterodyne detected signals that was not apparent in the homodyne studies. Many, but not all, tensor elements contain sign changes, or nodal planes, near the t_1 = 0 fs axis. Molecular dynamics (MD) simulations also show these nodes.

Figure 1 shows two measured and simulated tensor elements of the fifth-order signal of CS_2. The bottom panel of Figure 1a displays the measured R_{zzzzmm} surface. The signal extends to approximately 200 fs along the diagonal. There is significant amplitude along the t_2 = 0 fs slice, which is attributed to a hyperpolarizability signal not included in the simulation. There is also substantial (negative) signal along and around the t_1 = 0 fs slice, which extends to t_1 ~ 100 fs. The top panel of Figure 1a is the MD simulation by Saito and Ohmine of the same surface [4]. It shares several features with the measured surface, including the general shape and the presence of negative signal in the region of t_1 = 0 fs. In Figure 1b, the measured and simulated R_{zzmmzz} surfaces are shown. They again both show nodal planes in the region of t_1 = 0 fs. Here, however, the measured signal is larger and longer-lived than in the case of R_{zzzzmm}. The persistence of the signal along the diagonal slice of the surface comfirms the assignment of the lower limit on the inhomogeneous timescale of the liquid of several hundred femtoseconds [6]. Figure 1c shows the measured and simulated $R_{llzzl'l'}$ signal, a tensor element suggested to diminish cascade signals even further than the magic angle tensor elements. Again, both the measured and simulated signals show nodal planes.

The interesting and unexpected nodal plane in both the measured and simulated signals shows that the fifth-order experiment is a sensitive probe of details of intermolecular motions. Third-order spectroscopy of room temperature liquids has, to our knowledge, never shown changes of sign in the nuclear response. Preliminary analysis of this feature indicates that the node is neither due to the system's nonspherical polarizability nor the dipole-induced-dipole terms in the polarizibility. However, it does appear that the presence of the nodal plane is intimately related to anharmonic mode coupling. When expressing the fifth-order response as

$$R^{(5)}(t_1,t_2) = -\beta\left\langle \{\Pi(t_1+t_2),\Pi(t_1)\}_{PB},\dot{\Pi}(0)\right\rangle, \tag{1}$$

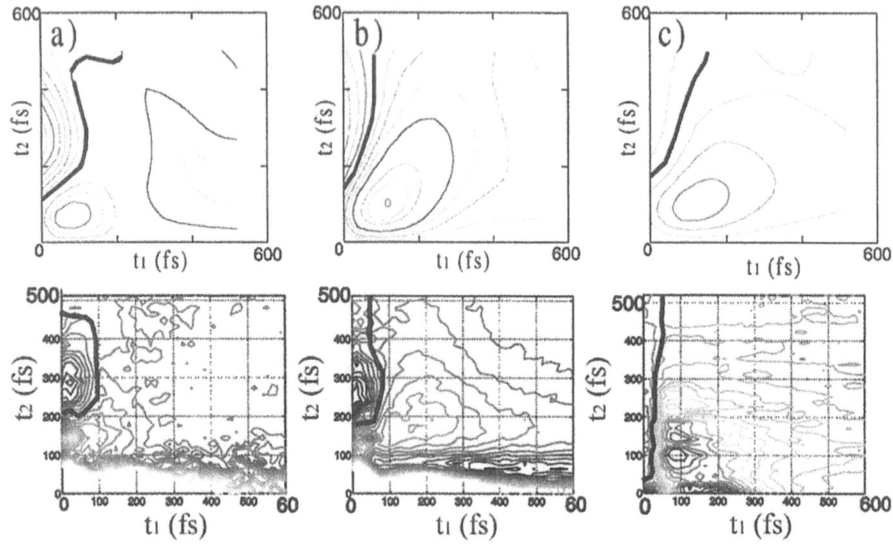

Fig. 1. a) Measured (bottom) and simulated (top) R_{zzzzmm} surface. b) Measured (bottom) and simulated (top) R_{zzmmzz} surface. . b) Measured (bottom) and simulated (top) $R_{llzzl'l'}$ surface.

where Π is the total polarizability of the system and PB denotes Poisson brackets, and analyzing this expression while *not* allowing for coupling between atomic coordinates via the following equation,

$$R(t_1,t_2) = \beta\left\langle \sum_{i=j} \frac{\partial\Pi(t_1+t_2)}{\partial q_i(t_1+t_2)} \frac{\partial q_i(t_1+t_2)}{\partial p_j(t_1)} \frac{\partial\Pi(t_1)}{\partial q_j(t_1)} \dot{\Pi}(0) \right\rangle \tag{2}$$

no nodes result [4]. By transforming into a "normal mode" basis set (in which the normal modes change in time as a result of anharmonicity), it is shown that rotational-rotational coupling largely determines the position of the nodal plane. Further analysis of the information contained in the measured signals is underway.

Acknowledgments. This work is supported by a grant from NSF.

References

1. Y. Tanimura and S. Mukamel, J. Chem. Phys., **99**, 9496 (1993).
2. D.A. Blank, L.J. Kaufman, and G.R. Fleming, J. Chem. Phys., **111**, 3105 (1999).
3. A. Ma and R.M. Stratt, J. Chem. Phys., **116**, 4972 (2002).
4. S. Saito and I. Ohmine, Phys. Rev. Lett., **88**, 207401 (2002).
5. L.J. Kaufman, J. Heo, L.D. Ziegler, and G.R. Fleming., Phys Rev. Lett, **88**, 207402 (2002); K.J. Kubarych, C.J. Milne, S. Lin, V. Astinov, and R.J.D. Miller, J. Chem. Phys., **116**, 2016 (2002).
6. D.A. Blank, L.J. Kaufman, and G.R. Fleming, J. Chem. Phys., **113**. 771 (2000); L.J. Kaufman, D.A Blank, and G.R. Fleming, J. Chem. Phys., **114**, 2312 (2001).

Femtosecond 2D Fourier transform study of electronic reorientation in silicon naphthalocyanine

Wei Qian, Allison Albrecht Ferro, Richard T. Treglio, and David M. Jonas

Department of Chemistry and Biochemistry, University of Colorado, Boulder CO 80309-0215 USA
E-mail: david.jonas@colorado.edu

Abstract. 26 fs excitation of a doubly degenerate transition in a molecule with square symmetry yielded a ~200 fs pump-probe anisotropy decay from 0.40(1) to 0.11(1). 2D spectra probe this rapid reorientation of the electronic wavefunction.

The breakdown of the adiabatic separation between fast electronic and slow nuclear motions is important throughout chemistry [1]. Such non-adiabatic effects are most important when the electronic motions are slowed by near electronic degeneracy. The near degenerate coupling between vibrational and electronic dynamics is probed by various four-wave mixing measurements using 26 fs pulses that cover the doubly degenerate Q(0-0) electronic transition of silicon 2,3-naphthalocyanine bis(trihexylsilyloxide) in benzonitrile solution [2]. Naphthalocyanine has the same point group symmetry as a square, and the Q(0-0) transition is analogous to the transition from the ground state (n_x=0, n_y=0) to the degenerate excited states (n_x=1, n_y=0) and (n_x=0, n_y=1) for a particle in a square two dimensional box. Upon electronic excitation to a degenerate state, a square molecule can lower its total energy by exciting vibrations which elongate the molecule perpendicular to the electronic node, thus breaking the square symmetry (Jahn-Teller effect). For randomly oriented molecules, the optical electric field prepares an excited state wavefunction with nodal line perpendicular to the projection of the field vector onto the molecular plane. When vibrations lower the molecular symmetry, the two electronic states have different energies and acquire phase at different rates so that the electronic wavefunction reorients. Topological phase shifts (Berry phase) which invert the electronic wavefunction can arise if the symmetric degeneracy point is circled [1]. Transient grating scattering was used to measure the molecular vibrations excited upon electronic excitation. The pump-probe polarization anisotropy, $r=(S_{\parallel}-S_{\perp})/(S_{\parallel}+2S_{\perp})$, was measured to characterize the decay of wavefunction alignment (it does not distinguish wavefunctions with opposite signs) in the excited state. Finally, phase resolved two dimensional Fourier transform (2D FT) spectra measure homogeneous electronic dephasing, which decays by both electronic wavefunction reorientation (the sign of the wavefunction matters) and vibrational motions which change the electronic transition frequency.

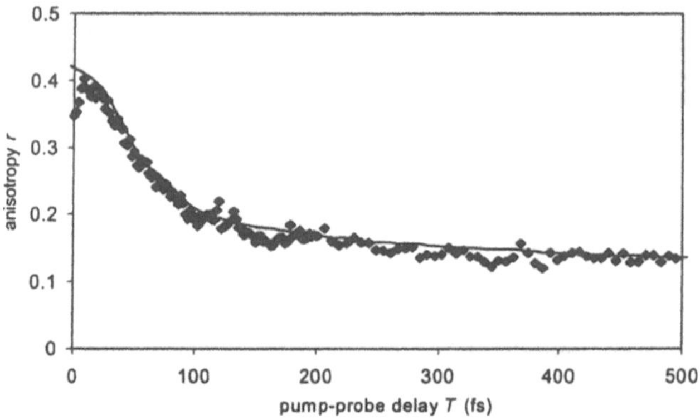

Fig. 1. Experimental pump-probe polarization anisotropy (diamonds) compared to a model calculation including intramolecular vibrations and uniform electronic reorientation (solid).

Transient grating signals recovered 6 underdamped vibrations and an overdamped 3 ps exponential decay which accounted for all but 8 cm^{-1} out of the 58 cm^{-1} reorganization energy. The initial pump-probe anisotropy of 0.40(1) matches the ratio $S_{\parallel} = 3S_{\perp}$ expected for impulsive excitation while the limiting anisotropy of 0.11(1) indicates orientational randomization within the molecular plane (r =1/10). The initially quadratic anisotropy decay near T=10 fs indicates that the nuclear motions which give rise to reorientation have been resolved so the anisotropy can be used to estimate the orientational correlation function $p(t)$ [2]:

$$r(T) = (1/10) + (3/10)p(2T) \qquad (1)$$

Modeling with correlation functions that decay quadratically near t=0 works well outside the pulse overlap region, but vibrations that dephase excited state absorption [to states analogous to (n_x=1, n_y=1) for a particle in a square 2D box] are not properly handled for overlapping pulses and require a new framework[3].

For impulsive excitation, the information content of four-wave mixing signals is maximized by using two Fourier transforms to separate the four-wave mixing signal across two frequency dimensions so that signal at each radiated frequency is dispersed according to excitation frequency. These transformations require measurement of the four-wave mixing signal electric field as a function of the time, t, after the third pulse for many delays, τ, between the first and second pulses. The four wave mixing signal field is detected by high precision Fourier transform spectral interferometry, which separately measures signal delays to 0.020 fs and phase shifts to 0.05 rad. Signal propagation delays and distortions caused by propagation from the sample to the detector are corrected using a tracer

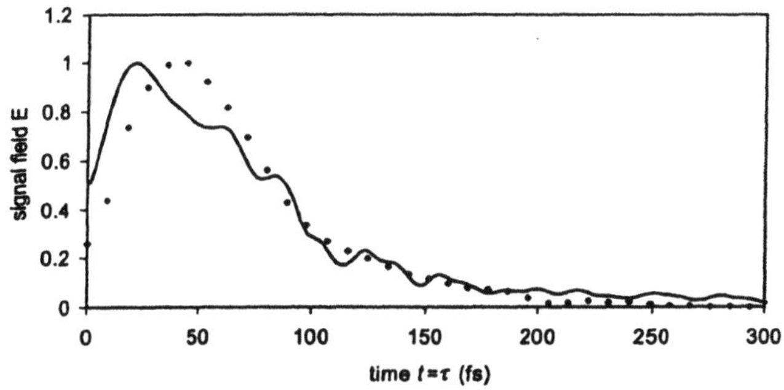

Fig. 2. Experimental four-wave mixing signal fields at the time of photon echo emission (diamonds) compared to a model calculation including both intramolecular vibrations and uniform electronic reorientation (solid).

pulse. The real (absorptive) and imaginary (refractive) 2D spectra are separated by comparing a projection of the 2D spectrum to spectrally resolved pump-probe [4]. The phased 2D FT spectrum was used to calculate the time domain "echo slice" fields at $t=\tau$ (when inhomogeneous broadening is eliminated by a photon echo pulse sequence) shown in Fig. 2.

There are at least three alternatives to explain the electronic reorientation correlation function: a static distribution of Jahn-Teller splittings caused by asymmetric solvent environments; electronic pseudo-rotation (a manifestation of Berry's topological phase); and electronic dephasing. The early part of the calculated echo slice field, dominated by electronic reorientation, rises too quickly, while the long time decay, determined by intramolecular vibrations, is correctly captured. This suggests degenerate electronic dephasing is slower than 100 fs. Electronic inhomogeneity alone can only yield anisotropy decay from 0.4 to 0.25, while adiabatic pseudo-rotation models miss the slow part of the anisotropy decay, so models incorporating both effects may be required to simulate the data.

References

[1] David R. Yarkony, "Diabolical conical intersections," Rev. Mod. Phys. **68**, 985-1013 (1996).
[2] Allison Albrecht Ferro, and David M. Jonas, "Pump-probe polarization anisotropy study of doubly degenerate electronic reorientation in silicon naphthalocyanine," J. Chem. Phys. **115**, 6281-6284 (2001).
[3] Jaeyoung Sung and Robert J. Silbey, "Four-wave mixing spectroscopy for a multi-level system," J. Chem. Phys. **115**, 9266-9287 (2001).
[4] John D. Hybl, Allison Albrecht Ferro, and David M. Jonas, "Two-dimensional Fourier transform electronic spectroscopy," J. Chem. Phys. **115**, 6606-6622 (2001).

Direct Measurement of Vibrational Relaxation and Dephasing of Molecules by Spectrally Resolved Two-Colour Three-Pulse Photon Echoes

Lap Van Dao, Craig Lincoln, Martin Lowe and Peter Hannaford

Centre for Atom Optics and Ultrafast Spectroscopy, School of Biophysical Sciences and Electrical Engineering, Swinburne University of Technology, Hawthorn, Australia 3122
E-mail: dvlap@swin.edu.au

Abstract: Spectrally resolved two-colour three-pulse stimulated photon echoes have been investigated on a femtosecond time scale for Rhodamine B and Rhodamine 101 in Methanol. The time evolution of the spectra of the PE signals and their dependence on the wavelength of the excitation pulses and the structure of the molecules are analysed. The dynamics of the vibrational relaxation and dephasing have been determined directly from the wavelength dependence of the two-colour three-pulse photon echo signals. The influence of rotational motion of the $N(C_2H_5)_2$ group on the echo intensity is discussed.

1. Introduction

The fundamental aim of ultrafast time-resolved spectroscopy is to observe various optical responses of a molecular system, which reflect the dynamics of the molecular interactions in the condensed phase, the collective motions in liquids, the solvent reorganisation motions, and the structural change of a protein in biological systems, in order to construct a physical picture of the underlying dynamical processes. For simple molecules with well-separated eigenstates conventional vibrational spectroscopies such as infrared absorption and Raman scattering give direct information on the vibrational energy levels and their properties. In complex systems the situation is very different; here the detailed information is often buried beneath broad featureless line shapes. Coherent nonlinear spectroscopy [1] such as photon echoes, four-wave mixing and Raman scattering has been used to overcome these difficulties and unravel the microscopic picture underlying complex spectra. Spectrally resolved photon echo signals have been reported for several systems. The spectrum of three-pulse photon echoes at 800 nm has been studied for the dye molecule DTTCI [2] and for GaAs [3]. However, both of these examples have been for one-colour photon echo experiments, in which the molecules can be modelled as two-level systems and thus are not expected to have multiple peaks in the spectrum. Wiersma *et al* [4] have used two-colour photon echoes as a versatile tool for the study of vibrational deactivation processes on a picosecond time scale.

In this paper we explore the use of spectrally resolved stimulated photon echoes (PE) to study the vibrational spectrum and the influence of molecular structures on a femtosecond time scale by considering two dye molecules, Rhodamine B (RhB) and Rhodamine 101 (Rh101). Using a two-colour three-pulse spectrally resolved PE technique we have carried out a detailed study of the dependence of the echo spectrum on the wavelength of the laser excitation pulses. We show that information about the dynamics of the dephasing and the vibrational relaxation can be deduced from spectrally resolved measurements of two-colour three-pulse stimulated photon echoes.

2. Theoretical Background

In a three-pulse PE experiment, three ultrashort pulses are used, two pump pulses with wave vectors $\mathbf{k}_1$ and $\mathbf{k}_2$ and a probe pulse with wavevector $\mathbf{k}_3$. The first pulse $\mathbf{k}_1$ excites the optical transition of the molecule coherently. After a time t_{12} the second pulse $\mathbf{k}_2$ interacts with the freely evolving system, resulting in a phase and population distribution (like a grating) in the ground and excited states. This phase and population grating in frequency space gives rise to a frequency-modulated transmission spectrum of the sample. The spacing of the grating in the ground and excited states of the molecule depends on the separation of the two pump pulses t_{12}, called the coherence time. The third pulse $\mathbf{k}_3$ interacts with this grating resulting in an echo at a time t (near t_{12}) after the third pulse when dephasing processes have occurred. The intramolecular energy and charge transfer and the interaction with other molecules (e.g., solvents) washes out the grating during the time t_{23}, called the population time. The population relaxation time can vary typically from picoseconds to hundreds of milliseconds.

The combination of pump and probe pulses in the sample creates a nonlinear polarization, which in turn radiates an electric field. Any nonlinear optical signal can be expressed as a convolution of the nonlinear electronic polarization $\mathbf{P}^{NL}$ with the temporal distribution of the excitation field. The experimental photon echo data can be analysed by comparing with the calculated signal using the third-order electronic polarization $P^{(3)}$. The third-order electronic polarization that generates the PE signal in the phase-matching direction $\mathbf{k}_4 = \mathbf{k}_3 + \mathbf{k}_2 - \mathbf{k}_1$ is given by [1]

$$P^{(3)}(t, t_{12}, t_{23}) \sim \int_0^\infty dt_3 \int_0^\infty dt_2 \int_0^\infty dt_1\, E_1^*(t+t_{12}-t_3-t_2-t_1)\, E_2(t+t_{23}-t_3-t_2)\, E_3(t-t_3)\, R^{(3)}(t_3, t_2, t_1) \quad (1)$$

where E_1, E_2, E_3 are the electric fields of the three excitation pulses and $R^{(3)}$ is the third-order nonlinear optical response function. For a time-integrated PE measurement the echo signal can be expressed as

$$S_{PE}(t_{12}, t_{23}) = \int_0^\infty |P^{(3)}(t, t_{12}, t_{23})|^2\, dt. \quad (2)$$

It has been pointed out that information about the temporal shape of the nonlinear polarization is lost by the time-integrated measurement [5]. A possible way to obtain detailed information about the nonlinear polarization is to record the PE signal as a function of wavelength using a monochromator. For a delta-function monochromator instrumental function the spectral response of the PE spectrum is given by [2]

$$S_{WPE}(\lambda_D, t_{12}, t_{23}) \sim (2\pi c/\lambda_D)^2\, |P^{(3)}(\lambda_D, t_{12}, t_{23})|^2 \quad (3)$$

where λ_D is the detection wavelength and c the speed of light. The time and wavelength dependence of $P^{(3)}(\lambda_D, t_{12}, t_{23})$ can be sensitive to inhomogeneous broadening and the vibrational energy spectrum because the different spectral regions can decay with different time constants [5].

3. Experimental Arrangement

The ultrafast pulses used in our experiment are generated by conversion of a 1 mJ amplified pulse (80 fs, 800 nm) in two independently tunable optical parametric amplifiers (OPA). The output of one OPA is split into two beams which act as pump pulses ($\mathbf{k}_1$ and $\mathbf{k}_2$) and the output of the second OPA acts as a probe pulse ($\mathbf{k}_3$). The PE signal in the phase-matched direction is analysed in a 0.25 m spectrometer (spectral resolution 1 nm) and detected with a photomultiplier. The wavelength of the pump pulses (FWHM ~ 15 nm) is set to 580, 550 and 520 nm while the wavelength of the probe beam is fixed at 640 nm. For measurements of spectra of the PE signal the monochromator is scanned while the delay times t_{12} (between pulses $\mathbf{k}_1$ and $\mathbf{k}_2$) and t_{23} (between $\mathbf{k}_2$ and $\mathbf{k}_3$) are fixed. For detailed spectral analysis the PE signal is measured as a function of t_{12} and t_{23} by holding the monochromator wavelength fixed and scanning the delay stages. Two laser dyes RhB and Rh101 (Lambda chrome laser dyes - Lambda Physik) at a concentration of 10^{-3} M in Methanol with absorption spectrum maximum at 552 nm and 568 nm, respectively, have been investigated.

4. Results and Discussion

Figure 1 shows the PE spectra taken at different population times t_{23} after excitation at 520 nm for RhB (fig. 1a) and 550 nm for Rh101 (fig. 1b) on the blue

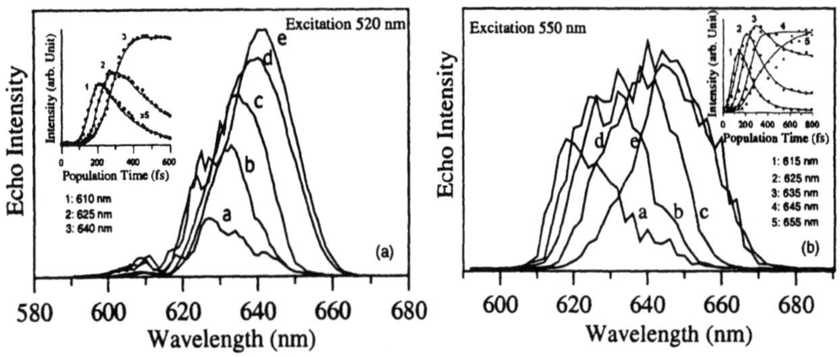

Fig. 1. Spectra of photon echo signals for (a) RhB and (b) Rh101 measured at different population times (t_{23}) (a) 100 fs, (b) 300 fs, (c) 350 fs, (d) 400 fs and (e) 450 fs. The wavelengths of the pump and probe beams were 520 nm and 640 nm for RhB 550 nm and 645 nm for Rh101.

wing of the absorption spectrum. The inset shows the time evolution of the PE spectral intensity in detail. At short population times the PE spectrum displays new bands around 600 nm to 625 nm. The short wavelength part decays rapidly with increasing population time T. For long population times the PE signal at 640 nm (i.e., the central probe wavelength) is dominant.

Components with fast decay time (300 fs) and slow decay time (>10 ps) are observed in the PE spectral intensity, as can be seen in the insets to fig. 1. The ratio of slow to fast decay components increases when the detected wavelength increases. The rise time of the PE intensity at different detection wavelengths for RhB and Rh101 (fig. 1a, 1b) is determined by the relaxation time from the initially excited vibrational level to the corresponding vibrational level in the excited electronic state. We can deduce values for the relaxation times from vibrational levels corresponding to transitions near 610 nm into levels corresponding to transitions near 625 nm to be about 80 fs and into levels corresponding to transitions near 640 nm to be about 200 fs for RhB and about 350 fs for Rh101 and up to about 800 fs for transitions into levels near 650 nm for Rh101.

Using a probe beam with fixed spectral profile and central wavelength about 640 nm, we find a significant dependence of the echo spectrum on excitation wavelength. Figure 2 shows the dependence on the excitation wavelength, where $\Delta\lambda=\lambda-\lambda_{max}$, $\Delta\lambda_0$ is the FWHM of the probe pulse, λ_{max} is the central wavelength of the probe pulse, and λ is the detected wavelength. Decreasing the wavelength of the pump pulse (from 580 nm to 550 nm for Rh101 and from 580 nm to 550 nm and 520 nm for RhB) results in a broadening of the echo spectrum (mostly on the short wavelength side).

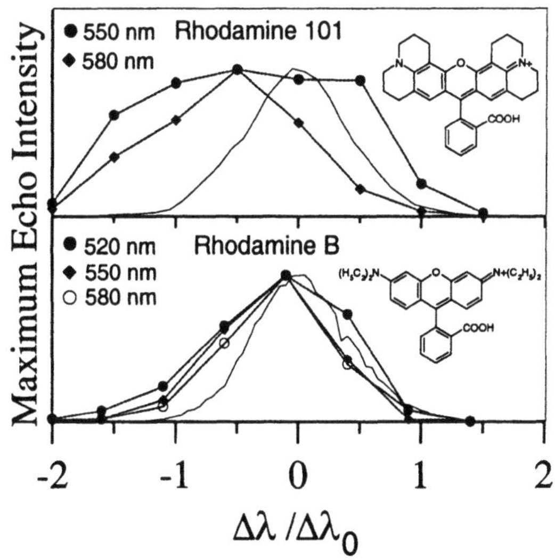

Fig. 2. Maximum echo intensity detected at different wavelengths for different pump wavelengths. The solid line without data points shows the spectrum of the probe pulse.

The new spectral features which differ from the spectral profile of the probe pulse and have a fast decay are associated with the "pure" echo signal. This fast decay time (300 fs) corresponds to the dephasing time of the electronic system S_1 and the slow decay (>10 ps) corresponds to the lifetime of the population grating created by the interaction of the two pump pulses.

With decreasing pump wavelength, more vibronic transitions within the electronic band S_1 are excited which can lead to a difference in the amplitude of the vibrational oscillations [6]. The intensity of the short wavelength part in comparison with the PE intensity at the probe maximum of the RhB sample is much weaker than for Rh101, as shown in fig. 1a and 1b and fig. 2. The structures of both Rh101 and RhB have a rotating Benzonic Acid group. The main difference between the two dyes lies in the substitution of the amino groups; Rh101 has nitrogen atoms rigidly linked to the xanthene skeleton, while RhB has free amine nitrogen with a di-alkyl substitution. Conformation of the N-alkyl group (3-N,N-dialkylimine group perfectly coplanar with the xanthene ring and a slightly twisted 6-N,N-diakylamine group) is considered to be important for the fluorescent and photochemical properties of RhB [7]. The 3-imino, 6-amino-xanthene cation radical is unstable in the ground electronic state unless it acquires a negative charge [8]. The latter is predicted to induce dramatic changes in the amino group such that the C-N bond elongates, the C-N charge separation increases, and the N,N-dialkyl group becomes perpendicular to the xanthene ring. The rotational motion of the $N(C_2H_5)_2$ group in RhB weakens the coupling of the vibrational mode and reduces coherent transfer between the vibrational levels.

5 Conclusions

We have investigated the spectra of two-colour three-pulse photon echo signals for Rhodamine B and Rhodamine 110 in Methanol and find the spectra of the echo signal are strongly dependent on the wavelength of the excitation pulses. Using a spectrally resolved PE technique we have been able to determine the vibrational relaxation times and the optical dephasing time of the molecules. We have shown that the echo spectrum is dependent on the energy-level structure of the molecule. This experimental technique will be used to study dephasing processes and other related chemical and physical processes involving nonequilibrium relaxation of vibrational excited states of molecules.

References

1 S. Mukamel, *Principles of Nonlinear Optical Spectroscopy*, Oxford University Press, New York, 1995.
2 L.D. Book and N.F. Schere, J. Chem. Phys. **111**, 792-795, 1999.
3 M.U. Wehner, D. Steinbach, M. Wegener, Phys. Rev. **B54**, R5211-R5214, 1996.
4 W. Kaiser, *Ultrashort Laser Pulses and Application*, Topics in Applied Physics, Vol 60, Springer-Verlag, Heidelberg, 1988, Ch. 6.
5 M.C. Asplund, M. Lim and R.M. Hochstrasser, Chem. Phys. Lett. **323**, 269-277, 2000.
6 K. Ohta, D.S. Larsen, M. Yang, and G.R. Fleming, J. Chem. Phys. **114**, 8020-8039, 2001.
7 F.P. Schafer, *Dye Lasers*, Springer-Verlag, Berlin, Heidelberg, 1990.
8 P. Ilich, P.K. Mishra, S. Macura and T.P. Burghardt, Spectrochim. Acta Part A: Molecular and Biomolecular Spectroscopy **52**, 1323-1330 (1996).

Correlated Line Broadening in Three Pulse Multidimensional Spectroscopy of Coupled Vibrations

Ravindra Venkatramani and Shaul Mukamel

University of Rochester, Dept of Chemistry and Dept of Physics and Astronomy, Rochester, NY 14627, USA
E-mail: mukamel@chem.rochester.edu
E-mail: ardnivar@pas.rochester.edu

Abstract: We present a theory for the three possible resonant, femtosecond, infrared four wave mixing techniques ($-\mathbf{k}_1+\mathbf{k}_2+\mathbf{k}_3$, $\mathbf{k}_1-\mathbf{k}_2+\mathbf{k}_3$, $\mathbf{k}_1+\mathbf{k}_2-\mathbf{k}_3$) in coupled, localized, vibrational systems which includes broadening effects with arbitrary time scales and degrees of correlation.

1. Introduction

Multidimensional nonlinear techniques provide a valuable tool for probing static and dynamic properties of molecules. Considerable effort has been devoted to identifying spectroscopic signatures of molecular properties such as couplings between energy levels, anharmonicities and dipole orientations [1]. Bath interactions with a multitude of time scales may result in complicated line shapes [2]. In this work we simulate line broadening effects with using a localized vibrational basis. The signals are directly analyzed in terms of the molecular structure and dynamical properties in real space.

2. Model System

We consider a model system of two, coupled local vibrations with a bilinear interaction and a quartic anharmonicity. The system is also linearly coupled to a harmonic bath with an overdamped brownian oscillator spectral density [3]. Only the first three energy levels for each vibration affect the signals and off diagonal couplings to the bath in the exciton (diagonalized) basis are neglected [4]. The second excited state anharmonicities and the couplings in the localized basis are held fixed so that there are two independent fluctuations i.e. the frequencies of the two oscillators. These fluctuations are allowed to have an arbitrary correlation through a parameter η, which takes values between -1 (anti correlated), through 0 (uncorrelated) to 1 (fully correlated). We have used the general expression for the third order response function derived in [2] and applied to electronic excitons in [4]. All possible two dimensional, four wave mixing signals for this system at $\mathbf{k}_I = -\mathbf{k}_1+\mathbf{k}_2+\mathbf{k}_3$, $\mathbf{k}_{II} = \mathbf{k}_1-\mathbf{k}_2+\mathbf{k}_3$ and $\mathbf{k}_{III} = \mathbf{k}_1+\mathbf{k}_2-\mathbf{k}_3$, which survive the rotating wave approximation (RWA) have been calculated. $\mathbf{k}_1$, $\mathbf{k}_2$ and $\mathbf{k}_3$ are the wave vectors of the three time ordered incident fields, separated by time intervals t_1 and t_2 . t_3 is the time interval between $\mathbf{k}_3$ and the signal field.

Our model has two single exciton transitions at 2015 and 2084 cm-1 and three double exciton states at 4015, 4078 and 4153 cm-1. Only transitions from ground

Table 1. All possible 2D resonances corresponding to the intervals (t_1,t_3) along with their peak intensities for each of the three techniques.

Transition Pairs (cm-1, cm-1)	Relative Peak Intensities			Transition Pairs (cm-1, cm-1)	Relative Peak Intensities		
	$\mathbf{k_I}$	$\mathbf{k_{II}}$	$\mathbf{k_{III}}$		$\mathbf{k_I}$	$\mathbf{k_{II}}$	$\mathbf{k_{III}}$
(ω_1, ω_1) (2015,2015)	0.971	1.000	1.000	(ω_2, ω_1) (2084,2015)	0.349	0.157	0.157
(ω_1, ω_2) (2015,2084)	0.343	0.156	0.156	(ω_2, ω_2) (2084,2084)	0.129	0.261	0.261
(ω_1, ϖ_1-ω_1) (2015,2000)	1.000	0.764	0.764	(ω_2, ϖ_1-ω_1) (2084,2000)	0.733	0.101	0.101
(ω_1, ϖ_1-ω_2) (2015,1931)	0.006	0.077	0.078	(ω_2, ϖ_1-ω_2) (2084,1931)	0.099	0.011	0.011
(ω_1, ϖ_2-ω_1) (2015,2063))	0.349	0.147	0.147	(ω_2, ϖ_2-ω_1) (2084,2063))	0.021	0.169	0.169
(ω_1, ϖ_1-ω_2) (2015,1994)	0.121	0.154	0.154	(ω_2, ϖ_1-ω_2) (2084,1994)	0.308	0.136	0.136
(ω_1, ϖ_3-ω_1) (2015, 2138)	0.071	0.015	0.015	(ω_2, ϖ_3-ω_1) (2084, 2138)	0.003	0.074	0.074
(ω_1, ϖ_3-ω_2) (2015, 2069)	0.087	0.037	0.037	(ω_2, ϖ_3-ω_2) (2084, 2069)	0.105	0.142	0.142

to single exciton states and single to double exciton states are allowed. We expect two possible resonances during t_1 and eight during t_3 (Table 1.).

The impulsive, time gated, homodyne signal is given by: $| S(\omega_3,t_2,\omega_1) |^2$ with $S(\omega_3,t_2,\omega_1) = \iint \exp(-i(\omega_1 t_1+\omega_3 t_3))\, R(t_3, t_2, t_1)$. R is the third order non-linear response function. The simulated $\log(| S(\omega_3,t_2{=}0,\omega_1) |^2)$ are presented in Fig 1. The first three rows represent technique $\mathbf{k_I}$. Each column corresponds to a given value of the parameter η and the three rows, represent different values of the parameter κ which measures the nature of broadening ($\kappa \ll 1$ (static bath), $\kappa \gg 1$ (fast bath)). As κ is increased the line shape crosses over from a Gaussian to a Lorentzian profile and all correlation effects begin to disappear. The same calculations are repeated for techniques $\mathbf{k_{II}}$ and $\mathbf{k_{III}}$ in the following two sets of three rows each.

2. Discussion of spectra

Correlation effects between fluctuations in energy levels manifest themselves in the orientation of the cross-peaks. Peaks with correlated fluctuations are oriented along the diagonal and those that are anticorrelated are oriented off-diagonally.

Anticorrelation between static fluctuations reverses the dephasing and rephasing of oscillators in the third time interval. For the photon echo technique $\mathbf{k_I}$ anti-correlated cross peaks are poorly resolved as compared to correlated diagonal peaks. The peak at (2015, 2138), for example, appears in the correlated spectra but is missing in the anticorrelated case. This is because anticorrelation effects between level fluctuations reverse the rephasing of the oscillators. The technique does not give a perfect echo in this case. Uncorrelated fluctuations reverse the rephasing process partly and hence the resolution of cross peaks is better than that in the anticorrelated case. Correlated fluctuations on the other hand lead to perfect rephasing of oscillators and hence the technique shows well resolved peaks. Techniques $\mathbf{k_{II}}$ and $\mathbf{k_{III}}$ are expected to show diffuse, unresolved signals as there is no echo and the signal is supposed to rapidly dephase in time. This is indeed the case in the correlated and uncorrelated spectra. However, if level fluctuations are anticorrelated the signal shows extremely well resolved peaks. Here again, anticorrelated effects reverse the dephasing of oscillators in the third time interval leading to an echo. As a result these techniques show the best resolution of anticorrelated cross peaks.

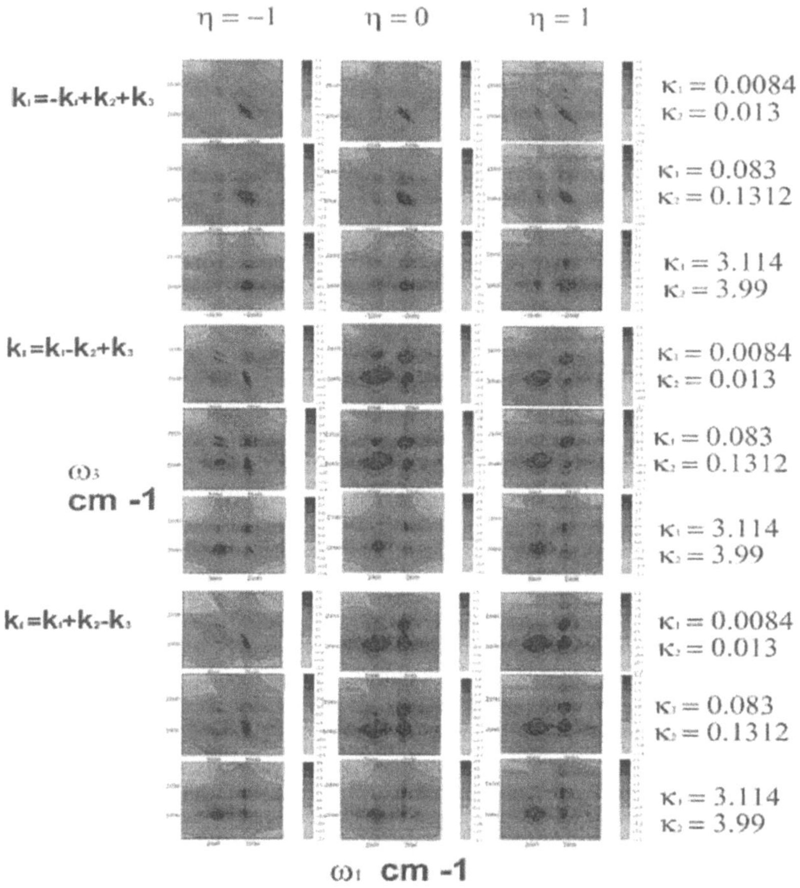

Fig. 1. 2D frequency Domain spectra for the three four wave mixing techniques as indicated. All levels in the localized basis are assumed to have the same inverse lifetime broadening of 1cm-1. The time interval t_2 is set to zero.

Acknowledgements. The support of the National Institutes Of Health (GM59230-01A2) and the National Science Foundation (CHE-0132571) is gratefully acknowledged.

References:

[1] S. Mukamel and R. Hochstrasser (edts), "Multidimensional spectroscopies", Special issue chemical Physics, **266**, 135-368 (2001).
[2] S. Mukamel, "Nonimpact Unified Theory of Four Wave Mixing and Two-Photon Processes",Phys. Rev. A. **28**, 3480-3492 (1983).
[3] S. Mukamel, "Principles of Nonlinear Optical Spectroscopy" (Oxford University Press, New York, 1995)
[4]W. M. Zhang, T. Meier, V. Chernyak, S. Mukamel , " Exciton-migration and three-pulse femtosecond optical spectroscopies of photosynthetic antenna complexes", J. Chem. Phys. **108** , 7763-7774 (1997).

Two Dimensional Analysis of Integrated Three Pulse Photon Echo

Yutaka Nagasawa, Kazushige Seike, Takayuki Muromoto, and Tadashi Okada

Department of Chemistry, Graduate School of Engineering Science, and
Research Center for Materials Science at Extreme Conditions,
Osaka University, 1-3 Machikaneyama, Toyonaka, Osaka 560-8531, Japan
E-mail: nagasawa@chem.es.osaka-u.ac.jp

Abstract. We have carried out two-dimensional measurement of the integrated three-pulse photon echo signal of an organic dye, Nile blue (NB), doped in poly(methyl methacrylate) (PMMA) at 30 K. Rephasing and echo formation was still clear even when the population period, *T*, was extended to >100 ps. Moreover, oscillation was present in the echo signal even when *T* was set to longer values than the dephasing time of the vibrational coherence.

1. Introduction

Recently three-pulse photon echo peak shift (3PEPS) measurement has been used to study solvation dynamics [1,2], energy transfer [3], and electron transfer [4]. However, 3PEPS measurement only monitors the peak position of the echo signal and ignores other valuable information like width and intensity of the echo signal. To extract all the information embraced in the signal, multi-dimensional measurement is necessary. Basically, 3PE is a three dimensional spectroscopy since there are three time periods to consider. During the first time period between the first and second pulses, τ, the system is in a coherent electronic superposition state and during the second time period between the second and third pulses, *T*, the system is in a diagonal population state. The third pulse creates the final coherent superposition state, which leads to rephasing and echo formation. Here we report two-dimensional (2D) measurement of integrated 3PE signal by controlling τ and *T*, and integrating the echo signal over the third period.

2. Experimental Methods

A very stable femtosecond cavity-dumped Kerr lens mode-locked Chromium-doped forsterite laser was developed for the experiment. The repetition rate of the cavity-dumping was 100 kHz and the second harmonic centered at 635 nm was generated in a LBO crystal. The second harmonic was divided into 3 beams and the autocorrelation traces between all three beams were measured. Gaussian pulse shape was assumed and the FWHM of the pulses were 26-28 fs. The sample was a film of NB/PMMA with a thickness of 100-200 μm. The temperature was controlled by a closed-cycle helium gas cryostat and the signals were detected by two pairs of photodiode and lock-in amplifier.

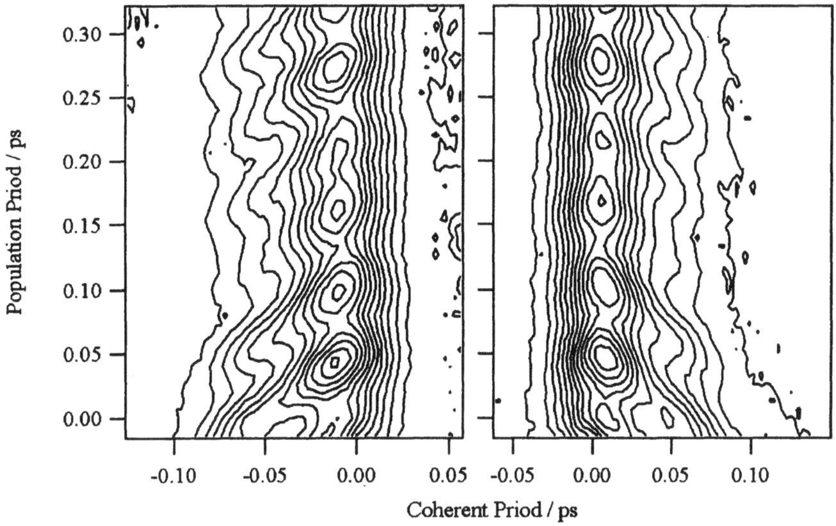

Fig. 1. 2D plots of the integrated three pulse photon echo signals of NB/PMMA at 30 K.

3. Results and Discussion

Figure 1 shows 2D plotted integrated 3PE signal of NB/PMMA film at 30 K. The left side represents the signal appearing in the phase-matching direction of $\boldsymbol{k}_1$ - $\boldsymbol{k}_2$ + $\boldsymbol{k}_3$ and the right side is for - $\boldsymbol{k}_1$ + $\boldsymbol{k}_2$ + $\boldsymbol{k}_3$. The symmetric feature of the two signals assures the accuracy of the measurement. The horizontal axis represents τ and the vertical axis represents T. Therefore, the electronic dephasing time can be attained from the horizontal axis, and the information on excited state lifetime and spectral diffusion dynamics can be attained from the vertical axis. In other words, the signal appearing along the horizontal direction is the echo signal, while the signal appearing at $\tau = 0$ along the vertical axis is the transient grating signal. The modulation of the signal is caused by the coherent intramolecular vibrations of NB.

Figure 2 shows the horizontal cross-section of the echo signal at various values of T. The asymmetric feature of the cross-section shows that rephasing is capable even at large values of T. Note the appearance of a bump or a shoulder in the signal, which is indicated by an arrow.

During the population period, the electronic phase information will be recorded in the interference pattern caused by pulses 1 and 2. Any type of spectral diffusion will lead to a loss of the phase memory and electronic rephasing will be prohibited. In such a case, echo formation will not occur and the horizontal cross-section of the signal will become symmetric. It can be seen from Figure 2 that rephasing and echo formation is still capable even when T was extended more than 100 ps. It can be concluded that spectral diffusion is absent in this timescale at 30 K. Interesting

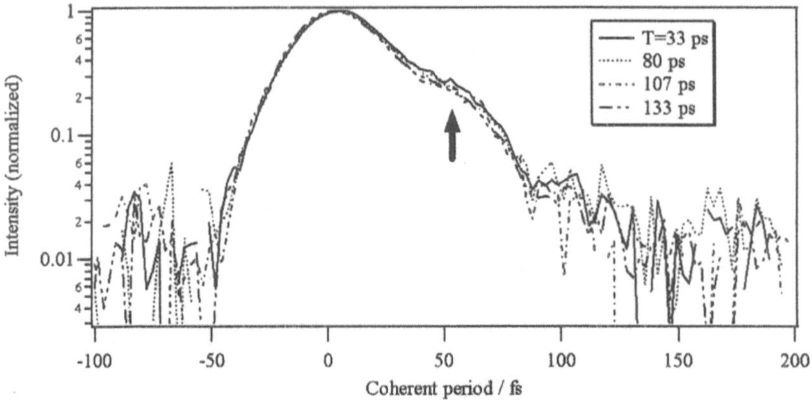

Fig. 2. The photon echo signals of NB/PMMA at 30 K with different values of T.

point is that all the coherent oscillations are not capable of erasing the phase memory. This means that the energy exhausted by the vibrational relaxation is not enough to alter the polymer configuration surrounding the dye molecule.

Even more interesting point is the appearance of the bump in the horizontal cross section of the signals at $T > 20$ ps shown in Figure 2. This bump or a shoulder in the signal must have arisen from vibrational oscillation of the dye molecule. However, it cannot be caused by the impulsive stimulated Raman scattering process, because at $T > 20$ ps, all the vibrational coherence must have been vanished. Usually the dephasing time of the vibrational coherence is less than a few picoseconds. We have concluded that integrated echo signal itself embraces the oscillatory structure. This means that electronic dephasing is not a simple process, which can be described by a single dephasing time. Similar bump can be observed in the previously reported mode-suppressed echo signal [5]. However, the origin of such an oscillation was not seriously discussed because the population period was set shorter than the vibrational dephasing time.

References

1 T. Joo, J. Jia, J.-Y. Yu, M. J. Lang, and G. R. Fleming, *J. Chem. Phys.* **104,** 6089, 1996.

2 W. P. de Boeij, M. S. Pshenichnikov, and D. A. Wiersma, *J. Phys. Chem.* **100,** 11806, 1996.

3 R. Agarwal, B. P. Krueger, G. D. Scholes, M. Yang, J. Yom, L. Mets, and G. R. Fleming, *J. Phys. Chem. B* **104**, 2908, 2000.

4 Q.-H. Xu, S. G. D., M. Yang, and G. R. Fleming, *J. Phys. Chem. A* **103,** 10348, 1999.

5 C. J. Bardeen and C. V. Shank, *Chem. Phys. Lett.* **203** 535, 1993.

Diffractive optics-based heterodyne detected three-pulse photon echo

J. P. Ogilvie, M. L. Cowan, M. Armstrong, A. Nagy, and R. J. D. Miller

Departments of Physics and Chemistry, University of Toronto, Toronto, Canada
Email: dmiller@lphys.chem.utoronto.ca

Abstract: A novel diffractive optics-based method for performing three pulse photon echoes is demonstrated and extends the inherent passive phase stabilization of diffractive optics to this class of experiments for complete recovery of the signal field.

Photon echo spectroscopy has proven to be an essential tool for studying fundamental problems in chemistry. In particular, the dynamics of solvation have been probed by a number of methods, including two and three pulse, time and frequency domain stimulated photon echoes (SPE), and more commonly, the three-pulse photon echo peak shift (3PEPS)[1-5]. Although more difficult to implement, heterodyne methods of detecting the SPE provide the maximum information about the echo signal. Frequency-resolved[2,3] heterodyne detection and actively phase-locked time domain[4] methods have both been demonstrated. Here we develop a novel heterodyne-detected diffractive optics-based method for the three pulse SPE, which we demonstrate with rhodamine6G (R6G) in ethanol. The methodology is similar to the spectral interferometry[6] approach developed by Jonas and co-workers[3], with the added benefit of passive phase-locking provided by the diffractive optic[7,8].

The laser source for the three pulse SPE experiment is a noncollinear OPA pumped by 775 nm, 350 μJ, 120 fs pulses at 1 kHz from a diode -pumped solid state chirped-pulse amplification system with a doubled Er-fiber oscillator. The output from the NOPA is compressed with a combination of chirped mirrors and a deformable mirror[9] to yield ~7 fs, 5 μJ pulses centred at 525 nm with ~200 nm bandwidth. These pulses are split at a 50% beamsplitter (BS) to give the pump and probe pulses which are focussed at a small angle by M1 (f=15cm) onto a diffractive optic element (NOI) designed for 30% diffraction into the first order beams. These four diffracted beams yield the boxcar geometry, which is imaged 1:1 onto the sample by two parabolic mirrors M2, M3 (f=17.5") used in an off-axis configuration. A mask blocks the unwanted beams and lenses collimate the signal and focus it into a spectrometer. To keep the advantage of passive phase-locking given by the DO, the τ delay was implemented using transmissive optics, through rotation of two ~1-mm fused silica flats (C) in the second pump (2) and the reference (r) beams. The flats are aligned so as to compensate for beam shifts due to refraction. Identical flats C_{13} are placed in beam 1 and the probe (p) for timing and dispersion compensation. The reference intensity is reduced using an ND filter, with compensating glass (C_3) placed in beam 3 ensuring that the weak reference precedes the pump pulses in time. The T delay is obtained via a delay stage as shown. The pulses are 20 fs in duration at the sample.

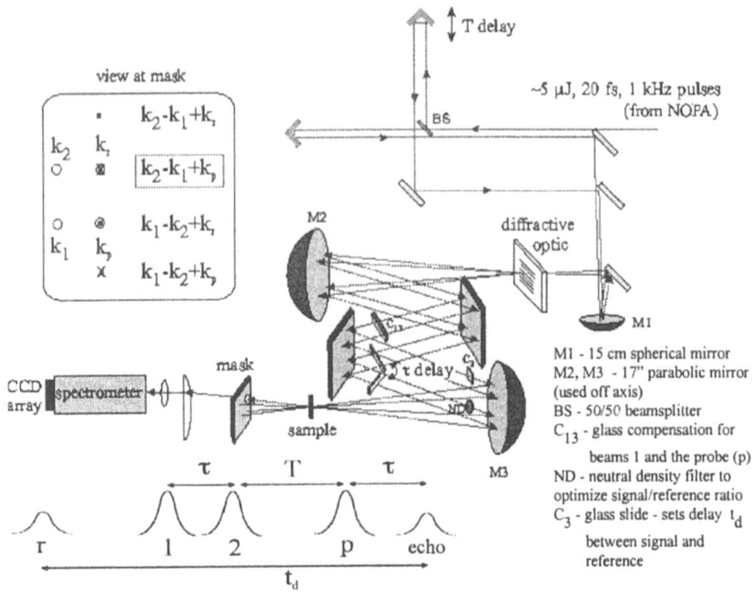

Figure 1: Experimental setup for the diffractive optics-based heterodyne detected 3 pulse echo. Echo signals generated by beam 3 are shown as , while signals from beam 4 are shown as X.

In order to fully determine the real and imaginary parts of the sample response, we measured spectral interferograms of the signal and reference fields, as well as of the 1-2 and p-r beam pairs. The diffractive optic provides automatic generation of the reference beam r for spectral interferometry with the signal, and determination of the phase difference between the signal and reference ($\Delta\phi_{sr}$). In order to measure spectral interferograms of the other two beam pairs a second diffractive optic was inserted in place of the sample, thereby recombining the beam pairs. This allowed us to measure the phase differences between beams 2 and 1 ($\Delta\phi_{21}$), and r and p ($\Delta\phi_{rp}$), to within a constant factor imprinted by the second diffractive optic (ϕ_{DO}). The phase of the signal response is then given by

$$\varphi_s(\omega) = \Delta\phi_{sr}(\omega) - \Delta\phi_{21}(\omega) + \Delta\phi_{rp}(\omega) - \pi/2. \tag{1}$$

Note that the constant phase ϕ_{DO} cancels, allowing us to determine the absolute phase of the sample response.

To demonstrate the feasibility of the method, a preliminary heterodyne detected photon echo spectrogram from R6G is shown in Fig. 2. We found phase stability of λ/90 over >20 minutes, demonstrating the excellent passive phase-locking provided by the diffractive optic. Delays of up to $\tau = 70$ fs have been achieved, with minimal phase distortion of the pulses, as measured using spectral interferometry with the second diffractive optic in place. Note that since the heterodyne detected phase depends on the difference between the reference and beam 2, if both beams strike the same reflective optic the resulting phase fluctuations should largely cancel. Thus the τ delay can be implemented using retroreflectors, while maintaining good phase stability.

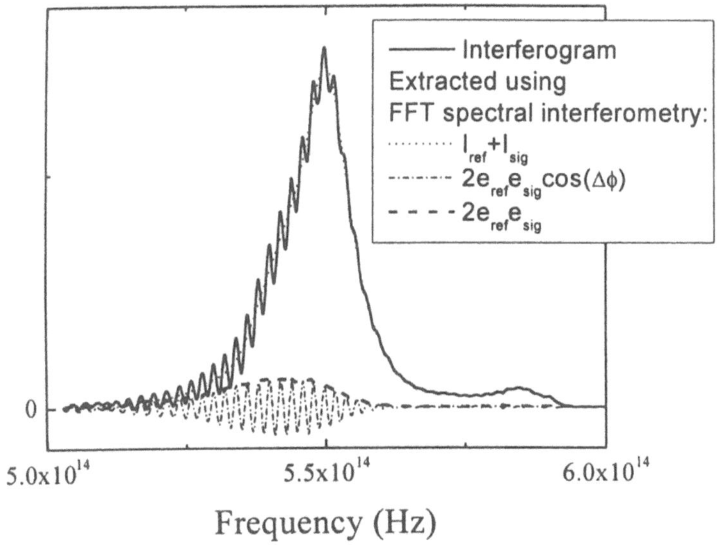

Figure 2: Heterodyne signal of rhodamine 6G in ethanol detected by spectral interferometry.

The great benefit of this method is the passive phase-locking provided by the diffractive optic[7,8] and accompanying information content in the real and imaginary components of the signal[1-4]. This method will allow full characterisation of photon echo responses in future studies of solvation processes and dynamics of heme proteins.

References

1. W.P. de Boeij *et al.*, Annu. Rev. Phys. Chem. **49,** 99-123 (1998).
2. M. F. Emde *et al.*, Opt. Lett. **22**, 1338-1340 (1997).
3. S. M. Gallagher *et al.*, J. Opt. Soc. Am. B **15,** 2338-2345 (1998).
4. W. P. de Boeij *et al.*, Chem. Phys. Lett. **238**, 1-8 (1995).
5. Q. Xu *et al.*, J. Phys. Chem. A **103**, 10348-10358, (1999).
6. L. Lepetit *et al.*, J. Opt. Soc. Am. B **12**, 2467-2474 (1995).
7. G. D. Goodno *et al.* J. Opt. Soc. Am. B **15**, 1791-1794 (1998).
8. J. A. Rogers *et al.* Physica B **219-220**, 562-564 (1996).
9. M. R. Armstrong *et al.*, Opt. Lett. **26**, 1152-1154 (2001).
10. A. A. Maznev *et al.*, Opt. Lett. **23**, 1378-1380 (1998).
11. G. Dadusc *et al.*, Isr. J. Chem. **38**, 191-206 (1998).

Structural Dynamics in Water Probed by Heterodyne-Detected Photon Echo

Sergey Yeremenko, Maxim S. Pshenichnikov, and Douwe A. Wiersma

Ultrafast Laser and Spectroscopy Laboratory, Materials Science Centre
University of Groningen, Nijenborgh 4, 9747 AG Groningen,
The Netherlands
E-mail: M.S.Pshenichnikov@chem.rug.nl

Abstract. Results of heterodyne-detected photon echo experiments on the OH stretching mode of HDO molecule in heavy water and acetonitrile are reported and discussed. Two vibrational dynamical processes with time constants of 130 fs and 900 fs were identified. The former is attributed to bond breaking dynamics of a single hydrogen bond, the latter to rearrangement of the hydrogen-bond network.

Despite the fact that structure and dynamics of water has been a subject of extensive research, there is still much to be learned [1]. In the liquid phase the structure of water is dynamical and changes on ultrafast time scales. Therefore, comprehension of liquid water dynamics is a key to understanding its structure.
Recent advances in infrared femtosecond technology enabled application of nonlinear optical methods to study vibrational dynamics [1-5]. Among them photon echo spectroscopy has proven itself as a powerful technique to investigate the liquid-state dynamics. It readily allows extracting characteristics of fast processes hidden in the background of slower ones.

In this contribution we report on heterodyne-detected photon echo experiments on the OH stretching mode of HDO molecules, diluted in heavy water. Combining extremely short 70-fs excitation pulses with heterodyne detection of the transient nonlinear polarization enabled us to obtain a detailed picture of the OH-vibrational dynamics.

IR pulses of 70 fs at a central wavelength of ~3 μm with ≥10 μJ energy are generated at a 1-kHz repetition rate in a home-built 3-stage optical parametrical amplifier (OPA). The OPA output is split into three parts. Two beams are focused into the sample to generate the photon echo signal at the phase-matched direction $\mathbf{2k_2\text{-}k_1}$. The third beam is aligned collinearly with the echo signal and serves as a local oscillator (LO) to heterodyne the photon echo.

A contour plot of the measured time-resolved photon echo signal is presented in Fig.1 as grayscale. This plot is generated as follows: for a fixed delay t_{12} between the excitation pulses, the delay of the LO is scanned and the heterodyned echo field recorded by a detector. The resulting interference pattern is then numerically processed using a Fourier bandpass filter to extract the echo transient profile.

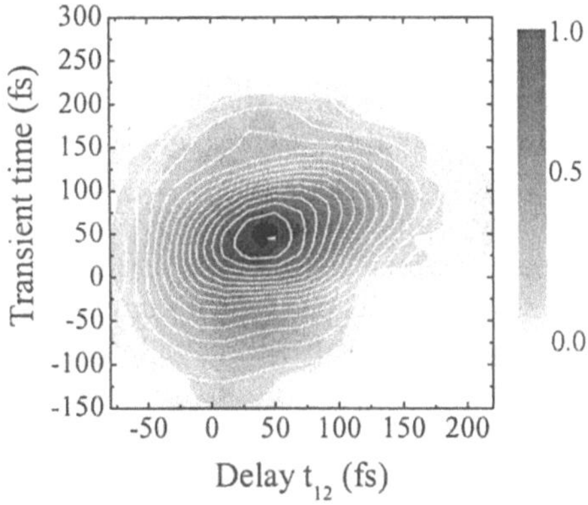

Fig.1. The experimental (grayscale) and simulated (white solid lines) time-resolved two-pulse photon echo profiles for HDO in D_2O.

The experiments are well described in the framework of the stochastic model of frequency fluctuations by a biexponential correlation function that exhibits a fast decay of 130 fs and a slower one of ~900 ps. The amplitudes of these two components are approximately equal and make 130 cm^{-1} and 115 cm^{-1} respectively. The simulated time-resolved photon echo signal is presented in Fig.1 as white contour lines.

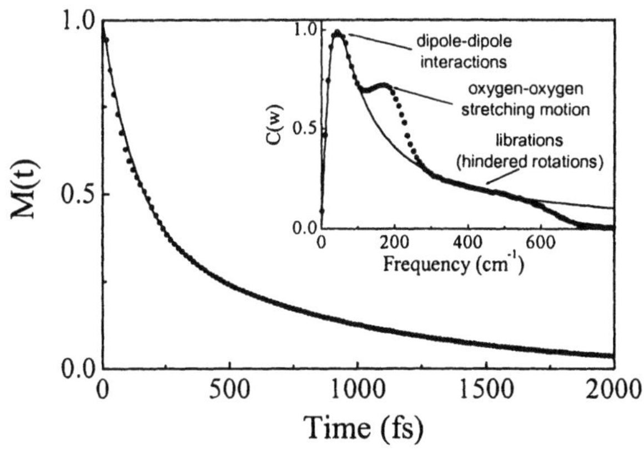

Fig.2. The model correlation function (solid line) and with the fast part obtained by Fourier-transformation of the spectral density (solid dots). The inset shows spectral densities of anisotropic Raman polarizability (solid dots) and of the fast part of the correlation function (solid line).

The corresponding correlation function is shown in the Fig. 2 as a solid line. The slow part of the correlation function has been observed before experimentally [3, 4] and in molecular dynamics simulations of water [6]. There is a general consensus that this time scale is related to breaking and making of hydrogen bonds and that it therefore reflects the mean lifetime of a hydrogen bond [5].

The fast part of the dynamics, however, has never been directly observed before. To elucidate the molecular dynamics at this time scale we calculated the spectral density generated by the experimental correlation function. The inset in the figure 2 depicts this spectral density as a solid line, which, as the figure shows, strongly resembles the Raman spectrum of heavy water (solid circles). Its spectral components have been identified as librations, collective oscillations due to the hydrogen-bond network, and intermolecular collisions of water molecules. Unlike any other probe species, the HDO molecule is chemically identical to bulk D_2O and, therefore, the OH-stretch vibration is expected to be fully involved in *all* molecular interactions. When the Raman spectral density of heavy water is used to calculate the fast part of the correlation function, while keeping the slow part unchanged we find a correlation function (Fig.2, solid circles) that is practically indistinguishable from the one derived from the experimental data.

Thus, the following picture of liquid water dynamics emerges from our study. The correlation time of ~130 fs corresponds to the time scale at which a single hydrogen bond of a four bonds connected molecule is disturbed by the local environment. A frequency excursion of 130 cm^{-1} associated with this time scale, is sufficient to break this bond with subsequent new bond formation or restoration the old one. Therefore, although there are local breakdowns of the hydrogen bond network at a 130-fs time scale, its global backbone remains intact. It takes almost 1 ps before the backbone loses its integrity as a consequence of several hydrogen bonds to have been broken, and a new network structure is formed.

References

1 For recent publications, see, e.g., *Liquid Water and Aqueous Solutions,* Chem. Phys. **258**, edited by. R.M. Hochstrasser, G.L. Hofacker, H.P. Trommsdorff (special issue), (2000).

2 J. Stenger, D. Madsen, P. Hamm, E. T.J. Nibbering, T. Elsaesser, Phys. Rev. Lett. **87**, 027401-1-4 (2001).

3 G. Gallot, G. M. Gale, F. Hache, N. Lascoux, S. Bratos, J-Cl. Leicknam, Phys. Rev. Lett. **82**, 1068 (1999).

4 H. J. Bakker, S. Woutersen, Phys. Rev. Lett **83**, 2077 (1999).

5 S. Woutersen, H. J. Bakker, H.-K. Nienhuys, Chem. Phys. **258**, 233 (2000).

6 F. Sciortino, S. L. Fornili, J. Chem. Phys. **90**, 2786 (1989).

A mid-infrared photon echo study of liquid water

J. Stenger[1], D. Madsen[1,2], P. Hamm[1,3], E. T. J. Nibbering[1], and T. Elsaesser[1]

[1] Max Born Institut für Nichtlineare Optik und Kurzzeitspektroskopie, Max Born Strasse 2A, D-12489 Berlin, Germany
[2] Current address: Department of Chemistry, Aarhus University, Langelandsgade 140, 8000 Aarhus C, Denmark
[3] New address: Physikalisch Chemisches Institut, Universität Zürich, Winterthurerstr. 190, 8057 Zürich, Switzerland
E-mail: stenger@mbi-berlin.de

Abstract. We present the first mid-infrared photon echo peak shift study of HOD/D_2O, demonstrating that spectral diffusion occurs on many time scales. The memory in the fluctuations of the local structure of the hydrogen bond network is relatively long-lived (5-15 ps).

Information on the structural fluctuations of the hydrogen bond network in water, that occur on ultrashort time scales, is a prerequisite for the understanding of its physical and chemical properties. It is well known that the O-H stretching vibration probes the local microscopic structure because its frequency shift due to hydrogen bonding scales roughly linearly with the OH...O distance [1]. Until now, ultrafast infrared spectroscopy of liquid water has concentrated on pump-probe studies, that probe the incoherent population dynamics of vibrational states. Transient hole burning experiments with 150 fs time resolution have revealed a spectral diffusion process with a time scale of 700 fs [2]. On the other hand, different structural components have been proposed based on hole burning results obtained with 1 ps pulses [3] and anisotropy measurements [4]. Furthermore, ultrafast energy redistribution of vibrational excitation towards the bending mode has been shown with infrared pump/Raman probe measurements [5].

Photon echo spectroscopy in the mid-infrared has the potential to give a complete picture of the dynamics of the hydrogen bond network in liquid water. It enables one to decipher the dephasing and spectral diffusion dynamics of vibrational transitions, giving direct access to the transition frequency fluctuation correlation function and hence to information on structural dynamics. We present the first photon echo study of the prototypical hydrogen bonded network, liquid water, by studying the O-H stretching vibration of HOD in deuterated water. We performed both 2-pulse [6] and 3-pulse photon echo measurements including transient grating and photon echo peak shift (3PEPS) experiments which provide detailed insight into the dynamics on time scales ranging from sub-100 fs up to several picoseconds.

Intense femtosecond mid-infrared laser pulses are generated by optical parametric amplification of a white light continuum and used in a three pulse echo set-up. The time separation between pulse 1 and 2 is called coherence time τ, and the temporal delay between pulse 2 and 3 is called population time T. We performed transient grating (TG, τ=0) and 3PEPS measurements at three excitation frequencies ν_{exc} at the red edge, the maximum and blue edge of the

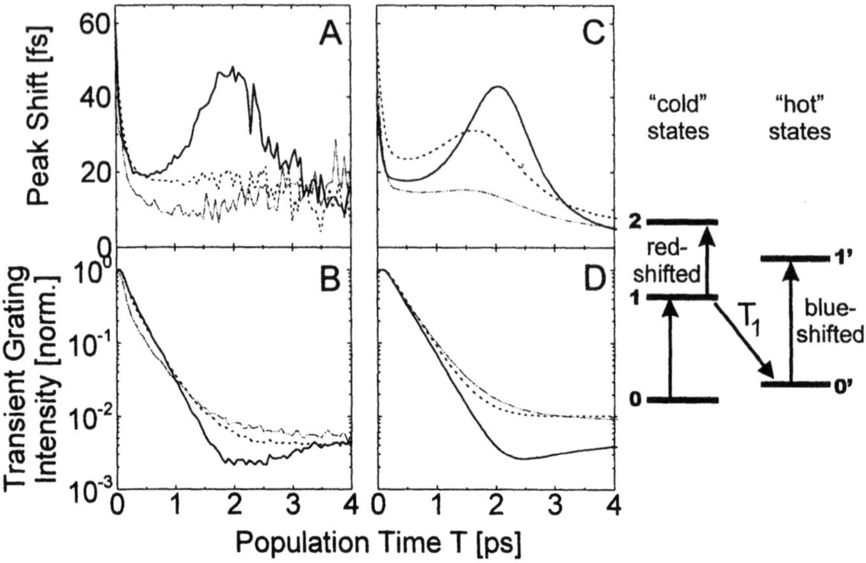

Fig. 1. Experimental data (A and B) and numerical simulation using Eqn.1 (C and D) of the three pulse echo experiment. In A and C the peak shift is shown for excitation frequencies 3300 cm^{-1} (solid), 3400 cm^{-1} (dash), and 3500 cm^{-1} (dots). The transient grating (τ = 0) is illustrated in B and D with the same symbols. On the right hand side the energy level scheme of the hydrogen bonded OH stretching transition is shown. T_1 relaxation leads the molecule into a vibrationally 'hot' state.

absorption band of the O-H stretching vibration of HOD. The experimental results are shown in the Fig.1 A and B. The three normalized TG traces (Fig. 1 B) show for short times T a decay behaviour close to exponential, that is due to population relaxation. The extracted T_1-value of 600 fs is somewhat faster than the T_1-value reported in [2], because the parallel polarization of the three beams makes the signal sensitive to reorientational dynamics. At longer delay times T the TG signal levels off for ν_{exc} = 3500 and 3400 cm^{-1} or has a local minimum before it reaches a constant offset (ν_{exc} = 3300 cm^{-1}). This unexpected absence of a complete decay of the TG signal is accompanied by an unusual behaviour of the peak shift.

After a common decay from an initial value of 50 fs the 3PEPS signals (Fig.1 A) exhibit a strong dependence on ν_{exc} for population times T larger than 500 fs. In the center and blue wing of the O-H band the 3PEPS signals remain constant at a value of about 10-20 fs. However, on the red side of the band the peak shift rises again at T ~ 1 ps, to reach a maximum value of 50 fs at T = 2 ps, followed by a decay to 10 fs at T = 4 ps.The observed experimental results can only be explained when taking into account an extended level scheme consisting of the v=0, v=1 and v=2 states of the O-H stretching transition, as well as the v'=0 and v'=1 states of a "hot" molecule that is reached after T_1-relaxation of the v=1 state. The v'=0 → v'=1 transition is blue-shifted with respect to the fundamental v=0 → v=1 transition, because the hydrogen bond is weakened after T_1-relaxation [7]. The complex interplay of these five states in the context of different frequency fluctuation correlation functions can be understood by comparing the experimental results with systematic model calculations [8]. The behavior of 3PEPS and TG data can only be modeled by including a relatively long lived contribution in the correlation function on the order of 5-15 ps. Simulated signals for a frequency correlation function of the form:

$$\left\langle \delta\omega_{10}(t)\delta\omega_{10}(0)\right\rangle = \Delta_1^2 e^{-t/\tau_1} + \Delta_2^2 e^{-t/\tau_2} + \delta(t)/T_2 \qquad (1)$$

with parameter values T_2 = 90 fs, τ_1 = 700 fs, Δ_1 = 54 cm^{-1}, τ_2 = 15 ps, Δ_2 = 38 cm^{-1} are shown in Fig. 1 C and D. The slow component indicates that there are long-lived local geometries in the hydrogen bond structure of water that exist in the picosecond regime. On the other hand, we have shown that the 2-pulse echo (T=0, [6]) reveals an ultrafast decay of the coherent polarization in the sub-100 fs domain. This ultrafast phase relaxation can be understood by taking into account the unusually large anharmonicity of the OH stretching mode that makes the oscillator more attackable to solvent fluctuations. Comparison with molecular dynamics simulations [9,10] showing 50 and 800 fs components suggest the same underlying mechanisms might apply, where the 50 fs component might be responsible for the ultrafast coherence 2-pulse echo decay. The 800 fs MD- and 700 fs 3PEPS- components seem to be close to the hydrogen bond lifetime.

In summary, we have performed the first mid-infrared photon echo study of the O-H stretching vibration of HOD in D_2O. The dynamics take place on many different time scales including a decay of phase coherence on a sub-100 fs time scale and spectral diffusion in the 5-15 ps regime.

References

1 D. Hadži, and S. Bratos, in *The Hydrogen Bond,* Edited by P. Schuster, G. Zundel, and C. Sandorfy, North Holland, Amsterdam, Vol. II, 565, 1976.
2 G. M. Gale, G. Gallot, F. Hache, and N. Lascoux, Phys. Rev. Lett. **82**, 1068, 1999.
3 R. Laenen, C. Rauscher, and A. Laubereau, Phys. Rev. Lett. **80**, 2622, 1998.
4 S. Woutersen, U. Emmerichs, and H. J. Bakker, Science **278,** 658, 1997.
5 J. C. Deak, S. T. Rhea, L. K. Iwaki, and D. D. Dlott, J. Phys. Chem. A **104**, 4866, 2000.
6 J. Stenger, D. Madsen, P. Hamm, E. T. J. Nibbering, and T. Elsaesser, Phys. Rev. Lett. **87**, 027401, 2001.
7 A. J. Lock, S. Woutersen, and H. J. Bakker, J. Phys. Chem. A **105**, 1238, 2001.
8 J. Stenger, D. Madsen, P. Hamm, E. T. J. Nibbering, and T. Elsaesser, J. Phys. Chem. A **106**, 2341, 2002.
9 M. Diraison, Y. Guissani, J.-Cl. Leicknam, and S. Bratos, Chem. Phys Lett. **61**, 5211, 1996.
10 A. Luzar, and D. Chandler, Nature **379**, 55, 1996.

Two-dimensional visible-infrared Fourier transform spectroscopy

Nadia Belabas and Manuel Joffre

Laboratoire d'Optique et Biosciences INSERM U451 - CNRS UMR 7645-
Ecole Polytechnique-ENSTA
91128 PALAISEAU CEDEX France
E-mail: nadia.belabas@polytechnique.org

Abstract. We demonstrate here two-dimensional spectroscopy of a phase-matched sample of known nonlinear response by means of coherent measurement of the emitted infrared field after a visible double-pulse excitation.

1. Principle

Multiple pulse nonlinear spectroscopy is the optical analog of nonlinear nuclear magnetic resonance. This technique serves to disentangle overlapping contributions to an emission by spreading it along one or more additionnal axes and thus characterize the emitting sample. The complex emission arising from a nth order non-linearity depends on the excitation and on the nonlinear response of the sample $\Xi^{(n)}$ ($\omega_{emission}$; $\pm\omega_2$, ..., $\pm\omega_{n+1}$). A coherent measurement of this field as a function of n independent variables will allow the determination of the desired $\Xi^{(n)}$. The retrieved multidimensional map (or its projection if less than n variables are used) of the nonlinear response for either noncentrosymmetric [1] or centrosymmetric media [2-5] gives precious insight on couplings between energy levels.

We present here a low-power two-dimensional (2D) Fourier transform visible excitation/infrared emitted field measurement. In contrast with infrared [3-5] or visible [1,2] experiments, in our visible-infrared configuration both electronic and vibrational couplings can be adressed (see Fig. 1). This technique could be useful to understand a molecular system such as a protein complex where electronic (visible) excitation of an optically active cofactor sets vibrational dipoles in a coherent oscillation [6]. It might also prove especially enlightening when studying quantum wells where the one dimension of visible frequencies is related to the interband transitions (hence to the wavevector) whereas emission due to intersubband transitions is along the other axis in the infrared spectral domain.
Figure 2 illustrates this principle in the experimental case studied here : the experimental signal $S(\tau,T)$ generated by difference frequency mixing (n=2) is a function of two independent temporal variables. Its 2D Fourier transform yields the 2D spectral map, $\Xi^{(2)}$ (ω_1 ; ω_2, $-\omega_3$).

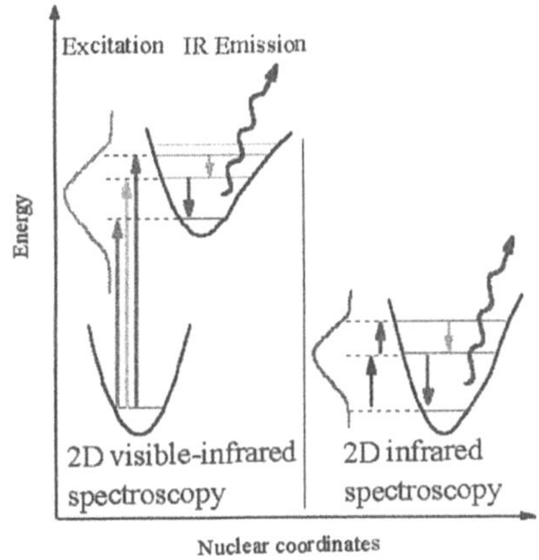

Fig. 1. Principle of 2D visible-infrared (electronic-vibrational) and infrared spectroscopy (vibrational).

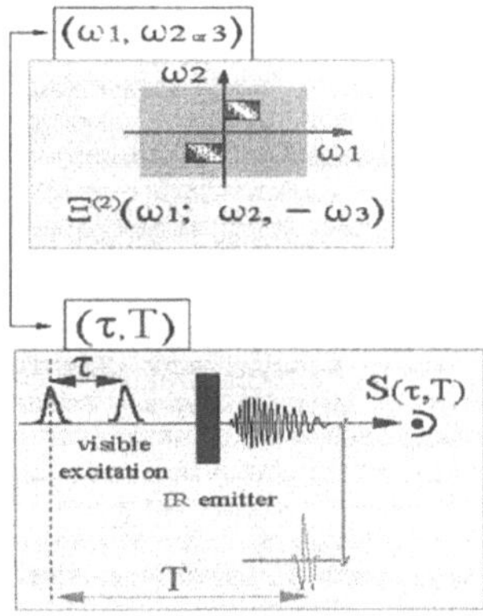

Fig. 2. Principle of 2D spectroscopy in the case of difference frequency generation $\omega_{emission} = \omega_1 = \omega_3 - \omega_2$.

2. Experimental Results

In practice, the infrared emitter is excited by a sequence of two 16-fs 0.2-nJ 800-nm pulses from a Ti:Sapphire oscillator. The time delay τ between the two 'visible' pulses is the Fourier conjugate of the visible spectral axis $\omega_{2\ or\ 3}$ of $\Xi^{(2)}$. The emitted field is measured in amplitude and phase by time-domain interferometry. The time delay T between the measured field and a broadband infrared reference pulse [7] transforms into the infrared spectral variable ω_1 of the 2D map. It must be noted that the validity of the $\Xi^{(2)}$ map yielded by a double Fourier transform of the interferometric signal $S(\tau,T)$ depends strongly on the accuracy of the time axes τ and T. The time delay τ, actively stabilized, is thus scanned using the time-domain interferences of a laser diode propagating along the interferometer allowing accurate and optimal sampling of the data.

As a test of the technique and of the experimental setup, we have measured the nonlinear response $\Xi^{(2)}(\omega_1;\omega_2,-\omega_3)$ of a birefringent crystal ($AgGaS_2$). The two visible exciting pulses are polarised each along one of the crystal axis so that the infrared emission results only from the cross term we are seeking and not from each individual pulse. The resulting map shown on Fig. 3, dominated by phase-matching, can be simulated with no adjustable parameter and depends on the crystal thickness and on the type of the nonlinear process involved. The good agreement between theory and experiment demonstrates the validity of the technique.

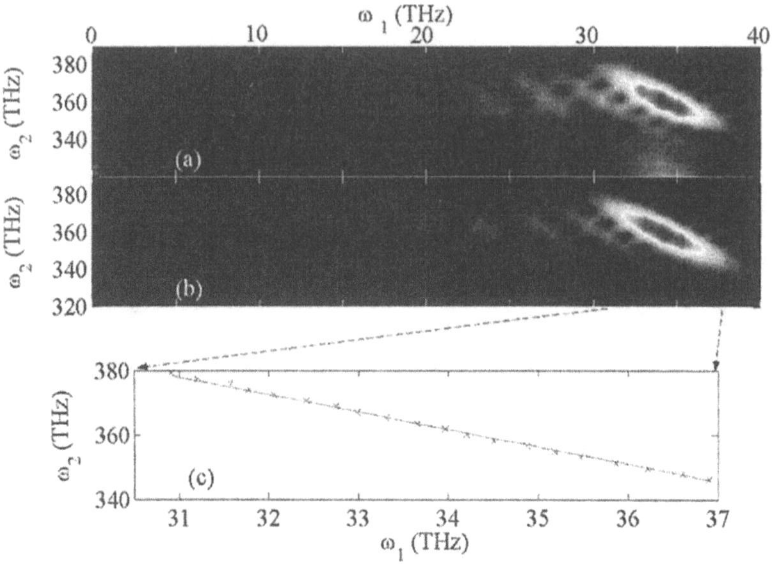

Fig. 3. . Top : Double Fourier transform of S(T,τ) emitted by a type I (ϕ=θ=45°) 500 µm thick $AgGaS_2$ crystal. Middle : Simulation. Bottom : Experimental maxima and theoretical slope (difference<2.6%).

References

1 L. Lepetit et al., "Two-dimensional nonlinear optics using Fourier-transform spectral interferometry", Opt. Lett. 21, 564 (1996)

2 J. D. Hybl et al., "Two-dimensional Fourier transform electronic spectroscopy", J. Chem. Phys. 115, 6606 (2001)

3 M. C. Asplund et al., "Two-dimensional infrared spectroscpy of paptide by phase-controlled femtosecond vibrational photon echoes", Proc. Natl. Acad. Sci. U.S.A. 97, 8219 (2000)

4 A. Tokmakoff et al., "Vibrational anharmonicities revealed by coherent two-dimensional infrared spectroscopy", Phys. Rev. Lett. 86, 2154 (2001)

5 A. Tokmakoff et al., "Two-dimensional Raman Spectroscopy of vibrational interactions in liquids", Phys. Rev. Lett. 79, 2702 (1997)

6 M-L Groot et al., "Coherent infrared emission from myoglobin crystals. an electric field measurement" Proc. Natl. Acad. Sci. U.S.A., in press.

7 A. Bonvalet et al., "Femtosecond infrared emission resulting from coherent charge oscillations in quantum wells", Phys. Rev. Lett. 76, 4392-4395 (1996).

Coherent Two-Dimensional Fourier Transform Infrared Spectroscopy

M. Khalil, N. Demirdöven and A. Tokmakoff

Department of Chemistry, Massachusetts Institute of Technology, Cambridge, MA 02139, USA
E-mail: tokmakof@mit.edu

Abstract. Absorptive two-dimensional (2D) infrared vibrational correlation spectra are obtained by adding together rephasing and non-rephasing 2D spectra. The modulation of cross-peaks and emergence of relaxation-induced peaks are demonstrated as a function of the waiting period.

Two-dimensional (2D) infrared (IR) vibrational spectroscopy has recently been employed to study the structure and correlated dynamics of strongly coupled C≡O vibrations in metal dicarbonyl compounds and to determine the structure of small peptides in solution [1-5]. Femtosecond heterodyne-detected third-order mid-infrared responses obtained as a function of an initial evolution period and a final detection period are Fourier transformed to give a 2D IR spectrum. Information on transient and evolving molecular structures is encoded in the position, amplitude, and shape of cross peaks in the 2D IR spectrum. From a practical perspective, the use of this method requires the ability to obtain intuitive and physically relevant spectra with absorptive lineshapes. This study demonstrates a method where two experiments sampling conjugate frequencies in the initial time are performed in a fixed phase-matched direction by exchanging the sequence of the first two pulses along the incident wave-vectors. The sum of the heterodyne-detected 2D spectra resulting from these two experiments with different pulse orderings gives a 2D IR correlation spectrum of dicarbonylacetylacetonato rhodium(I) (RDC) dissolved in hexane with absorptive lineshapes. The method is then used to explore the effects of relaxation on the various features in the 2D vibrational spectra by introducing a variable waiting period.

2D FT IR spectroscopy uses three field-matter interactions with the three input fields $E_a(k_a)$, $E_b(k_b)$ and $E_c(k_c)$ arranged in a box-car geometry to generate a third-order non-linear polarization in the phase-matched direction, $k_s = -k_a + k_b + k_c$. Two different signals are obtained at k_s which differ only in the time ordering of the k_a and k_b pulses: $S_I = -k_1 + k_2 + k_3$ (rephasing) and $S_{II} = + k_1 - k_2 + k_3$ (non-rephasing). This technique is experimentally realized using a five-beam IR interferometer with 90 fs pulses centered at 4.9 μm. The incoming IR beam is split into 4 equal parts, three of which are used as input beams while the fourth beam is further split into a tracer and a local oscillator (LO). The tracer beam is sent along the path of the signal and is blocked during data collection. After the sample, the signal beam is spatially and temporally overlapped with the LO and dispersed in a monochrometer onto a 64 channel MCT detector. Rephasing and non-rephasing heterodyned signals are collected as a function of τ_1 (delay between the first two pulses) for a particular waiting period, τ_2 (delay between the second and third pulses) by moving retroreflectors mounted on stepper-controlled translation stages. To determine the timing between pulses a and b to within $\lambda/10$, the pulses are recombined after the sample and dispersed in a second monochrometer. Interference fringes collected as the timing between the two pulses is varied are used to calibrate the τ_1 axis. Fourier transforming along the corrected τ_1 axis leads to symmetric features in the resulting 2D spectra.

It is known from NMR, that collecting only the rephasing or the non-rephasing signals yields lineshapes that have mixed-phase character resulting in both absorptive and dispersive features in the 2D spectrum. Absorptive lineshapes can be obtained by adding together the rephasing and non-rephasing 2D spectra. Figure 1 shows the rephasing and non-rephasing spectra with the phase-twisted lineshapes and their addition yielding absorptive features. The sum of the two responses shows ten features arising from the various ladder diagrams contributing to the third-order signals for this six-level system. We see that the diagonal and the cross-peaks consist of two or more signed features, separating the dynamics of the ground and first excited states. To correct for the phase mismatch between the signal and the LO fields arising due to pathlength differences, we phase the spectra by fitting the projection of the 2D absorption spectrum along ω_3 to a dispersed pump-probe signal taken with the tracer as the probe under the same conditions [6].

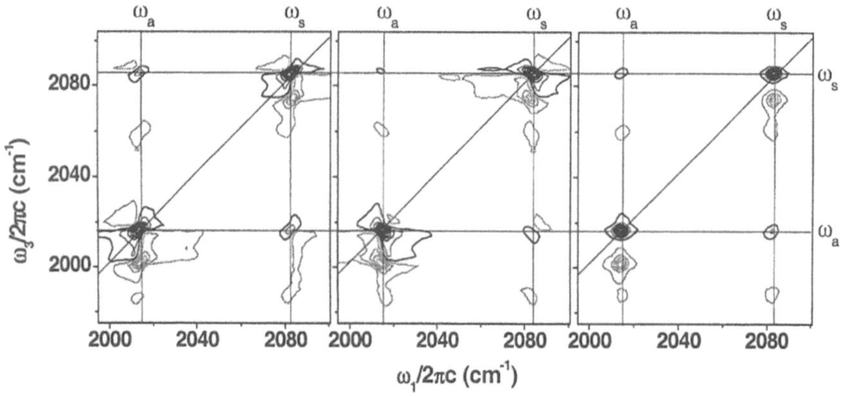

Fig. 1. Real part of 2D IR vibrational spectra at $\tau_2 = 0$ (a) S_I (b) S_{II} and (c) $S_I + S_{II}$.

Figure 2 demonstrates the effect of a finite τ_2 mixing time on the 2D rephasing spectra. The first two panels show the modulation of the cross-peak intensity as a function of τ_2. We see that the intensity of the cross-peak reaches the maximum in Fig. 2(a) when $\tau_2 = n\pi/2\omega_{as}$ where n is an even integer and ω_{as} is the frequency splitting between the symmetric and asymmetric carbonyl stretches. The cross-peaks disappear when n is an odd integer as shown in Fig. 2(b). The third panel obtained when $\tau_2 = 11.75$ ps indicates the emergence of four new features as a result of various coherent and incoherent population relaxation processes occurring during the mixing time. This results in the diagonal and cross-peaks splitting into three features instead of the usual two features obtained at smaller values of τ_2. A systematic study of the 2D rephasing and non-rephasing spectra as a function of τ_2 allows us to map out the complete dynamics of this multi-level system including the effects of solvent-induced relaxation and population relaxation.

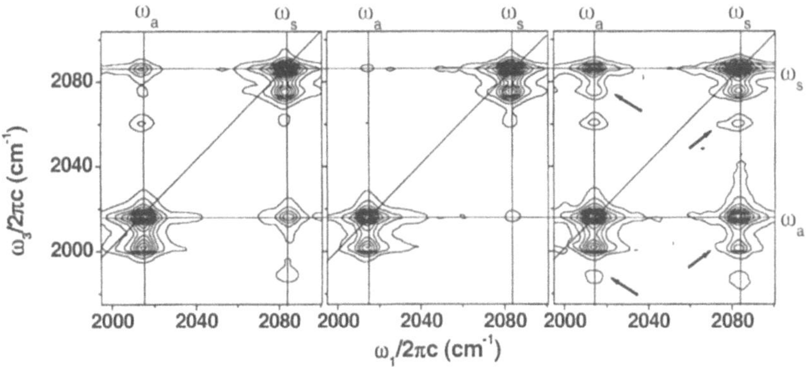

Fig. 2. Absolute value 2D IR rephasing spectra as a function of a variable mixing time. (a) $\tau_2 = 470$ fs (b) $\tau_2 = 705$ fs (c) $\tau_2 = 11.75$ ps.

References

1 O. Golonzka, M. Khalil, N. Demirdöven, and A. Tokmakoff, J. Chem. Phys., **115**, 10814 (2001).

2 N. Demirdöven, M. Khalil, O. Golonzka, and A. Tokmakoff, J. Phys. Chem. A, **105**, 8025 (2001).

3 D. E. Thompson, K. A. Merchant and M. D. Fayer, J. Chem. Phys., **115**, 317 (2001).

4 M. T. Zanni, S. Gnanakaran, J. Stenger, and R. M. Hochstrasser, J. Phys. Chem. B, **105**, 6520 (2001).

5 S. Woutersen, and P. Hamm, J. Phys. Chem. B, **104**, 11316-11320 (2000).

6 J. D. Hybl, A. Albrecht Ferro and D. M. Jonas, J. Chem. Phys., **115**, 6606 (2001).

Femtosecond Study of Vibrational Self-Trapping in Crystalline Acetanilide

J.Edler, P. Hamm

Max Born Institut für Nichtlineare Optik und Kurzzeitspektroskopie, Max Bornstr. 2a, 12489 Berlin, Germany
New Address: Physikalisch Chemisches Institut, Universität Zürich, Winterthurerstrasse 190, 8057 Zürich, Switzerland
Email: jedler@pci.unizh.ch

Abstract. Femtosecond infrared spectroscopy of NH and amide I bands of crystalline acetanilide confirms that vibrational self-trapping occurs in hydrogen-bonded model peptide crystals. The phonons which mediate self-trapping are identified.

1. Introduction

It has been suggested many years ago [1] that self-localization of vibrational states might play an important role for one of the most remarkable, yet not well understood, property of proteins, namely their capability of storing and transporting small quanta of energy. Acetanilide (ACN), a molecular crystal at room temperature, has been widely used as a model system to study this phenomenon. It consists of quasi-one-dimensional chains of hydrogen bonded peptide groups with structural properties that are comparable to those of an α-helix. Two coupling mechanisms have to be discussed to obtain self-trapping [2]: (i) Vibrational modes of the individual peptide units are known to be coupled by dipole-dipole-interaction and to form *delocalized* states, i.e. vibrational molecular excitons. (ii) These excitons are coupled to lattice phonons by nonlinear interaction mediated through the hydrogen bonds which stabilize the crystal structure. Excitation of an exciton hence leads to a deformation of the lattice, thereby localizing the excitation.

It has been recognized quite some time ago that the vibrational spectra of the amide I (i.e the C=O stretching) and NH stretching band of crystalline ACN exhibit 'anomalities' [3]. The amide I mode, which is observed at 1666 cm^{-1} at room temperature, splits into two bands at low temperatures with an additional 'anomalous band' at 1650 cm^{-1} (Fig. 1d). The NH stretching mode exhibits a main peak at 3295 cm^{-1} accompanied by an almost-regular sequence of 9 satellite peaks towards lower frequencies (Fig. 1a). It has been proposed that both phenomena can be explained by self-trapping theory [2], according to which the main peak represents the free exciton and the 'anomalous' side bands the self-trapped states.

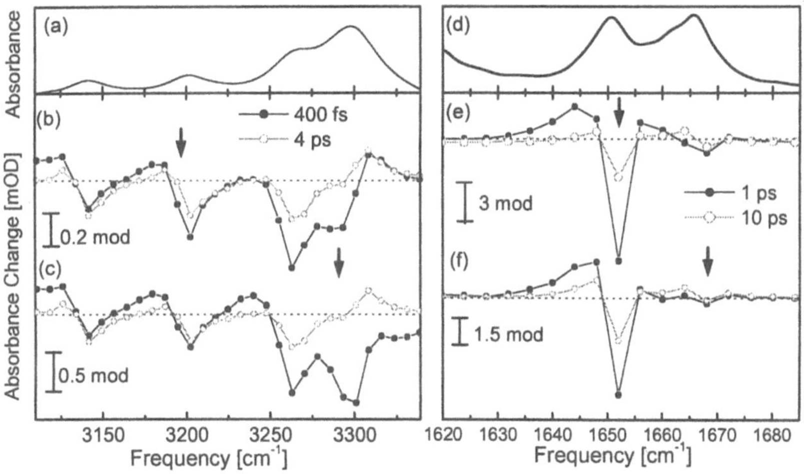

Fig 1: Part of the NH (a) and CO (d) band of crystalline ACN. (b, c) Response of the sample upon selective excitation of the self-trapped states (b,e) and the free exciton peak (c,f) for two different delay times. The position of the narrow band pump pulse is indicated by the arrow.

2. Results and Discussion

In a first experiment, short broadband pump pulses (spectral width 200 cm^{-1}, pulse duration 130 fs FWHM) were used to impulsively excite a section of the absorption spectrum which includes the NH free-exciton peak and the first three satellite peaks [4]. The transient absorbance change signal shows pronounced oscillations that persist up to about 15 ps and contain two distinct frequency components whose temperature dependence and frequencies match perfectly with two phonon bands in the non-resonant electronic Raman spectrum of ACN [3]. Therefore, the beating structure is assigned to a ground state phonon excited through a 'stimulated impulsive Raman effect', resonantly enhanced by the NH absorption band as a consequence of anharmonic coupling to lattice phonons. These are the phonons which modulate the hydrogen bond distance and thus mediate self-trapping.

In a second experiment, narrow band tunable pump pulses (spectral width 30 cm^{-1}, pulse duration 250 fs FWHM) were used to selectively excite individual sub-levels of the NH band (Fig. 1 b, c). On the sub-picosecond time scale, the free-exciton and the lower lying self-trapped states behave distinctly differently. When exciting the free-exciton (Fig. 1c), a strong bleach and stimulated emission signal is observed which recovers on a 400 fs time scale. Simultaneously, population is transferred into lower lying self-trapped states. On the other hand, when pumping one of the self-trapped states directly (Fig. 1b), population within all self-trapped states equilibrates essentially instantaneously, but the free exciton peak is not back-populated. This is the direct observation of ultrafast self-trapping: Excitation of the free-exciton leads to an *irreversible* population of self-trapped states, but not *vice versa*. The initially excited states relax on a 1ps time scale into states, that are either outside of our frequency window or spectroscopically dark, and return back to the ground state with a delay of 20 ps.

The narrow band pump experiment (spectral width 14 cm^{-1}, pulse duration 700 fs FWHM) was repeated on the amide-I band. Selective pumping of the free exciton state (1666 cm^{-1}) at low temperatures (90K) does not lead to a significant bleach of that band. Coherent broad band excitation experiments show that the lifetime of the normal amide-I mode is 1.5 ps [5]. The lack of an observable bleach can therefore only be explained by the cancellation of all contributions to the pump-probe signal (excited state absorption, stimulated emission and bleach), which is the case for a perfect harmonic state. It can be shown that the effective anharmonicity of a vibrational exciton is a direct measure of its degree of delocalization. Thus, we conclude that the free exciton state is almost perfectly delocalized at 90 K. As temperature increases, a bleach signal starts to be observed, pointing to a non-complete cancellation of the different contributions of the total pump-probe signal. Apparently, disorder induced by thermally excited phonons starts to localize the free exciton.

The anharmonicity of the self-trapped state (1650 cm^{-1}), on the other hand, originates from nonlinear interaction between the amide-I mode and the phonon system of the crystal. It is mediated through the hydrogen bonds and is nonzero even at low temperatures. With rising temperature, the bleach of the 1650 cm^{-1} band diminishes as it disappears in the absorption spectrum.

3. Conclusions

Unambiguous signature of self-trapping has been observed for both the amide I and the NH band of crystalline ACN. In both cases, the free exciton and the self-trapped states behave distinctly differently, supporting the previous assignments of the vibrational spectra [2,3]. Owing to the large self-trapping energy of the NH band, self-localization occurs even at room temperature. On the other hand, self-trapping of the CO band, the nonlinearity of which is significantly smaller, is observed only at low temperatures and breaks down at room temperature. The studies proof that vibrational excitations of regular protein structures reveal a rich variety of collective phenomena mediated by the large non-linearity of the hydrogen bonds stabilizing the secondary structure, that have yet to be explored.

References

1 A. S. Davidov, J. theor. Biol. **66**, 379 (1977).

2 A. C. Scott, Phys. Reports. **217** 1 (1992).

3 G. Careri, U. Buontempo, F. Galluzzi, A. C. Scott, E. Gratton, and E. Shyamsunder, Phys. Rev. B **30**, 4689 (1984).

4 J. Edler, P. Hamm, A.C. Scott, Phys. Rev. Lett **88**, 067403 (2002).

5 J. Edler, P. Hamm, J. Chem Phys., accepted for publication (2002).

Correlated vibrational dynamics in solution studied by two-dimensional Fourier transform infrared spectroscopy

N. Demirdöven, M. Khalil, and A. Tokmakoff

Department of Chemistry, Massachusetts Institute of Technology, Cambridge MA 02139, USA
E-mail: tokmakof@mit.edu

Abstract. We studied the ultrafast vibrational dynamics of a model system of two strongly coupled vibrations through characterization of the correlation between their transition energy fluctuations using two-dimensional Fourier transform infrared spectroscopy.

Femtosecond two-dimensional (2D) infrared (IR) spectroscopy allows the dynamics of transient molecular structures in solution to be probed by characterizing the correlated energy fluctuations of coupled vibrations. By Fourier transforming the third order nonlinear response as a function of an initial evolution period and final signal detection period, a 2D spectrum is obtained. Dynamical parameters that characterize the correlated system-solvent bath interactions are determined by the Fourier transform (FT) line shape analysis of the 2D IR correlation spectra.

2D IR spectroscopy is a heterodyned-detected four wave mixing (FWM) technique that monitors the propagation of vibrational coherences as a function of an initial evolution period τ_1, and a final signal detection period τ_3 for a fixed waiting time τ_2. This is done using three time-ordered mid-IR laser fields $\boldsymbol{E_a}$, $\boldsymbol{E_b}$ and $\boldsymbol{E_c}$ with wave vectors $\boldsymbol{k_a}$, $\boldsymbol{k_b}$ and $\boldsymbol{k_c}$ to generate the nonlinear signal field $\boldsymbol{E_s}$ along the wave vector matching direction $\boldsymbol{k_s}= -\boldsymbol{k_a}+\boldsymbol{k_b}+\boldsymbol{k_c}$. Experimentally controlled time delays are $\tau_1(=t_b-t_a)$ and $\tau_2(=\min(|t_a|,|t_b|))$, where t_a and t_b represent the relative delays between $\boldsymbol{E_a}$ and $\boldsymbol{E_b}$ with respect to $\boldsymbol{E_c}$ with t_c=0. The signal was spatially and temporally overlapped with a local oscillator field, focused into a monochromator and dispersed onto a 64-element array detector. Dispersing the signal acts to effectively Fourier transform with respect to the detection time τ_3. A set of interferograms recorded as a function of τ_1 are Fourier transformed to give the 2D FT spectrum. In general, the 2D FT spectrum of any particular FWM signal will have both absorptive and dispersive character. However, spectra with absorptive character must be obtained in order to measure meaningful and intuitive 2D FT line shapes. Such 2D correlation spectra were constructed from the sum of 2D spectra collected in the same signal direction by rephasing ($\boldsymbol{S_I}=-\boldsymbol{k_a}+\boldsymbol{k_b}+\boldsymbol{k_c}$, $\tau_1>0$) and non-rephasing ($\boldsymbol{S_{II}}=+\boldsymbol{k_a}-\boldsymbol{k_b}+\boldsymbol{k_c}$, $\tau_1<0$) experiments. The experiments were performed on the anharmonically coupled symmetric, s, (ω_s=2085 cm^{-1}) and asymmetric, a, (ω_a=2015 cm^{-1}) carbonyl (–C≡O) stretching modes of acetylacetonato dicarbonylrhodium (I) (RDC) using 90 fs 4.9-μm mid-IR pulses [1,2].

Figure 1 shows three 2D IR correlation spectra of RDC in $CHCl_3$ solution at three different waiting periods, τ_2=0, 1.5 and 3 ps. Each spectrum exhibits eight purely absorptive features – four diagonal and four off-diagonal cross peaks – each of which are two pairs of oppositely signed peaks.

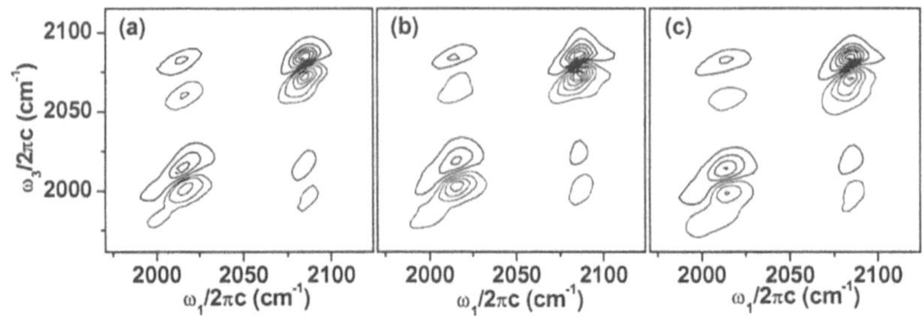

Fig. 1. Experimental 2D IR absorption spectra of RDC in $CHCl_3$ for different relaxation times (a) τ_2=0, (b) τ_2=1.5 ps, and (c) τ_2=3 ps.

Diagonal peaks arise from the processes involving only one fundamental (positive contours), and those involving one fundamental and its overtone (negative contours); whereas, cross peaks originate from the processes involving either both fundamentals (positive contours) or those involving one fundamental and the combination band (negative contours). The degree of elongation of 2D diagonal line shapes is an indication of inhomogeneous broadening, and 2D cross peak line shapes are sensitive to the correlated dynamics of coupled vibrations. As seen from Fig. 1, cross peaks are elongated and tilted parallel to the diagonal. As the system evolves with increasing τ_2, elongation becomes less pronounced and the tilt disappears. These features are expected for a system undergoing strongly correlated spectral diffusion in an inhomogeneous environment.

To model the correlated solvation dynamics of the coupled vibrational transitions, we propose an exponentially decaying form for the energy gap auto- and cross-correlation function

$$\zeta_{mn}(t) = \left\langle \delta\omega_m(t)\delta\omega_n(0) \right\rangle_B = \rho_{mn}\Delta_m\Delta_n \exp(-|t|/\tau_{mn}) \quad m,n = a,s \tag{1}$$

where $\delta\omega$(t) is the bath induced shift of the vibrational transition frequency, $\Delta_{m(n)}$ is the auto-correlation amplitude, and τ_{mn} is the correlation time of the vibrational frequency fluctuations. ρ_{mn} is the unitless correlation coefficient that reflects the degree of correlation between the fluctuations of ω_m and ω_n [2]. It is defined as the normalized zero time value of the energy gap cross correlation function. ρ_{mn} changes between +1 and -1 corresponding to fully anti-correlated and correlated cases, respectively. When m=n, Eq. 1 represents the energy gap auto-correlation function with ρ_{mn}=1 and gives the energy gap cross-correlation function for the coupled vibrational modes when $m \neq n$. It is this energy gap cross-correlation function that we are most interested in quantifying through 2D line shape analysis. The nonlinear response corresponding to certain Liouville pathways is calculated by using its corresponding dephasing function using a new FWM theory [3].

Fluctuation amplitudes for correlation functions involving doubly excited states are obtained using harmonic scaling relative to the corresponding fundamentals. The total 2D nonlinear response is then calculated by summing over all the accessible pathways, and double Fourier transformed to obtain the 2D spectrum.

Calculated 2D IR correlation spectra, which closely reproduce the experimental ones, are shown in Fig. 2. For the τ_2=0 spectrum, the immediate effect of the strong cross-correlation between the asymmetric and symmetric modes of RDC, ρ_{as}=0.9±0.1, is demonstrated as the tilt of the cross peaks parallel to diagonal. The loss of the tilt as a function of the evolution period τ_2=1.5 and 3 ps and the decrease in the overall intensities are also expected since energy gap correlation function loose amplitude and the rephasing spectrum approaches the non-rephasing one. A cross-correlation fluctuation amplitude Δ_{as} of 5.8±0.6 cm^{-1} and a cross-correlation fluctuation lifetime τ_{as} of 1.2±0.4 ps reproduce well the τ_2 dynamics of the 2D IR correlation spectra. The estimated auto-correlation fluctuation amplitudes and times are Δ_{aa}=7.4±0.6 cm^{-1}, Δ_{ss}=5.2±0.4 cm^{-1}, τ_{aa}=2.0±0.2 ps, and τ_{ss}=2.0±0.4 ps, respectively.

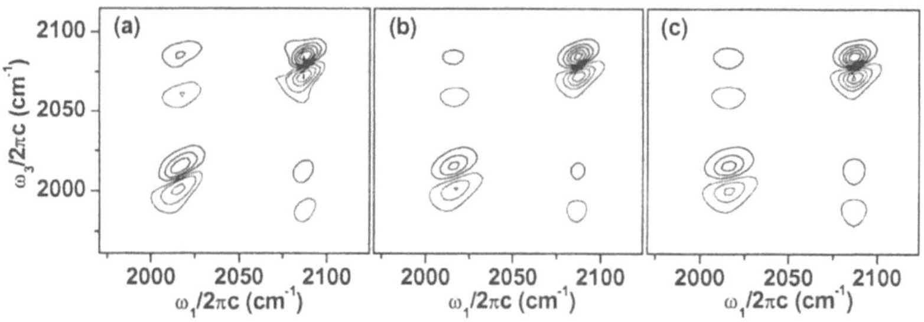

Fig. 2. Calculated 2D IR absorption spectra of RDC in $CHCl_3$ for different relaxation times (a) τ_2=0, (b) τ_2=1.5 ps, and (c) τ_2=3 ps with ρ_{as}=0.9±0.1.

RDC is a planar (d^8) coordination compound, and the Rh atom is exposed to solvent molecules in the primary solvation shell along the two axial positions. We suggest that correlated fluctuations in this system arise from modulation of the solvent density around these axial directions, which modulates the electron density of the Rh atom [2]. This in turn will modulate the strength of the d_π-π^* bonding between Rh and CO groups in a manner that symmetrically influences the individual CO anharmonicities. Ensemble averaging of either effect will appear as a correlated broadening of the transition frequencies.

References

1 O. Golonzka, M. Khalil, N. Demirdöven and A. Tokmakoff, J. Chem. Phys. **115**, 10814, 2001

2 N. Demirdöven, M. Khalil, O. Golonzka and A. Tokmakoff , J. Phys. Chem. A. **105**, 8025, 2001

3 J. Sung and R. J. Silbey, J. Chem. Phys, **115**, 9266, 2001

Local structure and dynamics of liquid acetone by heterodyned 2D IR spectroscopy

Nien-Hui Ge, Martin T. Zanni and Robin M. Hochstrasser

Department of Chemistry, University of Pennsylvania, Philadelphia, PA 19104, USA
E-mail: hochstra@sas.upenn.edu

Abstract. Vibrational energy flow and dynamical structures in neat liquid acetone and isotopic mixtures from 2D-IR spectra, photon echoes and transient gratings of its carbonyl modes, show ultrafast reorientation and energy transport.

Introduction. Multiple pulse two- dimensional IR spectroscopy promises to obtain snapshots of molecular structures in the condensed phase.[1] Here we show that it exposes the dynamics of the local structure of liquids: results for liquid acetone are obtained. The liquid responds like a molecular excitonic system undergoing structure fluctuations. Isotope dilution with acetone-2 -^{13}C allowed measurement of the coupling between molecules. New heterodyned 2D transient grating experiments reveal the dynamics of intermolecular energy transfer.

Experimental Methods. Femtosecond IR pulses (120 fs, 1 μJ, 1715 cm^{-1}) were generated by mixing the signal and idler beams of an 800 nm pumped OPA in an $AgGaS_2$ crystal. The output was split into three equal beams (wavevectors k_1, k_2, and k_3; each 300 nJ) and a local oscillator beam (k_{LO}; 3 nJ) which was combined with the signal field in the phase-matched direction $k_s = -k_1 + k_2 + k_3$, and focused into a single channel HgCdTe detector. The time interval between k_1 and k_2 is τ, T is the interval between k_2 and k_3, and t is the interval between k_3 and k_{LO}. Photon echo experiments are done by scanning τ and t with $T = 0$. In transient grating experiments, T and t are scanned while keeping $\tau = 0$.

Results and Discussion. For the 50:50 isotope mixture the 2D spectrum shows the ^{12}C=O and ^{13}C=O stretches of acetone molecules as diagonal peaks. The cross peaks between the two isotope distributions are present but quite weak. For the 3% mixture, the 2D spectrum, shown in Fig. 1(a), exhibits a diagonal peak corresponding to the ^{13}C=O stretch, while the corresponding ^{12}C=O diagonal peak shows a strongly attenuated echo field at the ^{12}C=O fundamental frequency. Pronounced off-diagonal peaks are now observed, characteristic of the existence of intermolecular coupling between ^{13}C=O acetone molecules with their surrounding ^{12}C=O acetone molecules. Figure 1(b) and (c) show slices through the magnitude and real part of the 2D spectrum at ω_t = 1671 cm^{-1}, taken at two polarization conditions (XXXX) and (XYYX) for (k_1, k_2,k_3,k_{LO}). The off-diagonal peak retains its phase and has similar intensity in the two polarizations, indicating that the coupling is from neighbors having nearly parallel or anti-parallel C=O groups.

A recent Monte Carlo simulation suggests that the nearest neighbors in liquid acetone can be classified into two groups: one is around 4 A separation in a side-by-side anti-parallel orientation. The other, near 5.3 A, is in a head- to-tail parallel

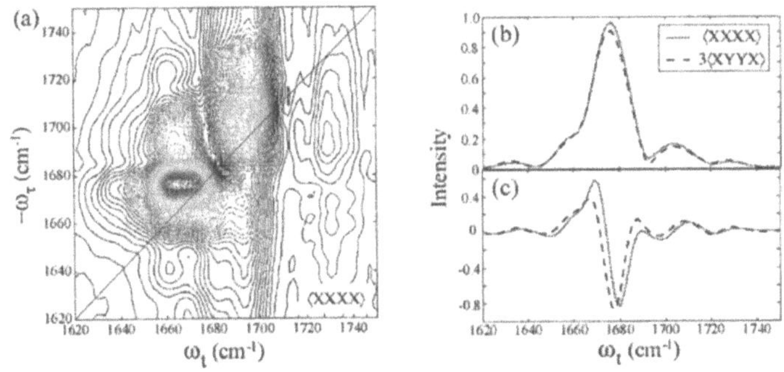

Fig. 1. (a) The absolute magnitude 2D photon echo spectrum of liquid acetone isotopic mixture containing 3% acetone-2-^{13}C. Sections though the magnitude (b) and real part (c) of the 2D spectra at ω_t = 1671 cm^{-1}, taken at (XXXX) and (XYYX) polarization conditions.

arrangement [2]. The transition dipole contributions to the intermolecular coupling are −3.3 and −4.4 cm^{-1} respectively for these two arrangements. Our measurements of the cross peaks are consistent with both these predictions. The polarization condition (-π/4,π/4,π/2,0) eliminates signal from aligned molecules and can be used to investigate coupling to molecules of different orientations .

The existence of intermolecular coupling is further verified by the results from heterodyned transient grating experiments. Figure 2(a) shows the absolute magnitude spectrum as a function of T for the 3% mixture. A slice through ω_t = 1656 cm^{-1} clearly shows oscillations with a period of 830 fs. These oscillations give rise to a cross peak in the 2D spectrum at ω_T = 40 cm^{-1}, corresponding to the isotope shift. Such a cross peak is expected to arise from coupling between isotopomers. Oscillations of 3.3 ps were also found: they were most pronounced in the ^{12}C=O regions and in the neat liquid. The possible influence of the strong attenuation of the pulse on the form of these oscillations is currently under study.

The transient grating signals in the time domain exhibit significant dependence on the isotope compositions. Figure 2(b) shows the T dependence of the heterodyned transient grating anisotropy for the ^{12}C=O peak in neat acetone and the ^{13}C=O peak in the 3% mixture. The anisotropy was obtained by taking the difference between the signals for (XXXX) and (XXYY) polarization conditions, and normalized by the isotropic signal. Clearly neat acetone exhibits faster anisotropy decay. Similar behavior is found in the integrated transient grating signals where neat liquid shows faster decay compared to the 50:50 mixture.

Coupling is also indicated by changes in the linear spectral linewidths as a function of isotopic composition: e.g. the bandwidths of ^{12}C=O in neat acetone and the ^{13}C=O in the 3% mixture are 16 cm^{-1} and 10 cm^{-1} , respectively. Integrated three pulse photon echoes of the 50:50 mixture exhibit oscillations and decay slower than those of neat acetone. The correlation function of the ^{13}C=O peak in the 3% isotope mixture can be measured by heterodyned three pulse photon echo. The extent of excitonic delocalization in the liquid is controllable by varying the isotope concentration. This effect on the correlation function is currently under investigation.

The current results suggest a slowed energy transfer between neighboring molecules in the isotopic liquid due to the energy mismatch. Also plotted in Fig. 2(b) is the expected anisotropy from the rotational correlation time of 0.75 ps

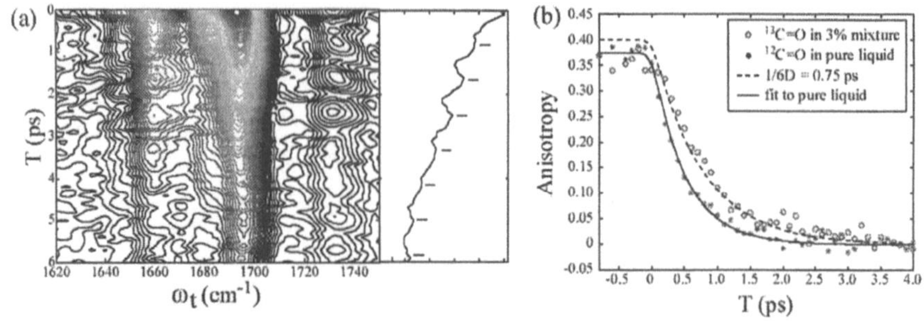

Fig. 2. (a) The absolute magnitude 2D transient grating spectrum of the 3% mixture for the (XXYY) polarization condition. The side panel shows a slice through the spectrum at ω_t = 1656 cm^{-1}. (b) Heterodyned transient grating anisotropy of the ^{12}C=O peak in neat acetone (*) and the ^{13}C=O peak in the 3% isotope mixture (o).

measured by Rayleigh scattering [3]. The anisotropy decay of the 3% mixture coincides with this curve very well, suggesting it is dominated by molecular reorientation. However in neat acetone the energy transfer (1.5 ps) between acetone molecules influences the decay. For a qualitative picture of the energy transfer, we consider an irreversible energy transfer process between two equivalent vibrators. In the presence of fast and uncorrelated fluctuations of the two levels, the ener gy transfer rate may be estimated by a Golden rule limit. For a coupling strength of 5 cm^{-1}, the rate is 1.7 ps. The rate slows down to 12 ps when there is a 40 cm^{-1} energy difference between the two levels. Although we do not expect such a simple picture to prevail, the possibility that only a few strongly coupled but not all nearest neighbors participate in the energy transfer process is suggested by this comparison to theory. This point will be addressed by simulations of the coupling and energy fluctuat ions.

Conclusions. We have measured the complex nonlinear response of liquid acetone in the region of 6 μm. Isotope dilution studies indicate the presence of intermolecular coupling between parallel or anti-parallel neighbors. Evidence is obtained for delocalization of the C=O mode and irreversible energy transfer among acetone molecules. This transport corresponds to relaxation amongst the partially delocalized states of the liquid. These new and exciting results demonstrate the power of nonlinear methods in revealing structure and dynamics of liquids.

Acknowledgements. This research was supported by NIH and NSF.

References

1 N.-H. Ge and R. M. Hochstrasser, *PhysChemComm*, 5, 17 (2002).
2 H. Torii, J. Mol. Struct. (Theochem) **311**, 199-203, (1994).
3 J. F. Dill, T. A. Litovitz and J. A. Bucaro, J. Chem. Phys. **62**, 3839-3850, (1975).

Part XII

Ultrafast Dynamics in Biology

Dynamics of peptide structures by multiple pulse 2D IR spectroscopy

R.M. Hochstrasser, J. Chang, J.-X. Chen, N.-H. Ge, Y.-S. Kim, J. Park, I. Rubtsov, and M.T. Zanni

Department of Chemistry and RLBL, University of Pennsylvania, Philadelphia, PA 19104
E-mail: hochstra@sas.upenn.edu

Abstract. The 2D IR COSY spectra of peptides is shown to yield novel information about structure and dynamics not available from linear IR methods. Three pulse correlation spectroscopy with heterodyning and passive phase control is carried out with rephasing and nonrephasing sequences yielding absolute value, real and purely absorptive 2D spectra for a variety of polarization conditions. The diagonal peaks are removable by polarization choices and improvements in resolution are obtained, especially for more complex systems such as α-helices. IR COSY studies are reported of amide-I, amide-A (N-H and N-D) and the coupling and angular relations between these modes are obtained through novel two color methods.

1. Introduction

An ideal approach to visualizing chemical and biological processes would involve snapshots of atomic positions of participating molecules throughout the course of chemical reactions or conformational changes. Although unlikely to be obtained from a single experimental technique, a combination of methods may well achieve this goal. New experiments have demonstrated the promise of using 2D IR methods to simplify IR spectroscopy of complex systems and thus obtain structural features with sufficient time resolution to resolve processes during chemical reactions and conformational dynamics. The 2D IR spectroscopy (1) naturally accesses the time scales that are hard to reach by 2D NMR and X-ray diffraction methods in solutions. Exciting biological applications include structural dynamics of nucleotides, peptides (2-4), and perhaps small proteins. Other applications in photochemistry, polymer dynamics, and liquids (see paper herein by Ge and Hochstrasser) are certainly possible and promising approaches.

2. The heterodyned 2D IR experiments

In IR COSY and THIRSTY (three pulse IR stimulated echo spectroscopy) experiments, three passively phase locked IR pulses with wavevectors k_1, k_2 and k_3 arrive at the sample after time intervals τ and T generating an electric field pulse $E(t;\tau,T)$ in the direction $-k_1 + k_2 + k_3$ which is heterodyne detected by the local oscillator pulse k_{LO}. The echo field is a sum of Bohr frequencies corresponding to

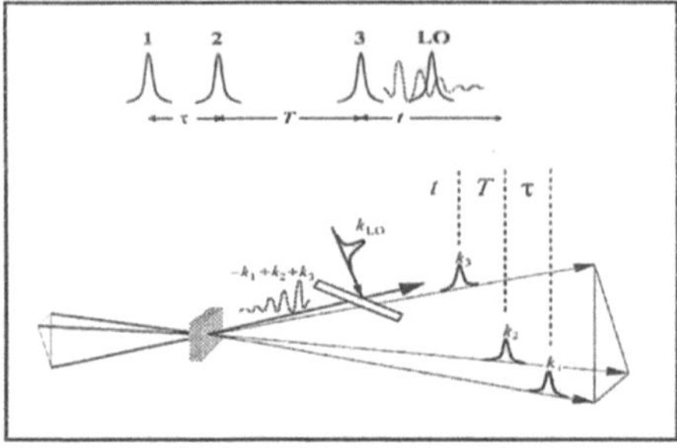

Fig. 1. 2D IR time domain interferometry.

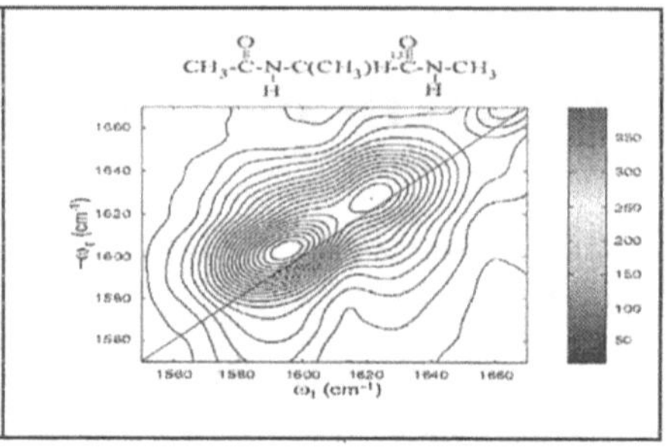

Fig. 2. 2D IR spectrum of alanine dipeptide

various Liouville pathways for the excitation and transfer between vibrational populations and coherences. Important structural constraints are also obtainable. For transition charge potentials the distance constraints are $<1/r^3>$from anharmonicity and $<1/r^6>$ from energy transfer. The cross peak polarizations yield averages $<P_1>$ or $<P_2>$ of angles between transition dipoles, which are not model dependent, and their distribution functions. A typical 2D IR peptide spectrum is shown in Fig. 2 for the alanine dipeptide, which shows the ^{12}C and ^{13}C elongated diagonal shapes from the inhomogeneous structure distributions, the cross peaks from coupling the two anharmonic amide units that are energetically separated by isotopic substitution (also being done with with $^{18}O{=}^{13}C$) and the structural differences between the ends of the molecule in agreement with MD simulations (5).

3. Structures and dynamics by 2D IR

Novel 2D IR phenomena have been demonstrated. **Polarization conditions.** Polarization conditions were found that eliminate diagonal peaks from the 2D IR spectra. For the polarization condition $(\pi/4,-\pi/4,\pi/2,0)$, the signal is $<\mu_1.\mu_3\mu_2.\mu_4 - \mu_1.\mu_4\mu_2.\mu_3>$ (6,7) which vanishes for a single dipole involvement (the symbols refer to the dipole directions at four times). An example of a diketone is given in Figure 3. Removing the intense diagonal peaks allows a clear observation of the phase of the cross peaks in the real 2D spectrum and represents a significant advance in simplifying complex spectra. The systematic use of polarization conditions considerably clarifies the 2D spectra of alpha helices because only the transitions between symmetric and antisymmetric excitonic levels of the α-helix need be exposed. The energy distributions and relaxation dynamics of delocalized states of the helix could be tested, free from interference by the diagonal peaks, based on the coupling between the amide oscillators. An example of the 20 unit helix Ac-AAAAKaaaaKAAAAKAAAAK-NH_2, where the lower case a's (alanines) contain $^{13}C{=}O$ groups down shifted by ca. 30 cm^{-1}, is shown in Fig. 3. The a-A coupling is clearly seen in these spectra (arrow), as is the broad distribution of two particle excitations of the •-helix.

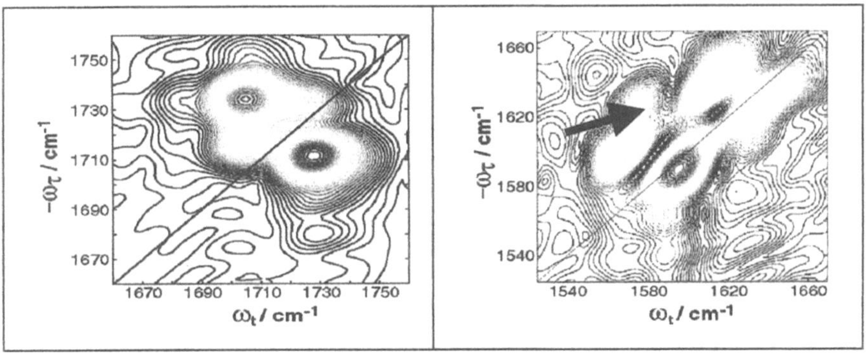

Fig. 3. 2D IR spectra of 1,3-cyclohexanedione (left) and the 20 unit helix (right)

The heterodyned photon echoes of amide-A modes. Heterodyned echoes of N-H modes of peptides have been obtained. These modes undergo very rapid (sub-picosecond) relaxation dependent on whether or not the N-H bond is hydrogen bonded and the frequency correlation functions signal the motions of nearby solvent molecules.

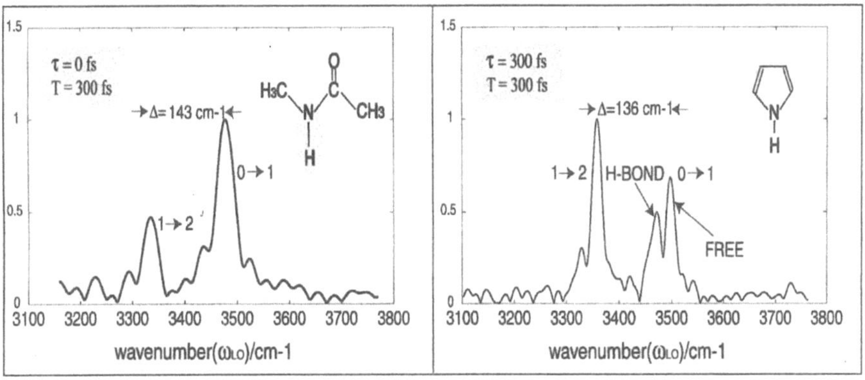

Fig. 4. Heterodyned photon echoes from N-methylacetamide (left) and pyrrole (right)

Examples of heterodyned N-H echoes are shown in Fig. 4 for some values of T and τ. For N-H the anharmonicity is ca. 140 cm-1 so self-localization of N-H excitation occurs in multiple unit peptides and the coupling scheme is quite different from that used for amide-I. **Two color experiments.** The coupling matrix, needed to interpret two color heterodyned echo signals for molecules with multiple N-H and amide-I modes incorporates the N-H/amide-I off-diagonal anharmonicity. We therefore used nonlinear IR methods to obtain these benchmarks for a number of peptide models. Typically the N-H/amide-I coupling is found to be 2 or 3 cm^{-1} (see paper by Rubtsov et al.). The use of two infrared "colors" in 2D IR experiments allows the measurement of coupling between modes of different frequencies. Experiments using 3 and 6 micron pulses provide the coupling between the N-H and amide-I modes and on the energy transfer between these modes, the TD directions and an evaluation of the coupling potential. **Correlations of frequency distributions.** When pulse k_1 interacts before k_2, the coherence has frequencies of opposite signs during the t and t periods, but if the ordering of pulses is reversed, the frequencies have the same sign. Acquiring 2D spectra using both pulse sequences either combined or separately provides information on correlations between the vibrational frequency distributions (3) and purely absorptive spectra (8) which may improve spectral resolution. The spectra each have significant inhomogeneous broadening so the nonrephasing signals are relatively small in the present examples. In addition to time ordering, the polarization of each laser pulse can be individually varied in all quadrants allowing linear combinations of Liouville paths to be observed and hence the exclusion or emphasis of particular pieces of the 2D IR spectra. Fig. 5. shows some recent results (the corresponding real parts of the 2D IR spectra were also presented).

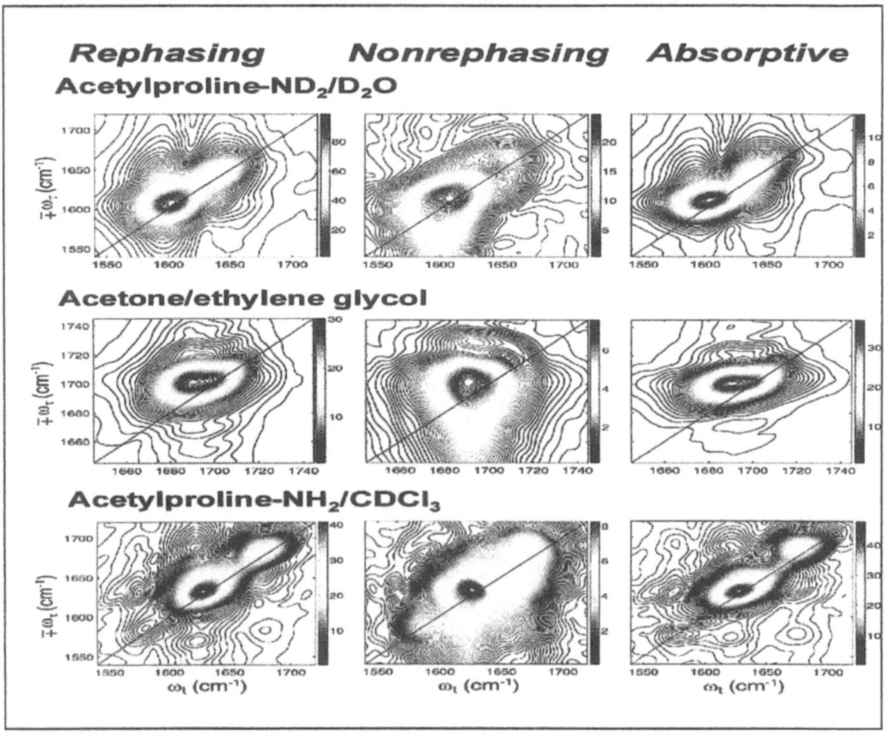

Fig. 5. Examples of absolute 2D IR spectra having oscillatory parts from the two times: left, FT(cos cos+sin sin); center, FT(cos cos –sin sin); and right, FT(left +right).

4. Conclusions

Multidimensional IR appears to be a promising method for simplifying the spectra of complex systems such as polypeptides, leading to sharper dynamical and structural information than obtainable from conventional IR spectra. This paper has focused on polarization properties as a means to remove the diagonal peaks which are dominating and obscuring useful information, the importance of combining rephasing and nonrephasing spectra in further simplifying spectra, the introduction of multicolor nonlinear IR methods to obtain the coupling between different types of modes. The first echoes of N-H and N-D modes were reported as were the first applications of the 2D IR method to liquids (Ge and Hochstrasser). 2D spectra were interpreted in inhomogeneous, aqueous environments by comparing to molecular dynamics simulations (5).

Acknowledgements. The work was supported by NIH and NSF. We are grateful to Dr. S. Decatur and his group for their essential contributions to the helix work.

References

1. M. C. Asplund, M. T. Zanni and R. M. Hochstrasser, Proc. Natl. Acad. Sci. U. S. A., 2000, 97, 8219.
2. M. T. Zanni, S. Gnanakaran, J. Stenger and R. M. Hochstrasser, J. Phys. Chem. B, 2001, 105, 6520.
3. N.-H. Ge, M. T. Zanni and R. M. Hochstrasser, J. Phys. Chem. A, 2002, 106, 962
4. N.-H. Ge and R. M. Hochstrasser, PhysChemComm, 2002,5,17
5. S. Gnanakaran and R. M. Hochstrasser, J. Am. Chem. Soc., 2001, 123, 12886."
6. R. M. Hochstrasser, Chem. Phys., 2001, 266, 273. 6
7. M. T. Zanni, N.-H. Ge, Y. S. Kim and R. M. Hochstrasser, Proc. Natl. Acad. Sci. USA, 2001, 98, 11265.
8. R.R. Ernst, G. Bodenhausen and A. Wokaun "Principles of nuclear magnetic resonance" Clarenden Press, Oxford 1987; p 309

Watching a protein as it functions: real-time observations using picosecond crystallography

Friedrich Schotte[1], Michael Wulff[2], and Philip Anfinrud[1]

[1] Laboratory of Chemical Physics, NIDDK, National Institutes of Health, Bethesda, MD 20892, USA
[2] European Synchrotron Radiation Facility, Grenoble Cedex 38043, FRANCE

Abstract. The structural changes associated with ligand translocation in myoglobin have been investigated with picosecond time-resolved x-ray crystallography. These studies, conducted on the ID09 beamline at the ESRF in Grenoble, France, unveil the protein's structural evolution with < 2 Å resolution.

1. Introduction

Proteins are engaged in a myriad of tasks that are essential to life. To understand in mechanistic detail of how proteins function, it is crucial to know the time ordering of events that give rise to their designed function. Myoglobin (Mb), a ligand-binding heme protein, has long served as a model system for investigating ligand transport and binding in proteins. Using femtosecond time-resolved polarized IR spectroscopy, the dynamics of CO motion after photodetachment from MbCO have been probed [1, 2]. Those studies revealed the time-dependent orientation of CO, the presence of a docking site that mediates the transport of ligands to and from the active binding site, as well as the dynamics of ligand binding and escape. In addition, site-specific mutations among the highly conserved residues circumscribing the ligand docking site were found to have a marked influence on the dynamics of ligand binding and escape. While much has been learned about ligand dynamics in Mb, much less is known about the structural evolution that accompanies ligand translocation. To probe the structural evolution of a protein as it executes its designed function, a multinational collaboration has been established to acquire time-resolved x-ray crystal structures on the ID09B beam line at the European Synchrotron and Radiation Facility [3]. Recent improvements in the experimental methodology have extended the time resolution down to 150 ps and improved the spatial resolution of the diffraction images to below 2 Å

2. Experimental Methods

Time-resolved x-ray diffraction images are acquired using the pump-probe method: a femtosecond or nanosecond laser pulse triggers ligand dissociation in a ~ 250 micron P6 MbCO crystal and a variably delayed 150 ps x-ray pulse probes its structure. The x-ray pulses, which are generated when a relativistic electron bunch (6 GeV at the ESRF) passes through the gap of an undulator, can be efficiently focused to the dimensions of the protein crystal. The diffracted x-ray photons are detected with high quantum efficiency using a MAR CCD. Reconstruction of the protein structure with atomic resolution requires a series of diffraction images with crystal orientations spanning 60 degrees (owing to the symmetry of the crystal). With undulator radiation, the x-ray bandwidth was sufficient to obtain redundant data with images collected every 2 degrees. To obtain high dynamic range diffraction images with the available x-ray flux, approximately 64 x-ray shots were integrated on the MAR CCD between readouts. Because the protein crystal requires sufficient time to recover between photolysis pulses, which are intense enough to excite a significant fraction of the protein molecules, the maximum repetition rate used was 3 Hz. Diffraction images were accumulated with and without photolysis to generate accurate difference-diffraction data.

To pursue these studies, the synchrotron is operated in a rare "single bunch" mode, where a single electron bunch is injected into the storage ring. Given the 844 m circumference of the ESRF storage ring, x-ray pulses are generated at 355 kHz. Single pulse selection from this pulse train is accomplished with a high speed synchronous x-ray chopper running at 900 Hz in combination with a single-shot operated millisecond shutter. Synchronization between the fs laser and x-ray pulses is accomplished with ps precision using a Synchro-Lock-equipped Coherent Mira Ti:sapphire laser; synchronization with a ns laser is accomplished with a digital delay generator.

3. Results and Discussion

A high-resolution electron difference density map of MbCO is shown in Fig. 1. This image is the first of a time series that reveal, with atomic resolution, the order of events that accompany ligand translocation. Numerous features are observed including the displacement of the heme iron toward the proximal histidine, tilting of the heme, the docking of CO in a site near the heme iron, and the correlated motion of protein side chains in the vicinity of the active binding site. For example, the distal histidine (residue 64) shifts toward the site once occupied by CO, possibly raising the barrier to geminate recombination. The photodetached CO, found in a heme pocket docking site only about 2 angstroms away from the binding site, causes a displacement of the nearby isoleucine (residue 107). The CO translocation and heme displacement are also evident in a time-resolved structure determined at 150 ps, suggesting that the protein response to the ligand translocation event is ultrafast, perhaps as fast as the 1.6 ps time constant assigned to protein relaxation in an earlier ultrafast time-resolved mid-IR study of photolyzed MbCO in solution [2]. On the time scale of a few hundred ns (not shown), the "docked" CO slips around to the proximal side of the heme and is found in the so-called Xe1 docking site, as observed in low temperature trapping studies [4,5] as well as in earlier time-resolved studies [6]. Within a few microseconds, the CO escapes from this site into the surrounding solvent. The structural changes that accompany ligand translocation help explain how the protein is able to excrete toxic CO with high efficiency, even though the CO is temporarily located so close to the active binding site.

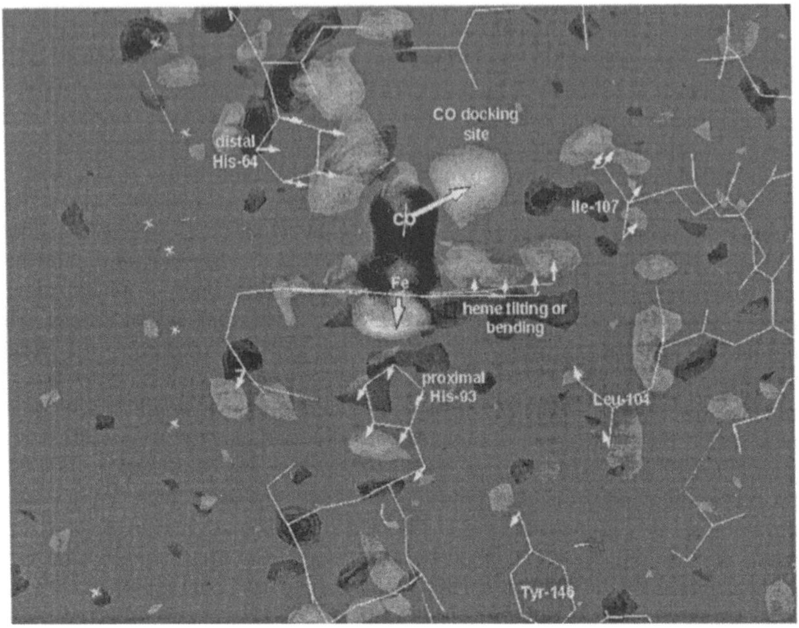

Fig. 1. Difference electron density map of MbCO determined 3 ns after photolysis of a P6 MbCO crystal at 10 °C. The dark (negative) and light (positive) regions correspond to photolysis-induced change in the electron density. The arrows indicate the direction of motion of the CO and the surrounding side chains.

4. Conclusions

Complementary time-resolved spectroscopic and crystallographic studies promise to unveil at a high level of structural detail the conformational changes that accompany protein function in this and other protein systems.

Acknowledgements. We thank Prof. J.S. Olson, Prof. G.N. Phillips, Jr., and Dr. J. Soman for supplying the P6 MbCO crystals used in this study, and thank Prof. K. Moffat and V. Srajer for participating in earlier P21 MbCO studies.

References

1 M. Lim, T.A. Jackson, and P.A. Anfinrud, in *Science*, Vol.269, 962, 1995.
2 M. Lim, T.A. Jackson, and P.A. Anfinrud, in *Nature Struct. Biol.*, Vol.4, 209, 1997.
3 F. Schotte et al., in *Third-Generation Hard X-ray Synchrotron Radiation Sources*, Edited by D.M. Mills, Wiley 2002.
4 K. Chu, et al., in *Nature*, Vol 403, 921, 2000
5 A. Osterman et al., in *Nature*, Vol. 404, 205, 2000.
6 V. Srajer, et al., in *Biochemistry* Vol. 40, 13802, 2001.

Primary Processes of Ligand Dissociation in Heme Proteins

M. Armstrong[1], J. P. Ogilvie[2], M. L. Cowan[2], A. M. Nagy[2], and R. J. D. Miller[2]

[1] Department of Physics and Astronomy, University of Rochester, Rochester, NY, USA
[2] Departments of Physics and Chemistry, University of Toronto, Toronto ON, Canada
E-mail: dmiller@lphys.chem.utoronto.ca

Abstract. Amplified few cycle pulses were used to observe the initial dynamics of the photodissociation process of carboxymyoglobin to specifically probe the issue of reaction driven modes in this system. Significant amplitude was observed at distinct modulation frequencies corresponding to deoxymyoglobin providing insight into the microscopic details of the reactive crossing.

Myoglobin (Mb) and hemoglobin (Hb) are, respectively, the oxygen storage and transport molecules found in animals, reversibly binding to diatomic ligands such as O_2, CO and NO at the heme site. Interest in these molecules is motivated by cooperativity in hemoglobin, where the binding affinity of hemes within the protein is modulated by their ligation state via quaternary structural reorientation. This modulation indicates a correlation between the localized ligation state of hemes and the large scale structural conformation of the protein. Initially, the structural changes with ligation must be localized at the heme site. This localized or small scale to large scale coupling mechanism in energy transduction is a central issue of the molecular dynamics of many biological systems [1]. In this study, we report the results of ultrafast pump-probe spectroscopy of carboxymyoglobin (MbCO) and deoxymyoglobin (deoxyMb) to discern the initial displacements coupled to ligand photodissociation.

The experimental observable is sensitive to pump-induced wavepacket motion, due to the absorptive $\chi^{(1)}$ character of the probe interaction in the well separated pulse limit [2]. The detected signal is thus composed of either Raman excited modes of unphotolyzed MbCO or photolysis induced wavepacket motion. These possibilities are mutually exclusive for a given molecule, since the quantum yield of dissociation upon photon absorption is unity for MbCO. If a pump photon is absorbed to create an excited electronic state, wavepacket motion must be due to the chemical reaction; if no pump photon is absorbed, then any wavepacket motion must be due to Raman excited ground state motion of MbCO. Although the actual signal will be due to some mixture of these two effects, we present several control studies favouring the reaction-driven case.

Amplified pulses of ~5μJ in the 500-600nm spectral region were generated using a noncollinear optical parametric amplifier, and precompensated with chirped mirrors and a deformable mirror. As described previously [3], the output of this system typically gives pulses with 70 nm bandwidth, and durations of less than 10 fs at the sample. The experimental setup is a standard noncollinear pump probe setup, and has been detailed elsewhere [3]. To determine the spectral content of the pump-probe signal, it was subjected to two kinds of analysis to

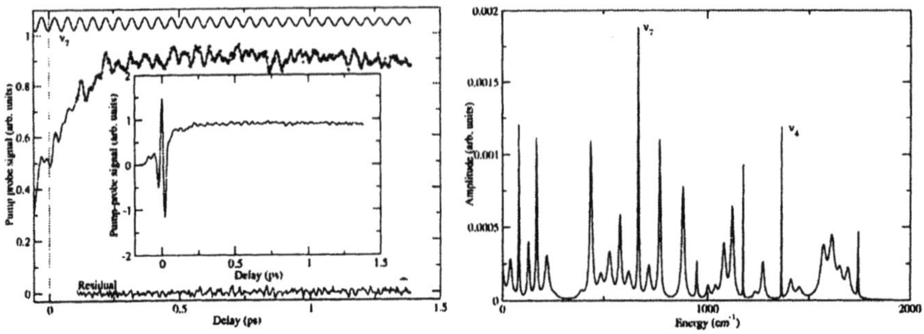

Figure 1. (left) Pump-probe signal and LPSVD fit of MbCO at 538 nm. The LPSVD derived ν_7 mode is shown along with residuals to the fit. (right) LPSVD amplitude spectrum.

extract the amplitude and phase of the molecular oscillations: standard discrete Fourier transform and linear predictive singular value decomposition (LPSVD) [4]. Typical processed data and fits along with LPSVD spectra for MbCO at 538 nm detection wavelength are shown in Figure 1. The ν_7 oscillation (670 cm^{-1}) is clearly the most significant contribution to the MbCO time domain signals. The observed frequency is significantly shifted from the equilibrium MbCO peak, in the direction corresponding to the deoxy species (within the first picosecond) as previously observed [5]. This indicates that ν_7 in the MbCO samples corresponds (at least in part) to an oscillation in just deligated deoxyMb. A second dominant frequency (1360cm^{-1}) was identified as the ν_4 mode, and is also shifted towards the equilibrium deoxyMb peak, indicating that it is also a reaction-driven mode. Interestingly, the pump probe data for deoxyMb did not show either of these modes with any significant amplitude, suggesting that our observations in MbCO are related to the dissociation process.

Equilibrium spontaneous Raman studies of MbCO and deoxyMb with CW excitation at 514 nm were also conducted. Unlike the pump-probe data that showed a much larger signal in MbCO than deoxyMb, the spontaneous Raman spectra show ν_4 signal in both species, but with amplitudes that are within a factor of two of each other. Neither species showed ν_7 signal. These differences between the spontaneous and impulsively driven Raman signal strongly suggest that the ν_7 and ν_4 modes are due to dissociation induced nuclear motion.

The experimentally derived phase of the ν_7 oscillation at 538nm detection wavelength is -160 +/- 30°. The phase of the actual physical vibration will be ambiguous, with a possible 180° phase shift [2], putting the actual phase of the vibrational motion at -160° or 20°. Using the method of Champion *et al.* [2], we calculate the phase for field induced Raman excitation of the ground state of MbCO to be 80°. Within the experimental error, this is inconsistent with the measured phase, and provides additional evidence that the observed oscillation is induced by the reaction.

As a final control study to confirm that the ν_7 and ν_4 modes are reaction-driven, we have conducted pump-probe experiments of cytochrome c. In this protein, the Fe-S bond between the heme and a methionine group is photolysed with ~unit quantum efficiency [8]. Preliminary results show no sign of either ν_7

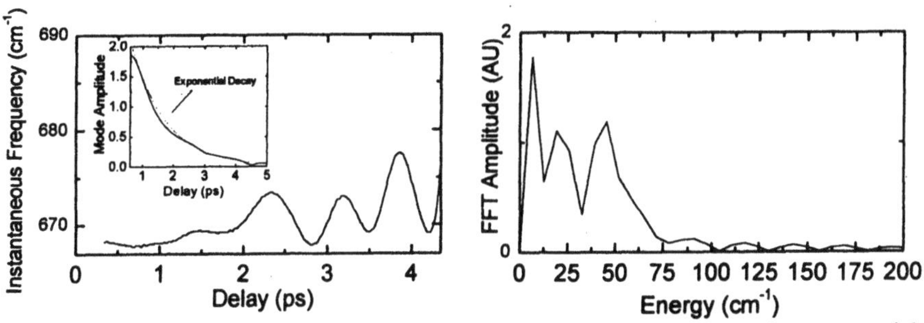

Figure 2: (a) Instantaneous frequency of the ν_7 mode. The inset shows the exponential decay of the amplitude of the mode. (b) Fourier spectrum of the instantaneous frequency.

or ν_4. This result can be understood within the context of reaction driven modes: the lower frequency of the Fe-S bond compared to the Fe-CO bond in myoglobin prevents impulsive excitation of either mode. Further experiments will be performed to confirm this point.

To gain an understanding of the relation between the reaction driven ν_7 mode and the subsequent protein motions along the dissociative pathway, analysis of the instantaneous frequency of the ν_7 mode was performed, revealing strong modulations as seen in Figure 2. FFT of the instantaneous frequency reveals a significant contribution at 50 cm^{-1}, similar to observations by Champion *et al.* [6] This mode has been identified with the doming motion of the heme [7], suggesting a pathway by which reaction driven displacements in ν_7 and ν_4 modes couple to low frequency modes in the protein, thereby distributing energy over a larger region, allowing for more effective displacement of the surrounding globin.

In conclusion, we find strong oscillations in the MbCO pump-probe data at the ν_7 and ν_4 modes of the porphyrin ring. Many control studies suggest that these modes are driven by the photodissociation process in myoglobin and form the first step in the energy transduction process between the ligation site and the surrounding protein. The picture that is emerging for the early time dynamics is that the initial crossing event in dissociation involves a superposition of displacements of ν_7 and ν_4 porphyrin modes that enable Fe to move out of plane and pick up the additional electron density with bond dissociation. Interesting modulation is also observed in the phase of these modes suggesting a mechanical coupling to longer length scale motions of the protein.

References

1 M. F. Perutz, Nature **228**, 726, 1970.
2 A. T. N. Kumar, *et al*,. J. Chem. Phys. **114**, 701, 2001.
3 M. R. Armstrong, *et al.* , Optics Letters **26**, 1152, 2001.
4 R. Kumaresan *et al.*, IEEE Trans. Ac., Sp., Signal Process. **ASSP-30**, 833, 1982.
5 Y. Mizutani and T. Kitagawa, Science **278**, 443, 1997.
6 L. Zhu *et al.*, Phys. Rev. Lett. **72**, 301, 1994.
7 X.-Y. Li, M. Z. Zgierski, Chem. Phys. Lett. **188**, 16, 1992.
8 W. Wang *et al.*, J. Phys. Chem. B **104**, 10789, 2000.

CARS Microscopy: 3D Vibrational Imaging of Proteins, DNA, and Lipids in Living Cells

Ji-Xin Cheng and X. Sunney Xie

Department of Chemistry and Chemical Biology, Harvard University, 12 Oxford Street, Cambridge, MA 02138, USA
Email: jxcheng@fas.harvard.edu, xie@chemistry.harvard.edu

Abstract. Vibrational mapping of intracellular proteins is achieved by polarization CARS microscopy. Fast imaging of lipids in live interphase cells, chromosomes in live mitotic cells, and lipid vesicles in apoptotic cells are carried out with a laser-scanning CARS microscope.

1. Introduction

Coherent anti-Stokes Raman scattering (CARS) microscopy is a 3D imaging technique with inherent chemical selectivity based on vibrational spectroscopy.[1] In CARS microscopy, the nonresonant electronic CARS background from the scatterer limits the spectral selectivity while that from the solvent limits the image contrast. Epi-detection (backward-detection) has been demonstrated as an efficient method to reject the solvent background.[2, 3] A CARS microscope using two synchronized near infrared picosecond pulse trains not only gives high spectral resolution, but also provides a high ratio of the resonant signal to the nonresonant background.[2] Recently, Cheng *et al.* developed polarization CARS (P-CARS) microscopy that effectively suppresses the nonresonant background by use of the polarization difference between the resonant and the nonresonant CARS fields.[4] Combination of picosecond excitation and a laser-scanning scheme allows fast imaging of living cells at an acquisition rate of several seconds per frame.[5] In this paper, we present CARS images of living cells as well as high-speed spectroscopic characterization of apoptosis with the methods described above.

2. Results and Discussion

Vibrational imaging of intracellular protein distribution with P-CARS microscopy is demonstrated in Figure 1. A P-CARS spectrum of N-methylacetamide liquid is displayed in Fig. 1A. The amide I band peaked at 1652 cm^{-1} shows a high signal to background ratio because of the effective suppression of the nonresonant background. The peak positions of the P-CARS bands coincide with the spontaneous Raman bands. Fig. 1B shows a P-CARS image of an epithelial cell. The nonresonant background from water and cellular organelles was effectively suppressed and the image contrast arose from the resonant CARS signal of the protein amide I band.

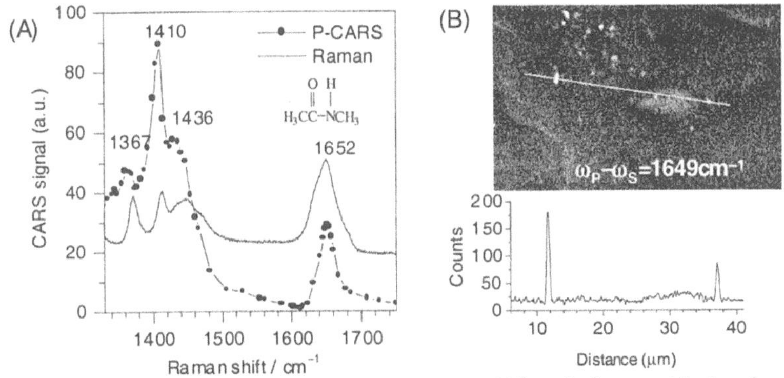

Fig.1. (A) Spontaneous Raman and P-CARS spectra of N-methylacetamide (see inset). (B) A P-CARS image of an epithelial cell based on the contrast from the protein amide I band. The intensity profile along the line in the image is shown below the image.

Laser-scanning CARS images of unstained live NIH3T3 fibroblast cells at different stages of a cell cycle are shown in Figure 2. In the forward (A) and backward (B) detected images of a cell in interphase, the pump and Stokes frequency difference ($\omega_p - \omega_s$) was tuned to the aliphatic C–H stretching vibration. The boundary of the cell nucleus is seen as a bright ring in the forward image as a consequence of the constructive interference of the CARS radiations from the nucleus lipid membrane. In contrast, a dark ring is seen in the backward image because of the destructive interference. The small lipid-rich organelles in the cytoplasm show a high contrast in the backward image owing to an effective rejection of the water background by epi-detection.[2, 3] By tuning $\omega_p - \omega_s$ to the DNA backbone vibration frequency, we could image the distribution of chromosomes in a live mitotic cell (Fig. 2C).

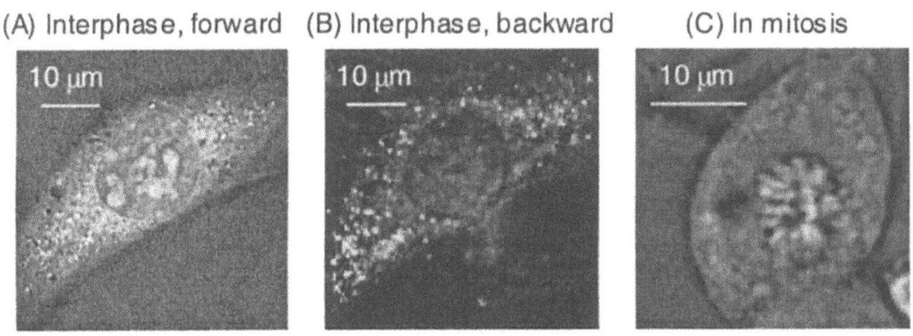

Fig.2. Laser-scanning CARS images of NIH3T3 cells acquired in 14 seconds. The average powers of the 5-ps pump and the 5-ps Stokes beams generated from two synchronized Ti:sapphire lasers were 40 and 20 mW, respectively, at a repetition rate of 80 MHz. (A) Forward-detected image of a cell in interphase with $\omega_p - \omega_s$ tuned to 2870 cm^{-1}. (B) Simultaneously epi-detected image of the same cell. (C) Forward-detected image of a cell in metaphase with $\omega_p - \omega_s$ tuned to 1090 cm^{-1}.

The high sensitivity of CARS microscopy permits visualization of intracellular dynamical processes. We carried out CARS imaging of NIH3T3 cells undergoing apoptosis, of which the early morphological change usually completes in several minutes. With the high acquisition speed of laser-scanning CARS microscopy, we were able to visualize the morphology changes at different stages of apoptosis based on C–H vibration. Figure 3 shows two CARS images of apoptotic cells in the same area taken at different depths (z) corresponding to two stages of apoptosis. The image contrast arose from lipids that are rich in C–H bonds. A lot of lipid vesicles are clearly seen in the early stage (Fig. 3A).

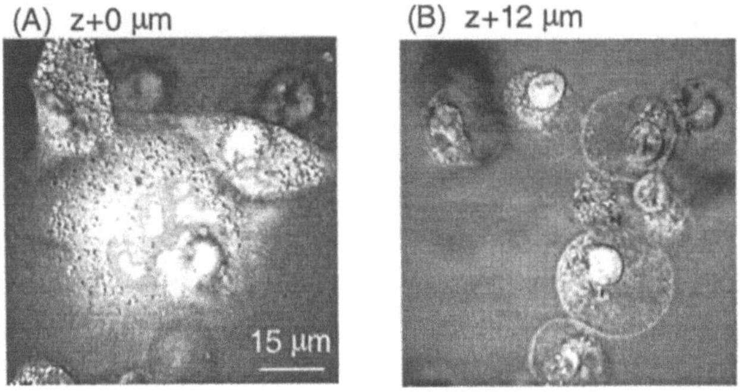

Fig. 3. Forward-detected CARS images of NIH3T3 cells treated with L-asparaginase for 20 hours. $\omega_p - \omega_s$ was tuned to the aliphatic C–H stretching vibration frequency at 2870 cm^{-1}. The pump and Stokes powers were 40 and 20 mW, respectively. The acquisition time was 8.5 seconds for each image of 512 × 512 pixels.

In conclusion, CARS microscopy provides an example of the usefulness of ultrafast technology to biology. Our results demonstrate that CARS microscopy is becoming a powerful tool for vibrational imaging of living cells.

Acknowledgements. This work was supported by an NIH grant (GM62536-01). We thank E. O. Potma, K. Jia, G. F. Zheng, X. L. Nan for their help.

References

1. A. Zumbusch, G. R. Holtom, and X. S. Xie, in *Phys. Rev. Lett.* Vol. 82, 4142, 1999.
2. J. X. Cheng, A. Volkmer, L. D. Book and X. S. Xie, in *J. Phys. Chem. B* Vol. 105, 1277, 2001.
3. A. Volkmer, J. X. Cheng, and X. S. Xie, in *Phys. Rev. Lett.* Vol. 87, 023901, 2001.
4. J. X. Cheng, L. D. Book, and X. S. Xie, in *Opt. Lett.* Vol. 26, 1341, 2001.
5. J. X. Cheng, K. Y. Jia, G. Zheng, and X. S. Xie, in *Biophys. J.* 2002, in press.

Ultrafast Elementary Events in the Excited State of Wild-Type Green-Fluorescent Protein

Kathrin Winkler, Marco Seidel, and Peter Vöhringer

Biomolecular and Chemical Dynamics Group, Max-Planck-Institute for biophysical Chemistry, Am Faßberg 11, 37077 Göttingen, Germany, E-mail: pvoehri@gwdg.de

Abstract. The optical response of wild-type (wt) green-fluorescent protein (GFP) has been measured at room temperature following 40-fs, 400-nm photoexcitation. In the spectral range of the stationary fluorescence of GFP, delayed emission due to excited-state proton transfer (ESPT) can be induced. The entire spectro-temporal response is simulated quantitatively using a novel dynamical model which includes an energy-dependent rate constant for ESPT, intra- and intermolecular flow of vibrational energy, and an additional non-radiative decay pathway leading to internal conversion.

1. Introduction

The primary steps leading to bioluminescence have recently attracted considerable interest [1,2]. An elucidation of the underlying molecular mechanisms and time scales will help establish fundamental principles involved in photon emission from protein environments. The photoactive protein, GFP, from *Aequoria Victoria* is prototypical for bioluminescence from β-barrel tertiary structures which can protect the chromophore from the aqueous surroundings and effectively prevent solvent-assisted quenching. Its chromophore (p-hydroxybenzylidene-imidazolidinone) is coupled to the protein backbone via an extended network of hydrogen-bonds which facilitates a primary ESPT followed by a characteristic green emission. Here, we present a detailed femtosecond time- and frequency-resolved pump-probe study to quantify the elementary dynamics accompanying ESPT in wt-GFP.

2. Results and Discussion

Wt-GFP was expressed in *E. coli* and purified as described in Ref. [3]. First, time-resolved pump-probe transients as a function of probe wavelengths were measured. This study was carried out with 40-fs, 400-nm pump pulses and probe pulses of 40 fs throughout the visible spectral range. Since photo-conversion strongly alters the GFP response, rapid scan data acquisition with pump and probe pulse energies of less than 1 nJ was performed while spinning the sample at 5000 rpm. Probe wavelength-dependent transients are shown in Fig. 1a for the spectral range covering the stationary emission spectrum of GFP. Here, a strong induced emission rises in a highly non-exponential fashion. Furthermore, a fraction of the emission appears within the time resolution. For wavelengths longer than 630 nm, a distinct transient absorption decays on similar time scales (see Fig. 1b). In a second study, transient differential transmission spectra (DTS) at various time delays were recorded (Fig. 1c). The following features are worth highlighting: (i) At the shortest delays, the stimulated emission is much broader than the stationary fluorescence of GFP but a shoulder reminiscent of the vibrational substructure of GFP is already visible. (ii) After about 7 ps, the red edges of the DTS and the stationary spectrum are identical. (iii) For delays larger than 1 ps, an isosbestic point at 496 nm is observed. (iv) A residual long-time transient absorption is detected for wavelengths shorter than 480 nm.

We discuss these results in terms of ESPT. The stable protonated form of the chromophore is denoted as A whose excited-state, A^*, is prepared by the pump pulse. Green emission originates from an intermediate, I^*, which is an environmentally unrelaxed form of the deprotonated excited-state, B^* [1]. Obviously, the build-up of the induced emission reflects the ESPT kinetics under I^* formation. The transient absorption for $\lambda \leq 480$ nm must also be due to I^* as it persists for long delays. Consequently, the decaying absorption for $\lambda \leq 630$ nm is assigned to A^*. Besides by ESPT, I^* is also accessed directly by the pump-pulse since the red edge of the prompt emission matches the stationary spectrum. The isosbestic point separates emission dominated either by A^* or I^* and implies a barrier for ESPT which is larger than thermal excitations. However, a simple 2-state interconversion is inconsistent with multi-exponential kinetics even if an ensemble of heterogeneous environments are invoked in which ESPT can occur.

Here, a refined dynamical model for GFP (Fig. 2a) is proposed which takes into account that 400-nm excitation prepares A^* with an excess energy of 2000 cm^{-1} as estimated from the 0-0-transition observed in hole-burning studies [3]. Therefore, A^* must be coarsely grained in energy to capture the intrinsic energy dependence of the rate coefficient for unimolecular barrier crossing leading to proton transfer. Vibrationally hot (index h), partially relaxed (p) and vibrationally cold (c) species of A^* are introduced which undergo ESPT with energy-specific rate constants, k_h, k_p, and k_c. Vibrational energy relaxation (VER) is characterized by k_r. Furthermore, measurements of the fluorescence quantum yield, Φ, give evidence for another decay channel of A^* leading to internal conversion via chromophore torsion [2]. The model includes this pathway by the rate coefficient, k_T, which is assumed to be energy-independent for simplicity. The precise value of k_T can be obtained independently from the time-resolved spontaneous fluorescence in the spectral reg-

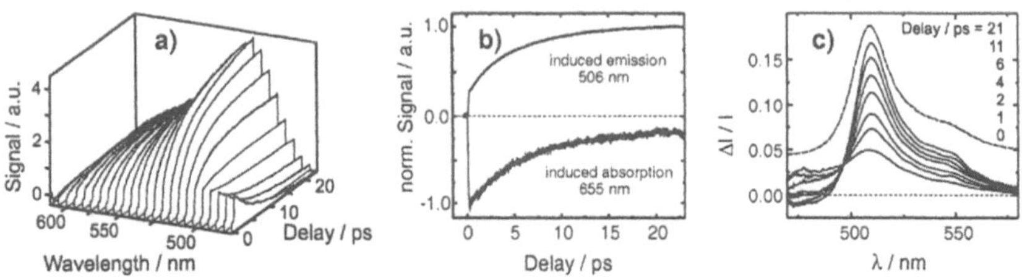

Fig. 1. a) Probe wavelength-dependent transients for GFP. b) Pump-probe transients emphasizing the matching behavior of induced emission and near-infrared induced absorption. (c). Comparison of DTS (solid) and stationary emission (dashed).

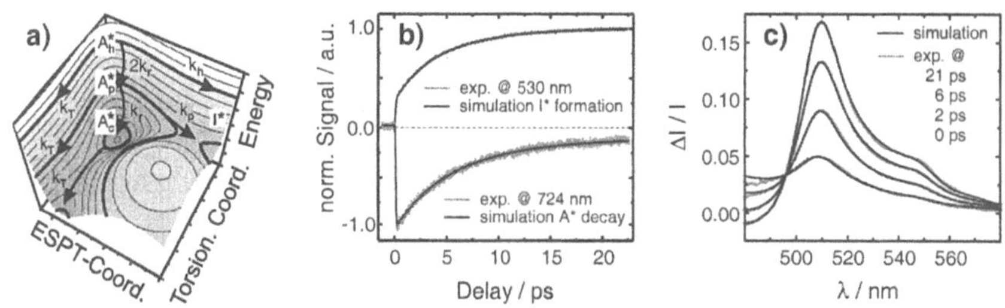

Fig. 1. Refined dynamical model for GFP (a) and comparison of the model with pump-probe transients (b) and DTS(c)

gion where only A^* can emit, i.e. λ<480 nm. We have performed time-correlated single-photon counting (TCSPC) which reveal a long-time decay with a time constant of 300 ps. When the build-up of I^* fluorescence is detected, a corresponding rise time cannot be found indicating that this component is related to the additional decay channel rather than ESPT. Deuteration studies confirm such an interpretation. This finding further implies that the ESPT channel is closed once the protonated chromophore is vibrationally cold, i.e $k_c \rightarrow 0$. Analytic expressions for the time-dependent populations, A^* and I^*, as well as for Φ are readily derived.

The model has to reproduce the pump-probe transients, the time-dependent DTS, the fluorescence quantum yield, the TCSPC data, and the deuteration effect simultaneously with only three adjustable parameters, namely k_h, k_p, and k_r. Model simulation are shown in Fig. 2b and c for the decay of A^* (probed at 724 nm), for the build-up of I^* (530 nm), and for various DTS. Both, the pump-probe data and the DTS can be reproduced quantitatively with the same set of parameters. Further, a value for Φ of 11% is obtained in agreement with the experimental value from [2]. Finally, TCSPC data and the deuteration effect can also be fully reproduced.

In conclusion, we were able to reproduce the entire optical response of GFP with a novel photodynamical model which includes VER, energy-dependent ESPT, and an additional decay pathway for the protonated chromophore of GFP leading to internal conversion via torsion. In particular, the non-exponential nature of the kinetics is traced back to a competition of VER and ESPT. Finally, we stress that within the framework of the model, VER is significantly slower than ESPT. This might be attributed to the highly rigid tertiary structure of the protein which protects the chromophore from inelastic collisions with the aqueous surroundings.

References

1 M. Chattoraj, B.A. King, G.U. Bublitz, and S.G. Boxer, PNAS, **93,** 8362, 1996.
2 A.D. Kummer et.al., Chem. Phys. **237**, 183, 1998.
3 T.M.H. Creemers, A.J. Lock, V. Subramaniam, T.M. Jovin, and S. Völker, Nature Struct. Biol. **6**, 557-560 (1999)
4 K. Winkler, J. Lindner, V. Subramaniam, T.M. Jovin, and P. Vöhringer, PCCP, **4**, 1072, 2002

Real-time Observation of Conformational Dynamics in Peptide Folding

W. Zinth[1], S. Spörlein[1], H. Carstens[1], C. Renner[2], R. Behrendt[2], L. Moroder[2], P. Tavan[1] and J. Wachtveitl[1,3]

[1] Lehrstuhl für BioMolekulare Optik, Oettingenstr. 67, Ludwig-Maximilians-Universität München, 80538 München, Germany
E-mail: zinth@physik.uni-muenchen.de

[2] Max-Planck-Institut für Biochemie, am Klopferspitz 18a, 82152 Martinsried, Germany

[3] Institut für Physikalische und Theoretische Chemie, Marie-Curie-Str. 11, Goethe Universität Frankfurt, 60439 Frankfurt am Main, Germany

Abstract. Femtosecond spectroscopy on optically triggered cyclic model-peptides combined with computer simulations reveals ultrafast events in protein folding. Important initial conformational dynamics occur within picoseconds and are quantitatively simulated with molecular dynamics calculations.

In a living organism, the synthesis of an amino acid chain in the ribosomes is followed by the folding of the chain into the native protein. This folding to the functional structure proceeds through a conformational dynamics on a nested hierarchy of temporal and spatial scales. The fastest processes comprise local rearrangements of the polypeptide chain and occur within picoseconds whereas slower reactions involving large parts of a protein may proceed on the time scale of seconds [1]. Experimentally, these events have been studied by classical stopped-flow or temperature jump techniques with a time-resolution of milli- or at best microseconds. Recently, the nanosecond time-scale has become accessible via laser-induced temperature-jump and by triplet-triplet energy transfer [2]. So far the speed limit of protein folding reactions was deduced from triplet-triplet quenching experiments of small peptides to be ca. 20 ns [3].

In order to investigate structural changes on the sub-nanosecond time scale we have developed a special molecular system: A photo-switchable peptide (see Fig. 1, top) is synthesized, which is backbone-cyclized with a suitably functionalized (4-aminophenyl)azobenzoic acid (APB). The APB molecule has been chosen as optical trigger, since azobenzene dyes are known to photoisomerize around the central diazene bond within fractions of picoseconds at high quantum yields [4] and to induce significant changes of the petide structure (see Fig. 1) [5,6]. Since the two isomeric forms of the chromophore retain pronounced spectral differences in the peptide construct, a selective trigger of the reaction combined with clear-cut spectroscopic distinction of the products is possible (see Fig. 1, bottom). Furthermore, the APB-peptide is simple enough as to allow a calculation of its conformations by molecular dynamics (MD) simulations. During the various steps of the protein folding process situations occur where parts of the protein undergo transitions be-

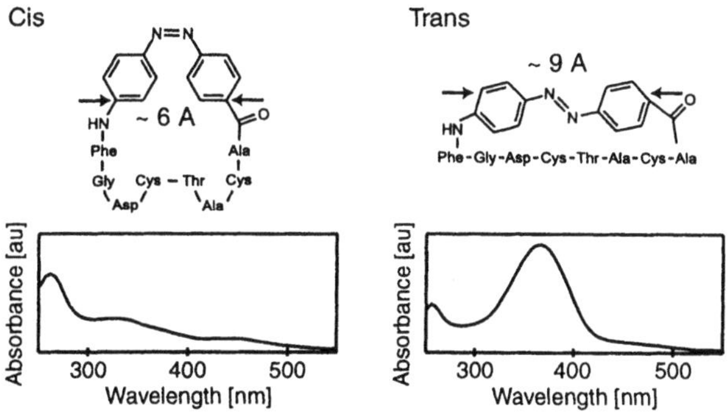

Fig. 1. Structure and absorption spectra of the two conformations of the investigated azopeptide. The isomerization leads to considerable elongation of the chromophore and to strongly changed spectra.

tween structures of well-defined geometries. Such a conformational dynamics is expected in our azo-peptide model system and is addressed in the present studies.

After excitation of the *cis*-conformers with a femtosecond light pulse, a number of reactions occur: the azobenzene moiety is expected to isomerize in the electronically excited state around the diazene double bond from *cis* to *trans* (step 1) and to relax into the ground state by dissipating excitation energy in the form of heat into the peptide chain and teh surrounding solvent (step 2). Subsequently, the strain exerted by the azobenzene in the *trans*-configuration with its changed geometry (elongation from 6 to 9 Å and planarization, Fig. 1) will force the peptide moiety through a series of conformational transitions into the structure of the *trans*-APB peptide (step 3). Without the ring restriction no conformational transitions within the peptide moiety are expected. Therefore, a corresponding linear APB-peptide is used as a reference system.

The experimental system for the femtosecond pump-probe experiment is based on a Ti-sapphire laser-amplifier set-up and a multichannel detection system [7]. The most important features in brief: Excitation wavelength 480 nm, white light probing in the range 300 nm to 700 nm. Width of the temporal instrumental response function ca. 80 fs. In our set-up the *cis-trans*-isomerization of APB is triggered by a 50 fs pulse and the reactions are probed by a delayed white-light pulse of the same duration. According to the static absorption spectra, the formation of the *trans*-APB ground state (in step 2) should lead to a strong increase of the absorption at about 370 nm. Furthermore, whenever absorption at wavelengths longer than 500 nm is observed, it has to be attributed to intermediate states, since the relaxed ground states of both isomers do not absorb in this spectral range. Such intermediates may represent the electronically excited state of APB (step 1), or a "hot" ground state of the *trans*-APB, whose vibrations decay upon dissipation of the excitation energy (step 2), or even a distorted cyclic *trans*-APB-peptide, whose internal strain relaxes during the conformational transitions (in step 3).

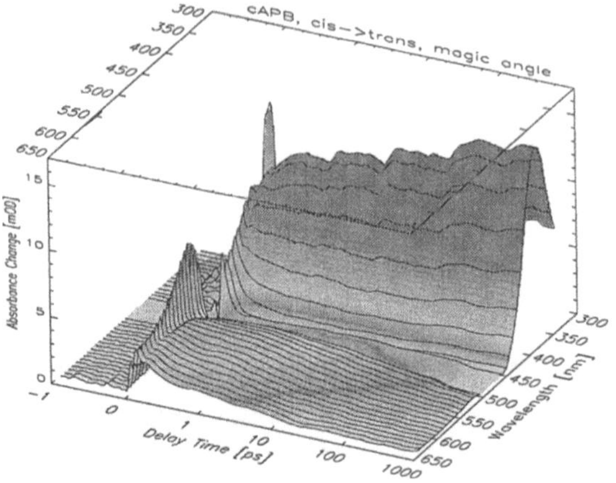

Fig. 2. Transient absorption change of cyclic APB-peptide as a function of wavelength after the photoexcitation of the *cis*-isomer.

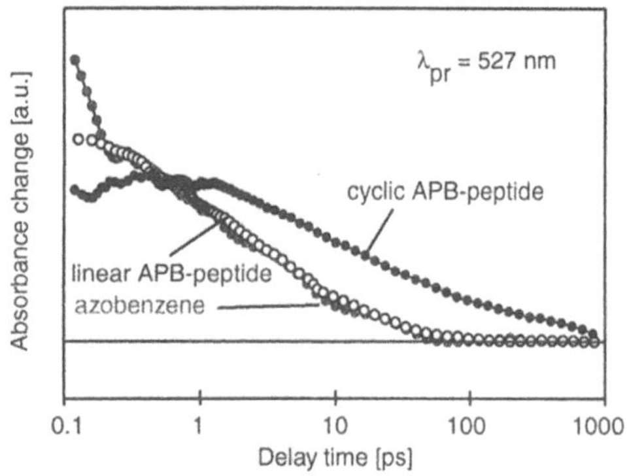

Fig. 3. Transient absorption change of different azo-compounds. While the linear peptide behaves similar to free azobenzene, the cyclic peptide shows much slower kinetics.

In Figure 2 shows transient absorption data of the cyclic APB-peptide obtained by photo-excitation of the *cis*-isomer. With increasing time-delays, the strong absorption of the cyclic *trans*-APB-peptide near 370 nm builds up and undergoes a blue-shift at increasing time-delay. This process occurs predominantly on the 10 to 50 ps time scale. Only weak absorption changes are found at later delay times. Interesting and similar features are seen in the range of the $n\pi^*$-transition at $\lambda > 500$ nm: At first, the excited electronic state shows up via its increased absorption. This excited state absorption decays within 3 ps. The following ca. 15 ps process can be attributed to the cooling of the vibrationally hot molecule. Later on, one observes a pronounced blue-shift of the remaining absorption band on the time scale of 10 to 50 ps. On the 100 ps time scale this absorption band starts to dis-

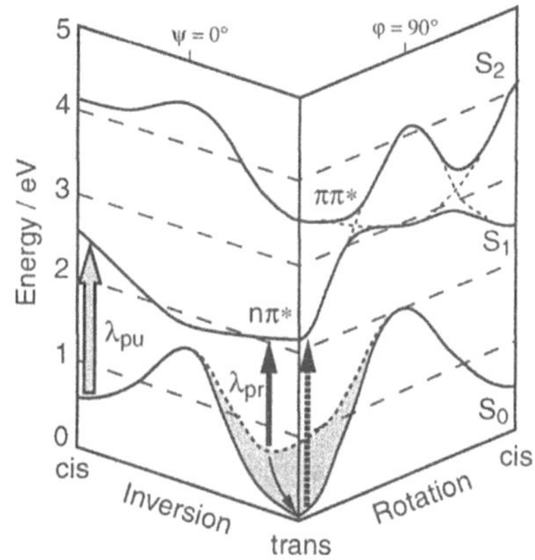

Fig. 4. Potential energy surfaces of azobenzene showing qualitatively the action of peptide strain on the absorption wavelength (surfaces according to ref. [8]).

appear. For the *linear* APB-peptide the picosecond reaction dynamics is considerably faster. This is evident from the traces at $\lambda_{pr} = 527$ nm shown in Fig. 3. While the linear ABP-peptide has a picosecond absorption dynamics very similar to that of azobenzene, the absorption changes of cyclic azopeptide persist for much longer times. Even after 1 ns the absorption has not recovered completely.

Interesting information is obtained from the analysis of the newly formed *trans*-band of the chromophore, which serves as a monitor for the molecular interactions on the azo-peptide chromophore. After photoisomerization of the APB-moiety, the incomplete relaxation of the peptide chain may exert strain on the chromophore. This results in a modification of the potential energy surfaces: When the APB molecule has reached the *trans* ground-state configuration, the dynamic interactions with the peptide prevents the molecule from reaching immediately the relaxed geometry (see Fig. 4, gray area). It is only after the disappearance of the strain on the chromophore, that this red-shift will disappear. In Figure 5 (circles) the temporal shift of the peak position of the ππ*-band around 370 nm is plotted together with the results of a molecular dynamics simulation (points, solid curve). The MD simulation calculate the energy content in the cyclic APB-peptide. Simulated energy content and observed absorption change are seen to match closely. This agreement demonstrates (i) that the peak shift actually senses all fast processes within the peptide, (ii) that it essentially reflects the energy content of the APB-peptide, and (iii) that the kinetics of the light-triggered folding of the cyclic peptide into its *trans*-conformation can be quantitatively described by MD-simulations. Unfortunately, due to the multidimensionality of the structural changes, a direct inspection of the MD-traces, calculated during the first nanosecond after the trigger, can not give information on the details of the average structural changes. However work is in progress to describe the motion in the multidimensional configuration space by elaborated statistical techniques. Preliminary data indicate that the relaxation toward the *trans*-configuration of the whole azopetide is about halfway completed. Preliminary transient IR data support this finding.

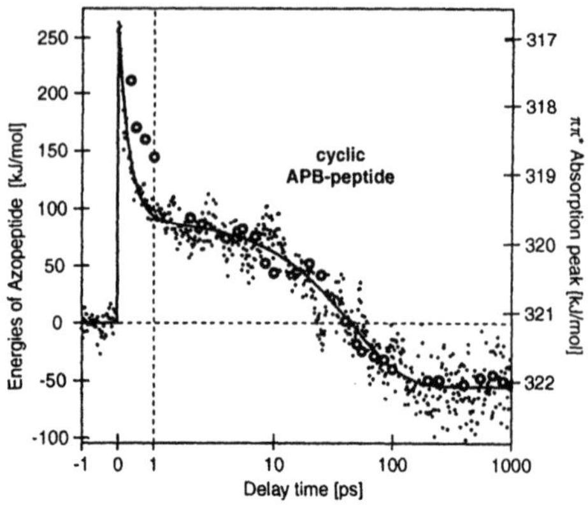

Fig. 5. Comparison of the energy content in the APB-peptide during MD simulations upon light-induced *cis/trans*-isomerization (small dots) compared with the peak position of the *trans*-band (circles).

In summary, we find that important conformational transitions in the cyclic azo-peptide occur on the 10 - 50 ps time scale. The surprisingly good agreement between spectroscopic and computer experiments shows that the virtual reality generated by high quality computer simulations faithfully models the molecular processes. We learn that substantial conformational transitions in the cyclic peptide proceed on the quite rapid time scale of a few picoseconds. This supports the view, that ultrafast conformational dynamics have to be included in descriptions of protein folding processes.

References

1 H. Frauenfelder, P.G. Wolynes and R.H. Austin, in *Reviews of Modern Physics*, Vol. 71, S419, 1999.

2 L.J. Lapidus, W.A. Eaton and J. Hofrichter, *Proceedings of the National Academy of Sciences of the United States of America*, Vol. 97, 7220, 2000.

3 O. Bieri, J. Wirz, B. Hellrung, M. Schutkowski, M. Drewello and T. Kiefhaber, in *Proceedings of the National Academy of Sciences of the United States of America*, Vol. 96, 9597, 1999.

4 T. Nägele, R. Hoche, W. Zinth and J. Wachtveitl, in *Chemical Physics Letters*, Vol. 272, 489, 1997.

5 R. Behrendt, C. Renner, M. Schenk, F.Q. Wang, J. Wachtveitl, D. Oesterhelt and L. Moroder, *Angewandte Chemie-International Edition*, Vol. 38, 2771, 1999.

6 C. Renner, R. Behrendt, S. Spörlein, J. Wachtveitl and L. Moroder, Biopolymers, Vol. 54, 489, 2000.

7 M. Seel, E. Wildermuth and W. Zinth, *Measurement Science and Technology*, Vol. 1997, 449-452, 1997.

8 S. Monti, G. Orlandi, P. Palmieri, *Chemical Physics*, Vol. 71, 87, 1982.

Multidimensional Nonlinear Spectroscopy of Conformational Relaxation in Azobenzene Peptide

Ravindra Venkatramani[a], Tomoyuki Hayashi[b] and Shaul Mukamel[a,b]

[a] Department of Physics and Astronomy, University of Rochester, Rochester, NY 14627 USA
E-mail: ardnivar@pas.rochester.edu
[b] Department of Chemistry, University of Rochester, Rochester, NY 14627, USA
E-mail: thayashi@chem.rochester.edu
E-mail: mukamel@chem.rochester.edu

Abstract: A simulation study of Azobenzene peptide is presented in which a ground state isomerization of the APB residue is simulated by applying dihedral angle constraints. A 750 ps post isomerization relaxation of the peptide chain is found. Schemes for probing the structural dynamics and the photoisomerization process using multidimensional, infrared, coherent techniques are outlined.

1. Introduction

Azobenzene peptide is an artificially synthesized molecule [1,3], comprising of the active site fragment of thioredoxin reductase, which is backbone-cyclized by the photochromic (4-Amino) phenylazobenzoic acid (APB). The peptide exists in two stable states corresponding to the *trans* and *cis* configurations of the APB residue. The peptide chain around APB shows different levels of conformational flexibility in these two states. According to NMR studies [2], the peptide chain is more flexible in the *cis* structure. In this work we construct and simulate a "ground state" isomerization process in APB peptide. Possible techniques, which can follow the relaxation of the peptide chain by coherent probing of vibrations are outlined. By extracting parameters for a vibrational hamiltonian from the simulation, it will be possible to follow the structural dynamics through ultrafast, infrared, three pulse techniques. The Collective Electronic Oscillator (CEO) formalism [4] may be used to simulate the photoisomerization process.

2. MD Simulation

The peptide structure was constructed using the CHARMM force field. The structure differs from the original given in [1] by excluding the *tert*-butyl groups. The starting geometries for the *cis* and *trans* structures were obtained from [1]. *cis*-APB peptide was embedded in a DMSO (Dimethyl Sulfoxide) box of side 40 A, containing 550 molecules (density – 1.12 g/cm^3). Solvent atoms overlapping with the peptide atoms were removed. The peptide was minimized and held fixed while the solvent was heated for 1ps and equilibrated at 300 K for 50 ps. The peptide-solvent system was then heated for 1ps and subsequently equilibrated at 300 K for 100 ps. During this stage and during the dynamics run, the dihedral angle of the APB residue was constrained at 360 degrees. The same procedure was repeated with a dihedral angle constraint of 180 degrees to get an equilibrated *trans*-APB structure in DMSO. Dynamics runs of 3 ns were carried out for the *cis* and *trans*

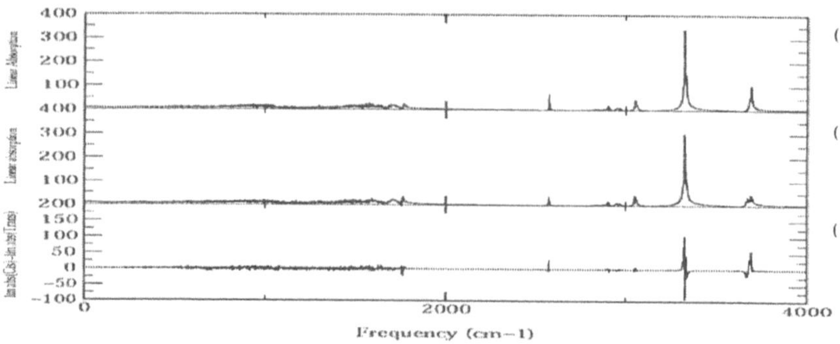

Fig. 1. Linear Absorption spectra for (a) *cis*-APB and (b) Trans-APB structures. Difference spectra Linear absorption (cis minus trans) is shown in (c).

structures with snapshots at every 1 ps. The *cis* structure was then subjected to a constrained rotation of the dihedral angle around the azo-group in steps of 5 degrees to change the configuration from *cis* to *trans*. The total time for rotation was 3.6 ps and resulting structures after each 5 degrees rotation were equilibrated at 300 K for 0.1 ps. A 3 ns dynamics run was carried out for the resultant *trans* structure with snapshots at every 1 ps.

2. Results and Discussion

A root mean square deviation (rmsd) calculation of the structures obtained during the 3ns dynamics run for the showed that both the *cis* and *trans*-APB peptides were stable in the last 1.5 ns. 500 structures were obtained from the last 1 ns of these dynamics runs. A normal mode analysis was performed and the averaged linear absorption spectra were calculated. These are plotted in Fig 1.

The linear absorption for both configurations is very similar with minor peak shifts and relative intensity changes for the vibrations. We expect the peptide chain to occupy different conformational subspaces for the *cis* and *trans* structures, but the linear absorption is not very sensitive to these differences. 2D non-linear spectroscopy shall be much more sensitive to the couplings and dipole orientations and hence to the structure. An rmsd calculation (fig 2) with respect to the average structure for the 3ns post-isomerization *trans*-APB peptide run shows evidence of

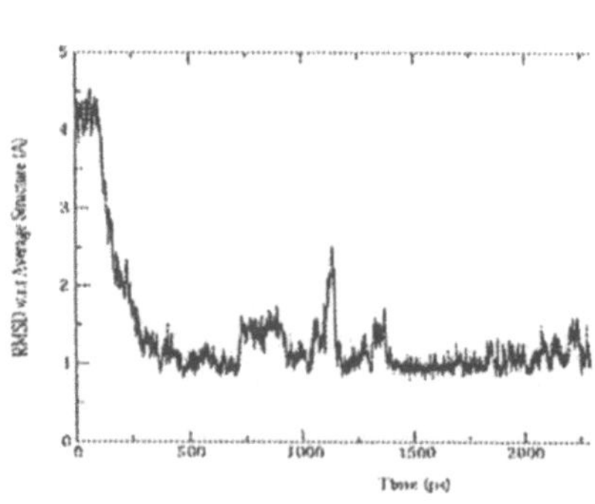

Fig. 2. RMSD with respect to the average structure from the 3ns post-isomerization run.

peptide chain relaxation lasting up to 750 ps The structure is stable for the remainder of the trajectory.

We have implemented a scheme for following the photoisomerization and subsequent peptide chain relaxation and using 2D infrared techniques:

In order to simulate the photoisomerization process, MD-CEO algorithm [5] is utilized. This is a molecular dynamics method based on the Collective Electronic Oscillator approach [4], which provides a fast, inexpensive algorithm to calculate excited state surfaces. The procedure is described below:

(i) Construction of the theoretical model for azobenzene peptide. Electronically excited state surface of azobenzene peptide is modeled using CEO formalism. Nuclear motions of azobenzene peptide are obtained by solving the classical equations of motion.

(ii) Implementation of non-adiabatic coupling, which is crucial in the simulation of photoisomerization, to the MD-CEO program.

(iii) Following the photoisomerization dynamics of azobenzene peptide.

(iv) Analyzing the spectral changes during the isomerization process and assigning these spectral changes to structural changes.

(v) Analyzing the coupling between the isomerization of azobenzene and the relaxation of the peptide.

An analysis of the trajectories of the pre and post isomerization runs for *cis* and *trans* structures is done in the following sequence:

(i) Vibrations, which are strongly coupled and are most sensitive to structural changes, are identified

(ii) The vibrational hamiltonian is constructed as a set of coupled localized vibrations

(iii) 2D and 3D spectra for all possible four wave mixing techniques are calculated.

(iv) Analyzing and comparing the spectra, allow to assess the sensitivity of each technique to various structural changes.

Acknowledgements. The support of the National Institutes of Health (GM59230-01A2) and the National Science Foundation (CHE-0132571) is gratefully acknowledged.

References:

1. C. Renner, R. Behrendt, S. Spörlein, J. Wachtveitl, L. Moroder, Biopolymers Vol. 54, 489, 2000.
2. J. Wachtveitl , S. Spörlein, W. Zinth, C. Renner, R. Behrendt, L. Mororder, Ultrafast Phenomena XII, Springer Series in Chem. Physics, 66, Eds. T. Elsaesser, S.Mukamel, M.M. Murnane, N.F. Scherer 695, 2000.
3. S. Spörlein, H. Carstens, C. Renner, R. Behrendt, L. Mororder, P. Tavan, W. Zinth, J. Wachtveitl , PNAS (In press).
4. E. V. Tsiper, V. Chernyak, V. Tretiak and S. Mukamel, J. Phys. Chem. Vol. 110, 8328, 1999.
5. M. Tommasini, G. Zerbi, V. Chernyak, and S. Mukamel, J. Phys. Chem. Vol. 105, 7057, 2001.

Coherent vibrational dynamics of green fluorescent proteins and blue coppers proteins

G. Cerullo[1], M. Zavelani-Rossi[1], S. De Silvestri[1], V. Pellegrini[2], F. Beltram[2], S. Cannistraro[3]

1 Istituto Nazionale per la Fisica della Materia, IFN-CNR, Dipartimento di Fisica, Politecnico di Milano, Italy, E-mail: sandro.desilvestri@fisi.polimi.it
[2] Istituto Nazionale per la Fisica della Materia, Scuola Normale Superiore, Pisa, Italy
[3] Istituto Nazionale per la Fisica della Materia, Università della Tuscia, Viterbo, Italy

Abstract. The vibrational motions coupled to the electronic transitions in optically active proteins are observed in real time by femtosecond spectroscopy. We present results on green fluorescent proteins and on blue copper proteins (azurin and plastocyanin).

1. Introduction

Femtosecond spectroscopy enables to follow in real time, in molecules and solids, vibrational motions coupled to the electronic transition. Time-domain vibrational spectroscopy is complementary to frequency-domain resonance Raman (RR) spectroscopy, providing information on nuclear dynamics in the excited state and on the shape of the potential energy surfaces. In addition, it allows to detect low-frequency modes which, in RR experiments, are often obscured by Rayleigh scattering. There have been only few reports of observation of coherent vibrational dynamics in biomolecules, such as proteins [1-3]; studies under normal operating conditions, i.e. at room temperature and in physiological solution, are challenging because of the fast dephasing times.

In this work we present results on the coherent vibrational dynamics in two important classes of proteins: green fluorescent proteins (GFPs) and blue copper proteins. GFP of the *Aequorea Victoria* jellyfish has an intrinsic green emitting chromophore, which makes it a unique fluorescent label. Blue copper proteins work as mobile electron carriers in a wide variety of biological processes such as bacterial respiration (azurin) and photosynthesis (plastocyanin).

2. Results and discussion

The pump-probe experiments were performed using a non-collinear optical parametric amplifier, pumped by the second harmonic of a Ti:sapphire laser at 1 kHz. The system generates pulses with ultrabroad bandwidth, extending from 500 to 650 nm, compressed to nearly transform-limited, sub-10-fs duration by multiple reflections onto chirped mirrors [4]. Samples are kept in a 0.5-mm cuvette at room temperature. Dynamics at different wavelengths are obtained by selecting after the sample spectral components of the broadband probe pulse with 10-nm bandwidth interference filters.

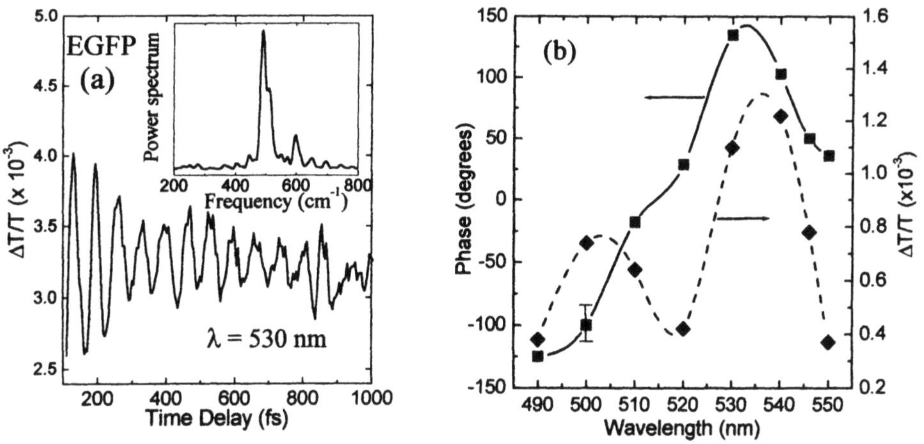

Figure 1: (a) Room temperature differential transmission of EGFP as a function of pump-probe delay. Inset shows the Fourier transform of the oscillatory component of the signal; (b) spectral dependence of amplitude (diamonds) and phase (squares) of the oscillatory component of the signal.

We first study two GFP mutants with enhanced brightness and photostability: F64L/S65T GFP [enhanced GFP (EGFP)], and T203Y EGFP (E^2GFP). Fig. 1(a) shows the differential transmission (ΔT/T) signal of EGFP: an oscillatory pattern is superimposed on a constant positive signal, due to stimulated emission (SE) from the long-lived excited state. A Fourier analysis of the data gives a dominant mode at 497 cm^{-1} and a weaker one at 593 cm^{-1}, both with damping time of about 1 ps. Comparable results are obtained for E^2GFP, with the dominant mode slightly blue-shifted to 510 cm^{-1} and the weaker mode red-shifted to 570 cm^{-1}.

A wavelength-dependent analysis of amplitude and phase of the 497 cm^{-1} mode in EGFP (see Fig. 1(b)) shows that oscillations are weaker on the peak and at the wings of the SE band and are larger on its steep sides, while a strong phase change, (approximately an inversion) centered around the wavelength of the emission maximum is observed. This analysis allows to associate unambiguously the 497 cm^{-1} mode to wave packet dynamics in the excited state. The 593 cm^{-1} mode is attributed to the ground state based on the analysis of steady state emission lineshape. To support this assignment we performed a*b initio* calculations of the vibrational properties of the GFP chromophore based on density functional theory [5]. The results show that, under appropriate nonthermally equilibrated conditions reproducing the atomic displacements induced by photoexcitation, two modes at frequencies around 600 cm^{-1} are coupled to the electronic excitation, both corresponding to angular in-plane deformations of the phenolic and heterocyclic rings of the chromophore. Simulations also show that energy is transferred to these modes from the carbon stretching modes, coupled to the electronic excitation, due to the geometric constraint introduced by the double-ring structure of the chromophore [6].

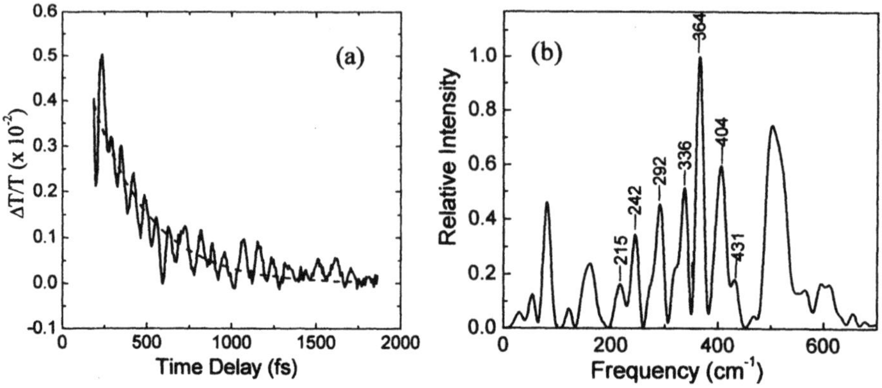

Fig. 2: (a) Room temperature ΔT/T dynamics of Azurin. Dashed line is an exponential fit. (b) Fourier transform of the oscillatory component of the signal. Bands labeled with their corresponding frequency have been found in the RR spectrum

We also studied blue copper proteins directly pumped in the ligand-to-metal charge transfer band. Fig. 2(a) shows the ΔT/T dynamics of Azurin from *Pseudomonas aeruginosa*. We observe a fast decay of ground state bleaching, with time constant of ≈ 270 fs, assigned to the reverse charge transfer process. Superimposed on this decay is a complex oscillatory pattern, as shown by the Fourier spectrum in Fig. 2(b), with a strikingly rich spectral content. Some of the frequencies are consistent with those measured in RR spectra, and assigned to Cu-S(Cys) and Cu-N(His) stretching modes. Three low-frequency bands are observed in the 30-80 cm^{-1} region and are attributed to collective "phonon" modes involving large portions of the protein. Finally, there is an intense mode around 510 cm^{-1}, not present in the RR spectrum; assignment of this mode to the excited state is supported by its fast damping time (≈ 300 fs). A similar mode was observed also in plastocyanin, in agreement with previous results [2]. Access to the vibrational motions coupled to the excited state in blue copper proteins should allow a more accurate simulation of the physiological electron transfer event.

References

[1] L. Zhu, P. Li, M. Huang, J.T. Sage, and P.M. Champion, Phys. Rev. Lett. **72,** 301 (1994).
[2] L.D. Book, D.C. Arnett, H. Hu, and N.F. Scherer. J. Phys. Chem. A **102,** 4350 (1998).
[3] U. Liebl, G. Lipowski, M. Negrerie, J.C. Lambry, J.L. Martin, and M. Vos, Nature **401,** 181 (1999).
[4] M. Zavelani-Rossi et al., Opt. Lett. **26,** 1155 (2001)
[5] R. A.G. Cinelli, V. Tozzini, V. Pellegrini, F. Beltram, G. Cerullo, M. Zavelani-Rossi, S. De Silvestri, M. Tyagi, and M. Giacca, Phys. Rev. Lett. **86,** 3439 (2001).
[6] K. Moritsugu, O. Miyashita, and A. Kidera, Phys. Rev. Lett. **85,** 3970 (2000).

Vibrational Dynamics, Mode Coupling and Structure of Acetylproline and other Dipeptides

Igor V. Rubtsov, Jianping Wang, and Robin M. Hochstrasser

Department of Chemistry, University of Pennsylvania, Philadelphia, PA 19104, USA
E-mail: hochstra@sas.upenn.edu

Abstract. Two-dimensional infrared transient absorption spectroscopy is applied to study vibrational relaxation, mode coupling and structure of several dipeptides in solution. Efficient vibrational energy transfer (EnT) between amide-I and amide-II bands of the same amide has been observed in $AcProNH_2$ (τ~3.1 ps). The energy transfer process from the amide-I band acetyl end to the amide-II band(amino) has been detected (~10 ps), but not that between the two amide-I bands. Interconversion of different Liouville pathways cased by vibrational energy transfer has been clearly demonstrated giving a reason for the nonexponential decay kinetics observed in the conventional pump-probe experiments. Coupling strengths of the amide-A mode with other modes (amide-I, amide-II, etc.) in dipeptides were measured for the first time, delivering essential structural information. Different coupling strengths were obtained in isolated amide and across the H-bond in the self-H-bonded conformation. Angles between transition moments of amide-A modes and other modes have been measured provides important benchmarks for the transition dipole directions of modes in peptides.

1. Introduction

The recent experimental developments in two dimensional infrared spectroscopy of peptides and proteins [1-4] have made clear the exciting possibilities for such methods in the determination of structures and conformational dynamics of biologically interesting molecules. Implementation of the full potential of these methods requires detailed information on vibrational dynamics, mode couplings and directions of the transition moments with respect to the molecular skeleton in different known peptide conformations.

The first part of this work is focused on determination of the energy relaxation pathways in the $AcProNH_2$ model dipeptide. Knowledge of the detailed vibrational energy relaxation pathways in peptides is essential for understanding biological mechanisms of vibrational energy transduction and chemical reactions in protein environments. The amide-I (Am-I) mode relaxes in *ca.* 1 ps in most cases [1] implying that it is strongly coupled to other internal modes. The mechanisms of vibrational energy transfer might be influenced by fluctuations in the energy gaps and couplings between the amide units. It is therefore necessary to determine experimentally not only the relaxation rates but also the couplings between the peptide units and their distributions.

In the second part we present the first experimental data on the coupling of Am-A (N-H stretch mode) and Am-I, Am-II modes in isolated amide and across the hydrogen bond. It is shown that the coupling strength and the Am-I mode transition moment direction are sensitive to the H-bond formation.

2. Two-Color Mid-IR Transient Absorption Spectrometer

The two-color mid-IR transient absorption spectrometer has been implemented based on a commercial Ti:S oscillator/regenerative amplifier laser system (Clark-MXR). The setup contains two OPAs, each followed by difference frequency generation, which allow tuning probe and pump frequencies independently over the broad spectral range. The pump pulses could be spectrally narrowed by a tunable Fabry-Perot interferometer or by interference filters. Wire-grid polarizers control the polarization of the beams. Transient spectra in the region of pumped bands are termed diagonal spectra.

3. Results and Discussion

Vibrational Energy Pathways. Evolution of the transient polarized spectra of $AcProNH_2$ dipeptide with time after selective excitation of each of the three modes (two Am-I modes and Am-II mode) was acquired (see example in Figure 1). The spectra were globally fitted taking into account diagonal anharmonicity of each mode, combination band anharmonicities, as well as energy transfer (EnT) pathways between different modes. An efficient EnT was observed between Am-I and Am-II modes of the same amide (Figure 2). No significant EnT was observed between two Am-I modes. Coupling strengths and angular relations between transition moments were determined. The dominant structure was determined to be C_7 self-H-bonded conformation. Interconversion between different Liouville paths was clearly demonstrated in both diagonal and cross peaks, giving a reason for the nonexponentiality of the time profiles.

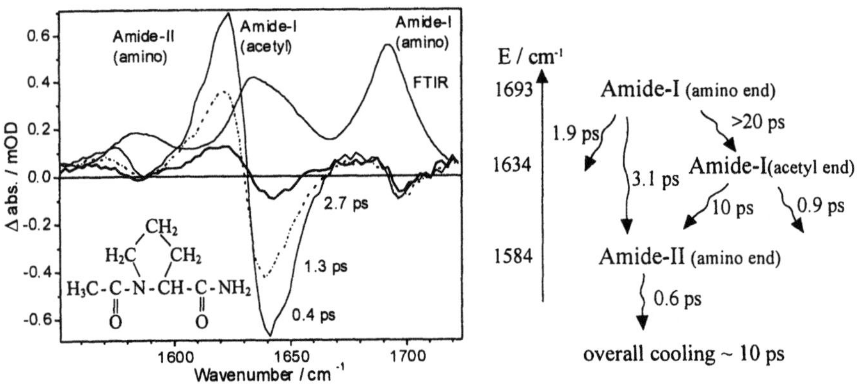

Fig. 1. Left panel: Transient absorption spectra at three delay times obtained for the Am-I(ac) band excitation at magic angle polarization along with FTIR spectrum. Right panel: Detailed energy relaxation pathways determined in this work.

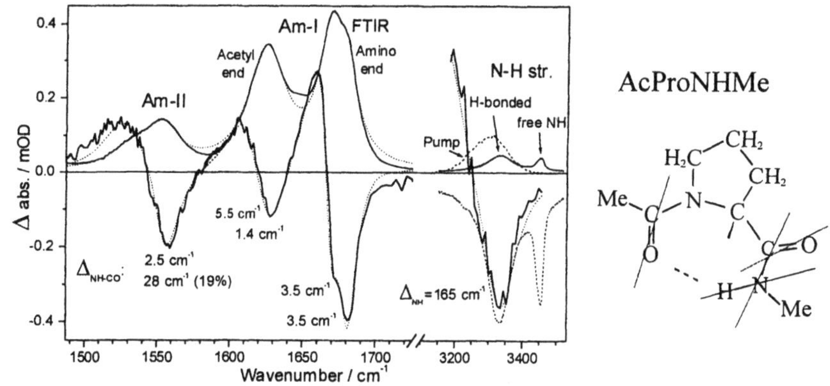

Fig. 2. Transient spectra for AcProNHMe after excitation of the H-bonded Am-A mode along with FTIR spectrum and pump spectrum. Simultaneous fit of the transient and FTIR spectra is shown.

Coupling of Am-A and Am-I / II modes. The Am-A mode, being the most localized mode in amide, can be used to obtain structural information on the H-bonding network as well as to characterize transition dipole directions of other modes in peptides. As a first step interaction of the Am-A and Am-I modes was detected in an isolated amide (the sample was AcAlaOMe, normal and deuterated). The coupling strength $\Delta_{Am\text{-}A/Am\text{-}I}$ is about 2 cm^{-1} similar for both -D and -H amides. The Am-I transition moment was determined to be at 21±3° to the C=O bond, in agreement with expectations [5]. However, in the deuterated amide this angle is 11±4°, which illustrate different PE distribution of the Am-I mode in the deuterated amide. This orientation and the coupling strength changes if the NH group is involved in H-bonding (AcProNHMe, Fig. 2). The Am-A / Am-I coupling across the H-bond is ~1.4 cm^{-1}, in C_7 conformation of AcProNHMe (Fig. 2). The Am-II(amine) transition dipole is at 75° to the Am-A mode dipole. Assuming that NH bond is almost parallel to the C=O bond in the C_7 conformation, the angle of 70-75° can be obtained for the Am-II dipole to the C=O bond, in agreement with expectations (73°) [5].

Interactions of Am-A mode with 7 different modes in the spectral range from 1300 to 1700 cm^{-1} including Am-I, Am-II, CCH_3 and NCH_3 symmetric and antisymmetric bending were observed in N-methylacetamide.

Acknowledgements. Research support through NIH and NSF and equipment from M. Therien is gratefully acknowledged.

References

1 P. Hamm, M. Lim, and R. M. Hochstrasser, J. Phys. Chem. B, **102**, 6123, 1998.
2 M. C. Asplund, M. T. Zanni, and R. M. Hochstrasser, Proc. Natl. Acad. Sci. USA, **97**, 8219, 2000.
3 S. Woutersen and P. Hamm J. Phys. Chem. B, **104**, 11316, 2000.
4 M. T. Zanni, S. Gnanakaran, J. Stenger, and R. M. Hochstrasser, J. Phys. Chem. B, **105**, 6520, 2001.
5 Krimm, S.; Bandekar, J. *Adv. Protein Chem.* **38**, 181, 1986.

Energy relaxation of carotenoids in solution and in LH2 complexes studied with sub-10-fs temporal resolution

Giulio Cerullo[1], Dario Polli[1], Guglielmo Lanzani[1], Sandro De Silvestri[1], H. Hashimoto[2], Richard Cogdell[2]

[1] Istituto Nazionale per la Fisica della Materia, IFN-C.N.R.,
Dipartimento di Fisica, Politecnico, Milano, Italy
Tel.: +39-02-23996164, Fax: +39-02-23996126, e-mail: giulio.cerullo@polimi.it
[2] Division of Biochemistry and Molecular Biology, University of Glasgow, Glasgow, UK

Abstract. The ultrafast internal conversion dynamics of carotenoids in solution and inside LH2 complex of purple bacteria are studied with sub-10-fs pulses. Energy is transferred from the S_2 state of carotenoids to bacteriochlorophylls with 40-60% efficiency.

1. Introduction

Carotenoids perform several vital functions in photosynthetic systems [1]: (i) light harvesting, by absorbing energy in the blue-green region and transferring it to the chlorophylls with very high efficiency; (ii) quenching of the chlorophyll triplet states; (iii) singlet oxygen scavenging. In carotenoids both the ground state (S_0) and the first excited state (S_1) possess A_g character, while the second excited state (S_2) possesses B_u symmetry. Therefore only the $S_0 \rightarrow S_2$ transition is optically allowed; however, after photoexcitation, an ultrafast internal conversion (IC) to S_1 takes place. The dynamics of this process is important for photosynthesis, because energy transfer from the carotenoids to the chlorophylls can take place either from S_2 or from S_1. The recent availability of sub-10-fs visible laser pulses has allowed to directly study IC in carotenoids with unprecedented time resolution [2].

In this work we compare the early steps of excited state dynamics of carotenoids in solution and inside the light harvesting complex 2 (LH2) of purple photosynthetic bacteria. Purple bacteria are anaerobic prokaryotes which, due to their simple structure and well separated spectral features, are excellent model organisms for investigating the basic mechanisms of photosynthetic light harvesting [3]. By comparing the IC dynamics and subsequent S_1 evolution inside and outside the LH2 complex, we are able to pinpoint the energy transfer pathways from the carotenoids to the bacteriochlorophylls (BChls).

2. Results and Discussion

The experiments are performed using a visible optical parametric amplifier [4] pumped by the second harmonic of an amplified Ti:sapphire laser at 1 kHz. Broadband amplification of a white light seed is achieved in a β-barium borate crystal using a non-collinear phase matching geometry; the amplified pulses (500-650 nm bandwidth) are then compressed to nearly transform-limited sub-10-fs duration by multiple reflections onto chirped dielectric mirrors. Samples are kept in a 0.2-mm cuvette at room temperature.

Fig. 1(a) shows ΔT/T spectra of the carotenoid rhodopin-glucoside in acetone solution, at different delays following photoexcitation; we observe the rapid formation of a photoinduced absorption band (PA) peaking at 560 nm, and a corresponding blue-shift of the isosbestic point. The PA band is assigned to $S_1 \rightarrow S_n$ absorption and provides a signature of the IC process. The PA formation is completed in ≈ 600 fs and a single exponential fitting of its risetime (see Fig. 2a) gives a time constant τ_1 = 145 fs.

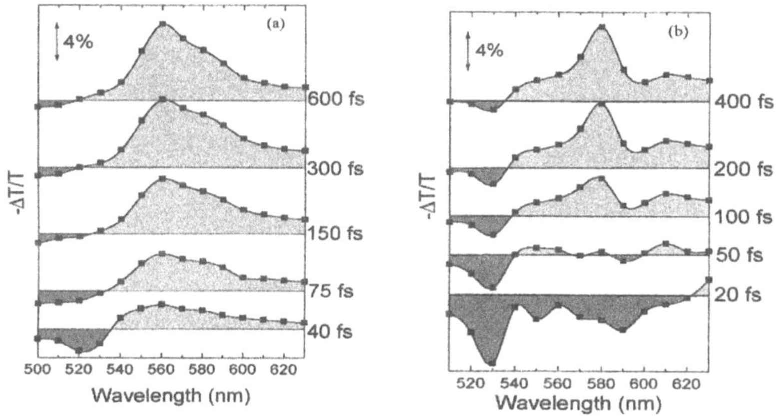

Fig. 1. (a) differential transmission spectra of rhodopin-glucoside in acetone, following photoexcitation by a sub-10-fs visible pulse. (b) same as (a), for the LH2 complex of Rhodopseudomonas acidophila.

Fig 1(b) shows ΔT/T spectra of the LH2 complex of the bacterium *Rhodopseudomonas Acidophila* [5], containing rhodopin-glucoside in addition to BChls. Our pulses excite essentially the carotenoids, because BChl absorption is at 800-850 nm (there is a very weak absorption from the Q_x band, and the corresponding bleaching at 590 nm can be seen as a dip in Fig. 1b). Initial photobleaching and stimulated emission rapidly give way to the buildup of PA. In this case however the process is much faster and is completed in ≈ 250 fs; exponential fitting of the risetime gives a time constant τ_1 = 61 fs. The shortening of S_2 lifetime in the LH2 complex is due to the opening of a new relaxation channel, in addition to IC to S_1: energy transfer to the BChls. Our measurements allow thus to evaluate an $S_2 \rightarrow$ BChl transfer efficiency of ≈ 58%. Fig. 2(b) shows, on a longer timescale, the decay of S_1 to S_0; the time constant τ_1 for this process, however, is nearly unchanged inside and outside the LH2 complex, indicating that there is negligible energy transfer from S_1 to the BChls.

The dynamics in Fig. 2(a) have a superimposed oscillatory pattern due to impulsively excited coherent vibrational motion in the carotenoids. The inset in Fig. 2(a) shows a Fourier analysis of the signal, after having subtracted the slowly varying component: several coupled modes are present, the strongest being at 1155 cm^{-1} and 1510 cm^{-1}, with frequencies not significantly changed by the protein environment; they are assigned to ground state (S_0) motion.

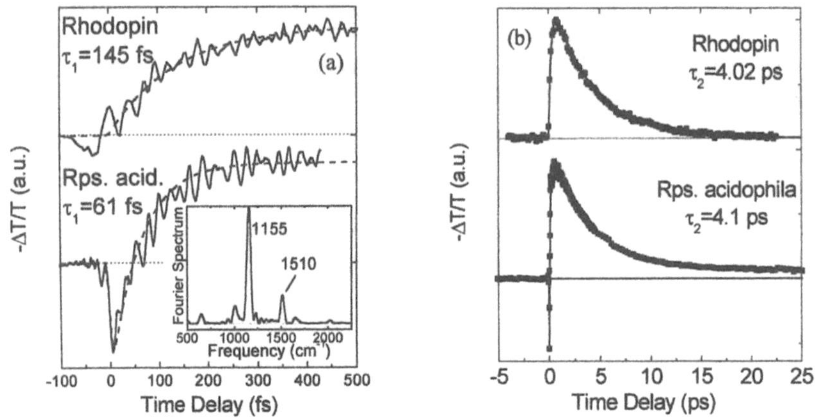

Fig. 2. (a) differential transmission dynamics at 560 nm (for rhodopin-glucoside) and 580 nm (for the LH2 complex of Rhodopseudomonas acidophila), with exponential fits; inset: Fourier analysis of the oscillating pattern in rhodopin-glucoside; (b) same as (a), over a longer timescale.

Similar results were obtained for the carotenoid okenone, found in the LH2 of Chromatium Purpuratum. In cyclohexane solution, the $S_2 \rightarrow S_1$ internal conversion occurs with a time constant $\tau_1 = 130$ fs and S_1 decays back to the ground state with $\tau_2 = 4.1$ ps. Inside the LH2 complex, τ_1 shortens to 80 fs, while τ_2 remains essentially unchanged. This indicates that energy transfer to the BChl takes place from S_2 with an efficiency of 38%, while virtually no energy is transferred from S_1.

References

1 H. A. Frank, and R.J. Cogdell, *Photochem. Photobiol.* **63**, 257 (1996).
2 G. Cerullo, G. Lanzani, M. Zavelani-Rossi, and S. De Silvestri, *Phys. Rev. B* **63**, R241104 (2001).
3 R.J. Cogdell, N.W. Isaacs, T.D. Howard, K. McLuskey, N.J. Fraser, and S.M. Prince, *J. Bacteriol.* **181**, 3869 (1999).
4 M. Zavelani-Rossi et al., *Opt. Lett.* **26**, 1155 (2001).
5 G. McDermott et al., *Nature* **374**, 517 (1995).

Early Events in Ligand Dissociation Dynamics in Myoglobin: an Infrared Study

T. Polack, M. H. Vos, J.-L. Martin, A. Alexandrou

Laboratoire d'Optique et Biosciences
CNRS UMR 7645 - INSERM U451- Ecole Polytechnique - ENSTA
Ecole Polytechnique, 91128 Palaiseau Cedex, France
Phone: (33)(1)69 33 36 52, fax: (33)(1)69 33 30 17,
E-mail: antigoni.alexandrou@polytechnique.fr

Abstract. We used a femtosecond visible-infrared pump-probe experiment to follow ligand dissociation in carbonmonoxy-myoglobin. The CO mid-infrared absorption was found to be significantly delayed with respect to the photodissiciation. Assuming a gradual vibrational frequency change, the mid-infrared differential transmission can be interpreted in terms of a 1.2 ps ligand transfer time from the heme to the docking site.

1. Introduction

Myoglobin is an oxygen storage protein found in muscle. Its active site, a planar prosthetic group (the heme), contains an iron atom which reversibly binds oxygen; it can also bind several other ligands such as NO and CO. Heme molecule are found in many other proteins playing a variety of roles, including ligand transport (hemoglobin) and catalysis (oxidases). Detailed knowledge of ligand transfer in heme proteins is a key point in the understanding of complex biochemical reactions [1]. CO has a large affinity for the heme and is IR active. In our experiment, a visible light pump pulse triggers the Fe-CO bond breaking and the transmission of a delayed IR probe pulse is recorded. In contrast to visible probe experiments, sensitive to the modification of the electronic transitions of the heme, mid-infrared probe experiments allow to directly follow the ligand dynamics via ligand vibrational frequency changes. Mid-infrared probe experiments have thus allowed the observation of the arrival of the ligand at a docking site in the heme pocket of myoglobin [2] after photodissociation of the ligand [3-5]. Here we report results on early events in the ligand photodissociation dynamics in myoglobin.

2. Visible pump-infrared probe experiment

Tunable mid-infrared probe pulses (3-18 μm) are obtained with a Ti:Sapphire oscillator followed by a 100-kHz regenerative amplifier (RegA, Coherent), optical parametric amplification and difference frequency mixing . We derive our 400-nm pump from the same amplifier by frequency doubling. We used a 5 mM solution of horse heart myoglobin (Sigma) in a 50 mM Tris buffer, pH=7.6. Carbonmonoxy-myoglobin (MbCO) is loaded between two 2-mm thick calcium fluoride windows using a 25-μm spacer. The cell is simultaneously rotated and translated, so that each shot illuminates a new spot of MbCO, with a 5% photodissociation rate. Pump and probe polarizations are perpendicular in order to maximize the signal. (The pump beam excites preferentially heme molecules with the heme plane parallel to the pump polarization while the CO ligand is initially nearly perpendicular to the heme plane [4].)

When bound to the heme, the CO vibrational frequency is 1945 cm^{-1}. This frequency is shifted to 2130 cm^{-1} for a CO at the initial docking site [2]. By tuning our infrared probe (200 fs duration, centered at 1900 cm^{-1}, spectral width : 150 cm^{-1}) close to the bound-to-heme ligand vibrational frequency, the docked CO frequency is out of the range of the probe. Our probe absorption thus discriminates between the two CO states as can be seen in Figure 1.

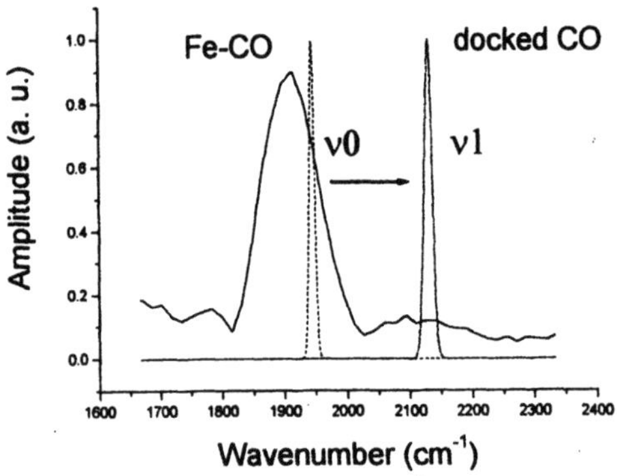

Fig. 1. Experimental probe spectrum and CO absorption lines simulated by gaussians (dashed: CO bound to heme or A-band, dash-dotted:"docked" CO or B-band). For the sake of simplicity, we do not show the structure of the different bands, nor their different absorption strengths.(A-band absorption is 30 times stronger.)

3. Results and Discussion

We have chosen a spectrally integrated configuration in which coherent artifact (perturbed free-induction decay) contributions cancel out. As a counterpart, we are confronted with the detection of very small signals (a few 10^{-5}).

Figure 2 shows the differential transmission as a function of pump-probe delay time in MbCO as well as in a direct gap semiconductor, GaAs, obtained in exactly the same experimental conditions (solid lines). In the case of GaAs, the differential transmission signal is due to the infrared absorption of free carriers created by the pump pulse. The rise of this signal is instantaneous and therefore reflects the instrument response function. The derivative of the GaAs signal (dotted line) corresponds to the pump-probe cross-correlation.

If we assume that the CO vibrational frequency "jumps" from the heme-bound value to that of docked CO, we can fit the MbCO differential transmission with an exponential taking into account the pump-probe temporal resolution. In this case, we find a characteristic transition time of 250 fs. However, in order to correctly fit the signal, a delay time of 200 fs is necessary (see dot-dashed line in Fig. 2 obtained without a delay time) which is not compatible with the quasi-instantaneous nature of the ligand photodissociation [6]. This suggests that the ligand vibrational frequency does not instantaneously change after dissociation. Using a simple model which takes into account a CO vibrational frequency, absorption strength and dipole orientation changing continuously in an exponential manner from the initial to the final value $[\nu(t)=(\nu0-\nu1)e^{-t/\tau}+\nu1]$ and incorporating the finite pump and probe duration, we can fit the MbCO differential transmission with a characteristic ligand transfer time of 1200 fs ± 300 fs (dashed line in Fig. 2). This gradual vibrational frequency change may be attributed to the effect of the local electric field felt by the ligand as it moves inside the adjusting heme pocket.

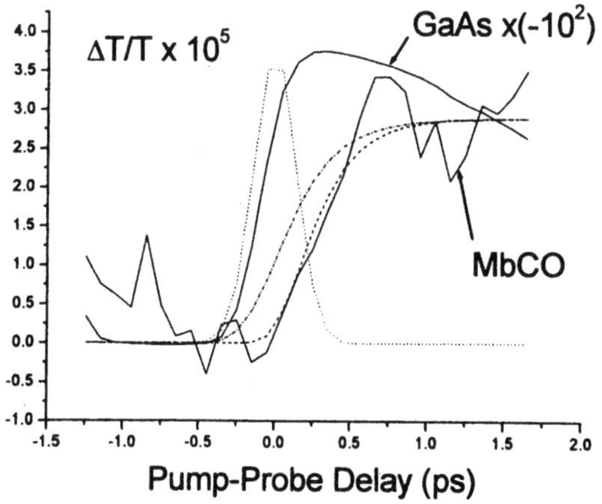

Fig. 2. Differential transmission of MbCO and of GaAs (solid lines). The dotted line shows the derivative of the rise time of the GaAs signal and corresponds to a pump-probe cross-correlation.

References

1 M. H. Vos and J.-L. Martin, Biochim. Biophys. Acta **1411**, 1-20, 1999.
2 T.Y. Teng, V. Srajer, and K. Moffat, Nature Struct. Biol. **1**, 701-705, 1994.
3 M. Lim, T. A. Jackson, and P.A. Anfinrud, Nature Struct. Biol. **4**, 209, 1997.
4 M. Lim, T. A. Jackson, and P.A. Anfinrud, Science **269**, 962, 1995.
5 M. Lim, T. A. Jackson, and P.A. Anfinrud, J. Chem. Phys. **102**, 11, 1995.
6 J. W. Petrich, C. Poyart, J.-L. Martin, Biochemistry, **27**, 4049, 1988.

Transition states in bacteriorhodopsin during retinal isomerization observed by sub-5-fs pulses

Takayoshi. Kobayashi[1], Takashi. Saito[1], and Hiroyuki Ohtani[2]

[1] Department of Physics, Graduate School of Science, University of Tokyo, 7-3-1 Hongo, Bunkyo, Tokyo 113-0033, Japan. E-mail: kobayashi@phys.s.u-tokyo.ac.jp
[2] Department of Biomolecular Engineering, Graduate School of Bioscience and Biotechnology, Tokyo Institute of Technology, 4259 Nagatsuta, Yokohama, 226-8501, Japan

Abstract. We report the first direct observation of structural changes in a complex biological system, the chromophore of bacteriorhodopsin, as it undergoes the *trans-cis* photoisomerization by monitoring changes in the vibrational spectra of the transition state.

1. Introduction

Real-time investigations of the rearrangement of chemical bonds during chemical transformations require femtosecond temporal resolution, so that the atomic vibrations within the reacting molecules can be observed. Following the development of lasers capable of emitting ultra-short laser flashes on this timescale, chemical reactions involving relatively simple molecules have been monitored in detail, revealing the transient existence of intermediate species as reactants are transformed into products.

Here we report the direct observation of transition state during photochemical reaction by probing the change in the electronic transition probability induced by nuclear motion. A complex biological system, the retinal chromophore of bacteriorhodopsin (bR_{568}), is studied as it undergoes the *trans-cis* photoisomerization that is fundamental to the vision process. bR_{568} undergoes the following photochemical cycle:

$$bR_{568} \xrightarrow{h\nu} H \rightarrow I_{460} \rightarrow J_{625} \rightarrow K_{610} \rightarrow L_{543} \rightarrow M_{412} \rightarrow N_{560} \rightarrow O_{640} \rightarrow bR_{568}$$

Here, H denotes the Franck-Condon excited state of bR_{568} with an all-*trans* retinal Schiff base. The subscripts represent the wavelengths(nm) of absorption maxima of the species. Several vibrational spectroscopy studies using ultrashort pulses have shown that H and I_{460} are excited-state species with a chromophore of an all-*trans* conformation, and that K_{610} is a ground-state species with the chromophore already converted to the 13-*cis* conformation.

2. Experimental, Results and Discussion

In order to elucidate the change in the geometrical configuration during the isomerization process, we measured the time-resolved transmittance change($\Delta T/T$), which we call the real-time signal, of bR_{568} using sub-5-fs pulses as both pump and probe at room temperature.

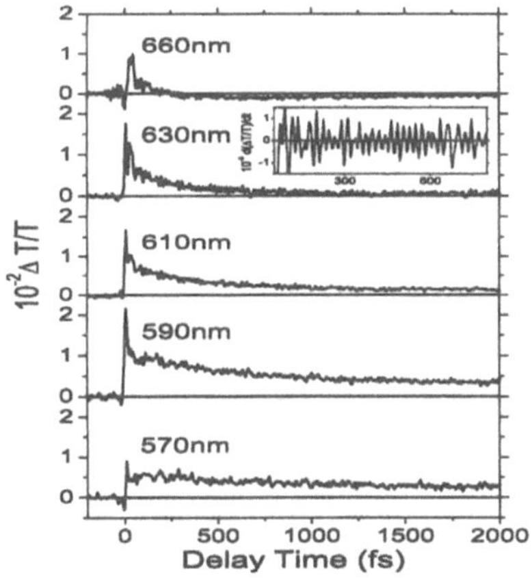

Fig. 1. Time-dependence of the transmittance change ($\Delta T/T$) following sub-5-fs pulse excitation of bR_{568} at room temperature. Sub-5-fs pump and probe pulses (520-750 nm) were generated by non-collinear optical parametric amplification. The inset shows an enlarged trace monitored at 630 nm.

Figure 1 shows the real-time signals of bR_{568} at several wavelengths. The measured $\Delta T/T$ is modulated by molecular vibrations, which correspond to Raman-active modes[1,2]. The Fourier transformation of $\Delta T/T$ around an arbitrary delay time t_d gives the amplitudes and frequencies of the molecular vibrations at t_d. Figure 2 shows spectrograms calculated for real-time signals probed at 610 nm. As shown in the spectrograms, instantaneous frequencies are modulated at a period of about 200 fs. This can be seen from the time-integrated Fourier power spectrum of the vibrationally induced transmittance modulation. We conclude that the temporal shifts in frequency are due to wave-packet motion of the excited-state rather than ground-state for the following three reasons. (1) The spectrogram calculated for the real-time signal at 660 nm due to excited-state absorption has the same features as those at other wavelengths. (2) The pulse width (sub-5-fs) is much shorter than the observed molecular vibration frequencies and is too short to enhance of the molecular vibration in the ground electronic state[1,2]. (3) The decay time of the molecular vibration amplitude is found to be shorter than that of $\Delta T/T$.

The first event occurs within 50-100 fs. The wave-packet reaches the flat potential region (I_{460} with a B_u-like electronic configuration) from H. Periodic out-of-plane distortion modes also starts to take place. An intense C=N stretching band from the protonated Schiff base (C_{15}=NH^+, 1620-1630 cm^{-1}) appears within 50 fs. The skeletal vibrations around 1250 cm^{-1} are activated in the first 100 fs as predicted theoretically[3,4]. The hydrogen-out-of-plane (HOOP) modes appear just after the excitation in the 800-1050 cm^{-1} region.

The second event is seen in the 100-200 fs region and involves splitting of the band centered around 1250 cm^{-1} into two peaks at the gate-delay time of about 100 fs after excitation of the chromophore. The frequency of one peak gradually increases, reaching 1520-1540 cm^{-1} at about 150 fs, while the frequency of the other peak gradually decreases, reaching a value as low as 1130 cm^{-1} at 200 fs after excitation.. These frequencies are modulated with 200-fs period of the C-C=C torsion. These mode frequencies approach each other and merge into a single peak, since there is neither an in-plane nor an out-of-plane mode in a distorted configuration.

The third event is the formation of the transition state in the 200-600 fs region. C=N stretching, C=C-H bending, and HOOP modes are also silent in this region. The weak intensity is partly due to the broad distribution of molecular conformations covered by the wave packet. Energy randomization occurs during the fast wave-packet motion along the potential curves of the excited B_u-like and A_g-like states and during the Landau-Zener tunneling between the potential curve B_u and A_g.

In conclusion, We report the first direct observation of structural changes in a complex biological system, the chromophore of bacteriorhodopsin, as it undergoes the *trans-cis* photoisomerization by monitoring changes in the vibrational spectra of the transition state.

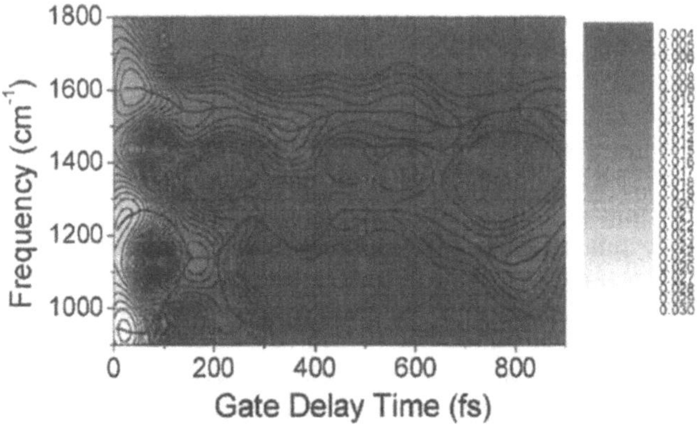

Fig. 2. Spectrogram calculated for the $\Delta T/T$ trace at 610 nm. The calculations used the Hanning window function (150-fs FWHM). The spectrogram shows the appearance of all significant vibrations[5-7] in the characteristic frequency ranges 1,500-1,650cm^{-1}(C=C and C=N stretching modes), 1,150-1,250cm^{-1}(in-plane C=C-H bending coupling with the C=C stretching modes) and 900-1,000cm^{-1}(the hydrogen-out-of plane (HOOP) mode).

References

1 W. T. Pollard, et al., *J. Phys. Chem.* **96,** 6147 (1992).
2 C. J. Bardeen, Q. Wang and C. V. Shank, *J. Phys. Chem. A* **102,** 2759 (1998).
3 T. Vreven, et al. *J. Am. Chem. Soc.* **119,** 12687 (1997).
4 M. Garavelli, F. Negri and M. Olivucci, *J. Am. Chem. Soc.* **121,** 1023 (1999).
5 S. J. Doig, P. J. Reid and R. A. Maties, *J. Phys. Chem.* **95,** 6372 (1991).
6 G. Eyring, B. Curry, A. Broek, J. Lugtenburg and R. A. Maties, *Biochemistry* **21,** 384 (1982).
7 A. B. Myers, R. A. Harris and R. A. Maties, *J. Chem. Phys.* **79,** 603 (1983).

Investigating excited state dynamics in bacteriorhodopsin by impulsive stimulated emission pumping

Sanford Ruhman[1], Bixue Hou[1], Michael Ottolenghi[1], Noga Friedman[2], Mudi Sheves[2]

[1] Department of Physical Chemistry, and the Farkas Center for Light Induced Processes, the Hebrew University, Jerusalem 91904, Israel. E-mail: sandy@fh.huji.ac.il.
[2] Department of Organic Chemistry, The Weizmann Institute of Science, Rehovot 76100, Israel.

Abstract. Femtosecond stimulated emission pumping experiments from bacteriorhodopsins fluorescent state show for the first time that it is a photocycle intermediate, with a constant nonisomerized structure throughout its 0.5 psec lifetime. Implications concerning internal conversion dynamics are discussed.

Bacteriorhodopsin (BR) is a light energized proton pumping protein, embedded in the membrane of the photosynthetic Halobacterium. The absorbing chromophore is a retinal molecule covalently linked to the protein. Early in the BR photocycle, the retinal selectively isomerizes from all trans to the *13-cis* form, and the protein undergoes a series of structural changes leading to ejection of one proton from the cytoplasm. Contrary to previous interpretations, recent studies in this lab prove that *all-trans* → *13-cis* retinal isomerization is not the primary consequence of light absorption in BR [1]. Furthermore, while appearing very rapidly, the fluorescent state (I_{460}) was shown to be the product of very rapid nuclear rearrangement along other retinal coordinates, giving rise to a red shifting in the emission, and periodical modulations which mirror impulsively induced wave packet motions [2]. Some very basic riddles concerning the appearance and decay of this state are, however, still unresolved. 1) While it is often assumed to be so, evidence that the fluorescent state is a photocycle intermediate is circumstantial. In view of the 60% quantum efficiency for proton pumping, this assumption needs direct experimental proof. 2) The cited studies suggest that the decay of I_{460} must reflects the stage of retinal isomerization. Despite this we, along with others, observe nearly no spectral evolution in the absorption and emission bands of I_{460} after the first few tens of femtoseconds [3]. In particular no continuous spectral shifts which would accompany dynamic wave packet motion along the active double bond twisting coordinate are observable. Does this mean that the structure of the fluorescent state is constant over its lifetime? If so how is the point of curve crossing to the ground state accessed, and how is this represented in the transient spectroscopy?

To address these questions 3 pulse experiments were conducted with a 1 kHz home made multipass amplified Ti;sapphire laser system equipped with a commercial NOPA followed by an 800 nm pumped booster producing 20-35 fsec pulses. First a 535 nm pulse excites (pump) a BR suspension to the fluorescent state. At various delays within the I_{460} lifetime a 1070 nm pulse impinges on the sample, stimulating emission back to the ground state (dump). Finally, probing from before, to ~10 psec after the pump was conducted at two wavelengths filtered from a continuum generated in sapphire-900 nm to follow rapid changes in the emission from I_{460}, and 650 nm to quantify relative yields of formation for the "K" photocycle intermediate, generated a few picoseconds after photoexcitation (and known to contain an isomerized retinal group).

The rationale behind this experiment is straightforward. A ~1 micron dump wavelength is chosen to avoid complicating stimulating emission with absorption to higher excited states demonstrated to exist at shorter wavelengths [4]. If I_{460} lies along the photocycle, and if the dumped population is nonreactive, a matching depletion should appear in concentrations of later intermediates, and can be tested by 650 nm probing. In deciphering I_{460} decay dynamics we consider two extreme scenarios. In the first, coined "dynamical", the full excited state population continuously changes leading to the observed exponential reduction in emission and absorption cross sections. In contrast in a "kinetic" scenario assuming proteins in I_{460} to have a well defined structure and cross section, the decay of "I_{460}" must reflect a gradual reduction in the number density of excited states. In the later case due to the cross section invariance, the probability of stimulating emission will be independent of delay, whereas in the former the fractional "dump" probability will diminish with the spectral features of "I_{460}". Testing the probability of SEP at various delays has the potential of differentiating between these scenarios, which are not distinguishable by pump-probe data alone.

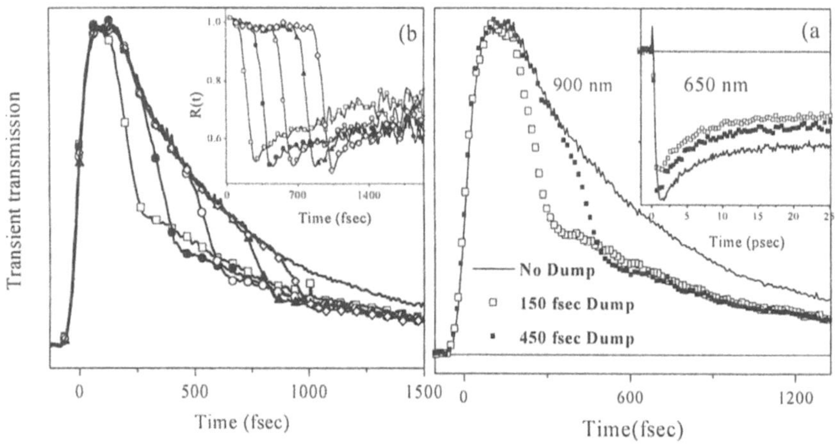

Results obtained provide clear answers to the questions posed above. The insert of (b presents *R(t)*, the ratio of stimulated emission signals with and without dumping for the various SEP delays. Accordingly, for SEP delays larger than ~200 fsec, the post dump portions of the scans converge at longer delay times. Also the ratio plots are identical within the noise margins, regardless of the dump timing, over the range from 200 to 1,000 fsec, in agreement with the kinetic scenario above. Stimulated emission scans exhibit an excess depletion immediately after the SEP which recovers within a few hundred femtoseconds. This stage of emission reconstitution is identical for all dumping delays beyond 200 fsec. As for the 650 nm probing, quantitative analysis of the data in panel (a shows that the depletion in the excited state (R^*e^{-2t}) leads to an equal depletion in production of "K" within experimental error.

The matching depletion in the later reactive intermediates proves that the dumped population *is short circuited from the photocycle.* For the first time this provides quantitatively proof that the fluorescent state lies along the multidimensional reaction coordinate involved in proton pumping! The invariance of the dumping probability over time in the fluorescent state agrees precisely with a kinetic scenario, which must indicate that its population is arrested in a constant structure throughout the emission lifetime, varying neither in absorption or emission spectra, nor in the emission cross section. The excess bleach observed at early times most likely reflects resonant ground state wave packet accumulation which most likely dissipates through vibrational relaxation within a few tenths of a picosecond as observed.

Having gained such significant new insights, one is tempted to consider their implications on the energetics and structure of $I_{460.}$ The fact that the dumped population is nonreactive, must indicate that the constant geometry of the fluorescent population is quite different from that of the reactive conical intersection. Bearing in mind that an equivalent state is produced even in a protein containing a rigidized retinal unable to isomerizes around the active coordinate [1], shows that reactive crossing to the ground state requires substantial twisting of the $C_{13}=C_{14}$ bond from the I_{460} geometry. The potential structure which stabilizes this geometry, and the mechanism by which internal conversion takes place are, however, not clear at all, and will be discussed further in our presentation.

Acknowledgements. This work was supported by the Human Frontier Science Program (HFSP), and the Israel Science Foundation (ISF). The Farkas Center is supported by the Minerva Gesellschaft, GmbH, Munich, Germany.

References

1 T. Ye, N. Friedman, Y. Gat, G. H. Atkinson, M. Sheves, M. Ottolenghi, S. Ruhman, J. Phys. Chem. B, 103, 5122 (1999).

2 T. Ye, E. Gershgoren, N. Friedman, M. Ottolenghi, M. Sheves, S. Ruhman, Chem. Phys. Lett., 314, 429-434 (1999).

3 G. Haran, K. Wynne, A. Xie, Q. He, M. Chance, R. M. Hochstrasser, Chem. Phys. Lett. **26**, 389 (1996).

4 F. Gai, C. Mcdonald, P. A. Anfinrud, *J. Am. Chem. Soc.*, **119**, 6201 (1997).

Vibrational dynamics upon photoexcitation of native bacteriorhodopsin (bR) and of the all-trans - locked bR5.12

Rolf Diller[1], Johannes Herbst[1], Karsten Heyne[2], Michael Ottolenghi[3], Noga Friedman[4], and Mudi Sheves[4]

[1] Fachbereich Physik, Freie Universität Berlin, Arnimallee 14, D-14195 Berlin, Germany
E-mail: diller@physik.fu-berlin.de
[2] Max-Born-Institut, Max-Born-Str. 2A, D-12489 Berlin, Germany
E-mail: heyne@mbi-berlin.de
[3] Department of Physical Chemistry, The Hebrew University, Jerusalem 91904, Israel
E-mail: micott@chem.ch.huji.ac.il
[4] Department of Organic Chemistry, The Weizmann Institute of Science, 76100 Rehovot, Israel
E-mail: mudi.sheves@weizmann.ac.il

Abstract. Femtosecond time-resolved infrared spectroscopy was used to study and compare the retinal all-trans to 13-cis isomerization dynamics in bacteriorhodopsin with the dynamics of bR reconstituted with an all-trans - locked retinal chromophore.

The all-trans to 13-cis photoisomerization of the retinal chromophore in bacteriorhodopsin (bR) acts as a strong steric and electronic perturbation for the protein environment, thereby driving the proton pump mechanism. Photoexcitation and torsional movement are expected to be paralleled by various intra- and intermolecular relaxation processes. According to femtosecond (fs) time-resolved optical absorption spectroscopy the electronic ground state is reached after about 500 fs (J) [1-4], which then relaxes on a 3 ps time-scale to the K-state with the chromophore in 13-cis configuration.

Transient infrared spectroscopy allows to study the structural dynamics along the photoreaction directly at femtosecond time- and sufficient spectral resolution. By this means it was demonstrated recently for the first time, that the isomerization occurs within 500 fs, on a similar time-scale as the decay of the excited electronic state [5]. Transient difference spectra between the light adapted all-trans parent state bR_{570} and the product states as well as respective kinetics were taken in spectral regions of C=NH - stretch, C=C - stretch and C-C - stretch vibrations of the chromophore. All product bands rise whithin ca. 500 fs, especially that around 1185 cm^{-1} (C-C - stretch) which indicates the formation of the 13-cis retinal configuration (Fig. 1). The partial decay of the negative band at 1200 cm^{-1}, indicating the partial recovery of the all-trans ground-state, occurs on a somewhat slower time-scale.

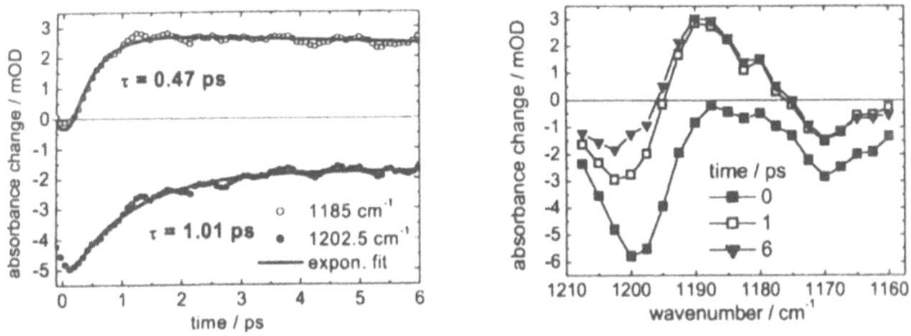

Fig. 1. IR transients (left) and difference spectra (right) of native bacteriorhodopsin in the fingerprint region. The experimants were performed at ambient temperature, excitation at 570 nm and an instrumental response function ofypically 250 fs used for deconvolution.

It was generally assumed that all protein conformational alterations of the bR photocycle are triggered by the retinal isomerization process. Recently it was proposed that the protein experiences changes due to retinal charge redistribution in the excited state [6]. To verify this possibility we have studied the vibrational dynamics upon photoexcitation of bR reconstituted with a chemically all-trans - locked retinal chromophore, bR5.12. This system shows a ca.15 ps excited state lifetime without all-trans to 13-cis isomerization.

Fig. 2 shows transient difference spectra of native bR (left) and bR5.12 (right) between 1475 and 1675 cm^{-1}, where C=NH - stretch, C=C - stretch vibrations of

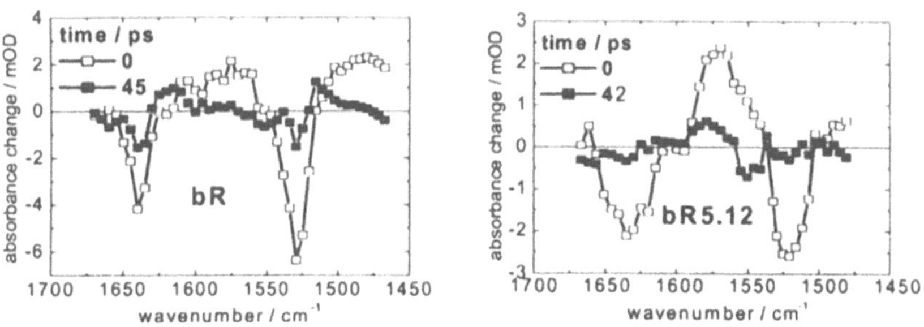

Fig. 2. IR difference spectra of native bacteriorhodopsin (left) and the all-trans – locked bR5.12 (right). Experimental condtions as given in Fig. 1.

the chromophore are found as well as amide I and amide II vibrations of the protein backbone and the bending vibration of protein bound water molecules. For bR5.12 the fingerprint region around 1200 cm^{-1} (data not shown) does not show significant signals, which confirms that isomerization does not occur around any double bond. The strong positive band at 1570 cm^{-1} decays with about 16 ps and is therefore attributed to a C=C - stretch vibration of the chromophore excited electronic state. Interestingly, the difference spectrum of bR5.12 at 42 ps shows residual negative difference signals at around 1640 and 1550 cm^{-1} and a positive signal at 1580 cm^{-1} that persist on longer time scales. These observations indicate that even without C=C - double bond photoisomerization but solely by photoexcitation (in bR5.12), i. e. fast change of the electric dipole, changes concerning the transition frequencies or absorption cross-section of non-chromophore vibrational transitions are induced. A similar effect might occur in native bR, however the respective changes in the vibrational spectrum are easily obscured by those due to the isomerization process.

Acknowledgements. We thank Prof. D. Stehlik for continuous support. This work was supported by Deutsche Forschungsgemeinschaft Sfb450.

References

1 J. Dobler, W. Zinth, W. Kaiser and D. Oesterhelt, Chem. Phys. Lett. **144**, 215, 1988.
2 R. Mathies, B. Cruz, W. Pollard and C. Shank, Science **240**, 777, 1988.
3 F. Gai, K. C. Hasson, J.Cooper McDonald and P.A. Anfinrud, Science **279**, 1886, 1998.
4 W. Zinth, A. O. Sieg, P. Huppmann, T. Blankenhorn, D. Oesterhelt and M. Nonella, in *Ultrafast Phenomena XII,* Edited by T. Elsaesser, S. Mukamel, M. M. Murnane, N. F. Scherer, Vol. 66 of Springer Series in Chem. Phys., 680, 2001.
5 J. Herbst, K. Heyne and R. Diller, in *Femtochemistry and Femtobiology,* Edited by A. Douhal, J. Santamaria, WorldScientific, to be pulished 2002.
6 A. Aharoni, L. Weiner, M. Ottolenghi and M. Sheves, J. Biol. Chem **275**, 21010, 2000.

Coherent Oscillations in Photoisomerization Reaction Dynamics of Photoactive Yellow Protein (PYP) and Related Systems

Haik Chosrowjan[1], Seiji Taniguchi[1], Noboru Mataga[1], Norio Hamada[2], Fumio Tokunaga[2], Yasushi Imamoto[3], and Mikio Kataoka[3]

[1] Institute for Laser Technology, c/o Technical Research Center, Kansai Electric Power Company 3-11-20, Nakoji, Amagasaki, Hyogo 661-0974, JAPAN
E-mail: haik@ile-osaka.u.ac.jp
[2] Department of Earth and Space Science, Graduate School of Science, Osaka University, Toyonaka, Osaka 560-0043, JAPAN
[3] Graduate School of Materials Science, Nara Institute of Science and Technology, Ikoma, Nara 630-0101, JAPAN

Abstract. Coherent oscillations in fluorescence dynamics of native PYP and it's several mutants have been observed. Two modes (~135 cm^{-1} and overdamped ~45 cm^{-1}) were identified. Both modes disappear in the dynamics of denatured PYP emphasizing the important role of protein nanospace (PNS) as the environment for photoreaction. Fluorescence dynamics in analogues show the familiar picture of *solvation effect* whereas the oscillations are almost negligible. The origin of oscillatory modes is discussed.

1. Introduction

In photosensory or photoactive proteins with chromophores absorbing in the visible region, such as rhodopsin (Rh) for vision, bacteriorhodopsin (bR), photoactive yellow protein (PYP) etc., light absorption leads to ultrafast and highly efficient reactions. Interestingly, despite quite different chromophores of these proteins, their behavior concerning the photoinduced primary processes in the protein nanospace in the course of the relaxation and twisting, including coherent processes seems to be similar. For instance, the fluorescence dynamics in all three proteins is nonexponential, no dynamic Stokes shift but a slight narrowing of the fluorescence spectra within first few picoseconds of the reaction were observed [1-3]. The spectral narrowing effect is most probably due to damping of coherently coupled intrachromophore high frequency modes and/or low frequency vibrations of environmental protein interacting with chromophore along the twisting coordinate. If this is the case, oscillatory behavior in fluorescence dynamics of photosensory proteins should be observed. To our knowledge, no such observation has been reported yet.

In this family of proteins, PYP, which functions as a blue light photoreceptor for the negative phototaxis of the purple sulfur bacterium *Ectothiorhodospira halophila* , is very stable, has the simplest structure and shows relatively simple photoreaction cycle, making it an ideal candidate for elucidation of the dynamics and mechanisms of the photoinduced primary processes in photosensory proteins.

2. Results and Discussion

The fluorescence dynamics measurements were carried out by up-conversion apparatus based on Ti:Sapphire laser (820 nm, 800 mW, 76 MHz, ~65 fs). The fwhm of the instrumental response was 110 fs.

Fluorescence decay profiles of native PYP at three characteristic wavelengths on blue, top and red sides of the spectrum are shown in Fig. 1A. The curves can be well reproduced by the sum of two exponentials (~300 fs and ~4 ps) and two sinusoidal decaying components. An underdamped mode (~135cm^{-1}) with ~550 fs damping time and an overdamped mode (~45cm^{-1}) with ~250 fs damping time were identified. The initial phases of both oscillations depend on observation wavelength and a phase shift of ~π from "blue" to "red" is clearly observed for both modes. The amplitudes of vibrations are rather small at the top compared with the "blue" and "red" edges of the spectrum. Hence, the observed oscillations can be explained by the transition frequency modulation mechanism.

In Fig. 1B fluorescence decay profiles at three characteristic wavelengths of denatured PYP are shown. In this case, PYP's chromophore (*p*-coumaric acid) is still bound to protein backbone but fully exposed to water environment. As expected, the dynamics is dominated by *solvation effect* and no oscillations in fluorescence decays can be observed. Two explanations are plausible. First, the oscillatory modes are of intrachromophore origin and are completely damped owing to fast energy dissipation due to the interaction with surrounding solvent molecules. Second, the vibrations originate from the chromophore-protein interaction due to the hydrogen bonding network in PNS. In this case, too,

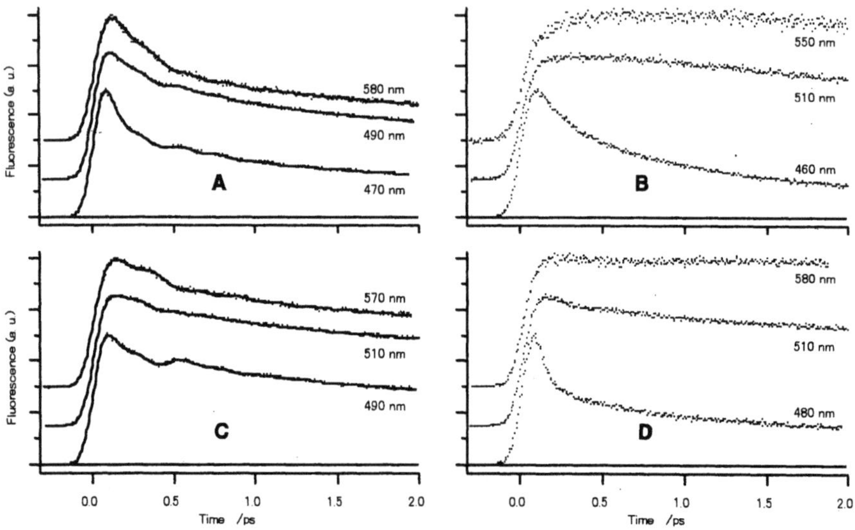

Fig. 1. Fluorescence decays at three characteristic wavelengths on blue, top and red sides of the spectrum for native PYP (A), denatured PYP (B), E46Q/T50V mutant (C) and ferulic acid analogue (D). Solid lines in A and C show the fits.

chromophore exposure to solvent will cause the disappearance of vibrations. Both explanations emphasize the important role of PNS as the environment for photoreaction.

To identify the origin of observed oscillatory modes and their role in photoisomerization reaction, seven site directed mutants (E46Q, T50V, R52Q, P68A, W119G, E46Q/R52Q, E46Q/T50V) were investigated in detail. The qualitative picture of fluorescence dynamics is very similar to the results observed for native PYP. In the mutants, too, two oscillatory modes with ~130-140cm^{-1} and ~40-60cm^{-1} frequencies together with double exponential decays have been observed. As an example, results of E46Q/T50V double mutant are shown in Fig. 1C. The most characteristic result seems to be the effect of mutation on the damping times. The damping time of high frequency mode seems to be very little affected by mutation whereas the damping time of low frequency mode further decreases as the PNS structure becomes looser and more disordered by mutation. Additionally, in some mutants (T50V, E46Q/T50V) the amplitudes of oscillatory modes are a little larger than in native PYP suggesting a stabilizing role for Thr50 amino acid in the hydrogen bonding network. We tentatively assume that the high frequency mode is due to intrachromophore deformation vibrations while the low frequency one is due to some low frequency vibrations in environmental protein interacting with chromophore in PNS. This picture is consistent also with the fact that no coherent oscillations were observed in denatured PYP, where the chromophore surrounding is fully disordered.

To further deepen our understanding concerning the origin and role of observed oscillations, three analogues with replaced chromophores (3-methoxy-4-hydroxy cinnamic acid (ferulic acid), 7-hydroxy-coumarin-3-carboxilic acid (locked chromophore) and 4-dimethyl amino cinnamic acid(dmac acid)) have been investigated. In the case of locked chromophore, where the bond twisting motions are hampered, fluorescence decay becomes very long (~60 ps) indicating that the fast photoisomerization does not occur. Nevertheless, a slight oscillatory feature on both short and long wavelength regions could be observed. It is not possible to reproduce the decay profiles with two oscillatory components. The picture is rather complex and requires additional analysis. On the other hand, in the case of dmac acid and ferulic acid analogues, fluorescence dynamics show the familiar picture of *solvation effect* whereas the oscillations are almost negligible (Fig. 1D). To our knowledge, such *intraprotein solvation effect* is observed for the first time and is a subject of further investigations.

Acknowledgements. The present work was partially supported by a Grant-in-Aid No. 13640519 from the JSPS.

References

1 N. Mataga, H. Chosrowjan, Y. Shibata, Y. Imamoto, and F. Tokunaga, *J. Phys. Chem. B* **104**, 5191, 2000.

2 M. Du, and G. R. Fleming, *Biophys. Chem.* **48**, 101, 1993.

3 H. Kandori, Y. Furutani, S. Nishimura, Y. Shichida, H. Chosrowjan, Y. Shibata, and N. Mataga, *Chem. Phys. Lett.* **334**, 271, 2001.

Protein Dynamics of Bacteriorhodopsin probed by Photon Echo and Transient Absorption Spectroscopy

John T.M Kennis[1], Delmar S. Larsen[1], Kaoru Ohta[1], Marc T. Facciotti[2], Robert M. Glaeser[2] and Graham R. Fleming[1,3]

[1] Department of Chemistry, University of California, Berkeley
E-mail: john@nat.vu.nl
[2] Department of Molecular & Cell Biology, University of California and Life Sciences Division, Lawrence Berkeley National Laboratory, Berkeley
[3] Physical Biosciences Division, Lawrence Berkeley National Laboratory, Berkeley, CA 94720, USA
E-mail: GRFleming@lbl.gov

Abstract. We report the results of a three-pulse echo peak shift (3PEPS), and transient absorption (TA) study of bacteriorhodopsin. The results are analyzed in the framework of the multimode Brownian oscillator (MBO) model. It is concluded that the protein environment of the retinal chromophore exhibits an extensive inertial solvation phase, which may be attributed to a dielectric response. Diffusive-type motions of the protein on a sub-ps to ps timescale do not occur.

1. Introduction

Bacteriorhodopsin (bR) is an efficient light-driven proton pump which shows a trans-cis isomerization reaction of its retinal chromophore upon absorption. The protein environment of retinal in bR not only serves to exert the light-triggered biological function, but also actively catalyzes the isomerization of retinal [1]. Despite the protein's defining role on bR's photochemistry, the protein solvation dynamics, i.e., the response of the protein matrix to photoexcitation of retinal, have largely remained unknown. To assess the protein solvation dynamics of bR, we have performed a three-pulse echo peak shift (3PEPS) and a transient absorption (TA) study.

2. Results and Discussion

Fig. 1A shows a 3PEPS measurement on bR recorded at 568 nm (circles). The peak shift profile is directly related to the transition frequency correlation function, M(t) [2], and thus provides for a direct probe of solvation timescales and amplitudes. The peak shift of bR has a low initial value of 10.5 fs, indicative of a large reorganization energy. The peak shift decays rapidly, within 80 fs, to a

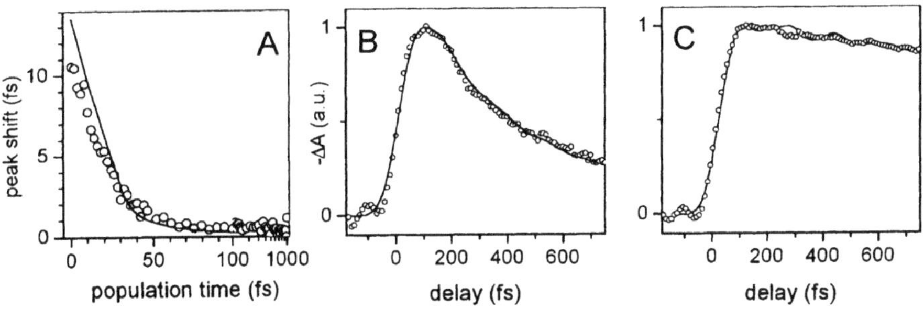

Fig. 1. A: 3PEPS profile of bR taken at 568 nm (circles); B: TA profile in bR excited at 606 nm, probed at 870 nm (circles); C: same as B for the D85S mutant. Solid lines denote the result of a MBO simulation as described in the text.

terminal value of 0.5 fs. After this, no discernible relaxations occur up to a population time of 1 ps.

Figure 1B shows a TA measurement on bR probed at 870 nm on excitation at 606 nm (circles). At this wavelength, stimulated emission (SE) at the far red edge of the emission spectrum is observed and thus the dynamic Stokes shift is monitored. The TA signal at 870 nm shows a rapid rise of the SE, followed by a decay in about 300 fs. The phenomenon that the SE from bR rises very rapidly has been well documented [1] and is consistent with the rapid decay of the peak shift. Fig.1C shows the TA signal in the D85S mutant at 870 nm on excitation at 606 nm. Again, a rapid rise of the SE is observed. The SE hardly decays during the first picosecond as a result of a 10-fold increased excited state lifetime. No slow rise components are observed, which indicates that protein solvation in the D85S mutant does not occur on a >100 fs timescale.

To quantitatively account for the excited state properties of bR, the data are simulated in the framework of the multimode Brownian oscillator (MBO) model [3]. We describe bR's lineshapes and nonlinear spectroscopic features with a minimal two-component model, consisting of a Gaussian protein relaxation with a decay of 50 fs and a λ of 1430 cm^{-1} (bR) or 1060 cm^{-1} (D85S mutant), and the intramolecular modes of retinal derived directly from resonance Raman spectroscopy [4], accounting for 1100 cm^{-1} of the total λ. The static absorption and fluorescence lineshapes of bR and the D85S mutant, are well reproduced with these parameters, as shown in Fig. 2. The results of the MBO simulation for the 3PEPS and TA measurements are shown as solid lines in Figs 1A, B and C. The agreement with the experimental data is satisfactory, and we conclude that this minimal nonreactive model is sufficient to reproduce the essential properties of the linear and nonlinear signals in bR.

The Gaussian component can, in analogy with solvation dynamics in liquids, be related to inertial motions, i.e. small scale, free and unrestricted motions of the individual groups of the protein scaffolding in the potential wells which they occupied before optical excitation. Our results strongly indicate that after the Gaussian, inertial relaxation, no diffusion-type motions take place up to a few ps

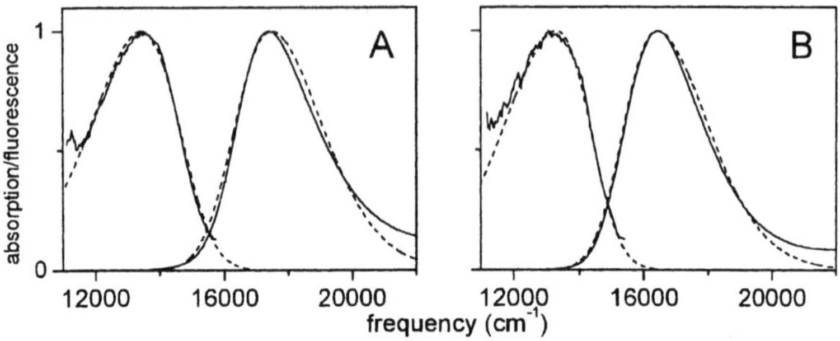

Fig. 2. A: Absorption and fluorescence lineshapes of bR (solid lines), MBO simulation (dashed line) B: same as A for the D85S mutant.

in bR. The absence of such diffusive motion is very likely due to the covalently constrained, polymeric nature of the protein.

Schulten and co-workers have studied the early light-induced events in bR by structure-based MD simulations [5]. The retinal chromophore in bR exhibits a very large difference dipole of 13.5 Debye on optical excitation, corresponding to the transfer of almost a quarter of an elementary charge towards the Schiff base terminus [6]. It was found that upon such a charge redistribution, the protein shows an extensive dielectric relaxation with a magnitude of ca. 1700 cm^{-1}, which occurs on a timescale of 100 fs, in striking quantitative agreement with our findings. The simulations on bR showed that the relaxations involve relatively small motions of many nuclei up to relatively long distances from the chromophore. The protein conformation was distorted only minimally due to large activation barriers, in agreement with our finding that 'diffusive' protein relaxations do not occur.

We conclude that on the sub-ps to ps timescale, on which the isomerization of retinal takes place, the protein's major energetic influence is electrostatic via a large number of small amplitude motions of charges and dipoles. Major structural rearrangement of the protein that might, for example, direct the reactive motion by repulsive interactions do not occur on the timescale of the isomerization.

References

1 F. Gai, K.C. Hasson, J. MacDonald, P.A. Anfinrud, *Science* 279, 1886, 1998.
2 G.R. Fleming and M.H. Cho, *Annu. Rev. Phys. Chem.* 47, 109, 1996.
3 K. Ohta, D.S. Larsen, M. Yang and G.R. Fleming, *J. Chem. Phys.* 114, 8020, 2001.
4 G.R. Loppnow, R. A. Mathies, T.R. Middendorf, D.S. Gottfried, S.G. Boxer, *J. Phys. Chem.* 96, 737, 1992.
5 D. Xu, C.H. Martin, K. Schulten, *Biophys. J.* 70, 453, 1996.
6 R.A. Mathies, L. Stryer, *Proc. Natl. Acad. Sci. (USA)* 1976, 73, 2169.

Two-photon standing wave patterned photobleaching to probe DNA dynamics in live cells

C. J. Bardeen and S. K. Davis

Department of Chemistry, University of Illinois, 600 S. Mathews Ave., Urbana, IL 61801, USA
E-mail: bardeen@uiuc.edu

Abstract. A combination of standing-wave two-photon excitation and patterned photobleaching has been used to study the diffusive motion of DNA in live cells on the lengthscale of 100 nm.

1. Introduction

Two-photon excitation, which avoids the harmful biological effects of ultraviolet exposure and provides intrinsic axial resolution, has been widely used in confocal microscopes to image static structures in biological samples . We have combined two-photon excitation [1] with patterned photobleaching [2] and a standing wave geometry [3] in order to study macromolecular dynamics in such samples [4]. The aim of this work is to measure the short-range diffusive motion of the DNA-containing chromatin in live cells, which is believed to be important in biological gene expression.

2. Experimental Methods

The experimental apparatus is shown in Figure 1. Briefly, a laser system consisting of a 90 MHz Ti:Sapphire femtosecond laser generates ~45 fs, 800 nm pulses which pass through a prism compressor (not shown) to compensate for linear chirp from the optics. The intensity of these pulses is varied by a Pockels cell. The beam is divided by a beam splitter upon entry into the microscope setup. One part is reflected to a mirror attached to a piezo-driven translation stage whose movement is controlled to better than 10 nm precision. This beam is reflected upward through the inverted microscope, while the other beam reflected downward. The two counterpropagating beams are focused into the sample, spatially overlapped using adjustable mirror mounts, and temporally overlapped using a translation stage. In situ measurements show that the standing wave interference pattern has a visibility close to one. For the photobleaching experiments, a brief pulse of high intensity light is used to destroy molecules at the peaks of the standing wave pattern, leaving a periodic distribution of unbleached molecules which washes out due to diffusion. The decay of the fringe

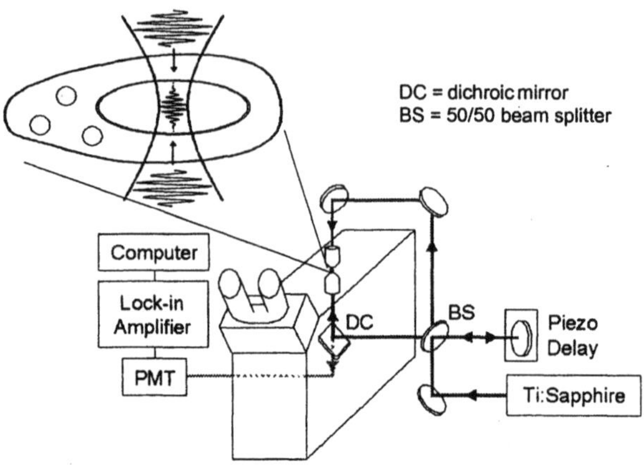

Fig. 1. Experimental set-up for focusing two counterpropagating laser beams to make a two-photon standing wave in a localized region of a cell. The sample area in the fluorescence microscope is enlarged. BS = beamsplitter, DC = dichroic mirror.

pattern is monitored by varying the phase delay of one of the beams, which scans the standing wave across the bleach pattern. The resulting fluorescence signal has both an oscillatory component that decays due to motion across the axial fringes, and a non-oscillatory component that recovers due to motion of unbleached molecules from outside the focal spot.

3. Results and Discussion

This experiment is highly sensitive to small-scale spatial displacements on the order of 100 nm or less. Given the laser parameters used in this experiment, the 1/e decay of the oscillations corresponds to a three dimensional root-mean-square displacement of 115 nm. This increased sensitivity is required for measuring chromatin movements inside the nuclei of live cells. Figure 2 shows data from a single spot inside the nucleus of a *Xenopus laevis* frog cell, showing the oscillations due to the standing wave pattern decaying on a timescale of seconds. The average diffusion constant obtained from the oscillation decays is 3×10^{-12} cm^2/s. It is also constrained – the unmodulated bleach component never recovers fully, since chromatin motion on a lengthscale of 1-2 microns does not occur. Such subdiffusive behavior on larger lengthscales is expected in environments characterized by a high degree of molecular crowding, as expected in the cell nucleus. By moving the probe spot within the cell, we have found considerable variation in the diffusion depending on location within the nucleus. Attempts to correlate the diffusion rate with local chromatin density or proximity to larger scale structures like the nucleolus or the nuclear membrane were unsuccessful.

Experiments on isolated nuclei confirm that the environment of a live cell is necessary for the observed diffusion. The most consistent explanation for our data is that dynamic protein binding events are responsible for the *in vivo* flexibility of the chromatin. We are currently testing this hypothesis using photo-crosslinking experiments and *in vitro* studies.

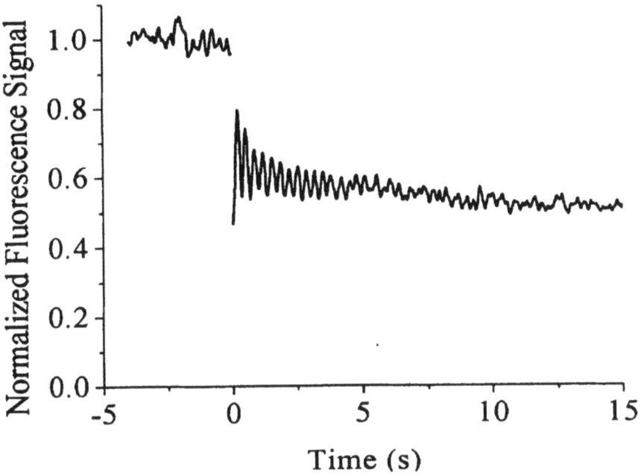

Fig. 2. Two-photon fluorescence recovery after patterned photobleaching in a live *Xenopus laevis* cell. Before time = 0, scanning the probe standing wave across the sample does not generate oscillations in the fluorescence signal. After the high intensity bleaching pulse, the bleached fringes decay on a timescale of seconds due to chromatin motion in the nucleus.

4. Conclusions

The combination of two-photon excitation and the standing wave geometry has enabled us to measure very small-scale diffusive motion in cells, providing new insights into macromolecular dynamics in living media.

References

1 W. Denk, J. H. Strickler, and W. W. Webb, Science **248**, 73, 1990.
2 M. T. Cicerone, F. R. Blackburn, and M. D. Ediger, Macromolecules 28, 8224, 1995.
3 B. Bailey, D. L. Farkas, D. L. Taylor, and F. Lanni, Nature **366**, 44, 1993.
4 S. K. Davis and C. J. Bardeen, Rev. Sci. Inst. **73**, 2128, 2002.

Part XIII

Applications of Ultrashort Pulses

Ultrahigh-Resolution Distance Meter Using a Frequency Comb of a Femtosecond Mode-Locked Laser

K. Minoshima, T. Tomita, Y. Yamaoka, and H. Matsumoto

Metrology Institute of Japan, National Institute of Advanced Industrial Science and Technology (AIST), Central 3, 1-1-1 Umezono, Tsukuba, Ibaraki 305-8563, Japan
E-mail: k.minoshima@aist.go.jp

Abstract. An optical distance meter is developed using intermode beats of a femtosecond frequency comb with a simple setup and high reliability for open-field measurement. Phase measurements of 10-GHz beat components realize one-micrometer resolution up to 240 m distance. The distance meter follows the variation of the environmental parameters with an accuracy of 2.3×10^{-8}, and then realizes automatic refractive index correction with an accuracy of 1.3×10^{-6} by using two-color measurements.

1. Introduction

High-resolution long-distance measurements are required for applications in a wide range of fields, such as for optical geodesy, shape measurement of parabola antennae, and sensors. Conventionally, several methods have been used for length measurements. Optical length-measuring interferometry and time-of-flight measurement of optical pulses have been used for high-resolution measurements, but these methods are not suitable for long distance, because they require continuous scanning of the optical path as a reference. On the other hand, phase measurement of a sinusoidally modulated cw optical wave has been used for long-distance measurement because it allows real-time measurement without a scanning reference [1]. In this method, either a high-resolution phase meter or high-frequency modulation is required to realize high resolution in the distance measurement. And although high-frequency cw modulation of as high as 28 GHz has recently been reported using this method [2], the requirement of an external modulator with a complex setup inevitably leads to large cyclic error that originates from the electric and optical crosstalk and that is independent of the distance. To overcome this problem, we have developed an optical distance meter based on phase measurement of an intermode beat of a frequency comb using a mode-locked femtosecond laser [3]. Because this method does not require an external modulator, we can significantly suppress the cyclic error. Moreover, the setup is simple, which is important for practical application. In addition, the femtosecond distance meter has marked advantages over cw methods in that it is suitable for multicolor measurement and is less sensitive to sharp absorption lines in air. These qualities are important for reliability and accuracy. In the previous study, we achieved 50 μm resolution at a 240 m distance using 1 GHz beat frequency. Without a modulator, it is easy to utilize high frequency by simply selecting one of the beat components in the broad and stable frequency comb. In this study, we achieved ultrahigh one-micrometer resolution in the femtosecond distance meter by using up to 10 GHz components of beat frequencies.

2. Experimental Methods

The experimental setup is shown in Fig. 1. A femtosecond mode-locked laser generates pulses whose temporal widths are 80 fs and which appear with a 20 ns repetition rate in the time domain (Femtolite-C-20-SP, IMRA America, Inc.). In contrast, in the frequency domain, the pulse train is an optical frequency comb with optical frequency modes that appear every 50 MHz precisely over THz wide. The intermode beats between all possible pairs in the comb give a wide range of stable electric frequency components if the pulse train is detected by a photo-electronic detector. This electric frequency comb is a series of electric waves whose phase provides the information on the pulse propagation. The femtosecond mode-locked pulse train thus is used as a series of sinusoidally modulated optical waves that can be applied to the distance measurement.

The laser emits both fundamental (center wavelength, 1560 nm) and second harmonic (780 nm) components that are coaxial to each other. The main measurements in this study were done using the second harmonic component. The beam is separated into two, and one of the resulting beams is detected by a photodetector as a reference (PD1). The other beam propagates the target distance and is detected by a photodetector as a probe (PD2). From the series of electric beat frequency components, specified frequency components, f and f-f_{rep}, are selected by bandpass filters (BPF) for the probe and reference, respectively. Here f_{rep} is the repetition frequency of the laser, 50 MHz (48.1543 MHz). In this study, we performed measurements using 10 GHz (10.0643 GHz, 209-th harmonic of the repetition rate) as the measurement frequency, f. A double balanced mixer (DBM) selects a component at a frequency, f_{rep}, derived as the difference between the frequencies of the probe and the reference, which provides the information on the phase difference. The phase difference is measured by a phase meter when the fundamental beat at f_{rep} from the reference signal is used as the reference.

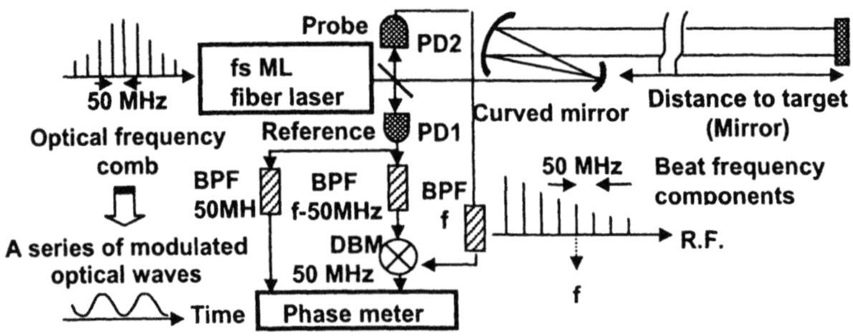

Fig. 1. Experimental setup for the femtosecond comb distance meter.

Table 1. Femtosecond comb distance measurement at 10 GHz in the optical tunnel.

distance (m)	positioning resolution (μm)	nonlinearity in scanning (μm) residuals of linear fitting	nonlinearity in scanning (μm) after cyclic error correction
0	1	10 (0.24 deg)	4
10	2	---	2
25	2	---	4
50	2	---	7
240	2	14 (0.33 deg)	2

We performed distance measurements in a 310-m-long semi-underground optical tunnel covered by soil. The annual temperature fluctuation at this site is about 2 °C. The temperature is monitored every 6 m, and the pressure and humidity are monitored at three points inside the tunnel. These facilities are suitable for evaluating the effect of the environmental fluctuations on the distance measurement.

3. Results and Discussion

We performed phase measurements at several different distances in the optical tunnel. The results are summarized in Table 1.

In order to evaluate the positioning resolution, we monitored the phase fluctuation for 200 s and calculated the standard deviation for an average time of 10 s. An ultrahigh resolution of 1.6 μm at a 240 m distance was obtained (Fig. 1).

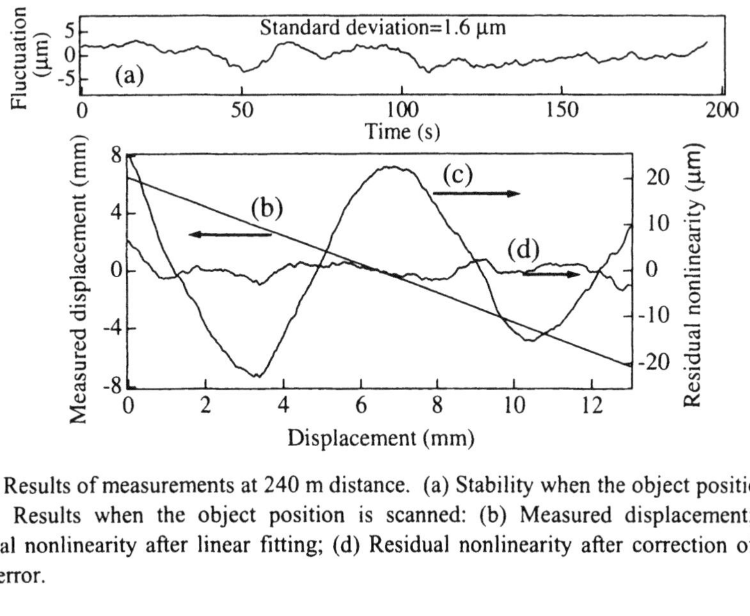

Fig. 2. Results of measurements at 240 m distance. (a) Stability when the object position is fixed. Results when the object position is scanned: (b) Measured displacement; (c) Residual nonlinearity after linear fitting; (d) Residual nonlinearity after correction of the cyclic error.

In order to evaluate the linearity of the distance measurement, the phase shift was measured while varying the object position from 240 m, although scanning is not necessary in the actual distance measurement. The measured displacement clearly shows a linear relationship with the applied displacement (Fig. 2 (a)). After linear fitting, the residual nonlinearity is obtained (Fig. 2 (b)). The residual nonlinearity shows a sinusoidal dependence at f (corresponding to a cycle of about 15 mm) or higher harmonics, which is called a cyclic error. Here we see a small cyclic error of 0.3 deg, while in conventional methods, the cyclic error may be more than several degrees for similar measurements. This small cyclic error is due to the simplicity of the setup, and particularly to the absence of a modulator. Cyclic error is thought to originate from electric and optical cross-talk, such as an optical multi path. If the optical multi path is dominant, the cyclic error is expected to be more significant for shorter distance. However, the results show comparable cyclic errors for the distances of 0 m and 240 m, which suggests that the optical multi path is negligible. Because the current system is only a mock-up, it is rather sensitive to electric noise. We expect that the cyclic error will be further reduced by making a robust electric system.

Moreover, we performed fitting for sinusoidal functions at f and 2f for this nonlinear error because it is expected to be systematic. The standard deviation of the final residuals (Fig. 2 (c)) was 1.6 μm after 10 s time averaging, which gives the resolution of the distance measurement with scanning. That this result was consistent with the values of the resolution without scanning supports the validity of the fitting procedures described above.

In long-distance measurements, the refractive index of air is a significant parameter in terms of precision because it fluctuates with environmental conditions. In order to evaluate the ability of our femtosecond comb distance meter to follow the environmental fluctuation, we measured a fixed distance, 240 m, for a long term, and extracted the temporal variation of the group refractive index. Simultaneously, the environmental temperature, pressure, and humidity in the optical tunnel were monitored, and the group refractive index of air was calculated [4]. The temporal variations of the group refractive index are plotted in Fig. 3 for both the measured (a) and calculated (b) cases. The measured and calculated results show good agreement. The difference between the two sets of values is also plotted in Fig. 3(c). The standard deviation is $2.3\text{x}10^{-8}$, which indicates a high ability of our femtosecond comb distance meter to follow the variation of the environmental conditions.

In practical application, the environmental parameters cannot be fully obtained, especially because they vary with respect to both time and space. By a simultaneous measurement using pulses of two different colors, which is called a two-color method [3,5], the effect of the refractive index variation is cancelled out. This is because the dispersion relation of the refractive index of air does not change although the environmental condition changes. A femtosecond mode-locked laser is suitable for generating different colors for pulses.

In the two-color method, the geometrical distance (D), which is without the ambiguity of the refractive index, is obtained as Eq. 1. Here the group refractive indices of air for each color component are denoted as n_1 and n_2, respectively. The heart of the method is that the value A is constant under practical environmental conditions. By using the optical distances for two colors, D_1 and D_2, which are

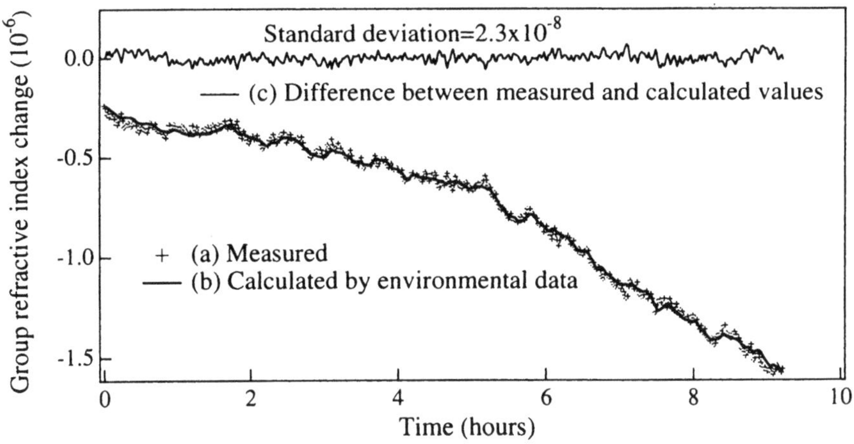

Fig. 3. Temporal variation of the group refractive index of air, which is measured by the femtosecond comb distance meter and calculated using environmental data.

obtained by the phase measurements using the femtosecond comb distance meter, the geometrical distance D is obtained.

$$D = D_2 - A(D_2 - D_1), \quad \text{where } A = \frac{n_2 - 1}{n_2 - n_1} \text{ and } D_{1,2} = n_{1,2} \cdot D. \tag{1}$$

We performed two-color measurement using 780 nm and 1560 nm components which are emitted coaxially from the laser. Phase measurement for each color was performed using the same setup only changing the color filters to select each component. We continuously measured the fixed distance of 240 m for 15 hours. As a result, D was constant within a relative accuracy of 6×10^{-7}, which corresponds to 140 μm, independent of the environmental condition change up to 2×10^{-6}.

In conclusion, an ultrahigh resolution distance meter with a simple setup and high reliability for an open-field measurement was developed using a broad and stable femtosecond frequency comb. This distance meter is expected to enable new real-world applications of ultrafast technology.

Acknowledgements. This work is supported by the Research Project on Femtosecond Technology (FST), METI, Japan. T. Tomita was also with Tokyo Science University, Japan. We thank Dr. H. Fujimoto for fruitful discussions.

References

1 J. Rueger, *Electronic Distance Measurement*, Springer-Verlag, 1990.
2 I. Fujima, S. Iwasaki, and K. Seta, Meas. Sci. Technol. **9**, 1049, 1998.
3 K. Minoshima and H. Matsumoto, Appl. Opt. **39**, 5512, 2000.
4 P.E. Ciddor and R.J. Hill, Appl. Opt. **38**, 1663, 1999.
5 P. Bender and J. Owens, J. Geophys. Res. **70**, 2461, 1965.

Ultrahigh Resolution Optical Coherence Tomography using Novel Femtosecond Lasers

I. Hartl[1], A. M. Kowalevicz[1], P.-L. Hsiung[1], T. H. Ko[1], T. Schibli[1], F. X. Kärtner[1], J. G. Fujimoto[1], T. A. Birks[2], W. J. Wadsworth[2], U. Bünting[3], and D. Kopf[3]

[1] Department of Electrical Engineering and Computer Science and Research Laboratory of Electronics, Massachusetts Institute of Technology, Cambridge, MA 02139, USA
[2] Optoelectronics Group, Department of Physics, University of Bath, UK
[3] High Q Laser Production GmbH, Hohenems, Austria

Abstract. We describe new light sources for ultrahigh resolution OCT. A compact Nd:Glass femtosecond laser, spectrally broadened in a tapered fiber and an ultralow threshold, low cost, Ti:Al_2O_3 laser can achieve image resolutions of <5 μm at 1300 nm and 800 nm.

Optical coherence tomography (OCT) is an emerging technology for micron-scale cross-sectional imaging of biological tissue and materials [1]. High resolution detection of backscattered light is performed using low coherence interferometry. One of the key problems in OCT is the lack of compact, high performance, low coherence light sources with sufficient bandwidth and power to enable high resolution, real time imaging. A Ti:Al_2O_3 laser generating ~5 fs pulses and bandwidths of 300 nm at 800 nm has been demonstrated for 1.5 μm axial resolution OCT imaging [2,3]. High nonlinearity, air-silica microstructure fibers [4] or tapered fibers [5] can generate broadband continua using low energy femtosecond pulses and have been used to achieve 2.5 μm image resolutions at 1300 nm wavelengths [6]. While these approaches achieve high resolutions in the important 800 nm and 1300 nm wavelength ranges, these lasers are relative expensive and bulky.

In this paper we describe new femtosecond laser light sources for ultrahigh resolution OCT imaging which are compact, robust, and suitable for clinical applications. Ultrahigh resolution OCT at 1.3 μm was demonstrated using a compact Nd:Glass femtosecond laser generating ~110-150 fs pulses which are spectrally broadened in a tapered single mode fiber. Bandwidths of 132 nm can be generated near 1300 mm, corresponding to image resolutions of <5 μm. This system is based on a commercially available Nd:Glass laser which is compact and portable. Short coherence length light at 800 nm is generated using a new, ultra-low threshold modelocked Ti:Al_2O_3 laser. This laser operates with pump powers in the 200 mW range, generating output powers of 15 mW with pulse durations of 10-15 fs and bandwidths of ~90-130 nm. By reducing the pump power requirements, this laser can cost up to 5x less than standard Ti:Al_2O_3 lasers. These sources will enable ultrahigh resolution OCT imaging at 1300 nm and 800 nm.

Spectrally Broadened Femtosecond Nd:Glass Laser at 1300 nm. Figure 1 shows the experimental setup for 1300 nm imaging. A compact diode pumped femtosecond Nd:Glass laser (High Q Laser Production) generates pulses with 110-

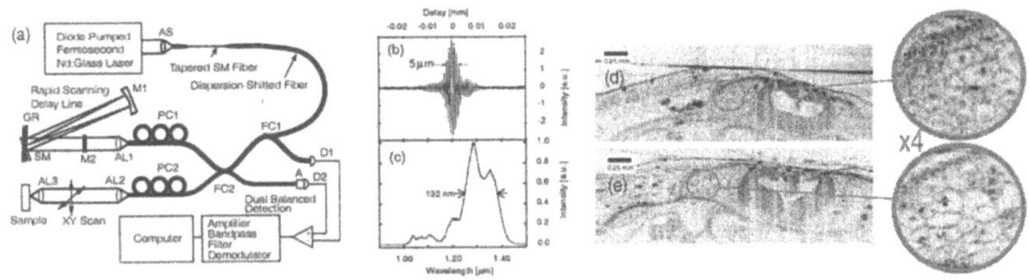

Fig. 1. Left: OCT system using a femtosecond diode pumped Nd:Glass laser with continuum generation in a tapered single mode fiber and a broadband rapid scanning delay line. Right: High resolution in vivo OCT image (bottom) of an African frog tadpole (Xenopus laevis) at 1300 nm with 5 μm resolution compared to conventional 20 μm resolution image (top).

150 fs duration and 130 mW average power at 75 MHz repetition rate and 1.06 μm wavelength. The pulses are coupled into a single mode fiber (Corning SMF-28). The fiber was tapered by stretching in a flame so that, after a short length (~20 mm) of normal fiber, the fiber tapers down to a uniform waist with a diameter of 2 μm and a length of 90 mm, before tapering up again to normal fiber. This taper results in efficient generation of continuum [5]. The output of the tapered fiber is fusion spliced to a 10 m length of dispersion shifted fiber (zero dispersion at longer wavelengths) to reduce parasitic contributions from four wave mixing. This fiber is then connected to the fiber optic OCT system which consists of broadband 3 dB fiber couplers (FC), designed for 1300 nm center wavelength, together with custom designed achromatic lenses.

In order to accommodate the broad bandwidths and enable real time imaging, delay scanning is performed using a rapid scanning delay line with reflective optics. The group delay is scanned at 1 m/s and the heterodyne frequency is set to 1 MHz. Dispersion is matched in the two arms of the interferometer in order to maintain resolution. Dual balanced detection with two InGaAs photodiodes (D1, D2) is used to cancel excess noise.

Figure 1 shows a typical optical spectrum of the continuum as well as the measured interferometer output from a single reflection, giving the axial point spread function. The OCT axial resolution was measured to be 5 μm in air (or <4 um in tissue) and the sensitivity was 90 dB. To demonstrate the performance of the system for ultrahigh resolution in vivo imaging, an African frog tadpole (Xenopus laevis) was imaged using the continuum light and a conventional semiconductor light source (AFC Technologies). The continuum light source yields an axial resolution of 5 μm in air or ~4 μm in biological tissue, which is four times higher than a conventional semiconductor light source.

Ultralow Threshold Ti:Sapphire Laser at 800 nm. Figure 2 shows a schematic of the ultra-low-threshold KLM Ti:Al_2O_3 laser. The pump source is a frequency doubled, diode pumped Nd:Vanadate laser. The cavity is a standard folded X configuration with a 2 mm thick Ti:Al_2O_3 laser crystal with $\alpha = 5.0\ cm^{-1}$. The focusing mirrors are 7.5 cm radius of curvature. The output coupler has a transmission of ~2% between 700 nm and 900 nm and is low dispersion.

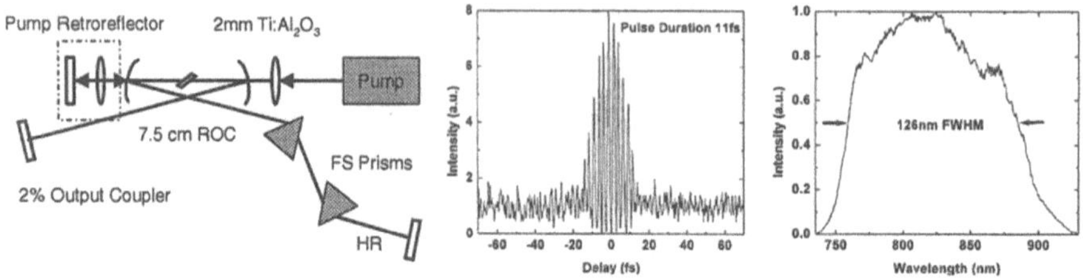

Fig. 2. a: Schematic diagram of the ultra low threshold Ti:Al_2O_3 laser. b: Output pulse duration of 11 fs and spectrum with bandwidth of 126 nm obtained using DCMs.

Intracavity dispersion compensation and tuning is provided by a pair of fused silica Brewster prisms. To achieve high performance KLM operation at low thresholds the laser cavity is extended and has a 50 MHz repetition rate. In addition to increasing the energy per pulse by a factor of 2 over the standard 100 MHz configuration, the increased arm lengths also reduces the laser mode size in the crystal to about 8 μm radius. Increasing the pulse energy by cavity length scaling enables stronger KLM and shorter pulses at low pump powers. The tight focusing of the laser mode also reduces the laser threshold. To more efficiently use pump power, a double pass pump configuration is used.

Using double chirped mirrors (DCMs) the pump threshold for starting KLM is <260 mW and once initiated, modelocking can be sustained with pump powers of 170 mW. Pumping at 260 mW, the laser generates 13 mW with 11 fs pulse durations with bandwidth of ~125 nm. Using standard commercial mirrors, the KLM threshold was reduced to ~156 mW. Pumping at 200 mW, the laser generates output powers of 16 mW with 14 fs pulses and 90 nm bandwidths. These bandwidths yield OCT resolutions of <5 μm. The ability to achieve ultrashort pulses and broad bandwidths using pump powers of 200 mW, more than one order of magnitude less than standard lasers, enables this laser to be built at a fraction of the cost of conventional Ti:Al_2O_3 laser technology.

References

1 D. Huang, E. Swanson, C. P. Lin, J. S. Schuman, W. G. stinson, W. Chang, M. R. Hee, T. Flotte, K. Gregory, C. A. Puliafito, and J. G. Fujimoto, Science **254**, 1178, 1991.
2 U. Morgner, F. X. Kärtner, S. H. Cho, Y. Chen, H. A. Haus, J. G. Fujimoto, E. P. Ippen, V. Scheuer, G. Angelow, and T. Tschudi, Optics Letters **24**, 417
3 W. Drexler, U. Morgner, F. X. Kärtner, C. Pitris, S. A. Boppart, X. D. Li, E. P. Ippen, and J. G. Fujimoto, Optics Letters **24**, 1221, 1999.
4 J. K. Ranka, R. S. Windeler, and A. J. Stentz, Optics Letters **25**, 25, 2000.
5 T. A. Birks, W. J. Wadsworth, and P. S. J. Russell, Optics Letters **25**, 1415, 2000.
6 I. Hartl, X. D. Li, C. Chudoba, R. Ghanta, T. Ko, J. G. Fujimoto, J. K. Ranka, R. S. Windeler, and A. J. Stentz, Optics Letters **26**, 608, 2001.

Functional device fabrication by nonlinear material processing of glass with a femtosecond laser oscillator

Kaoru Minoshima, Andrew M. Kowalevicz, Erich P. Ippen, and James G. Fujimoto

Department of Electrical Engineering and Computer Science and Research Laboratory of Electronics, Massachusetts Institute of Technology, Cambridge, MA 02139, USA

Abstract. Functional photonic devices are fabricated in glass using high-energy laser pulses from an unamplified laser oscillator. Devices including directional couplers and interferometers are fabricated. Interaction length and spectral dependences are evaluated.

Photonic device fabrication using nonlinear material processing with near-IR femtosecond pulses can generate localized, clean, three-dimensional (3D) structures in many materials [1-3]. Recently, fabrication of glass waveguides using pulses directly from femtosecond lasers oscillators, without the need for amplification, has been reported [4-6]. Unamplified laser oscillators have several advantages over amplified systems for device fabrication including lower cost and complexity as well as greater control of exposure parameters. Because of their high repetition rates, multiple low intensity shots interact within the focal volume, leading to well controlled cumulative effects. Several photonic devices have been demonstrated including single-mode waveguides [5], X-couplers [6], and directional couplers [4]. Our group has developed a high-power femtosecond Ti:Sapphire laser oscillator using a novel long cavity design [7]. Because of its high pulse energies and high repetition rate, this laser enables the versatile fabrication of a wide range of photonic devices.

In this paper we demonstrate the fabrication and characterization of directional couplers and what is, to our knowledge, the first Mach-Zehnder interferometer. Interferometers are of interest because of their use in practical devices such as sensors and switches. These devices are characterized using a variety of techniques including near field and far field mode measurements, interaction length dependence analysis, and wavelength dependent device properties. Laser processing enables rapid prototyping and fabrication of 3D device geometries.

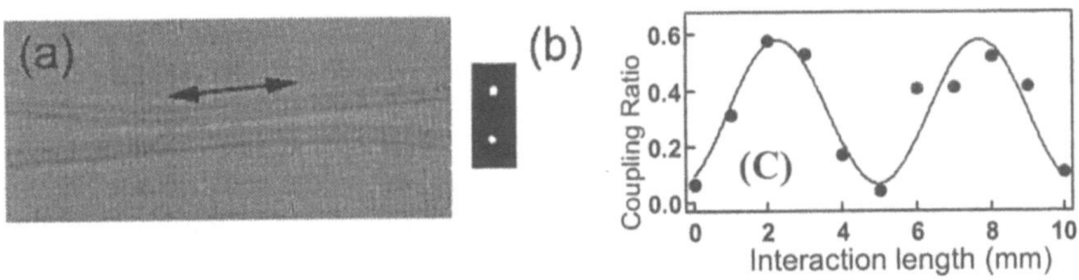

Fig. 1. (a) Microscopic view of a directional coupler with 8 μm waveguide separation; (b) output image of a He-Ne laser coupled into one of the waveguides; (c) coupling ratio as a function of the interaction length between two waveguides branches.

The laser used to perform the nonlinear material processing is a high pulse energy Ti:Sapphire laser oscillator. The laser cavity has been extended using a Herriott type multipass cell to reduce the repetition rate to 4MHz and increase the pulse energy. Operating in net negative dispersion with a saturable bragg reflector (SBR) to stabilize operation, 80 fs pulses with up to 100 nJ energies can be generated [7]. The laser is focused into a 1 mm thick glass plate using a 100X oil-immersion microscope objective with an effective 0.6 NA. Device structures were fabricated by 2D translation of the sample perpendicular to the incident light using precision computer controlled stages.

Couplers such as X couplers as well as coupled mode devices have been demonstrated, however detailed characterization requires more than only output spot observation. Here we analyzed mode-coupling behavior. Fig. 1(a) shows a two waveguide directional coupler fabricated inside a glass plate. Waveguides were fabricated by scanning the plate in the X-Y plane at 10 mm/s with an incident pulse energy of ~10 nJ. Evaluation of the mode profile at the output of the coupler shows that the waveguides have diameters of 4-5 μm and are single mode in the near-IR. We fabricated a series of couplers with several different interaction lengths (indicated with an arrow in Fig 1(a)). Fig. 1(c) shows the coupling ratio as a function of the interaction length with 8 μm separation. The data shows that the coupling ratio oscillates as a function of interaction length. The period of oscillation changes as a function of the separation. The oscillatory behavior is the first clear evidence of mode coupling between two waveguides fabricated by the nonlinear laser processing.

As an example of a more complex device, we demonstrate, what we believe is for the first time, the fabrication and characterization of a Mach-Zehnder interferometer by the nonlinear laser processing. Fig. 2 shows the interferometer, consisting of two X-couplers placed back-to-back, with crossing angles of 2 deg. The path length difference between the two arms is approximately 10 μm.

In order to characterize the operation of the interferometer, we used the output of a KLM Ti:Sapphire laser which can generate ~5 fs pulses and bandwidths of ~300 nm [8]. This laser source provides a high brightness, single mode beam, which can be easily coupled into the waveguides by using a microscope objective. The edges of the glass substrate were polished to optical quality in order to enable efficient coupling. The output spectrum from the interferometer waveguide was collimated using a second microscope objective. The spectral data was measured with an optical multichannel analyzer (OMA).

If a given frequency of light is coupled into the interferometer, it will be split into the two arms of the interferometer at the first X-coupler, travel different path lengths, and will either constructively or destructively interfere after being recombined at the second X-coupler. Thus, the unbalanced path length Mach-Zehnder functions as a filter. The frequency or wavelength dependence of the interferometer can be measured using a broadband light source. Fig. 3(a) shows the input spectrum with a FWHM of 130 nm. Fig. 3(b) shows the wavelength transfer function in the crossed interferometer arm. The transfer function is constructed by normalizing the output spectrum by the input spectrum. With an arm length difference of 10 μm and a spectrum centered at 800 nm, oscillations in the output should occur with periodicity of ~65 nm. We observe a full modulation of the transfer function with a periodicity of approximately 50 nm, in good agreement with theory.

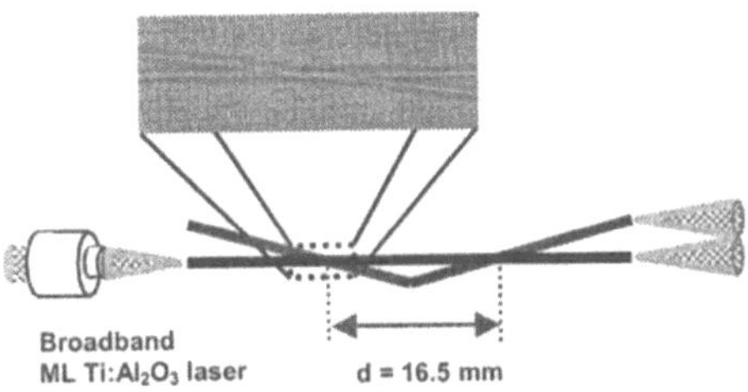

Fig. 2. Schematic representation of Mach-Zehnder Interferometer with phase-contrast microscope images of waveguides.

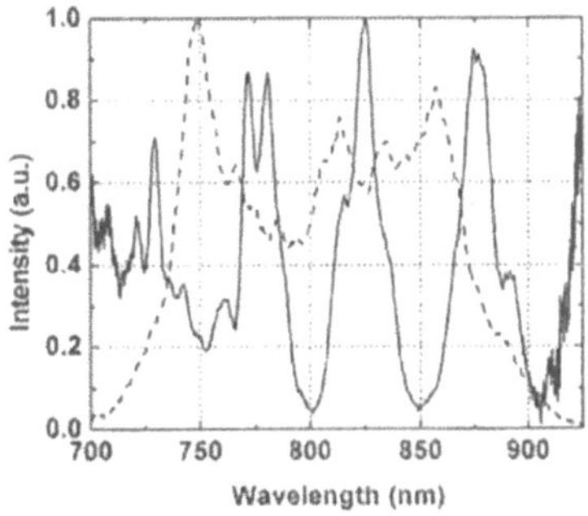

Fig. 3. (a) Input spectrum and, (b) normalized output demonstrating the filtering effect of the interferometer as oscillations with a period of 50 nm.

In conclusion, we demonstrate photonic device fabrication using nonlinear materials processing in glass with a high power femtosecond laser oscillator. The building blocks of photonic devices such as X couplers, coupled mode devices, as well as interferometers can be fabricated. Interaction length dependence and wavelength dependent characteristics, using femtosecond broadband lasers as light sources, can be measured. Femtosecond materials processing will enable rapid and versatile device prototyping as well as the fabrication of three-dimensional structures not possible with planer techniques.

Dr. Minoshima was visiting from the National Institute of Advanced Industrial Science and Technology (AIST) in Tsukuba, Japan. Andrew Kowalevicz is also with the Division of Engineering and Applied Sciences at Harvard University.

References

1 K.M. Davis, K. Miura, N. Sugimoto, and K. Hirao, Opt. Lett. **21**, 1729, 1996.
2 D. Homoelle, S. Wielandy, A.L. Gaeta, N.F. Borrelli, and C. Smith, Opt. Lett. **24**, 1311, 1999.
3 Y. Sikorski, A.A. Said, P. Bado, R. Maynard, C. Florea, and K.A. Winick, Elect. Lett. **36**, 226, 2000.
4 A.M. Streltsov and N.F. Borrelli, Opt. Lett. **26**, 42, 2001.
5 C.B. Schaffer, A. Brodeur, J.F. Garcia, and E. Mazur, Opt. Lett. **26**, 93, 2001.
6 K. Minoshima, A. M. Kowalevicz, I. Hartl, E. P. Ippen, and J. G. Fujimoto, Opt. Lett. **26**, 1516, 2001.
7 S. H. Cho, F. X. Kärtner, U. Morgner, E. P. Ippen, J. G. Fujimoto, J. E. Cunningham, W. H. Knox, Opt. Lett. **26**, 560, 2001.
8 U. Morgner, F.X. Kärtner, S.H. Cho, Y. Chen, H.A. Haus, J.G. Fujimoto, E.P. Ippen, V. Scheurer, G. Angelow, and T. Tschudi, Opt. Lett. **24**, 411, 1999.

Nanostructuring of transparent materials with femtosecond laser pulses

F. Korte[1], J. Koch[1], S. Nolte[2], C. Fallnich[1], and B.N. Chichkov[1]

[1] Laser Zentrum Hannover e.V., Hollerithallee 8, D-30419 Hannover, Germany
Phone: +49-511-2788-211, fax: +49-511-2788-100
E-mail: fk@lzh.de
[2] Institut für Angewandte Physik, Friedrich-Schiller-Universität Jena, Max-Wien-Platz 1, D-07743 Jena, Germany

Abstract. Detailed investigations of possibilities to use femtosecond lasers for the fabrication of periodic structures in transparent materials with the scale length of the order of several hundreds nanometers are reported.

Femtosecond lasers have established themselves as excellent and universal tools for direct ablative microstructuring of solid materials [1-6]. At present, the challenge is to develop a simple laser-based technology for the fabrication of periodic nanostructures in high refractive index transparent materials with a structure size of the order of several hundreds nanometers (100 – 500 nm). This is required for many applications in photonics, for fabrication of photonic crystals, LEDs and laser diodes, microoptical devices, data storage etc.

In this paper investigations of possibilities to use femtosecond lasers for direct ablative nanostructuring of transparent materials are reported. Our goals are to understand how small structures can be fabricated with femtosecond laser pulses and how practicable and reliable this technique can be. Due to the nonlinear nature of interaction of femtosecond laser pulses with transparent materials, to initiate ablation simultaneous absorption of several photons is required. Multiphoton absorption produces initial free electrons which are further accelerated by a femtosecond laser field. These electrons induce avalanche ionization, optical breakdown, and generate microplasma. Subsequent microplasma expansion results in the fabrication of a small structure at the target surface. It is reasonable to expect that the minimum structure size, which can be produced by this way, is determined by the following equation $d = k\lambda/NA\, q^{1/2}$, where λ is the radiation wavelength, NA is the numerical aperture of the focusing optics, k is a proportionality constant ($k = 0.5$ - 1), and q is the number of photons required to overcome the energy band gap.

To test this prediction we investigated the dependence of the minimum structure size, which can be fabricated in different transparent materials, on their band gap energy. The results are shown in Fig. 1. In our experiments we used a commercial kilohertz femtosecond laser system [7] delivering 1 mJ, 30 fs laser pulses at 800 nm. To focus the laser radiation a 36× Schwarzschild objective with $NA = 0.5$ was used. As can be seen in Fig. 1, for transparent materials the minimum structure size is in good agreement with the above equation.

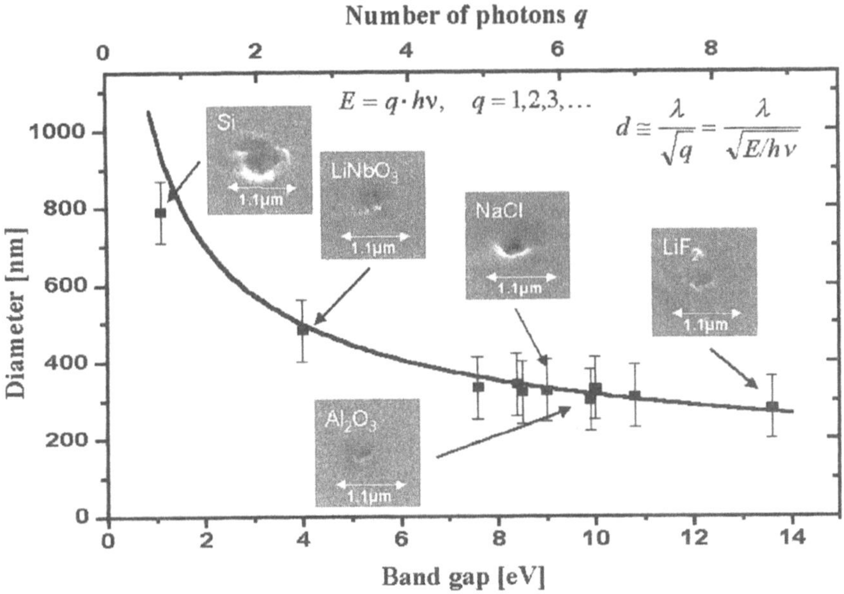

Fig. 1. Dependence of the minimum structure size for different transparent materials on their band gap energy. The solid line represents the structure size calculated due to the expression $d \cong \lambda/q^{1/2}$.

To perform nanostructuring one should carefully control the laser intensity distribution in the focus area. This distribution was measured by imaging the corresponding position in focus onto a CCD camera with a high numerical aperture (NA=0.95) 100× microscope objective. In Fig. 2, the measured laser intensity distribution in the focus of a Schwarzschild objective and an example of a hole drilled in glass are shown.

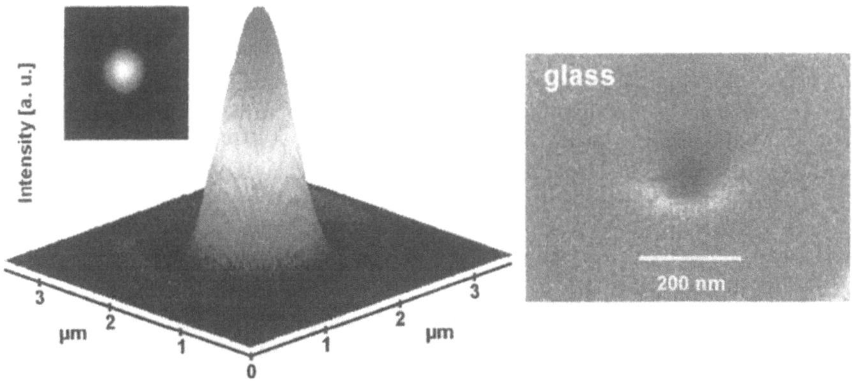

Fig. 2. Measured laser intensity distribution in the focus of a Schwarzschild objective and an example of a 200 nm hole drilled in glass with femtosecond laser pulses.

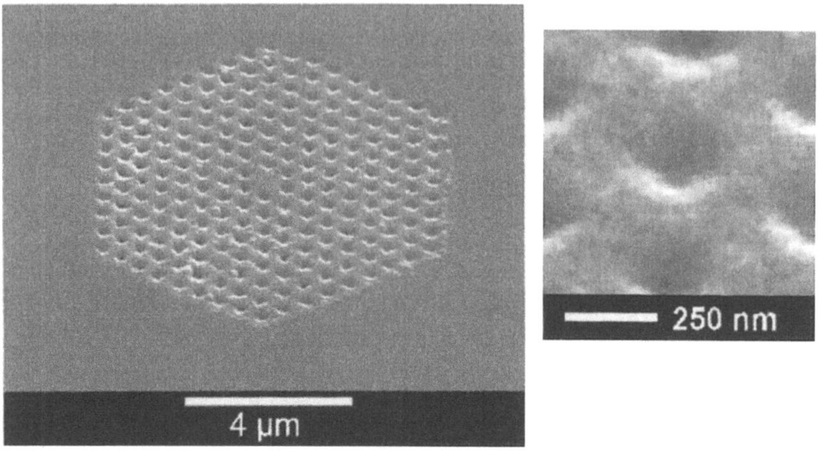

Fig. 3. An example of a periodic nanostructure with a defect cavity fabricated in a sapphire crystal with femtosecond laser pulses. On right side, an enlarged fragment of a single hole is shown.

In Fig.3, first periodic nanostructures fabricated in a sapphire crystal with femtosecond lasers are shown. For each hole ten lasers pulses have been used. The total processing time for 216 holes shown in Fig.3 was 30 seconds (including the time required for the sample positioning).

In conclusion, with tightly focused 800 nm femtosecond laser pulses periodic structures with the structure size down to 100 nm can be fabricated by direct ablative writing at atmospheric pressure. It has been observed that nanostructuring with femtosecond lasers is material dependent. Identification of materials which can be nanostructured with femtosecond laser pulses has been performed.

References

[1] P. P. Pronko, S. K. Dutta, J. Squier, J. V. Rudd, D. Du, and G. Mourou, Opt. Commun. **114**, 106 (1995).

[2] C. Momma, B. N. Chichkov, S. Nolte, F. von Alvensleben, A. Tünnermann, and H. Welling, Opt. Commun. **129**, 134 (1996); B. N. Chichkov, C. Momma, S. Nolte, F. von Alvensleben, and A. Tünnermann, Appl. Phys. A **63**, 109 (1996).

[3] B. C. Stuart, M. D. Feit, S. Herman, A. M. Rubenchik, B. W. Shore, and M. D. Perry, J. Opt. Soc. Am. B **13**, 459 (1996).

[4] P. Simon and J. Ihlemann, Appl. Phys. A **63**, 505 (1996).

[5] W. Kautek, J. Krüger, M. Lenzner, S. Sartania, C. Spielmann, and F. Krausz, Appl. Phys. Lett. **69**, 3146 (1996).

[6] S. Nolte, C. Momma, H. Jacobs, A. Tünnermann, B. N. Chichkov, B. Wellegehausen, and H. Welling, J. Opt. Soc. Am. B **14**, 2716 (1997).

[7] FEMTOLASERS Produktions GmbH, Vienna, www.femtolasers.com.

Demonstration of the high temporal and spatial resolution in junction-mixing ultrafast scanning tunneling microscopy

Dzmitry A. Yarotski, Antoinette J. Taylor

Material Science and Technology Division, MST-10, Los Alamos National Laboratory, MS K764, Los Alamos, NM 87545
E-mail: dzmitry@lanl.gov

Abstract. We present results of junction-mixing ultrafast scanning tunneling microscopy that demonstrate a combined spatial/temporal resolution of 1 nm/13 ps, yielding nearly a factor of 3 improvement in the temporal resolution over previous results.

1. Introduction

The growing field of nanotechnology requires special tools capable of probing ultrafast dynamics of nanostructures on the atomic scales. One of the possible practical realizations of such technique is junction-mixing STM (JM-STM), which combines scanning tunneling microscopy (STM) and nonlinear optical spectroscopy [1, 2]. Using JM-STM, the spatial resolution of 1 nm and the temporal resolution of more than 35 ps (full width at half maximum (FWHM)) have been demonstrated [3]. However, this temporal resolution is not yet sufficient for the ultrafast dynamics studies.

Here we present recent results for JM-STM with a greatly improved temporal resolution of less than 13 ps FWHM.

2. Results and Discussions

The experimental setup used in these measurements is described in detail elsewhere [4]. To investigate the temporal resolution of JM-STM we used voltage pulses generated on a Pt transmission line deposited on the low temperature grown GaAs (LT-GaAs) with a 2 ps carrier lifetime. The LT-GaAs wafer was polished to 75 μm thickness in order to lower the line impedance. To determine the spatial resolution of the JM-STM, Ti dots with 2.5 μm diameter and 20 nm height were deposited on top of the Pt transmission line. The time-dependent I-V relation for the STM is of the form: $I(t) = \beta[V(t) + \gamma V^3(t)]$, where $I(t)$ $[V(t)]$ is the tunneling current [bias voltage], β - the linear tunneling conductance, and γ – the nonlinear tunneling coefficient. Because of a difference in the value of γ between Ti and Pt, the amplitude of the transient tunneling current induced by voltage pulses is strongly dependent on the position of the tip.

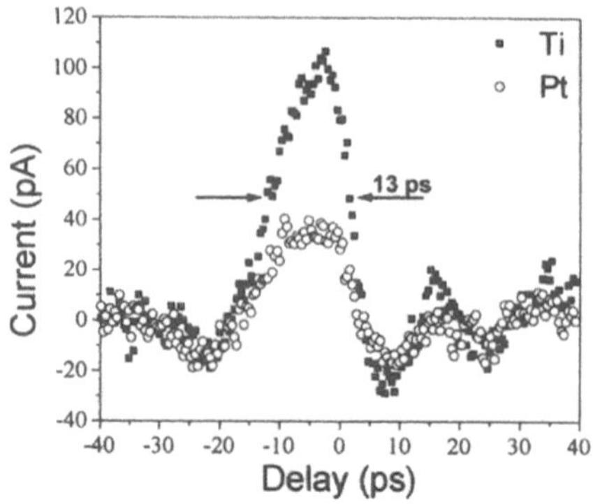

Fig. 1. Transient tunneling signals taken on top of the titanium dot (black squares) and on top of the platinum substrate (open circles) at V_b=0.1 V, I_t=3nA.

Figure 1 presents the transient tunneling current I(τ) as a function of delay τ between the two optical pulses with the tip placed above the Pt stripline and above the Ti dot. FWHM of the transient tunneling signal is 13 ps, limited partially by the long carrier lifetime in LT-GaAs photoswitch, as well as by the stripline impedance. By scanning the tip through the border between Pt and Ti and recording the time-resolved tunneling signal at every point of this scan, a spatio-temporal picture of the surface was acquired which is presented in Fig. 2.

Now a spatial scan of the transient tunneling current amplitude (Figure 3, solid line) taken at a fixed (-3.6 ps) delay can be extracted and compared with a topographic scan, acquired simultaneously (Figure 3, dashed line). This scan

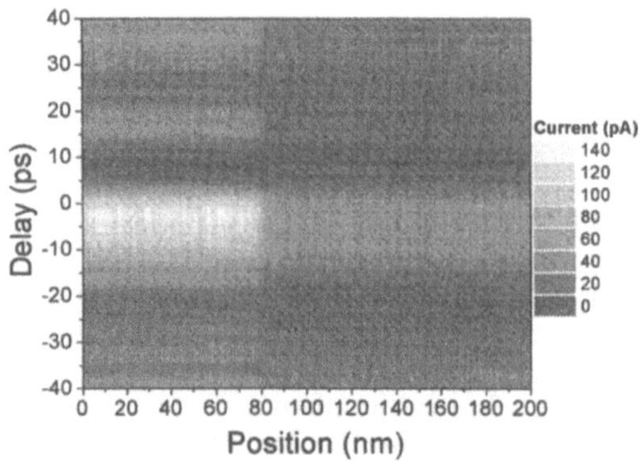

Fig. 2. Spatio-temporal scan of the patterned sample across the Pt/Ti border.

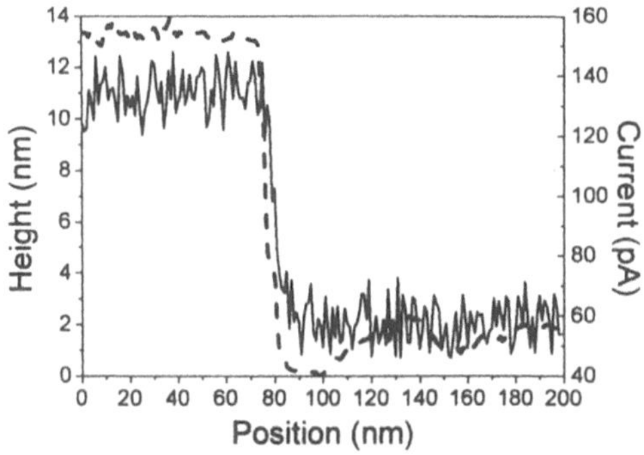

Fig. 3. Individual scan line extracted from the spatio-temporal scan (solid line) along with a topographic scan (dashed line) across the Ti/Pt border.

reveals that the amplitude of the signal changes significantly as the tip scans 1 nm, yielding a 1 nm spatial resolution for our JM-USTM set-up. This 1 nm spatial resolution is limited by the fabrication of the Ti dots since the topographic scan exhibits the same spatial resolution across that border.

3. Summary

In summary, we have demonstrated 1 nm spatial and 13 ps FWHM temporal resolution using JM-USTM, where the spatial resolution is limited by the fabrication of the structure studied. This result shows comparable spatial and greatly improved temporal resolution compared to the 1 nm/35 ps temporal resolution previously reported [3].

References

1 G. Nunes, Jr. and M.R Freeman, Science **262**, 1029 (1993).
2 G.M. Steeves, A.Y. Elezzabi, M.R. Freeman, Appl. Phys. Lett. **72**(4), 504 (1998).
3 N.N. Khusnatdinov, T.J. Nagle, and G. Nunes, Appl. Phys. Lett. **77**, 4434 (2000).
4 G. P. Donati, G. Rodriguez, and A. J.Taylor, J. Opt. Soc. Am. B **17**, 1077 (2000).

Gouy Phase Shift of Single-Cycle Picosecond Acoustic Pulses

N.C.R. Holme, M.T. Myaing, and T.B. Norris

Center for Ultrafast Optical Science
University of Michigan
2200 Bonisteel Blvd.
Ann Arbor, Mi 48109-2099
Phone 734-764-9269
Fax 734-763-4876
Email: tnorris@eecs.umich.edu

Abstract. We have investigated the effect of diffraction on single-cycle picosecond acoustic transients generated using femtosecond lasers. The Gouy phase shift has been observed, indicating the feasibility of new nanometer-scale acoustic imaging schemes.

The generation of terahertz (THz) radiation via photoconductive switching or optical rectification with femtosecond optical pulses has enabled the investigation of diffractive effects with single-cycle pulses. The far-field diffraction of a broadband pulse from an aperture can be described by the time-domain Huygens-Fresnel integral:

$$u(P_0,t)=\iint_{\Sigma}\frac{\cos\theta}{2\pi c r_{01}}\frac{\partial}{\partial t}u\left(P_1,t-\frac{r_{01}}{c}\right)d\sigma$$

That the far-field pulse shape is the temporal derivative of the input (near field) pulse is a manifestation of the Gouy phase shift, which has been observed for THz electromagnetic pulses by Ruffin et al. [1]. Because THz measurements can directly yield the diffracted field (not just the intensity), a new form of imaging is possible, in which the measured far-field THz waveforms are input into a *time-reversed* Huygens-Fresnel integral to reconstruct the object. Both one- and two-dimensional imaging have recently been demonstrated by our group [2].

Femtosecond optical pulses are also used to generate single-cycle *acoustic* pulses with up to THz bandwidth [3]. Such pulses are now often used to perform 1-D tomographic imaging of thin films (e.g. in commercial integrated circuit fabrication lines) [4]. Given the very strong analogies between picosecond electromagnetic and acoustic (phonon) pulses, it is natural to ask whether the 2-D time-reversal imaging technique might be adaptable to picosecond acoustics. This is an intriguing possibility, since THz bandwidth implies few-nanometer resolution in the acoustics case. The first step in developing such a technique is highly nontrivial: we must first demonstrate that THz acoustic pulses can be diffracted into the far field and successfully detected. The direct manifestation of this would be to observe the Gouy phase shift.

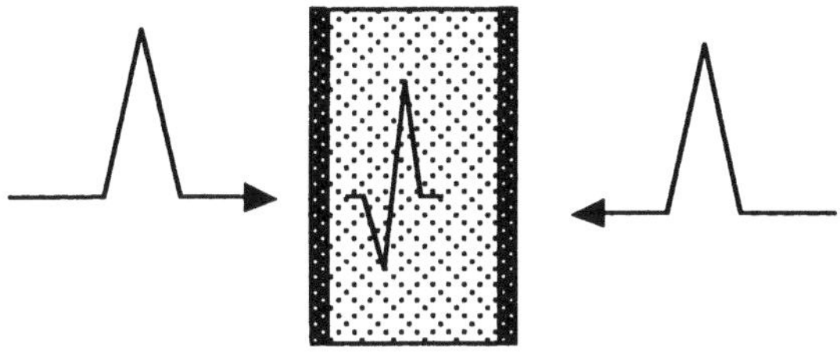

Figure 1. Acoustic pulse excitation and detection scheme.

The basic scheme of our experiment is shown in figure 1. A 100-fs optical pulse from an 80-MHz Ti:sapphire oscillator is incident on a 15-nm thick Al film deposited on a 500-um high purity single-crystal <100> Si wafer, launching a single-cycle acoustic pulse into the silicon. The acoustic wave propagates through the Si crystal, and is detected on the other side by the strain-induced modulation of the reflectivity of a thin Al film probed by a delayed fs optical pulse. Because the large propagation delay of the acoustic pulse would require a 15-meter optical delay, a probe delay line the length of the oscillator was built with a self-aligning beam stabilization system to allow arbitrary pulse delays to be obtained. The principal challenge of the experiment was to find the picosecond-duration 10^{-6} reflectivity signal within the 12-ns time window.

The peak frequency of the single-cycle pulse spectrum for this structure is 150 GHz, corresponding to a wavelength of 67 nm. For a beam waist of 3 um, the acoustic Rayleigh range is 360 um, so after 500 um of propagation the pulse shape should evolve from a single cycle to its derivative; the calculated pulse shapes are shown in figure 2. If the optical beam is not tightly focused, however, so that the Rayleigh range becomes much longer than the 500-um crystal thickness, then the acoustic wave suffers negligible diffraction and does not change its shape (i.e. it remains in the near field). (To date all picosecond acoustics experiments of which we are aware have been performed in this regime.)

Some representative measured acoustic waveforms are shown in figure 3. In 3(a), the optical pulse is not tightly focussed, and the propagated waveform is

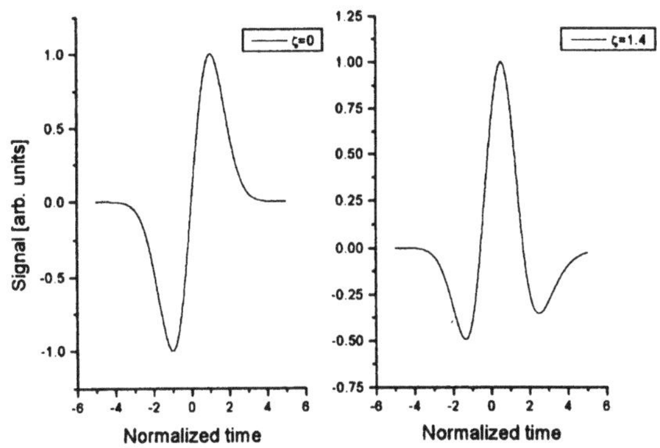

Figure 2. Calculated acoustic pulse shapes (left) in the near field, and (right) in the far field.

essentially the same as the input waveform (measured on a separate sample with only 70 nm propagation length in Si). In 3(b), the excitation beam is focussed to a waist of about 3 um, and the shape has clearly evolved towards the derivative shape. The data are somewhat complicated by a reflection appearing near 20 ps, but no deconvolution has been attempted with this data.

The observation of the waveform change associated with the Gouy phase shift which occurs as the excitation spot size is varied indicates that diffractive effects of picosecond acoustic transients can be observed with sufficient signal-to-noise ratio, and can ultimately be exploited. The diffractive spreading associated with the 3-um excitation is not large; much larger effects are expected as the acoustic source size is reduced below 1 micron. Such effects will be investigated in the near future.

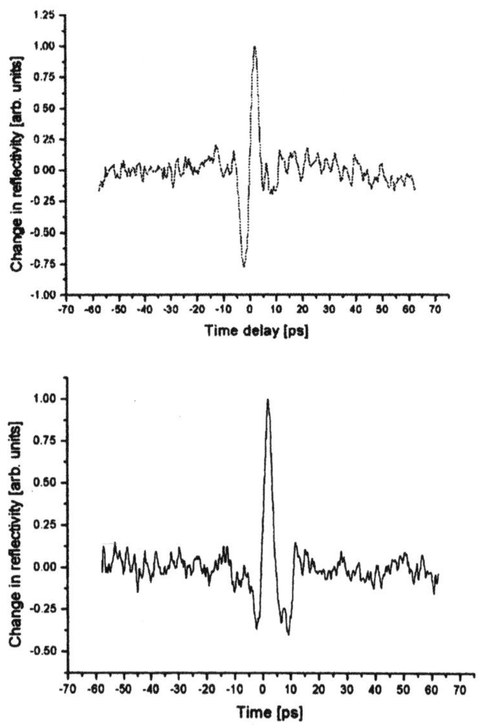

Figure 3. Measured acoustic waveforms.

References.

1. A.B. Ruffin, et al., Phys. Rev. Lett. 83, 3410 (1999)
2. A.B. Ruffin, et al., Opt. Lett. 26, 681 (2001)
3. H.T. Grahn, et al., IEEE J. Quant. Electron. 25, 2562 (1989).
4. H. Maris, Scientific American, 86 (1998).

Ultrashort-pulse-laser ablation of metals: Significant changes in ablation rates with depth

Rune Lausten[1], Jakob A. Olesen[1], Kasper Vestentoft[1], and Peter Balling[1,2]

[1] Institute of Physics and Astronomy, University of Aarhus, DK-8000 Aarhus C, Denmark
[2] E-mail: balling@phys.au.dk

Abstract. Femtosecond-laser ablation rates for a broad range of metals have been investigated and general trends are discussed. A newly developed technique allows the accurate measurement of the ablation rates at all times during the formation of holes.

1. Introduction

The interest in ultrashort-pulse-laser ablation of materials is of both fundamental and applied nature. On the fundamental level, ultrashort-pulse-laser excitation of materials can provide information about the timescales for energy transfer in solids. On the other hand, materials processing with ultrashort-pulse lasers has been shown to exhibit very distinct characteristics that may prove these laser types superior for certain applications in micromachining and laser surgery.

This paper describes a study of ablation rates in a range of amorphous metals. The investigation uses a newly developed technique [1], which applies time-gated imaging to measure the depth of the structures formed *during* the ablation process, see Sec. 2. The depth versus time thus obtained provides the ablation rate at all times during the ablation, and some of the general trends observed in this study are discussed in Sec. 3.

2. On-the-fly depth profiling

The short pulse duration of femtosecond lasers allows three-dimensional imaging, since the selection of a specific time of flight for the light corresponds to a specific depth. When combined with laser ablation, this technique allows for the extraction of geometrical information about the region subjected to ablation *during* the ablation process.

A specific implementation of time-gated imaging, which is very convenient for ablation or micromachining applications, was recently developed [1]. The technique uses one and the same laser pulse to perform the ablation and the geometrical measurement. As shown in Fig. 1, the principle resembles that behind a single-shot autocorrelator: The output of a femtosecond laser is split in two parts by a beam splitter. One part is focussed onto a sample and performs ablation, while the second part enters a delay line and is used for gating. Part of the backscattered light is collected by the lens in front of the sample and after being picked off by the beam-splitter it is sent through an imaging system designed to image the ablation region of the sample onto the non-linear crystal that forms the optical gate. Inside the non-linear crystal, the backscattered, re-imaged light is crossed by the gate beam. As shown in the insert of Fig. 1, part of the image is up-converted to the second harmonic, and the resulting pattern reflects in one direction the time (or depth) coordinate of the backscattered light and in the perpendicular direction an image across the laser-ablated region. Consequently, the pattern represents a depth profile across the ablation region.

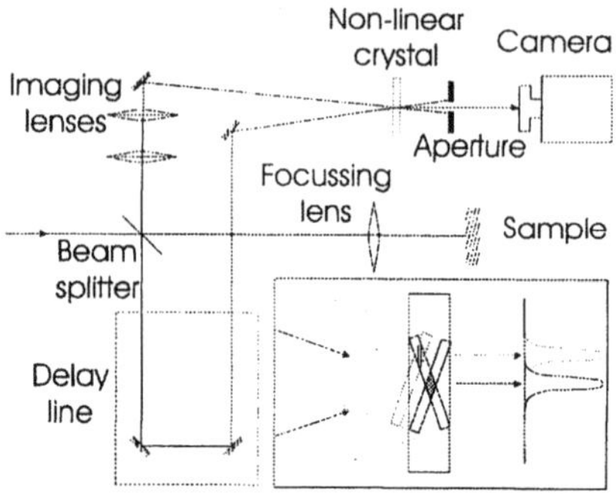

Fig. 1. The setup for on-the-fly depth profiling. The insert shows how the temporal information (depth) is converted to a spatial pattern

Such on-line depth profiles are useful for certain applications of femtosecond lasers in micromachining and laser surgery since they can be used as a feedback signal to control the ablation process. The depth-resolution depends on the laser pulseduration and the material subjected to ablation. For example for a 100-femtosecond pulseduration, an accuracy of 1 μm is obtained in steel, increasing to ~5 μm in soft biological materials.

3. Depth-resolved ablation rates

Using the setup described above, the depth versus time during ablation can be readily obtained, and hence also the ablation rate. For this study, a 5-mm-diameter aperture is inserted in the laser beam and imaged (demagnified ×40) onto the sample by a 50-mm-focal-length lens to produce a roughly constant intensity across the ablation region. Furthermore, the imaging system allows the selection of the central ~Ø30 μm of the 120 μm circular spot, which represents to good accuracy a constant-intensity region.

Figure 2 shows the measured depth versus number of laser pulses for a copper sample. At all fluences the ablation rate drops significantly with increasing depth. The same behavior has been observed in various degrees for many of the preliminary measurements on a broad range of metals. The drop in ablation rate has important consequences for the reproducibility of ultrashort-pulse-laser machining and for the efficiency of the formation of deep structures.

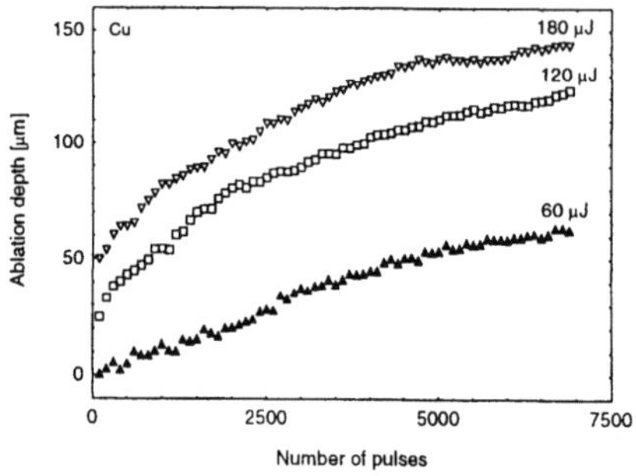

Fig. 2. Measured hole depths versus time for a copper sample at three different fluences corresponding to pulse energies of 60 μJ, 120 μJ and 180 μJ. Note that the curves have been offset for clarity.

Recent data, focusing on the development of ablation during the first few hundreds of laser pulses, show that also in this regime the ablation process

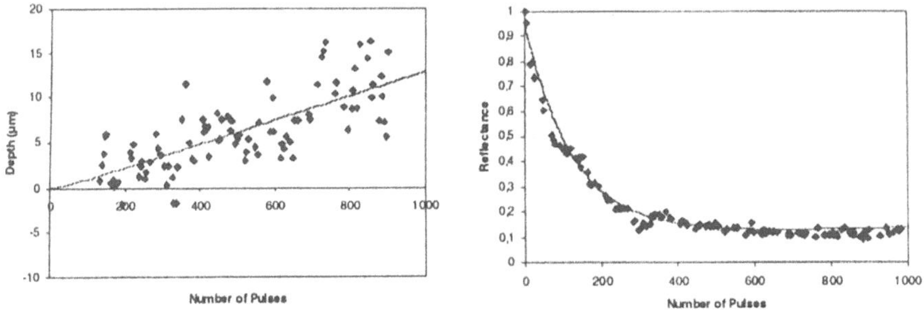

Fig. 3. (a) Measured hole depths versus the number of pulses for a copper target at a pulse energy of 120 µJ. (b) The intensity of the backscattered light, or the effective reflectance. The solid line is the result of a model calculation, as described in the text.

changes with the number of pulses. For example, the amount of backscattered light drops significantly during the first ~100 shots, see Fig. 3 (b). This could indicate that a larger fraction of the energy is gradually coupled to the surface during this interval leading to an initial increase of the ablation rate. Figure 3 (a) shows a preliminary measurement in this regime, which may show this tendency.

In order to illuminate the few-pulse behavior of the ablation process, a series of scanning-electron microscope (SEM) images were acquired. During the first hundreds of pulses, the ablated area changes: a complicated surface morphology, including ripples perpendicularly to the laser polarization with a spacing of roughly the laser wavelength appear. This phenomenon, which was first observed with nanosecond pulses [2], is due to interference between the incoming laser light and a surface-scattered wave. The occurrence of the surface corrugations and in particular ripples influences the energy coupling to the copper target: the laser is allowed to excite surface plasmons in the metal. Presumably, these collective electronic excitations decay rapidly to electronic heat, offering a new excitation mechanism of the metal. The observed change in reflectivity and the ablation rate is reproduced by a simple iterative model, which invokes an exponential growth of the height of surface ripples and assumes that the energy coupling to the sample is proportional to the ripple height. The ablation rate is assumed to follow the behavior $\ln(F / F_{th})$, which has been confirmed by many experiments.

The drop in ablation rate observed at larger depth is caused by a combination of effects, one of which may be the surface morphology: SEM images show that at later stages of the ablation, the rippled areas develop domains and other secondary structures that could alter the energy-coupling mechanisms.

References

1 R. Lausten and P. Balling, "On-the fly depth profiling during ablation with ultrashort laser pulses: A tool for accurate micromachining and laser surgery", Appl. Phys. Lett. **79**, 884-886 (2001).

2 Z. Guosheng, P. M. Fauchet, and A. E. Siegman, "Growth of spontaneous periodic structures on solids during laser illumination", Phys. Rev. B **26**, 5366-5381 (1982).

Sub-Micrometer Pulsetrain Machining of Diamond-Like Amorphous Carbon Films with Femtosecond Pulses of a Ti:sapphire Laser Oscillator

S. Lakó[1], P. Apai[1], I. Pócsik[1], E. Wintner[2], and R. Szipőcs[1]

[1] Research Institute for Solid State Physics and Optics
H-1525 Budapest P.O.Box 49, Hungary
E-mail: szipoecs@sunserv.kfki.hu
[2] Institut für Photonik, Technische Universitat Wien
Gusshauss strasse 27/387 A-1040 Wien, Austria

Abstract: Fabrication results of sub-micrometer holes on diamond-like amorphous carbon (DLC) coated surfaces produced by ablation with femtosecond laser pulses are reported. This technologically important coating material has not been studied in laser ablation experiments so far. In contrast to earlier experiments with metallic coatings on glass substrates, even the damaged area does not exceed the sub-micrometer size scale on DLC coatings deposited on Si wafers due to the basically different mechanical and ion diffusion properties and nonlinearity playing role in the ablation process.

1. Introduction

Although the production of sub-micron structures has a great importance [1,2,3], the modification of material properties in areas smaller than 1 micrometer is still a challenge to modern technologies [4]. Laser-based optical methods face serious difficulties working on the border of diffraction-limited resolution and focal spot size. Using deep-ultraviolet light sources and lasers the diffraction limit can be reduced, but the laser systems and the necessary optical components of the ultimate high performance are extremely expensive in this regime. Beside the diffraction limit another fundamental problem is the heat diffusion which acts against the spatial confinement of the laser energy deposition. Because of these factors, only very special laser systems and materials can be used in sub-micron material processing. Recently, a few research groups reported on experimental results indicating that utilizing non-linear processes and/or laser pulses of sub-picosecond duration, the difficulties mentioned above can be overcome and structures smaller than the laser focal spot can be generated [5,6]. Unfortunately, both in the nanojoule and in the microjoule pulse regime, surface deformations occur at the edge of the produced sub-micrometer structures. These deformations of at least 1 micrometer diameter limit most of the possible applications, e.g. repairing the photolitographic masks. Our aim was to overcome this limitation on the size of the damaged area with an arrangement based on the widely used femtosecond pulse Ti:sapphire oscillator and by using diamond like carbon (DLC)

[7] film as target. Unlike the metallic films used in the above cited experiments [5,6], DLC is a hard and covalently bound network of atoms, and this is why we expected a basically different behaviour during the ablation process.

2. Experimental setup and targets

The laser source we used in our experiments was a tunable, fs mode-locked Ti:sapphire laser (FemtoRose 100 TUN, product of R&D Ultrafast Lasers Ltd.). The laser emitted 85 fs long pulses at a 76 MHz repetition rate and 0.7 W output power at around 800 nm in our experiments.

Considering the possible applications of fs pulse ablation with a 76 MHz laser source, such as DVD mastering at MHz repetition rates, the shortest ablation process is desirable. In order to determine the number of laser pulses needed to drill through a thin film, we had to apply variable but well defined exposure times (and energy densities) at each experimental spot. The shortest exposure time we generated in our experimental setup was a single pulse train of 700 ns duration.

We used one beamsplitter in order to observe the reflection from the surface of the sample on a screen and to determine the proper focusing position. The focusing optics was an aspheric lens with N.A. = 0.6. The diameter of the focal spot was measured by the knife-edge method. The focal spots were elliptical and 50% of the power concentrated in an area 2 μm x 4 μm in our experiments.

The DLC targets were prepared by radio frequency aimed glow discharge CVD decomposition of methane. Single crystal silicon wafer was used as a substrate. Our preparation conditions were optimal for the formation of so called „diamond-like" carbon [7], a glassy matrix with mechanical parameters usually comparable to the hardness of diamond, and this deposition technique is known as being adequate to prepare a smooth surface.

3. Machining, evaluation and interpretation

Our focusing lens was assembled on a translation stage and its position was adjusted along its optical axis according to the observed reflection. The samples were fixed on an X-Y translation stage and they could be moved in the plane perpendicular to the beam propagation. Exposures were done in regular spatial sequences, so the length-scale was self-calibrated. After each exposure series we inserted some space and while controlling and correcting the focusing by long exposure, we observed the change appearing in the reflected light spot. The exposure time was set by varying the number of exposures. Usually we applied 1, 2, 3, ... 10 exposures of single pulse train of 700 ns duration in a series.

The traces of the laser shots were investigated by optical microscopy. We found that there was no visible deformation outside the drilled holes (see Figure 1.). In the worst case, some fine carbon dust fall-out could be seen in a circle of 2 μm

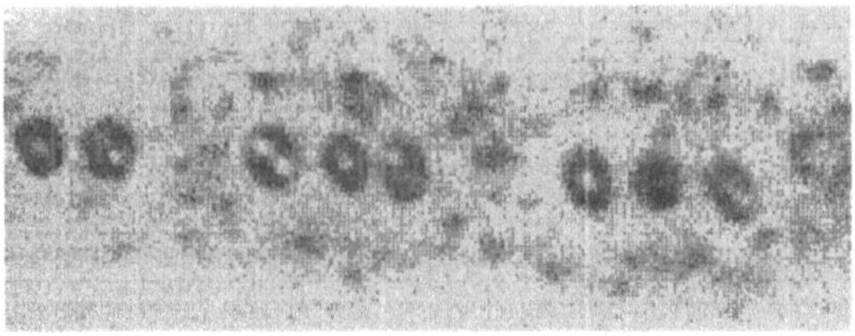

Fig. 1. A sequence of ablated holes on a DLC film with white light illumination. The distance between the hole centres is 2 micrometers in the hole triplets.

diameter, wich is probably the consequence of the agglomerated unburned carbon particles created during the ablation. The fall-out dust of the formally drilled hole is not present in the fallout region of the next hole, proving that these fine dust particles are wiped away easily by the shock wave of air of the next ablation experiment, so they can be eliminated mechanically from the film surface.

4. Conclusions

Holes with sub-micrometer diameter were produced on DLC coatings deposited on Si wafers. One order of magnitude less than the focal spot, sub-micrometer damaged area could be achieved due to nonlinear processes. We are convinced that DLC coatings are excellent candidates for submicron fabrication applications.

Acknowledgements. This work was supported by the Hungarian Science Foundation under contracts OTKA-T026073 and OTKA-T029578.

References

1 Y. Kaneda and S. Kubota in *IEEE Quant. El.* Vol. 3, 35, 1997.
2 Y. Wu, H. Khoo and T. Kogure in *Appl. Phys. Lett.* Vol. 64, 3225, 1994.
3 E. N. Glezer, M. Milosavljevic, L. Huang, R. J. Finlay, T.-H. Her, J. P. Callan and E. Mazur in *Opt. Lett.* Vol. 21, 2023, 1996.
4 Ch. Schaffer, A. Brodeur, J. F. Garcia, and E. Mazur in *Opt. Lett.* Vol. 26, 93, 2001.
5 P. P. Pronko, S. K. Dutta, J. Squier, J. V. Rudd, D. Du and G. Mourou in *Opt. Comm.* Vol. 114, 106, 1995.
6 F. Korte, S. Adams, A. Egbert, C. Fallnich, A. Ostendorf, S. Nolte, M. Will, J.-P. Ruske, B. N. Chichkov and A. Tünnermann in *Opt. Express* Vol. 7 ,41, 2000.
7 W. I. Milne, S. R. P. Sylva, J. Robertson and G. A. J. Amaratunga in *Amorphus Carbon: State of Art* World Scientific Singapore 1998.

Two-photon polymerization of photosensitive resins with femtosecond pulses and fabrication of microoptical devices and photonic structures

J. Serbin, C. Fallnich, and B.N. Chichkov

Laser Zentrum Hannover e.V., Hollerithallee 8, D-30419 Hannover, Germany
Phone:+49-511-2788-237, fax: +49-511-2788-100
E-mail: js@lzh.de

Abstract. Investigations on two-photon polymerization of photosensitive resins initiated by femtosecond laser pulses are performed. Applications of this technique for the fabrication of three-dimensional microoptical devices, photonic structures and micro-models are demonstrated.

Two-photon polymerization (2PP) of photosensitive resins initiated by femtosecond laser pulses has recently been demonstrated as a very powerful technique for the fabrication of three-dimensional microstructures [1,2]. When tightly focused into the volume of a liquid resin, which is transparent in the infrared, femtosecond laser pulses can initiate two-photon-polymerization and produce 3D microstructures with a resolution better than 0.2 μm [2].

In this paper, investigations of the two-photon polymerization technique and its potential for the fabrication of 3D microstructures and low-loss, high-performance microoptical devices are reported. The resins we use are designed for ultraviolet photo-lithography and contain photo-initiators sensitive to 390 nm radiation. These resins are exposed to tightly focused 80-fs Ti:sapphire laser pulses (at 780 nm and a repetition rate of 80 MHz) initiating two-photon polymerization in a small focal volume.

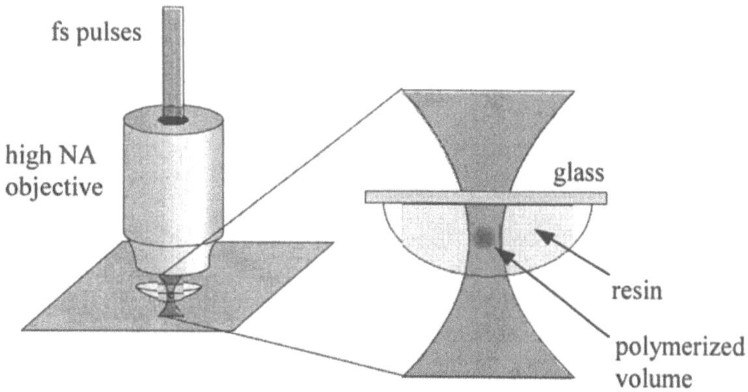

Fig. 1. Experimental Setup for 2PP

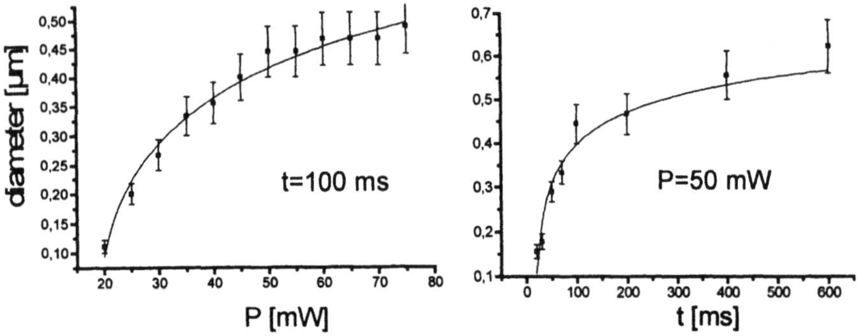

Fig. 2. Calculated and measured minimum structure sizes fabricated by 2PP in a hybrid polymer as a function of the average laser power (left) and the irradiation time (right).

Since these resins are transparent to the incident laser light, the polymerization can be initiated everywhere in the volume of a liquid droplet.

The experimental setup is shown in Fig. 1. By moving the laser focus through the resin, complicated three-dimensional structures can be produced along the trace of the focus. To avoid the generation of filaments, high NA (NA=1.3) microscope objectives are used for focussing of femtosecond laser pulses. To perform three-dimensional micro structuring, the focused laser beam was scanned in the x- and y-planes with a galvo-scanner and the sample was moved in the z-direction with a translation stage. To guarantee a flat surface for the incident laser beam, the resin was irradiated through a 170 μm thick glass substrate. Non-irradiated liquid resin was washed out with alcohol.

As shown in Fig. 2, the minimum structure sizes of the polymerized resin depend on the average laser power (at a fixed irradiation time) and on the irradiation time (at a fixed laser power). By introducing a threshold density of radicalized starter molecules required for solidification, one can estimate the minimum structure sizes, which can be produced by tightly focussed Gaussian laser pulses, by

$$d = r_0 \sqrt{\ln \frac{t \sigma_2 I_0^2}{C}} \quad \text{with} \quad C = \ln\left(\frac{\rho_0}{\rho_0 - \rho_{th}} \right), \tag{1}$$

where r_0 is the focus radius, σ_2 is the two-photon absorption cross section for the starter molecules, ρ_0 and ρ_{th} are the initial and the threshold starter densities, I_0 is the peak laser intensity, and t is the irradiation time.

In these experiments, the formation of bubbles in the liquid resin under laser irradiation was not a limiting factor for the 2PP. However, when the pulse energy and/or the irradiation time exceed a certain value, the polymerization process starts also in the regions above and below the focal volume corresponding to the intensity distribution near the focus. This fact restricts the suitable working range for microstructuring to rather small pulse energies and/or irradiation times. For example, with a laser irradiation time of 100 ms the optimum process window for high resolution microstructuring is between 15 mW and 30 mW.

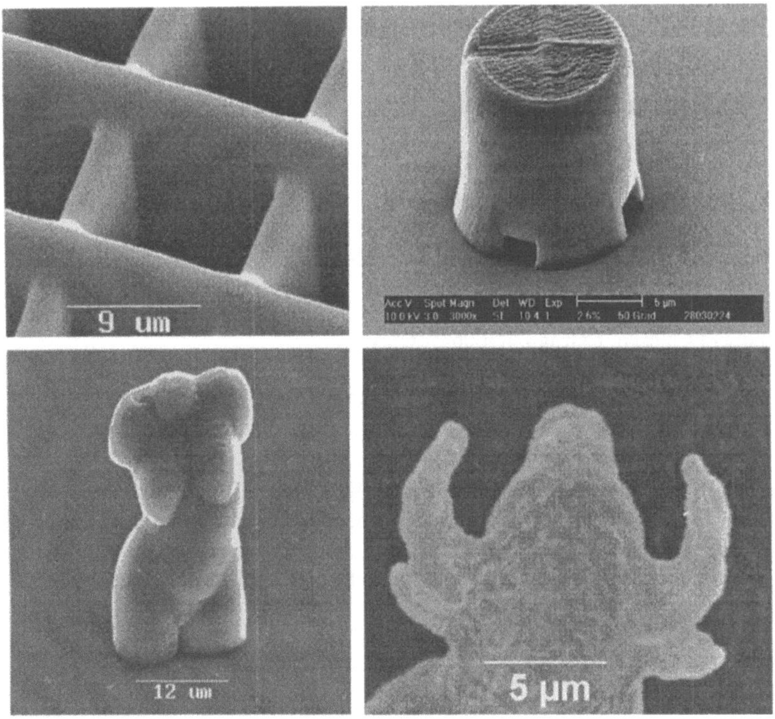

Fig. 3. Examples of different 3D microstructures fabricated by 2PP in a hybrid polymer using femtosecond laser pulses.

In Fig. 3, some examples of 3D microstructures produced by the two-photon-polymerization technique are shown.

In conclusion, it has been demonstrated for the first time that two-photon polymerization of hybrid polymers with femtosecond laser pulses is possible. This technique allows to fabricate different microoptical devices like three dimensional photonic structures, waveguide couplers, light deflectors, photonic bandgap structures, etc. Functional characteristics of the fabricated structures are at present under investigations.

References

1 B. H. Cumpston, S. P. Ananthavel, S. Barlow, D. L. Dyer, J. E. Ehrlich, L. L. Erskine, A. A. Heikal, S. M. Kuebler, I.-Y. Sandy Lee, D. McCord-Maughon, J. Qin, H. Röckel, M. Rumi, X.-L. Wu, S. R. Marder, and J. W. Perry, "Two-photon polymerization initiators for three-dimensional optical data storage and microfabrication," Nature **398**, 51-54 (1999).

2 S. Kawata, H-B. Sun, T. Tanaka, and K. Takada, "Finer features for functional microdevices," Nature **412**, 697 (2001).

Widely tunable femtosecond pulses from a tapered fiber for ultrafast microscopy and multiphoton applications

Justin E. Jureller[1] and Norbert F. Scherer[1]

T.A. Birks[2], W.J. Wadsworth[2] , and P. St. J. Russell[2]

[1] Department of Chemistry, James Franck Institute, and Institute for Biophysical Dynamics, University of Chicago, Chicago, IL 60637, USA
E-mail: nfschere@uchicago.edu
[2] Department of Physics, University of Bath, Claverton Down, Bath BA2 7AY, UK
E-mail: pystab@bath.ac.uk

Abstract. Sub-50fs transform limited pulses are produced in the 670-1050nm range by compressing sections of the ultrabroadband supercontinuum generated in a tapered fiber injected with pulses from a cavity dumped Ti:Sapphire laser. 2-Photon imaging of fluorescent beads is demonstrated. Biological and chemical applications are discussed.

1. Introduction

The recent introduction of new types of optical fibers with novel nonlinear properties has caused much excitement in part because of their ability to generate an ultrabroadband continuum spanning the visible to the near IR. Most of the attention has been focused on microstructured, so called "holey," photonic crystal fibers (PCF's) that consist of a 2-D hexagonal array of airholes surrounding a central 1-2μm diameter fused silica core.[1] Through an effective index of refraction model, the airholes act as an air cladding and shift the zero GVD-wavelength of the fiber into the visible, thereby allowing very long nonlinear interaction lengths for propagating pulses of the appropriate wavelength. Unamplified Ti:Sapphire (Ti:S) lasers operating around 800nm have been used as the primary sources to demonstrate continuum generation as well as other nonlinear processes and optical meteorologies.[2-4] Tapered fibers, or standard single-mode fibers that have had a length of their midsection uniformly reduced to 1-2μm in diameter, have been used in a similar manner to generate broad continua.[5] Tapered fibers are superior to PCF's as stable continuum sources because of their more robust coupling conditions as well as their higher damage threshold. Even though the spectral output of both types of fibers have been well documented, the time domain output still remains relatively uninvestigated. Furthermore, few applications that utilize short duration pulses from microstructured fiber sources have been reported.

2. Experimental Methods

Experiments were conducted with a home-built cavity dumped Ti:S laser that delivered 30nJ, 16fs pulses centered at 800nm at a 125kHz repetition rate.[6] 10nJ of this light was sent through a prism-pair precompensation line and then focused into the tapered fiber through a 10x plan-achromat objective. The remaining light was diverted and used as a gate pulse for cross-correlation analyses of the continuum performed by sum frequency generation in a thin Type I BBO crystal. The tapered fiber was 17.5cm overall in length with a midsection uniformly narrowed to a diameter of 2.5µm (zero GVD point ~800nm) that was approximately 9cm in length. The output of the fiber was recollimated with a 20x plan-achromat objective or aspheric lens and recompressed in a prism-pair postcompensation line. Spectral selection was achieved with an adjustable pair of razor blades on the back mirror plane of the prism line, and the pulses then sent either to the cross-correlation apparatus or directly to the 2-photon microscope.

The 2-photon microscope consisted of a 100x oil immersion objective (Zeiss) mounted horizontally on a piezo actuated XYZ stage. Epi-detected fluorescence was separated from the excitation pulses with a dichroic mirror (Chroma), detected with a fiber-coupled PMT, and acquired with a photon counter.

3. Results and Discussion

Figure 1 shows the cross-correlation of a 35nm FWHM continuum slice centered at 930nm. The gate-deconvoluted near-IR pulse is a 38fs FWHM nearly transform limited Gaussian. Pulses of comparable duration (<50fs) have been obtained at other wavelengths in the 670-1050nm range. Since the full continuum extends beyond our detector range (650-1250nm), various applications beyond the spectral coverage of Ti:S lasers are envisioned.

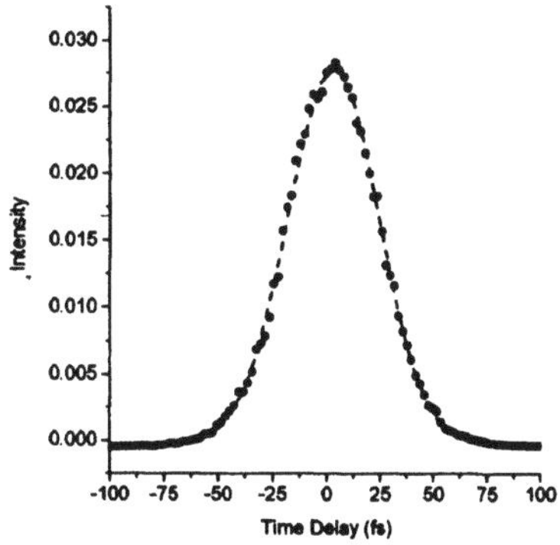

Fig. 1. Cross-correlation of imaging pulse

2-photon imaging was demonstrated with yellow-green fluorescent beads (Molecular Probes) with a 1-photon absorption maximum at 500nm and emission maximum at 510nm. Bead samples were placed between coverslips and mounted on an XY piezo scanstage (Queensgate) that was raster scanned to acquire an image. 2-photon excitation was achieved with the 38fs transform limited 930nm pulses characterized in Figure 1. Fig. 2(b) shows the image obtained from raster scanning a roughly 13µm x 13µm area. Individual beads are well resolved. Fig. 2(a) shows a cross-sectional slice of one 500nm bead with a fitted FWHM of 380nm, implying some bead sectioning.

These short duration tunable pulses are intended for microscopy applications that allow novel functional imaging and study of dynamics from sub-micron regions and nanostructures. Extension of 2-photon interferometry measurements of individual particles [7] are in progress. The multi-spectral potential of the tapered

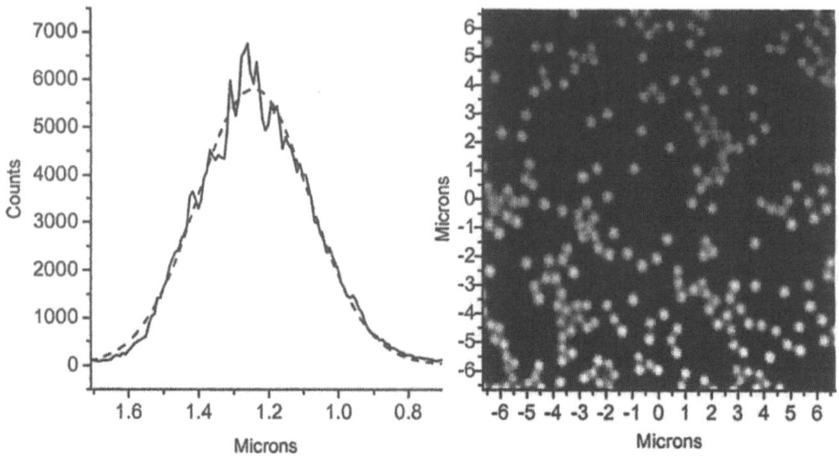

Fig. 1. (a) Cross-section of a 500nm diameter bead (solid) with a Gaussian fit (dotted). (b) 13μm x 13μm 2-photon image of 500nm diameter fluorescent beads.

fiber source can be employed for sub-diffractive imaging of molecular relaxation by 2-color stimulated emission depletion (STED).[8] The time-resolved chemical dynamics of processes like solvation can be studied within 100 attoliter focal volumes in heterogeneous systems like argarose gels and intracellular environments. Similarly, the extension of 2-photon microscopy [9] to a 2-color 2-photon variant for the ratiometric study of multiple (indigenous or exogenous) fluorophores becomes more straightforward.

We wish to thank the W.M. Keck Foundation, the Consortium for Nanoscience Research at the University of Chicago and Argonne National Labs, the NSF, and the UK EPSRC funding. WJW is a Royal Society University Research Fellow.

References

1 T.A. Birks, J.C. Knight, and P. St. J. Russell, *Opt. Lett.* 22, 961-963 (1997).
2 J.K. Ranka, R.S. Windeler, and A.J. Stentz, *Opt. Lett.* 25, 25-27 (2000).
3 J.E. Sharping, M. Fiorentino, A. Coker, P. Kumar, and R. Windeler, *Opt. Lett.* 26, 1048-1050 (2001).
4 S.A. Diddams, et al, *Phys. Rev. Lett.* 84, 5102-5105 (2000).
5 T.A. Birks, W.J. Wadsworth, and P. St. J. Russell, *Opt. Lett.* 25, 1415-1417 (2000).
6 Y.-H. Liau, A.N. Unterreiner, D.C. Arnett, and N.F. Scherer, *Appl. Opt.* 38, 7386-7392 (1999).
7 Y.-H. Liau, A.N. Unterreiner, Q. Chang, and N.F. Scherer, *J. Phys. Chem. B.* 105, 2135-2142 (2001).
8 M. Dyba, T.A. Klar, S. Jakobs, and S. W. Hell, *Appl. Phys. Lett.* 77, 597-599 (2000).
9 W. Denk, J.H. Strickler, W.W. Webb, *Science* 248, 73-76 (1990).

Femtosecond pump-probe spectroscopy using a two-dimensional smart pixel detector array

S. Bourquin[1], R. P. Prasankumar[1], U. Morgner[2], F. Kärtner[1], T. Lasser[3], R. P. Salathé[3], and J. G. Fujimoto[1]

[1] Department of Electrical Engineering and Computer Science and Research Laboratory of Electronics, Massachusetts Institute of Technology, Cambridge, MA 02139, USA
E-mail: bourquin@mit.edu
[2] High Frequency and Quantum Electronics Laboratory, University of Karlsruhe, D-76128 Karlsruhe, Germany
[3] Institute of Applied Optics, Swiss Federal Institute of Technology, CH-1015 Lausanne, Switzerland

Abstract. A smart pixel detector array combined with a broadband $Ti:Al_2O_3$ laser enables parallel sub-40-fs resolution, spectrally resolved pump-probe measurement. This smart pixel array can simultaneously demodulate a 58x58 pump probe data set.

1. Introduction

Femtosecond pump-probe spectroscopy is a well-known method used to investigate ultrafast excited state dynamics in condensed matter, chemical, and biological systems. The standard technique uses a narrowband pump and continuum probe to excite the sample, and employs a spectrometer to acquire the wavelength dependent absorption saturation dynamics. This technique often requires the use of a femtosecond amplifier to achieve sufficient intensities for continuum generation and sufficient pump-probe signal magnitudes. With the development of 5 fs lasers, which can generate spectra spanning one octave, spectrally resolved pump-probe measurements can be performed without the need for amplifiers [1,2,3]. Although the pulse repetition rate from a laser oscillator is extremely high, pulse energies are low and signal levels are small. Standard CCD detectors cannot detect modulated signals and thus it is not possible to take advantage of high signal to noise measurements that are possible with high repetition rate laser sources. Here, we present a new technique that enables the parallel acquisition of pump-probe measurements for multiple wavelengths. This is made possible using a novel, two-dimensional smart pixel detector array, which was originally developed for high speed optical coherence tomography (OCT) [4,5]. Each pixel performs amplitude demodulation, and in combination with a diffraction grating, probe transmission signals can be acquired in parallel for multiple wavelengths. The smart pixel array can achieve sensitivities comparable to lock-in amplification but can perform demodulation of probe transmission signals at multiple wavelengths simultaneously, enabling time and wavelength resolved femtosecond pump-probe spectroscopy.

2. Experimental Method

The experimental setup uses the smart pixel detector array in combination with a 100 MHz, 5 fs, 250 nm bandwidth Ti:sapphire laser and a femtosecond pump-probe system. The output of the laser is separated into a 90 mW pump beam and 0.4 mW probe beam. Reflective optics are used to preserve pulse duration. A grating spectrally spreads the probe beam after the sample and a lens focuses the beam onto one row of the detector array. The pump beam is chopped at 3.8 kHz and scanned. A row of the smart pixel array is read out multiple times for each pump-probe time delay and averaging is performed to increase the signal to noise ratio. The probe beam is subsequently chopped and the same row is read to acquire the linear transmission of the sample for data normalization.

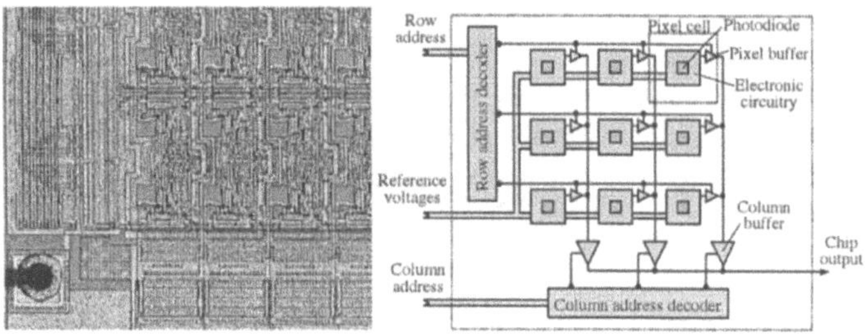

Fig. 1. Photograph of a section of the smart pixel detector array and corresponding array layout.

Figure 1. shows a photograph of a section of the detector array and a schematic diagram. The silicon detector chip is realized with a 2 μm complementary metal-oxide semiconductor (CMOS) process with a bipolar transistor option. The die size is 7.2 mm x 7.2 mm, which allows a 58 x 58 pixel array. Each pixel is 110 μm x 110 μm and contains a 35 μm x 35 μm photodiode and electronic circuitry for performing amplification, band-pass filtering centered at the modulation frequency, rectification and low-pass filtering. The filter cut-off frequencies are adjusted by off-chip reference voltages. The generated amplitude modulation signals are selected sequentially by a row and a column address decoder, via the pixel and the column buffers. The analog output signal is read out of the chip, digitized by a 12-bit data acquisition card and transferred to a computer.

3. Results and Discussion

We demonstrate spectrally resolved measurement of carrier dynamics in a thin sample of GaAs to show the functionality of this novel system. A 3D data set acquired by the smart pixel detector array is shown in Figure 2. It can be seen that the dynamics are a strong function of wavelength. Measurements performed far above the band edge of 870 nm have a rapid relaxation due to carrier-carrier and carrier-phonon scattering, which remove carriers from their initial optically excited

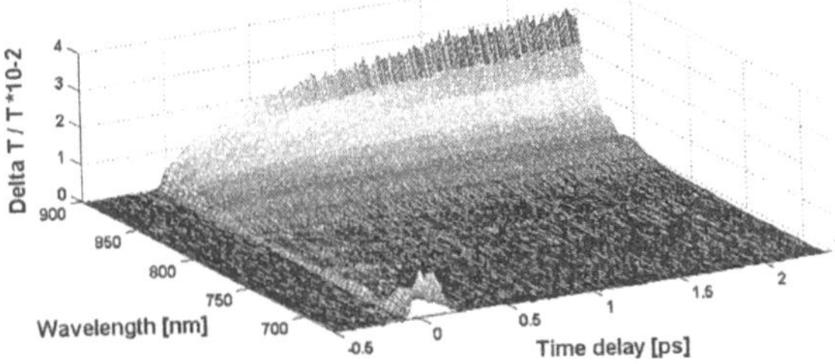

Fig. 2. Femtosecond spectrally resolved pump-probe measurements acquired in parallel by the smart pixel detector array.

states. Measurements closer to the band edge show increased absorption saturation as a function of time from carrier relaxation and state filling.

These results demonstrate the feasibility of performing parallel spectrally resolved pump-probe measurements over 58 simultaneous wavelengths with a sensitivity of $5 \cdot 10^{-4}$ when averaging each pixel 5000 times, corresponding to 200 seconds of measurement time for a 2.8 ps scan. Even higher sensitivities are possible with increased averaging. This technology can be applied to a wide range of pump-probe measurements in condensed matter, chemistry, and biology. Using the two dimensional imaging capabilities of the smart pixel array, an additional dimension of detection is also possible, enabling for example, one-dimensional spatial imaging combined with spectrally and time resolved measurement.

4. Conclusions

A new method for simultaneously acquiring pump-probe signals at several wavelengths using a smart pixel detector array and broadband femtosecond laser has been demonstrated and promises to be a powerful technology for performing rapid, high sensitivity, wavelength-dependent pump-probe spectroscopy.

References

1 U. Morgner, F. X. Kärtner, S. H. Cho, Y. Chen, H. A. Haus, J. G. Fujimoto, E. P. Ippen, V. Scheurer, G. Angelow, and T. Tschudi, in *Optics Letters,* Vol. 24, 411, 1999.

2 D. H. Sutter, G. Steinmeyer, L. Gallmann, N. Matuschek, F. Morier-Genoud, U. Keller, V. Scheuer, G. Angelow, T. Tschudi, in *Optics Letters*, Vol. 24, 631, 1999.

3 R. Ell, U. Morgner, F. X. Kärtner, J. G. Fujimoto, E. P. Ippen, V. Scheuer, G. Angelow, T. Tschudi, M. J. Lederer, A. Boiko, and B. Luther-Davies, in *Optics Letters*, Vol. 26, 373, 2001.

4 S. Bourquin, P. Seitz and R. P. Salathé, in *Electronics Letters*, Vol. 37, 975, 2001.

5 S. Bourquin, P. Seitz and R. P. Salathé, in *Optics Letters*, Vol. 26, 512, 2001.

The Control of Optical Damage Threshold by the Chirp of Ultraviolet Femtosecond Pulses

Hidetoshi Murakami[1], Shingo Ono[1], Yuji Suzuki[1], Toshimasa Kozeki[1], Hideyuki Ohtake[1], Shin-ya Koshihara[2, 3], and Nobuhiko Sarukura[1, 2]

[1] Institute for Molecular Science (IMS), Myodaiji, Okazaki 444-8585, Japan
E-mail: murakami@ims.ac.jp

[2] Kanagawa Academy of Science and Technology (KAST)

[3] Tokyo Institute of Technology

Abstract. The damage process of the aluminum coating for the ultraviolet-femtosecond pulses is investigated. It is found that the damage threshold becomes lower for the positively-chirped pulse irradiation case even the pulse duration becomes longer.

1. Introduction

The optical damage process of the material by femtosecond laser pulses is still attractive and important target of research both for the material processing and for the component design of high-peak-power laser systems. The damage of the optical material for the infrared ultrashort pulses was well studied [1, 2]. On the other hand, the material processing and the high-peak-power laser development trend have been moving into the ultraviolet region [3], owing to the successful development of new categories of ultraviolet high-peak-power lasers including KrF laser amplifier system operating at 248-nm and Ce:LiCAF chirped pulse amplifier system at 290-nm [4-6]. However, such optical damage information for ultraviolet femtosecond lasers is not investigated well. For the further expansion of ultraviolet ultrashort laser applications, detailed research of the basic damage process in ultraviolet region is inevitable. In this presentation, we will report the pulse duration and chirp dependence of the aluminum coating damage. The damage threshold was found to be lower for the positively chirped pulse irradiation case.

2. Experimental and Results

The experimental setup for the damage threshold measurement is shown in Fig. 1. The femtosecond seed pulses at 290 nm with 1-kHz repetition rate were generated by frequency tripling of the Ti:sapphire regenerative amplifier output. The durations of 290-nm femtosecond pulses were evaluated by a third-order autocorrelator using the visible fluorescence of the XeF C-A transition [7]. The resolution of this system was confirmed with a test pattern to be less than 10 μm, which corresponded to a pulsewidth resolution down to 50 fs. The chirp of the ultraviolet femtosecond pulse was controlled by the closely placed Brewster silica

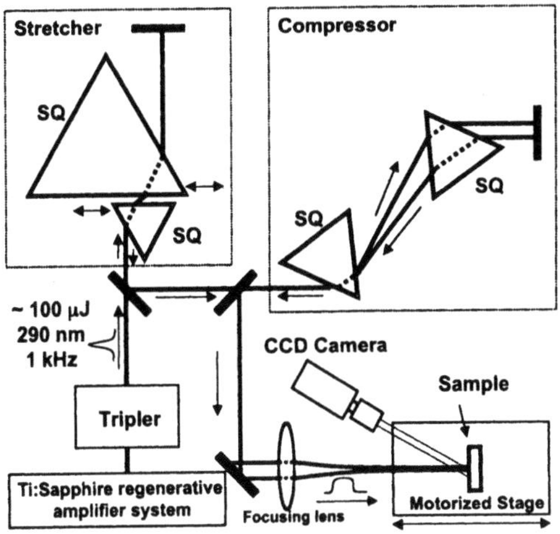

Fig. 1. Experimental setup of the damage evaluation system for the different chirp condition of ultraviolet femtosecond pulses.

prism pair to provide large adjustable positive dispersion and 1.5-m separated Brewster silica prism pair to provide fixed negative dispersion. Using this optics, positively or negatively chirped pulses with 700-fsec pulse duration and nearly transform limited 140-fsec pulses were provided as shown Fig.2 (a). For the accurate control of the relative fluence, the sample was mounted on the motorized stage after the fixed focusing lens. The damage was monitor by the CCD camera. The sample was slowly moved to the beam waist from the upward until the damage process would start up. In this way, the damage threshold dependence of different pulse duration and chirp conditions were relatively compared. The damage threshold was found to be lower for the positively-chirped pulse irradiation case as shown Fig.2 (b). The larger gap from the beam waist position means lower

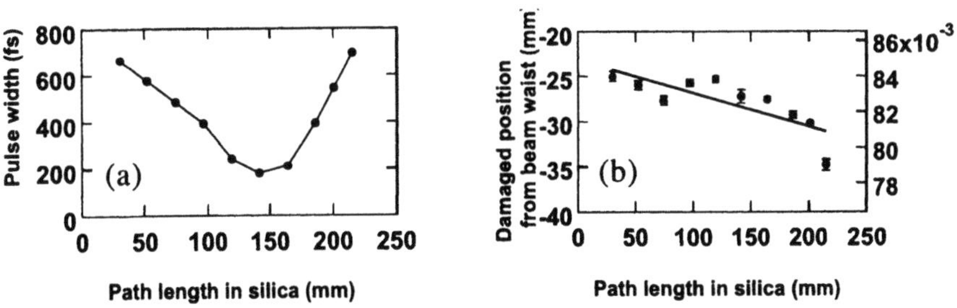

Fig. 2. (a) The pulse duration (FWHM) dependence on different insertion depth of the optical path. The shortest pulse duration corresponds to 140-fs assuming a $sech^2$ pulse shape considering the deconvolution factor of 1.29. (b) The relative damage starting position for the different chirp condition. The larger gap from the beam waist position means lower optical fluence. The amount of the chirping can be controlled by the silica optical path length. The larger amount of material resulted in the larger amount of positive chirp. Within the accuracy of the experimental data, clear drop of the damage threshold was observed for the larger amount of positive chirp.

optical fluence. The amount of the chirping can be controlled by the silica optical path length. The larger amount of material resulted in the larger amount of positive chirp. Within the accuracy of the experimental data, a clear drop of the damage threshold was observed for the larger amount of positive chirp. The major point of this suggestive new finding is that the damage threshold was found to be lower for the positively-chirped pulse irradiation case even the pulse duration became longer for such pulses. This should be noticed for the design of the high-peak-power laser system. This experimental results also suggest that lower photon-cost for material processing will be achieved with the positively-chirped ultraviolet femtosecond pulse. Further detailed research including much larger span of the pulse duration should be needed to clarify the mechanism of this interesting chirping dependence.

3. Conclusions

In conclusion, the damage of the aluminum coating for the ultraviolet femtosecond pulses was investigated for the basic study of material processing. The damage threshold was found to be lower for the positively-chirped pulse irradiation case not for the shortest pulse case. This new finding suggests that the control of the chirping will reduce the effective photon cost for the material processing. Additionally there should be rich and interesting physics to be explored behind this anomalous dependence.

References

1 B. C. Stuart, M. D. Feit, A. M. Rubenchik, B. W. Shore, and M. P. Perry, Phys. Rev. Lett. **74**, 2248 (1995).
2 M. Lenzner, J. Kruger, S. Sartania, Z. Cheng, Ch. Spielmann, G. Mourou, W. Kautek, and F. Krausz, Phys. Rev. Lett. **80**, 4076 (1998).
3 P. Simon, J. Bekesi, C. Dolle, J. -H. Klein-Wiele, G. Marowsky, S. Szatmari, B. Wellegehausen, Ultrafast Optics, Quebec, Canada, 75 (2001).
4 Y. Nabekawa, K. Kondo, N. Sarukura, K. Sajiki, S. Watanabe, Opt. Lett. **18**, 1922 (1993).
5 S. Szatmari, Appl. Phys. B **58**, 211 (1994).
6 Z. Liu, T. Kozeki, Y. Suzuki, N. Sarukura, K. Shimamura, T. Fukuda, M. Hirano, H. Hosono, Opt. Lett. **26**, 301 (2001).
7 N. Sarukura, M. Watanabe, A. Endoh, and S. Watanabe, Opt. Lett. **13**, 996 (1988).

Index of Contributors

Springer Series in Chemical Physics

Editors: Vitalii I. Goldanskii Fritz P. Schäfer J. Peter Toennies

Managing Editor: H. K.V. Lotsch

40 **High-Resolution Spectroscopy of Transient Molecules**
By E. Hirota

41 **High Resolution Spectral Atlas of Nitrogen Dioxide 559–597 nm**
By K. Uehara and H. Sasada

42 **Antennas and Reaction Centers of Photosynthetic Bacteria**
Structure, Interactions, and Dynamics
Editor: M. E. Michel-Beyerle

43 **The Atom-Atom Potential Method**
Applications to Organic Molecular Solids
By A. J. Pertsin and A. I. Kitaigorodsky

44 **Secondary Ion Mass Spectrometry SIMS V**
Editors: A. Benninghoven, R. J. Colton, D. S. Simons, and H. W. Werner

45 **Thermotropic Liquid Crystals, Fundamentals**
By G. Vertogen and W. H. de Jeu

46 **Ultrafast Phenomena V**
Editors: G. R. Fleming and A. E. Siegman

47 **Complex Chemical Reaction Systems**
Mathematical Modelling and Simulation
Editors: J. Warnatz and W. Jäger

48 **Ultrafast Phenomena VI**
Editors: T. Yajima, K. Yoshihara, C. B. Harris, and S. Shionoya

49 **Vibronic Interactions in Molecules and Crystals**
By I. B. Bersuker and V. Z. Polinger

50 **Molecular and Laser Spectroscopy**
By Zu-Geng Wang and Hui-Rong Xia

51 **Space-Time Organization in Macro molecular Fluids**
Editors: F. Tanaka, M. Doi, and T. Ohta

52 **Multiple-Photon Laser Chemistry**
By. R. V. Ambartzumian, C. D. Cantrell, and A. Puretzky

53 **Ultrafast Phenomena VII**
Editors: C. B. Harris, E. P. Ippen, G. A. Mourou, and A. H. Zewail

54 **Physics of Ion Impact Phenomena**
Editor: D. Mathur

55 **Ultrafast Phenomena VIII**
Editors: J.-L. Martin, A. Migus, G. A. Mourou, and A. H. Zewail

56 **Clusters of Atoms and Molecules**
Solvation and Chemistry of Free Clusters, and Embedded, Supported and Compressed Clusters
Editor: H. Haberland

57 **Radiationless Transitions in Polyatomic Molecules**
By E. S. Medvedev and V. I. Osherov

58 **Positron Annihilation in Chemistry**
By O. E. Mogensen

59 **Soot Formation in Combustion**
Mechanisms and Models
Editor: H. Bockhorn

60 **Ultrafast Phenomena IX**
Editors: P. Barbara, W. H. Knox, G. A. Mourou, and A. H. Zewail

61 **Gas Phase Chemical Reaction Systems**
Experiments and Models 100 Years After Max Bodenstein
Editors: J. Wolfrum, H.-R. Volpp, R. Rannacher, and J. Warnatz

62 **Ultrafast Phenomena X**
Editors: P. F. Barbara, J. G. Fujimoto, W. H. Knox, and W. Zinth

MIX
Papier aus verantwortungsvollen Quellen
Paper from responsible sources
FSC® C105338
FSC
www.fsc.org

If you have any concerns about our products,
you can contact us on
ProductSafety@springernature.com

In case Publisher is established outside the EU,
the EU authorized representative is:
Springer Nature Customer Service Center GmbH
Europaplatz 3, 69115 Heidelberg, Germany

Printed by Libri Plureos GmbH
in Hamburg, Germany